Handbook of Neurochemistry and Molecular Neurobiology

Amino Acids and Peptides in the Nervous System

Abel Lajtha (Ed.)

Handbook of Neurochemistry and Molecular Neurobiology
Amino Acids and Peptides in the Nervous System

Simo S. Oja, Pirjo Saransaari and Arne Schousboe

With 58 Figures and 23 Tables

Springer

Editor
Abel Lajtha
Director
Center for Neurochemistry
Nathan S. Kline Institute for Psychiatric Research
140 Old Orangeburg Road
Orangeburg
New York, 10962
USA

Volume Editors

Prof. Simo S. Oja
The Centre for Laboratory Medicine
Tampere University Hospital
PB 2000
FI-33520 Tampere
Finland
E-mail: Simo.Oja@uta.fi

Prof. Arne Schousboe
Inst. for Pharmacology
Royal Danish School of Pharmacy
Universitetsparken 2
DK-2100 Copenhagen
Denmark
E-mail: as@dfuni.dk

Prof. Pirjo Saransaari
Brain Research Center
Medical School
FI-33014 University of Tampere
Finland
E-mail: pirjo.saransaari@uta.fi

Library of Congress Control Number: 2006922553

ISBN: 978-0-387-30342-0

Additionally, the whole set will be available upon completion under ISBN: 978-0-387-35443-9
The electronic version of the whole set will be available under ISBN: 978-0-387-30426-7
The print and electronic bundle of the whole set will be available under ISBN: 978-0-387-35478-1

© 2007 Springer Science+Business Media, LLC.

All rights reserved. This work may not be translated or copied in whole or in part without the written permission
of the publisher (Springer Science+Business Media, LLC., 233 Spring Street, New York, NY 10013, USA),
except for brief excerpts in connection with reviews or scholarly analysis. Use in connection with any form
of information storage and retrieval, electronic adaptation, computer software, or by similar or dissimilar
methodology now known or hereafter developed is forbidden.
The use in this publication of trade names, trademarks, service marks, and similar terms, even if they are not
identified as such, is not to be taken as an expression of opinion as to whether or not they are subject to
proprietary rights.

springer.com

Printed on acid-free paper   SPIN: 11416661 2109 - 5 4 3 2 1 0

Preface

This volume deals with the role of amino acids and small peptides in the central nervous system (CNS). The various chapters describe individually the metabolism and functions of the different compounds. Amino acids are essential building blocks of proteins but they serve also other important functions in the CNS, for instance, as neurotransmitters or neuromodulators in their own right or as precursors for neurotransmitters. The neurotransmitter actions are mostly described in other volumes of this handbook, as the articles in this volume emphasize the other roles of amino acids in brain intermediary metabolism. For instance, glutamate and γ-aminobutyric acid (GABA) constitute together with glutamine and alanine an important metabolic cluster involved in brain ammonia homeostasis. Glycine and serine are closely metabolically intertwined. Histidine is the parent molecule for histamine with less limited distribution in nervous tissue. Aromatic amino acids are likewise neurotransmitter precursors, and their normal concentrations essential for healthy brain development and functions. D-Amino acids are more recently recognized compounds of significance. The precise role of taurine is still somewhat obscure. It is not broken down in the organism but has been postulated to function in cell volume regulation, together with other amino acids and organic and inorganic elements. It has also been suggested to be an inhibitory neuromodulator. Arginine, citrulline, and ornithine are constituents of the urea cycle, which in the liver is important in the elimination of ammonia. However, this cycle is not operating in the brain, and therefore ammonia accumulation is a significant factor in the development of encephalopathy upon failure of liver functions (hepatic encephalopathy). Disturbances in the urea cycle are life threatening for affected individuals. So are also the disturbances in the metabolism of aromatic and branched-chain aliphatic amino acids. Inborn errors in amino acid metabolism are generally not common but they always result in severe diseases. A great number of small-molecular oligopeptides have been detected in brain tissue. Their functions and significance are still largely unknown, except in case of N-acetylaspartate and glutathione.

The editors thank the authors for their sincere efforts to produce individual up-to-date articles in this volume. We think that this volume well covers the most important aspects of metabolism and functions of amino acids and small peptides in the CNS. We hope that the readers will share our opinion.

Simo S. Oja
Pirjo Saransaari
Arne Schousboe

Table of Contents

Preface v

Contributors ix

1 Glutamine, Glutamate, and GABA: Metabolic Aspects 1
    H. S. Waagepetersen · U. Sonnewald · A. Schousboe

2 Serine, Glycine, and Threonine 23
    T. J. de Koning · S. A. Fuchs · L. W. J. Klomp

3 Histidine 47
    P. Panula

4 Aromatic Amino Acids in the Brain 59
    M. Cansev · R. J. Wurtman

5 Arginine, Citrulline, and Ornithine 99
    H. Wiesinger

6 Branched Chain Amino Acids (BCAAs) in Brain 117
    S. M. Hutson · A. J. Sweatt · K. F. LaNoue

7 Sulfur-Containing Amino Acids 133
    G. J. McBean

8 Taurine 155
    S. S. Oja · P. Saransaari

9 Neurobiology of D-Amino Acids 207
    E. Dumin · H. Wolosker

**10 Amino Acids and Brain Volume Regulation: Contribution
     and Mechanisms** 225
    H. Pasantes-Morales

11 Urea Cycle Enzymopathies 249
    R. Butterworth

© Springer-Verlag Berlin Heidelberg 2007

| 12 | **Ammonia Toxicity in the Central Nervous System** 261
J. Albrecht |
|---|---|
| 13 | **Disorders of Amino Acid Metabolism** 277
M. Yudkoff |
| 14 | ***N*-Acetylaspartate and *N*-Acetylaspartylglutamate** 305
M. H. Baslow |
| 15 | **Glutathione in the Nervous System: Roles in Neural Function and Health and Implications for Neurological Disease** 347
R. Janáky · R. Cruz-Aguado · S. S. Oja · C. A. Shaw |
| 16 | **Low Molecular Weight Peptides** 401
K. L. Reichelt |
| | **Index** .... 413 |

Contributors

J. Albrecht
Department of Neurotoxicology, M. Mossakowski Medical Research Centre, Polish Academy of Sciences, Pawińskiego St. 5, 02-106 Warsaw, Poland

M. H. Baslow
Nathan S. Kline Institute for Psychiatric Research, Center for Neurochemistry, 140 Old Orangeburg Road, Orangeburg, New York 10962, USA

R. Butterworth
Neuroscience Research Unit, Hôpital Saint-Luc (CHUM), Montréal, Que, Canada

M. Cansev
Department of Brain and Cognitive Sciences, Massachusetts Institute of Technology, Cambridge MA, 02139, USA and Department of Pharmacology and Clinical Pharmacology, Uludag University Medical School, Gorukle, Bursa, 16059, Turkey

R. Cruz-Aguado
Department of Ophthalmology, Physiology, and Experimental Medicine and Neuroscience Program, University of British Columbia, Vancouver, British Columbia, Canada

T. J. de Koning
Department of Metabolic Diseases and the Laboratory for Metabolic Diseases and Endocrinology, University Medical Centre Utrecht, The Netherlands

E. Dumin
Department of Biochemistry, B. Rappaport Faculty of Medicine, Technion-Israel Institute of Technology, Haifa 31096, Israel

S. A. Fuchs
Department of Metabolic Diseases and the Laboratory for Metabolic Diseases and Endocrinology, University Medical Centre Utrecht, The Netherlands

S. M. Hutson
Wake Forest University School of Medicine, Winston-Salem, NC 27157 and Department of Biochemistry, Wake Forest University School of Medicine, Medical Center Boulevard, Winston-Salem, NC 27157, USA

R. Janáky
Department of Physiology, University of Tampere Medical School, FIN-33014, Tampere, Finland

L. W. J. Klomp
Department of Metabolic Diseases and the Laboratory for Metabolic Diseases and Endocrinology, University Medical Centre Utrecht, The Netherlands

K. F. LaNoue
Penn State College of Medicine, Hershey, PA 17033, USA

G. J. McBean
Department of Biochemistry, Conway Institute of Biomolecular and Biomedical Research, University College Dublin, Belfield, Dublin 4, Ireland

S. S. Oja
Centre for Laboratory Medicine, Tampere University Hospital, Tampere, Finland

P. Panula
Neuroscience Center, Institute of Biomedicine, POB 63 (Haartmaninkatu 8), FI-00014, University of Helsinki, Finland

H. Pasantes-Morales
Department of Biophysics, Institute of Cell Physiology, National University of Mexico, UNAM, Mexico City, Mexico

P. Saransaari
Brain Research Center, Tampere University Medical School, Tampere, Finland

© Springer-Verlag Berlin Heidelberg 2007

A. Schousboe
Department of Pharmacology and Pharmacotherapy,
Danish University of Pharmaceutical Sciences,
DK-2100 Copenhagen, Denmark

C. A. Shaw
Department of Ophthalmology, Physiology,
and Experimental Medicine and Neuroscience
Program, University of British Columbia,
Research Pavilion 828 W. 10th Ave. Vancouver,
British Columbia, V5Z 1L8, Canada

U. Sonnewald
Department of Neuroscience, Norwegian University
of Science and Technology (NTNU),
N-7489 Trondheim, Norway

A. J. Sweatt
Wake Forest University School of Medicine,
Winston-Salem, NC 27157, USA

H.S. Waagepetersen
Department of Pharmacology and Pharmacotherapy,
Danish University of Pharmaceutical Sciences,
DK-2100 Copenhagen, Denmark

H. Wiesinger
Physiologisch-chemisches Institut, Universität
Tübingen, Hoppe-Seyler-Str. 4, D-72076 Tübingen,
Germany

H. Wolosker
Department of Biochemistry, B. Rappaport Faculty of
Medicine, Technion-Israel Institute of Technology,
Haifa 31096, Israel

R. J. Wurtman
Department of Pharmacology and Clinical
Pharmacology, Uludag University Medical School,
Gorukle, Bursa, 16059, Turkey

M. Yudkoff
Department of Pediatrics, University of Pennsylvania
School of Medicine, Children's Hospital of Philadelphia,
34th Street & Civic Center Boulevard, Philadelphia,
PA 19104, USA

1 Glutamine, Glutamate, and GABA: Metabolic Aspects

H. S. Waagepetersen · U. Sonnewald · A. Schousboe

1	Introduction	2
2	Content of Glutamate, Glutamine, and GABA in Different Brain Regions	2
3	Metabolic Compartmentation at the Multicellular Level	3
4	GABA and Glutamate Homeostasis	3
4.1	Glutamate	3
4.2	GABA	4
4.3	GABA–Glutamate–Glutamine Cycle	5
4.3.1	Glutamine	5
4.3.2	Ammonia Homeostasis in the Glutamate–Glutamine Cycle	8
4.4	Alternative Glutamate Precursors	9
4.5	Ammonia Toxicity	9
5	Glutamate-Metabolizing Enzymes	10
5.1	Glutamate Dehydrogenase	10
5.2	Aminotransferases	10
5.3	Aminotransferase Versus GDH	11
6	Anaplerosis	11
6.1	Pyruvate Carboxylase	11
7	Enzymes Involved in Glutamine Metabolism	12
7.1	Glutamine Synthetase	12
7.2	Phosphate-Activated Glutaminase	12
8	Enzymes Involved in GABA Metabolism	13
8.1	Glutamate Decarboxylase	13
8.2	GABA aminotransferase	13
9	Metabolic Compartmentation at the Single-Cell Level	14

© Springer-Verlag Berlin Heidelberg 2007

Abstract: Glutamine synthesized exclusively in astrocytes is the quantitatively most significant precursor for the excitatory and inhibitory neurotransmitters, glutamate and GABA (γ-aminobutyrate), respectively. Glutamate subsequent to its release and receptor interaction is taken up by surrounding cellular elements, predominantly astrocytes. In astrocytes, glutamate is directly converted to glutamine or it enters the tricarboxylic acid (TCA) cycle. Glutamine is transferred to glutamatergic neurons to serve as precursor for glutamate or for TCA cycle intermediates. The latter is important due to the lack of anaplerosis in neurons. GABA is also taken up by surrounding cellular elements following receptor interaction. However, neuronal reuptake seems to be most important in case of GABA. GABA is either reutilized as neurotransmitter or metabolized via the GABA shunt. To the extent that GABA is lost to the astrocytic compartment, glutamine serves as the precursor for its de novo synthesis. This occurs by deamidation to glutamate and further decarboxylation to GABA. This chapter describes the neuronal and astrocytic enzymatic machineries active in sustaining GABA and glutamate homeostasis involving glutamine and ammonia. The transporters that are also important participants in this scenario are briefly mentioned and for a more detailed description the reader is referred to other chapters.

List of Abbreviations: AAT, aspartate aminotransferase; ALAT, alanine aminotransferase; BCAT, branched-chain aminotransferase; GABA, γ-aminobutyrate; GABA-T, GABA aminotransferase; GAD, l-glutamate decarboxylase; GDH, glutamate dehydrogenase; GS, glutamine synthetase; PAG, phosphate-activated glutaminase; PC, pyruvate carboxylase; PLP, pyridoxal phosphate; TCA, tricarboxylic acid

1 Introduction

Glutamate, the major excitatory, and γ-aminobutyrate (GABA), the major inhibitory neurotransmitter in the central nervous system (CNS), only differ by a single carboxyl group. The mechanisms involved in the homeostasis of glutamate and GABA, i.e., synthesis, degradation, and transport, are all instrumental in the maintenance of the balance between excitation and inhibition which, in turn, is essential for brain function. The two amino acids are metabolically closely related since GABA is synthesized directly from glutamate, and the amino group of GABA is transaminated into glutamate in GABA degradation. Glutamate and GABA metabolism is intimately associated with energy metabolism via the tricarboxylic acid (TCA) cycle, and glutamate has a prominent role in general amino acid metabolism. Both transmitters are important for cellular differentiation in the developing brain (Meier et al., 1984; Belhage et al., 1990; Hack and Balazs, 1994). An interesting and complicating feature of brain amino acid metabolism is the lack of anaplerosis in neurons, making them incapable of de novo synthesis of glutamate and GABA from glucose. This together with the fact that the majority of glutamatergic and a part of GABAergic neurotransmission are terminated by astrocytic uptake of the neurotransmitters result in the obligatory involvement of glutamine as a key player. Glutamine is exclusively synthesized in astrocytes (Norenberg and Martinez-Hernandez, 1979) and since it is a non-neuroactive amino acid, it can be transferred from astrocytes to neurons, and thus function as precursor for synthesis of neurotransmitter glutamate and GABA without interfering with neuronal signaling.

2 Content of Glutamate, Glutamine, and GABA in Different Brain Regions

The brain contains a very high concentration of glutamate ranging from 80 to 100 nmol/mg protein (i.e., approximately 10 mM) depending on the region (Erecinska and Silver, 1990). The concentration of glutamate has been estimated to be 1–10 µM in the CSF and 0.5–4 µM in the extracellular fluid of the brain. The concentration in the synaptic cleft varies from 2–1000 µM with the extent of neuronal activity. In contrast to these low concentrations of glutamate in the extracellular space, the concentration of glutamate is thought to be approximately 100 mM within the vesicles (Meldrum, 2000; Danbolt, 2001). The concentration of aspartate ranges from 20 to 30 nmol/mg protein whereas that of glutamine ranges from 50 to 60 nmol/mg protein depending on the brain region. The level of GABA is approximately 10–30 nmol/mg protein (Erecinska and Silver, 1990).

3 Metabolic Compartmentation at the Multicellular Level

Berl and Clarke (1969) defined metabolic compartmentation as the presence in a tissue of more than one distinct pool of a given metabolite. These separate pools of a metabolite are not in rapid equilibrium with each other but maintain their own integrity and turnover rates. Compartmentation of metabolites and cellular functions exists at different levels in an organism. It seems self-evident that different organs are optimized for different functions. Within one organ, distinct cell types may have their own properties and characteristics. Berl and coworkers (1961) demonstrated that intracerebral injection of radioactive glutamate leads to a higher specific radioactivity of glutamine than that of glutamate. This finding was the first indication of metabolic compartmentation in the brain. The product, in this case glutamine, having a higher specific radioactivity than its precursor, in this case glutamate, is a classical indication of compartmentation. A small glutamate pool that is precursor for glutamine and a large glutamate pool with a slow turnover characterized the two compartments of glutamate in the brain (van den Berg and Garfinkel, 1971). Injection of several substrates confirmed compartmentalized glutamate metabolism in the brain. Radioactively labeled glucose, pyruvate, and lactate injected into the brain predominantly labeled the large glutamate pool, since the specific radioactivity of glutamine was found to be lower than that of glutamate (Gaitonde, 1965; O'Neal and Koeppe, 1966; van den Berg et al., 1969; Minchin and Beart, 1975; Berl and Clarke, 1983). However, acetate and bicarbonate preferentially labeled glutamine over glutamate, showing that these substrates mainly entered the compartment of the small glutamate pool (Berl et al., 1962; Garfinkel, 1966; Balazs et al., 1970; Minchin and Beart, 1975; Berl and Clarke, 1983). The observation by Norenberg and Martinez-Hernandez (1979), that glutamine synthetase (GS) (EC 6.3.1.2) exclusively is located in astrocytes, was instrumental for the elucidation of the results obtained with radioactively labeled precursors. The origin of the two glutamate compartments was thereby defined, i.e., the neuronal compartment contains the large glutamate pool with a slow turnover and the glial compartment contains the small glutamate pool, which is precursor for glutamine. In addition to the large amount of glutamate in the brain, aspartate which is metabolically closely related to glutamate is also present in a significant amount; aspartate is assigned to the neuronal compartment (Gundersen et al., 1998; Oz et al., 2004). Considering our current knowledge, the two-compartment model of the brain is an oversimplification since the neuronal as well as the glial compartment consist of heterogeneous populations of cells. Moreover, evidence of intracellular metabolic compartmentation, which partly might be the consequence of the dynamics of multienzyme complexes and mitochondrial heterogeneity, complicates the metabolic map of the brain even more (McKenna et al., 1996a, 2000a, 2006a; Waagepetersen et al., 2001a, 2006). For further discussion of intracellular compartmentation see below.

4 GABA and Glutamate Homeostasis

4.1 Glutamate

Glutamate is an unbranched five-carbon α-amino acid having a variety of functions in general metabolism in addition to being the key neurotransmitter of excitatory signaling in the brain. On top of this, glutamate is excitotoxic and it is essential to keep the extracellular concentration below a critical concentration in the low micromolar range (Danbolt, 2001; Schousboe and Waagepetersen, 2004). Glutamate is an important intermediate in energy metabolism, being a link between the TCA cycle and amino acid synthesis (Erecinska and Silver, 1990). Glutamate is also part of the malate–aspartate shuttle transferring reducing equivalents from the cytosol to the mitochondria for reoxidation and for generation of ATP in the electron transport chain (McKenna et al., 2006b). The malate–aspartate shuttle is of particular importance in a highly aerobic organ as the brain, which uses nearly exclusively glucose as energy substrate. Interestingly, the glutamate–aspartate carrier (ARALAR1) might be absent from astrocytes (Ramos et al., 2003). Glutamate is synthesized from α-ketoglutarate via transamination or reductive amination, the latter catalyzed by glutamate dehydrogenase (GDH) (EC 1.4.1.3). The transaminases of importance for maintenance of glutamate homeostasis in the brain are mainly aspartate aminotransferase (AAT) (EC 2.6.1.1), branched-chain aminotransferase (BCAT) (EC 2.6.1.42), and alanine aminotransferase (ALAT) (EC 2.6.1.2). Glutamate is a precursor for GABA, glutamine, proline, and arginine and is also involved in the degradation of the carbon skeletons of these amino acids as well as histidine, before conversion to α-ketoglutarate and oxidative degradation in the TCA cycle. The synthesis and degradation pathways of GABA and glutamine are discussed in detail below. Glutamate is also part of the peptide glutathione, which additionally consists of one molecule of cysteine and one of glycine (*see Chapter* by McBean, this volume). Glutathione functions as a major intracellular antioxidant in the brain and is synthesized in both astrocytes and neurons, in the latter preferentially by a supply of the precursor CysGly from astrocytes (Dringen et al., 1999). The content of glutathione varies greatly in different brain regions, and the distribution among neurons and astrocytes differs also within these regions (Langeveld et al., 1996).

Excitatory glutamatergic neurotransmission involves filling of vesicles in the nerve endings by the vesicular glutamate transporters and exocytotic release of glutamate from the presynaptic neuron (Fremeau Jr et al., 2001). Astrocytes might also have the capability of releasing glutamate from vesicles in an exocytotic manner (Parpura et al., 2004). Glutamate activates a variety of glutamate receptors, ionotropic as well as metabotropic, in the synaptic cleft including the astrocytic entity (Hansson and Rönnbäck, 2004; Kew and Kemp, 2005). This activation may be terminated by desensitization of the receptor and/or binding to and uptake via high-affinity glutamate transporters present in the plasma membrane of surrounding astrocytes as well as of the pre- or postsynaptic neurons (Danbolt, 2001; Schousboe et al., 2004). The astrocytic glutamate transporters seem to constitute the quantitatively most important uptake site in the synaptic cleft (Gegelashvili and Schousboe, 1997, 1998; Danbolt, 2001). This might, however, vary between different neuronal phenotypes. Thus, it should be noted that in cultured glutamatergic cerebellar neurons glutamate uptake appears to be important (Waagepetersen et al., 2005; Olstad et al., 2006). Five different glutamate transporters have been cloned namely EAAT1/GLAST, EAAT2/GLT-1, EAAT3/EAAC1, EAAT4, and EAAT5 (Wadiche et al., 1995; Lehre and Danbolt, 1998; Levy, 2002).

4.2 GABA

GABA is the most abundant inhibitory neurotransmitter in the CNS. It was not until the 1960s that GABA was suggested to be an inhibitory neurotransmitter in the CNS (Curtis and Watkins, 1960). GABA is also present in other organs than the brain, for example, in pancreatic β-cells where it appears to act as a signaling molecule between cells within the pancreatic islets. GABA functions in the reproductive system by regulating release of estradiol and progesterone from the ovaries and spermatogenesis in the testes (for review on the peripheral GABAergic system, see Gladkevich et al., 2006). L-Glutamate decarboxylase (GAD) (EC 4.1.1.15) is the enzyme catalyzing the biosynthesis of GABA from glutamate (❯ *Figure 1-1*). It has been suggested that during early postnatal development, GABA might be synthesized through an alternative pathway involving ornithine (Seiler, 1980). Classically, GAD has been considered to be a marker for GABAergic neurons in the CNS. However, recent studies have indicated that GAD might be present in at least a subpopulation of excitatory neurons as well (Schwarzer and Sperk, 1995; Sloviter et al., 1996). This localization in excitatory neurons might reflect the possible role of GABA as a neurodifferentiative agent during early postnatal development in the CNS (Belhage et al., 1998).

GABA acts in an inhibitory fashion via postsynaptically located GABA$_A$ or GABA$_B$ receptors. GABA$_A$ receptor activation causes an influx of Cl$^-$, whereas stimulation of GABA$_B$ receptors causes an efflux of K$^+$. Consequently, both receptors induce a hyperpolarization of the plasma membrane. GABA$_B$ receptors are also located presynaptically, regulating Ca^{2+} channels, and thereby causing inhibition of transmitter release (Olsen and Betz, 2006). Termination of the action of GABA in the synaptic cleft is mediated by high-affinity GABA uptake carriers. These carriers are localized on presynaptic GABAergic neurons and on surrounding astrocytes, and the prevalent view is that the majority of transmitter GABA is taken up back into the presynaptic neuron (❯ *Figure 1-1*) (Krogsgaard-Larsen et al., 1987; Borden, 1996). However, the fraction of GABA taken up by the presynaptic neuron compared with that taken up by surrounding astrocytes has only been estimated from indirect measurements. Using nuclear magnetic resonance spectroscopy (NMRS), it has been estimated that the GABA–glutamine–glutamate cycle accounts for approximately 23% of the total cycling of glutamine from astrocytes to neurons (Patel et al., 2005b). GABA is metabolized by GABA

Figure 1-1

Key metabolic processes and release and uptake of neurotransmitters in glutamatergic and GABAergic synapses interacting with a surrounding astrocyte. Glutamate (GLU) is released from the glutamatergic neuron, and subsequently, predominantly taken up into the astrocyte. The quantitative relationship between neuronal and astrocytic uptake is indicated by the thickness of the *arrows*. In the neuronal compartment, glutamate may reenter the vesicular pool or enter the TCA cycle via α-ketoglutarate (α-KG), catalyzed by either glutamate dehydrogenase (GDH) or aspartate aminotransferase (AAT). In the astrocyte, glutamate is amidated to glutamine (GLN) via glutamine synthetase (GS) or enters the TCA cycle, as presented in detail in ❯ *Figure 1-2*. Glutamine is released and acts as precursor for glutamate in the glutamatergic neuron. GABA is released from the GABAergic neuron, the transmission predominantly terminated by presynaptic uptake. In the nerve ending, GABA may reenter the vesicular pool or be metabolized in the mitochondria to succinate (Succ) via succinate semialdehyde (SSA), catalyzed by GABA transaminase (GABA-T) and succinate semialdehyde dehydrogenase (SSADH). The degradation pathway also occurs with regard to the fraction of GABA taken up by the astrocyte. α-Ketoglutarate is generated after successive steps in the TCA cycle and may eventually in the astrocyte be the precursor for glutamine, which is released and may be taken up by the GABAergic neurons, being deamidated to glutamate via glutaminase (PAG) and subsequently either directly or after metabolism in the TCA cycle glutamate acts as immediate precursor for GABA catalyzed by glutamate decarboxylase (GAD)

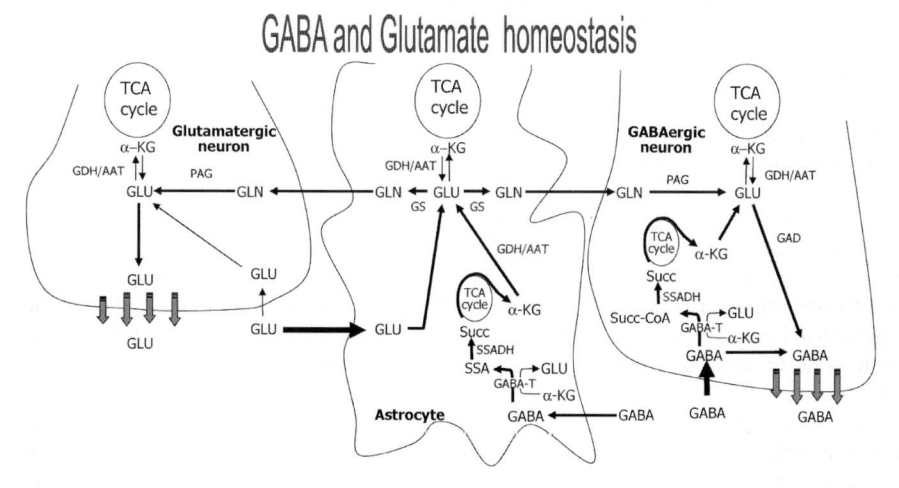

aminotransferase (GABA-T) (EC 2.6.1.19) in the GABA shunt (Balazs et al., 1970), illustrated in ❯ *Figure 1-1*. The GABA shunt consists of both the synthetic and the degradative machinery for GABA involving GAD and GABA-T, respectively. Large differences ranging from 2% to 17% exist in the reported fluxes through the GABA shunt compared with TCA cycle flux, with the highest value obtained from analysis of mouse brains (Balazs et al., 1970; Machiyama et al., 1970; Martin and Rimvall, 1993; Hassel et al., 1998). In contrast to GAD, GABA-T, which is localized in the mitochondrial matrix (Schousboe et al., 1977b; Chan-Palay et al., 1979), is an enzyme ubiquitously expressed not only in the brain but also in other organs (Roberts and Bregoff, 1953). For detailed description of the enzymes GAD and GABA-T, see below.

4.3 GABA–Glutamate–Glutamine Cycle

4.3.1 Glutamine

A flow of glutamate occurs from the neuronal to the astrocytic compartment during glutamatergic neurotransmission (❯ *Figure 1-1*). This flow needs to be compensated for by a flow of a glutamate precursor in the opposite direction owing to the lack of a neuronal anaplerotic pathway. Although it has recently been shown that carboxylation of pyruvate, presumably via mitochondrial malic enzyme, can occur in cultured cerebellar granule cells (Hassel and Bråthe, 2000), the extent to which this occurs at physiological concentrations of pyruvate is disputable (Lieth et al., 2001). The closely related nonneuroactive amino acid, glutamine, has been shown to be the predominant glutamate precursor, being rapidly and extensively metabolized to glutamate in neurons (Berl and Clark, 1969; Bradford et al., 1978; Rothstein and Tabakoff, 1984; Szerb and O'Regan, 1985; Shank et al., 1989; Waagepetersen et al., 2005). Glutamine is exclusively synthesized in astrocytes by an ATP-dependent amidation of glutamate, catalyzed by the cytosolic enzyme GS (Norenberg and Martinez-Hernandez, 1979). Thus, the net flow of glutamate from neurons to astrocytes is counteracted by a glutamine flow from astrocytes to neurons, a concept referred to as the glutamate–glutamine cycle (❷ *Figure 1-1*) (Berl and Clarke, 1983). Efflux of glutamine from astrocytes has been suggested to be mediated by system N, which is a Na^+–glutamine symporter and a H^+ antiporter (Broer and Brookes, 2001; Chaudhry et al., 2002). With regard to neuronal uptake of glutamine, GlnT or SNAT1 has been suggested to fulfill this role. It is a high-affinity Na^+–glutamine symporter and it was found to be preferentially located on glutamatergic neurons (Varoqui et al., 2000), supporting the importance of glutamine as a glutamate precursor. The metabolic flux of the glutamate–glutamine cycle has been estimated from 0.01 µmol/g/min in pentobarbital-treated rats to 0.51 µmol/g/min and 0.3 µmol/g/min in the awake rat and in the awake human brain, respectively (Gruetter et al., 1998; Oz et al., 2004; Hyder et al., 2006). Glutamine also functions as a precursor for the inhibitory neurotransmitter GABA via glutamate (Reubi et al., 1978; Battaglioli and Martin, 1990, 1991; Sonnewald et al., 1993c; Westergaard et al., 1995), thereby expanding the glutamate–glutamine cycle to a GABA–glutamate–glutamine cycle (van den Berg and Garfinkel, 1971; Berl and Clarke, 1983), as illustrated in ❷ *Figure 1-1*. The metabolic flux of the GABA–glutamate–glutamine cycle has been estimated to 0.14 µmol/g/min in the anesthetized rat cerebral cortex (Hyder et al., 2006). The GABA–glutamine–glutamate cycle and the glucose oxidation rate of GABAergic neurons increase with cortical activity, although to a smaller extent, compared with that observed for glutamatergic neurons and the rate of the glutamate–glutamine cycle (Patel et al., 2005b).

The stoichiometry of the glutamate–glutamine cycle is hampered by the considerable extent of oxidative metabolism of glutamate in astrocytes, and that of glutamine in neurons (Yu et al., 1984; Yudkoff et al., 1988; Hertz et al., 1992; Westergaard et al., 1995). Entrance of the amino acid carbon skeletons into the TCA cycle can accelerate metabolism of acetyl CoA or be the first step in the total degradation of the carbon skeleton via pyruvate recycling, both events ensuing an elevated energy production. The hydrolysis of glutamine to glutamate is catalyzed by phosphate-activated glutaminase (PAG) (EC 3.5.1.2) located in the inner mitochondrial membrane (Kvamme et al., 1988; Laake et al., 1999), a location which might accelerate the oxidative consumption of glutamine. GABA synthesis from glutamine has been shown to involve TCA cycle metabolism by entrance of the carbon skeleton of glutamine into the TCA cycle before reformation of glutamate from α-ketoglutarate and eventually GABA synthesis, a pathway which seems to be more important for synthesis of GABA during vesicular release compared with release from the cytoplasmic pool (Waagepetersen et al., 1999a, 2001b). As seen for glutamine in the neuronal compartment, glutamate, in addition to being amidated to glutamine, is oxidatively metabolized in astrocytes (Yu et al., 1982; Qu et al., 2001). To what extent glutamate enters the TCA cycle compared with the amount converted to glutamine is dependent on the extracellular concentration of glutamate, the amount entering the TCA cycle being the concentration-dependent pathway (McKenna et al., 1996b). Lactate is formed from glutamate via the TCA cycle and malic enzyme (ME) (EC 1.1.1.39) or phosphoenolpyruvate carboxykinase (PEPCK) (EC 4.1.1.49) in astrocytes (Sonnewald et al., 1993b). The amount of lactate generated via these pathways is, however, quantitatively negligible compared with that obtained from glucose via glycolysis (McKenna et al., 1996b).

Cultured neurons as well as synaptosomes synthesize considerable amounts of aspartate from exogenously added glutamine via a "truncated" TCA cycle producing 9–12 molecules of ATP. This pathway is upregulated during hypoglycemic conditions (Hertz et al., 1992; Yudkoff et al., 1994; Waagepetersen et al., 2005). In neurons and synaptosomes in which PAG is predominantly localized, glutamine seems to be a superior energy substrate compared to glutamate (Westergaard et al., 1995). In astrocytes, on the other hand, glutamate seems to be the preferred precursor in contrast with glutamine for net synthesis of aspartate via a truncated TCA cycle, a process which is also upregulated during hypoglycemic conditions (Bakken et al., 1998).

As mentioned above, complete oxidative degradation of the glutamate or glutamine carbon skeleton requires pyruvate recycling, i.e., entrance of the amino acid carbon skeleton into the TCA cycle and exit via decarboxylation of malate to pyruvate catalyzed by malic enzyme or the concerted action of phosphoenolpyruvate carboxykinase and pyruvate kinase converting oxaloacetate into pyruvate and subsequent oxidative decarboxylation to acetyl CoA before re-entrance into the TCA cycle (❯ *Figure 1-2*). Pyruvate recycling was observed in the brain by ex vivo NMRS and found to be a neuronal phenomenon since the characteristic labeling pattern of pyruvate recycling was observable in glutamate and not in glutamine (Cerdan et al., 1990). Shortly after, several investigations using cultured brain cells have indicated that the primary localization of this pathway is in astrocytes (Sonnewald et al., 1996; Bakken et al., 1997; Alves et al., 2000; Waagepetersen et al., 2002). One of the key enzymes in this pathway, malic enzyme, is differentially located in neurons and astrocytes, i.e., the mitochondrial form is present primarily in neurons whereas the cytosolic form is in astrocytes (Vogel et al., 1998a, b). This might be one of the clues to the elucidation of the localization of this pathway; however, diverging results exist regarding the importance of the cytosolic contra

◘ Figure 1-2

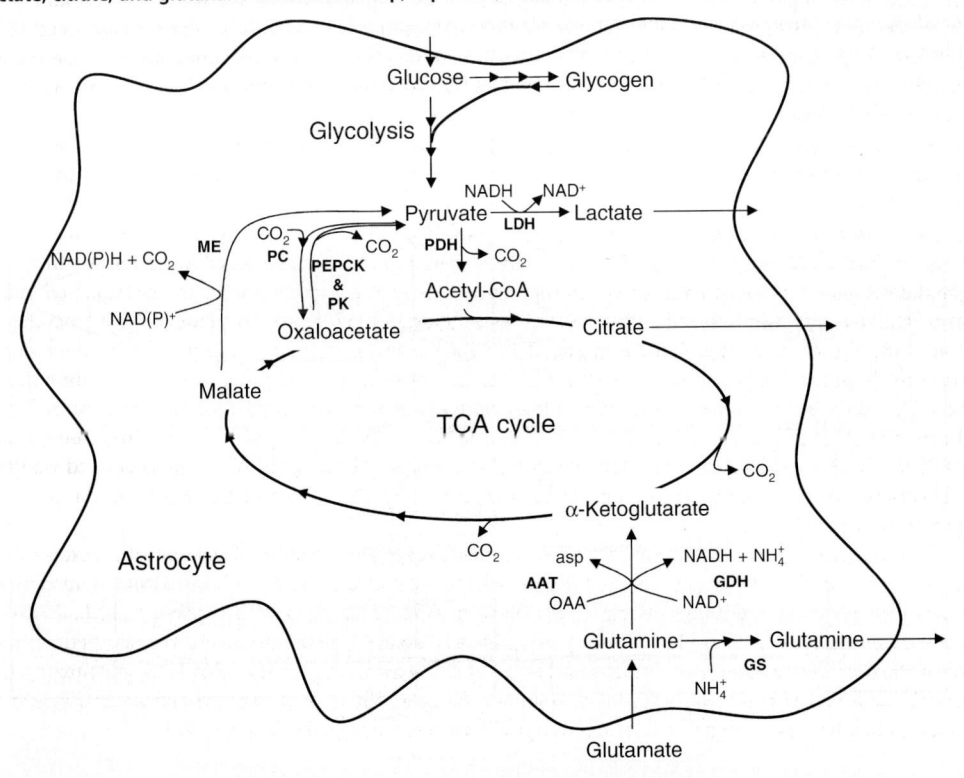

Metabolism of glutamate and glucose in astrocytes. Glucose is metabolized to pyruvate via glycolysis or used in the synthesis of glycogen. Pyruvate may either be oxidatively decarboxylated to acetyl CoA catalyzed by pyruvate dehydrogenase (PDH) for oxidation in the TCA cycle, carboxylated to oxaloacetate by pyruvate carboxylase (PC), or be reduced to lactate by the operation of lactate dehydrogenase (LDH). Pyruvate recycling is the conversion of malate to pyruvate via malic enzyme (ME) and subsequent oxidation of pyruvate in the TCA cycle. Alternatively, pyruvate recycling is initiated by the concerted action of phosphoenolpyruvate carboxykinase (PEPCK) and pyruvate kinase (PK) converting oxaloacetate to pyruvate. The operation of ME or PEPCK plus PK in the direction of pyruvate is necessary for complete oxidation of glutamate as well as glutamine. Analogously, the operation of PC is required to obtain net formation of glutamate and glutamine. Glutamate may either be metabolized in the TCA cycle after conversion to α-ketoglutarate by aspartate aminotransferase (AAT) or glutamate dehydrogenase (GDH) or be amidated to glutamine by glutamine synthetase (GS). Release of lactate, citrate, and glutamine is indicated. Asp, aspartate; OAA, oxaloacetate the mitochondrial form (Vogel et al., 1998a; McKenna et al., 2000b and references therein). The extent to which the pyruvate recycling pathway occurs in vivo has not been completely elucidated, and also the cellular localization is at present still unclear. The results of investigations employing ^{13}C-labeled isotopes in order to obtain a quantitative measurement of pyruvate recycling are complicated by a considerable scrambling of labeling because of the reversible steps in the TCA cycle between oxaloacetate and the symmetrical molecule succinate. It might seem somewhat puzzling that the extent of pyruvate recycling does not parallel the extent of pyruvate carboxylation; however, efflux of glutamine from the brain might to some extent account for the anaplerosis. The rate of pyruvate carboxylation occurring in glia seems to be activity dependent since it was significantly higher in awake rats compared with rats under deep pentobarbital anesthesia (Choi et al., 2002; Oz et al., 2004). However, bicuculline-induced seizures had no accelerating effect on the pyruvate carboxylase (PC) pathway in halothane-anesthetized rats, a treatment leading to increased activity of the glutamate–glutamine cycle (Patel et al., 2004, 2005a). The rate of PC has been estimated to be 28% of the glial TCA cycle rate and the glutamate–glutamine cycle rate, both estimated to be 0.5 μmol/g/min (Oz et al., 2004).

4.3.2 Ammonia Homeostasis in the Glutamate–Glutamine Cycle

Ammonia is an important component of the glutamate–glutamine cycle. It is formed in the neuronal compartment by hydrolysis of glutamine and utilized in the synthesis of glutamine in the astrocytic compartment (▶ *Figure 1-3*). This flux of ammonia via glutamine from the astrocytic to the neuronal

Figure 1-3
Ammonia homeostasis associated with the glutamate (GLU)–glutamine (GLN) cycle between glutamatergic neurons and astrocytes. The transfer of glutamate from the neuronal to the astrocytic compartment following glutamatergic transmission is accompanied by a flow of glutamine from the astrocyte to the neuron, creating a net flow of nitrogen from the astrocytes to the neurons. This flow has been suggested to be counteracted either by (A) diffusion of ammonia or (B) by transport of an amino acid (AA). In case of the latter, the ammonia generated in the glutaminase (PAG) reaction has to be fixed in glutamate by the functioning of glutamate dehydrogenase (GDH) and subsequent transamination (AT) with a keto acid (KA), which is suggested to be either pyruvate or one of the branched-chain keto acids. In the astrocyte, the concerted action of a transaminase and an oxidative deamination liberates ammonia for the glutamine synthetase (GS) reaction. Abbreviations: α-ketoglutarate (α-KG)

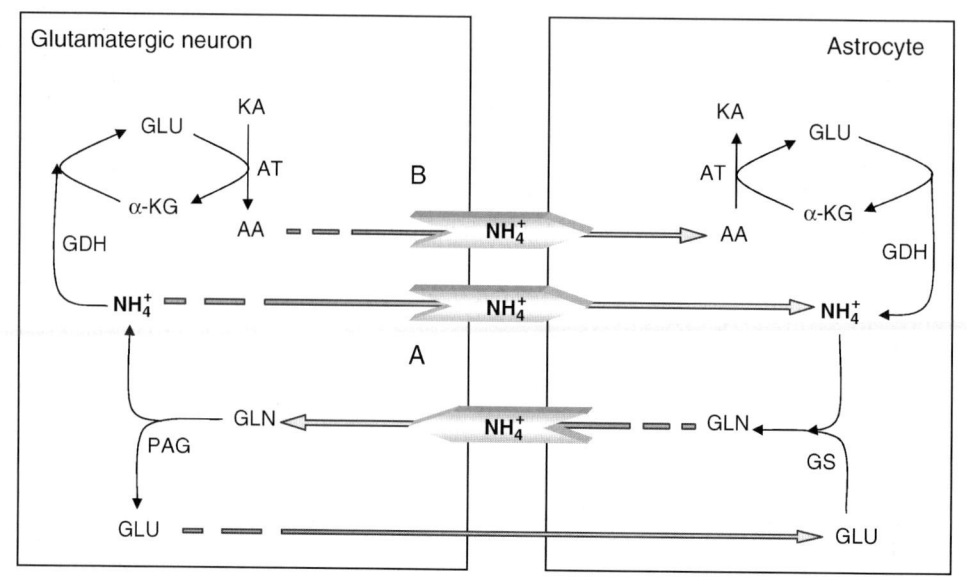

compartment has been suggested to be counteracted by transfer of ammonia from the neuronal to the astrocytic compartment either by diffusion of ammonia, transport of ammonia by K^+/Cl^- cotransporters, or by an amino acid (Yudkoff et al., 1996; Waagepetersen et al., 2000; Lieth et al., 2001; Marcaggi and Coles, 2001). If ammonia is to be carried by an amino acid, it has to be fixed in glutamate via the action of GDH subsequent to the deamidation of glutamine. Glutamate may then be transaminated, transferring the amino group to a nonneuroactive amino acid, which has been suggested to be either alanine or one of the BCAAs, via ALAT or BCAT, respectively. These amino acids are transferred to the astrocytic compartment in which ammonia is liberated via oxidative deamination of glutamate in the GDH reaction subsequent to the action of ALAT or BCAT. Compensation for the loss of carbon skeletons from the neuron has been proposed to be achieved by transfer of lactate or one of the branched-chain ketoacids from the astrocytes (Yudkoff et al., 1996; Waagepetersen et al., 2000; Lieth et al., 2001). The importance of these shuttles in sustaining ammonia homeostasis related to glutamatergic neurotransmission is at present not clear; however, no activity-dependent increase of alanine shuttling could be observed when glutamatergic neurotransmission was induced in cerebellar cultures (Bak et al., 2005). The major obstacle for the proof of the operation of these shuttles is the lack of evidence for the functioning of the GDH reaction in the direction of reductive amination. The thermodynamic equilibrium constant of the GDH reaction favors reductive amination but the K_m value of ammonia is in the range of 10–28 mM (Chee et al., 1979; Colon et al., 1986; Filla et al., 1986), which probably under normal physiological conditions (0.1–0.2 mM ammonia) will direct the reaction toward formation of α-ketoglutarate. For comparison, the K_m value of ammonia in the reaction of GS has been reported to be 180 μM (Cooper and Lai, 1987). Also the ratio of $NAD^+/NADH$ influences deamination of glutamate to α-ketoglutarate in which a simultaneous reduction of NAD^+ to NADH occurs.

4.4 Alternative Glutamate Precursors

The glutamate–glutamine cycle does not operate in a stoichiometric fashion, which appears to necessitate compensating pathways or substrates such as TCA cycle intermediates. Alternatively, de novo synthesis of glutamine via pyruvate carboxylation in the astrocytes may provide the necessary amount of glutamate precursor (see above). α-Ketoglutarate may to some extent serve as an additional precursor for the releasable neurotransmitter pool of glutamate (Kihara and Kubo, 1989; Shank et al., 1989; Peng et al., 1991). Utilization of TCA cycle intermediates for synthesis of neurotransmitter glutamate necessitates an amino donor, and the nitrogen of alanine has been shown to be incorporated into glutamate of cultured cortical neurons and cerebellar granule neurons (Yudkoff et al., 1990; Peng et al., 1991). α-Ketoglutarate as well as citrate, malate, and succinate are all released to a larger extent from astrocytes than from neurons, with citrate being released much more intensely than the others (Sonnewald et al., 1991; Westergaard et al., 1994a, b). The rate of release is dependent on the concentration of bicarbonate, suggesting that pyruvate carboxylation plays an important role in the synthesis of releasable TCA cycle intermediates (Westergaard et al., 1994a; Waagepetersen et al., 2001a). Investigating the neuronal compartment, however, has not provided compelling evidence that citrate, malate, or succinate has a significant supplementary role as glutamate precursor (Shank and Campbell, 1984; Hertz et al., 1992; Westergaard et al., 1994a, b). None of the mentioned TCA cycle intermediates have been found to serve as a GABA precursor either (Hertz et al., 1992). This finding might be in concord with a much lower quantitative demand for precursors in GABAergic neurons, since the net flow of glutamate to astrocytes from neurons is noticeably higher than that of GABA (Hertz and Schousboe, 1987), illustrated in ❷ *Figure 1-1*.

4.5 Ammonia Toxicity

An elevated ammonia concentration is toxic for astrocytes as well as neurons, although astrocytes have been suggested to be the primary target. The exact mechanisms behind ammonia toxicity are at present not understood. However, ammonia interferes with energy metabolism, including inhibition of the pyruvate and α-ketoglutarate dehydrogenase complexes as well as the malate–aspartate shuttle (Cooper and Lai, 1987). Moreover, acute exposure of cultured astrocytes to ammonia (5 mM) has been shown to induce transient increase in intracellular calcium inducing glutamate release, the underlying mechanism being alkalinization of the cytosol (Rose et al., 2005). Because of the deleterious effects of high ammonia, it is essential to keep brain ammonia concentration at a low level. Ammonia removal in the brain depends almost exclusively on glutamine synthesis since the brain lacks two enzymes of the urea cycle namely carbamoyl-phosphate synthase I and ornithine transcarbamylase. Injection of ammonium chloride into the brain or the blood efficiently labels the amide position of glutamine whereas the amino position is labeled at a much lower rate (Cooper et al., 1979). The preferential labeling of the amide position shows the importance of the GS reaction in the removal of ammonia compared to the reductive amination of α-ketoglutarate catalyzed by GDH (Duffy et al., 1983; Cooper and Plum, 1987). The brain cannot continuously remove ammonia via the synthesis of glutamine without a complementary anaplerosis via pyruvate carboxylation, a pathway which is upregulated during hyperammonemia (Sibson et al., 2001).

5 Glutamate-Metabolizing Enzymes

5.1 Glutamate Dehydrogenase

GDH and several different aminotransferases, such as AAT and ALAT, are capable of catalyzing the reversible conversion of glutamate to its corresponding α-ketoacid, the TCA cycle intermediate α-ketoglutarate. The action of GDH in the direction of α-ketoglutarate does provide an immediate net synthesis of TCA cycle intermediates that is in contrast to the action of AAT. GDH is a strictly mitochondrial enzyme, presumably located in both the inner membrane and in the matrix, located predominantly in astrocytes in the brain; however, the neuronal compartment is not totally devoid of activity (Aoki et al., 1987; Rothe et al., 1994, Plaitakis and Zaganas, 2001 and references therein). GDH is a highly regulated enzyme that catalyzes the oxidative deamination of glutamate into α-ketoglutarate using either NAD^+ or $NADP^+$, and this process therefore is energy producing (Chee et al., 1979; Colon et al., 1986; Filla et al., 1986). In humans, GDH exists as two different isoforms encoded for by two separate genes, GLUD1 being the housekeeping gene and GLUD2 the nervous tissue-specific gene (Plaitakis and Zaganas, 2001). Both isoforms are allosterically activated by ADP in the range of 0.1–1 mM and by L-leucine in the range of 1.0–10 mM. The activating effect is tenfold higher for the nervous tissue-specific isoform than for the housekeeping form. The two activators act synergistically in certain low concentration ranges. It has been suggested that the nervous tissue-specific form of GDH is only active in the presence of ADP (Plaitakis et al., 2003). GTP acts inhibitory only on the housekeeping isoform with an IC_{50} of 0.2 mM. Surprisingly, a genetic basis of two different GLUD genes has not been found in any other mammalian species than humans. However, Cho and coworkers (1995) found two soluble fractions from bovine brain containing distinct forms of GDH, and Colon and coworkers (1986) have characterized two different portions of GDH: one solubilized and one particulate bound, the latter two differing in their allosterical regulation and heat stability.

5.2 Aminotransferases

The aminotransferases, which are pyridoxal phosphate (PLP)-dependent enzymes, transfer the amino group of glutamate to an amino group accepting α-ketoacid generating α-ketoglutarate and a new α-amino acid. The aminotransferases with highest activity in the brain are AAT, ALAT, GABA-T, and BCAT. AAT, BCAT, and ALAT exist in a cytosolic as well as a mitochondrial form. In case of BCAT, the cytosolic form is restricted to the neuronal compartment and the mitochondrial to the astrocytes (Fonnum 1968; Hutson, 2001). AAT operates in the brain at a very high rate, which is several-fold higher than the flux through the TCA cycle (Mason et al., 1992; Yudkoff et al., 1994) and at near equilibrium in both nerve endings and whole brain. In contrast, the activity of ALAT in the brain, being 10–20 times lower than in the liver, is too low to maintain a near-equilibrium situation under in vivo conditions (Erecinska and Silver, 1990).

The rate of ALAT in brain homogenate has been determined to 13 nmol/min/mg protein, and it differs 2.5-fold among different brain regions (Erecinska and Silver, 1990). The two isoforms of ALAT are very similar, a sixfold higher K_m for alanine of the mitochondrial form being the only difference (Erecinska and Silver, 1990). There is highly diverging information in the literature concerning the proportion of AAT activity comprised by the mitochondrial and cytosolic forms (for references see Erecinska and Silver, 1990). The K_m of α-ketoglutarate is tenfold higher for the mitochondrial form compared with the cytosolic form, directing the reaction toward aspartate formation in the mitochondria (Erecinska and Silver, 1990). This supports the action of a truncated TCA cycle operating from glutamate to aspartate via the TCA cycle, leading to energy production even during limited availability of acetyl CoA (see below). The activity of AAT differs among different cells and cellular compartments; however, it is impossible to conclude from enzyme activity measurements and immunocytochemistry whether cellular specificities exist with regard to this enzyme (Erecinska et al., 1993). The two isoforms of AAT together with the cytosolic and mitochondrial isoforms of malate dehydrogenase constitute the enzymes of the malate–aspartate shuttle (see above). It should be noted that the action of AAT alone does not provide an immediate net synthesis of TCA cycle intermediates. This notion is important since neurons are generally considered to lack an anaplerotic pathway, providing a net synthesis of TCA cycle intermediates and subsequently amino acids from glucose (Yu et al., 1983).

5.3 Aminotransferase Versus GDH

Whether exogenous glutamate and endogenous glutamate generated from glutamine enter the TCA cycle via the action of GDH or a transaminase seems unclear. The transaminase involved, preferentially, is most likely AAT since this transaminase is the one with the highest activity in brain tissue (Cooper, 1988). McKenna and coworkers (1993, 1996b) demonstrated in a synaptosomal preparation that the oxidative metabolism of endogenous glutamate was inhibited only to a limited extent by aminooxyacetic acid (AOAA), a transaminase inhibitor. This finding indicates that glutamate is converted to α-ketoglutarate by GDH. However, glutamate release from cultured glutamatergic neurons was inhibited by AOAA, indicating the importance of AAT in the synthesis of neurotransmitter glutamate (Palaiologos et al., 1989; Peng et al., 1991). Results obtained from cultured astrocytes seem even more complex and to some extent controversial. With regard to oxidative metabolism of endogenous glutamate in astrocytes, McKenna and coworkers (1993, 1996b) demonstrated the importance of AAT. In contrast, AOAA did not affect the production of CO_2 from glutamine in a study by Schousboe and coworkers (1993) and in line with this Yu and coworkers (1982) showed that oxidation of glutamate in astrocytes could not be inhibited by AOAA indicating that GDH is active in the oxidation of glutamate. However, Farinelli and Nicklas (1992) have shown inhibition by AOAA of oxidative metabolism of exogenous glutamate in cultured astrocytes. Using ^{13}C NMRS and cultured astrocytes it was elucidated by Westergaard and coworkers (1996) that the entrance of exogenous glutamate into the TCA cycle occurred via the action of GDH, whereas the production of glutamate from α-ketoglutarate was catalyzed by AAT. This is in line with a low capacity for reductive amination of α-ketoglutarate in neural cells (see above).

6 Anaplerosis

6.1 Pyruvate Carboxylase

PC (EC 6.4.1.1) is the main enzyme for carbon dioxide fixation in the brain that converts pyruvate to oxaloacetate (❯ *Figure 1-2*), and it is thought to be exclusively localized in the mitochondrial matrix of the glial compartment (Yu et al., 1983; Shank et al., 1985; Kaufman and Driscoll, 1992; Cesar and Hamprecht, 1995; Jitrapakdee and Wallace, 1999). The reaction is energy dependent in the form of ATP; it uses biotin and Mg^{2+} as cofactors and is allosterically activated by acetyl-CoA. The role of PC in brain is mainly anaplerotic, and it is thought to be essential for maintenance of glutamine, glutamate, and GABA homeostasis (Gamberino et al., 1997). The activity of PC was found to be 0.25 nmol/min/mg of protein in the adult brain (Yu et al., 1983). K^+ stimulates the CO_2 fixation rate in cultured astrocytes, which was determined to be 3.4 nmol of fixed CO_2/min/mg of protein at 2 mM K^+ and 6.8 nmol at 25 mM K^+ (Kaufman and Driscoll, 1992). Astrocytes take up K^+ released from neurons during neuronal activity, and thus K^+ might be the link between neuronal activity and the operation of PC in astrocytes (Hertz and Schousboe, 1975; Hertz et al., 2006).

7 Enzymes Involved in Glutamine Metabolism

7.1 Glutamine Synthetase

GS is a cytosolic enzyme located exclusively in astrocytes in the brain and catalyzes the ATP-dependent amidation of glutamate into glutamine (Norenberg and Martinez-Hernandez, 1979; Cooper and Lai, 1987). The rate of glutamine synthesis has been estimated to be 0.21 μmol/min/g wet wt. in an in vivo NMR study, a value significantly lower than the GS activity of 26 nmol/min/mg of protein (3 μmol/min/g wet wt.) obtained in a homogenate of cultured astrocytes (Schousboe et al., 1977a; Sibson et al., 1997). Evaluation of reported K_m values indicates that GS is not saturated under in vivo conditions. Thus, substrate alterations may influence enzyme activity (Erecinska and Silver, 1990). Methionine sulfoximine inhibits GS nonselectively; however, it has been used for elucidation of the role of GS in glutamate and GABA homeostasis (Sonnewald et al., 1993c). GS activity has been shown to be increased in response to activation of AMPA receptors in cultured astrocytes (Fleischer-Lambropoulos et al., 1996). Others have shown that GS and the glutamine content in brain might be modulated by neuronal NMDA receptor activation via neuronal nitric oxide release, affecting GS in an inhibitory manner in surrounding astrocytes (Kosenko et al., 2003).

7.2 Phosphate-Activated Glutaminase

PAG catalyzes the deamidation of glutamine to glutamate and is totally dependent on the P_i concentration, which acts as an allosteric effector. The P_i concentration that saturates PAG is higher the physiological concentration of P_i in the brain; hence, P_i is likely a prominent regulator of PAG activity in vivo (Erecinska and Silver, 1990). Different activities of PAG measured in various preparations range from 0.2 nmol/min/mg protein at 5 mM P_i in bulk isolated astrocytes to 150 nmol/min/mg protein at 50 mM Pi in whole-brain homogenate, illustrating the importance of the P_i concentration (for references see Erecinska and Silver, (1990)). The low rate determined in the bulk isolated astrocytes by Patel and coworkers (1982) is in line with a predominant neuronal localization of PAG with highest immunoreactivity observed in glutamatergic neurons (Laake et al., 1999; Kvamme et al., 2000). However, poor immunoreactivity was also observed in some glutamatergic fibres of the cerebellum, indicating high extent of heterogeneity with regard to PAG (Laake et al., 1999). In cultured cells, the activity in neurons is approximately twofold higher than that obtained in astrocytes using the same concentration of P_i (Schousboe et al., 1979; Drejer et al., 1985). PAG activity is upregulated by TCA cycle intermediates, fatty acyl CoA derivatives and Ca^{2+}. The latter affects PAG activity by decreasing the apparent K_m for P_i activation in a physiological concentration, making Ca^{2+} a realistic candidate for in vivo regulation (Erecinska and Silver, 1990). Ca^{2+} operates most likely in an indirect manner since it does not affect activity in a preparation of purified enzyme. The main inhibitors of PAG activity are protons and one of the products of the reaction, glutamate, both being present in relevant amounts for being functionally important inhibitors in vivo (for references see Kvamme et al., 2001).

PAG has a preferential mitochondrial localization, evident from both subcellular fractionation and immunogold labeling (Salganicoff and De Robertis, 1965; Laake et al., 1999; Kvamme et al., 2001). Contradictory observations exist regarding the functional localization of PAG: whether it is localized on the inner or outer face of the inner mitochondrial membrane (Kvamme et al., 2000; Zieminska et al., 2004). The localization is important with regard to the concentration of regulators that the enzyme is exposed to in the cytosol versus the mitochondrial matrix (Kvamme et al., 2000, 2001).

8 Enzymes Involved in GABA Metabolism

8.1 Glutamate Decarboxylase

The GABA-synthesizing enzyme GAD exists in two isoforms, known as GAD_{65} and GAD_{67}, according to their molecular masses (i.e., approximately 65 and 67 kDa), and they are encoded for by two independent genes (Kaufman et al., 1986, 1991; Kobayashi et al., 1987; Erlander et al., 1991; Martin et al., 1991a, b; Bu et al., 1992). GAD_{67} seems to be a cytosolic enzyme distributed throughout the GABAergic neurons in the cell body as well as in processes. In contrast, GAD_{65} is predominantly found to be located in nerve terminals and it might be anchored in or associated with the membrane of the vesicles containing the neurotransmitter (Kaufman et al., 1991; Christgau et al., 1991, 1992). GAD_{67} is not restricted to GABAergic neurons but is localized also in glutamatergic granule cells of the dentate gyrus. Interestingly, kainic acid-induced seizures cause a marked but transient increase in the expression of GAD_{67} in the hippocampus (Sperk et al., 2003). The regulation of GAD is complex and has not been fully elucidated. The association and dissociation of the cofactor PLP plays a major role in the short-term regulation of GAD. Fifty percent of GAD is present in the brain as the apoenzyme, thereby providing a large "dormant" catalytic capacity, which may be readily recruited whenever necessary (Miller et al., 1978; Itoh and Uchimura, 1981; Battaglioli et al., 2003). GAD_{65} is the predominant contributor to this large pool of apoenzyme, i.e., 93% of GAD_{65} exists as apoenzyme whereas GAD_{67} exists as holoenzyme approximately 72%. The apoenzyme is produced by an alternative transamination reaction catalyzed by GAD that converts the normally tightly bound PLP to pyridoxamine-5′-phosphate, which dissociates readily from the enzyme (Battaglioli et al., 2003). The interconversion between apo- and holoGAD is regulated by physiological concentrations of ATP and inorganic phosphate (Martin and Rimvall, 1993; Battaglioli et al., 2003). GAD_{65} is more sensitive than GAD_{67} toward the inactivation by aspartate (Porter and Martin, 1987; Battaglioli et al., 2003). Soluble GAD (presumably GAD_{67}) is inhibited by protein kinase A-mediated protein phosphorylation (Bao et al., 1995), whereas membrane-associated GAD (presumably GAD_{65}) is activated by ATP-mediated phosphorylation (Hsu et al., 1999).

The rationale for the existence of two enzymes, GAD_{65} and GAD_{67}, expressed by independently regulated genes, that are distinct with regard to regulation and localization is at present not understood. It has been speculated that this might be connected to the fact that GABA exists in at least two separate compartments, namely the cytoplasmic and the vesicular pool, which might have distinct functions (Soghomonian and Martin, 1998). The neurotransmitter function of the vesicular pool seems obvious. However, the function of the metabolic pool is more uncertain but it may to some extent consist of GABA acting as a neurodifferentiative agent during development (Meier et al., 1991; Belhage et al., 1998). As further evidence in support of multiple pools of GABA and their functional significance, antiepileptic drugs were found both in vitro and in vivo to affect distinct pools of GABA (Iadarola and Gale, 1980, 1982; Gram et al., 1988; Preece and Cerdan, 1996). It has been proposed that particularly GAD_{65} may be associated with synthesis of neurotransmitter GABA during synaptic activity, which is supported by its localization in the nerve terminal (Christgau et al., 1991, 1992; Kaufman et al., 1991). The importance of GAD_{65} for the synthesis of GABA during seizure was investigated using in vivo ^{1}H NMRS on vigabatrin-treated rats, a treatment leading to selective inhibition of GAD_{67} expression (Sheikh and Martin, 1998), and GAD_{65} was shown to account for the majority of GABA synthesis during seizure (Patel et al., 2006).

8.2 GABA aminotransferase

GABA-T catalyzes the first step in the degradation of GABA and is together with GAD a central element in the regulation of the GABA level in the brain. GABA-T is restricted to a mitochondrial localization but exhibits a ubiquitous distribution in both neurons and astrocytes (Hyde and Robinson, 1978). GABA-T transaminates GABA to succinate semialdehyde, converting the pyridoxal 5′-phophate (PLP) cofactor to its pyridoxamine 5′-phosphate form. The enzyme and cofactor are regenerated by the conversion of α-ketoglutarate to glutamate (Baxter and Roberts, 1958; Schousboe et al., 1973, 1974). Succinate semialdehyde is subsequently oxidized by succinate semialdehyde dehydrogenase, using NAD$^+$ as cofactor, to succinate, which enters the TCA cycle as illustrated in ❷ *Figure 1-1*. Inhibition of GABA-T has been used in the treatment of certain forms of epilepsy. Vigabatrin or γ-vinyl GABA, an active-site-directed suicide inhibitor of GABA-T, elevates the level of synaptic GABA, and consequently increases the efficacy of GABAergic neurotransmission (Iadarola and Gale, 1980; Wood et al., 1981; Gram et al., 1988). Vigabatrin inhibits predominantly the neuronal fraction of GABA-T (Schousboe et al., 1986), and evidence has been provided suggesting that the astrocytic fraction is upregulated due to a rise in the cellular GABA content upon acute vigabatrin treatment. Such explanation requires that GABA-T is not saturated in the astrocytic compartment. These considerations might explain an observed unchanged total GABA level and GABA-T activity in vigabatrin-treated animals (de Graaf et al., 2006).

9 Metabolic Compartmentation at the Single-Cell Level

Using ^{13}C NMRS to study extracts of relatively homogenous cell cultures, metabolic compartmentation has been observed in primary cultures of neurons as well as astrocytes (Schousboe et al., 1993; Sonnewald et al., 1993a, 1998; Westergaard et al., 1995; Bakken et al., 1997; Waagepetersen et al., 1998a, b, 1999b, 2003). These findings indicate that compartmentation exists within one cell and it is mostly referred to as mitochondrial heterogeneity or intramitochondrial compartmentation. Indeed this might be the case as confirmed by Lopez-Beltran and coworkers (1996) who showed that the viscosity of the mitochondria is such that free diffusion within the mitochondria is unlikely. Furthermore, studies of the three-dimensional structure of mitochondria using electron microscope tomography have suggested that inner membrane proteins might be compartmentalized (Perkins et al., 2001). Additionally, α-ketoglutarate dehydrogenase, the rate-limiting enzyme of the TCA cycle, has been shown to be heterogeneously distributed among mitochondria in cultured astrocytes at the single-cell level (Waagepetersen et al., 2006). The functional importance of these indications of metabolic compartmentation at the single-cell level might be illustrated by the finding that the main intracellular pools of glutamate and glutamine are compartmentalized from the glutamate pool, which is taken up from the extracellular space, and the glutamine pool from which glutamine is released into the extracellular space, suggesting that different pools of a metabolite may serve different functional purposes (Schousboe et al., 1993; Waagepetersen et al., 2001a).

References

Alves PM, Nunes R, Zhang C, Maycock CD, Sonnewald U, et al. 2000. Metabolism of 3-^{13}C-malate in primary cultures of mouse astrocytes. Dev Neurosci 22: 456-462.

Aoki C, Milner TA, Berger SB, Sheu KF, Blass JP, et al. 1987. Glial glutamate dehydrogenase: Ultrastructural localization and regional distribution in relation to the mitochondrial enzyme, cytochrome oxidase. J Neurosci Res 18: 305-318.

Bak LK, Sickmann HM, Schousboe A, Waagepetersen HS. 2005. Activity of the lactate–alanine shuttle is independent of glutamate–glutamine cycle activity in cerebellar neuronal–astrocytic cultures. J Neurosci Res 79: 88-96.

Bakken IJ, White LR, Aasly J, Unsgard G, Sonnewald U. 1997. Lactate formation from [U-^{13}C]aspartate in cultured astrocytes: Compartmentation of pyruvate metabolism. Neurosci Lett 237: 117-120.

Bakken IJ, White LR, Unsgard G, Aasly J, Sonnewald U. 1998. [U-^{13}C]glutamate metabolism in astrocytes during hypoglycemia and hypoxia. J Neurosci Res 51: 636-645.

Balazs R, Machiyama Y, Hammond BJ, Julian T, Richter D. 1970. The operation of the γ-aminobutyrate bypath of the tricarboxylic acid cycle in brain tissue in vitro. Biochem J 116: 445-461.

Bao J, Cheung WY, Wu JY. 1995. Brain L-glutamate decarboxylase. Inhibition by phosphorylation and activation by dephosphorylation. J Biol Chem 270: 6464-6467.

Battaglioli G, Martin DL. 1990. Stimulation of synaptosomal γ-aminobutyric acid synthesis by glutamate and glutamine. J Neurochem 54: 1179-1187.

Battaglioli G, Martin DL. 1991. GABA synthesis in brain slices is dependent on glutamine produced in astrocytes. Neurochem Res 16: 151-156.

Battaglioli G, Liu H, Martin DL. 2003. Kinetic differences between the isoforms of glutamate decarboxylase: Implications for the regulation of GABA synthesis. J Neurochem 86: 879-887.

Baxter CF, Roberts E. 1958. The γ-aminobutyric acid-α-ketoglutaric acid transaminase of beef brain. J Biol Chem 233: 1135-1139.

Belhage B, Hansen GH, Elster L, Schousboe A. 1998. Effects of γ-aminobutyric acid (GABA) on synaptogenesis and synaptic function. Perspect Dev Neurobiol 5: 235-246.

Belhage B, Hansen GH, Meier E, Schousboe A. 1990. Effects of inhibitors of protein synthesis and intracellular transport on the γ-aminobutyric acid agonist-induced functional differentiation of cultured cerebellar granule cells. J Neurochem 55: 1107-1113.

Berl S, Clarke DD. 1969. Compartmentation of amino acid metabolism. Handbook of Neurochemistry, Vol. 2. Lajtha A, editor. New York: Plenum Press, pp. 447-472.

Berl S, Clarke DD. 1983. The metabolic compartmentation concept. Glutamine, Glutamate and GABA in The Central Nervous System. Hertz L, Kvamme E, McGeer EG, Schousboe A, editors. New York: Alan R. Liss, Inc., pp. 205-217.

Berl S, Lajtha A, Waeelsch H. 1961. Amino acid and protein metabolism—IV. Cerebral compartments of glutamic acid metabolism. J Neurochem 7: 186-197.

Berl S, Takagaki G, Clarke DD, Waelsch H. 1962. Metabolic compartments in vivo. Ammonia and glutamic acid metabolism in brain and liver. J Biol Chem 237: 2562-2569.

Borden LA. 1996. GABA transporter heterogeneity: Pharmacology and cellular localization. Neurochem Int 29: 335-356.

Bradford HF, Ward HK, Thomas AJ. 1978. Glutamine—a major substrate for nerve endings. J Neurochem 30: 1453-1459.

Broer S, Brookes N. 2001. Transfer of glutamine between astrocytes and neurons. J Neurochem 77: 705-719.

Bu DF, Erlander MG, Hitz BC, Tillakaratne NJ, Kaufman DL, et al. 1992. Two human glutamate decarboxylases, 65-kDa GAD and 67-kDa GAD, are each encoded by a single gene. Proc Natl Acad Sci USA 89: 2115-2119.

Cerdan S, Kunnecke B, Seelig J. 1990. Cerebral metabolism of [1,2-^{13}C2]acetate as detected by in vivo and in vitro ^{13}C NMR. J Biol Chem 265: 12916-12926.

Cesar M, Hamprecht B. 1995. Immunocytochemical examination of neural rat and mouse primary cultures using monoclonal antibodies raised against pyruvate carboxylase. J Neurochem 64: 2312-2318.

Chan-Palay V, Wu JY, Palay SL. 1979. Immunocytochemical localization of γ-aminobutyric acid transaminase at cellular and ultrastructural levels. Proc Natl Acad Sci USA 76: 2067-2071.

Chaudhry FA, Reimer RJ, Edwards RH. 2002. The glutamine commute: Take the N line and transfer to the A. J Cell Biol 157: 349-355.

Chee PY, Dahl JL, Fahien LA. 1979. The purification and properties of rat brain glutamate dehydrogenase. J Neurochem 33: 53-60.

Cho SW, Lee J, Choi SY. 1995. Two soluble forms of glutamate dehydrogenase isoproteins from bovine brain. Eur J Biochem 233: 340-346.

Choi IY, Lei H, Gruetter R. 2002. Effect of deep pentobarbital anesthesia on neurotransmitter metabolism in vivo: On the correlation of total glucose consumption with glutamatergic action. J Cereb Blood Flow Metab 22: 1343-1351.

Christgau S, Aanstoot HJ, Schierbeck H, Begley K, Tullin S, et al. 1992. Membrane anchoring of the autoantigen GAD65 to microvesicles in pancreatic β-cells by palmitoylation in the NH$_2$-terminal domain. J Cell Biol 118: 309-320.

Christgau S, Schierbeck H, Aanstoot HJ, Aagaard L, Begley K, et al. 1991. Pancreatic β cells express two autoantigenic forms of glutamic acid decarboxylase, a 65-kDa hydrophilic form and a 64-kDa amphiphilic form which can be both membrane-bound and soluble. J Biol Chem 266: 21257-21264.

Colon AD, Plaitakis A, Perakis A, Berl S, Clarke DD. 1986. Purification and characterization of a soluble and a particulate glutamate dehydrogenase from rat brain. J Neurochem 46: 1811-1819.

Cooper AJ. 1988. L-Glutamate(2-oxoglutarate) aminotransferases. Glutamine and glutamate in mammals, Vol. 2. Kvamme E, editor. Florida: CRC Press, pp 123-152.

Cooper AJ, Lai JC. 1987. Cerebral ammonia metabolism in normal and hyperammonemic rats. Neurochem Pathol 6: 67-95.

Cooper AJ, Plum F. 1987. Biochemistry and physiology of brain ammonia. Physiol Rev 67: 440-519.

Cooper AJ, McDonald JM, Gelbard AS, Gledhill RF, Duffy TE. 1979. The metabolic fate of ^{13}N-labeled ammonia in rat brain. J Biol Chem 254: 4982-4992.

Curtis DR, Watkins JC. 1960. The excitation and depression of spinal neurones by structurally related amino acids. J Neurochem 6: 117-141.

Danbolt NC. 2001. Glutamate uptake. Prog Neurobiol 65: 1-105.

de Graaf RA, Patel AB, Rothman DL, Behar KL. 2006. Acute regulation of steady-state GABA levels following GABA-transaminase inhibition in rat cerebral cortex. Neurochem Int 48: 508-514.

Drejer J, Larsson OM, Kvamme E, Svenneby G, Hertz L, et al. 1985. Ontogenetic development of glutamate metabolizing enzymes in cultured cerebellar granule cells and in cerebellum in vivo. Neurochem Res 10: 49-62.

Dringen R, Pfeiffer B, Hamprecht B. 1999. Synthesis of the antioxidant glutathione in neurons: Supply by astrocytes of CysGly as precursor for neuronal glutathione. J Neurosci 19: 562-569.

Duffy TE, Plum F, Cooper AJL. 1983. Cerebral ammonia metabolism in vivo. Glutamine, Glutamate and GABA in The Central Nervous System. Hertz L, Kvamme E, McGeer EG, Schousboe A. editors. New York: Alan R. Liss, Inc., pp 371-388

Erecinska M, Silver IA. 1990. Metabolism and role of glutamate in mammalian brain. Prog Neurobiol 35: 245-296.

Erecinska M, Pleasure D, Nelson D, Nissim I, Yudkoff M. 1993. Cerebral aspartate utilization: Near-equilibrium relationships in aspartate aminotransferase reaction. J Neurochem 60: 1696-1706.

Erlander MG, Tillakaratne NJ, Feldblum S, Patel N, Tobin AJ. 1991. Two genes encode distinct glutamate decarboxylases. Neuron 7: 91-100.

Farinelli SE, Nicklas WJ. 1992. Glutamate metabolism in rat cortical astrocyte cultures. J Neurochem 58: 1905-1915.

Filla A, De MG, Brescia M, Palma V, Di LA, Di GG, Campanella G, 1986. Glutamate dehydrogenase in human brain: Regional distribution and properties. J Neurochem 46: 422-424.

Fleischer-Lambropoulos E, Kazazoglou T, Geladopoulos T, Kentroti S, Stefanis C, et al. 1996. Stimulation of glutamine synthetase activity by excitatory amino acids in astrocyte cultures derived from aged mouse cerebral hemispheres may be associated with non-*N*-methyl-D-aspartate receptor activation. Int J Dev Neurosci 14: 523-530.

Fonnum F. 1968. The distribution of glutamate decarboxylase and aspartate transaminase in subcellular fractions of rat and guinea-pig brain. Biochem J 106: 401-412.

Fremeau RT Jr, Troyer MD, Pahner I, Nygaard GO, Tran CH, et al. 2001. The expression of vesicular glutamate transporters defines two classes of excitatory synapse. Neuron 31: 247-260.

Gaitonde MK, 1965. Rate of utilization of glucose and compartmentation of α-oxoglutarate and glutamate in rat brain. Biochem J 95: 803-810.

Gamberino WC, Berkich DA, Lynch CJ, Xu B, LaNoue KF. 1997. Role of pyruvate carboxylase in facilitation of synthesis of glutamate and glutamine in cultured astrocytes. J Neurochem 69: 2312-2325.

Garfinkel D. 1966. A simulation study of the metabolism and compartmentation in brain of glutamate, aspartate, the Krebs cycle, and related metabolites. J Biol Biochem 17: 3918-3929.

Gegelashvili G, Schousboe A. 1997. High affinity glutamate transporters: Regulation of expression and activity. Mol Pharmacol 52: 6-15.

Gegelashvili G, Schousboe A. 1998. Cellular distribution and kinetic properties of high-affinity glutamate transporters. Brain Res Bull 45: 233-238.

Gladkevich A, Korf J, Hakobyan VP, Melkonyan KV. 2006. The peripheral GABAergic system as a target in endocrine disorders. Auton Neurosci 124: 1-8.

Gram L, Larsson OM, Johnsen AH, Schousboe A. 1988. Effects of valproate, vigabatrin and aminooxyacetic acid on release of endogenous and exogenous GABA from cultured neurons. Epilepsy Res 2: 87-95.

Gruetter R, Seaquist ER, Kim S, Ugurbil K. 1998. Localized in vivo ^{13}C-NMR of glutamate metabolism in the human brain: Initial results at 4 Tesla. Dev Neurosci 20: 380-388.

Gundersen V, Chaudhry FA, Bjaalie JG, Fonnum F, Ottersen OP, et al. 1998. Synaptic vesicular localization and exocytosis of L-aspartate in excitatory nerve terminals: A quantitative immunogold analysis in rat hippocampus. J Neurosci 18: 6059-6070.

Hack N, Balazs R. 1994. Selective stimulation of excitatory amino acid receptor subtypes and the survival of granule cells in culture: Effect of quisqualate and AMPA. Neurochem Int 25: 235-241.

Hansson E, Rönnbäck L. 2004. Astrocytic receptors and second messenger systems. Advances in Molecular and Cell Biology. Hertz L, editor. The Netherlands: Elsevier, Amsterdam, pp. 475–501.

Hassel B, Bråthe A. 2000. Neuronal pyruvate carboxylation supports formation of transmitter glutamate. J Neurosci 20: 1342-1347.

Hassel B, Johannessen CU, Sonnewald U, Fonnum F. 1998. Quantification of the GABA shunt and the importance of the GABA shunt versus the 2-oxoglutarate dehydrogenase pathway in GABAergic neurons. J Neurochem 71: 1511-1518.

Hertz L, Schousboe A. 1975. Ion and energy metabolism of the brain at the cellular level. Int Rev Neurobiol 18: 141-211.

Hertz L, Schousboe A. 1987. Primary cultures of GABAergic and glutamatergic neurons as model systems to study neurotransmitter functions. I. Differentiated cells. Model systems of development and aging of the nervous system. Vernadakis A, Privat A, Lauder JM, Timiras PS, Giacobini E, editors. Boston: Martinus Nijhoff publishing, pp 19-31.

Hertz L, Peng L, Dienel GA. 2007. Energy metabolism in astrocytes: High rate of oxidative metabolism and spatiotemporal dependence on glycolysis/glycogenolysis. J Cereb Blood Flow Metab. Epub.

Hertz L, Peng L, Westergaard N, Yudkoff M, Schousboe A. 1992. Neuronal–astrocytic interactions in metabolism of transmitter amino acids of the glutamate family. Drug Research Related to Neuroactive Amino Acids, Alfred

Benzon Symposium 32. Schousboe A, Diemer NH, Kofod H, editors. Copenhagen: Munksgaard, pp. 30-48.

Hsu CC, Thomas C, Chen W, Davis KM, Foos T, et al. 1999. Role of synaptic vesicle proton gradient and protein phosphorylation on ATP-mediated activation of membrane-associated brain glutamate decarboxylase. J Biol Chem 274: 24366-24371.

Hutson S. 2001. Structure and function of branched chain aminotransferases. Prog Nucleic Acid Res Mol Biol 70: 175-206.

Hyde JC, Robinson N. 1978. Electron cytochemical localization of γ-aminobutyric acid catabolism in rat cerebellar cortex. Histochemistry 49: 51-65.

Hyder F, Patel AB, Gjedde A, Rothman DL, Behar KL, et al. 2006. Neuronal–glial glucose oxidation and glutamatergic-GABAergic function. J Cereb Blood Flow Metab 26: 865-877.

Iadarola MJ, Gale K. 1980. Evaluation of increases in nerve terminal-dependent vs nerve terminal-independent compartments of GABA in vivo. Brain Res Bull 5: 13-19.

Iadarola MJ, Gale K. 1982. Substantia nigra: Site of anticonvulsant activity mediated by γ-aminobutyric acid. Science 218: 1237-1240.

Itoh M, Uchimura H. 1981. Regional differences in cofactor saturation of glutamate decarboxylase (GAD) in discrete brain nuclei of the rat. Effect of repeated administration of haloperidol on GAD activity in the substantia nigra. Neurochem Res 6: 1283-1289.

Jitrapakdee S, Wallace JC. 1999. Structure, function and regulation of pyruvate carboxylase. Biochem J 340 (Pt 1): 1-16.

Kaufman DL, Houser CR, Tobin AJ. 1991. Two forms of the γ-aminobutyric acid synthetic enzyme glutamate decarboxylase have distinct intraneuronal distributions and cofactor interactions. J Neurochem 56: 720-723.

Kaufman DL, McGinnis JF, Krieger NR, Tobin AJ. 1986. Brain glutamate decarboxylase cloned in λgt-11: Fusion protein produces γ-aminobutyric acid. Science 232: 1138-1140.

Kaufman EE, Driscoll BF. 1992. Carbon dioxide fixation in neuronal and astroglial cells in culture. J Neurochem 58: 258-262.

Kew JN, Kemp JA. 2005. Ionotropic and metabotropic glutamate receptor structure and pharmacology. Psychopharmacology (Berl) 179: 4-29.

Kihara M, Kubo T. 1989. Aspartate aminotransferase for synthesis of transmitter glutamate in the medulla oblongata: Effect of aminooxyacetic acid and 2-oxoglutarate. J Neurochem 52: 1127-1134.

Kobayashi Y, Kaufman DL, Tobin AJ. 1987. Glutamic acid decarboxylase cDNA: Nucleotide sequence encoding an enzymatically active fusion protein. J Neurosci 7: 2768-2772.

Kosenko E, Llansola M, Montoliu C, Monfort P, Rodrigo R, et al. 2003. Glutamine synthetase activity and glutamine content in brain: Modulation by NMDA receptors and nitric oxide. Neurochem Int 43: 493-499.

Krogsgaard-Larsen P, Falch E, Larsson OM, Schousboe A. 1987. GABA uptake inhibitors: Relevance to antiepileptic drug research. Epilepsy Res 1: 77-93.

Kvamme E, Roberg B, Johansen L, Torgner IA. 1988. Interrelated effects of calcium and sulfhydryl reagents on renal phosphate-activated glutaminase. Contrib Nephrol 63: 156-160.

Kvamme E, Roberg B, Torgner IA. 2000. Phosphate-activated glutaminase and mitochondrial glutamine transport in the brain. Neurochem Res 25: 1407-1419.

Kvamme E, Torgner IA, Roberg B. 2001. Kinetics and localization of brain phosphate activated glutaminase. J Neurosci Res 66: 951-958.

Laake JH, Takumi Y, Eidet J, Torgner IA, Roberg B, et al. 1999. Postembedding immunogold labelling reveals subcellular localization and pathway-specific enrichment of phosphate activated glutaminase in rat cerebellum. Neuroscience 88: 1137-1151.

Langeveld CH, Schepens E, Jongenelen CA, Stoof JC, Hjelle OP, et al. 1996. Presence of glutathione immunoreactivity in cultured neurones and astrocytes. Neuroreport 7: 1833-1836.

Lehre KP, Danbolt NC. 1998. The number of glutamate transporter subtype molecules at glutamatergic synapses: Chemical and stereological quantification in young adult rat brain. J Neurosci 18: 8751-8757.

Levy, LM. 2002. Structure, function and regulation of glutamate transporters. Glutamate and GABA Receptors and Transporters. Structure, Function and Pharmacology. Egebjerg J, Schousboe A, Krogsgaard-Larsen P, editors. London: Taylor and Francis, pp. 307-336.

Lieth E, LaNoue KF, Berkich DA, Xu B, Ratz M, et al. 2001. Nitrogen shuttling between neurons and glial cells during glutamate synthesis. J Neurochem 76: 1712-1723.

Lopez-Beltran EA, Mate MJ, Cerdan S. 1996. Dynamics and environment of mitochondrial water as detected by ^{1}H NMR. J Biol Chem 271: 10648-10653.

Machiyama Y, Balazs R, Hammond BJ, Julian T, Richter D. 1970. The metabolism of γ-aminobutyrate and glucose in potassium ion-stimulated brain tissue in vitro. Biochem J 116: 469-481.

Marcaggi P, Coles JA. 2001. Ammonium in nervous tissue: Transport across cell membranes, fluxes from neurons to glial cells, and role in signalling. Prog Neurobiol 64: 157-183.

Martin DL, Rimvall K. 1993. Regulation of γ-aminobutyric acid synthesis in the brain. J Neurochem 60: 395-407.

Martin DL, Martin SB, Wu SJ, Espina N. 1991a. Cofactor interactions and the regulation of glutamate decarboxylase activity. Neurochem Res 16: 243-249.

Martin DL, Martin SB, Wu SJ, Espina N. 1991b. Regulatory properties of brain glutamate decarboxylase (GAD): The apoenzyme of GAD is present principally as the smaller of two molecular forms of GAD in brain. J Neurosci 11: 2725-2731.

Mason GF, Rothman DL, Behar KL, Shulman RG. 1992. NMR determination of the TCA cycle rate and α-ketoglutarate/glutamate exchange rate in rat brain. J Cereb Blood Flow Metab 12: 434-447.

McKenna MC, Hopkins IB, Lindauer SL, Bamford P. 2006a. Aspartate aminotransferase in synaptic and nonsynaptic mitochondria: Differential effect of compounds that influence transient hetero-enzyme complex (metabolon) formation. Neurochem Int 48: 629-636.

McKenna MC, Waagepetersen HS, Schousboe A, Sonnewald U. 2006b. Neuronal and astrocytic shuttle mechanisms for cytosolic-mitochondrial transfer of reducing equivalents: Current evidence and pharmacological tools. Biochem Pharmacol 71: 399-407.

McKenna MC, Stevenson JH, Huang X, Hopkins IB. 2000a. Differential distribution of the enzymes glutamate dehydrogenase and aspartate aminotransferase in cortical synaptic mitochondria contributes to metabolic compartmentation in cortical synaptic terminals. Neurochem Int 37: 229-241.

McKenna MC, Stevenson JH, Huang X, Tildon JT, Zielke CL, et al. 2000b. Mitochondrial malic enzyme activity is much higher in mitochondria from cortical synaptic terminals compared with mitochondria from primary cultures of cortical neurons or cerebellar granule cells. Neurochem Int 36: 451-459.

McKenna MC, Tildon JT, Stevenson JH, Boatright R, Huang S. 1993. Regulation of energy metabolism in synaptic terminals and cultured rat brain astrocytes: Differences revealed using aminooxyacetate. Dev Neurosci 15: 320-329.

McKenna MC, Tildon JT, Stevenson JH, Huang X. 1996a. New insights into the compartmentation of glutamate and glutamine in cultured rat brain astrocytes. Dev Neurosci 18: 380-390.

McKenna MC, Sonnewald U, Huang X, Stevenson J, Zielke HR. 1996b. Exogenous glutamate concentration regulates the metabolic fate of glutamate in astrocytes. J Neurochem 66: 386-393.

Meier E, Drejer J, Schousboe A. 1984. GABA induces functionally active low-affinity GABA receptors on cultured cerebellar granule cells. J Neurochem 43: 1737-1744.

Meier E, Hertz L, Schousboe A. 1991. Neurotransmitters as developmental signals. Neurochem Int 19: 1-15.

Meldrum BS. 2000. Glutamate as a neurotransmitter in the brain: Review of physiology and pathology. J Nutr 130: 1007S-1015S.

Miller LP, Martin DL, Mazumder A, Walters JR. 1978. Studies on the regulation of GABA synthesis: Substrate-promoted dissociation of pyridoxal-5′-phosphate from GAD. J Neurochem 30: 361-369.

Minchin MC, Beart PM. 1975. Compartmentation of amino acid metabolism in the rat posterior pituitary. J Neurochem 24: 881-884.

Norenberg MD, Martinez-Hernandez A. 1979. Fine structural localization of glutamine synthetase in astrocytes of rat brain. Brain Res 161: 303-310.

Olsen RW, Betz H. 2006. Energy metabolism of the brain. Basic Neurochemistry: Molecular, Cellular, and Medical Aspects, 7th ed. Siegel GJ, Agranoff BW, Albers RW, Fisher SK, Uhler MD, editors. Philadelphia: Lippincott-Raven, pp. 531-557.

Olstad E, Qu H, Sonnewald U. 2007. Glutamate is preferred over glutamine for intermediary metabolism in cultured cerebellar neurons. J Cereb Blood Flow Metab.

O'Neal RM, Koeppe RE. 1966. Precursors in vivo of glutamate, aspartate and their derivatives of rat brain. J Neurochem 13: 835-847.

Oz G, Berkich DA, Henry PG, Xu Y, La Noue K, et al. 2004. Neuroglial metabolism in the awake rat brain: CO_2 fixation increases with brain activity. J Neurosci 24: 11273-11279.

Palaiologos G, Hertz L, Schousboe A. 1989. Role of aspartate aminotransferase and mitochondrial dicarboxylate transport for release of endogenously and exogenously supplied neurotransmitter in glutamatergic neurons. Neurochem Res 14: 359-366.

Parpura V, Scemes E, Spray DC. 2004. Mechanisms of glutamate release from astrocytes: Gap junction "hemichannels", purinergic receptors and exocytotic release. Neurochem Int 45: 259-264.

Patel AB, Chowdhury GM, de Graaf RA, Rothman DL, Shulman RG, et al. 2005a. Cerebral pyruvate carboxylase flux is unaltered during bicuculline-seizures. J Neurosci Res 79: 128-138.

Patel AB, de Graaf RA, Mason GF, Rothman DL, Shulman RG, et al. 2005b. The contribution of GABA to glutamate/glutamine cycling and energy metabolism in the rat cortex in vivo. Proc Natl Acad Sci USA 102: 5588-5593.

Patel AB, de Graaf RA, Martin DL, Battaglioli G, Behar KL. 2006. Evidence that GAD65 mediates increased GABA synthesis during intense neuronal activity in vivo. J Neurochem 97: 385-396.

Patel AB, de Graaf RA, Mason GF, Kanamatsu T, Rothman DL, et al. 2004. Glutamatergic neurotransmission and neuronal glucose oxidation are coupled during intense neuronal activation. J Cereb Blood Flow Metab 24: 972-985.

Patel AJ, Hunt A, Gordon RD, Balazs R. 1982. The activities in different neural cell types of certain enzymes associated with the metabolic compartmentation glutamate. Brain Res 256: 3-11.

Peng LA, Schousboe A, Hertz L. 1991. Utilization of α-ketoglutarate as a precursor for transmitter glutamate in cultured cerebellar granule cells. Neurochem Res 16: 29-34.

Perkins GA, Renken CW, Frey TG, Ellisman MH. 2001. Membrane architecture of mitochondria in neurons of the central nervous system. J Neurosci Res 66: 857-865.

Plaitakis A, Zaganas I. 2001. Regulation of human glutamate dehydrogenases: Implications for glutamate, ammonia and energy metabolism in brain. J Neurosci Res 66: 899-908.

Plaitakis A, Spanaki C, Mastorodemos V, Zaganas I. 2003. Study of structure–function relationships in human glutamate dehydrogenases reveals novel molecular mechanisms for the regulation of the nerve tissue-specific (GLUD2) isoenzyme. Neurochem Int 43: 401-410.

Porter TG, Martin DL. 1987. Rapid inactivation of brain glutamate decarboxylase by aspartate. J Neurochem 48: 67-72.

Preece NE, Cerdan S. 1996. Metabolic precursors and compartmentation of cerebral GABA in vigabatrin-treated rats. J Neurochem 67: 1718-1725.

Qu H, Konradsen JR, van Hengel M, Wolt S, Sonnewald U. 2001. Effect of glutamine and GABA on [U-^{13}C]glutamate metabolism in cerebellar astrocytes and granule neurons. J Neurosci Res 66: 885-890.

Ramos M, del Arco A, Pardo B, Martinez-Serrano A, Martinez-Morales JR, et al. 2003. Developmental changes in the Ca^{2+}-regulated mitochondrial aspartate–glutamate carrier aralar1 in brain and prominent expression in the spinal cord. Brain Res Dev Brain Res 143: 33-46.

Reubi JC, Van den Berg CJ, Cuénod M. 1978. Glutamine as precursor for the GABA and glutamate transmitter pools. Neurosci Lett 10: 171-174.

Roberts E, Bregoff HM. 1953. Transamination of γ-aminobutyric acid and β-alanine in brain and liver. J Biol Chem 201: 393-398.

Rose C, Kresse W, Kettenmann H. 2005. Acute insult of ammonia leads to calcium-dependent glutamate release from cultured astrocytes, an effect of pH. J Biol Chem 280: 20937-20944.

Rothe F, Brosz M, Storm-Mathisen J. 1994. Quantitative ultrastructural localization of glutamate dehydrogenase in the rat cerebellar cortex. Neuroscience 62: 1133-1146.

Rothstein JD, Tabakoff B. 1984. Alteration of striatal glutamate release after glutamine synthetase inhibition. J Neurochem 43: 1438-1446.

Salganicoff L, De Robertis E. 1965. Subcellular distribution of the enzymes of the glutamic acid, glutamine and γ-amino-butyric acid cycles in rat brain. J Neurochem 12: 287-309.

Schousboe A, Waagepetersen HS. 2004. Role of astrocytes in homeostasis of glutamate and GABA during physiological and pathophysiological conditions. Adv Mol Cell Biol 31: 461-474.

Schousboe A, Hertz L, Svenneby G, Kvamme E. 1979. Phosphate-activated glutaminase activity and glutamine uptake in primary cultures of astrocytes. J Neurochem 32: 943-950.

Schousboe A, Larsson OM, Seiler N. 1986. Stereoselective uptake of the GABA-transaminase inhibitors γ-vinyl GABA and γ-acetylenic GABA into neurons and astrocytes. Neurochem Res 11: 1497-1505.

Schousboe A, Sarup A, Bak LK, Waagepetersen HS, Larsson OM. 2004. Role of astrocytic transport processes in glutamatergic and GABAergic neurotransmission. Neurochem Int 45: 521-527.

Schousboe A, Svenneby G, Hertz L. 1977a. Uptake and metabolism of glutamate in astrocytes cultured from dissociated mouse brain hemispheres. J Neurochem 29: 999-1005.

Schousboe I, Bro B, Schousboe A. 1977b. Intramitochondrial localization of the 4-aminobutyrate-2-oxoglutarate transaminase from ox brain. Biochem J 162: 303-307.

Schousboe A, Westergaard N, Sonnewald U, Petersen SB, Huang R, et al. 1993. Glutamate and glutamine metabolism and compartmentation in astrocytes. Dev Neurosci 15: 359-366.

Schousboe A, Wu JY, Roberts E. 1973. Purification and characterization of the 4-aminobutyrate-2-ketoglutarate transaminase from mouse brain. Biochemistry 12: 2868-2873.

Schousboe A, Wu JY, Roberts E. 1974. Subunit structure and kinetic properties of 4-aminobutyrate-2-ketoglutarate transaminase purified from mouse brain. J Neurochem 23: 1189-1195.

Schwarzer C, Sperk G. 1995. Hippocampal granule cells express glutamic acid decarboxylase-67 after limbic seizures in the rat. Neuroscience 69: 705-709.

Seiler N. 1980. On the role of GABA in vertebrate polyamine metabolism. Physiol Chem Phys 12: 411-429.

Shank RP, Campbell GL. 1984. α-Ketoglutarate and malate uptake and metabolism by synaptosomes: Further evidence for an astrocyte-to-neuron metabolic shuttle. J Neurochem 42: 1153-1161.

Shank RP, Baldy WJ, Ash CW. 1989. Glutamine and 2-oxoglutarate as metabolic precursors of the transmitter pools of glutamate and GABA: Correlation of regional uptake by rat brain synaptosomes. Neurochem Res 14: 371-376.

Shank RP, Bennett GS, Freytag SO, Campbell GL. 1985. Pyruvate carboxylase: An astrocyte-specific enzyme implicated in the replenishment of amino acid neurotransmitter pools. Brain Res 329: 364-367.

Sheikh SN, Martin DL. 1998. Elevation of brain GABA levels with vigabatrin (γ-vinylGABA) differentially affects GAD65 and GAD67 expression in various regions of rat brain. J Neurosci Res 52: 736-741.

Sibson NR, Dhankhar A, Mason GF, Behar KL, Rothman DL, et al. 1997. In vivo ^{13}C NMR measurements of cerebral glutamine synthesis as evidence for glutamate–glutamine cycling. Proc Natl Acad Sci USA 94: 2699-2704.

Sibson NR, Mason GF, Shen J, Cline GW, Herskovits AZ, et al. 2001. In vivo ^{13}C NMR measurement of neurotransmitter glutamate cycling, anaplerosis and TCA cycle flux in rat brain during [2-^{13}C]glucose infusion. J Neurochem 76: 975-989.

Sloviter RS, Dichter MA, Rachinsky TL, Dean E, Goodman JH, et al. 1996. Basal expression and induction of glutamate decarboxylase and GABA in excitatory granule cells of the rat and monkey hippocampal dentate gyrus. J Comp Neurol 373: 593-618.

Soghomonian JJ, Martin DL. 1998. Two isoforms of glutamate decarboxylase: Why? Trends Pharmacol Sci 19: 500-505.

Sonnewald U, Hertz L, Schousboe A. 1998. Mitochondrial heterogeneity in the brain at the cellular level. J Cereb Blood Flow Metab 18: 231-237.

Sonnewald U, Westergaard N, Hassel B, Muller TB, Unsgard G, et al. 1993a. NMR spectroscopic studies of ^{13}C acetate and ^{13}C glucose metabolism in neocortical astrocytes: Evidence for mitochondrial heterogeneity. Dev Neurosci 15: 351-358.

Sonnewald U, Westergaard N, Petersen SB, Unsgard G, Schousboe A. 1993b. Metabolism of [U-^{13}C]glutamate in astrocytes studied by ^{13}C NMR spectroscopy: Incorporation of more label into lactate than into glutamine demonstrates the importance of the tricarboxylic acid cycle. J Neurochem 61: 1179-1182.

Sonnewald U, Westergaard N, Schousboe A, Svendsen JS, Unsgard G, et al. 1993c. Direct demonstration by [^{13}C] NMR spectroscopy that glutamine from astrocytes is a precursor for GABA synthesis in neurons. Neurochem Int 22: 19-29.

Sonnewald U, Westergaard N, Jones P, Taylor A, Bachelard HS, et al. 1996. Metabolism of [U-^{13}C5]glutamine in cultured astrocytes studied by NMR spectroscopy: First evidence of astrocytic pyruvate recycling. J Neurochem 67: 2566-2572.

Sonnewald U, Westergaard N, Krane J, Unsgard G, Petersen SB, et al. 1991. First direct demonstration of preferential release of citrate from astrocytes using [^{13}C]NMR spectroscopy of cultured neurons and astrocytes. Neurosci Lett 128: 235-239.

Sperk G, Schwarzer C, Heilman J, Furtinger S, Reimer RJ, et al. 2003. Expression of plasma membrane GABA transporters but not of the vesicular GABA transporter in dentate granule cells after kainic acid seizures. Hippocampus 13: 806-815.

Szerb JC, O'Regan PA. 1985. Effect of glutamine on glutamate release from hippocampal slices induced by high K$^+$ or by electrical stimulation: Interaction with different Ca^{2+} concentrations. J Neurochem 44: 1724-1731.

van den Berg CJ, Garfinkel D. 1971. A stimulation study of brain compartments. Metabolism of glutamate and related substances in mouse brain. Biochem J 123: 211-218.

van den Berg CJ, Krzalic L, Mela P, Waelsch H. 1969. Compartmentation of glutamate metabolism in brain. Evidence for the existence of two different tricarboxylic acid cycles in brain. Biochem J 113: 281-290.

Varoqui H, Zhu H, Yao D, Ming H, Erickson JD. 2000. Cloning and functional identification of a neuronal glutamine transporter. J Biol Chem 275: 4049-4054.

Vogel R, Hamprecht B, Wiesinger H. 1998a. Malic enzyme isoforms in astrocytes: Comparative study on activities in rat brain tissue and astroglia-rich primary cultures. Neurosci Lett 247: 123-126.

Vogel R, Jennemann G, Seitz J, Wiesinger H, Hamprecht B. 1998b. Mitochondrial malic enzyme: Purification from bovine brain, generation of an antiserum, and immunocytochemical localization in neurons of rat brain. J Neurochem 71: 844-852.

Waagepetersen HS, Bakken IJ, Larsson OM, Sonnewald U, Schousboe A. 1998a. Comparison of lactate and glucose metabolism in cultured neocortical neurons and astrocytes using ^{13}C-NMR spectroscopy. Dev Neurosci 20: 310-320.

Waagepetersen HS, Bakken IJ, Larsson OM, Sonnewald U, Schousboe A. 1998b. Metabolism of lactate in cultured GABAergic neurons studied by ^{13}C nuclear magnetic resonance spectroscopy. J Cereb Blood Flow Metab 18: 109-117.

Waagepetersen HS, Hansen GH, Fenger K, Lindsay JG, Gibson G, et al. 2006. Cellular mitochondrial heterogeneity in cultured astrocytes as demonstrated by immunogold labeling of α-ketoglutarate dehydrogenase. Glia 53: 225-231.

Waagepetersen HS, Qu H, Hertz L, Sonnewald U, Schousboe A. 2002. Demonstration of pyruvate recycling in primary cultures of neocortical astrocytes but not in neurons. Neurochem Res 27: 1431-1437.

Waagepetersen HS, Qu H, Sonnewald U, Shimamoto K, Schousboe A. 2005. Role of glutamine and neuronal glutamate uptake in glutamate homeostasis and synthesis during vesicular release in cultured glutamatergic neurons. Neurochem Int 47: 92-102

Waagepetersen HS, Sonnewald U, Larsson OM, Schousboe A. 1999a. Synthesis of vesicular GABA from glutamine involves TCA cycle metabolism in neocortical neurons. J Neurosci Res 57: 342-349.

Waagepetersen HS, Sonnewald U, Qu H, Schousboe A. 1999b. Mitochondrial compartmentation at the cellular level: Astrocytes and neurons. Ann NY Acad Sci 893: 421-425.

Waagepetersen HS, Sonnewald U, Larsson OM, Schousboe A. 2000. A possible role of alanine for ammonia transfer between astrocytes and glutamatergic neurons. J Neurochem 75: 471-479.

Waagepetersen HS, Sonnewald U, Larsson OM, Schousboe A. 2001a. Multiple compartments with different metabolic characteristics are involved in biosynthesis of intracellular and released glutamine and citrate in astrocytes. Glia 35: 246-252.

Waagepetersen HS, Sonnewald U, Gegelashvili G, Larsson OM, Schousboe A. 2001b. Metabolic distinction between vesicular and cytosolic GABA in cultured GABAergic neurons using ^{13}C magnetic resonance spectroscopy. J Neurosci Res 63: 347-355.

Waagepetersen HS, Sonnewald U, Schousboe A. 2003. Compartmentation of glutamine, glutamate, and GABA metabolism in neurons and astrocytes: Functional implications. Neuroscientist 9: 398-403.

Wadiche JI, Arriza JL, Amara SG, Kavanaugh MP. 1995. Kinetics of a human glutamate transporter. Neuron 14: 1019-1027.

Westergaard N, Drejer J, Schousboe A, Sonnewald U. 1996. Evaluation of the importance of transamination versus deamination in astrocytic metabolism of [U-^{13}C]glutamate. Glia 17: 160-168.

Westergaard N, Sonnewald U, Petersen SB, Schousboe A. 1995. Glutamate and glutamine metabolism in cultured GABAergic neurons studied by ^{13}C NMR spectroscopy may indicate compartmentation and mitochondrial heterogeneity. Neurosci Lett 185: 24-28.

Westergaard N, Sonnewald U, Schousboe A. 1994a. Release of α-ketoglutarate, malate and succinate from cultured astrocytes: Possible role in amino acid neurotransmitter homeostasis. Neurosci Lett 176: 105-109.

Westergaard N, Sonnewald U, Unsgard G, Peng L, Hertz L, et al. 1994b. Uptake, release, and metabolism of citrate in neurons and astrocytes in primary cultures. J Neurochem 62: 1727-1733.

Wood JD, Kurylo E, Tsui SK. 1981. Interactions of di-*n*-propylacetate, gabaculine, and aminooxyacetic acid: Anticonvulsant activity and the γ-aminobutyrate system. J Neurochem 37: 1440-1447.

Yu AC, Drejer J, Hertz L, Schousboe A. 1983. Pyruvate carboxylase activity in primary cultures of astrocytes and neurons. J Neurochem 41: 1484-1487.

Yu AC, Fisher TE, Hertz E, Tildon JT, Schousboe A, et al. 1984. Metabolic fate of [^{14}C]-glutamine in mouse cerebral neurons in primary cultures. J Neurosci Res 11: 351-357.

Yu AC, Schousboe A, Hertz L. 1982. Metabolic fate of ^{14}C-labeled glutamate in astrocytes in primary cultures. J Neurochem 39: 954-960.

Yudkoff M, Daikhin Y, Grunstein L, Nissim I, Stern J, et al. 1996. Astrocyte leucine metabolism: Significance of branched-chain amino acid transamination. J Neurochem 66: 378-385.

Yudkoff M, Nelson D, Daikhin Y, Erecinska M. 1994. Tricarboxylic acid cycle in rat brain synaptosomes. Fluxes and interactions with aspartate aminotransferase and malate/aspartate shuttle. J Biol Chem 269: 27414-27420.

Yudkoff M, Nissim I, Hertz L. 1990. Precursors of glutamic acid nitrogen in primary neuronal cultures: Studies with ^{15}N. Neurochem Res 15: 1191-1196.

Yudkoff M, Nissim I, Pleasure D, 1988. Astrocyte metabolism of [^{15}N]glutamine: Implications for the glutamine–glutamate cycle. J Neurochem 51: 843-850.

Zieminska E, Hilgier W, Waagepetersen HS, Hertz L, Sonnewald U, et al. 2004. Analysis of glutamine accumulation in rat brain mitochondria in the presence of a glutamine uptake inhibitor, histidine, reveals glutamine pools with a distinct access to deamidation. Neurochem Res 29: 2121-2123.

2 Serine, Glycine, and Threonine

T. J. de Koning · S. A. Fuchs · L. W. J. Klomp

1	**Introduction**	**25**
2	**Metabolism of L-Serine, Glycine, and Threonine**	**25**
2.1	L-Serine Biosynthesis	25
2.2	L-Serine Utilization	26
2.2.1	Glycine and One-Carbon Group Production by L-Serine Hydroxymethyltransferases	26
2.2.2	L-Serine and Phospholipid Synthesis	27
2.2.3	L-Serine and Synthesis of Glycine and D-Serine	27
2.3	Glycine Biosynthesis	28
2.4	Glycine Utilization	28
2.5	L-Threonine Utilization	28
3	**L-Serine in the Central Nervous System**	**29**
3.1	L-Serine During Central Nervous System Development	29
3.2	Functions of L-Serine-Derived Metabolites in Central Nervous System	31
3.2.1	Phospholipids and Sphingolipids	31
3.2.2	D-Serine and Glycine	31
3.3	Serine Transport	32
3.4	Clinical Relevance of L-Serine	32
3.4.1	3-Phosphoglycerate Dehydrogenase Deficiency	32
3.4.2	3-Phosphoserine Phosphatase Deficiency	33
3.4.3	Disorders Associated with Alterations in Serine Metabolism	33
3.4.4	Clinical Use of L-Serine or L-Serine-Derived Molecules	34
4	**Glycine in the Central Nervous System**	**34**
4.1	Glycine During Central Nervous System Development	34
4.2	Functions of Glycine in Central Nervous System	35
4.2.1	Glycine Receptors	35
4.2.2	Glycine and N-Methyl-D-Aspartate Neurotransmission	36
4.2.3	Glycine Transporters	37
4.3	Clinical Relevance of Glycine	37
4.3.1	Hyperekplexia	37
4.3.2	Nonketotic Hyperglycinemia	38
4.3.3	Disease States with a Secondary Elevation of Glycine	38
4.3.4	Glycine Deficiency	39
4.3.5	Clinical Use of Glycine	39
5	**L-Threonine in the Central Nervous System**	**39**
5.1	L-Threonine During Development of the Central Nervous System	40

© Springer-Verlag Berlin Heidelberg 2007

5.2	Clinical Relevance of L-Threonine	40
5.2.1	Hyperthreoninemia	40
5.2.2	Disease States with a Secondary Elevation of Threonine	40
5.2.3	Clinical Use of L-Threonine	40
6	***Conclusions***	***41***

Abstract: Amino acids are not only indispensable for protein synthesis but also play important cellular functions. In this chapter, the central nervous system (CNS) functions of serine, glycine, and threonine are discussed. Both the specific functions of the aforementioned amino acids and the functions of their derivatives are reviewed in relation to brain development and neurotransmission. In particular, for serine and glycine exciting new functions were unravelled recently.

List of Abbreviations: CNS, central nervous system; GCS, glycine cleavage system; SHMTs, serine hydroxymethyltransferases; 3-PGDH, 3-phosphoglycerate dehydrogenase; 5,10-MTHF, 5,10-methylenetetrahydrofolate; mSHMT, mitochondrial isoenzyme; cSHMT, cytosolic isoenzyme; SAM, S-adenosylmethionine; HSN I, hereditary sensory neuropathy type I; GlyT1, glycine transporter subtype 1; EDG, endothelial differentiation gene; AMPA, 2-amino-3-hydroxy-5-methyl-4-isoxazolepropionate; MRI, magnetic resonance imaging; CSF, cerebrospinal fluid; 3-PSP, 3-phosphoserine phosphatase; DAO, D-amino acid oxidase; VIAAT, vesicular inhibitory amino acid transporter; NKH, nonketotic hyperglycinemia; PLP, pyridoxal phosphate; ALS, amyotrophic lateral sclerosis; GRs, glycine receptors

1 Introduction

L-Serine, glycine, and L-threonine are three closely related amino acids that share common biochemical pathways and therefore are discussed together in this chapter. However, L-threonine is classified as an indispensable or essential amino acid, whereas both L-serine and glycine are dispensable or nonessential amino acids. Classifying the latter two as nonessential can be rather misleading because these amino acids have very important functions in both development and function of the central nervous system (CNS). In recent years, a vast amount of data on these amino acids has appeared in the literature, and at present it has become difficult to maintain a comprehensive view on their functions in brain tissue. This is particularly true for L-serine and L-serine-derived metabolites. In this chapter, we focus on the synthesis and utilization of L-serine, glycine, and threonine, and subsequently on some specific functions of the amino acids and their metabolites in the CNS. Finally, clinical disorders related to biochemical abnormalities in the pathways, transport, or the receptors of L-serine, glycine, and L-threonine are discussed.

2 Metabolism of L-Serine, Glycine, and Threonine

2.1 L-Serine Biosynthesis

L-Serine can be derived from four possible sources: dietary intake; biosynthesis from the glycolytic intermediate 3-phosphoglycerate; from glycine; and by protein and phospholipid degradation. Few data are available on the relative contribution of each of these four sources of L-serine to overall L-serine homeostasis. However, it is very likely that the predominant source of L-serine varies in different tissues and during different stages of human development. For instance, the majority of L-serine synthesized by the fetal liver comes from glycine by the combined action of the glycine cleavage system (GCS) (EC 1.4.4.2, EC 1.8.1.4, EC 2.1.2.10) and serine hydroxymethyltransferases (SHMTs) (EC 2.1.2.1) (Narkewicz et al., 1996). By contrast in the adult, the kidney synthesizes most of its L-serine from 3-phosphoglycerate (Lowry et al., 1987). In the CNS, it is likely that synthesis from 3-phosphoglycerate is a predominant source of L-serine and that dietary intake of L-serine contributes little or nothing to serine metabolism, and this is discussed in more detail in later sections.

In the L-serine synthetic pathway, 3-phosphoglycerate is converted to L-serine in three consecutive steps (❶ Figure 2-1). Although the presence of the L-serine synthetic pathway in brain tissue was established a long time ago, little is known about the regulation of this pathway in brain tissue. In contrast, regulation of the pathway in the liver has been studied in more detail where, for instance, protein intake, and in particular, that of sulphur-containing amino acids such as methionine and cysteine influences hepatic

◘ **Figure 2-1**
Metabolic pathways of L-serine and glycine. Enzymes involved: I, 3-phosphoglycerate dehydrogenase (3-PGDH); II, 3-phosphohydroxypyruvate aminotransferase; III, 3-phosphoserine phosphatase (3-PSP); IV, serine hydroxymethyltransferases (SHMTs); V, cystathionine b synthase; VI, serine racemase; VII, the synthesis of sphingolipids starts with the condensation of L-serine and palmitoyl-CoA via serine palmitoyltransferase; VIII, phosphatidylserine is derived from phosphatidylcholine and phosphatidylethanolamine by the enzymes phosphatidylserine synthase I and II; IX, glycine cleavage system (GCS). Abbreviations: THF, tetrahydrofolate; 5, 10-MTHF, 5,10-methylenetetrahydrofolate; 5-MTHF, 5-methyltetrahydrofolate

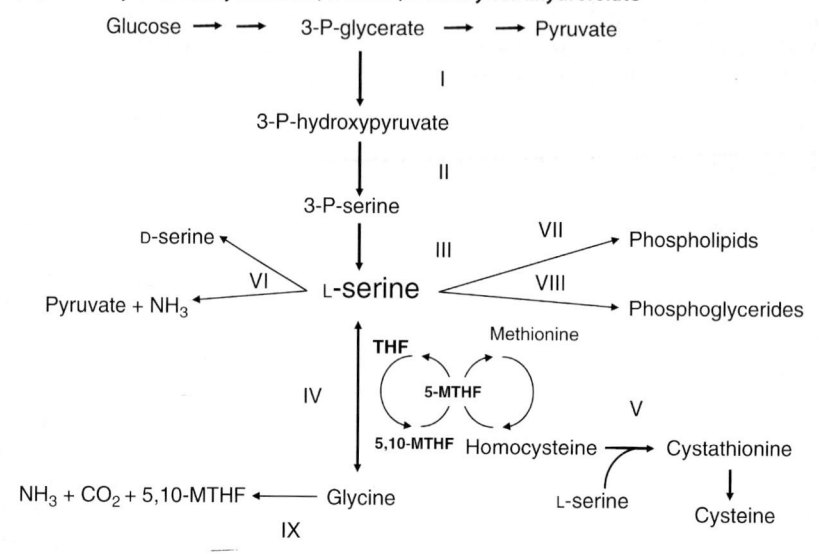

L-serine synthesis (Achouri et al., 1999). The data available suggest that cerebral expression of 3-phosphoglycerate dehydrogenase (3-PGDH) (EC 1.1.1.95) is not regulated by dietary protein intake (Achouri et al., 1997).

2.2 L-Serine Utilization

L-Serine has a well-recognized role in cellular proliferation, being a precursor for nucleotide synthesis (Snell, 1984). However, one should realize that multiple pathways of L-serine utilization are present and that these pathways do not relate only to the supply of nucleotide precursors. The pathways of L-serine utilization known to have a role in brain metabolism are discussed next.

2.2.1 Glycine and One-Carbon Group Production by L-Serine Hydroxymethyltransferases

This pathway of L-serine utilization is a major source of one-carbon groups, providing formyl groups for purine synthesis and methyl groups for pyrimidine synthesis and the remethylation of homocysteine. It further provides methyl groups for many other methylation reactions in cellular homeostasis. SHMTs catalyze the formation of glycine from serine, thereby generating 5,10-methylenetetrahydrofolate (5,10-MTHF) (◙ *Figure 2-1*). Two isoenzymes of SHMT exist: a mitochondrial isoenzyme (mSHMT) and a cytosolic isoenzyme (cSHMT). The reversible interconversion from L-serine to glycine by the different isoenzymes is likely to play a role in maintaining the intracellular concentrations of one-carbon groups in the different cellular compartments. The exchange of L-serine, glycine, and formate through the mitochondrial membrane is an important pathway for the equilibrium of activated intramitochondrial one-carbon groups such as 5,10-MTHF. Mitochondria are the major site for the production of one-carbon groups. L-Serine is the donor of one-carbon groups via mSHMT, subsequently forming formate that exits the mitochondria (Fu et al., 2001). On the other hand in the cytosol, cSHMT is an important regulator of S-adenosylmethionine (SAM) synthesis. In addition, cSHMT is also involved in the supply of one-carbon units for thymidylate synthesis (Herbig et al., 2002).

These functions of the SHMT isoforms have been investigated in mammalian cell cultures, and therefore do not necessarily reflect the functions of SHMTs in the CNS. Very little information is available on SHMTs in brain tissues. Early publications (Daly and Aprison, 1974) gave indirect evidence for the presence of only the mitochondrial isoform in the rat brain, and this was confirmed by the more recent study of Verleysdonk and coworkers (1999) suggesting that the aforementioned functions of the two isoforms of SHMT do not take place in mammalian brain tissue.

2.2.2 L-Serine and Phospholipid Synthesis

In addition to being involved in one-carbon metabolism, L-serine is an important precursor for the synthesis of phosphoglycerides and complex macromolecules such as sphingolipids and glycolipids. In *Escherichia coli* and yeast, the phosphoglyceride phosphatidylserine is synthesized from L-serine and cytidine diphosphodiacylglycerol (CDP-diacylglycerol), whereas in mammalian cells phosphatidylserine is derived from phosphatidylcholine and phosphatidylethanolamine by phosphatidylserine synthases (EC 2.7.8.8) (▶ *Figure 2-1*) (Kuge and Nishijima, 1997). Phosphatidylserine is an important lipid messenger and a key molecule in multiple cellular mechanisms such as the apoptosis-signaling pathways and hemostasis in which externalization of phosphatidylserine is crucial (Tyurina et al., 2000; Zwaal et al., 2004).

The synthesis of sphingolipids starts with the condensation of L-serine and palmitoyl-CoA via serine palmitoyltransferase (EC 2.3.1.50), leading to the formation of ketosphinganine, which is converted to ceramide in three consecutive steps. Ceramide, sphingosine, and sphingosine-1-phosphate are interconvertible sphingolipid metabolites and form the precursors for sphingomyelin and glycosphingolipids such as gangliosides. These L-serine-derived sphingolipids are important membrane components and myelin constituents, and further play a role in cellular differentiation, proliferation, migration, and apoptosis. Moreover, cells with null mutations in serine palmitoyltransferase are not viable, indicating the essential role of sphingolipid synthesis from L-serine for cell survival (Kuge and Nishijima, 1997). Mutations in subunit 1 of serine palmitoyltransferase are a cause of the nonlethal human disorder hereditary sensory neuropathy type I (HSN I), leading to sensory loss and subsequent injuries in, particularly, the lower limbs (Bejaoui et al., 2001).

2.2.3 L-Serine and Synthesis of Glycine and D-Serine

The formation of glycine from L-serine via SHMT is an important reaction since it not only results in the transfer of a one-carbon group to folates but also that glycine itself has important functions particularly in the CNS. In its role as a neurotransmitter, glycine is involved in both inhibitory neurotransmission via the glycine receptor (GR) and excitatory N-methyl-D-aspartate (NMDA) receptor-mediated neurotransmission. D-Serine has recently been discovered to function as neuromodulator in the CNS. D-Serine cannot be qualified as a classical neurotransmitter, because D-serine is not excreted via presynaptic vesicles, but instead released from astrocytes and perhaps also from neurons in the synaptic cleft. Although the presence of significant amounts of D-serine in the human brain was detected in the 1990s, its origin and functions remained unknown for a long time. Recently, the existence of the enzyme DL-serine racemase (EC 5.1.1.10) has been reported that directly converts L-serine into D-serine (Wolosker et al., 1999). When compared to glycine, it was demonstrated that D-serine is a selective and at least equally potent ligand for the "glycine site" of the NMDA receptor. According to some studies, D-serine is the predominant endogenous ligand for NMDA receptors in most mammalian brain regions and the retina, whereas on the other hand, glycine appears to be the principal NMDA receptor ligand in the brain stem, spinal cord, and cerebellum (Fuchs et al., 2005). These specific affinities of glycine and D-serine on NMDA receptor activity underscore the importance of L-serine synthesis and transport in brain tissues, with L-serine being the precursor of both amino acids.

2.3 Glycine Biosynthesis

Glycine is the smallest amino acid and in contrast with all other amino acids glycine is a nonchiral molecule. As is the case for L-serine, glycine can also be obtained from different sources such as the diet, protein breakdown, or synthesis from different precursors such as L-serine, glyoxylate, sarcosine (N-methylglycine) or from condensation of CO_2, NH_3, and 5,10-MTHF. Direct synthesis of glycine from L-serine via SHMTs is discussed earlier (● *Figure 2-1*). Another source of glycine is synthesis from glyoxylate via glycine transaminase (EC 2.6.1.4) or synthesis from sarcosine via sarcosine dehydrogenase (EC 1.5.99.1 and EC 2.1.1.20). However, there is little evidence to support a role of significance for these latter pathways in glycine synthesis.

2.4 Glycine Utilization

Glycine catabolism can take place via the GCS yielding 5,10-MTHF and by the serine hydromethyltransferase resulting in L-serine and tetrahydrofolate. Most publications cite the GCS as the major degradative enzyme complex in glycine catabolism, but it is likely that in the CNS glycine degradation occurs by the combined action of the GCS and mSHMT (Verleysdonk et al., 1999). In this latter study it was also demonstrated that glycine degradation in an astrocyte-rich cell culture results not only in L-serine but also in lactate formation. Another minor pathway of glycine degradation includes the formation of amino acetone (or L-threonine) via 2-amino-3-ketobutyrate ligase (EC 2.3.1.29). However, the latter pathway, namely the interconversion of L-threonine and glycine, is almost nonfunctional in humans and is discussed later in the section on L-threonine catabolism.

Glycine can serve as an amino acid precursor in multiple biochemical reactions. For instance, glycine, glutamate, and cysteine are the precursors for the synthesis of glutathione. At least in cell cultures glycine incorporation from the culture medium into glutathione appears to be significant (Dringen et al., 1998). Insufficient availability of glutathione causes hemolytic anaemia and neurological abnormalities as are observed in patients with glutathione synthesis defects (Ristoff et al., 2001). In another reaction, glycine and L-arginine are the precursors of creatine synthesis and form guanidinoacetic acid by the action of the enzyme arginine/glycine amidinotransferase (EC 2.1.4.1). Defects in creatine synthesis lead to psychomotor retardation, movement and behavioral disorders, and seizures (Schulze, 2003). Furthermore, glycine can be converted to sarcosine via glycine N-methyltransferase (EC 2.1.1.20), of which recently a defect was reported in two siblings with mild hepatomegaly, elevation of serum transaminases, and hypermethioninemia (Luka et al., 2002). Interestingly, glycine and sarcosine display an opposite effect on the glycine transporter subtype 1 (GlyT1), with sarcosine inhibiting the glycine transporter and glycine activating it. In addition, glycine plays a role as a precursor in many other reactions such as the formation of collagen, d-aminolevulinate, heme, and glycocholate. Finally, glycine is used in conjugation reactions involved in detoxification.

2.5 L-Threonine Utilization

In contrast to L-serine and glycine, L-threonine is an indispensable or essential amino acid and uptake from the diet is therefore required. There are no data to support the existence of L-threonine synthesis routes of any significance in humans. In mammals, at least two enzymes are involved in L-threonine utilization, namely L-threonine 3-dehydrogenase (EC 1.1.1.103) and L-threonine dehydratase (EC 4.3.1.19) (● *Figure 2-2*).

◘ Figure 2-2
Metabolic pathways of L-threonine. Enzymes involved: I, L-threonine dehydratase; II, L-threonine 3-dehydrogenase (pseudogene in humans); III, 2-amino-3-ketobutyrate ligase

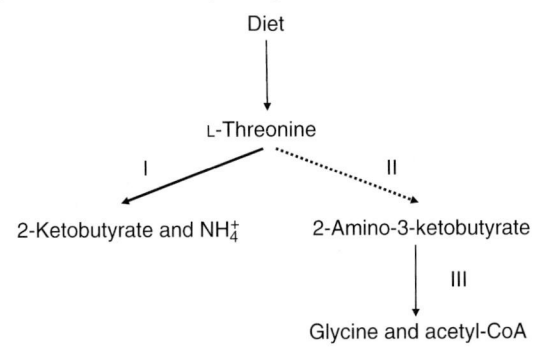

```
                        Diet
                          │
                          ▼
                      L-Threonine
                    I ╱         ╲ II
                    ╱             ╲
     2-Ketobutyrate and NH₄⁺    2-Amino-3-ketobutyrate
                                          │
                                         III
                                          ▼
                                Glycine and acetyl-CoA
```

Theoretically, L-threonine aldolase (EC 4.1.2.5) might play a role, but it is doubtful whether it exists at all in mammals. In humans, the major catabolic pathway of L-threonine is through L-threonine dehydratase, which converts L-threonine to 2-ketobutyrate and NH_4^+. The other catabolic enzyme L-threonine 3-dehydrogenase catalyzes the formation of 2-amino-3-ketobutyrate. The latter can be converted to glycine and acetyl-CoA via 2-amino-3-ketobutyrate ligase (EC 2.3.1.29) or alternatively 2-amino-3-ketobutyrate can be decarboxylated nonenzymatically to aminoacetone. However, only a minority of L-threonine is catabolized via the dehydrogenase pathway, and it was recently demonstrated that in humans L-threonine 3-dehydrogenase is an expressed pseudogene (Edgar, 2002). This limited function of the human dehydrogenase is in contrast with its significant contribution to L-threonine catabolism in other mammals and rodents, where it can serve as a precursor for glycine. Surprisingly, transcripts in human tissues of the other enzyme in the pathway from L-threonine to glycine, namely 2-amino-3-ketobutyrate ligase, are present in significant amounts in multiple tissues including brain (Edgar and Polak, 2000). Human L-threonine aldolase is a nontranscribed pseudogene and is therefore not involved in L-threonine metabolism; so, other putative L-threonine-catabolizing enzymes may well exist (Edgar, 2005).

3 L-Serine in the Central Nervous System

3.1 L-Serine During Central Nervous System Development

During development, the fetus itself is probably responsible for the majority of L-serine and glycine syntheses. Transport of both L-serine and glycine from the mother to the fetus seems to be limited. In fetal sheep, it has been shown that neither L-serine nor glycine is transported across the placenta in significant amounts from mother to fetus (Cetin et al., 1992; Moores et al., 1993). This limited placental transport of amino acids was confirmed for glycine in human pregnancies (Cetin et al., 1995). The human placenta has low SHMT activity, again suggestive of the insignificance of the conversion of serine to glycine, further implying that glycine is produced by the fetus itself (Lewis et al., 2005).

Serine concentrations are high in all body fluids during early fetal development when measured by standard amino acid analysis, thus comprising both D-and L-serine. Once a blood–brain barrier has been established, differences in serine concentration between plasma and CSF emerge (Huether and Lajtha, 1991).

Of all the enzymes involved in L-serine metabolism, 3-PGDH has been studied most extensively during fetal brain development. 3-PGDH is responsible for the first step in the synthesis of L-serine, and 3-PGDH is highly expressed in all fetal tissues, including brain tissue. Data from in situ hybridization of 3-PGDH during development of the CNS show a very strong expression of 3-PGDH mRNA during early fetal development, in particular in the ventricular and subventricular zone of the fetal brain (Yamasaki et al., 2001). At later stages of development 3-PGDH is no longer expressed in neuronal–glial precursors and neurons, but only in glial cells, and mainly in astrocytes. In the adult murine brain, 3-PGDH is highly expressed in the olfactory bulb and the cerebellum, whereas 3-PGDH again is not expressed in neurons but only in the Bergmann glia.

The importance of 3-PGDH for fetal brain development was clearly demonstrated by Yoshida and associates (2004), who showed that a *3-PGDH* gene-knockout mouse resulted in a lethal phenotype with severe neurodevelopmental abnormalities. The 3-PGDH$^{-/-}$ embryos died after embryonic day 13.5. They were much smaller than normal littermates and showed smaller brains with hypoplasia of the telencephalon, diencephalon, mesencephalon, and the complete absence of certain brain structures such as the cerebellum. A subset of embryos had exencephaly. The CNS abnormalities in the 3-PGDH-knockout mice were most likely caused by a diminished proliferation of neuronal–glial precursor cells in the ventricular zone. The 3-PGDH-knockout mouse establishes the importance of this particular L-serine biosynthetic pathway for brain development, despite the fact that L-serine can be obtained through other biochemical pathways. Later in gestation, other factors besides local L-serine synthesis may play a role. Sakai and coworkers (2003) have shown in animals that toward the end of gestation the ASCT-1 amino acid transporter is upregulated in brain capillaries, facilitating L-serine transport over the blood–brain barrier. This suggests a specific CNS requirement or function of L-serine obtained from the circulation during this stage of development.

In addition to L-serine itself, L-serine-derived ligands such as D-serine, glycine, and their respective receptors have important functions during development of the CNS. These complex functions are only briefly summarized in this chapter and the reader is referred to recent reviews on this subject.

Limited information is available on the actual concentrations of D-serine during brain development. D-Serine has been localized predominantly to the rodent and human forebrain, with highest levels in the cerebral cortex, hippocampus, and striatum, followed by the limbic forebrain, diencephalon, and midbrain and low levels in the pons, medulla, cerebellum, and spinal cord (Hashimoto et al., 1993a, b; Nagata et al., 1994). So, low concentrations of D-serine are observed in areas where glycine is an important neurotransmitter and D-serine concentrations are high in areas known to have low glycine concentrations. Immunohistochemical localization of D-serine has suggested a selective localization to protoplasmic type II astrocytes, a subtype of glial cells that ensheathes nerve terminals and is particularly enriched in cortical gray matter. However, very recently, another study showed in contrast to all earlier studies that D-serine is present and produced in neurons as well (Kartvelishvily et al., 2006). These latter experiments shed a new light on possible mechanisms by which D-serine exerts its functions on neurotransmission.

The anatomical distribution of D-serine in the CNS closely mimics that of the NR2 A/B subtypes of the NMDA receptor, and therefore since some years a functional relationship has been eminent from different kinds of studies between D-serine metabolism and NMDA receptor functions (Schell et al., 1997).

The NMDA receptor complex has well recognized functions during fetal brain development (Sugiura et al., 2001). In the immature brain, synaptic transmission is considered to be weak and plastic. Neurotransmission is mediated in large part by NMDA receptors; this is in contrast to adults. Its functions in the fetal brain have been related to activity-dependent plasticity and synaptic refinement. NMDA receptor activation can promote survival of neuronal populations as well as their migration and dendritic outgrowth. Insufficient NMDA receptor activation will result in a loss of cell populations by apoptosis during development (Ikonomidou et al., 1999).

In both NMDA and glycine receptors (GRs), changes in the predominant subunit composition occur. During development NMDA receptors switch from predominantly NR2B subtype to NR2A, thereby supposedly enabling rapid synaptic transmission. Other subunits such as the NR3A have their highest expression shortly after birth (Brody et al., 2005). Interestingly, the NR3 subunit in combination with NR1 can form a GR of which glycine shows an excitatory and D-serine an inhibitory effect (Chatterton et al., 2002). This illustrates again the strong relationships in brain tissue between the L-serine-derived metabolites D-serine and glycine and their respective receptors.

3.2 Functions of L-Serine-Derived Metabolites in Central Nervous System

3.2.1 Phospholipids and Sphingolipids

Savoca and coworkers (1995) were the first to show that L-serine clearly has trophic effects on neurons in culture. They demonstrated in cultured neuronal cells that the addition of L-serine in physiological concentrations had a marked enhancing effect on dendritogenesis and axon length in these cells. Others have confirmed these observations in later experiments (Mitoma et al., 1998a; Furuya et al., 2000). The mechanism through which L-serine influences neuronal function is unknown, but this might be through sphingolipid synthesis. The trophic effects reported on neurons in culture in the presence of L-serine have also been reported for L-serine-derived sphingolipids, and impaired neuronal survival and increased apoptosis were observed in cultured cerebellar Purkinje cells after inhibition of ceramide synthesis (Furuya et al., 1998; Mitoma et al., 1998b).

Astrocytes are the key cells that metabolize L-serine. After synthesis of L-serine they release a significant amount to supply surrounding neurons with the necessary precursors for phospholipid and sphingolipid synthesis. Neurons in culture need exogenous L-serine for the synthesis of L-serine-derived phospholipids and sphingolipids. The biosynthesis of phosphatidylserine and sphingolipids is severely reduced when exogenous L-serine is absent (Mitoma et al., 1998c).

The mechanisms by which sphingolipids exert their function in the CNS at the cellular level were extensively reviewed by Colombaioni and Garcia-Gil (2004) and include many complex effects such as the functioning of lipid rafts and caveolae in cell membranes. Other mechanisms include the activation of a subgroup of G-proteins called the endothelial differentiation gene (EDG)-related receptors by sphingosine-1-phosphate, which results in activation of multiple intracellular signaling pathways. Furthermore, regulation of protein phosphorylation by ceramide through activating several kinases (such as JNK, PKC, and DAPK) and finally sphingosine, sphingosine-1-phosphate, and ceramide also play a role in excitability and neurotransmitter release through several postulated mechanisms. This complexity of functions of sphingolipids is further exemplified by the fact that the composition and roles of sphingolipids are different among the different cell types in brain tissues and at different stages of development (Colombaioni et al., 2004).

Finally, the role of sphingolipids, in particular, the role of ceramide in apoptosis is well recognized, and ceramide is involved in both intrinsic and extrinsic pathways of cell death. The cellular mechanisms involved again are complex and reviewed elsewhere (Luberto et al., 2002).

3.2.2 D-Serine and Glycine

Both D-serine and glycine are synthesized from the same amino acid precursor L-serine, discussed in earlier sections. Until very recently there was only evidence that the metabolism of D-serine and glycine is, similar to that of glutamate, restricted to glial cells and these cells provide the microenvironment for optimal neuronal development and neuronal functioning. Snyder and Kim (2000) proposed a model of the interaction between astrocytes and neurons, including D-serine release. They suggest that glutamate after release from the presynaptic neuron not only binds to the NMDA receptor complex but also triggers the release of D-serine from nearby astrocytes via activation of non-NMDA receptors (possibly 2-amino-3-hydroxy-5-methyl-4-isoxazolepropionate (AMPA) receptors). Subsequently, both glutamate and D-serine activate the postsynaptic NMDA receptor complex. However, Kartvelishvily and coworkers (2006) recently demonstrated that D-serine is also present in neurons and produced indeed in significant amounts via serine racemase. It was also shown that D-serine release was mediated after activation of different types of ionotropic glutamate receptors through a yet unidentified mechanism. The mechanism by which D-serine subsequently is removed from the synaptic cleft is unclear. One likely possibility involves uptake in neurons through the asc1 amino acid transporters (Helboe et al., 2003). Recently, a murine asc1-knockout model was generated and the asc1$^{-/-}$ mice showed a severe and lethal phenotype with tremor, ataxia, and seizures

(Xie et al., 2005). According to the authors, the phenotype resembles NMDA receptor overexcitation and demonstrates the importance of asc1 in regulating synaptic D-serine concentrations.

With the exception of the olfactory bulb in rodents where a high immunoreactivity for both D-serine and glycine has been observed, all other data suggest that either glycine or D-serine is present to activate the NMDA receptor. In part, these differences can be explained by the presence of D-amino acid oxidase (DAO) (EC 1.4.3.3), the enzyme degrading D-serine and not glycine in areas with low D-serine concentrations and high glycine concentrations (Schell et al., 1995; Wang and Zhu, 2003). Similarly, the concentrations of glycine in different areas of the brain have been linked to SHMT activity (Daly and Aprison, 1974).

3.3 Serine Transport

Uptake and transport of L-serine occurs through several transport systems, including the Na^+-dependent neutral amino acid transporter ASCT, system A amino acid transporters, and the Na^+-independent transporter system L asc transporters. The neutral amino acid transporters ASCT1 and ASCT2 transport L-alanine, L-serine, L-cysteine, and L-threonine, whereas ASCT2 in addition also transports L-glutamine, L-asparagine, L-methionine, L-leucine, and glycine albeit with different affinities for the later amino acids. ASCT 1 and -2 do also transport D-amino acids like D-serine, but with much lower affinity compared with that of L-amino acids.

A recent report suggests that L-serine is differently transported into neurons and astrocytes. However, the results from cell cultures contrast with the results from in situ hybridization and immunohistochemistry. In primary cultures of astrocytes and neurons, Yamamoto and coworkers (2003) reported that ASCT1 is the primary neuronal L-serine transporter, whereas in astrocytes both ASCT1 and ASCT2 are involved. L-Serine transport into neurons was faster than in astrocytes. However, Sakai and coworkers (2003) investigated ASCT1 during development and found an expression pattern identical to that of 3-PGDH with ubiquitous expression in neuroepithelial cells in the ventricular zone during early fetal development. At later stages of development, immunostaining for ASCT1 was low in neurons and high in astrocytes, although ASCT1 mRNA remained detectable in neurons. In their studies, ASCT1 and 3-PGDH were clearly coexpressed in astrocytes, suggesting release of L-serine via ASCT1 to meet the metabolic demands of the neurons. They hypothesize that other L-serine transporters, namely the Na^+-dependent system A transporters SAT1 and SAT2 are responsible for neuronal uptake of L-serine. Indeed, SAT 1 and SAT 2 are expressed in neurons (Armano et al., 2002). However, in the experiments of Yamamoto and coworkers (2003) this hypothesis in cultured cells appeared less likely because L-serine transport into cultured neurons occurred in the presence of a system A transporter inhibitor.

The system L, Na^+-independent neutral amino acid transporter consists of two transporters asc1 and asc 2, respectively. Asc 2 is not present in brain tissues and only expressed in peripheral tissues like the kidney. In the CNS, asc1 is found only in neurons and not in glia and white matter. Asc1 is involved in transport of D-serine, but also in glycine transport. Asc1 might play a role in amino acid mobilization and synaptic clearance of D-serine (Helboe et al., 2003; Matsuo et al., 2004). We already discussed in a previous section that very recently a murine asc1 knockout was generated and the animals showed a lethal phenotype with tremors, ataxia, and seizures (Xie et al., 2005), underscoring the functional role of asc1 and suggesting specific biological functions of D-serine.

3.4 Clinical Relevance of L-Serine

3.4.1 3-Phosphoglycerate Dehydrogenase Deficiency

Patients with serine deficiency, due to the autosomal recessive disorder 3-PGDH deficiency, demonstrate severe neurological abnormalities, and again this underscores the importance of the L-serine synthetic pathway in the CNS. In the majority of patients, congenital microcephaly has been observed with severe psychomotor retardation in the first months of life followed by the onset of intractable seizures (de Koning and Klomp, 2004). Cranial magnetic resonance imaging (MRI) of the patients showed severe attenuation of the white matter volume in combination with hypomyelination. Low concentrations of the amino acid serine and to a variable degree of glycine are the biochemical hallmarks of this disorder and are more pronounced in the cerebrospinal fluid (CSF) than in plasma. Oral supplementation of the deficient amino acids L-serine and glycine proved to be very effective in the treatment of seizures (de Koning et al., 1998). Control of seizures was even observed in patients with longstanding seizures unresponsive to many antiepileptic treatment regimens.

Long-term follow-up of patients on amino acid therapy demonstrates that not all remain free of seizures. Furthermore, patients with a relatively late onset of treatment, i.e., after their first year of life, showed no or little progress in psychomotor development during treatment (de Koning et al., 2002). Conceivably, early treatment would improve psychomotor development. This was confirmed by the case of a female baby diagnosed prenatally with 3-PGDH deficiency in whom L-serine therapy to the mother was started at 27 weeks of gestation. With this maternal L-serine therapy a normalization of the initially decelerating head growth was observed by fetal ultrasound and the girl child was born normocephalic. Treatment was continued on the second day of life, and until the age of 4 years the girl has demonstrated a normal psychomotor development (de Koning et al., 2004).

The 3-PGDH gene is located on chromosome band 1q12. So far, only missense mutations encoding the C-terminal part of the protein have been identified. Expression studies of these mutations as well as expression of a mutant protein lacking the last 209 amino acids of the C-terminal region have demonstrated a considerable residual enzyme activity, similar to the 3-PGDH residual activity in fibroblasts from patients with 3-PGDH deficiency (Klomp et al., 2000). These data suggest that only mild mutations are compatible with life and that more severe mutations like in the 3-PGDH-knockout mouse will result in a lethal phenotype.

3.4.2 3-Phosphoserine Phosphatase Deficiency

The second disorder of serine biosynthesis, 3-phosphoserine phosphatase (3-PSP) (EC 3.1.3.3) deficiency, has been reported only in a single case (Jaeken et al., 1997). This patient, a boy who also suffered from Williams–Beuren syndrome, was found to have a low CSF serine concentration, but less pronounced than those observed in 3-PGDH deficiency. This boy was also treated with L-serine and a favorable response to L-serine therapy, most notably an increase in head circumference, was reported. Unfortunately, the patient was lost in follow-up and there are no data on long-term effects of therapy. Recently, the molecular defect in this patient with 3-PSP deficiency was confirmed (Veiga-da-Cunha et al., 2004). To date, no other patients with 3-PSP have been reported neither in isolated form nor in combination with Williams–Beuren syndrome.

3.4.3 Disorders Associated with Alterations in Serine Metabolism

Serine deficiency of unknown cause has been described in another single case report of a girl with progressive polyneuropathy, ichthyosis, growth retardation, and delayed puberty. No deficiency in any of the three L-serine biosynthetic enzymes could be identified and the basic defect remains to be resolved. Nevertheless, treatment with L-serine resulted in a clear and sustained improvement of muscle strength and ichthyosis.

In the past, elevated concentrations of serine and glycine have been associated with schizophrenia in some studies and a possible role of the NMDA receptor complex has been suggested. However, these elevations of serine and glycine concentrations in schizophrenia were not confirmed by others and more recently a different hypothesis suggesting a hypofunction of the NMDA receptor was proposed and linked to D-serine metabolism (Tsai and Coyle, 2002). A recent study demonstrated a genetic association between schizophrenia and DAO and G72, a DAO-activator protein, in a large series of patients (Chumakov et al., 2002). This study suggests that D-amino acids and most likely D-serine metabolism is indeed associated with schizophrenia. Plasma D-serine concentrations have been investigated in a limited number of patients with schizophrenia. Hashimoto and coworkers (2003) found that plasma D-serine concentrations were lower in a group of schizophrenia patients when compared with healthy controls. Although this study favors the hypothesis that D-serine is involved in the pathogenesis of schizophrenia, it is doubtful whether plasma D-serine concentrations actually reflect D-serine concentrations in the brain, given the role of the kidney in the metabolism of D-serine. The same authors more recently reported CSF D- and L-serine concentrations in patients with schizophrenia and showed that the D-serine/total serine ratio was lower in schizophrenia compared with healthy controls. However, the observed differences can also be explained by altered L-serine concentrations in patients with schizophrenia (Hashimoto et al., 2005).

3.4.4 Clinical Use of L-Serine or L-Serine-Derived Molecules

In the previous sections, the clinical use of L-serine and L-serine-derived metabolites such as glycine and D-serine has been briefly mentioned. Obviously, L-serine and glycine have been used to treat deficiencies in the serine biosynthesis disorders. Some L-serine-derived molecules have also been used in groups of patients without evidence of deficiency but as a pharmacological compound.

Some evidence for this came from clinical studies in which adding glycine, D-serine, or D-cycloserine was indeed beneficial for the negative symptoms in some patients with schizophrenia. However, a systematic review on the pharmacological use of D-serine and glycine demonstrated an effect only on the negative symptoms of schizophrenia and not on the positive symptoms, with the magnitude of the effect being rather moderate (Tuominen et al., 2005).

Phosphatidylserine has been used as a "brain specific" nutrient in the elderly to prevent or improve declining memory. Use of phosphatidylserine in aging animals has shown promising some results (Casamenti et al., 1991). However, the limited number of studies in humans showed at best only modest effects and the clinical relevance of phosphatidylserine needs to be proven (McDaniel et al., 2003).

4 Glycine in the Central Nervous System

4.1 Glycine During Central Nervous System Development

For many years glycine gained most attention being an inhibitory neurotransmitter, in particular in the spinal cord and brain stem. Nowadays, a surprising amount of specific functions of glycine are known in the CNS and new and exciting functions of glycine are still being reported. Glycine exerts its functions during fetal development through GRs, glycine transporters, and the NMDA receptor. The first two are briefly discussed here and in more detail in subsequent sections; the NMDA receptor was briefly discussed in ❯ Sect. 3.

During fetal development, functional GRs are observed as early as the neural stem/progenitor cell lineages (Nguyen et al., 2002). Functional GRs are present on immature migrating neurons, and several studies have demonstrated that glycinergic neurotransmission actually occurs at an earlier developmental stage than glutaminergic neurotransmission (Tapia et al., 2001).

In contrast to postnatal development, GRs cause depolarization in fetal neurons, and glycine functions as an excitatory neurotransmitter during development. The same is true for $GABA_A$ receptors, which also operate as excitatory receptors during development. It has been suggested that glycine and the activation of GRs play a role in cellular signaling for neuronal development and the maturation of synapses. Tapia and coworkers (2001) demonstrated that synaptic GRs can indeed regulate neurite outgrowth in developing spinal cord neurons. A reduction of glycinergic neurotransmission enhanced neurite outgrowth effectuated via increased calcium influx. The developmental switch from an excitatory to an inhibitory GR is mediated by the expression of a cotransporter KCC2, a neuronal K^+/Cl^- cotransporter. KCC2 lowers the cellular chloride concentration, thereby shifting the chloride equilibrium toward more negative values, and thus converting the GR from excitatory to inhibitory. An identical mechanism during development is observed for $GABA_A$ receptors where the switch between excitatory and inhibitory function is also mediated by KCC2 expression (Stein and Nicoll, 2003).

Not all GRs are synaptically located and in distinct areas neuronal GRs are present, without evidence for glycinergic neurotransmission. These extrasynaptic GRs are also activated by taurine and possibly play a role in neurite outgrowth like the synaptic GRs (Mori et al., 2002). Like other neurotransmitter receptors during development, fetal GR composition differs from adult GR subunit composition and consists of a monomeric α or heteromeric α2β subunit, and during development this changes to the adult α1β subunit in the GR.

Limited data are available on glycine concentrations during brain development. Glycine concentrations are higher in the spinal cord than in the cortex (Aprison et al., 1969). The cortical glycine concentrations vary in different developmental stages. When compared to GABA concentrations glycine and GABA show different developmental profiles in the murine cortex. Cortical glycine concentrations are relatively high and higher than those of GABA during the proliferative stage of fetal brain development and show a decrease toward the end of gestation, with a further reduction postnatally. This is in contrast with GABA, the concentrations of which increase during prenatal and postnatal development. However, when compared to other amino acids the actual concentrations of glycine and GABA are not very high. Taurine concentrations are several times higher than those of both glycine and GABA (Benitez-Diaz et al., 2003). Taurine is a ligand for both extrasynaptic GRs and $GABA_A$ receptors, and the fact that it is abundant during fetal development suggests specific functions in corticogenesis. Accordingly, taurine is the principal ligand for GRs on immature migrating and differentiating neurons during corticogenesis, given the relatively low concentrations of glycine and the abundance of taurine (Flint et al., 1998).

4.2 Functions of Glycine in Central Nervous System

Glycine is a classical neurotransmitter since it is released from synaptic vesicles and removed in the synaptic cleft by specific glycine transporters. As we discuss in more detail later, the neurotransmitter functions comprise both inhibitory and excitatory neurotransmission through GRs and NMDA receptors.

L-Serine is likely to be the most important precursor for the synthesis of glycine in the CNS and this occurs through the mSHMT (Shank and Aprison, 1970; Daly et al., 1976). Similar to the production of its precursor L-serine, glycine is likely to be produced within the CNS given the low affinity for glycine of the neutral amino acid transporter at the blood–brain barrier (Smith et al., 1987). Given the functions of glycine, its synthesis and breakdown will be tightly regulated and together with glycine transporters will determine glycine concentrations.

Glycine and GABA share a common vesicular inhibitory amino acid transporter (VIAAT), which can handle glycine, GABA, or both. The ability to transport glycine or GABA may depend on the relative extravesicular concentration; according to some studies glycine inhibits GABA uptake and vice versa (Christensen and Fonnum, 1991). It is suggested that uptake mainly depends on the synthesis of GABA and glycine and their transport at the plasma membrane. Glycine and GABA are coreleased in a considerable number of synapses, for instance in brain stem motor neurons, spinal cord, superior olivary complex, and cerebellar Golgi cells (Aragon and Lopez-Corcuera, 2003). However, the VIAAT is not present in all nerve endings containing glycine and GABA; so, additional transporters must exist or alternative modes of neurotransmitter release take place.

Glycine has been linked to trophic effects on neurons both in vivo and in vitro. Trophic effects of both glycine and L-serine were reported in cultured cerebellar Purkinje neurons. However, it is possible that these effects relate to L-serine-derived phospholipids such as ceramide. Trophic effects on cerebellar Purkinje neurons were also reported for ceramide and the synthesis of phospholipids is severely decreased when glycine or L-serine is absent (Mitoma et al., 1998c).

4.2.1 Glycine Receptors

Glycine acts as an inhibitory neurotransmitter at specific GRs, predominantly in the brain stem and spinal cord. Mammalian structures with a high density of inhibitory GRs are the medulla and spinal cord and lower levels of expression are observed in the pons, thalamus, and hypothalamus with low expression in higher brain regions. For detailed information on expression of GRs, we refer to excellent reviews on this subject (Legendre, 2001; Lynch, 2004). Besides this clustering of GRs in spinal cord and brain stem neurons, GRs are present in other brain areas such as the cortex and hippocampus. However, glycinergic neurotransmission was only observed in cerebellar Golgi cells, retinal ganglions, and neurons of the ventral tegmental area, indicating that cortical and hippocampal GRs might be extrasynaptically located and have a somewhat different function. It was shown that some glial cells contain GRs together with the capacity of glia cells to release glycine, and therefore glycine may also act as a neuromodulator.

The GR is a member of the neurotransmitter-gated ion channel superfamily. Other members of this family are the nicotinic acetylcholine receptor and $GABA_A$ and $GABA_C$ receptors. These ligand-gated ion channels are involved in mediating fast neurotransmission in the CNS. The GR is a ligand-gated chloride channel activated by glycine and antagonized by strychnine. For this reason GRs are also called strychnine-sensitive GRs. They are clustered on the postsynaptic membrane and upon binding of glycine the ion channel opens, leading to an influx of chloride ions. The influx of chloride results in hyperpolarization, thereby increasing the threshold for neuronal activation and subsequently leading to synaptic inhibition.

The GR consists of three subunits: $\alpha 1$, $\alpha 2$, and β that form a pentameric complex mediating chloride channel function (Betz et al., 1999). In a minority of GRs a $\alpha 3$ subunit is found. Two GR isoforms have been identified, a fetal form consisting predominantly of $\alpha 2$ homomers and an adult form consisting of $\alpha 1 \beta$ heteromers.

4.2.2 Glycine and *N*-Methyl-D-Aspartate Neurotransmission

NMDA receptors have a high affinity for binding glycine, much higher than displayed by GRs discussed in the previous section. NMDA receptors are important for excitatory glutamatergic neurotransmission. They are members of a class of ionotropic receptor channels, organized as heteromeric assemblies composed of an NR1 subunit, combined with at least one of four NR2 (A–D) subunits. NMDA receptors are unique in requiring simultaneous ligand binding at two sites for activation. Currently, it is generally agreed that glutamate molecules bind to the NR2 subunit, while the "glycine binding site" is located on the NR1 subunit. In addition to glutamate, glycine was assumed to be the necessary physiological co-agonist, reacting with the strychnine-insensitive "glycine site" of the NMDA receptor. However, immunohistochemical localization of glycine failed to show a specific pattern of colocalization with the NMDA receptor, except in the brain stem. More recent data strongly support a role for D-serine as the predominant endogenous ligand for NMDA receptors in most mammalian brain regions (and the retina), whereas glycine is the principal ligand in the brain stem, spinal cord, and cerebellum. In the past, it was suggested that given the high affinity of NMDA receptors for glycine the glycine-binding sites were saturated at the physiological concentrations observed for glycine in tissue and CSF. However, this is not the case in the synapse because exogenously added glycine potentiates NMDA receptor activation at low glycine concentrations, and high glycine concentrations prime NMDA receptors for internalization. Indeed, similar to AMPA receptors NMDA receptors are not static on the synaptic surface but are also internalized after agonist binding. Glycine (but likely also D-serine or perhaps both amino acids) plays a particular role in this internalization process in that it primes the NMDA receptor for internalization independent of glutamate, and that internalization only takes place after agonist binding of both ligands, e.g., glutamate or NMDA and glycine or D-serine (Nong et al., 2003). Regulation of NMDA receptor internalization by glycine priming might play important roles determining the receptor numbers during development, subunit composition, or modulation of NMDA receptor currents.

Finally, a new NMDA receptor subunit expressed in the spinal cord was recently reported, extending the properties of glycine as an excitatory neurotransmitter. This NR3 subunit can form a complex with NR1 subunits and produces a unique receptor that is excited by glycine and inhibited by D-serine (Chatterton et al., 2002).

4.2.3 Glycine Transporters

After release in the synaptic cleft, glycine like other neurotransmitters is promptly removed by specific glycine transporters to stop signaling. Two specific glycine transporters are known to date, namely GlyT1 and GlyT2, and furthermore, glycine can also be transported by other amino acid transporters such as system A family SNAT (small neutral amino acid transporters) transporters, but the latter is not discussed in this chapter. Different transcripts of GlyT1 are recognized, GlyT1 (a, b, c, e, f), (Aragon and Lopez-Corcuera, 2003) of which not all isoforms have been isolated in humans. Of GlyT2, three isoforms are found namely GlyT2a, -b, and -c. GlyT1 and GlyT2 both have a considerable degree of homology and a high substrate specificity for glycine, but can be distinguished pharmacologically because only GlyT1 is inhibited by sarcosine.

The two glycine transporters have distinct functions in the CNS and spinal cord. GlyT1 is expressed in regions of the brain not associated with glycinergic inhibition such as the hippocampus, cerebellum, and brain hemispheres. GlyT1 is present mainly in glial cells (and in particular in astrocytes) and selected neuronal populations in the spinal cord and forebrain (Cubelos et al., 2005). It has been suggested that GlyT1 is involved in regulating excitatory NMDA synapses by controlling glycine levels. In contrast to GlyT1, GlyT2 is exclusively present in glycinergic neurons (Zafra et al., 1995). GlyT2 is predominantly colocalized in axons and presynaptic terminals of glycinergic neurons in the brain stem and spinal cord, and therefore it was suggested that it might play a role in removing glycine from the synaptic cleft (Zafra et al., 1995). However, recently important information about the specific roles of the glycine transporters was obtained from GlyT1- and GlyT2-knockout mice, which demonstrated surprisingly a role for GlyT1 in lowering extracellular glycine levels at glycinergic synapses. This is in contrast to what was suggested by previous studies. GlyT2 was shown to be essential for glycine uptake and recycling in the presynaptic cytosol (Gomeza et al., 2003a, b). These elegant studies reveal complementary the functions of GlyT1 and GlyT2 in removing glycine from the synapse and providing glycine for neurotransmitter release. The role of GlyT1 in glutamatergic NMDA receptor neurotransmission needs to be established further, but at present GlyT1 might be related to prevention of saturation of the NMDA receptor glycine-binding site and in the regulation of efficacy of the receptors through controlling glycine concentrations (Eulenburg et al., 2005).

What was very striking was that the phenotype and electrophysiological studies in the GlyT1- and GlyT2 knockouts resembled very much known glycine disease entities. The GlyT1 knockout displayed a lethal phenotype and showed features of glycine encephalopathy or nonketotic hyperglycinemia, and the GlyT2-knockout mice showed a phenotype comparable with hyperekplexia or startle disease. Indeed mutations in human GlyT2 (*SLC6A5*) were subsequently found in patients with hyperekplexia based on the observations in the knockout mice (Rees et al., 2006).

4.3 Clinical Relevance of Glycine

4.3.1 Hyperekplexia

Hyperekplexia or hereditary startle disease is a predominantly autosomal dominant disorder caused by a defective function of the GR. In rare cases, the disorder is caused by autosomal recessive defects of GLRA1, the β-subunit of the GR, mutations in the GR-clustering proteins gephyrin and collybistin, or by defects in the glycine transporter GlyT2 (Rees et al., 2006). The main features of the disorder are disturbances of muscle tone and an exaggerated startle response usually to auditory stimuli. In the newborn period a generalized stiffness or "stiff baby syndrome" is present from birth onward. Hypertonia, hyperreflexia, and exaggerated startle response to noise or handling complicated by feeding difficulties, aspiration pneumonia, inguinal hernias, and an increased risk of sudden death are features in newborns. Nose tapping in a newborn with hyperekplexia typically induces a startle (head retraction) response and provides an important diagnostic clue. After the first years of life the generalized stiffness diminishes. Mental development is usually normal. In adults, an exaggerated startle response to auditory stimuli occurs with unprotected falling with normal consciousness, which may lead to serious head injuries (Bakker et al., 2006). Mutations in autosomal dominant hyperekplexia are predominantly observed in the α1 subunit (GLRA1) of the GR leading to uncoupling of ligand binding and chloride channel function (Shiang et al., 1993), either by disrupting GR surface expression or by reducing the ability to conduct chloride ions. Mutations in GLRA1 impair glycinergic inhibitory neurotransmission in reflex circuits of the spinal cord and brain stem, increasing the general level of excitability of motor neurons.

Treatment of hyperekplexia patients consists of clonazepam administration. The drug enhances the GABA-gated chloride channel function, and this most likely compensates the impaired glycine-gated function.

4.3.2 Nonketotic Hyperglycinemia

Glycine encephalopathy or nonketotic hyperglycinemia (NKH) is characterized by accumulation of glycine in body fluids. It is a severe autosomal recessive disorder within the neonatal form or classical NKH, many patients not surviving the newborn period (Hoover-Fong et al., 2004). In a typical case, a severe encephalopathy develops after a short symptom-free interval with lethargy, hypotonia, abnormal eye movements, hypoventilation, and apnea. The symptoms progress to coma with myoclonus and seizures. In the second to third week, spontaneous breathing may resume. However, finally a very severe psychomotor retardation with spastic quadriplegia and intractable seizures develops, with some patients who will be able to stand and communicate. A striking difference was observed in the outcome between boys and girls with an overall outcome being more favorable in boys (Hoover-Fong et al., 2004). In patients with NKH, brain malformations can be found indicating altered early fetal brain development. Dysgenesis of the corpus callosum is most frequently observed, suggesting an early insult to the developing brain at around 11–20 weeks of gestation.

Biochemical investigations show markedly elevated glycine concentrations in all body fluids. CSF glycine concentrations are elevated to a greater extent than plasma concentrations. Therefore, an elevated glycine CSF/plasma ratio is diagnostic of the disorder and CSF glycine concentrations can be found to be up to 30 times the normal values.

In rare cases, less severe late-onset and transient forms of NKH have been reported, and these patients certainly have a better prognosis. Seizures are usually the presenting symptoms in the less severe forms, but psychomotor retardation and movement disorders have also been reported as presenting symptoms. Presentation of late/atypical NKH has been reported up to adulthood (39 years). In these milder-affected patients an abnormal CSF/plasma glycine ratio is also present albeit that lower concentrations of glycine are observed than in classical cases. In a few cases, a transient form of NKH was diagnosed with a favorable outcome in some and a poor outcome in other patients (Aliefendioglu et al., 2003). Milder mutations with considerable residual activity were found in these patients.

NKH is caused by defects in the GCS. The GCS comprises different subunits: P-protein or glycine decarboxylase, T-protein or aminomethyltransferase, and H-protein or the hydrogen carrier protein. The majority of mutations are observed in the P-protein, less in the T-protein (Applegarth and Toone, 2004), and only in a single case, an abnormality in the H-protein.

4.3.3 Disease States with a Secondary Elevation of Glycine

Elevated concentrations of glycine in plasma, urine, and CSF are frequently observed in patients using the antiepileptic drug sodium valproate, and this is important when a diagnosis of NKH is suspected.

Recently, a new epileptic encephalopathy in newborns was reported (Clayton et al., 2003) in which glycine in both plasma and CSF was also highly elevated, although the values in the CSF were lower than usually in NKH discussed in the previous section. Furthermore, the elevation in glycine is accompanied by an elevation in threonine and low values for biogenic amines in the CSF. The seizures responded well to oral administered pyridoxal phosphate (PLP). PLP-responsive seizures are caused by mutations in pyridox(am)ine 5′-phosphate oxidase (EC 1.4.3.5) causing a defect in PLP-dependent enzymes (Mills et al., 2005). Elevations in glycine are therefore supposed to be related to an inhibition of the GCS for which PLP is a required cofactor. The significance of the elevated glycine in the pathogenesis of the disorder is yet unclear.

Glycine concentrations are secondarily elevated in a number of other inborn errors of metabolism. For instance, in ketotic forms of hyperglycinemia such as propionic acidemia and methylmalonic acidemia glycine concentrations are very high and most likely caused by an inhibition of the GCS by the toxic accumulation of propionyl-CoA.

Furthermore, plasma glycine is elevated in D-glyceric acidemia owing to D-glycerate kinase (EC 2.7.1.31) deficiency. Only a limited number of patients reported with this disorder have been associated with psychomotor retardation and seizures. Elevated CSF glycine has also been reported in vanishing white matter disease (van der Knaap et al., 1999). In these patients with a severe and progressive white matter disorder, glycine concentrations are moderately elevated with an abnormal CSF/plasma ratio. No defects in GCS are present in these patients, and as in all disorders mentioned in this section, glycine can be used as an additional diagnostic biochemical marker.

4.3.4 Glycine Deficiency

To date, only one case has been reported with glycine deficiency (Quackenbush et al., 1999). In a boy with xeroderma pigmentosum consistently low concentrations of glycine were found. No enzymatic or molecular defect was established. A good clinical response was noted to oral glycine therapy.

4.3.5 Clinical Use of Glycine

Glycine is used in the treatment of inborn errors of metabolism such as isovaleric acidemia, glycine and serine deficiency disorders. Glycine is used in these disorders to conjugate abnormal organic acids or to treat deficiencies. Identically to D-serine, glycine has been used in psychiatric disorders based on the hypothesis that the NMDA receptor under stimulation is a factor in the pathogenesis of psychiatric diseases. A systematic review has shown that only small numbers of patients have been treated with glycine added to regular medication and small and not always consistent effects were observed mainly on the negative symptoms of the disorder (Tuominen et al., 2006).

Apart from the CNS, glycine has been reported to have immunomodulatory properties, and clinical application of glycine has been attempted in a wide range of disorders, including ischemic stroke and effects of alcoholism. In most reports the results were promising but not systematically investigated (Gundersen et al., 2005).

5   L-Threonine in the Central Nervous System

It is obvious that L-threonine, being an essential amino acid, is necessary for protein synthesis in the CNS. Recent data show that the CNS senses changes in essential amino acid concentrations very rapidly, in particular in response to dietary sources deficient of essential amino acids such as L-threonine, and this must have played an important role in human evolution (Hao et al., 2005). In contrast, there is surprisingly little data available on specific functions of L-threonine in the CNS. In animals, it most likely serves as an important precursor for glycine. But as already discussed earlier, the human L-threonine dehydrogenase is a pseudogene and displays hardly any enzyme activity, and therefore L-threonine is unlikely to be a precursor for glycine in humans.

L-Threonine has gained most of its attention in relation to infant formulas. Infants and in particular preterm infants fed a whey-protein formula develop mild hyperthreoninemia. This is caused by the high L-threonine content of bovine whey protein. The plasma threonine concentrations of premature infants on an infant-formula diet can be up to twice as high as what is observed in children receiving human milk. The clinical significance of the elevated plasma threonine levels remains unclear. In several studies a correlation was observed between plasma threonine and brain threonine concentrations, causing concern about the elevated plasma concentrations in premature infants (Boehm et al., 1998). However, animal data suggest that L-threonine is a relatively nontoxic amino acid. Elevated dietary threonine intake in rats resulted in a reduced growth and body weight and was likely to be caused by an amino acid imbalance and not by a direct toxic effect (Castagne et al., 1996). In the same study, high L-threonine intake did not result in significant changes in behavior nor was neurological deficits observed. When interpreting such animal data, concerning safety issues and translating the consequences to humans, one need to bear in mind that L-threonine catabolism is different in humans and that there is a limited capacity of L-threonine catabolic pathways.

5.1 L-Threonine During Development of the Central Nervous System

There are no data available about specific functions of L-threonine in the CNS and the same is true for L-threonine during development of the CNS. L-Threonine is transported by neutral amino acid transporters such as the Na^+-dependent ASCT1 and -2. Further research is warranted on functions of L-threonine in the CNS. It is very likely, given the many known and specific functions of other amino acids in the CNS, that L-threonine is not only indispensable for protein synthesis but also has specific functions itself.

5.2 Clinical Relevance of L-Threonine

5.2.1 Hyperthreoninemia

Hyperthreoninemia as a possible inborn error of metabolism has been reported only in two case reports. Reddi (1978) was the first to report hyperthreoninemia in an 8-month-old boy with seizures and growth retardation. High concentrations of threonine were found in the plasma and urine of this boy. No elevations of other amino acids were detected. A threonine-loading test led to a further increase of plasma threonine. No enzyme studies were performed in this patient and follow-up was not reported.

Another case report involves two sibs with Leber's congenital amaurosis and hyperthreoninemia. Extensive ophthalmological abnormalities were present with additional features being anemia, hepatomegaly, elevated transaminases, and pericardial effusions. In one of the children other amino acids such as serine, glycine, and hydroxyproline were elevated as well (Hayasaka et al., 1986).

5.2.2 Disease States with a Secondary Elevation of Threonine

The most common cause of high threonine appears to be related to dietary intake of the amino acid, as already discussed in a previous section. In PLP-responsive seizures not only an elevated CSF glycine is observed but also a marked increase of threonine in the plasma and CSF (Clayton et al., 2003). Given the fact that elevated threonine concentrations in the CSF are rarely observed, threonine is an important diagnostic marker for this disorder. L-Threonine dehydratase is a PLP-dependent enzyme and dysfunction of the dehydratase is therefore suggested to cause an elevation in threonine. The clinical relevance of the elevated CSF threonine in itself is not known.

5.2.3 Clinical Use of L-Threonine

L-Threonine has been used as a potential therapy in a number of neurological disorders. For instance, L-threonine has been used in the treatment of spasticity, and several studies have been conducted albeit with small numbers of patients. These studies suggested an effect of oral L-threonine on spinal spasticity. In two such trials, a double-blind controlled study was performed that showed that this treatment improved motor functions and gave some relief of spasticity, but that improvement was modest at best (Growdon et al., 1991; Lee and Patterson, 1993). Unfortunately, no recent studies are available to confirm these interesting results and to shed some light on the possible mechanisms by which L-threonine exerts a function on the muscle tone.

L-Threonine has also been used as a possible therapy for amyotrophic lateral sclerosis (ALS). A systematic review of the data available on this subject revealed no evidence that L-threonine therapy influences beneficially the natural course of this severe disorder (Parton et al., 2003). A similar systematic review failed to reveal clear effects of L-threonine in the treatment of multiple sclerosis.

6 Conclusions

In recent years the list of functions of L-serine and glycine in the CNS has expended rapidly and we have seen many exciting developments related to these two amino acids. Although L-serine and glycine have much in common, a major difference between the two is the fact that L-serine operates as the precursor for many CNS metabolites and glycine is predominantly a neurotransmitter. In contrast to L-serine and glycine, there is surprisingly little information on L-threonine. The sparse data available on the clinical use of L-threonine in spinal spasticity suggest that it may have specific functions and warrants further research.

References

Achouri Y, Rider MH, Schaftingen EV, Robbi M. 1997. Cloning, sequencing and expression of rat liver 3-phosphoglycerate dehydrogenase. Biochem J 323: 365-370.

Achouri Y, Robbi M, Van Schaftingen E. 1999. Role of cysteine in the dietary control of the expression of 3-phosphoglycerate dehydrogenase in rat liver. Biochem J 344: 15-21.

Aliefendioglu D, Tana Aslan A, Coskun T, Dursun A, Cakmak FN, et al. 2003. Transient nonketotic hyperglycinemia: Two case reports and literature review. Pediatr Neurol 28: 151-155.

Applegarth DA, Toone JR. 2004. Glycine encephalopathy (nonketotic hyperglycinaemia): Review and update. J Inherit Metab Dis 27: 417-422.

Aprison MH, Shank RP, Davidoff RA. 1969. A comparison of the concentration of glycine, a transmitter suspect, in different areas of the brain and spinal cord in seven different vertebrates. Comp Biochem Physiol 28: 1345-1355.

Aragon C, Lopez-Corcuera B. 2003. Structure, function and regulation of glycine neurotransporters. Eur J Pharmacol 479: 249-262.

Armano S, Coco S, Bacci A, Pravettoni E, Schenk U, et al. 2002. Localization and functional relevance of system A neutral amino acid transporters in cultured hippocampal neurons. J Biol Chem 277: 10467-10473.

Bakker MJ, van Dijk JG, van den Maagdenberg AM, Tijssen MA. 2006. Startle syndromes. Lancet Neurol 5: 513-524.

Bejaoui K, Wu C, Scheffler MD, Haan G, Ashby P, et al. 2001. SPTLC1 is mutated in hereditary sensory neuropathy, type 1. Nat Genet 27: 261-262.

Benitez-Diaz P, Miranda-Contreras L, Mendoza-Briceno RV, Pena-Contreras Z, Palacios-Pru E. 2003. Prenatal and postnatal contents of amino acid neurotransmitters in mouse parietal cortex. Dev Neurosci 25: 366-374.

Betz H, Kuhse J, Schmieden V, Laube B, Kirsch J, et al. 1999. Structure and functions of inhibitory and excitatory glycine receptors. Ann N Y Acad Sci 868: 667-676.

Boehm G, Cervantes H, Georgi G, Jelinek J, Sawatzki G, et al. 1998. Effect of increasing dietary threonine intakes on amino acid metabolism of the central nervous system and peripheral tissues in growing rats. Pediatr Res 44: 900-906.

Brody SA, Nakanishi N, Tu S, Lipton SA, Geyer MA. 2005. A developmental influence of the N-methyl-D-aspartate receptor NR3A subunit on prepulse inhibition of startle. Biol Psychiatry 57: 1147-1152.

Casamenti F, Scali C, Pepeu G. 1991. Phosphatidylserine reverses the age-dependent decrease in cortical acetylcholine release: A microdialysis study. Eur J Pharmacol 194: 11-16.

Castagne V, Maire JC, Gyger M. 1996. Neurotoxicology and amino acid intake during development: The case of threonine. Pharmacol Biochem Behav 55: 653-662.

Cetin I, Fennessey PV, Sparks JW, Meschia G, Battaglia FC. 1992. Fetal serine fluxes across fetal liver, hindlimb, and placenta in late gestation. Am J Physiol 263: E786-E793.

Cetin I, Marconi AM, Baggiani AM, Buscaglia M, Pardi G, et al. 1995. In vivo placental transport of glycine and leucine in human pregnancies. Pediatr Res 37: 571-575.

Chatterton JE, Awobuluyi M, Premkumar LS, Takahashi H, Talantova M, et al. 2002. Excitatory glycine receptors containing the NR3 family of NMDA receptor subunits. Nature 415: 793-798.

Christensen H, Fonnum F. 1991. Uptake of glycine, GABA and glutamate by synaptic vesicles isolated from different regions of rat CNS. Neurosci Lett 129: 217-220.

Chumakov I, Blumenfeld M, Guerassimenko O, Cavarec L, Palicio M, et al. 2002. Genetic and physiological data implicating the new human gene G72 and the gene for D-amino acid oxidase in schizophrenia. Proc Natl Acad Sci USA 99: 13675-13680.

Clayton PT, Surtees RA, De Vile C, Hyland K, Heales SJ. 2003. Neonatal epileptic encephalopathy. Lancet 361: 1614.

Colombaioni L, Garcia-Gil M. 2004. Sphingolipid metabolites in neural signalling and function. Brain Res Brain Res Rev 46: 328-355.

Cubelos B, Gimenez C, Zafra F. 2005. Localization of the GLYT1 glycine transporter at glutamatergic synapses in the rat brain. Cereb Cortex 15: 448-459.

Daly EC, Aprison MH. 1974. Distribution of serine hydroxymethyltransferase and glycine transaminase in several areas of the central nervous system of the rat. J Neurochem 22: 877-885.

Daly EC, Nadi NS, Aprison MH. 1976. Regional distribution and properties of the glycine cleavage system within the central nervous system of the rat: Evidence for an endogenous inhibitor during in vitro assay. J Neurochem 26: 179-185.

de Koning TJ, Duran M, Dorland L, Gooskens R, Van Schaftingen E, Jaeken J, Blau N, Berger R, Poll-The BT. 1998. Beneficial effects of L-serine and glycine in the management of seizures in 3-phosphoglycerate dehydrogenase deficiency. Ann Neurol 44: 261-265.

de Koning TJ, Klomp LW. 2004. Serine-deficiency syndromes. Curr Opin Neurol 17: 197-204.

de Koning TJ, Klomp LW, van Oppen AC, Beemer FA, Dorland L, et al. 2004. Prenatal and early postnatal treatment in 3-phosphoglycerate-dehydrogenase deficiency. Lancet 364: 2221-2222.

de Koning TJ, Klomp LW, van Oppen AC, Beemer FA, Dorland L, et al. 2004. Prenatal and early postnatal treatment in 3-phosphoglycerate-dehydrogenase deficiency. Lancet 364: 2221-2222.

Dringen R, Verleysdonk S, Hamprecht B, Willker W, Leibfritz D, et al. 1998. Metabolism of glycine in primary astroglial cells: Synthesis of creatine, serine, and glutathione. J Neurochem 70: 835-840.

Edgar AJ. 2002. The human L-threonine 3-dehydrogenase gene is an expressed pseudogene. BMC Genet 3: 18.

Edgar AJ. 2005. Mice have a transcribed L-threonine aldolase/GLY1 gene, but the human GLY1 gene is a non-processed pseudogene. BMC Genomics 6: 32.

Edgar AJ, Polak JM. 2000. Molecular cloning of the human and murine 2-amino-3-ketobutyrate coenzyme A ligase cDNAs. Eur J Biochem 267: 1805-1812.

Eulenburg V, Armsen W, Betz H, Gomeza J. 2005. Glycine transporters: Essential regulators of neurotransmission. Trends Biochem Sci 30: 325-333.

Flint AC, Liu X, Kriegstein AR. 1998. Nonsynaptic glycine receptor activation during early neocortical development. Neuron 20: 43-53.

Fu TF, Rife JP, Schirch V. 2001. The role of serine hydroxymethyltransferase isozymes in one-carbon metabolism in MCF-7 cells as determined by ^{13}C NMR. Arch Biochem Biophys 393: 42-50.

Fuchs SA, Berger R, Klomp LW, de Koning TJ. 2005. D-Amino acids in the central nervous system in health and disease. Mol Genet Metab 85: 168-180.

Furuya S, Mitoma J, Makino A, Hirabayashi Y. 1998. Ceramide and its interconvertible metabolite sphingosine function as indispensable lipid factors involved in survival and dendritic differentiation of cerebellar Purkinje cells. J Neurochem 71: 366-377.

Furuya S, Tabata T, Mitoma J, Yamada K, Yamasaki M, et al. 2000. L-Serine and glycine serve as major astroglia-derived trophic factors for cerebellar Purkinje neurons. Proc Natl Acad Sci USA 97: 11528-11533.

Gomeza J, Hulsmann S, Ohno K, Eulenburg V, Szoke K, et al. 2003a. Inactivation of the glycine transporter 1 gene discloses vital role of glial glycine uptake in glycinergic inhibition. Neuron 40: 785-796.

Gomeza J, Ohno K, Hulsmann S, Armsen W, Eulenburg V, et al. 2003b. Deletion of the mouse glycine transporter 2 results in a hyperekplexia phenotype and postnatal lethality. Neuron 40: 797-806.

Growdon JH, Nader TM, Schoenfeld J, Wurtman RJ. 1991. L-Threonine in the treatment of spasticity. Clin Neuropharmacol 14: 403-412.

Gundersen RY, Vaagenes P, Breivik T, Fonnum F, Opstad PK. 2005. Glycine—an important neurotransmitter and cytoprotective agent. Acta Anaesthesiol Scand 49: 1108-1116.

Hao S, Sharp JW, Ross-Inta CM, McDaniel BJ, Anthony TG, et al. 2005. Uncharged tRNA and sensing of amino acid deficiency in mammalian piriform cortex. Science 307: 1776-1778.

Hashimoto A, Kumashiro S, Nishikawa T, Oka T, Takahashi K, et al. 1993a. Embryonic development and postnatal changes in free D-aspartate and D-serine in the human prefrontal cortex. J Neurochem 61: 348-351.

Hashimoto A, Nishikawa T, Oka T, Takahashi K. 1993b. Endogenous D-serine in rat brain: N-methyl-D-aspartate receptor-related distribution and aging. J Neurochem 60: 783-786.

Hashimoto K, Engberg G, Shimizu E, Nordin C, Lindstrom LH, et al. 2005. Reduced D-serine to total serine ratio in the cerebrospinal fluid of drug naïve schizophrenic patients. Prog Neuropsychopharmacol Biol Psychiatry 29: 767-769.

Hashimoto K, Fukushima T, Shimizu E, Komatsu N, Watanabe H, et al. 2003. Decreased serum levels of D-serine in patients with schizophrenia: Evidence in support of the N-methyl-D-aspartate receptor hypofunction hypothesis of schizophrenia. Arch Gen Psychiatry 60: 572-576.

Hayasaka S, Hara S, Mizuno K, Narisawa K, Tada K. 1986. Leber's congenital amaurosis associated with hyperthreoninemia. Am J Ophthalmol 101: 475-479.

Helboe L, Egebjerg J, Moller M, Thomsen C. 2003. Distribution and pharmacology of alanine-serine-cysteine transporter 1 (asc-1) in rodent brain. Eur J Neurosci 18: 2227-2238.

Herbig K, Chiang EP, Lee LR, Hills J, Shane B, et al. 2002. Cytoplasmic serine hydroxymethyltransferase mediates competition between folate-dependent deoxyribonucleotide and S-adenosylmethionine biosyntheses. J Biol Chem 277: 38381-38389.

Hoover-Fong JE, Shah S, Van Hove JL, Applegarth D, Toone J, et al. 2004. Natural history of nonketotic hyperglycinemia in 65 patients. Neurology 63: 1847-1853.

Huether G, Lajtha A. 1991. Changes in free amino acid concentrations in serum, brain, and CSF throughout embryogenesis. Neurochem Res 16: 145-150.

Ikonomidou C, Bosch F, Miksa M, Bittigau P, Vockler J, et al. 1999. Blockade of NMDA receptors and apoptotic neurodegeneration in the developing brain. Science 283: 70-74.

Jaeken J, Detheux M, Fryns JP, Collet JF, Alliet P, et al. 1997. Phosphoserine phosphatase deficiency in a patient with Williams syndrome. J Med Genet 34: 594-596.

Kartvelishvily E, Shleper M, Balan L, Dumin E, Wolosker H. 2006. Neuron-derived D-serine release provides a novel means to activate N-methyl-D-aspartate receptors. J Biol Chem 281: 14151-14162.

Klomp LW, de Koning TJ, Malingre HE, van Beurden EA, Brink M, et al. 2000. Molecular characterization of 3-phosphoglycerate dehydrogenase deficiency—a neurometabolic disorder associated with reduced L-serine biosynthesis. Am J Hum Genet 67: 1389-1399.

Kuge O, Nishijima M. 1997. Phosphatidylserine synthase I and II of mammalian cells. Biochim Biophys Acta 1348: 151-156.

Lee A, Patterson V. 1993. A double-blind study of L-threonine in patients with spinal spasticity. Acta Neurol Scand 88: 334-338.

Legendre P. 2001. The glycinergic inhibitory synapse. Cell Mol Life Sci 58: 760-793.

Lewis RM, Godfrey KM, Jackson AA, Cameron IT, Hanson MA. 2005. Low serine hydroxymethyltransferase activity in the human placenta has important implications for fetal glycine supply. J Clin Endocrinol Metab 90: 1594-1598.

Lowry M, Hall DE, Hall MS, Brosnan JT. 1987. Renal metabolism of amino acids in vivo: Studies on serine and glycine fluxes. Am J Physiol 252: F304-F309.

Luberto C, Kraveka JM, Hannun YA. 2002. Ceramide regulation of apoptosis versus differentiation: A walk on a fine line. Lessons from neurobiology. Neurochem Res 27: 609-617.

Luka Z, Cerone R, Phillips JA, Mudd HS, Wagner C. 2002. Mutations in human glycine N-methyltransferase give insights into its role in methionine metabolism. Hum Genet 110: 68-74.

Lynch JW. 2004. Molecular structure and function of the glycine receptor chloride channel. Physiol Rev 84: 1051-1095.

Matsuo H, Kanai Y, Tokunaga M, Nakata T, Chairoungdua A, et al. 2004. High affinity D- and L-serine transporter Asc-1: Cloning and dendritic localization in the rat cerebral and cerebellar cortices. Neurosci Lett 358: 123-126.

McDaniel MA, Maier SF, Einstein GO. 2003. "Brain-specific" nutrients: A memory cure? Nutrition 19: 957-975.

Mills PB, Surtees RA, Champion MP, Beesley CE, Dalton N, et al. 2005. Neonatal epileptic encephalopathy caused by mutations in the PNPO gene encoding pyridox(am)ine 5′-phosphate oxidase. Hum Mol Genet 14: 1077-1086.

Mitoma J, Furuya S, Hirabayashi Y. 1998a. A novel metabolic communication between neurons and astrocytes: Nonessential amino acid L-serine released from astrocytes is essential for developing hippocampal neurons. Neurosci Res 30: 195-199.

Mitoma J, Ito M, Furuya S, Hirabayashi Y. 1998b. Bipotential roles of ceramide in the growth of hippocampal neurons: Promotion of cell survival and dendritic outgrowth in dose- and developmental stage-dependent manners. J Neurosci Res 51: 712-722.

Mitoma J, Kasama T, Furuya S, Hirabayashi Y. 1998c. Occurrence of an unusual phospholipid, phosphatidyl-L-threonine, in cultured hippocampal neurons. Exogenous L-serine is required for the synthesis of neuronal phosphatidyl-L-serine and sphingolipids. J Biol Chem 273: 19363-19366.

Moores RR, Rietberg CC, Battaglia FC, Fennessey PV, Meschia G. 1993. Metabolism and transport of maternal serine by the ovine placenta: Glycine production and absence of serine transport into the fetus. Pediatr Res 33: 590-594.

Mori M, Gähwiler BH, Gerber U. 2002. β-Alanine and taurine as endogenous agonists at glycine receptors in rat hippocampus in vitro. J Physiol 539: 191-200.

Nagata Y, Horiike K, Maeda T. 1994. Distribution of free D-serine in vertebrate brains. Brain Res 634: 291-295.

Narkewicz MR, Thureen PJ, Sauls SD, Tjoa S, Nikolayevsky N, et al. 1996. Serine and glycine metabolism in hepatocytes from mid gestation fetal lambs. Pediatr Res 39: 1085-1090.

Nguyen L, Malgrange B, Belachew S, Rogister B, Rocher V, et al. 2002. Functional glycine receptors are expressed by postnatal nestin-positive neural stem/progenitor cells. Eur J Neurosci 15: 1299-1305.

Nong Y, Huang YQ, Ju W, Kalia LV, Ahmadian G, et al. 2003. Glycine binding primes NMDA receptor internalization. Nature 422: 302-307.

Parton M, Mitsumoto H, Leigh PN. 2003. Amino acids for amyotrophic lateral sclerosis/motor neuron disease. Cochrane Database Syst Rev 4: CD003457.

Quackenbush EJ, Kraemer KH, Gahl WA, Schirch V, Whiteman DA, et al. 1999. Hypoglycinaemia and psychomotor delay in a child with xeroderma pigmentosum. J Inherit Metab Dis 22: 915-924.

Reddi OS. 1978. Threoninemia—a new metabolic defect. J Pediatr 93: 814-816.

Rees MI, Harvey K, Pearce BR, Chung SK, Duguid IC, et al. 2006. Mutations in the gene encoding GlyT2 (SLC6A5) define a presynaptic component of human startle disease. Nat Genet 38: 801-806.

Ristoff E, Mayatepek E, Larsson A. 2001. Long-term clinical outcome in patients with glutathione synthetase deficiency. J Pediatr 139: 79-84.

Sakai K, Shimizu H, Koike T, Furuya S, Watanabe M. 2003. Neutral amino acid transporter ASCT1 is preferentially expressed in L-Ser-synthetic/storing glial cells in the mouse brain with transient expression in developing capillaries. J Neurosci 23: 550-560.

Savoca R, Ziegler U, Sonderegger P. 1995. Effects of L-serine on neurons in vitro. J Neurosci Methods 61: 159-167.

Schell MJ, Brady RO, Molliver ME, Snyder SH. 1997. D-Serine as a neuromodulator: Regional and developmental localizations in rat brain glia resemble NMDA receptors. J Neurosci 17: 1604-1615.

Schell MJ, Molliver ME, Snyder SH. 1995. D-Serine, an endogenous synaptic modulator: Localization to astrocytes and glutamate-stimulated release. Proc Natl Acad Sci USA 92: 3948-3952.

Schulze A. 2003. Creatine deficiency syndromes. Mol Cell Biochem 244: 143-150.

Shank RP, Aprison MH. 1970. The metabolism in vivo of glycine and serine in eight areas of the rat central nervous system. J Neurochem 17: 1461-1475.

Shiang R, Ryan SG, Zhu YZ, Hahn AF, O'Connell P, et al. 1993. Mutations in the α1 subunit of the inhibitory glycine receptor cause the dominant neurologic disorder, hyperekplexia. Nat Genet 5: 351-358.

Smith QR, Momma S, Aoyagi M, Rapoport SI. 1987. Kinetics of neutral amino acid transport across the blood–brain barrier. J Neurochem 49: 1651-1658.

Snell K. 1984. Enzymes of serine metabolism in normal, developing and neoplastic rat tissues. Adv Enzyme Regul 22: 325-400.

Snyder SH, Kim PM. 2000. D-Amino acids as putative neurotransmitters: Focus on D-serine. Neurochem Res 25: 553-560.

Stein V, Nicoll RA. 2003. GABA generates excitement. Neuron 37: 375-378.

Sugiura N, Patel RG, Corriveau RA. 2001. N-Methyl-D-aspartate receptors regulate a group of transiently expressed genes in the developing brain. J Biol Chem 276: 14257-14263.

Tapia JC, Mentis GZ, Navarrete R, Nualart F, Figueroa E, et al. 2001. Early expression of glycine and GABA$_A$ receptors in developing spinal cord neurons. Effects on neurite outgrowth. Neuroscience 108: 493-506.

Tsai G, Coyle JT. 2002. Glutamatergic mechanisms in schizophrenia. Annu Rev Pharmacol Toxicol 42: 165-179.

Tuominen HJ, Tiihonen J, Wahlbeck K. 2005. Glutamatergic drugs for schizophrenia: A systematic review and meta-analysis. Schizophr Res 72: 225-234.

Tuominen HJ, Tiihonen J, Wahlbeck K. 2006. Glutamatergic drugs for schizophrenia. Cochrane Database Syst Rev 19: CD003730.

Tyurina YY, Shvedova AA, Kawai K, Tyurin VA, Kommineni C, et al. 2000. Phospholipid signaling in apoptosis: Peroxidation and externalization of phosphatidylserine. Toxicology 148: 93-101.

van der Knaap MS, Wevers RA, Kure S, Gabreels FJ, Verhoeven NM, et al. 1999. Increased cerebrospinal fluid glycine: A biochemical marker for a leukoencephalopathy with vanishing white matter. J Child Neurol 14: 728-731.

Veiga-da-Cunha M, Collet JF, Prieur B, Jaeken J, Peeraer Y, et al. 2004. Mutations responsible for 3-phosphoserine phosphatase deficiency. Eur J Hum Genet 12: 163-166.

Verleysdonk S, Martin H, Willker W, Leibfritz D, Hamprecht B. 1999. Rapid uptake and degradation of glycine by astroglial cells in culture: Synthesis and release of serine and lactate. Glia 27: 239-248.

Wang LZ, Zhu XZ. 2003. Spatiotemporal relationships among D-serine, serine racemase, and D-amino acid oxidase during mouse postnatal development. Acta Pharmacol Sin 24: 965-974.

Wolosker H, Sheth KN, Takahashi M, Mothet JP, Brady RO Jr, et al. 1999. Purification of serine racemase: Biosynthesis of the neuromodulator D-serine. Proc Natl Acad Sci USA 96: 721-725.

Xie X, Dumas T, Tang L, Brennan T, Reeder T, et al. 2005. Lack of the alanine-serine-cysteine transporter 1 causes tremors, seizures, and early postnatal death in mice. Brain Res 1052: 212-221.

Yamamoto T, Nishizaki I, Furuya S, Hirabayashi Y, Takahashi K, et al. 2003. Characterization of rapid and high-affinity uptake of L-serine in neurons and astrocytes in primary culture. FEBS Lett 548: 69-73.

Yamasaki M, Yamada K, Furuya S, Mitoma J, Hirabayashi Y, et al. 2001. 3-Phosphoglycerate dehydrogenase, a key enzyme for L-serine biosynthesis, is preferentially expressed in the radial glia/astrocyte lineage and olfactory ensheathing glia in the mouse brain. J Neurosci 21: 7691-7704.

Yoshida K, Furuya S, Osuka S, Mitoma J, Shinoda Y, et al. 2004. Targeted disruption of the mouse 3-phosphoglycerate dehydrogenase gene causes severe neurodevelopmental defects and results in embryonic lethality. J Biol Chem 279: 3573-3577.

Zafra F, Aragon C, Olivares L Danbolt NC, Gimenez C, et al. 1995. Glycine transporters are differentially expressed among CNS cells. J Neurosci 15: 3952-3969.

Zwaal RF, Comfurius P, Bevers EM. 2004. Scott syndrome, a bleeding disorder caused by defective scrambling of membrane phospholipids. Biochim Biophys Acta 1636: 119-128.

3 Histidine

P. Panula

1	Introduction	48
2	Histidine Transport and Uptake in Brain	48
3	Metabolism of Histidine	50
4	Histidine as Histamine Precursor	50
5	Histidinemia	51
5.1	Mechanism and Consequences of Histidinemia	51
5.2	Brain Disorders Related to Histidinemia	52
5.3	Portocaval Anastomosis and Histidinemia	53
6	Summary and Conclusions	54

© Springer-Verlag Berlin Heidelberg 2007

Abstract: Histidine is metabolized through three different pathways, of which decarboxylation to histamine yields the active neurotransmitter histamine in the tuberomamillary neurons of the brain. L-Histidine is transported into the brain through the blood–brain barrier (BBB) by a system that recognizes several amino acids. Lack of functional histidase activity because of an autosomal recessive disorder leads to elevated histidine levels in the blood, which is occasionally associated with neurological symptoms, although the disorder is generally considered benign. Experimental histidinemia is associated with increased levels of histamine, which accumulates in brain histaminergic neurons and their axons. Histamine is the major biologically active product derived from L-histidine in histidinemia, and it can potentially have functional significance because of the actions of histamine on the four known G protein-coupled histamine receptors, of which three are abundantly expressed in the brain.

List of Abbreviations: AP2/KER1, transcription factor AP2/KER1; BBB, blood-brain barrier; BCH, 2-aminobicyclo[2.2.1]heptane-2-carboxylic acid; BMEC, Brain microvascular endothelial cells; C/EBP, transcription factor C/EBP; DQ, developmental quotient; g17, human amino acid transporter g17; HDC, L-Histidine decarboxylase; HE, hepatic encephalopathy; HNF5, liver-specific DNA-binding protein HNF5; IQ, intelligence quotient; LAT1, system L transporter LAT1; mNAT, murine N-system amino acid transporter; NFIL6, nuclear factor interleukin-6; MNF, transcription factor MNF; NMDA, N-methyl D-aspartate; NSD-1015, L-amino acid decarboxylase inhibitor NSD-1015; PCA, portocaval anastomosis; SCC, cytochrome P-450 specific for cholesterol side chain cleavage; SN1, rat amino acid transporter SN1

1 Introduction

L-Histidine is an essential amino acid necessary for protein synthesis and various other functions in cells and tissues, e.g., as a precursor for histamine. The mammalian brain is a complex environment for amino acid transport, storage, and metabolism, as many amino acids and their derivatives activate specific cell-surface receptors, and thus participate in nervous transmission. Thus, the presence of amino acids in the limited extracellular space in the brain needs to be carefully controlled. Both neurons and glial cells need a continuous supply of nutrients. For example, glutamatergic, GABAergic, and glycinergic neurons need precursor amino acids for neurotransmitter synthesis, and released amino acid transmitters are taken up and sequestered by glial cells and neurons for repeated use. All cells in the brain need L-histidine for protein synthesis, and histaminergic neurons in the posterior hypothalamus (Panula et al., 1984; Watanabe et al., 1984) need L-histidine for the synthesis of the neurotransmitter histamine. Whereas L-histidine does not activate known specific receptors, histamine activates three (H_1, H_2, and H_3) of the four known G protein-coupled histamine receptors, which are widespread in the brain (Haas and Panula, 2003). In addition, histamine may modulate the polyamine site of the N-methyl-D-aspartate (NMDA) glutamate receptor (Bekkers, 1993; Vorobjev et al., 1993). Thus, the effects of histidinemia in the brain can derive from the overload of L-histidine in cells, interference of L-histidine with the transport and storage of other amino acids, or overproduction of decarboxylated histidine (histamine) which is highly active at different receptors on both neurons and glial cells.

2 Histidine Transport and Uptake in Brain

L-Histidine is taken up by brain slices more efficiently than many other amino acids (Neame, 1961, 1962), and the uptake of isomers of histidine is more efficient than that of many other imidazole compounds (Neame, 1964). However, uptake in vitro does not directly lend support to the availability of L-histidine in the extracellular space in the brain protected by the blood–brain barrier (BBB). The cell type responsible for the uptake is also difficult to analyze in slices. Uptake in cultured cells has been studied, and neuroblastoma cells accumulate more L-histidine than astrocytoma cells (Hannuniemi et al., 1985), and they are also more sensitive to the enhancement of uptake of other large neutral amino acids by histidine. The complex transport of amino acids across biological membranes and through the BBB has been dealt with in several reviews (Oldendorf, 1971; Christensen, 1982, 1990; Abumrad et al., 1989). The complexity of transport systems derives in part from the low specificity of the transport (Christensen, 1990), in which a given amino acid can be transported by a number of different systems, which are shared with several other substances. The basic properties, nomenclature, and specificity of systems for different amino acids have been described in detail (Christensen, 1990), and the additional complexity that emerged with gene cloning and nomenclature has been clarified (Christensen et al., 1994). Both Na^+-dependent (e.g., system N) and Na^+-independent (e.g., system L) transport systems are involved in histidine transport (Christensen, 1990).

Histidine enters the brain through the endothelial cells of brain capillaries, which are characterized by tight junctions and few pinocytotic vesicles. These cells can be isolated and maintained in culture (Panula et al., 1978), which allows analysis of uptake kinetics of various substances and comparisons to other cell types. However, the polarity of the cells may be lost in part or completely, since the interactions of the endothelial cells with the basal lamina and circulating substances are lost. More sophisticated multi-compartment systems have also been developed for use in permeability and uptake studies (Audus and Borchardt, 1986), although the polarity and distribution of major transport proteins has not been possible to assess in full. There is evidence for differential expression of transporters on the luminal and abluminal surfaces of the brain vascular endothelial cells, suggesting that Na^+-dependent transport is predominant on the abluminal surface while Na^+-independent transport is predominant on the luminal surface (Oldendorf and Szabo, 1976; Sershen and Lajtha, 1976; Christensen, 1979; O'Kane and Hawkins, 2003; O'Kane et al., 2004). Cultured rat brain microvascular endothelial cells (BMECs) take up L-histidine more efficiently than D-histidine, suggesting stereoselectivity of this saturable process (Yamakami et al., 1998). Ouabain, an inhibitor of Na^+/K^+-ATPase, reduces the uptake of L-histidine. Moreover, the substitution of Na^+ with choline chloride and choline bicarbonate in the incubation buffer decreased the uptake rates. Thus, L-histidine is actively taken up by a carrier-mediated mechanism into BMECs, with energy supplied by Na^+. However, L-histidine uptake at 0°C was not completely inhibited, and it was reduced in the presence of a Na^+-independent transport system L substrate, 2-aminobicyclo[2.2.1]heptane-2-carboxylic acid (BCH), suggesting facilitated diffusion (the Na^+-independent process) by a carrier-mediated mechanism into BMECs. L-Histidine uptake in rat BMECs appears to be system N-mediated since it was inhibited by glutamine, asparagine, and L-glutamic acid γ-monohydroxamate. System N-mediated transport is not pH sensitive. D-Histidine transport was also studied in rat BMECs. However, system N transport did not play a role in D-histidine uptake.

An in vivo model system of carotid artery injections was used by Oldendorf and coworkers (1988) who found that cationic amino acid uptake was increasingly inhibited by unlabeled histidine as the pH of the injection solution decreases, whereas the inhibitory effect of unlabeled histidine on neutral amino acid uptake decreased with decreasing pH. They found that the neutral form of histidine inhibited phenylalanine uptake whereas the cationic form competed with arginine uptake. Because phenylalanine decreases [^{14}C]histidine uptake at all pH values and arginine does not, it appears that cationic histidine has affinity for the cationic transporter, but is not transported by it (Oldendorf et al., 1988). Thus, L-histidine shows lower affinity for the system L transporter and inhibits cationic amino acid transport as histidine gets protonated, although the cationic system does not transport it. The saturable entry of L-histidine seems to be mediated exclusively by the carrier of neutral amino acids (Oldendorf et al., 1988).

Recent identification of the genes encoding for the major Na^+-dependent (A and N) (Chaudhry et al., 1999; Reimer et al., 2000; Yao et al., 2000; Nakanishi et al., 2001) transport system proteins enables detailed studies on the expression and function of these systems in tissues and cells. The mouse mNAT transporter corresponds to rat SN1 and human g17 transporters (Chaudhry et al., 1999; Fei et al., 2000; Gu et al., 2000). Some of the transporters, e.g., the second identified mouse N system transporter mNAT2, expressed specifically in the brain and retina (Gu et al., 2001), also show highest specificity for glutamine and L-histidine. System N transport has been originally characterized and limited to the liver (Christensen, 1990), but several other tissues including the nervous system express these transporters (Gu et al., 2001). During recent years, a large number of transporters have been identified and attempts to rationalize the complex nomenclature have been made (Christensen et al., 1994).

Expression cloning of the system L transporter LAT1 (Kanai et al., 1998) allowed detailed studies on the tissue-specific expression and function of this important carrier of neutral amino acids. The expression of the LAT gene is highly enriched in the BBB (Boado et al., 1999), which supports the role of this transporter in the endothelium rather than in neurons or glial cells. It is currently not known which transporter is responsible for L-histidine transport into histaminergic neurons, in which a high rate of influx is needed to provide substrate for diurnally regulated histamine synthesis.

3 Metabolism of Histidine

Histidine catabolism occurs through transamination to imodazole pyruvate (Lin et al., 1958; Spolter and Baldridge, 1964), deamination to urocanate (Mehler and Tabor, 1953), or decarboxylation to histamine (Schayer, 1952). Histidase (EC 4.3.1.3) is a cytosolic enzyme that catalyzes the nonoxidative deamination of histidine to urocanic acid. The enzyme is expressed in a tissue-specific manner in the liver and epidermis in the rat (Bhargava and Feigelson, 1976; Hryb and Feigelson, 1983). cDNA clones encoding human histidase (histidine ammonia-lyase) were isolated from a human λgt10 library (Suchi et al., 1993). The cDNA predicted a 657-amino-acid protein of 72,651 Da. The human histidase amino acid sequence appeared to be 93% conserved with both rat and mouse histidase sequences, including four N-glycosylation consensus sites (Suchi et al., 1993). The deduced amino acid sequence of chicken histidase has greater than 85% identity with the amino acid sequences of rat, mouse, and human histidases (Chendrimada and Davis, 2005).

The histidase gene was isolated from a human genomic library using the human histidase cDNA as a probe (Suchi et al., 1995). Restriction mapping and Southern blot analysis of the isolated clones revealed a single-copy gene spanning approximately 25 kb consisting 21 exons. Exon 1 encodes only the 5′-untranslated sequence of the liver histidase mRNA, with protein coding beginning in exon 2. A rarely observed 5′-GC, similar to that reported in the human P-450 (SCC) gene, is present in intron 20. All other splicing junctions adhere to the canonical GT/AG rule. A TATA box sequence is located 25 bp upstream of the liver histidase transcription initiation site, as determined by S1 nuclease protection analysis. Several liver- and epidermis-specific transcription factor-binding sites, including C/EBP, NFIL6, HNF5, AP2/KER1, MNF, and others, are also identified in the 5′-flanking region. A polymorphism (A to G transition) in the histidase-coding region of exon 16 has also been identified (Suchi et al., 1995).

4 Histidine as Histamine Precursor

L-Histidine was found to be converted to histamine in the cat brain in vitro (White, 1959) and in vivo (White, 1960). The enzymatic activity specific for this conversion in biological tissues (Schayer et al., 1959; Hakanson, 1967a) was found in developing tissues (Burkhalter, 1962; Hakanson, 1967a, b) and stomach (Aures et al., 1968). In the brain, the apparent K_m was 4×10^{-4} M, whereas the free L-histidine concentration in the brain is approximately 5×10^{-5} M, suggesting that the enzyme is not saturated by physiological concentrations of L-histidine (Schwartz et al., 1970). Under these circumstances, it would be feasible that any increase in brain histidine would lead to increased brain histamine and all subsequent consequences. This is indeed what happens, as has been repeatedly shown with high doses of L-histidine (Schwartz et al., 1972; Lozeva et al., 2003a).

L-Histidine decarboxylase (HDC) belongs to the pyridoxal-5′-phosphate-dependent decarboxylase family together with dopa decarboxylases and glutamate decarboxylases, for instance. It has been estimated that the HDC activity in the brain is very low, 0.76 ± 0.23 pmol/min/mg protein (Yamada et al., 1984; Watanabe et al., 1991) corresponding to 1/26000 of that of aromatic amino acid decarboxylase (EC 4.1.1.28) or 1/2400 of that of glutamate decarboxylase (EC 4.1.1.15) (Watanabe et al., 1991), which renders it very difficult to purify the unstable enzyme from the brain. The HDC activity in other mature organs is generally also low, with a few exceptions. The highest activities have been found in fetal rat tissues (Kahlson et al., 1960), hamster placenta, mouse mastocytoma, and gastric mucosa of the rat and mouse kidney (for a review, see Watanabe et al. (1991)). In the rat and hamster brains, the HDC activity is about fourfold higher in the hypothalamus than in the striatum and midbrain, which display higher activity than most other brain areas (Watanabe et al., 1991). Cloning of the HDC-encoding cDNA revealed a putative M_r 74,000 (655 amino acid residues) protein (Joseph et al., 1990; Yamamoto, 1990), whereas the purified liver protein had been characterized as a smaller ($M_r = 54\,000$) protein. Experiments on a rat basophilic/mast cell line (RBL-2H3) suggest that the 74-kDa form of HDC, synthesized in the cytosol, is translocated into the lumen of the endoplasmic reticulum (ER), where it is converted to the 53-kDa form (Tanaka et al., 1998). Similar studies have not been carried out with brain neurons.

Systemic injections into rats with L-histidine lead to clearly increased brain histidine and histamine concentrations provided that the dose is sufficiently high. Early studies with low doses (e.g., 100 mg/kg i.p. in rats) have given somewhat inconsistent results (Green and Erickson, 1967; Medina et al., 1969; Boissier et al., 1970), whereas administration of 500 mg/kg leads to a rapid and transient (peak at 1 h) increase in brain histidine levels and a more sustained (peak at 3 h) increase in the brain histamine concentration (Schwartz et al., 1972).

Decarboxylation of L-histidine to histamine occurs at least in tuberomamillary neurons, which express high levels of HDC mRNA (Castren and Panula, 1990), HDC protein (Watanabe et al., 1984), and histamine (Panula et al., 1984), and in mast cells of developing brain (Kinnunen et al., 1998; Karlstedt et al., 2001), whereas it has not been shown that brain vascular endothelial cells in vivo express significant levels of HDC mRNA (Karlstedt et al., 2001). This may be due to methodological reasons, like low sensitivity of the in situ hybridization technique or to a true lack of functional HDC translation in the brain vessels of intact animals. However, histamine has not been visualized in endothelial cells in vivo using antibody techniques, whereas neurons and mast cells show strong fluorescence, suggesting that endothelial cells do not store histamine.

The presence of HDC has been reported in rat BMECs (Sakurai et al., 2003) using RT-PCR and antiserum against HDC. HDC activity was also reported. On the other hand, an immortalized rat brain endothelial cell line did not display HDC mRNA with PCR or in situ hybridization, and histamine was not detected in these cells under normal culture conditions (Karlstedt et al., 1999; Panula et al., 2000). Thus, the significance of possible decarboxylation of L-histidine to histamine in brain vascular endothelial cells is unclear at the moment.

5 Histidinemia

5.1 Mechanism and Consequences of Histidinemia

Histidinemia is a result of histidase deficiency (Ghadimi, 1961). It appears to be inherited as an autosomal recessive disorder that affects the liver (Selden et al., 1995). Recently, mutations in the histidase gene were identified (Kawai et al., 2005). Four missense mutations (R322P, P259L, R206T, and R208L), two exonic polymorphisms (T141T c.423A→T and P259P c.777A→G), and two intronic polymorphisms (IVS6-5T→C and IVS9 + 25A→G) have been identified (Kawai et al., 2005). It is frequently but not always (Levy et al., 1974) associated with mental retardation, infantile spasms, epilepsy, speech defects, and motor incoordination. The long-term outcome of patients diagnosed by newborn screening in the north-west of England between 1966 and 1990 showed that in most cases the disorder is benign: neither DQ nor IQ correlated with plasma histidine at diagnosis or with the mean plasma histidine throughout life (Lam et al., 1996). Growth has been normal in all patients. There was no apparent benefit from a low-histidine diet in early childhood. In contrast to some earlier studies, there was no excess of clinical symptoms.

The findings of increased heterozygosity for histidinemia among schizophrenic patients make the histidase gene a plausible candidate for genetic studies in schizophrenia. In one study, a tetranucleotide repeat polymorphism in intron 8 of the histidase gene was used to examine the possibility that the histidase gene contributes to the genetic component of schizophrenia. In the first sample of 161 patients and 128 controls, the four-repeat allele was found to be in excess in the patients. In contrast, the three-repeat allele was less frequent in patients. The second sample of 95 patients and 93 controls was utilized to test these hypotheses. The observations were not replicated in this sample. The results thus do not support an involvement of the histidase gene in the development of schizophrenia (Nobile et al., 1997).

Histidinemia can be induced in animals by including excessive amounts of L-histidine to the diet (Woodworth and Baldridge, 1970) or by inhibition of the histidase by nitromethane (Lee and Wang, 1975). Inclusion of 5% histidine in the diet induces an over tenfold increase in histidine level in blood plasma, whereas nitromethane can produce an almost fivefold increase (Lee and Wang, 1975). Nitromethane has been found to bind to the active center of the enzyme (Givot et al., 1969; Givot and Abeles, 1970). Use of enzyme inhibitors can be considered a more relevant model of genetic lack of histidase activity, although even that does not reproduce the compensatory metabolic changes associated with the congenital lack of enzyme activity. Furthermore, unknown effects on other systems are difficult to exclude with chemically induced enzyme inhibition. Nitromethane together with L-histidine loading can also produce behavioral abnormalities including lower motor activity in open field tests (Dutra-Filho et al., 1989). Once again, the possible role of brain histamine was not investigated, which renders it difficult to estimate the role of L-histidine not decarboxylated and distinguish it from that of central histamine, which is likely to affect behavior. Treatment of rats with DL-α-hydrazinoimidazolylpropionic acid also produces histidinemia by inhibition of histidine ammonia-lyase (Brand and Harper, 1976), but the treatment also inactivates hepatic aspartate aminotransferase.

L-Histidine has been injected either systemically or directly into the brain. Many of the central nervous system (CNS) effects observed seem to be caused by decarboxylation of L-histidine to histamine within the CNS. However, experimental administration of histidine does not always seem to have the same effects as the physiological increase. Lozeva and coworkers (2003a) have studied the effect of L-histidine loading on both histamine and its metabolites. Brain and plasma histamine and tele-methylhistamine levels and binding properties and regional distribution of H_3 receptors in the brain were studied. In the L-histidine-loaded rats, tissue histamine levels were significantly increased by 40%–70% in the cortex, hypothalamus, and the rest of the brain. Histamine concentrations in the cerebellum and plasma and tele-methylhistamine concentrations in the cortex and hypothalamus did not change. The binding properties of H_3 receptors in the cerebral cortex were unchanged. Changes were seen in the regional distribution of R-α-[^{3}H]methylhistamine-binding sites, suggesting region-selective changes in the histamine H_3 receptor or isoforms. These results imply that following repeated L-histidine administration in the rat brain histamine synthesis is increased, but little if any change in its functional release can be detected. The excess of histamine may be stored in a nonreleasable pool, the identity of which remains unknown. After loading with L-histidine, histamine receptor binding properties are not altered, whereas the receptor density is changed in selected regions (Lozeva et al., 2003a).

5.2 Brain Disorders Related to Histidinemia

Early reports on clinical symptoms associated with histidinemia or consequences of histidinemia included a number of neurological and psychiatric abnormalities. However, the majority of the patients seem to be asymptomatic (Tada et al., 1982, 1984; Sano et al., 1997). In one prospective study on 16 families, the IQ scores, visiomotor integration, and skills in reading and mathematics were analyzed in probands and siblings (mean age 9.5 years) with histidinemia (none treated with a low-histidine diet) (Coulombe et al., 1983). CNS development in children with histidinemia was normal, and outlier score values did not correlate with biochemical values. Scriver and Levy (1983) conclude that histidinemia in the typical form (autosomal recessive impairment of L-histidine ammonia-lyase activity (EC 4.3.1.3)) is not a disease in humans, because both retrospective and prospective studies (Coulombe et al., 1983; Rosenmann et al., 1983) together indicate that the prevalence of disadaptive phenotypes (e.g., impaired intellectual or speech development, seizures, behavioral or learning disorder) in the histidinemia population is not higher than the frequency of these functional disorders in the nonhistidinemia population. The possibility remains that histidinemia is a risk factor for development of an unfavorable CNS phenotype, in particular in individuals under specific circumstances. It is also possible that histidinemia is a phenotype that may derive from a more than one type of histidase enzyme deficiency, and the cases with more severe CNS symptoms may have defects beyond those now found in the histidase gene. Furthermore, no published data is available on the association between the histidase gene and genes related to histamine synthesis and metabolism, e.g., histidine decarboxylase and histamine N-methyltransferase. The latter gene (Girard et al., 1994) shows polymorphism associated with significant differences in activity (Preuss et al., 1998). As such, this polymorphism lacks association with schizophrenia (Yan et al., 2000).

Preischemic administration of histamine (i.c.v.) suppresses ischemic release of glutamate and ameliorates neuronal damage, whereas the blockade of central histamine H_2 receptors aggravates ischemic injury. These findings suggest that histamine provides beneficial effects against ischemic damage through histamine H_2 receptors, when administered before induction of ischemia. Postischemic loading with L-histidine, a precursor of histamine, alleviates both brain infarction and delayed neuronal death. Since the alleviation is abolished by the blockade of central histamine H_2 receptors, facilitation of central histamine H_2 action caused by histidine may prevent reperfusion injury after ischemic events. Because the ischemia-induced increase in the glutamate level rapidly resumes after reperfusion of cerebral blood flow, beneficial effects caused by postischemic loading with histidine may be due to other mechanisms besides suppression of excitatory neurotransmitter release. Antiinflammatory action by histamine H_2 receptor stimulation is a likely mechanism responsible for the improvement (Adachi et al., 1993; Adachi, 2005).

5.3 Portocaval Anastomosis and Histidinemia

In portocaval shunted rats, the histamine concentration in the brain is significantly increased (Fogel et al., 1991), and the increase is due to the availability of L-histidine and increased synthesis rather than slow degradation (Fogel et al., 1991). The increase in histamine concentration is 2.4- to 13-fold higher in the hypothalamus and 1.5- to 2.5-fold higher in the rest of the brain as compared to sham-operated control animals. The consequences include the formation of morphologically distinct enlarged varicosities along the axons of histaminergic neurons, which contain large amounts of histamine (Fogel et al., 2002). The effects on the histaminergic system remain for at least 8 months after the operation, as histamine levels are elevated throughout the brain with the greatest increase found in the hypothalamus and striatum, and tele-methylhistamine levels are significantly elevated in the cortex and hypothalamus (Lozeva et al., 1998b). The spontaneous and K^+-evoked overflow of histamine from anterior hypothalamus and the H_3 inverse agonist thioperamide-induced overflow from both anterior hypothalamus and cortex are increased after chronic portocaval anastomosis (PCA). In spite of the significantly elevated brain concentrations and the moderate increase in histamine release, the binding properties of R-α-[^{3}H]methylhistamine to cortical membranes are not significantly changed (Lozeva et al., 1998a). In the brains of male Wistar rats, which had been 8 months previously subjected to PCA, the catecholamine levels were unchanged in all brain regions. In contrast, tryptophan was evenly increased throughout the brain. The accumulation of 5-hydroxytryptophan after decarboxylase inhibition (NSD-1015; 100 mg/kg, i.p.) and the endogenous levels of 5-hydroxyindoleacetic acid were significantly higher in PCA rats, particularly in the hypothalamus and midbrain, whereas 5-hydroxytryptamine concentrations were unchanged. It thus seems that 8 months after PCA, the catecholaminergic systems have been reestablished, whereas serotonergic and histaminergic systems still show disturbances in their function (Lozeva et al., 1998b). Lozeva et al. (1999b) also found that the levels of histamine, tele-methylhistamine, and 5-hydroxyindole-3-acetic acid (5-HIAA) but not that of 5-hydroxytryptamine (5-HT) (serotonin), show sustained elevation in the frontal cortex 6 months after surgery. The tissue levels of both histamine and metabolites in the frontal cortex remained increased. PCA led to changes in the synchronized, low-frequency, high-amplitude frontal cortex EEG activity during the light phase. Delta-wave amplitude but not delta time was significantly decreased, whereas spindle amplitude and spindling time were significantly decreased. PCA-operated rats showed a change in the pattern of EEG activity with increasing age, similar to sham-operated rats. Lozeva and her coauthors concluded that once established, the resetting of the systems regulating the sleep-waking behavior is being maintained with time. Interestingly, there was a significant negative correlation between the spindling time and the tissue histamine levels. Histamine, which participates in the control of vigilance, sleep, and wakefulness (Lin et al., 1988; Parmentier et al., 2002; Haas and Panula, 2003), as well as in the modulation of circadian rhythmicity, may thus play a role in the development of sleep disturbances in rats with PCA. The metabolic and endocrine effects of PCA are also important, as histamine is a known regulator of several hypothalamic and pituitary hormones. In one experiment (Lozeva et al., 1999a), Wistar rats after PCA exhibited marked growth retardation (weight gain of 20 g versus 140 g for the sham rats), increased plasma levels of prolactin (9.7 ± 2.4 versus 3.6 ± 0.6) and unaltered growth hormone levels (6.2 ± 0.5 versus 8.1 ± 1.0). A sixfold elevation of histamine concentration (29.5 ± 3.9 versus 4.8 ± 0.4) and a twofold increase of tele-methylhistamine levels (1.8 ± 0.1 versus 0.8 ± 0.02) were found in the hypothalamus. Although increased histaminergic activity in the hypothalamus may be involved in the development of growth retardation and in the enhanced basal secretion of prolactin in male rats with long-term PCA, other mechanisms associated with PCA may also contribute. These observations may also be relevant in human liver cirrhosis patients suffering from hepatic encephalopathy (HE). In HE patients, a significant fourfold increase of histamine in the caudate-putamen, and a significant increase in the cerebral cortex has been reported (Lozeva et al., 2003a). Tele-methylhistamine was also increased and the densities of histamine H_3 receptor sites were significantly decreased in the patient material, suggesting activation of the histaminergic system in HE. Given that histamine participates in the regulation of arousal and circadian rhythmicity, the induction of central histamine mechanisms may contribute to the development of neuropsychiatric symptoms, such as sleep disturbances and altered circadian rhythms in chronic HE in humans. Thus, pharmacological manipulation of the histaminergic system could be beneficial in the treatment of HE in chronic liver failure (Lozeva et al., 2003b).

6 Summary and Conclusions

L-Histidine is an essential amino acid needed by all cells. Its transport across the cell membrane and through the BBB (capillary endothelial cells, basal lamina, and astrocyte processes) is complex and pH-dependent. In addition to being one of the amino acids needed for protein synthesis, L-histidine can be transaminated to imidazole pyruvate, deaminated to urocanate, or decarboxylated to histamine. Of these, only histamine has well-characterized effects in the CNS. Although the decarboxylation of L-histidine to histamine is a minor metabolic pathway in the whole body, the potent site- and receptor-specific effects of histamine render to consider this pathway as an important one when the consequences of histidinemia are evaluated. Histidinemia, which as such was previously thought to associate with several severe CNS disorders, is now considered a mostly benign disorder. The cases with major CNS problems may be associated with other, hitherto unknown associated defects.

Acknowledgments

The author's original research has been supported by the Academy of Finland, the Sigrid Juselius Foundation, the Finnish Foundation for Alcohol Studies, and the Magnus Ehrnrooth's Foundation.

References

Abumrad NN, Williams P, Frexes-Steed M, Geer R, Flakoll P, et al. 1989. Inter-organ metabolism of amino acids in vivo. Diabetes Metab Rev 5: 213-226.

Adachi N. 2005. Cerebral ischemia and brain histamine. Brain Res Rev 50: 275-286.

Adachi N, Oishi R, Itano Y, Yamada T, Hirakawa M, et al. 1993. Aggravation of ischemic neuronal damage in the rat hippocampus by impairment of histaminergic neurotransmission. Brain Res 602: 165-168.

Audus KL, Borchardt RT. 1986. Characteristics of the large neutral amino acid transport system of bovine brain microvessel endothelial cell monolayers. J Neurochem 47: 484-488.

Aures D, Hakanson R, Schauer A. 1968. Histidine decarboxylase and DOPA decarboxylase in the rat stomach.

Properties and cellular localization. Eur J Pharmacol 3: 217-234.

Bekkers JM. 1993. Enhancement by histamine of NMDA-mediated synaptic transmission in the hippocampus. Science 261: 104-106.

Bhargava MM, Feigelson M. 1976. Studies on the mechanisms of histidase development in rat skin and liver. I. Basis for tissue specific developmental changes in catalytic activity. Dev Biol 48: 212-225.

Boado RJ, Li JY, Nagaya M, Zhang C, Partridge WM, 1999. Selective expression of the large neutral amino acid transporter at the blood–brain barrier. Proc Natl Acad Sci USA 96: 12079-12084.

Boissier JR, Guernet M, Tillement JP, Blanco I, Blanco M. 1970. Variations in brain histamine levels caused by diphenhydramine and L-histidine in the rat. Life Sci 9: 249-256.

Brand LM, Harper AE. 1976. Studies on the production and assessment of experimental histidinemia in the rat. Biochim Biophys Acta 444: 294-306.

Burkhalter A. 1962. The formation of histamine by fetal rat liver. Biochem Pharmacol 11: 315-322.

Castren E, Panula P. 1990. The distribution of histidine decarboxylase mRNA in the rat brain: An in situ hybridization study using synthetic oligonucleotide probes. Neurosci Lett 120: 113-116.

Chaudhry FA, Reimer RJ, Krizaj D, Barber D, Storm-Mathisen J, et al. 1999. Molecular analysis of system N suggests novel physiological roles in nitrogen metabolism and synaptic transmission. Cell 99: 769-780.

Chendrimada TP, Davis AJ. 2005. Molecular cloning of chicken hepatic histidase and the regulation of histidase mRNA expression by dietary protein. J Nutr Biochem 16: 114-120.

Christensen HN. 1979. Developments in amino acid transport, illustrated for the blood–brain barrier. Biochem Pharmacol 28: 1989-1992.

Christensen HN. 1982. Interorgan amino acid nutrition. Physiol Rev 62: 1193-1233.

Christensen HN. 1990. Role of amino acid transport and countertransport in nutrition and metabolism. Physiol Rev 70: 43-77.

Christensen HN, Albritton LM, Kakuda DK, Mac Leod CL. 1994. Gene-product designations for amino acid transporters. J Exp Biol 196: 51-57.

Coulombe JT, Kammerer BL, Levy HL, Hirsch BZ, Scriver CR. 1983. Histidinaemia. Part III: Impact; a prospective study. J Inherit Metab Dis 6: 58-61.

Dutra-Filho CS, Wannmacher CM, Pires RF, Gus G, Kalil AM, et al. 1989. Reduced locomotor activity of rats made histidinemic by injection of histidine. J Nutr 119: 1223-1227.

Fei YJ, Sugawara M, Nakanishi T, Huang W, Wang H, et al. 2000. Primary structure, genomic organization, and functional and electrogenic characteristics of human system N 1, a Na^+- and H^+-coupled glutamine transporter. J Biol Chem 275: 23707-23717.

Fogel WA, Andrzejewski W, Maslinski C. 1991. Brain histamine in rats with hepatic encephalopathy. J Neurochem 56: 38-43.

Fogel WA, Michelsen KA, Granerus G, Sasiak K, Andrzejewski W, Panula P, et al. 2002. Neuronal storage of histamine in the brain and tele-methylimidazoleacetic acid excretion in portocaval shunted rats. J Neurochem 80: 375-382.

Ghadimi H, Partington MW, Hunter A. 1961. A familial disturbance of histidine metabolism. N Engl J Med 265: 221-224.

Girard B, Otterness DM, Wood TC, Honchel R, Wieben ED, et al. 1994. Human histamine N-methyltransferase pharmacogenetics: Cloning and expression of kidney cDNA. Mol Pharmacol 45: 461-468.

Givot IL, Abeles RH. 1970. Mammalian histidine ammonia lyase. In vivo inactivation and presence of an electrophilic center at the active site. J Biol Chem 245: 3271-3273.

Givot IL, Smith TA, Abeles RH. 1969. Studies on the mechanism of action and the structure of the electrophilic center of histidine ammonia lyase. J Biol Chem 244: 6341-6353.

Green H, Erickson RW. 1967. Effect of some drugs upon the histamine concentration of guinea pig brain. Arch Int Pharmacod T 166: 121-126.

Gu S, Roderick HL, Camacho P, Jiang JX. 2000. Identification and characterization of an amino acid transporter expressed differentially in liver. Proc Natl Acad Sci USA 97: 3230-3235.

Gu S, Roderick HL, Camacho P, Jiang JX. 2001. Characterization of an N-system amino acid transporter expressed in retina and its involvement in glutamine transport. J Biol Chem 276: 24137-24144.

Haas H, Panula P. 2003. The role of histamine and the tuberomamillary nucleus in the nervous system. Nature Rev Neurosci 4: 121-130.

Hakanson R. 1967a. Kinetic properties of mammalian histidine decarboxylase. Eur J Pharmacol 1: 42-46.

Hakanson R. 1967b. Mammalian histidine decarboxylase: Interaction between apoenzyme and pyridoxal-5'-phosphate. Eur J Pharmacol 1: 381-390.

Hannuniemi R, Holopainen I, Korpi ER, Oja SS. 1985. Stimulation of amino acid accumulation in neuroblastoma and astrocytoma cells by L-histidine. Neurochem Res 10: 483-489.

Hryb DJ, Feigelson M. 1983. Histidase mRNA. Nature of translational products, tissue specificity, and differential development in male and female rat liver. J Biol Chem 258: 11377-11383.

Joseph DR, Sullivan PM, WangYM, Kozak C, Fenstermacher DA, et al. 1990. Characterization and expression of the complementary DNA encoding rat histidine decarboxylase. Proc Natl Acad Sci USA 87: 733-737.

Kahlson G, Rosengren E, White T. 1960. The formation of histamine in the rat foetus. J Physiol 151: 131-138.

Kanai Y, Segawa H, Miyamoto K, Uchino H, Takeda E, et al. 1998. Expression cloning and characterization of a transporter for large neutral amino acids activated by the heavy chain of 4F2 antigen (CD98). J Biol Chem 273: 23629-23632.

Karlstedt K, Nissinen M, Michelsen KA, Panula P. 2001. Multiple sites of L-histidine decarboxylase expression in mouse suggest novel developmental functions for histamine. Dev Dyn 221: 81-91.

Karlstedt K, Sallmen T, Eriksson KS, Lintunen M, Couraud PO, et al. 1999. Lack of histamine synthesis and down-regulation of H_1 and H_2 receptor mRNA levels by dexamethasone in cerebral endothelial cells. J Cereb Blood Flow Metab 19: 321-330.

Kawai Y, Moriyama A, Asai K, Coleman-Campbell CM, Sumi S, et al. 2005. Molecular characterization of histidinemia: Identification of four missense mutations in the histidase gene. Hum Genet 116: 340-346.

Kinnunen A, Lintunen M, Karlstedt K, Fukui H, Panula P. 1998. In situ detection of H_1-receptor mRNA and absence of apoptosis in the transient histamine system of the embryonic rat brain. J Comp Neurol 394: 127-137.

Lam WK, Cleary MA, Wraith JE, Walter JH. 1996. Histidinaemia: A benign metabolic disorder. Arch Dis Child 74: 343-346.

Lee SC, Wang ML. 1975. Histidinemia produced in the rat by treatment with nitromethane 1. Nutr Metab 18: 79-88.

Levy HL, Shih VE, Madigan PM. 1974. Routine newborn screening for histidinemia. Clinical and biochemical results. N Engl J Med 291: 1214-1219.

Lin EC, Pitt BM, Civen M, Knox WE. 1958. The assay of aromatic amino acid transaminations and keto acid oxidation by the enol borate-tautomerase method. J Biol Chem 233: 668-673.

Lin JS, Sakai K, Jouvet M. 1988. Evidence for histaminergic arousal mechanisms in the hypothalamus of cat. Neuropharmacology 27: 111-122.

Lozeva V, Anttila E, Tuominen RK, Hippeläinen M, Männistö PT, et al. 1999a. Hypothalamic histamine, growth rate, plasma prolactin, and growth hormone levels in rats with long-term portacaval anastomosis. Inflamm Res 48: 81-85.

Lozeva V, Valjakka A, Anttila E, MacDonald E, Hippeläinen M, et al. 1999b. Brain histamine levels and neocortical slow-wave activity in rats with portacaval anastomosis. Hepatology 29: 340-346.

Lozeva V, Attila M, Anttila E, Laitinen K, Hippeläinen M, et al. 1998a. Brain histamine H_3 receptors in rats with portacaval anastomosis: in vitro and in vivo studies. Naunyn-Schmiedebergs Arch Pharmacol 358: 574-581.

Lozeva V, MacDonald E, Belcheva A, Hippeläinen M, Kosunen H, et al. 1998b. Long-term effects of portacaval anastomosis on the 5-hydroxytryptamine, histamine, and catecholamine neurotransmitter systems in rat brain. J Neurochem 71: 1450-1456.

Lozeva V, Tarhanen J, Attila M, Männistö PT, Tuomisto L. 2003a. Brain histamine and histamine H_3 receptors following repeated L-histidine administration in rats. Life Sci 73: 1491-1503.

Lozeva V, Tuomisto L, Tarhanen J, Butterworth RF. 2003b. Increased concentrations of histamine and its metabolite, tele-methylhistamine and down-regulation of histamine H_3 receptor sites in autopsied brain tissue from cirrhotic patients who died in hepatic coma. J Hepatol 39: 522-527.

Medina MA, Landez JH, Foster LL. 1969. Inhibition of tissue histamine formation by decaborane. J Pharmacol Exp Ther 169: 132-137.

Mehler AH, Tabor H. 1953. Deamination of histidine to form urocanic acid in liver. J Biol Chem 201: 775-784.

Nakanishi T, Kekuda R, Fei YJ, Hatanaka T, Sugawara M, et al. 2001. Cloning and functional characterization of a new subtype of the amino acid transport system N. Am J Physiol Cell Physiol 281: C1757-C1768.

Neame KD. 1961. Uptake of amino acids by mouse brain slices. J Neurochem 6: 358-366.

Neame KD. 1962. Uptake of L-histidine, L-proline, L-ornithine, L-lysine, and L-methionine by brain tissue in vitro: A comparison with uptake by sarcoma RD3 and other tissues. J Neurochem 9: 321-324.

Neame KD. 1964. Uptake of histidine, histamine, and other imidazole derivatives by brain slices. J Neurochem 11: 655-662.

Nobile M, Maffei P, Nothen MM, Rietschel M, Smeraldi E, et al. 1997. Association study of schizophrenia and the histidase gene. Psychiatr Genet 7: 107-109.

O'Kane RL, Hawkins RA. 2003. Na^+-dependent transport of large neutral amino acids occurs at the abluminal membrane of the blood–brain barrier. Am J Physiol Endocrinol Metab 285: E1167-E1173.

O'Kane RL, Vina JR, Simpson I, Hawkins RA. 2004. Na^+-dependent neutral amino acid transporters A, ASC, and N of the blood–brain barrier: Mechanisms for neutral amino acid removal. Am J Physiol Endocrinol Metab 287: E622-E629.

Oldendorf WH. 1971. Brain uptake of metabolites and drugs following carotid arterial injections. T Am Neurol Assoc 96: 46-50.

Oldendorf WH, Szabo J. 1976. Amino acid assignment to one of three blood–brain barrier amino acid carriers. Am J Physiol 230: 94-98.

Oldendorf WH, Crane PD, Braun LD, Gosschalk EA, Diamond JM. 1988. pH dependence of histidine affinity for blood–brain barrier carrier transport systems for neutral and cationic amino acids. J Neurochem 50: 857-861.

Panula P, Joo F, Rechardt L. 1978. Evidence for the presence of viable endothelial cells in cultures derived from dissociated rat brain. Experientia 34: 95-97.

Panula P, Lintunen M, Karlstedt K. 2000. Histamine in brain development and tumors. Semin Cancer Biol 10: 11-14.

Panula P, Yang HY, Costa E. 1984. Histamine-containing neurons in the rat hypothalamus. Proc Natl Acad Sci USA 81: 2572-2576.

Parmentier R, Ohtsu H, Djebbara-Hannas Z, Valatx JL, Watanabe T, et al. 2002. Anatomical, physiological, and pharmacological characteristics of histidine decarboxylase knockout mice: Evidence for the role of brain histamine in behavioral and sleep–wake control. J Neurosci 22: 7695-7711.

Preuss CV, Wood TC, Szumlanski CL, Raftogianis RB, Otterness DM, et al. 1998. Human histamine N-methyltransferase pharmacogenetics: Common genetic polymorphisms that alter activity. Mol Pharmacol 53: 708-717.

Reimer RJ, Chaudhry FA, Gray AT, Edwards RH. 2000. Amino acid transport system A resembles system N in sequence but differs in mechanism. Proc Natl Acad Sci USA 97: 7715-7720.

Rosenmann A, Scriver CR, Clow CL, Levy HL. 1983. Histidinemia. Part II: Impact; a retrospective study. J Inherit Metab Dis 6: 54-57.

Sakurai E, Yamakami I, Sakurada T, Ochiai Y, Tanaka Y. 2003. Blood–brain barrier carrier-mediated transport and metabolism of L-histidine. Biogenic Amines 17: 335-348.

Sano H, Tada T, Moriyama A, Ogawa H, Asai K, et al. 1997. Isolation of a rat histidase cDNA sequence and expression in *Escherichia coli*—Evidence of extrahepatic/epidermal distribution. Eur J Biochem 250: 212-221.

Schayer RW. 1952. Biogenesis of histamine. J Biol Chem 199: 245-250.

Schayer RW, Rothschild Z, Bizony P. 1959. Increase in histidine decarboxylase activity of rat skin following treatment with compound 48/80. Am J Physiol 196: 295-298.

Schwartz JC, Lampart C, Rose C. 1970. Properties and regional distribution of histidine decarboxylase in rat brain. J Neurochem 17: 1527-1534.

Schwartz JC, Lampart C, Rose C. 1972. Histamine formation in rat brain in vivo: Effects of histidine loads. J Neurochem 19: 801-810.

Scriver CR, Levy HL. 1983. Histidinemia. Part I: Reconciling retrospective and prospective findings. J Inherit Metab Dis 6: 51-53.

Selden C, Calnan D, Morgan N, Wilcox H, Carr E, et al. 1995. Histidinemia in mice: A metabolic defect treated using a novel approach to hepatocellular transplantation. Hepatology 21: 1405-1412.

Sershen H, Lajtha A. 1976. Capillary transport of amino acids in the developing brain. Exp Neurol 53: 465-474.

Spolter H, Baldridge RC. 1964. Multiple forms of histidine-pyruvate transaminase in rat liver. Biochim Biophys Acta 90: 287-290.

Suchi M, Harada N, Wada Y, Takagi Y. 1993. Molecular cloning of a cDNA encoding human histidase. Biochim Biophys Acta 1216: 293-295.

Suchi M, Sano H, Mizuno H, Wada Y. 1995. Molecular cloning and structural characterization of the human histidase gene (HAL). Genomics 29: 98-104.

Tada K, Tateda H, Arashima S, Sakai K, Kitagawa T, et al. 1982. Intellectual development in patients with untreated histidinemia. A collaborative study group of neonatal screening for inborn errors of metabolism in Japan. J Pediatr 101: 562-563.

Tada K, Tateda H, Arashima S, Sakai K, Kitagawa T, et al. 1984. Follow-up study of a nation-wide neonatal metabolic screening program in Japan. A collaborative study group of neonatal screening for inborn errors of metabolism in Japan. Eur J Pediatr 142: 204-207.

Tanaka S, Nemoto K, Yamamura E, Ichikawa A. 1998. Intracellular localization of the 74- and 53-kDa forms of L-histidine decarboxylase in a rat basophilic/mast cell line, RBL-2H3. J Biol Chem 273: 8177-8182.

Vorobjev VS, Sharonova IN, Walsh IB, Haas HL. 1993. Histamine potentiates N-methyl-D-aspartate responses in acutely isolated hippocampal neurons. Neuron 11: 837-844.

Watanabe T, Taguchi Y, Maeyama K, Wada H. 1991. Formation of histamine: Histidine decarboxylase. Histamine and Histamine Antagonists. Uvnäs B, editor. Berlin: Springer-Verlag; pp. 145-163.

Watanabe T, Taguchi Y, Shiosaka S, Tanaka J, Kubota H, et al. 1984. Distribution of the histaminergic neuron system in the central nervous system of rats; a fluorescent immunohistochemical analysis with histidine decarboxylase as a marker. Brain Res 295: 13-25.

White T. 1959. Formation and catabolism of histamine in brain tissue in vitro. J Physiol 149: 34-42.

White T. 1960. Formation and catabolism of histamine in cat brain in vivo. J Physiol 152: 299-308.

Woodworth ME, Baldridge RC. 1970. Metabolic effects of an experimental histidinemia. Biochem Med 4: 425-434.

Yamada M, Watanabe T, Fukui H, Taguchi Y, Wada H. 1984. Comparison of histidine decarboxylases from rat stomach and brain with that from whole bodies of rat fetus. Agents Actions 14: 143-152.

Yamakami J, Sakurai E, Sakurada T, Maeda K, Hikichi N. 1998. Stereoselective blood–brain barrier transport of histidine in rats. Brain Res 812: 105-112.

Yamamoto J, Yatsunami K, Ohmori E, Sugimoto Y, Fukui T, et al. 1990. cDNA-derived amino acid sequence of L-histidine decarboxylase from mouse mastocytoma P-815 cells. FEBS Lett 276: 214-218.

Yan L, Szumlanski CL, Rice SR, Sobell JL, Lachman HM, et al. 2000. Histamine N-methyltransferase functional polymorphism: Lack of association with schizophrenia. Am J Med Genet 96: 404-406.

Yao D, Mackenzie B, Ming H, Varoqui H, Zhu H, et al. 2000. A novel system A isoform mediating Na^+/neutral amino acid cotransport. J Biol Chem 275: 22790-22797.

4 Aromatic Amino Acids in the Brain

M. Cansev · R. J. Wurtman

1	Introduction	60
2	Sources of Aromatic Amino Acids	61
3	Plasma Concentrations of the Aromatic Amino Acids	62
3.1	Plasma Tryptophan	66
3.1.1	Tryptophan Dioxygenase and Indoleamine Dioxygenase	66
3.1.2	Eosinophilia-Myalgia Syndrome	69
3.2	Plasma Tyrosine	69
3.2.1	Tyrosine Aminotransferase	70
3.3	Plasma Phenylalanine	72
3.3.1	Phenylalanine Hydroxylase	72
4	Brain Tryptophan and Tyrosine	73
4.1	Transport of Plasma Tryptophan and Tyrosine into the Brain	74
4.2	Brain Tryptophan	75
4.2.1	Tryptophan Hydroxylase	77
4.2.2	5-Hydroxytryptophan and l-DOPA	78
4.3	Brain Tyrosine	78
4.3.1	Tyrosine Hydroxylase	79
5	Consequences of Changing Brain Tryptophan and Tyrosine Levels	80
5.1	Precursor Availability and Neurotransmission	80
5.1.1	Neurons That Lack Multisynaptic or Autoreceptor-Based Feedback Loops	81
5.1.2	Neurons That Are Components of Positive Feedback Loops	82
5.1.3	Neurons That Normally Release Variable Quantities of Neurotransmitter Per Firing Without Engaging Feedback Responses	82
5.1.4	Physiologic Situations in Which Neurons Undergo Sustained Increases in Firing Frequency	83
5.1.5	Neurologic Diseases That Cause Either a Decreased Number of Synapses of Decreased Transmitter Release Per Unit Time	84
5.2	Brain Tryptophan and Serotonin	85
5.3	Brain Tyrosine and the Catecholamines	87

© Springer-Verlag Berlin Heidelberg 2007

Abstract: This chapter describes the aromatic L-amino acids tryptophan and tyrosine and the effects on tyrosine metabolism of phenylalanine. Tryptophan and phenylalanine are essential amino acids and must ultimately be derived from dietary proteins; tyrosine is obtained both from dietary proteins and from the hydroxylation of phenylalanine by phenylalanine hydroxylase (PAH). The proportions of dietary tryptophan, tyrosine, and phenylalanine that enter the systemic circulation are limited by three hepatic enzymes—tryptophan dioxygenase, tyrosine aminotransferase, and phenylalanine hydroxylase—that destroy them. These enzymes all have high substrate K_m's, hence they have little effect on their amino acid substrates present in systemic blood but major, concentration-dependent effects on the elevated concentrations, present postprandially, in portal venous blood.

All of the large, neutral amino acids (LNAA)—e.g., the three aromatic amino acids; the three branched-chain amino acids, leucine, isoleucine, and valine—across from the brain's capillaries into its substance through the action of a single transport molecule, LAT1. The kinetic properties of this molecule are such that it is saturated with LNAA at normal concentrations in systemic blood so that the individual LNAA compete with each other for blood–brain barrier transport. Hence the effect of any treatment on, for example, brain tryptophan, will depend not on plasma tryptophan, per se, but on the ratio of the plasma tryptophan concentration to the summed concentrations of the other, competing LNAA. Small quantities of LNAA molecules also enter the brain via choroid plexus transport into the cerebrospinal fluid.

The levels of tryptophan in the brain determine the substrate-saturation of tryptophan hydroxylase, and thus the rate at which tryptophan is converted to 5-hydroxytryptophan and subsequently to serotonin or melatonin. Brain tyrosine levels may or may not affect the rate at which tyrosine is hydroxylated, and converted to the catecholamines dopamine and norepinephrine, depending on the firing frequency of the particular catecholaminergic neuron. If the neuron is firing with high frequency, the tyrosine hydroxylase enzyme becomes multiply phosphorylated; this markedly increases its affinity for its otherwise-limiting cofactor (tetrahydrobiopterin) so that local tyrosine concentrations become limiting (several groups of prefrontal dopaminergic neurons normally fire unusually frequently, and are thus always susceptible to precursor control by available tyrosine levels). The abilities of the precursor amino acids, tryptophan and tyrosine, to control the rates at which neurons can produce and release their neurotransmitter products underlie a number of physiological processes, and also constitute a potential tool for amplifying or decreasing synaptic neurotransmission.

List of Abbreviations: AMP, adenosine monophosphate; BBB, blood-brain barrier; BH_4, tetrahydrobiopterin; CSF, cerebrospinal fluid; DOCA, deoxycorticosterone; DOPA, dihydroxyphenylalanine; DOPAC, dihydrophenylacetic acid; EMS, Eosinophilia-Myalgia syndrome; 5-HIAA, 5-hydroxyindole acetic acid; 5-HT, 5-hydroxytriptamine; 5-HTP, 5-hydroxytryptophan; HVA, homovanillic acid; IDO, indoleamine 2,3-dioxygenase; INF-gamma, interferon-gamma; L-AAAD, aromatic-L-amino acid decarboxylase; L-DOPA, L-dihydroxyphenylalanine; LAT1, Large Neutral Amino Acid Transporter 1; LNAA, Large Neutral Amino Acid; MAO, monoamine oxidase; MOPEG-SO4, 3-methoxy-4-hydroxyphenylethyleneglycol-Sulphate; NAD, nicotinamide adenine dinucleotide; NEFA, nonesterified fatty acids; NMDA, N-methyl D-aspartate; PAH, phenylalanine hydroxylase; PKU, phenylketonuria; SOD, Superoxide dismutase; TAT, tyrosine aminotransferase; TDO, tryptophan dioxygenase; TH, tyrosine hydroxylase; TNF-alpha, tumor necrosis factor-alpha; TPH, tryptophan hydroxylase

1 Introduction

This chapter describes the aromatic L-amino acids tryptophan and tyrosine, as well as the utilization of a third aromatic L-amino acid, phenylalanine, to produce tyrosine (the metabolism of phenylalanine in phenylketonuria [PKU] is described elsewhere in this volume). Like all dietary amino acids, each of these three compounds is used ubiquitously to synthesize proteins. But also, within some cell types, tryptophan is converted to the neurotransmitter serotonin (5-hydroxytryptamine; 5-HT) (● *Figure 4-1*), or tyrosine is converted to the catecholamines—the neurotransmitters dopamine and norepinephrine and the hormone epinephrine (● *Figure 4-2*) (tryptophan is also used in the pineal gland to make the hormone

Figure 4-1
Biosynthesis of serotonin from tryptophan

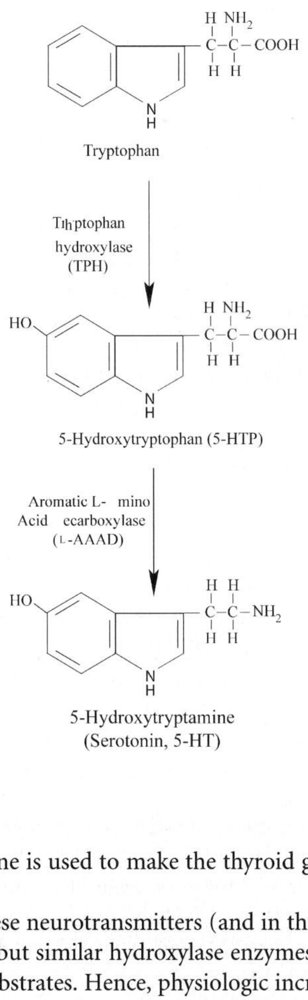

Tryptophan

↓ Tryptophan hydroxylase (TPH)

5-Hydroxytryptophan (5-HTP)

↓ Aromatic L-Amino Acid Decarboxylase (L-AAAD)

5-Hydroxytryptamine (Serotonin, 5-HT)

melatonin [❯ Figure 4-3], and tyrosine is used to make the thyroid gland's hormones, and the melanin in skin and brain).

The initial steps in producing these neurotransmitters (and in the process of converting phenylalanine to tyrosine) are catalyzed by specific but similar hydroxylase enzymes which, under certain conditions, are unsaturated with their amino acid substrates. Hence, physiologic increases in brain tryptophan or tyrosine levels can, by enhancing the saturation of their respective hydroxylases, control the rates at which serotoninergic or catecholaminergic cells form the intermediates 5-hydroxytryptophan (5-HTP) and dihydroxyphenylalanine (DOPA) and, ultimately, their neurotransmitter products. Similarly, changes in phenylalanine availability can affect tyrosine synthesis in the liver (❯ Figure 4-4) or DOPA formation in catecholaminergic neurons. This ability of precursor levels to control the syntheses of their biologically active products is unusual in the body: consumption of cholesterol, a precursor of testosterone or estrogens, in no way affects the syntheses of these gonadal steroids. This ability requires that plasma amino acid levels be allowed to vary (for example, in response to the macronutrient composition of the foods most recently consumed); that these variations be allowed to affect brain tryptophan or tyrosine levels; and that, as above, changes in these levels be sufficient to affect the rates at which the amino acids are hydroxylated.

2   Sources of Aromatic Amino Acids

Humans and other mammals are incapable of synthesizing tryptophan or phenylalanine de novo, and must ultimately obtain these essential amino acids by consuming proteins. The liver is able to make tyrosine from phenylalanine through the action of phenylalanine hydroxylase (PAH), hence mammals normally obtain

◘ Figure 4-2
Biosynthesis of the catecholamines dopamine, norepinephrine, and epinephrine from tyrosine

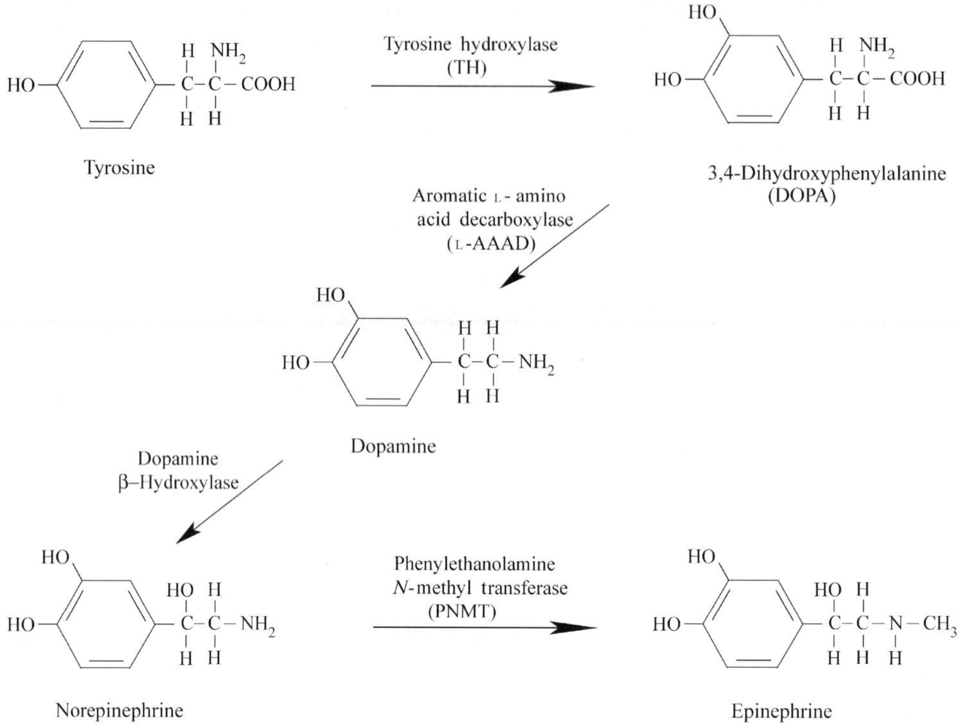

tyrosine from both an exogenous source, dietary protein, and endogenous synthesis (which provides about 15–20% of the tyrosine in human plasma [Barazzoni et al., 1998]). Tryptophan is usually the least abundant amino acid in most dietary proteins, constituting, for example, only 1–1.5% of the amino acids in casein, ovalbumin, and most meats (Orr and Watt, 1968), however, a few proteins—notably α-lactalbumin, a minor milk protein which is 6% tryptophan (Markus et al., 2002)—contain substantially more. Phenylalanine and tyrosine generally account for 3–4% of the amino acids in most dietary proteins. All three aromatic amino acids are primarily metabolized in the liver, by tryptophan dioxygenase (TDO), PAH, and tyrosine aminotransferase (TAT), respectively. Hence, only a portion of each actually enters the systemic circulation after a meal. The three aromatic amino acids can also be released from reservoirs in tissue or circulating proteins, however, this contribution is minor except in starvation.

3 Plasma Concentrations of the Aromatic Amino Acids

Aromatic amino acid concentrations in the systemic blood principally reflect the composition of the most recently consumed meal or snack (Fernstrom et al., 1979; Maher et al., 1984), and whether that food is still being digested and absorbed (▶ *Figure 4-5*). Consumption of carbohydrates with high glycemic indices (e.g., sucrose and starches, but not fructose) lowers plasma levels of most amino acids, principally via insulin-mediated facilitation of their uptake into skeletal muscle for conversion to protein (and, for the branched-chain amino acids leucine, isoleucine, and valine, for transamination and oxidation). In contrast, protein consumption raises plasma amino acid levels by directly contributing molecules which pass unmetabolized from the portal to the systemic circulations (i.e., virtually all of the leucine, isoleucine, and valine; a portion of each of the aromatic amino acids). In humans, a meal containing about 25-g protein

◘ **Figure 4-3**
Biosynthesis of melatonin from serotonin

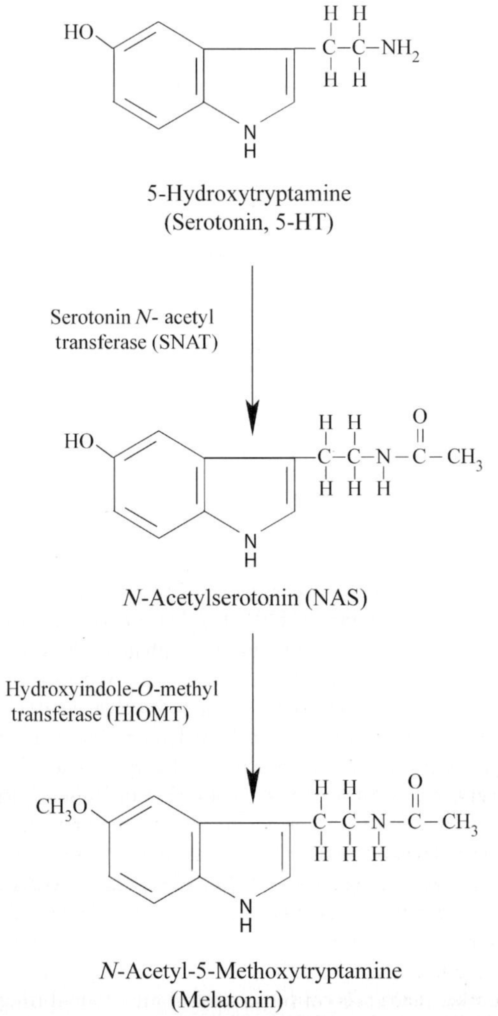

5-Hydroxytryptamine
(Serotonin, 5-HT)

↓ Serotonin N-acetyl transferase (SNAT)

N-Acetylserotonin (NAS)

↓ Hydroxyindole-O-methyl transferase (HIOMT)

N-Acetyl-5-Methoxytryptamine
(Melatonin)

and 180-g carbohydrate will neither raise nor lower plasma levels of tryptophan, phenylalanine, or tyrosine (Fernstrom et al., 1979), because the insulin-mediated passage of these amino acids into tissues is compensated by their entry from the splanchnic system; in rats, the null-effect proportion of proteins to carbohydrates is somewhat less (Yokogoshi and Wurtman, 1986).

As described below, enzymes exist in the liver, which function as "gates" to control the proportions of dietary tryptophan, tyrosine, and phenylalanine that are allowed to gain access to the systemic circulation. These enzymes are characterized by having a substrate K_m that is appreciably higher than the concentrations of their amino acid substrates in systemic blood, but lower than the concentrations that may be present postprandially in portal venous blood. This kinetic property allows the enzymes to metabolize only small proportions of the aromatic amino acid molecules reaching the liver by the hepatic arteries, but half or more of those arriving via the portal venous blood—and to vary the rates at which they destroy these substrates depending on the amounts that were consumed in the most recent meal.

◘ Figure 4-4
Conversion of phenylalanine to tyrosine

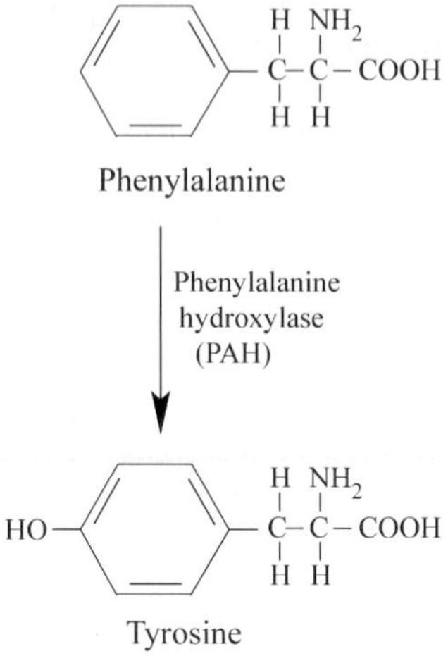

Phenylalanine

↓ Phenylalanine hydroxylase (PAH)

Tyrosine

The uptakes of plasma tryptophan, tyrosine, and phenylalanine into the brain depend not only on their own concentrations, but also on the plasma concentrations of other large neutral amino acids (LNAA) that compete with them for attachment to an LNAA carrier protein in brain capillaries. Hence, insulin—which profoundly lowers the plasma concentrations of the LNAA leucine, isoleucine, and valine—increases brain tryptophan uptake without increasing human plasma tryptophan levels because it decreases the competition for uptake generated by the other LNAA (Fernstrom and Wurtman, 1972b). Brain tyrosine and phenylalanine uptakes are not similarly increased by insulin, because insulin decreases their plasma levels by almost as much as it decreases those of the branched-chain amino acids (◉ Figure 4-5). The basis of plasma tryptophan's unique response to insulin, discussed below, is its also-unique ability to bind to circulating albumin: as much as 75–80% of the tryptophan in human plasma travels loosely bound to albumin (McMenamy and Oncley, 1958), but still largely able to enter the brain as shown by Pardridge (1977). Nonesterified fatty acids (NEFA)—which bind to a different site on the albumin molecule as shown by Goodman (1958)—inhibit this binding, hence insulin, which causes NEFA to strip off the albumin and enter adipocyes, increases albumin-bound tryptophan. This increase largely compensates for the reduction in "free" tryptophan that results from its insulin-mediated entry into skeletal muscle (◉ Table 4-1) (Lipsett et al., 1973; Madras et al., 1973).

Since people and most other mammals consume most of their food during either the day or the night depending on when their species sleeps, food consumption generates circadian rhythms in plasma amino acid levels (◉ Figure 4-5) (Wurtman et al., 1968b; Fernstrom et al., 1979), acting via insulin's effects and the passage of dietary amino acids from the portal to systemic circulations. These rhythms tend to disappear when people have been deprived of dietary carbohydrates and proteins for a day or two (Marliss et al., 1970).

When plasma glucose levels are above or below an "allowable" range, homeostatic feedback mechanisms are engaged to restore them to within that range, for example, insulin secretion in hyperglycemia, epinephrine secretion and glycogen breakdown in hypoglycemia. Similarly, when body temperature is above or below its allowable range, sweating or shivering are activated to restore it to normal. No such mechanisms regulate plasma amino acid levels: these levels are under "open-loop" control and, as described above, principally reflect the protein and carbohydrate contents of the meal or snack most recently consumed (◉ Figure 4-5). A behavioral feedback mechanism does exist through which a carbohydrate-rich, protein-poor snack can, by increasing brain serotonin, decrease the likelihood of continuing to eat carbohydrates (Wurtman et al., 1983). However, this mechanism does not "defend" allowable ranges for plasma

Aromatic amino acids in the brain   4   65

◘ Figure 4-5
Diurnal variations in plasma aromatic amino acid concentrations (*top*) and ratios (*bottom*) in normal human subjects consuming different levels of dietary protein. Each diet was consumed for five consecutive days and blood samples were drawn on the 4th and 5th days of each period. Plasma amino acid concentrations are expressed in nmol/ml. Vertical bars represent SD. Abbreviations: P,L,I,V,M,T + T are phenylalanine, leucine, isoleucine, valine, methionine, and tyrosine and/or tryptophan, respectively. Data from Fernstrom et al. (1979)

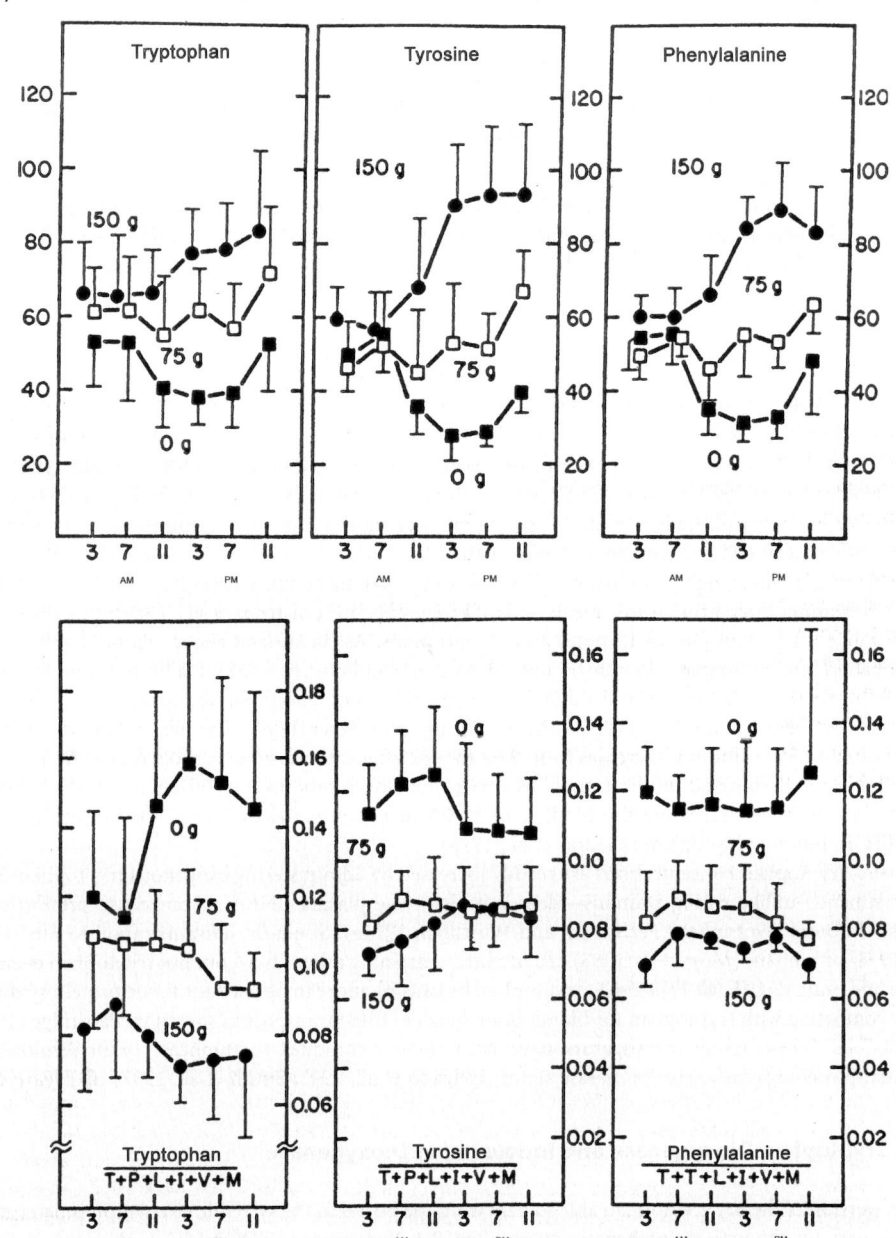

◘ Table 4-1

Effects of glucose ingestion on brain tryptophan and on serum free and albumin-bound tryptophan

	Control	Glucose (1 h)	Glucose (2 h)
Serum total tryptophan (μg/ml)	16.2 ± 0.2	19.6 ± 0.6	19.9 ± 0.4***
Serum free tryptophan (μg/ml)	5.5 ± 0.1	4.8 ± 0.3*	4.2 ± 0.2***
Free (% of total)	34	25	21
Serum bound tryptophan (μg/ml)	10.7 ± 0.3	14.8 ± 0.6**	15.7 ± 0.5***
NEFA (meq/L)	1.147 ± 0.034	0.648 ± 0.077***	0.604 ± 0.044***
Brain typtophan (μg/ml)	4.16 ± 0.42	6.42 ± 0.56**	5.93 ± 0.72**

Rats received D-glucose (2 g/4 ml tap water) by stomach tube; control animals received tap water. Values in all tables are given as mean ± SEM. *$p < 0.05$; **$p < 0.01$; ***$p < 0.001$, differs from controls. Data from Madras et al. (1973)

amino acid levels in the same sense the body defends blood glucose levels or temperatures. Food-induced changes in the plasma amino acid pattern are thus able, as described below, to affect neurotransmitter synthesis, as well as appetitive and other behaviors mediated by affected neurotransmitters.

3.1 Plasma Tryptophan

Plasma tryptophan concentrations among fasting normal humans vary between 55 and 65 μM (❯ Figure 4-5), depending in part on the individual's prior protein intake (i.e., higher after consuming a high-protein diet for a few days) (Fernstrom et al., 1979); prior caloric intake (i.e., dieting [Goodwin et al., 1990]); age (lower in older men [Caballero et al., 1991]); gender (higher in males [Demling et al., 1996]); and body mass index (lower in obesity [Caballero et al., 1988]). In rats, fasting tryptophan concentrations reportedly vary between 80 and 150 μM (Fernstrom and Wurtman, 1971a; Madras et al., 1973). Maximal levels among people consuming three high-protein meals (50 g/meal) per day are about twice as high as minimal levels in people consuming three protein-free meals daily (❯ Figure 4-5) (Fernstrom et al., 1979); this defines the normal range for human plasma tryptophan concentrations. As mentioned above, about 75–80% of the tryptophan in human plasma is loosely bound (McMenamy and Oncley, 1958) to albumin. This binding is of low affinity: the Kd's for rats and rabbits under pentobarbital anesthesia are greater than 1 mM (as compared with in vitro estimates of 0.13 mM) (Pardridge and Fierer, 1990). The proportion of circulating tryptophan bound to albumin increases from 0.62 to 0.82 after rats consume carbohydrates (❯ Table 4-1) (Madras et al., 1973) because insulin causes "free" (i.e., nonalbumin-bound) tryptophan to decline (like the other LNAA) but, by decreasing the binding of NEFA to albumin (Madras et al., 1973), enhances the albumin's affinity for tryptophan (Madras et al., 1973).

Plasma tryptophan concentrations are readily increased by administering exogenous tryptophan. Since this treatment—unlike eating proteins—does not also raise plasma LNAA, it can cause proportionate increases in brain tryptophan (Fernstrom and Wurtman, 1971a). Similarly, administration to rats (Gessa et al., 1974) or humans (Moja et al., 1988) of a mixture containing other LNAA but not tryptophan decreases plasma and brain or CSF (❯ Figure 4-6) tryptophan by causing more to be used for tissue protein synthesis, and by competing with tryptophan for blood–brain barrier (BBB) transport as shown by Pardridge (1977). As described below, many investigators have used these techniques to implicate brain serotonin in particular physiologic processes or disease states (Delgado et al., 1991; Smith et al., 1997) (❯ Figure 4-7).

3.1.1 Tryptophan Dioxygenase and Indoleamine Dioxygenase

The proportion of dietary tryptophan able to pass from the portal to the systemic circulation is determined in large part by the activity of hepatic tryptophan 2,3-dioxygenase (TDO) (EC 1.13.11.11), a heme-containing enzyme that irreversibly cleaves tryptophan's indolic nucleus to form N-formylkynurenine. This enzyme, found only in the liver, probably has only minor effects on the breakdown of tryptophan

⬛ Figure 4-6
Effect of consuming a tryptophan-free drink containing other LNAA on plasma and CSF tryptophan concentrations. Data on four or five subjects were obtained from Carpenter et al. (1998) and Delgado et al. (1991)

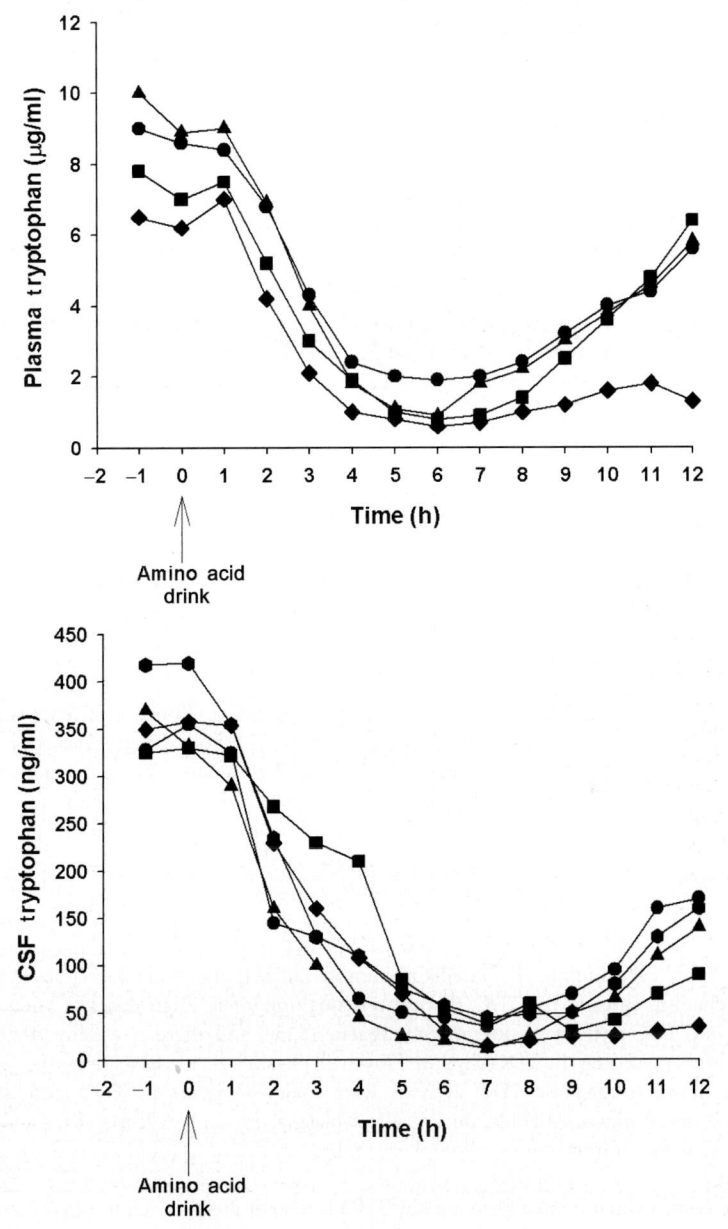

reaching the liver via the systemic circulation, because of its kinetic properties, i.e., a very high K_m—0.5×10^{-3} M (Schimke et al., 1965)—considerably greater than systemic, but not portal, venous tryptophan concentrations. In contrast, indoleamine 2,3-dioxygenase (IDO) (EC 1.13.11.17), the other enzyme that destroys tryptophan's indole nucleus, has a K_m for tryptophan at least an order of magnitude lower than

◘ Figure 4-7
Effect of consuming a tryptophan-free amino acid mixture on HAM-D (Hamilton rating scale for depression) in women with a prior history of depression. Data from Smith et al. (1997)

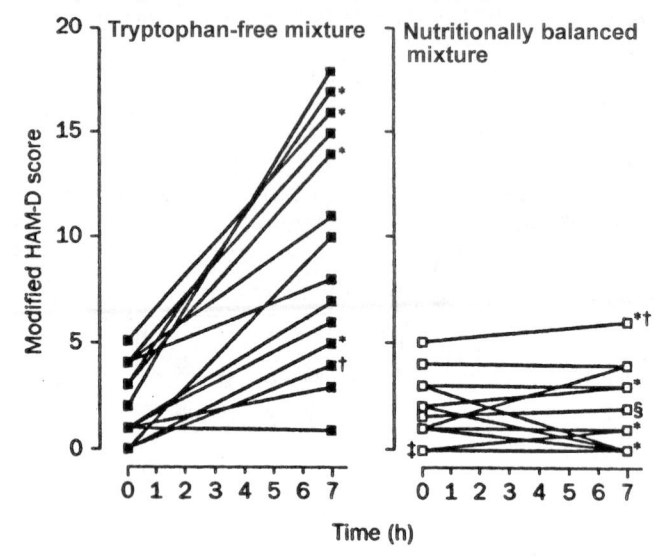

that of TDO (i.e., 26 μM [Yamazaki et al., 1985]; 45 μM [Shimizu et al., 1978]), and is active at the tryptophan concentrations found in systemic blood.

Tryptophan dioxygenase was initially called tryptophan pyrrolase (Kotake and Masayama, 1936), but renamed TDO after Hayaishi and coworkers (Hayaishi et al., 1957) showed that it incorporates two atoms of molecular oxygen into tryptophan to form the N-formylkynurenine. TDO, a tetrameric protein, is specific for the L-isomer of tryptophan (Tanaka and Knox, 1959), while IDO, a monomer, metabolizes a broad range of substrates (L-and D-tryptophan; serotonin; melatonin), and is found in all extrahepatic tissues (Yamazaki et al., 1985), including brain (Kwidzinski et al., 2005).

The human TDO gene is located on chromosome 4 (Comings et al., 1991). In rats, TDO activity exhibits a characteristic daily rhythm, peaking several hours after the onset of darkness, when the animals consume most of their food (Rapoport et al., 1966). This rhythm persists when animals consume protein-free diets (Ross et al., 1973), unlike the parallel rhythm in TAT activity (Wurtman et al., 1968b) described below. The mechanism causing the TDO rhythm remains unknown. High doses of glucocorticoid hormones can elevate TDO activity (Knox and Auerbach, 1955), and there is a daily rhythm in plasma glucocorticoid levels preceding the TDO rhythm. However, physiologic increases in plasma glucocorticoids have not been shown to increase TDO activity. Even major increases in TDO, produced by giving pharmacologic doses of glucocorticoids, do not affect plasma tryptophan levels (Kim and Miller, 1969) supporting the view that portal venous tryptophan, and not systemic tryptophan, is the normal substrate for this enzyme.

The hepatic N-formylkynurenine generated by TDO is further metabolized to L-kynurenine, kynurenic acid, xanthurenic acid, quinolinic acid, nicotinamide adenine dinucleotide (NAD), and, ultimately, to CO_2 and water. A kynurenine-producing pathway exists in rat brain (Guidetti et al., 1995) and several of its intermediates may have significant biological activity. Thus neurodegenerative effects have been attributed to quinolinic acid, which acts as an agonist for NMDA type glutamatergic receptors (Stone and Perkins, 1981); neuroprotective effects to the glutamate receptor antagonist kynurenic acid (Perkins and Stone, 1982); inhibition of striatal dopamine release by kynurenic acid (which blocks α-7 nicotinic receptors [Rassoulpour et al., 2005]); and an ability to stimulate neuronal growth and development to kynurenine (which is conjectured to stimulate nerve growth factor production [Dong-Ruyl et al., 1998]). NAD formed from this pathway is, of course, a cofactor in various enzymatic reactions.

The gene for human IDO is located on chromosome 8 (Burkin et al., 1993). The enzyme uses reduced molecular oxygen and superoxide (O^-_2) as substrates; it is the only enzyme known to do so besides superoxide dismutase (SOD), suggesting a role for it as an antioxidant. Whereas TDO activity can be induced by tryptophan, tyrosine, phenylalanine, histidine, and kynurenine (Taylor and Feng, 1991), IDO activity is induced by viruses, lipopolysaccharides, TNF-alpha, and interferons such as INF-gamma and INF-alpha. Its induction leads to intracellular depletion of tryptophan and to inhibition of the proliferation of various cancer cells, viruses, bacteria (Carlin et al., 1989; MacKenzie et al., 1999) and parasites as shown by Pfefferkorn (1984). In agreement with the in vitro finding that T cell proliferation could be inhibited by IDO activation (Munn et al., 1999), IDO expression in placenta was shown to prevent fetal allograft rejection by depleting tryptophan and suppressing maternal T cell responses in pregnant mice (Munn et al., 1998).

3.1.2 Eosinophilia-Myalgia Syndrome

Prior to 1990, L-tryptophan was freely available as a dietary supplement within the USA, purchased principally for self-treatment of insomnia. Then, in 1989, a manufacturer, the Showa Denka Company, began to market a new tryptophan preparation, the synthesis of which involved fermentation using a newly engineered strain of Bacillus amyloliquefacien (Yamaoka et al., 1994). Soon thereafter a new disease, the "Eosinophilia-Myalgia Syndrome" (EMS) was identified (Centers for Disease Control, 1989), initially among users of this preparation (Slutsker et al., 1990), who lived in New Mexico (Eidson et al., 1990). The syndrome was characterized by muscle pain and weakness, striking eosinophilia, dyspnea, skin rash, and various abnormal laboratory findings. Subsequent chemical analysis of this preparation revealed that it contained a variety of novel impurities, including "Peak E," or 1,1-ethylidenebis[tryptophan]—a compound later shown to activate human eosinophils and to enhance cytokine production from T lymphocytes (Yamaoka et al., 1994).

Excellent medical detective work led to the rapid removal of tryptophan containing toxic impurities from the market. It has also led, in the USA, but not elsewhere, to the continuing unavailability of pure tryptophan for medical uses, other than as a constituent of enteral and parenteral preparations. Had tryptophan been regulated as a drug, any new preparation would have had to undergo Phase I safety testing, during which the propensity of the new Showa Denka preparation to cause severe eosinophilia would have been noted, and the product would probably not have been approved for medical use. Unfortunately, because of the Dietary Supplement Act of 1994—which exempts "amino acids and their products" (!) from having to undergo FDA-regulated testing prior to sale—other amino acid "dietary supplements" continue to be sold in the USA without prior Phase I testing.

3.2 Plasma Tyrosine

Plasma tyrosine concentrations among fasting normal humans vary between 50 and 80 µM (❯ *Figure 4-5*), (Fernstrom et al., 1979; Glaeser et al., 1979; Maher et al., 1984) depending in part on the individual's prior protein intake (Fernstrom et al., 1979), and age (higher in older than younger women [Caballero et al., 1991]). In rats, fasting tyrosine levels vary between 90 and 120 µM (Fernstrom and Faller, 1978; Agharanya and Wurtman, 1982a). Maximal concentrations among people consuming three high-protein meals (50 g/meal) per day are about 3.5 times as great as minimal levels observed in people consuming three protein-free meals daily (Fernstrom et al., 1979). Unlike tryptophan, neither tyrosine nor phenylalanine in blood is appreciably bound to albumin.

Plasma tyrosine levels are also readily increased by administering exogenous tyrosine (❯ *Figures 4-8* and ❯ *4-9*) (Glaeser et al., 1979; Melamed et al., 1980a), which also raises tyrosine levels in human CSF (Growdon et al., 1982) and rat brain (Morre et al., 1980). A single oral dose of 100 mg/kg increased human

Figure 4-8

Accumulation of MOPEG-SO4 in brains of cold-stressed rats treated with neutral amino acids. Rats received valine (200 mg/kg, i.p.) or tyrosine (125 mg/kg, i.p.), or saline; 30 min later they were placed in single cages in a cold (40 C) environment. After 1 h, all animals were killed, and their whole brains were analyzed for tyrosine and MOPEG-SO4. Each point represents the tyrosine and MOPEG-SO4 present in a single brain. Brain tyrosine and MOPEG-SO4 levels in animals kept at room temperature were 14.4 µg/g and 80 ng/g, respectively. Symbols: closed circles, animals pretreated with valine; open circles, animals pretreated with saline; closed squares, animals pretreated with tyrosine. Data from Gibson and Wurtman (1978)

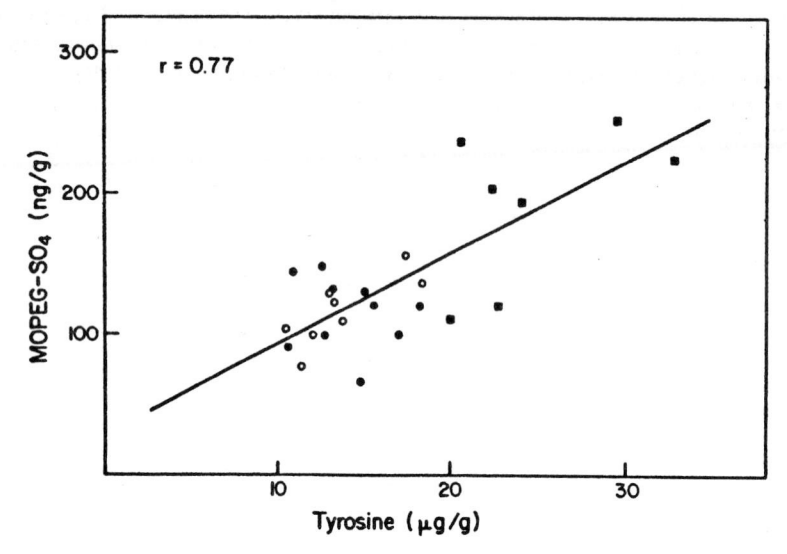

plasma tyrosine concentrations, after 2 h, from 69 to 154 µM, a 150-mg/kg dose increased this level to 203 µM. Both treatments reduced plasma levels of the other LNAA (Glaeser et al., 1979), probably by enhancing their utilization for tissue protein synthesis. A single intraperitoneal dose of 100 mg/kg given to rats increased tyrosine levels throughout the brain, but effects were greatest in hippocampus and cortex (Morre et al., 1980).

Similarly, administration to rats (Biggio et al., 1976) or humans (Sheehan et al., 1996; McTavish et al., 2005) of an LNAA mixture lacking tyrosine or its precursor phenylalanine lowers plasma tyrosine and, in rats, can be shown to deplete brain tyrosine as well (Biggio et al., 1976). Since brain tyrosine levels can, like those of tryptophan, control the rates of synthesis of its neurotransmitter products (the catecholamines dopamine and norepinephrine [Wurtman et al., 1974]), investigators are starting to use this technique to implicate brain catecholamines in particular physiologic processes or disease states (McTavish et al., 2005).

3.2.1 Tyrosine Aminotransferase

The proportion of dietary tyrosine, or tyrosine synthesized from phenylalanine by hepatic PAH, which can pass from the portal to the systemic circulations is determined in large part by the activity of hepatic TAT (EC 2.6.1.5). This enzyme catalyzes the conversion of tyrosine to p-hydroxyphenylpyruvate, the initial intermediate in tyrosine's complete degradation to fumarate and acetoacetate. Like TDO for tryptophan, TAT probably has only minor effects on the breakdown of the tyrosine that reaches the liver via the systemic circulation, because its K_m for tyrosine—1.7×10^{-3} M (Hayashi et al., 1967)—is even greater than TDO's for tryptophan, and also much greater than plasma tyrosine levels.

Figure 4-9

Effect of tyrosine administration on the accumulation of HVA in corpora striata of rats given haloperidol or probenecid. Rats received tyrosine (100 mg/kg) or its diluent followed in 20 min by haloperidol (2 mg/kg) or probenecid (200 mg/kg); they were sacrificed 70 min after the second injection. Data from individual animals receiving haloperidol are indicated by open circles; data from rats receiving haloperidol plus tyrosine are indicated by closed circles. Striatal HVA levels were highly correlated with brain tyrosine levels in all animals receiving haloperidol ($r = 0.70$, $p < 0.01$). In contrast, the striatal HVA levels of animals receiving probenecid alone did not differ from those of rats receiving probenecid plus tyrosine. Brain tyrosine and striatal HVA concentrations in each group were (respectively): probenecid, 17.65 ± 1.33 and 1.30 ± 0.10 µg/g; probenecid plus tyrosine, 44.06 ± 3.91 and 1.31 ± 0.11 µg/g; haloperidol, 17.03 ± 0.97 and 2.00 ± 0.10 µg/g; and haloperidol plus tyrosine, 36.02 ± 2.50 and 3.19 ± 0.20 µg/g. Data from Scally et al. (1977)

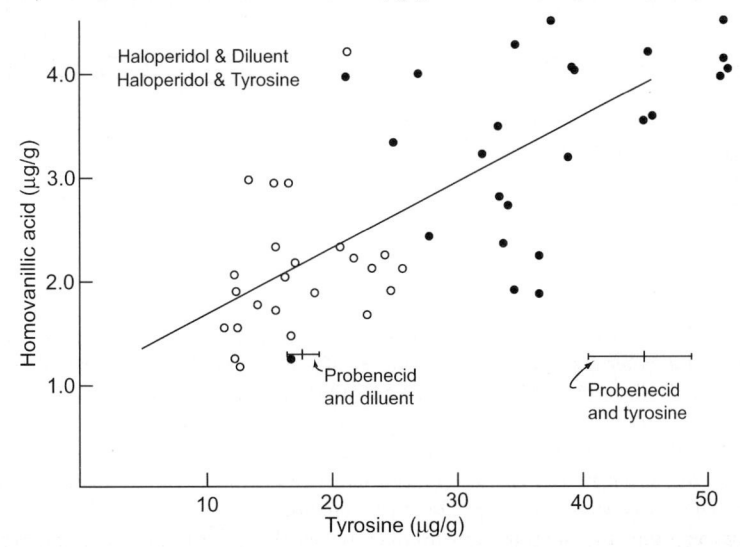

The gene for TAT is located on human chromosome 16 (Natt et al., 1986); mutations in this gene cause an inherited disorder, tyrosinemia type II (Richner-Hanhart syndrome). TAT was thought to have several chromatographically distinguishable forms (Johnson et al., 1973; Belarbi et al., 1979), however, these were subsequently shown to be generated by inappropriate purification methods: when the complete cDNA sequence coding for the rat gene was cloned (Grange et al., 1985), homogenous enzyme was obtained by Dietrich (1992). TAT is a relatively short-lived enzyme, with a half-life of less than 3 h in vivo and a rapid turnover rate as shown by Kenney (1967). It uses pyridoxal and pyridoxamine phosphates as cofactors and alpha-ketoglutarate as cosubstrate (Hayashi et al., 1967). Very low activities of an uninduceable form of TAT, about 1/50 of those present in liver, have been described in kidney and heart (Lin and Knox, 1958).

TAT activity in rats exhibits marked daily periodicity, increasing by fourfold or more in the evening, when the animal initiates rapid food consumption (Wurtman and Axelrod, 1967). This rhythm is, in fact, generated by the cyclic consumption of protein, because proteins contribute tryptophan, the limiting amino acid in hepatic protein synthesis (Wurtman et al., 1968b). Apparently, consumption of the tryptophan in protein allows the aggregation of long-lived messenger RNA coded for TAT into polyribosomes which synthesize TAT as shown by Munro (1968) and others (Fishman et al., 1969). The rhythm is rapidly extinguished in starved rats or in rats fed with a protein-free diet, and exhibits temporal shifts as soon as the feeding schedule is modified (Fuller and Snoddy, 1968). It is not generated by the daily rhythm in plasma glucocorticoids, even though high doses of these hormones can induce TAT activity, since it persists following adrenalectomy (Wurtman and Axelrod, 1967). Exogenous tyrosine, tryptophan, insulin, glucagon, dibutyrylcyclic AMP, and numerous other compounds can, in high concentrations, enhance TAT synthesis (Lin and Knox, 1957; Kenney and Flora, 1961; Holten and Kenney, 1967; Wicks et al., 1969; Kroger and Gratz, 1980).

3.3 Plasma Phenylalanine

This chapter considers plasma phenylalanine only in relation to tyrosine and tryptophan, i.e., as a precursor for hepatic tyrosine and brain catechols; and as a competitor for their transport across the BBB. Phenylalanine's metabolism in PKU and related diseases is described elsewhere in this volume.

Plasma phenylalanine concentrations among fasting normal humans vary between 45 and 60 µM (❍ *Figure 4-5*) (Fernstrom et al., 1979; Maher et al., 1984), depending in part on the individual's prior protein intake (i.e., higher after consuming a high-protein meal for several days [Fernstrom et al., 1979]), age (higher in older than younger women [Caballero et al., 1991]), and body mass index (higher in obese subjects, with less of a fall in response to insulin [Caballero et al., 1988]). In rats, fasting phenylalanine concentrations vary between 75 and 100 µM (Fernstrom and Faller, 1978). Maximal levels among normal people consuming three high-protein meals (50 g/meal) per day are about three times as high as minimal levels in people consuming three protein-free meals daily (Fernstrom et al., 1979).

3.3.1 Phenylalanine Hydroxylase

Properties PAH (E.C. 1.14.16.1), principally a hepatic enzyme, limits the proportion of dietary phenylalanine that is allowed to enter the systemic circulation. Its K_m for phenylalanine, estimated as $2.5–8.3 \times 10^{-4}$ M (Ayling et al., 1974; Abita et al., 1976) is much higher than systemic plasma phenylalanine concentrations (0.8×10^{-4} M [Fernstrom and Faller, 1978]) suggesting that this enzyme has little role in metabolizing phenylalanine outside the portal vascular system, except perhaps in patients with untreated PKU. The enzyme catalyzes the initial step in the metabolism of phenylalanine, its hydroxylation at the 4-position of the benzene ring to generate tyrosine. This hydroxylation, in mammals, is the obligatory and rate-limiting step in the complete oxidation of phenylalanine to CO_2 and water; no other pathway exists which destroys phenylalanine's benzene ring (Milstein and Kaufman, 1975). In PKU, the lack of PAH or its natural cofactor BH_4 causes the accumulation of phenylalanine, which, in this circumstance, is decarboxylated to phenylethylamine, or transaminated to phenylpyruvic acid. Accumulation of these metabolites in brain, and the relative depletion of PAH's product, tyrosine, causes the clinical findings of this disease (Kim et al., 2004).

Like the tyrosine and tryptophan hydroxylases (TPHs), PAH uses ferrous iron as a cofactor, and molecular oxygen and tetrahydrobiopterin (BH_4) as cosubstrates with phenylalanine. Phenylalanine can also be a substrate for tyrosine hydroxylase (TH), which transforms it to tyrosine and then DOPA. This supplemental pathway for catechol synthesis has been demonstrated in preparations of bovine adrenal gland and guinea pig heart (Ikeda et al., 1967), rat brain synaptosomes (Katz et al., 1976), striatal slices (Milner et al., 1986), PC12 cells (DePietro and Fernstrom, 1998), and brains of rats receiving phenylalanine (During et al., 1988). At high concentrations, phenylalanine becomes an inhibitor of PAH; in vitro studies with PC12 cells showed that this effect represents "substrate inhibition" by phenylalanine itself (DePietro and Fernstrom, 1998). High concentrations of phenylalanine can also suppress DOPA synthesis and release in vivo (Wurtman et al., 1974; During et al., 1988) partly by competing with tyrosine for binding to the common LNAA transporter (Fernstrom and Faller, 1978) at the BBB.

The gene that encodes PAH has been located on human chromosome 12 (Lidsky et al., 1985). Rat liver PAH exists in two oligomeric forms, a tetramer which accounts for 75–80% and a dimer for the remainder (Parniak and Kaufman, 1985). Recombinant human and rat liver PAH's, and enzyme isolated from rat liver (Kowlessur et al., 1996), share similar oligomeric composition, whereas recombinant human PAH has different regulatory properties; the PAH's from livers of Sprague-Dawley rats are composed of identical subunits (Iwaki et al., 1985).

Most of the PAH in the body is in liver; small amounts are also found in kidney and pancreas (Tourian et al., 1969). However, the contribution of human kidney to total in vivo tyrosine synthesis may be substantial (Garibotto et al., 2002).

Regulation Evidence exists that PAH activity can be modulated at three sites; phenylalanine itself also attaches to a fourth "catalytic site" at which it is converted to tyrosine. The three noncatalytic sites include a serine residue (Wretborn et al., 1980) that becomes phosphorylated; an "activator" site to which physiologic concentrations of phenylalanine attach; and an additional site at which very high-phenylalanine concentrations (i.e., 2 mM or greater) inhibit enzyme activity (Dhondt et al., 1978). PAH's phosphorylation and the attachment of phenylalanine to its activator site both have the effect of increasing the hydroxylation of phenylalanine to tyrosine, phosphorylation increasing the enzyme's Vmax without changing its K_m for phenylalanine (Abita et al., 1976). The mechanism by which the activator site enhances phenylalanine's hydroxylation is not known. Phosphorylation of PAH doubles the affinity of the regulatory site for phenylalanine (Døskeland et al., 1984), thus further enhancing the enzyme's activity, and attachment of phenylalanine to the activator site similarly enhances the enzyme's susceptibility to phosphorylation. Hence, the two mechanisms manifest a positive feedback relationship. Activation of the rat hepatic enzyme by phosphorylation (i.e., after giving glucagon, in vivo [Kaufman, 1986]) reportedly increases plasma tyrosine and decreases phenylalanine, while activation of the regulatory site has been shown to increase tyrosine release from the isolated perfused liver (Shiman et al., 1982). It should be noted that the rat is considerably better at hydroxylating phenylalanine than the human. In fasted rats about 75 μmol/kg/h are converted to tyrosine, contributing about 20% of the tyrosine entering the circulation (Moldawer et al., 1983). If plasma phenylalanine concentrations are increased eightfold, the conversion of phenylalanine to tyrosine also increases, now contributing 70% of the tyrosine entering the circulation. In humans, PAH is much less responsive to a phenylalanine load: the basal hydroxylation rate is only 6 μmol/kg/h (Clarke and Bier, 1982), and a phenylalanine load preferentially elevates plasma phenylananine, not tyrosine (Caballero and Wurtman, 1988).

The proportion of PAH that is phosphorylated is diminished by a specific PAH phosphatase enzyme (Jedlicki et al., 1977). Moreover, phosphorylation of PAH can be inhibited by its own cosubstrate BH_4, which also stabilizes the low-activity, unphosphorylated form of the enzyme; this inhibition can be blocked by phenylalanine (Døskeland et al., 1984). A number of endogenous compounds, including lysolecithin and α-chymotrypsin, have been shown to modify PAH activity in vitro; moreover, numerous other amino acids, including methionine, norleucine, and tryptophan, can serve as substrates for PAH in vitro as shown by Kaufman (1986). None of these compounds has been shown to affect the enzyme, nor hydroxylated by it, in vivo.

4 Brain Tryptophan and Tyrosine

The levels of tryptophan and tyrosine in the brain can be of major importance in controlling the rates at which neurons synthesize and release their neurotransmitter products, serotonin, dopamine, and norepinephrine: all serotonin-producing neurons and some dopamine-producing neurons invariably synthesize more or less of their transmitter when tryptophan or tyrosine levels rise or fall; the other dopaminergic and noradrenergic neurons can become tyrosine dependent when they are firing frequently. Brain levels of tryptophan and tyrosine are controlled in part by their plasma concentrations (and share with these concentrations the property of not being subject to feedback control). However, they are even more dependent on plasma concentrations of other LNAA—particularly phenylalanine and the branched-chain compounds, leucine, isoleucine, and valine—which compete with tryptophan or tyrosine for transport across the BBB. Thus, for example, a treatment—consumption of a carbohydrate that elicits insulin secretion—which lowers plasma LNAA concentrations will raise brain tryptophan, even though it does not raise plasma tryptophan in humans. Or one that raises plasma tryptophan—consumption of a protein—can lower brain tryptophan by contributing larger amounts of the other LNAA than tryptophan to the blood (❷ *Figure 4-10*).

◘ Figure 4-10

Effect of consuming a protein-free (CARBO) or protein-containing (CHOW; 18%) meal on brain and plasma tryptophan levels in overnight-fasted rats. Rats were killed 2 h after diet presentation. Two-hour plasma tryptophan levels were significantly greater in rats consuming either diet than in fasting controls (CHOW: $p < 0.001$; CARBO: $p < 0.01$). Two-hour brain tryptophan levels were significantly elevated above control only in rats consuming the carbohydrate-plus-fat diet ($p < 0.001$). Data from Fernstrom et al. (1973)

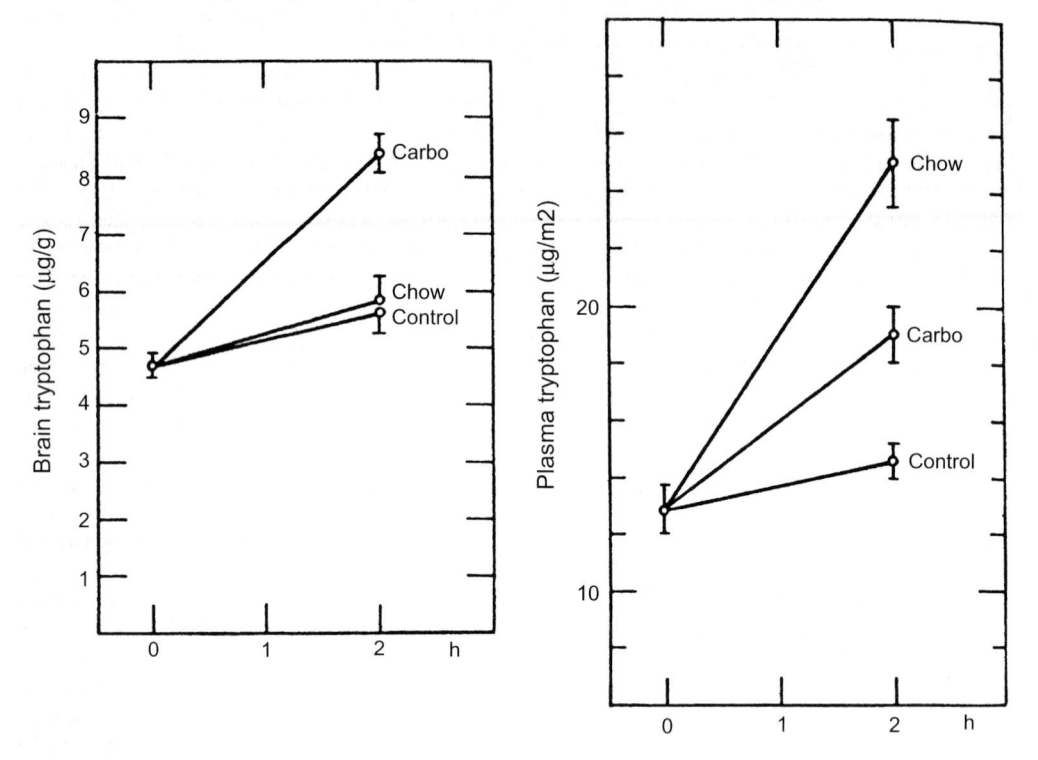

Brain tryptophan and tyrosine are utilized in all brain cells for protein synthesis, and serve as substrates for TPH or TH in monoaminergic neurons. Smaller quantities of brain tryptophan apparently are metabolized by IDO to form kynurenine and its products (Stone and Darlington, 2002) or, conceivably, by aromatic L-amino acid decarboxylase to form tryptamine (Saavedra and Axelrod, 1974). In the pineal organ serotonin formed from the hydroxylation and decarboxylation of tryptophan is further transformed to the hormone melatonin by N-acetylation followed by O-methylation (◉ Figure 4-3) (Axelrod et al., 1969). This process—which is accelerated each night (i.e., during the hours of darkness [Wurtman and Axelrod, 1965; Lynch et al., 1975])—is associated with increases in the activities of the enzymes (serotonin N-acetyltransferase and hydroxyindole-O-methyltransferase) that catalyze these two reactions, however, it probably depends more on the liberation of "bound" serotonin within pinealocytes, making the serotonin accessible both to serotonin N-acetyltransferase and to monoamine oxidase (MAO) (Wurtman, 2005).

4.1 Transport of Plasma Tryptophan and Tyrosine into the Brain

All amino acids are ionized at physiological pH and would not be able to cross membrane bilayers to gain access to the brain were it not for two highly specialized sets of transport molecules.

The most important set is located within the endothelial cells that line brain capillaries. It includes three different types of macromolecules: those that allow LNAA to enter the brain, by facilitated diffusion;

those that do the same for basic amino acids (e.g., lysine; arginine); and those that actively transport acidic amino acids (e.g., glutamate, aspartate) in the opposite direction—from the brain's extracellular fluid to the intravascular space as shown by Pardridge (1977). It should be noted that many of the LNAA and the basic amino acid lysine are "essential," and cannot be made by brain tissue; the acidic amino acids, in contrast, are readily synthesized from glucose. The other set of macromolecules that allow some circulating amino acids to enter the brain are in the cells that line the choroid plexus as shown by Lorenzo (1974). Because the surface they cover is very much smaller than the brain's capillaries, they transport only about 1/1000 as many molecules per unit time as shown by Pardridge (2001).

Carrier-mediated transport of LNAA at the BBB and in other tissues is affected by a family of transport proteins called "L-System," which contain LAT1, a catalytic subunit (also known as the light chain) and a type II glycoprotein subunit (4F2hc, also known as heavy chain) (Kanai et al., 1998). The LAT1-4F2hc heterodimer is connected by a single cysteine residue (Mastroberardino et al., 1998) in a disulfide linkage. LAT1-4F2hc is selective for LNAA transport and is essential for BBB transport (Boado et al., 1999) of the LNAA as well as for these compounds to enter tissues with comparable anatomic barriers (e.g., placenta and testis as reviewed by Verrey [2003]). This transport is bidirectional, Na-independent, and nonenergy-requiring (facilitated diffusion). The affinity of BBB LAT1 for LNAA is extremely high compared with that of peripheral LNAA transporters and the carrier is highly saturated at physiological plasma LNAA concentrations (i.e., the K_m of BBB LNAA transport approximates the plasma concentration of these LNAA), which causes the competition among LNAA for entering the brain as shown by Pardridge (1977). Two alkylating agents: melphalan and DL-2-amino-7-bis[(2-chloroethyl)amino]-1,2,3,4-tetrahydro-2-naphtoic acid (DL-NAM) inhibit BBB LNAA transport by damaging the disulfide bridge between LAT1 and 4F2hc in the heterodimer.

Once tryptophan, tyrosine, and the other LNAA molecules have passed from the plasma to the brain's extracellular fluid, they are able to enter all brain cells, to be used for synthesizing new protein molecules (chiefly in perikarya of neurons). Monoaminergic neurons also need large quantities of tryptophan and tyrosine in their nerve terminals to make serotonin and the catecholamines, and have additional mechanisms for obtaining such quantities. Amino acids are well transported into brain slices, by both saturable uptake and unidirectional influx (Vahvelainen and Oja, 1975), and it was using such slices that the competition for transport among LNAA was initially noted (Blasberg and Lajtha, 1966). Tryptophan (Grahame-Smith and Parfitt, 1970) and tyrosine (Morre and Wurtman, 1981) are also concentrated within synaptosomes, in competition with other LNAA. The proportions of tryptophan molecules used in serotoninergic neurons to synthesize serotonin versus proteins apparently are not known, however, it can safely be assumed that much more is used for the former purpose: among pineal cells, which also make both serotonin (and melatonin) and proteins, this ratio is greater than 100:1 (Wurtman et al., 1968a; Wurtman et al., 1969). Similarly, although acetylcholine-producing neurons constitute only a tiny fraction (about 1%) of brain cells, they are estimated to utilize at least 60% of the choline entering the brain to make acetylcholine (Farber et al., 1996), the rest being used in all brain cells to generate phospholipids.

4.2 Brain Tryptophan

The quantities of tryptophan that enter the brain, and sustain or elevate brain tryptophan levels, depend on three sets of nutrients in the plasma: tryptophan itself; the other LNAA; and NEFA, which, by binding loosely to albumin, diminish the binding of tryptophan and change the proportions of plasma tryptophan that are albumin-bound and free. (As described below, the actual effects of the fatty acids on brain tryptophan levels are minimal [Fernstrom et al., 1976; Pardridge and Fierer, 1990].) Brain tyrosine is similarly affected by plasma tyrosine and LNAA levels; however, tyrosine does not bind appreciably to albumin.

That giving high-tryptophan doses to rats could increase brain tryptophan (and serotonin) levels was first shown in 1962 for dietary tryptophan (Green et al., 1962; Wang et al., 1962) and in 1965 for injected tryptophan (800 mg/kg: Ashcroft et al., 1965). By 1971, it had been shown that brain tryptophan levels in rats normally vary within a twofold range, and that giving the animals as little as 12.5 mg/kg could

Figure 4-11

Dose-response curve relating brain tryptophan and brain serotonin. Rats received tryptophan (12.5, 25, 50, or 125 mg/kg, i.p.) at noon and were killed 1 h later. Horizontal bars represent standard errors of the mean for brain tryptophan; vertical bars represent standard errors of the mean for brain serotonin. All brain tryptophan levels were significantly higher than control values ($p < 0.001$). All brain serotonin levels were significantly higher than control values ($p < 0.01$). Data from Fernstrom and Wurtman (1971a)

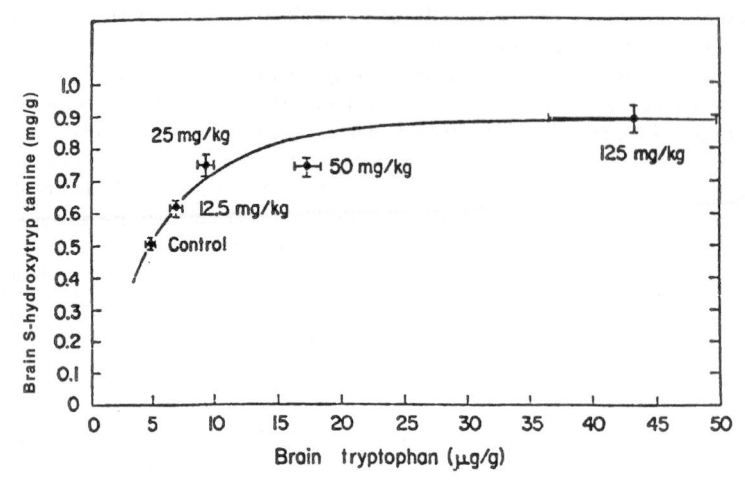

significantly increase brain tryptophan but keep it within this normal range (▶ Figure 4-11) (Fernstrom and Wurtman, 1971a). The rise in brain tryptophan occurs because it displaces other LNAA from the LAT1-4F2hc transport molecule in brain capillaries. The magnitude of this increase is predicted by the change in the "plasma tryptophan ratio"—the ratio of the plasma tryptophan concentration to the summed concentrations of other LNAA which exhibit high affinities for the transport carrier (usually taken as tyrosine, phenylalanine, and the branched-chain amino acids, but sometimes also methionine) (Fernstrom and Wurtman, 1971b). Since these affinities are not equal, one can theoretically improve the correlation between brain level and plasma ratio by correcting the summed LNAA concentrations for each amino acid's K_m, but operationally this correction is usually unnecessary (Fernstrom and Faller, 1978).

After giving tryptophan, the increase in brain tryptophan can also be predicted from plasma tryptophan levels alone, but more often this is not the case. Thus, a glucose-rich snack can increase brain tryptophan without increasing plasma tryptophan or increasing it only slightly in rats (▶ Figure 4-10), because the resulting insulin secretion lowers plasma levels of the other LNAA, while a protein-rich meal raises plasma tryptophan without raising brain tryptophan, because most proteins contain only 1.0–1.5% tryptophan, and thus contribute very little of this amino acid to the plasma, in comparison with the other LNAA (Fernstrom and Wurtman, 1972b). Individual LNAA (e.g., isoleucine) can be administered along with tryptophan, as experimental controls to block a tryptophan-induced rise in brain tryptophan, and groups of amino acids lacking tryptophan are often used to lower brain tryptophan levels (Gessa et al., 1974) (▶ Figure 4-6).

As noted above, about 75–80% of the tryptophan in the plasma travels loosely bound to albumin (McMenamy and Oncley, 1958). Initially it was anticipated that this binding would substantially retard the passage of tryptophan across the BBB, and scientists considered measuring plasma "free" (nonalbumin-bound) tryptophan, or the ratio of the "free tryptophan" to the other LNAA, as the best predictor of brain tryptophan levels (Knott and Curzon, 1972). However, subsequent studies often described treatment-induced changes in plasma "free" tryptophan which were opposite in direction to those in brain tryptophan. For example, rats consuming glucose (▶ Table 4-1) exhibited increases in brain tryptophan and albumin-bound plasma tryptophan (because the resulting secretion of insulin caused NEFA to dissociate from albumin and enter adipocytes), but a decrease in "free" tryptophan (Madras et al., 1973) (because the uptake of the "free" amino acid into muscle—like that of all other LNAA—was enhanced by insulin). Or rats consuming high-carbohydrate or high-protein meals that also did or did not contain 40% fat exhibited no fat-dependent changes in brain tryptophan, even though "free" plasma tryptophan levels were markedly elevated by the fat (Fernstrom et al., 1976). At present few investigators differentiate between "free" and "total" (free plus albumin-bound) tryptophan in calculating plasma tryptophan ratios, nor does there seem any reason to do so. In actuality, albumin-bound tryptophan has been shown by Pardridge and his associates (Pardridge and Fierer, 1990) to be "... readily available for transport into the brain, secondary to enhanced dissociation within the cerebral microcirculation of amino acid from the albumin binding site, as represented by an increased Kd in vivo ..." (i.e., an in vivo change in albumin's affinity for tryptophan rather than a "stripping" of tryptophan off the albumin molecule).

4.2.1 Tryptophan Hydroxylase

Properties TPH (E.C. 1.14.16.4) catalyzes the initial and rate-limiting step in serotonin biosynthesis, the hydroxylation of tryptophan at the 5-position to form 5-HTP (❷ *Figure 4-1*). This product, like the DOPA formed from tyrosine's hydroxylation, is readily decarboxylated to the corresponding monoamine, serotonin or dopamine, by the action of aromatic L-amino acid decarboxylase (L-AAAD), a widely distributed pyridoxine-dependent enzyme. TPH is in fact two distinct enzyme proteins, TPH1 and TPH2, which are encoded by genes on human chromosomes 11 and 12, respectively (Walther and Bader, 2003). TPH1 is localized within peripheral tissues (e.g., the enterochromaffin cells and certain intrinsic neurons of the gut) and the pineal organ, while TPH2 is a brain enzyme, concentrated within the perikarya and terminals of serotoninergic neurons (Walther and Bader, 2003; Patel et al., 2004). Both enzymes use ferrous iron as a cofactor and molecular oxygen and tetrahydrobiopterin (BH_4) as cosubstrates.

The kinetic properties of the two enzymes differ: the K_m of TPH2 for its major substrate, tryptophan, has been estimated to be several-fold (40.3 μM versus 22.8 μM [McKinney et al., 2005]) to tenfold (142 μM versus 13–23 μM [Kowlessur and Kaufman, 1999]) higher than that of TPH1, while its K_m for BH_4 is lower (20 μM versus 39 μM; McKinney et al., 2005). Since brain tryptophan and BH_4 concentrations in rats are around 5–10 μg/g brain tissue (Fernstrom and Wurtman, 1971a) and 3 μM (Nagatsu, 1983), respectively, it would be anticipated that—as is actually observed—the in vivo activity of TPH2 and the overall rate of serotonin synthesis in brain, both vary broadly with brain tryptophan levels (Fernstrom and Wurtman, 1971a), while peripheral serotonin synthesis, as reflected by blood serotonin, is much less responsive to changes in plasma tryptophan (Colmenares and Wurtman, 1979). And since, as described above, the principal factor that normally controls *brain* tryptophan levels is not plasma tryptophan, per se, but, rather, plasma concentrations of the other LNAA, the effects of any meal on brain serotonin synthesis will depend on the LNAA in that meal's protein and on the meal's ability to elicit insulin secretion (which lowers plasma LNAA), but only to a minor extent on its tryptophan content (Fernstrom and Wurtman, 1972b). On the other hand, the synthesis of serotonin in peripheral tissues, which lack a BBB, would be expected to depend principally on the meal's tryptophan content (Colmenares and Wurtman, 1979).

TPH enzymes are, along with PAH and TH, members of a protein superfamily (Hufton et al., 1995). They share considerable amino acid homology (71% for TPH1 and TPH2 [Walther and Bader, 2003]; 52% between TPH1 and PAH [Kappock and Caradonna, 1996]); utilize the same cofactors and cosubstrates; and to a variable extent, can hydroxylate all three amino acid substrates to variable extents (Kappock and Caradonna, 1996).

Regulation Both TPH and TH are readily phosphorylated on specific residues (e.g., serine-58 and 260 for TPH1 [Walther and Bader, 2003]; these plus serine-19 for TPH2 [McKinney et al., 2005]). This phosphorylation causes major changes in TH's kinetic properties, enhancing its saturation with BH_4 and making its net activity more dependent on available tyrosine levels. However, phosphorylation does not appear to cause major changes in the properties of TPH, whether the phosphorylated enzyme also subsequently binds to a 14-3-3 protein (McKinney et al., 2005). Experimental procedures that block serotonin autoreceptors or activate neuronal firing can increase brain serotonin synthesis (Stenfors and Ross, 2002), however, such procedures have not been shown to affect the enzyme's Vmax or its K_m's for tryptophan or BH$_4$. A major reduction in TPH2 activity does diminish brain serotonin synthesis, as shown in studies comparing this rate in a wild mouse strain and one containing a TPH2 allele with a single nucleotide polymorphism (C1473G). The 5-HTP synthesis and levels were markedly reduced in the animals with the polymorphism (Zhang et al., 2005).

4.2.2 5-Hydroxytryptophan and L-DOPA

The biosynthesis of serotonin can also be accelerated by administering 5-HTP, the amino acid intermediate in its physiologic synthesis from tryptophan (❍ *Figure 4-1*). Given orally, this compound readily enters the blood stream of humans and, via the LNAA transport carrier, the brains of rats (Amer et al., 2004). It can be decarboxylated to serotonin in any of the numerous CNS and peripheral cell types that contain the enzyme aromatic-L-amino acid decarboxylase (L-AAAD), for example, monoaminergic brain neurons, kidney, and gut. Presumably, only authentic serotoninergic cells have the capacity to store the serotonin thus formed—which protects that serotonin from immediate destruction by MAO—and to recapture intrasynaptic serotonin by serotonin-uptake molecules. But for a period of time after 5HTP's administration, many nonserotoninergic neurons and other cells produce and release serotonin as a "false neurotransmitter."

Similar caveats affect the use of oral L-DOPA, e.g., in treating Parkinson's disease: this catechol amino acid, an intermediate in dopamine's synthesis from tyrosine (❍ *Figure 4-2*), does enter the brain via the LNAA transport system—hence its efficacy can be enhanced by dietary carbohydrates (Berry et al., 1991) or suppressed by concurrently eating proteins (Mena and Cotzias, 1975). However, in both the brain and periphery the L-DOPA is decarboxylated to dopamine in many cell types besides its targeted nigrostriatal neurons. Both L-DOPA and 5-HTP (Van Woert and Rosenbaum, 1979) are usually given along with peripheral decarboxylase inhibitors; since only the brain dopamine or serotonin is desired. This lowers the required therapeutic dose. 5-HTP has been used experimentally to treat depression by van Praag (1981) and to suppress stress-induced eating (Amer et al., 2004); it not uncommonly causes sleepiness as a side effect. The brain uptakes of amino acid drugs like the antihypertensive agent α-methyldopa similarly depend on corresponding plasma ratios (to concentrations of the LNAA in dietary proteins [Zavisca and Wurtman, 1978; Pardridge et al., 1986]).

4.3 Brain Tyrosine

Brain tyrosine levels in fasted rats vary between about 60–80 micromolar (Milner and Wurtman, 1986). Administration of tyrosine increases these levels, a 100-mg/kg intraperitoneal dose causing peak elevations of about 150–200% (Morre et al., 1980). Consumption of a protein-free, carbohydrate-rich meal may or may not change rat brain tyrosine levels significantly (Gibson and Wurtman, 1977), probably because the insulin-induced fall in plasma tyrosine is comparable to that in the other LNAA (Fernstrom and Fernstrom, 1995). However, if the meal contains 8–40% protein, brain tyrosine rises, roughly in proportion to the protein content (Gibson and Wurtman, 1977). Chronic consumption (14 days) of meals containing 10% protein was associated with cortical tyrosine levels approximately double those seen among rats consuming 2% protein (Fernstrom and Fernstrom, 1995). As with tryptophan, brain tyrosine levels vary not with plasma tyrosine, per se but with the ratio of the tyrosine concentration to the summed concentrations of the other principal LNAA (Fernstrom and Faller, 1978).

CSF tyrosine levels were significantly elevated, by about 70%, among Parkinsonian patients receiving tyrosine (100 mg/kg/day, in 6 divided doses; the last dose was administered 2 h before the second lumbar puncture [Growdon et al., 1982]). CSF levels of the dopamine metabolite homovanillic acid (HVA) were also significantly elevated (by 36%), while those of the serotonin metabolite 5HIAA were, as expected, unchanged.

4.3.1 Tyrosine Hydroxylase

Properties TH (E.C. 1.14.16.2) catalyzes the initial and rate-limiting step in dopamine biosynthesis (Levitt et al., 1965), and thereby also affects the rates of formation of dopamine's biologically active products: norepinephrine and epinephrine (❷ *Figure 4-2*). This step involves the hydroxylation of p-tyrosine at the m-position to generate the catechol amino acid 3,4-dihydroxyphenylalanine, or DOPA. Like the tryptophan and PAHs, TH uses ferrous iron as a cofactor, and molecular oxygen and tetrahydrobiopterin (BH_4) as well as tyrosine as substrates.

In humans, there are four distinct TH enzyme proteins, exhibiting perhaps twofold variations in catalytic activity (Kappock and Caradonna, 1996); all are products of mRNAs arising from a single gene on chromosome 11 (Kaneda et al., 1987). In the rat (Grima et al., 1985) and mouse (Ichikawa et al., 1991), this gene encodes a single TH protein. TH is present in all of the cells that synthesize catecholamines, e.g., dopaminergic nigrostriatal neurons, mesocortical and mesolimbic tracts, tuberohypophyseal neurons, and neurons in the retina and olfactory bulbs; noradrenergic neurons originating in the locus coeruleus and lateral tegmentum; epinephrine-containing neurons in the brainstem; postganglionic noradrenergic sympathetic neurons; and adrenomedullary chromaffin cells.

Regulation It has been recognized for decades that the rate at which catecholamine-producing cells produce catecholamines from tyrosine is highly regulated, and coupled to the rates at which the cells are releasing these compounds. Thus, even major increases in release caused by prolonged neuronal firing tend not to lower the quantities of catecholamine remaining in the cells, as first shown by Elliott (1912). At least three regulatory processes can enhance tyrosine's hydroxylation when the need for additional catecholamine molecules arises: the phosphorylation of TH on specific serine residues—which activates the enzyme and markedly increases its affinity for its BH_4 cosubstrate; enhanced de novo synthesis of the enzyme protein; and development of the ability to respond to increased tyrosine levels, once the TH has been phosphorylated. A fourth regulatory process—end-product inhibition by cytoplasmic catecholamines—inhibits the enzyme's activity and catecholamine synthesis.

Like TPH, TH is readily phosphorylated. But unlike TPH, TH's phosphorylation leads to major changes in its kinetic properties, increasing its affinity for BH_4 (Lloyd and Kaufman, 1975; Wang et al., 1991) and thereby increasing the extent to which its net activity is regulated by local levels of tyrosine, its primary substrate (Wurtman et al., 1974). Four of the enzyme's serine residues (Ser8, Ser19, Ser31, and Ser40) are substrates for phosphorylation reactions, which are catalyzed, in human cells, by cAMP- and Calmodulin-dependent protein kinases (Harris et al., 1974). Phosphorylation at the Ser19 position sensitizes the TH to the subsequent phosphorylation of its Ser40 moiety, which increases the enzyme's net activity, both in vitro and in vivo (Dunkley et al., 2004). The phosphorylated enzyme, like TPH, is able to combine with a 14–3–3 "activator" protein, however, this step is not required in order for the TH to be activated by the phosphorylation reactions (Kleppe et al., 2001). The increase in TH's affinity for BH_4 as a consequence of its phosphorylation can be considerable: In one study, the K_m for various BH_4 analogs fell from 300 µM in unphosphorylated enzyme to 0.8–14.0 µM (Bailey et al., 1989). Other investigators, using BH_4, described decreases in its K_m of 2- to 12-fold (Kappock and Caradonna, 1996).

TH's Vmax apparently is not affected by its phosphorylation (Le Bourdelles et al., 1991), however, major increases in the rate at which it produces catechols do occur in vivo, possibly caused by the increases in the enzyme's saturation with BH4 (Zigmond et al., 1989) and by consequent decreases in its susceptibility to end-product inhibition by cytoplasmic catecholamines (Ames et al., 1978). Moreover, once the enzyme has been phosphorylated the rate at which it produces DOPA can readily be enhanced by administering tyrosine, as described below. Phosphorylated TH is rapidly dephosphorylated, with an initial half-life approximating 5 min (Yamauchi and Fujisawa, 1979); this suggests that one or more phosphatase enzymes are colocalized in the cell with TH (Yamauchi and Fujisawa, 1979).

TH is inhibited by catechols, particularly by its catecholamine end-products dopamine, norepinephrine, and epinephrine (Nagatsu et al., 1964). Most of the catecholamine molecules in neurons and adrenomedullary cells are sequestered within synaptic vesicles and thus unable to interact with TH and affect its activity. However not all of them are "free": cytoplasmic dopamine continues to be formed from the decarboxylation of DOPA, and dopamine also enters presynaptic terminals from the synaptic cleft via high-affinity catecholamine uptake system. And until such molecules are sequestered in vesicles, or destroyed by the mitochondrial enzyme MAO, they are able to bind to the TH enzyme protein and to diminish its activity—probably by inhibiting its phosphorylation (Almas et al., 1992) and by competing with BH_4 for binding to the ferric iron in the TH (Andersson et al., 1988). The levels of cytoplasmic dopamine (or norepinephrine, or epinephrine) in cells are, as might be expected, increased by drugs that inhibit MAO activity, hence such drugs can also act as potent inhibitors of tyrosine's hydroxylation in vivo (Spector et al., 1967).

The major changes in TH's kinetic properties caused by its phosphorylation allow its net activity to increase rapidly when neurons or adrenomedullary cells are physiologically activated and are releasing catecholamines at a more rapid rate (Weiner and Rabadjija, 1968). This has been shown by in vivo studies using electrical stimulation or potassium-induced depolarization of neurons. Moreover, sustained increases in in vivo TH activity, caused by enhanced synthesis of the enzyme protein (Silberstein et al., 1972) occur when the accelerated firing of catecholaminergic neurons (or of the cholinergic neurons that innervate the adrenal medulla) is sustained, for example, in sympathetic neurons of animals exposed to hemorrhagic shock (Conlay et al., 1981); or in adrenal medullas of rats receiving drugs that cause hypotension (Thoenen et al., 1969) or hypoglycemia (Viveros et al., 1969), or that destroy postganglionic terminals (Thoenen et al., 1969); or in animals exposed to cold or to immobilization stress (Kvetnansky et al., 1992); or in retinas of rats exposed to light (Witkovsky et al., 2004). TH synthesis can also be induced in vitro by culturing adrenals in a medium containing depolarizing concentrations of potassium (Silberstein et al., 1972). As described below, in most of these experimental systems, elevation of tissue tyrosine has been shown to further increase the synthesis and release of the catecholamines. "If the Km of neuronal tyrosine hydroxylase is, as described (Wurtman et al., 1974) 100–140 micromolar, then the enzyme may be only 25–50% saturated with this substrate under basal conditions, in as much as basal brain tyrosine levels reportedly are about 60–80 micromolar (Milner and Wurtman, 1986). Giving exogenous tyrosine could raise neuronal tyrosine levels substantially, however dopamine synthesis would still be limited by the poor saturation of tyrosine hydroxylase with its cofactor, BH_4, until the neuron began firing. Then tyrosine levels would affect the rate of dopamine synthesis."

Certain neurons, for example, the mesocortical dopaminergic neurons projecting to the rat's medial prefrontal and cingulate cortices, produce and release more catecholamine whenever tyrosine levels have been increased, even in the absence of an additional treatment to accelerate their firing (and, presumably, TH's phosphorylation) (Tam et al., 1990). This ability may derive from their lack of somatodendritic autoreceptors which would modulate impulse flow, and of nerve-terminal autoreceptors which would modulate dopamine synthesis (Chiodo et al., 1984). This lack causes the neurons to exhibit faster dopamine turnover, faster basal firing rates, and more bursting activity than other midbrain dopamine neurons. Insofar as these neurons couple dopamine synthesis to tyrosine levels under basal conditions, they are similar to brain serotoninergic neurons, in which changes in substrate (i.e., tryptophan) levels will always affect the net activity of the hydroxylase enzyme (TPH). Presumably, most of the TH in these mesocortical neurons is phosphorylated under basal conditions; however this has not yet been demonstrated.

5 Consequences of Changing Brain Tryptophan and Tyrosine Levels

5.1 Precursor Availability and Neurotransmission

The total amount of information that a group of neurons can transmit during any particular interval depends, in large part, on the number of neurotransmitter molecules that their presynaptic terminals release during that interval. This, in turn, depends on the total number of synapses that the neurons make, the average frequency with which the neurons happen to be firing, and the average amount of transmitter released at each synapse per firing. If the changes in neurotransmitter *synthesis* caused by elevating brain tryptophan or tyrosine levels are to be of physiologic relevance, they must be associated with increases in the amount of transmitter *released*, per depolarization and per unit time. This may or may not modify postsynaptic responses depending on, for example, whether unoccupied postsynaptic receptors are available to respond to the additional neurotransmitter molecules, i.e., increased transmitter release is necessary but not yet sufficient in order for the precursor effect to have functional significance. If the firing rates of most brain neurons are, as is generally believed, constrained by mechanisms, involving presynaptic receptors or multisynaptic reflex arcs, programed to keep total neurotransmitter release constant despite fluctuations in transmitter levels within presynaptic terminals, then when does precursor availability actually affect neurotransmission? There are a number of situations in which this seems probable.

5.1.1 Neurons That Lack Multisynaptic or Autoreceptor-Based Feedback Loops

Peripheral sympathetic neurons and chromaffin cells in humans (Agharanya et al., 1981) and experimental animals (Alonso et al., 1980) release more catecholamines after tyrosine has been administered or a protein-rich meal has been consumed. That the resulting increase in urinary catecholamine levels (❯ *Figure 4-12*) represents increased catecholamine release and not, for example, alterations in catecholamine metabolism, is indicated by the fact that levels of catecholamine metabolites in the urine also rise; that this reflects accelerated catecholamine synthesis and not simply release of stored material is indicated by the failure of tissue catecholamine levels to decline.

◘ Figure 4-12
Urinary levels of catecholamines in normal humans after tyrosine administration. Thirteen subjects fasted overnight; the next morning eight received a single oral dose (100 or 150 mg/kg) of tyrosine mixed in water; while five controls received only water. Urinary samples were obtained 0, 2, 4, and 8 h after tyrosine or water injection. Data are given as μg excreted/h (mean ± SEM). •$p < 0.01$ by ANOVA for repeated measurements; ••$p < 0.005$. Data from Alonso et al. (1982)

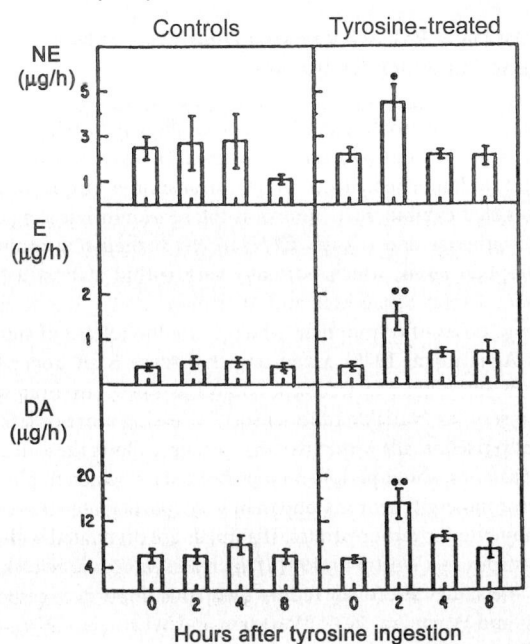

Similarly, rat mesocortical dopaminergic neurons lacking impulse-regulating somatodendritic and synthesis-modulating nerve terminal autoreceptors (White and Wang, 1984; Tam et al., 1990) respond without additional treatment to physiological tyrosine doses by synthesizing more dopamine (Tam et al., 1990) (❱ Table 4-2).

◼ Table 4-2
Effect of tyrosine administration on prefrontal cortex DOPA and dopamine levels after a decarboxylase inhibitor

	Time after tyrosine treatment (min)			
	0	30	40	60
DOPA accumulation (% control)	100 ± 5	117 ± 13	124 ± 7*	113 ± 5
Dopamine (% control)	100 ± 10	104 ± 12	115 ± 16	160 ± 15*

Rats received tyrosine (50 mg/kg, i.p.) 30, 40, or 60 min before sacrifice, and to measure DOPA accumulation, a decarboxylase inhibitor, NSD-1015 (100 mg/kg, i.p.), 30 min before sacrifice. Values are expressed as percentages of levels in saline-treated control animals at each respective time point. *Differs significantly from saline controls ($p < 0.05$). Data from Tam et al. (1990)

5.1.2 Neurons That Are Components of Positive Feedback Loops

If a neuron releases a precursor-dependent *excitatory* neurotransmitter directly onto its own receptors, or if its depolarization, acting transsynaptically, causes it to receive greater quantities of excitatory transmitters from other neurons and thus to fire more frequently, then the initial increase in transmitter release after precursor administration could enhance subsequent responses to the precursor.

5.1.3 Neurons That Normally Release Variable Quantities of Neurotransmitter Per Firing Without Engaging Feedback Responses

If the mechanisms controlling a neuron's firing frequencies allow it to release widely varying amounts of neurotransmitter per unit time without undergoing feedback changes in firing, then precursor availability might be expected to exert undampened effects on neurotransmitter output within this broad range.

One group of neurons that exhibits such control is the serotonin-releasing cells of the raphe nucleus. The rate at which they synthesize and release (❱ Figure 4-13) their neurotransmitter apparently varies directly with brain tryptophan levels, which normally vary within at least a twofold physiologic range (Fernstrom and Wurtman, 1971a; Schaechter and Wurtman, 1989). Raphe firing does decrease when animals are given *very large* doses of tryptophan, which cause the release of supraphysiologic amounts of serotonin (Gallager and Aghajanian, 1976), indicating that there *is* an upper limit to serotonin release beyond which the neurons are subject to feedback control as shown by Bramwell (1974). The ability of serotoninergic neurons to serve as "variable ratio sensors," releasing more or less of their transmitter when the plasma tryptophan ratio rises or falls within its normal range, allows these neurons to provide the rest of the brain with useful information about peripheral metabolic state, which might then be used to formulate behavioral strategies. Serotoninergic neurons apparently do participate in a complex neural-behavioral mechanism controlling appetite for carbohydrates. If animals are pretreated with a drug that, like carbohydrate consumption (Fernstrom and Wurtman, 1971a), increases serotonin release, and if they are then given a choice between various diets, they selectively reduce their consumption of carbohydrates while sustaining protein intake (Wurtman and Wurtman, 1977; Wurtman and Wurtman, 1979). This effect is independent of whether the carbohydrates in the test foods happen to be sweet (Wurtman and Wurtman, 1979).

◘ Figure 4-13
Effect of tryptophan availability on serotonin content of, and release from, rat brain slices. The superfusion media contained the tryptophan concentrations indicated. Both spontaneous and electrically evoked release were measured. Data from Schaechter and Wurtman (1989)

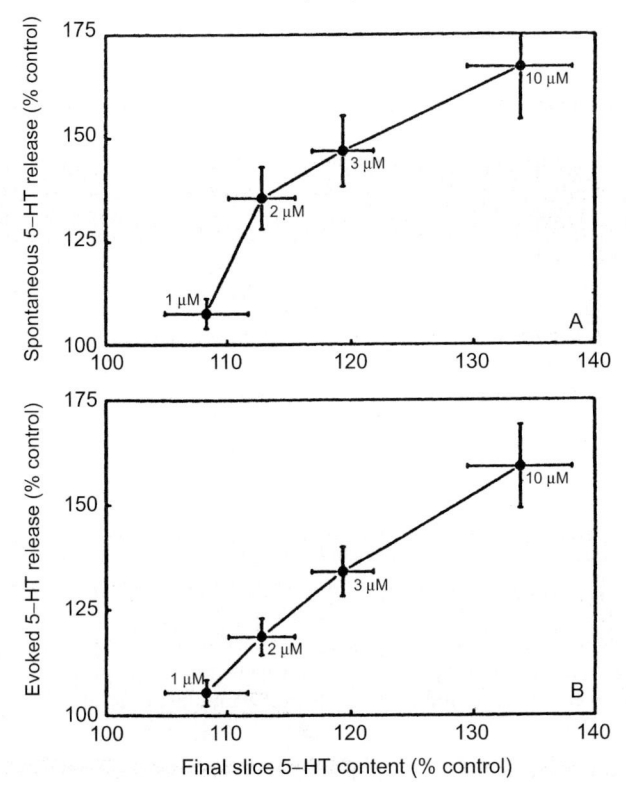

5.1.4 Physiologic Situations in Which Neurons Undergo Sustained Increases in Firing Frequency

The relation between sustained neuronal firing and precursor responsiveness is well illustrated by the ability of exogenous tyrosine to raise *or* lower blood pressure when the starting blood pressure is too low or too high, depending on which of the animal's noradrenergic neurons happen to be most active at the time of its administration. Noradrenergic neurons at several loci participate in the control of blood pressure (Palkovits and Zaborszky, 1977): Norepinephrine release from peripheral sympathetic nerves tends to elevate blood pressure, while its application to or release from certain brainstem sites tends to lower blood pressure (presumably by diminishing sympathetic outflow [DeJong et al., 1975]). If normotensive rats (mean systolic blood pressure = 120–130 mm Hg) receive a given dose of tyrosine (100–200 mg/kg, i.p.), blood pressure changes only slightly or not all (Sved et al., 1979a); blood pressure also fails to change in normotensive human subjects (Glaeser et al., 1979; Melamed et al., 1980b). If the same tyrosine dose is given to a spontaneously *hypertensive* rat (mean systolic blood pressure = 170–210 mm Hg), blood pressure *falls* by 28–46 mm Hg for several hours (❯ Figure 4-14) (Sved et al., 1979a); however, if that dose is given to a *hypotensive* animal (mean blood pressure = 63 mm Hg, 45 min after a hemorrhage of 20% of its calculated blood volume), systolic pressure *rises* by 31 mm Hg (Conlay et al., 1981). Treatments that cause hypertension accelerate norepinephrine release from brainstem neurons terminating in the anterior hypothalamus (presumably activating compensatory mechanisms), while those that cause hypotension

Figure 4-14
Effect of tyrosine on blood pressure in spontaneously hypertensive rats (SHR). The amino acid was given intraperitoneally (i.p.) and blood pressure was measured at 0 time and 90 min after. Data from Sved et al. (1979a)

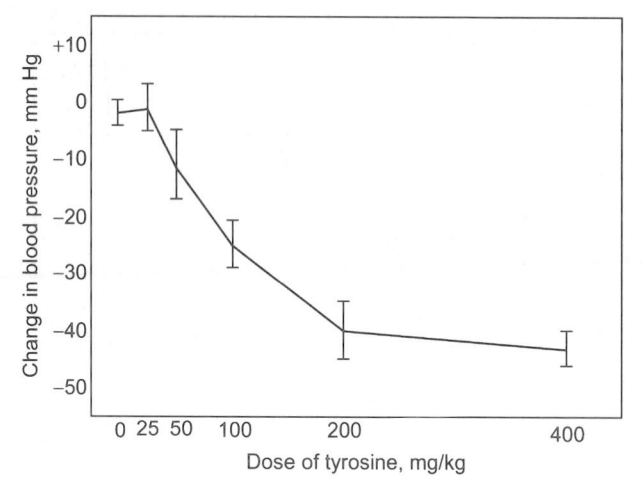

activate sympathoadrenal structures and catecholaminergic terminals in the posterior hypothalamus (Philippu et al., 1980). Hence, the most economical explanation for the paradoxical ability of tyrosine to lower or raise blood pressure, depending on whether it is chronically elevated or depressed, is that in the hypertensive animal, the precursor enhances norepinephrine release selectively within one set of brainstem neurons (because these happen to be firing frequently), while in shock peripheral sympathetic neurons and adrenomedullary cells, and posterior hypothalamic neurons, are activated and thus become tyrosine-sensitive.

Intravenous tyrosine markedly reduces blood pressure in animals with renovascular or DOCA-salt hypertension. These changes are associated with reductions in heart rate as shown by Bramwell (1974). Tryptophan has about half the blood-pressure-lowering activity of tyrosine, probably acting by increasing serotonin release from bulbospinal neurons, while branched-chain LNAA lack any effect on blood pressure and, when coadministered with tyrosine, block its effect (Sved et al., 1979a).

Similar relationships between precursor-dependence and chronic changes in firing frequency have also been described in nigrostriatal dopaminergic neurons (Melamed et al., 1980b), where unilateral destruction of most of the neurons renders the surviving ipsilateral neurons, but not the contralateral dopaminergic neurons, tyrosine-sensitive. Thus, physiologic conditions that accelerate the firing of precursor-depending neurons may overcome the feedback mechanisms that would otherwise maintain the constancy of neurotransmitter *release*, and allow the neuron to couple precursor availability to transmitter synthesis.

5.1.5 Neurologic Diseases That Cause Either a Decreased Number of Synapses of Decreased Transmitter Release Per Unit Time

Neurodegenerative disorders that diminish the number of presynaptic terminals issuing from a precursor-dependent brain nucleus may be associated with accelerations in the average firing rates of the surviving neurons (Bernheimer et al., 1973), enhancing their sensitivity to precursor control. This formulation provides a theoretical basis for testing neurotransmitter precursors in, for example, Parkinson's disease (Growdon et al., 1982). It also offers an explanation for the physiological specificity that seems to be associated with the therapeutic use of neurotransmitter precursors. If the brain contains, for example, ten groups of catecholaminergic neurons, and if all but one of these groups are intact and functioning normally, it might be anticipated that only this one group would be substantially affected by giving tyrosine. An elevation in brain tyrosine levels would initially enhance dopamine or norepinephrine release from all ten groups, but in the normal nuclei, this effect would rapidly be dampened by presynaptic or multisynaptic feedback mechanisms.

Brains of normal women produce significantly less serotonin than those of men (Nishizawa et al., 1997). This might explain why, as described below, women not infrequently develop "carbohydrate craving" in serotonin-related disorders like the premenstrual syndrome (or "late-luteal phase dysphoric syndrome"), the behavioral syndrome that can follow nicotine withdrawal, and the seasonal affective disorder. Eating the insulin-secreting carbohydrates will increase brain serotonin release, as described above, and improve serotonin-dependent feelings (Wurtman and Wurtman, 1989).

5.2 Brain Tryptophan and Serotonin

Basal brain serotonin levels in fasting rats vary between 0.47 and 0.70 µg/g tissue (Fernstrom and Wurtman, 1971a, b). In such animals, brain tryptophan levels are 3.3–6.8 µg/g. Administration of large doses of tryptophan via the diet (Green et al., 1962; Wang et al., 1962) or by injection (800 mg/kg: Ashcroft et al., 1965) were first shown in 1962 and 1965, respectively, to increase brain serotonin levels. A few years later (Fernstrom and Wurtman, 1971a), it was demonstrated that even much lower doses (12.5 mg/kg)—which increased brain tryptophan but kept it within its normal range—could elevate brain serotonin significantly. Doubling that dose approximately doubled the increment in brain serotonin (Fernstrom and Wurtman, 1971a), however, increasing the dose further caused only minor increments in serotonin levels, instead increasing those of serotonin's deaminated metabolite 5-hydroxyindole acetic acid (5-HIAA). This suggested that a perhaps-twofold variation in brain serotonin levels, resulting from parallel changes in the rate of serotonin synthesis, is "allowed" to exist in serotoninergic neurons, but that beyond that range the excess serotonin cannot be stored (and protected from intracellular MAO), nor, probably, released.

Follow-up studies on rats attempted to determine whether a short-term treatment that lowered brain tryptophan levels would similarly decrease those of serotonin. Toward this end, rats received insulin—a treatment known to lower plasma levels of most amino acids; surprisingly, insulin failed to lower plasma or brain tryptophan, actually raising them in rats (Fernstrom and Wurtman, 1972a) (insulin secreted in response to carbohydrate consumption does lower plasma tryptophan in people but by less than it lowers plasma concentrations of other LNAA [Martin-Du-Pan et al., 1982]). This response—shown later to result from tryptophan's unique propensity, described above, to bind to circulating albumin (❯ Table 4-1)—raised the question of whether insulin, secreted physiologically in response to an insulin-releasing mix of dextrose, sucrose, and dextrin would similarly affect plasma and brain tryptophan (Fernstrom and Wurtman, 1971b). Insulin did, and also elevated brain serotonin, providing the first evidence that a particular macronutrient, certain carbohydrates, could cause characteristic changes in a brain neurotransmitter.

If, indeed, plasma tryptophan concentrations determined brain tryptophan, then giving animals a high-protein meal would be expected to raise brain tryptophan and serotonin levels by even more than the carbohydrate, since even though tryptophan is scarce in proteins (1–1.5%), protein consumption still would elevate plasma tryptophan concentrations. But surprisingly, though plasma tryptophan concentrations did rise, this change was not accompanied by parallel elevations in brain tryptophan—or serotonin (Fernstrom and Wurtman, 1972b). Prior studies had shown that other large, neutral amino acids can suppress the transport of tryptophan into perfused slices (Blasberg and Lajtha, 1966), and that the LNAA compete with each other for transport into brain (Guroff and Udenfriend, 1962). These findings raised the possibility that the failure of dietary proteins to elevate brain tryptophan or serotonin levels resulted from the larger increases the proteins would produce in plasma concentrations of these other, more abundant LNAA. Thus an experiment was performed comparing the changes in brain tryptophan and serotonin occurring after animals received either tryptophan alone, or an amino acid mixture containing tryptophan plus the other two aromatic amino acids and the three branched-chain amino acids. As anticipated, brain levels of the indoles rose after tryptophan alone but not after animals consumed the mixture. This led to recognition of the paradoxical effects of macronutrients on brain serotonin: a meal that lacks tryptophan (but contains carbohydrates and perhaps fats) maximally *elevates* brain tryptophan (because it lowers plasma levels of the LNAA competitors), while one that contains the most tryptophan (i.e., a high-protein meal) *fails to elevate* brain tryptophan (❯ *Figure 4-10*). Hence, serotoninergic brain neurons are "variable ratio sensors" of the plasma amino acid pattern—specifically the ratio of tryptophan to other LNAA—producing more serotonin when protein-poor carbohydrate-rich foods are eaten, and less after consumption of protein-rich foods.

That such changes in serotonin production and tissue levels following carbohydrate consumption can be sufficient to affect serotonin *release* was initially shown in studies using rat brain slices (❯ *Figure 4-13*) (Schaechter and Wurtman, 1989). Superfusion of the slices with a physiologic medium containing 1–10 µM tryptophan caused dose-dependent elevations in tissue tryptophan levels which, below 5 µM tryptophan, were in the physiologic range for brain tryptophan in rats. Serotonin and 5-HIAA levels also rose significantly, proportionately greatest effects being observed with the lowest tryptophan concentrations. Both the spontaneous release of serotonin from the slices and the release caused by neuronal depolarization were also significantly increased (Schaechter and Wurtman, 1989). Moreover, the effects of providing tryptophan and of increasing the frequency of field stimulation were additive, and tryptophan's effects were blocked by adding leucine, another LNAA, to the medium (Schaechter and Wurtman, 1990).

That such changes in brain serotonin release also occur in response to meals which are, for Americans, normal, was demonstrated indirectly in a study on the effects of a high-carbohydrate versus a high-protein breakfast on plasma tryptophan ratios (Wurtman et al., 2003). The carbohydrate-rich breakfast (waffles, maple syrup, orange juice, coffee with sugar—containing 69.9 g of carbohydrate and 5.2 g of protein) generated plasma tryptophan ratios that were 54% higher than the protein-rich meal (turkey ham, eggs, cheese, butter) (Wurtman et al., 2003). Since a 50% increase in the plasma tryptophan ratio in rats can elevate brain tryptophan concentrations by 39% (Fernstrom and Faller, 1978; Fernstom et al., 1973), this increase, in turn, would be expected to elevate brain serotonin levels by 25% (Schaechter and Wurtman, 1989) and cause significant increases in the spontaneous and evoked release of serotonin (by 28 and 14%, respectively) (Schaechter and Wurtman, 1989). It seems probable that the 50% difference generated by the two breakfast meals also affects brain serotonin release in humans. The two meals also caused smaller (30%) but significant differences in the plasma tyrosine ratio.

A number of diseases and disorders are characterized by both mood disturbances—sadness, anger, anxiety—and weight gain associated with "carbohydrate craving," i.e., a strong desire, usually not associated with subjective hunger, to consume carbohydrate-rich foods which may or may not be sweet, usually at a characteristic time of day (Wurtman et al., 1993). The weight gain and the excess caloric load thus consumed relate not so much to the carbohydrates eaten but to the fats, which usually accompany them in the foods that the patients tend to select. Examples of such disorders include the seasonal affective disorder syndrome ("winter blues": Rosenthal et al., 1984), the premenstrual syndrome (Brzezinski et al., 1990), the behavioral syndrome sometimes following nicotine withdrawal (Spring et al., 1991), and, in some cases, obesity itself (Wurtman et al., 1981; Lieberman et al., 1986). Inasmuch as the consumption of carbohydrate-rich, protein-poor meals or, especially, snacks can, like antidepressant drugs, increase brain serotonin, and since many such patients volunteer that they choose these snacks because they have learned that the snacks make them "feel better—less tense, more motivated, less socially withdrawn, less sad"—it can be conjectured that the carbohydrate-craving represents a usually unrecognized attempt at self-medication directed at increasing intrasynaptic serotonin levels (Wurtman and Wurtman, 1989). All of these syndromes can be treated with serotoninergic drugs; some apparently also can be treated by giving patients a mixture of insulin-secreting carbohydrates (Sayegh et al., 1995) formulated for rapid absorption and sustained release.

When oral tryptophan was available in the USA for human use it was widely used to diminish symptoms of depression (Coppen et al., 1972) (a use which has continued in Canada, the UK, and elsewhere) and to treat insomnia (Demisch et al., 1987). Its administration can also decrease appetite (Wurtman et al., 1981) and caloric intake (Hrboticky et al., 1985); decrease sleep latency (Hartmann and Greenwald, 1984) and increase subjective drowsiness and fatigue (Lieberman et al., 1985); and decrease pain sensitivity (Lieberman et al., 1982/1983; Seltzer et al., 1982). Conversely, administration of amino acid mixtures containing LNAA but lacking tryptophan reportedly can bring on depressive symptoms and even full-scale depressive episodes in patients with histories of depression (❯ *Figure 4-7*) (Delgado et al., 1991; Smith et al., 1997), and can exacerbate aggressive tendencies in subjects with histories of aggressive behavior (Bjork et al., 2000). Interestingly, the presence of a less-active allele of the gene for hTPH2, the brain enzyme that initiates the conversion of tryptophan to serotonin, is also highly associated with unipolar major depression (Zhang et al., 2005).

5.3 Brain Tyrosine and the Catecholamines

Basal brain catecholamine levels in fasting rats vary widely from region to region; indeed, it was this variation that initially suggested to Marthe Vogt (1954) that norepinephrine, concentrated within the hypothalamus, and to Arvid Carlsson (Carlsson et al., 1958) that dopamine, concentrated within the corpus striatum, might function as neurotransmitters. Administration of even large doses of tyrosine generally does not affect brain norepinephrine or dopamine levels (although it does elevate urinary catecholamines in humans [❯ *Figure 4-12*; Alonso et al., 1982] and rats [Agharanya and Wurtman, 1982a], and does correct the depletion of brain norepinephrine observed in stressed rats [Reinstein et al., 1984]). But if the conversion of tyrosine's hydroxylated product DOPA to a brain catecholamine is suppressed pharmacologically, and catecholamine release is thus diminished, then giving tyrosine will enhance DOPA synthesis (Wurtman et al., 1974; Carlsson and Lindqvist, 1978; Badawy and Williams, 1982) and giving the LNAA leucine has the opposite effect (Wurtman et al., 1974)—observations that first suggested that, under particular circumstances, tyrosine availability might affect catecholamine production, just as tryptophan availability affects that of serotonin.

Those circumstances relate to the physiological activity of the neuron (or adrenomedullary chromaffin cell [Agharanya and Wurtman, 1982b]) that is producing the catecholamine. If the neuron's firing rate is sufficiently high, much of its TH enzyme protein is polyphosphorylated, causing conformational changes which increase its affinity for its BH_4 cofactor and allowing its activity to become limited by the extent to which it is saturated with its amino acid substrate (Lloyd and Kaufman, 1975). In general—as described above—this happens, for example, when the amounts of dopamine or norepinephrine released into brain synapses are diminished (e.g., in Parkinson's Disease), or when physiologic circumstances require a greater-than-usual release of the transmitter (as in hemorrhagic hypotension), or among neurons that normally fire frequently (mesocortical dopaminergic neurons that lack impulse-regulating somatodendritic and synthesis-modulating nerve terminal autoreceptors [❯ *Table 4-2*: Tam et al., 1990]), or in peripheral sympathoadrenal cells that are not components of polysynaptic feedback systems.

These relationships are well illustrated in rats, described above, that have been subjected to a unilateral lesion which destroys about 80% of a dopaminergic nigrostriatal tract, causing the surviving neurons to sustain rapid firing frequencies (Hefti et al., 1980): If such animals receive tyrosine systemically, no changes are noted in markers of dopamine release on the intact side of the brain, but major increases (which can be blocked by the LNAA valine) are observed on the lesioned side (Melamed et al., 1980b). The relationship between tyrosine availability and catecholamine synthesis is also evident when dopaminergic transmission has been blocked pharmacologically with haloperidol (Scally and Wurtman, 1977) or chronically administered reserpine (Sved et al., 1979a); when dopaminergic firing is enhanced by giving gamma-buytrolactone (Sved and Fernstrom, 1981) or amfonelic acid (Fuller and Snoddy, 1982), or when brain norepinephrine release is enhanced by cold stress (Gibson and Wurtman, 1978).

That brain tyrosine levels can also control catecholamine release has been shown directly using the technique of in vivo microdialysis: Giving tyrosine caused a short-term increase in striatal dopamine release (Acworth et al., 1988), which was enhanced in animals previously subjected to partial destruction of a nigrostriatal tract by 6-hydroxydopamine (During et al., 1989), and was prolonged if animals that also received the dopamine receptor antagonist haloperidol (During et al., 1989). Giving rats 250 mg/kg of phenylalanine enhanced striatal dopamine release by 59%, peaking after 75 min; a 500-mg/kg dose had no effect and a larger (1,000 mg/kg) dose reduced dopamine release by 26%—suggesting inhibition of TH (During et al., 1988). In none of these situations were changes noted in the release of dopamine's metabolites homovanillic acid (HVA) or dihydrophenylacetic acid (DOPAC), suggesting that the metabolites are a less sensitive marker of dopamine release than the transmitter itself. In patients with Parkinson's disease given tyrosine chronically, however, significant increases have been noted in CSF levels of HVA (Growdon et al., 1982). Tyrosine also increased the levels of dopamine metabolites in retinas of dark-adapted rats exposed to light (▶ *Table 4-3*) (Gibson et al., 1983).

◻ Table 4-3
Effect of tyrosine on increase in retinal catechols caused by light exposure

		Dopamine (ng/pair)	DOPAC (ng/pair)
Dark	Saline	2.9 ± 0.2	1.5 ± 0.3
	Tyrosine	2.6 ± 0.3	1.6 ± 0.3
Light	Saline	4.7 ± 0.4*	2.8 ± 0.1*
	Tyrosine	4.3 ± 0.5*	4.0 ± 0.5**

Rats were dark-adapted for 11 h and then exposed to light (1 h; 350 lux) or continued in darkness before killing. Thirty minutes before decapitation animals received tyrosine (100 mg/kg, i.p.) or its vehicle. *$p < 0.05$, differs significantly from either dark group. **$p < 0.01$, differs significantly from dark groups and from light-exposed controls. Data from Gibson et al. (1983)

The changes in catecholamine release caused by giving tyrosine are associated with some of the known physiological and behavioral effects of dopamine and norepinephrine. Thus, tyrosine can restore blood pressure in hemorrhage-induced hypotension (Conlay et al., 1981); this effect is not mediated by tyramine (Conlay et al., 1984) and is diminished if animals have been deprived of their adrenal medullas (Conlay et al., 1981). Tyrosine also lowers blood pressure in spontaneously hypertensive rats (Sved et al., 1979a)—an effect which also is blocked by other LNAA and is associated with an increase in brain levels of norepinephrine metabolites. Moreover, tyrosine diminishes serum prolactin levels in rats following their elevation by chronic treatment with reserpine (Sved et al., 1979b), and suppresses the rise in plasma corticosterone that follows acute stress in rats (Reinstein et al., 1985). In contrast, amino acid mixtures lacking tyrosine and phenylalanine increased plasma prolactin and decreased performance on a neurophysiological task sensitive to impaired dopaminergic function in humans (Gijsman et al., 2002). Giving tyrosine potentiated the brain-mediated anorexic effect of sympathomimetic drugs (phenylpropanolamine; ephedrine; amphetamine) (Hull and Maher, 1990), but not peripheral, sympathoadrenal-mediated as changes in gastric transit, thermogenesis (Hull and Maher, 1991), or blood pressure (Hull and Maher, 1992). Tyrosine administration also prevented hypoxia-induced decrements in learning and memory in rats (Shukitt-Hale et al., 1996).

Of potential significance for the development of treatments for psychiatric disturbances, tyrosine enhanced the release of dopamine from medial prefrontal cortical neurons of rats given clozapine, an atypical antipsychotic drug, but had no effect on striatal dopamine release in clozapine-treated animals. In contrast, tyrosine did potentiate the increase on striatal dopamine released caused by haloperidol, a "typical" antipsychotic agent, but had no effect on mesocortical dopamine release after haloperidol (Jaskiw et al., 2001). In otherwise-untreated animals, a low dose of tyrosine (25 mg/kg) enhanced dopamine synthesis in rapidly firing mesoprefrontal neurons. Higher doses initially also enhanced dopamine synthesis, but then stopped doing so, suggesting the operation of feedback processes (Tam et al., 1990).

Acknowledgements

Studies described in this review were supported in part by research grants from the National Institutes of Mental Health (grant MH-28783) and the Center for Brain Sciences and Metabolism Charitable Trust, and by an NIH grant to the Massachusetts Institute of Technology Clinical Research Center (grant M01-RR-01066).

References

Abita J-P, Milstein S, Chang N, Kaufman S. 1976. In vitro activation of rat liver phenylalanine hydroxylase by phosphorylation. J Biol Chem 251: 5310-5314.

Acworth IN, During MJ, Wurtman RJ. 1988. Tyrosine: Effects on catecholamine release. Brain Res Bull 21: 473-477.

Agharanya JC, Wurtman RJ. 1982a. Effect of acute administration of neutral and other amino acids on urinary excretion of catecholamines. Life Sci 30: 739-746.

Agharanya JC, Wurtman RJ. 1982b. Studies on the mechanism by which tyrosine raises urinary catecholamines. Biochem Pharm 31: 3577-3580.

Agharanya JC, Alonso R, Wurtman RJ. 1981. Changes in catecholamine excretion after short-term tyrosine ingestion in normally fed human subjects. Am J Clin Nutr 34: 82-87.

Almas B, Le Bourdelles B, Flatmark T, Mallet J, Haavik J. 1992. Regulation of recombinant human tyrosine hydroxylase isozymes by catecholamine binding and phosphorylation structure/activity studies and mechanistic implications. Eur J Biochem 209: 249-255.

Alonso R, Agharanya JC, Wurtman RJ. 1980. Tyrosine loading enhances catecholamine excretion by rats. J Neural Transm 49: 31-43.

Alonso R, Gibson C, Wurtman RJ, Agharanya JC, Prieto L. 1982. Elevation of urinary catecholamines and their metabolites following tyrosine administration in humans. Biol Psych 17: 781-790.

Amer A, Breu J, McDermott J, Wurtman RJ, Maher TJ. 2004. 5–Hydroxy-tryptophan, raises plasma 5–HTP levels in humans and rats and suppresses food intake by hypophagic and stressed rats. Pharm Biochem Behav 77: 137-143.

Ames MM, Lerner P, Lovenberg W. 1978. Tyrosine hydroxylase activation by protein phosphorylation and end product inhibition. J Biol Chem 253: 27-31.

Andersson KK, Cox DD, Que L Jr, Flatmark T, Haavik J. 1988. Resonance Raman studies on the blue-green-colored bovine adrenal tyrosine 3–monooxygenase (tyrosine hydroxylase). Evidence that the feedback inhibitors adrenaline and noradrenaline are coordinated to iron. J Biol Chem 263: 18621-18626.

Ashcroft GW, Eccleston D, Crawford TBB. 1965. 5–Hydroxyindole metabolism in rat brain A study of intermediate metabolism using the technique of tryptophan loading-I. J Neurochem 12: 483-492.

Axelrod J, Shein HM, Wurtman RJ. 1969. Stimulation of C^{14}–melatonin synthesis from C^{14}–tryptophan by noradrenaline in rat pineal in organ culture. Proc Natl Acad Sci USA 62: 544-549.

Ayling JE, Pirson WD, Al-Janabi JM, Helfand GD. 1974. Kidney phenylalanine hydroxylase from man and rat comparison with the liver enzyme. Biochemistry 13: 78-85.

Badawy AA-B, Williams DL. 1982. Enhancement of rat brain catecholamine synthesis by administration of small doses of tyrosine and evidence for substrate inhibition of tyrosine hydroxylase activity by large doses of the amino acid. Biochem J 206: 165-168.

Bailey SW, Dillard SB, Bradford-Thomas K, Ayling JE. 1989. Changes in the cofactor binding domain of bovine striatal tyrosine hydroxylase at physiological pH upon cAMP-dependent phosphorylation mapped with tetrahydrobiopterin analogues. Biochemistry 28: 494-504.

Barazzoni R, Zanetti M, Vettore M, Tessari P. 1998. Relationships between phenylalanine hydroxylation and plasma aromatic amino acid concentrations in humans. Metabolism 47: 669-674.

Belarbi A, Beck G, Bergmann C, Bollack C. 1979. Intrinsic forms of soluble and mitochondrial tyrosine aminotransferase from rat tissues. FEBS Lett 104: 59-65.

Bernheimer H, Birkmayer W, Hornykiewicz O, Jellinger K, Seitelberger F. 1973. Brain dopamine and the syndromes of Parkinson and Huntington Clinical, morphological and neurochemical correlations. J Neurol Sci 20: 415-455.

Berry EM, Growdon JH, Wurtman JJ, Caballero B, Wurtman RJ. 1991. A balanced carbohydrate: Protein diet in the management of Parkinson's disease. Neurology 41: 1295-1297.

Biggio G, Porceddu ML, Gessa GL. 1976. Decrease of homovanillic, dihydroxyphenylacetic acid and cyclic-adenosine-3,5′-monophosphate content in the rat caudate nucleus induced by the acute administration of an amino acid mixture lacking tyrosine and phenylalanine. J Neurochem 26: 1253-1255.

Bjork JM, Dougherty DM, Moeller G, Swann AC. 2000. Differential behavioral effects of plasma tryptophan depletion and loading in aggressive and nonaggressive men. Neuropsychopharmacology 22: 357-369.

Blasberg R, Lajtha A. 1966. Heterogeneity of the mediated transport systems of amino acid uptake in brain. Brain Res 1: 86-104.

Boado RJ, Li JY, Nagaya M, Zhang C, Pardridge WM. 1999. Selective expression of the large neutral amino acid transport at the blood-brain barrier. Proc Natl Acad Sci USA 96: 12079-12084.

Bramwell GJ. 1974. Factors affecting the activity of 5–HT-containing neurones. Brain Res 79: 515-519.

Brzezinski A, Shalitin N, Ever-Hadani P, Schenker JG. 1990. Plasma concentrations of tryptophan and dieting. Brit Med J 301: 183

Burkin DJ, Kimbro KS, Barr BL, Jones C, Taylor MW, et al. 1993. Localization of the human indoleamine 2,3–dioxygenase (IDO) gene to the pericentromeric region of human chromosome 8. Genomics 17: 262-263.

Caballero B, Wurtman RJ. 1988. Control of plasma phenylalanine. Dietary Phenylalanine and Brain Function. Wurtman RJ, editor. Boston/Basel: Birkhauser; pp. 3-12.

Caballero B, Finer N, Wurtman RJ. 1988. Plasma amino acids and insulin levels in obesity: Response to carbohydrate intake and tryptophan supplements. Metabolism 37: 672-676.

Caballero B, Gleason RE, Wurtman RJ. 1991. Plasma amino acid concentrations in healthy elderly men and women. Am J Clin Nutr 53: 1249-1252.

Carlin JM, Ozaki Y, Byrne GI, Brown RR, Borden EC. 1989. Interferons and indoleamine 2,3-dioxygenase: Role in antimicrobial and antitumor effects. Experientia 45: 535-541.

Carlsson A, Lindqvist M. 1978. Effects of antidepressant agents on the synthesis of brain monoamines. J Neural Transm 43: 73-91.

Carlsson A, Lindquist M, Magnusson T, Waldeck B. 1958. On the presence of 3-hydroxytyramine in brain. Science 127: 471

Carpenter LL, Anderson GM, Pelton GH, Gudin JA, Kirwin PDS, Price LH, Heninger GR, McDougle CJ. 1998. Tryptophan depletion during continuous CSF sampling in healthy human subjects. Neuropsychopharmacology 19: 26-35.

Centers for Disease Control (CDC). 1989. Eosinophilia-myalgia syndrome: New Mexico. Morb Mortal Wkly Rep 38: 765–767.

Chiodo LA, Bannon MJ, Grace AA, Roth RH, Bunney BS. 1984. Evidence for the absence of impulse-regulating somatodendritic and synthesis-modulating nerve terminal autoreceptors on subpopulations of mesocortical dopamine neurons. Neuroscience 12: 1-16.

Clarke JTR, Bier DM. 1982. The conversion of phenylalanine to tyrosine in man. Direct measurement by continuous intravenous tracer infusions of L-[$ring$-$^2H_5$]Phenylalanine and L-[1–^{13}C]Tyrosine in the postabsorptive state. Metabolism 31: 999-1005.

Colmenares JL, Wurtman RJ. 1979. The relation between urinary 5-hydroxyindoleacetic acid levels and the ratio of tryptophan to other large neutral amino acids placed in the stomach. Metabolism 28: 820-827.

Comings DE, Muhleman D, Dietz GW Jr, Donlon T. 1991. Human tryptophan oxygenase localized to 4q31: Possible implications for alcoholism and other behavioral disorders. Genomics 9: 301-308.

Conlay LA, Maher TJ, Wurtman RJ. 1981. Tyrosine increases blood pressure in hypotensive rats. Science 212: 559-560.

Conlay LA, Maher TJ, Wurtman RJ. 1984. Tyrosine's pressor effect in hypotensive rats is not mediated by tyramine. Life Sci 35: 1207-1212.

Coppen A, Whybrow PC, Noguera R, Maggs R, Prange AJ Jr. 1972. The comparative antidepressant value of L-tryptophan and imipramine with and without attempted potentiation by liothyronine. Arch Gen Psychiatry 26: 234-241.

De Jong W, Zandberg P, Bohus B. 1975. Role of noradrenaline in the central control of blood pressure in normotensive and spontaneously hypertensive rats. Arch Int Pharmacodyn Ther 213: 272-284.

Delgado PD, Price LH, Miller HL, Salomon RM, Licinio J, et al. 1991. Rapid serotonin depletion as a provocative challenge test for patients with major depression: Relevance to antidepressant action and the neurobiology of depression. Psychopharmacol Bull 27: 321-330.

Demisch K, Bauer J, Georgi K, Demisch L. 1987. Treatment of severe chronic insomnia with L-tryptophan: Results of a double blind cross-over study. Pharmacopsychiatry 20: 242-244.

Demling J, Langer K, Mehr MQ. 1996. Age dependence of large neutral amino acid levels in plasma. Focus on tryptophan. Advances in Experimental Medicine and Biology, Vol. 398. Filippini GA, Costa CVL, Bertazzo A, editors. New York: Plenum Press; pp. 579-582.

De Pietro FR, Fernstrom JD. 1998. The effect of phenylalanine on DOPA synthesis in PC12 cells. Neurochem Res 23: 1011-1020.

Dhondt JL, Dautrevaux M, Biserte G, Farriaux JP. 1978. Phenylalanine analogues as inhibitors of phenylalaninehydroxylase from rat liver. New conclusions concerning kinetic behaviors of the enzyme. Biochimie 60: 787-794.

Dietrich J-B. 1992. Tyrosine aminotransferase: A transaminase among others? Cell Mol Biol 38: 95-114.

Dong-Ruyl L, Sawada M, Nakano K. 1998. Tryptophan and its metabolite, kynurenine, stimulate expression of nerve growth factor in cultured mouse astroglial cells. Neurosci Lett 244: 17-20.

Døskeland AP, Døeskland SO, Ogreid D, Flatmark T. 1984. The effect of ligands of phenylalanine 4-monooxygenase on the cAMP-dependent phosphorylation of the enzyme. J Biol Chem 259: 11242-11248.

Dunkley PR, Bobrovskaya L, Graham ME, von Nagy-Felsobuki EI, Dickson PW. 2004. Tyrosine hydroxylase phosphorylation: Regulation and consequences. J Neurochem 91: 1025-1043.

During MJ, Acworth IN, Wurtman RJ. 1988. Phenylalanine administration influences dopamine release in the rat's corpus striatum. Neurosci Lett 93: 91-95.

During MJ, Acworth IN, Wurtman RJ. 1989. Dopamine release in rat striatum: Physiological coupling to tyrosine supply. J Neurochem 52: 1449-1454.

Eidson M, Philen RM, Sewell CM, Voorhees R, Kilbourne EM. 1990. L-Tryptophan and eosinophilia-myalgia syndrome in New Mexico. Lancet 335: 645-648.

Elliott TR. 1912. The control of the suprarenal glands by the splanchnic nerves. J Physiol (Lond) 44: 374-409.

Farber SA, Savci V, Wei A, Slack BE, Wurtman RJ. 1996. Choline's phosphorylation in rat striatal slices is regulated by the activity of cholinergic neurons. Brain Res 723: 90-99.

Fernstrom JD, Faller DV. 1978. Neutral amino acids in the brain: Changes in response to food ingestion. J Neurochem 30: 1531-1538.

Fernstrom JD, Wurtman RJ. 1971a. Brain serotonin content: Physiological dependence on plasma tryptophan levels. Science 173: 149-152.

Fernstrom JD, Wurtman RJ. 1971b. Brain serotonin content: Increase following ingestion of carbohydrate diet. Science 174: 1023-1025.

Fernstrom JD, Wurtman RJ. 1972a. Elevation of plasma tryptophan by insulin in rat. Metabolism 21: 337-342.

Fernstrom JD, Wurtman RJ. 1972b. Brain serotonin content: Physiological regulation by plasma neutral amino acids. Science 178: 414-416.

Fernstrom JD, Hirsch MJ, Faller DV. 1976. Tryptophan concentrations in rat brain failure to correlate with free serum tryptophan or its ratio to the sum of other serum neutral amino acids. Biochem J 160: 589-595.

Fernstrom JD, Wurtman RJ, Hammarstrom-Wiklund Rand WM, Munro HN, et al. 1979. Diurnal variations in plasma concentrations of tryptophan, tyrosine, and other neutral amino acids: Effect of dietary protein intake. Am J Clin Nutr 32: 1912-1922.

Fernstrom MH, Fernstrom JD. 1995. Effect of chronic protein ingestion on rat central nervous system tyrosine levels and in vivo tyrosine hydroxylation rate. Brain Res 672: 97-103.

Fishman B, Wurtman RJ, Munro HN. 1969. Daily rhythms in hepatic polysome profiles and tyrosine transaminase activity: Role of dietary protein. Proc Natl Acad Sci USA 64: 677-682.

Fuller RW, Snoddy HD. 1968. Feeding schedule alteration of daily rhythm in tyrosine alpha-ketoglutarate transaminase of rat liver. Science 159: 738

Fuller RW, Snoddy HD. 1982. L-Tyrosine enhancement of the elevation of 3,4–dihydroxyphenylacetic acid concentration in rat brain by spiperone and amfonelic acid. J Pharm Pharmacol 34: 117-118.

Gallager DW, Aghajanian GK. 1976. Inhibition of firing of raphe neurones by tryptophan and 5–hydroxytryptophan: Blockade by inhibiting serotonin synthesis with Ro-4–4602. Neuropharmacology 15: 149-156.

Garibotto G, Tessari P, Verzola D, Dertenois L. 2002. The metabolic conversion of phenylalanine into tyrosine in the human kidney: Does it have nutritional implications in renal patients? J Ren Nutr 12: 8-16.

Gessa FL, Biggio G, Fadda F, Corsini GU, Tagliamonte A. 1974. Effect of the oral administration of tryptophan-free amino acid mixtures on serum tyrptophan, brain tryptrophan and serotonin metabolism. J Neurochem 22: 869-870.

Gibson CJ, Wurtman RJ. 1977. Physiological control of brain catechol synthesis by brain tyrosine concentration. Biochem Pharm 26: 1137-1142.

Gibson CJ, Wurtman RJ. 1978. Physiological control of brain norepinephrine synthesis by brain tyrosine concentration. Life Sci 22: 1399-1406.

Gibson CJ, Watkins CJ, Wurtman RJ. 1983. Tyrosine administration enhances dopamine synthesis and release in light-activated rat retina. J Neural Trans 56: 153-160.

Gijsman HJ, Scarnà A, Harmer CJ, McTavish SFB, Odontiadis J, et al. 2002. A dose-finding study on the effects of branch chain amino acids on surrogate markers of brain dopamine function. Psychopharmacology (Berl) 160: 192-197.

Glaeser BS, Melamed E, Growdon JH, Wurtman RJ. 1979. Elevation of plasma tyrosine after a single oral dose of L-tyrosine. Life Sci 25: 265-271.

Goodman DS. 1958. The interaction of human serum albumin with long-chain fatty acid anions. J Am Chem Soc 80: 3892-3898.

Goodwin GM, Cowen PJ, Fairburn CG, Parry-Billings M, Calder PC, et al. 1990. Plasma concentrations of tryptophan and dieting. Br Med J 300: 1499-1500.

Grahame-Smith DG, Parfitt AG. 1970. Tryptophan transport across the synaptosomal membrane. J Neurochem 17: 1339-1353.

Grange T, Guenet C, Dietrich JB, Chasserot S, Fromont M, et al. 1985. Complete complementary DNA of rat tyrosine aminotransferase messenger RNA. Deduction of the primary structure of the enzyme. J Mol Biol 184: 347-350.

Green H, Greenberg SM, Erickson RW, Sawyer JL, Ellison T. 1962. Effect of dietary phenylalanine and tryptophan upon rat brain amine levels. J Pharmacol Exp Ther 136: 174-178.

Grima B, Lamouroux A, Blanot F, Biguet NF, Malli J. 1985. Complete coding sequence of rat tyrosine hydroxylase mRNA. Proc Natl Acad Sci USA 82: 617-621.

Growdon JH, Melamed E, Logue M, Hefti F, Wurtman RJ. 1982. Effects of oral L-tyrosine administration on CSF tyrosine and homovanillic acid levels in patients with Parkinson's disease. Life Sci 30: 827-832.

Guidetti P, Eastman CL, Schwarcz R. 1995. Metabolism of [5–^{3}H]kynurenine in the rat brain in vivo: Evidence for the existence of a functional kynruenine pathway. J Neurochem 65: 2621-2632.

Guroff G, Udenfriend S. 1962. Studies on aromatic amino acid uptake by rat brain in vivo. Uptake of phenylalanine and of tryptophan; Inhibition and stereoselectivity in the uptake of tyrosine by brain and muscle. J Biol Chem 237: 803-806.

Harris JE, Morgenroth VH, Baldessarini RJ, Roth RH. 1974. Regulation of catecholamine synthesis in the rat brain in vitro by cyclic AMP. Nature 252: 156-158.

Hartmann E, Greenwald D. 1984. Tryptophan and human sleep: An analysis of 43 studies. Progress in Tryptophan and Serotonin Research. Schlossberger W, Kochen B, Steonhart E, editors. Berlin: Walter de Gruyter; pp. 297-304.

Hayaishi O, Rothberg S, Mehler AH, Saito Y. 1957. Studies on oxygenases; enzymatic formation of kynurenine from tryptophan. J Biol Chem 229: 889-896.

Hayashi S-I, Granner DK, Tomkins GM. 1967. Tyrosine aminotransferase. J Biol Chem 242: 3998-4006.

Hefti F, Melamed E, Wurtman RJ. 1980. The decarboxylation of DOPA in the Parkinsonian brain: In vivo studies on an animal model. J Neur Trans Suppl 16: 95-101.

Holten D, Kenney FT. 1967. Regulation of tyrosine α-ketoglutarate transaminase in rat liver VI. Induction by pancreatic hormones. J Biol Chem 19: 4372-4377.

Hrboticky N, Leiter LA, Anderson GH. 1985. Effects of L-tryptophan on short term food intake in lean men. Nutr Res (Los Angel) 5: 595-607.

Hufton SE, Jennings IG, Cotton RGH. 1995. Structure and function of the aromatic amino acid hydroxylases. Biochem J 311: 353-366.

Hull KM, Maher TJ. 1990. L-Tyrosine potentiates the anorexia induced by mixed-acting sympathomimetic drugs in hyperphagic rats. J Pharmacol Exp Ther 255: 403-409.

Hull KM, Maher TJ. 1991. L-Tyrosine fails to potentiate several peripheral actions of the sympathomimetics. Pharmacol Biochem Behav 39: 755-759.

Hull KM, Maher TJ. 1992. Effects of L-tyrosine on mixed-acting sympathomimetic-induced pressor actions. Pharmacol Biochem Behav 43: 1047-1052.

Ichikawa S, Sasaoka T, Nagatsu T. 1991. Primary structure of mouse tyrosine hydroxylase deduced from its cDNA. Biochem Biophys Res Commun 176: 1610-1616.

Ikeda M, Levitt M, Udenfriend S. 1967. Phenylalanine as substrate and inhibitor of tyrosine hydroxylase. Arch Biochem Biophys 120: 420-427.

Iwaki M, Parniak MA, Kaufman S. 1985. Studies on the primary structure of rat liver phenylalanine hydroxylase. Biochem Biophys Res Commun 126: 922-932.

Jaskiw GE, Collins KA, Pehek EA, Yamamoto BK. 2001. Tyrosine augments acute clozapine – but not haloperidol-induced dopamine release in the medial prefrontal cortex of the rat: an in vivo microdialysis study. Neuropsychopharmacology 25: 149-156.

Jedlicki E, Kaufman S, Milstein S. 1977. Partial purification and characterization of rat liver phenylalanine hydroxylase phosphatase. J Biol Chem 252: 7711-7714.

Johnson RW, Roberson LE, Kenney FT. 1973. Regulation of tyrosine aminotransferase in rat liver X characterization and interconversion of the multiple enzyme forms. J Biol Chem 248: 4521-4527.

Kanai Y, Segawa H, Miyamoto K-I, Uchino H, Takeda E, et al. 1998. Expression cloning and characterization of a transporter for large neutral amino acids activated by the heavy chain 4F2 antigen (CD98). J Biol Chem 273: 23629-23632.

Kaneda N, Kobayashi K, Ichinose H, Kishi F, Nakazawa A, et al. 1987. Isolation of a novel cDNA clone for human tyrosine hydroxylase: Alternative RNA splicing produces four kinds of mRNA from a single gene. Biochem Biophys Res Commun 146: 971-975.

Kappock TJ, Caradonna JP. 1996. Pterin-dependent amino acid hydroxylases. Chem Rev 96: 2659-2756.

Katz I, Lloyd T, Kaufman S. 1976. Studies on phenylalanine and tyrosine hydroxylation by rat brain tyrosine hydroxylase. Biochim Biophys Acta 445: 567-578.

Kaufman S. 1986. Regulation of the activity of hepatic phenylalanine hydroxylase. Advances in Enzyme Regulation, Vol. 25. Weber G, editor. Elmsford, NY: Pergamon Press; pp. 37-65.

Kenney FT. 1967. Turnover of rat liver tyrosine transaminase: Stabilization after inhibition of protein synthesis. Science 156: 525-528.

Kenney FT, Flora RM. 1961. Induction of tyrosine-α-ketoglutarate transaminase in rat liver I. Hormonal nature. J Biol Chem 236: 2699-2702.

Kim JH, Miller LL. 1969. The functional significance of changes in activity of the enzymes, tryptophan pyrrolase and tyrosine transaminase, after induction in intact rats and in the isolated, perfused rat liver. J Biol Chem 244: 1410-1416.

Kim W, Erlandsen H, Surendran S, Stevens RC, Gamez A, et al. 2004. Trends in enzyme therapy for phenylketonuria. Mol Ther 10: 220-224.

Kleppe R, Toska K, Haavik J. 2001. Interaction of phosphorylated tyrosine hydroxylase with 14–3–3 proteins: Evidence for a phosphoserine 40–dependent association. J Neurochem 77: 1097-1107.

Knott PJ, Curzon G. 1972. Free tryptophan in plasma and brain tryptophan metabolism. Nature 239: 452-453.

Knox WE, Auerbach VH. 1955. The hormonal control of tryptrophan peroxidase in the rat. J Biol Chem 214: 307-313.

Kotake Y, Masayama T. 1936. Uber den mechanismus der kynurenin-bildung aus tryptophan Z. Physiol Chem 243: 237-244.

Kowlessur D, Kaufman S. 1999. Cloning and expression of recombinant human pineal tryptophan hydroxylase in

Escherischia coli: Purification and characterization of the cloned enzyme. Biochim Biophys Acta. 1434: 317-330.

Kowlessur D, Citron BA, Kaufman S. 1996. Recombinant human phenylalanine hydroxylase: Novel regulatory and structural properties. Arch Biochem Biophys 333: 85-95.

Kroger H, Gratz R. 1980. On the induction of tyrosine aminotransferase (TAT) by tryptophan and methionine. Int J Biochem 12: 575-582.

Kvetnansky R, Armando I, Weise VK, Holmes C, Fukuhara K, et al. 1992. Plasma dopa responses during stress: Dependence on sympathoneural activity and tyrosine hydroxylation. J Pharmacol Exp Ther 261: 899-909.

Kwidzinski E, Bunse J, Aktas O, Richter D, Mutlu L, 2005. Indolamine 2,3–dioxygenase is expressed in the CNS and down-regulates autoimmune inflammation. FASEB J 19: 1347-1349.

Le Bourdelles B, Horellou P, Le Caer J-P, Denèfle P, Latta M, et al. 1991. Phosphorylation of human recombinant tyrosine hydroxylase isoforms 1 and 2: An additional phosphorylated residue in isoform 2, generated through alternative splicing. J Biol Chem 266: 17124-17130.

Levitt M, Spector S, Sjoerdsma A, Udenfiend S. 1965. Elucidation of the rate-limiting step in norepinephrine biosynthesis in the perfused guinea pig heart. J Pharmacol Exp Ther 148: 1-8.

Lidsky AS, Law ML, Morse HG, Kao F-T, Rabin M, et al. 1985. Regional mapping of the phenylalanine hydroxylase gene and the Phenylketonuria locus in the human genome. Proc Natl Acad Sci USA 82: 6221-6225.

Lieberman HR, Corkin S, Spring BJ, Growdon JH, Wurtman RJ. 1982/1983. Mood, performance, and pain sensitivity: Changes induced by food constituents. J Psychiat Res 17: 135-145.

Lieberman HR, Corkin S, Spring BJ, Wurtman RJ, Growdon JH. 1985. The effects of dietary neurotransmitter precursors on human behavior. Am J Clin Nutr 42: 366-370.

Lieberman HR, Wurtman JJ, Chew B. 1986. Changes in mood after carbohydrate consumption among obese individuals. Am J Clin Nutr 44: 772-778.

Lin ECC, Knox WE. 1957. Adaptation of the rat liver tyrosine-α-ketoglutarate transaminase. Biochim Biophysic Acta 26: 85-88.

Lin ECC, Knox WE. 1958. Specificity of the adaptive response of tyrosine-α-ketoglutarate transaminase in the rat. J Biol Chem 233: 1186-1189.

Lipsett D, Madras BK, Wurtman RJ, Munro HM. 1973. Serum tryptophan level after carbohydrate ingestion: Selective decline in non-albumin-bound tryptophan coincident with reduction in serum free fatty acids. Life Sci 12: 57-64.

Lloyd T, Kaufman S. 1975. Evidence for the lack of direct phosphorylation of bovine caudate tyrosine hydroxylase following activation by exposure to enzymatic phosphorylating conditions. Biochem Biophys Res Commun 66: 907-913.

Lorenzo AV. 1974. Amino acid transport mechanisms of the cerebrospinal fluid. Fed Proc 33: 2079-2085.

Lynch HJ, Wurtman RJ, Moskowitz MA, Archer MC, Ho HM. 1975. Daily rhythm in human urinary melatonin. Science 187: 169-171.

Mac Kenzie CR, Hucke C, Müller D, Seidel K, Takikawa O, et al. 1999. Growth inhibition of multiresistant enterococci by interferon-γ-activated human uro-epithelial cells. J Med Microbiol 48: 935-941.

Madras BK, Cohen EL, Fernstrom JD, Larin F, Munro HN, et al. 1973. Dietary carbohydrate increases brain tryptophan and decreases serum-free tryptophan. Nature 244: 34-35.

Maher TJ, Glaeser BS, Wurtman RJ. 1984. Diurnal variations in plasma concentrations in plasma concentrations of basic and neutral amino acids and in red cell concentrations of aspartate and glutamate: Effects of dietary protein intake. Am J Clin Nutr 39: 722-729.

Markus CR, Olivier B, de Haan EHF. 2002. Whey protein rich in α-lactalbumin increases the ratio of plasma tryptophan to the sum of the other large neutral amino acids and improves cognitive performance in stress-vulnerable subjects. Am J Clin Nutr 75: 1051-1056.

Marliss EB, Aoki TT, Unger RH, Soeldner JS, Cahill GF Jr. 1970. Glucagon levels and metabolic effects in fasting man. J Clin Invest 49: 2256-2270.

Martin-Du-Pan R, Mauron C, Glaeser B, Wurtman RJ. 1982. Effect of various oral glucose doses on plasma neutral amino acid levels. Metabolism 31: 937-943.

Mastroberardino L, Spindler B, Pfeiffer R, Skelly PJ, Loffing J, et al. 1998. Amino-acid transport by heterodimers of 4F2hc/CD98 and members of a permease family. Nature 395: 288-291.

McKinney J, Knappskog PM, Haavik J. 2005. Different properties of the central and peripheral forms of human tryptophan hydroxylase. J Neurochem 92: 311-320.

McMenamy RH, Oncley JL. 1958. The specific binding of L-tryptophan to serum albumin. J Biol Chem 233: 1436-1447.

McTavish SFB, Mannie ZN, Harmer CJ, Cowen PJ. 2005. Lack of effect of tyrosine depletion on mood in recovered depressed women. Neuropsychopharmacology 30: 786-791.

Melamed E, Glaeser B, Growdon JH, Wurtman RJ. 1980a. Plasma tyrosine in normal humans: Effects of oral tyrosine and protein-containing meals. J Neural Transm 47: 299-306.

Melamed E, Hefti F, Wurtman RJ. 1980b. Tyrosine administration increases striatal dopamine release in rats with partial nigrostriatal lesions. Proc Natl Acad Sci USA 77: 4305-4309.

Mena I, Cotzias GC. 1975. Protein intake and treatment of Parkinson's disease with levodopa. N Engl J Med 292: 181-184.

Milner JD, Wurtman RJ. 1986. Commentary: Catecholamine synthesis: Physiological coupling to precursor supply. Biochem Pharmacol 35: 875-881.

Milner JD, Irie K, Wurtman RJ. 1986. Effects of phenylalanine on the release of endogenous dopamine from rat striatal slices. J Neurochem 47: 1444-1448.

Milstein S, Kaufman S. 1975. Studies on the phenylalanine hydroxylase system in vivo. An in vivo assay based on the liberation of deuterium or tritium into the body water from ring-labeled L-phenylalanine. J Biol Chem 250: 4782-4785.

Moja EA, Stoff DM, Gessa GL, Castoldi D, Assereto R, et al. 1988. Decrease in plasma tryptophan after tryptophan-free amino acid mixtures in man. Life Sci 42: 1551-1556.

Moldawer LL, Kawamura I, Bistrian BR, Blackburn GR. 1983. The contribution of phenylalanine to tyrosine metabolism in vivo. Studies in the post-absorptive and phenylalanine-loaded rat. Biochem J 210: 811-817.

Morre MC, Wurtman RJ. 1981. Characteristics of synaptosomal tyrosine uptake in various brain regions: Effect of other amino acids. Life Sci 28: 65-75.

Morre MC, Hefti F, Wurtman RJ. 1980. Regional tyrosine levels in rat brain after tyrosine administration. J Neural Trans 49: 45-50.

Munn DH, Zhou M, Attwood JT, Bondarev I, Conway SJ, et al. 1998. Prevention of allogeneic fetal rejection by tryptophan catabolism. Science 281: 1191-1193.

Munn DH, Shafizadeh E, Attwood JT, Bondarev I, Pashine A, et al. 1999. Inhibition of T cell proliferation by macrophage tryptophan catabolism. J Exp Med 189: 1363-1372.

Munro HN. 1968. Role of amino acid supply in regulating ribosome function. Fed Proc 27: 1231-1237.

Nagatsu T. 1983. Biopterin cofactor and monoamine-synthesizing monooxygenases. Neurochem Int 5: 27-38.

Nagatsu T, Levitt M, Udenfriend S. 1964. Tyrosine hydroxylase. The initial step in norepinephrine biosynthesis. J Biol Chem 239: 2910-2917.

Natt E, Kao F-T, Rettenmeier R, Scherer G. 1986. Assignment of the human tyrosine aminotransferase gene to chromosome 16. Hum Genet 72: 225-228.

Nishizawa S, Benkelfat C, Young SN, Leyton M, Mzengeza S, et al. 1997. Differences between males and females in rates of serotonin synthesis in human brain. Proc Natl Acad Sci USA 94: 5308-5313.

Orr ML, Watt BK. 1968. Amino acid content of foods USDA. Home economics research report No 4.

Palkovits M, Zaborszky L. 1977. Neuroanatomy of central cardiovascular control nucleus tractus solitarii: Afferent and efferent neuronal connections in relation to the baroreceptor reflex arc. Prog Brain Res 47: 9-34.

Pardridge WM. 1977. Regulation of amino acid availability to the brain. Nutrition and brain, Vol. 1. Wurtman RJ, Wurtman JJ, editors. New York: Raven Press; pp. 141-204.

Pardridge WM. 2001. Invasive brain drug delivery. Brain Drug Targeting: The Future of Brain Drug Development. Pardridge WM, editor. Cambridge: Cambridge University Press; pp. 13-35.

Pardridge WM, Fierer G. 1990. Transport of tryptophan into brain from the circulating, albumin-bound pool in rats and in rabbits. J Neurochem 54: 971-976.

Pardridge WM, Oldendorf WH, Cancilla P, Frank HJ. 1986. Blood-brain barrier: Interface between internal medicine and the brain. Ann Intern Med 105: 82-95.

Parniak MA, Kaufman S. 1985. Catalytically active oligomeric species of phenylalanine hydroxylase. Biochemistry 24: 3379A.

Patel PD, Pontrello C, Burke S. 2004. Robust and tissue-specific expression of TPH2 versus TPH1 in rat raphe and pineal gland. Biol Psychiatry 55: 428-433.

Perkins MN, Stone TW. 1982. An iontophoretic investigation of the actions of convulsant kynurenines and their interaction with the endogenous excitant quinolinic acid. Brain Res 247: 184-187.

Pfefferkorn ER. 1984. Interferon γ blocks the growth of *Toxoplasma gondii* in human fibroblasts by inducing the host cells to degrade tryptophan. Proc Natl Acad Sci USA 87: 908-912.

Philippu A, Dietl H, Sinha JN. 1980. Rise in blood pressure increases the release of endogenous catecholamines in the anterior hypothalamus of the cat Naunyn Schimiedeberg's. Arch Pharmacol 310: 237-240.

Rapoport MI, Feigin RD, Bruton J, Beisel WR. 1966. Circadian rhythm for tryptophan pyrrolase activity and its circulating substrate. Science 153: 1642-1644.

Rassoulpour A, Wu H-Q, Ferre S, Schwarcz R. 2005. Nanomolar concentrations of kynurenic acid reduce extracellular dopamine levels in the striatum. J Neurochem 93: 762-765.

Reinstein DK, Lehnert H, Scott NA, Wurtman RJ. 1984. Tyrosine prevents behavioral and neurochemical correlates of an acute stress in rats. Life Sci 34: 2225-2232.

Reinstein DK, Lehnert H, Wurtman RJ. 1985. Dietary tyrosine suppresses the rise in plasma corticosterone following acute stress in rats. Life Sci 37: 2157-2163.

Rosenthal NE, Sack DA, Gillin JC, Lewy AJ, Goodwin FK, et al. 1984. Seasonal affective disorder. Arch Gen Psychiatry 41: 72-80.

Ross DS, Fernstrom JD, Wurtman RJ. 1973. The role of dietary protein in generating daily rhythms in rat liver tryptophan pyrrolase and tyrosine transaminase. Metabolism 22: 1175-1184.

Saavedra JM, Axelrod J. 1974. Brain tryptamine and the effects of drugs. Adv Biochem Psychopharmacol 10: 135-139.

Sayegh R, Schiff I, Wurtman J, Spiers P, McDermott J, et al. 1995. The effect of a carbohydrate-rich beverage on mood, appetite, and cognitive function in women with premenstrual syndrome. Obstet Gynecol 86: 520-528.

Scally MC, Wurtman RJ. 1977. Brain tyrosine level controls striatal dopamine synthesis in haloperidol-treated rats. J Neural Trans 41: 1-6.

Schaechter JD, Wurtman RJ. 1989. Tryptophan availability modulates serotonin release from rat hypothalamic slices. J Neurochem 53: 1925-1933.

Schaechter JD, Wurtman RJ. 1990. Serotonin release varies with brain tryptophan levels. Brain Res 532: 203-210.

Schimke RT, Sweeney EW, Berlin CM. 1965. Studies of the stability in vivo and in vitro of rat liver tryptophan pyrrolase. J Biol Chem 240: 4609-4620.

Seltzer S, Stoch R, Marcus R, Jackson E. 1982. Alteration of human pain thresholds by nutritional manipulation and L-tryptophan supplementation. Pain 13: 385-393.

Sheehan BD, Tharyan P, McTavish SFB, Campling GM, Cowen PJ. 1996. Use of a dietary manipulation to deplete plasma tyrosine and phenylalanine in healthy subjects. J Psychopharmacol 10: 231-234.

Shiman R, Mortimore GE, Schworer CM, Gray DW. 1982. Regulation of phenylalanine hydroxylase activity by phenylalanine in vivo, in vitro, and in perfused rat liver. J Biol Chem 257: 11213-11216.

Shimizu T, Nomiyama S, Hirata F, Hayaishi O. 1978. Indoleamine 2,3 dioxygenase. Purification and some properties. J Biol Chem 253: 4700-4706.

Shukitt-Hale B, Stillman MJ, Lieberman HR. 1996. Tyrosine administration prevents hypoxia-induced decrements in learning and memory. Physiol Behav 59: 867-871.

Silberstein SD, Lemberger L, Klein DC, Axelrod J, Kopin IJ. 1972. Induction of adrenal tyrosine hydroxylase in organ culture. Neuropharmacology 11: 721-726.

Slutsker L, Hoesly FC, Miller L, Williams P, Watson JC, et al. 1990. Eosinophilia-myalgia syndrome associated with exposure to tryptophan from a single manufacturer. JAMA 264: 213-217.

Smith KA, Fairburn CG, Cowen PJ. 1997. Relapse of depression after rapid depletion of tryptophan. Lancet 349: 915-919.

Spector S, Gordon R, Sjoerdsma A, Udenfriend S. 1967. End-product inhibition of tyrosine hydroxylase as a possible mechanism for regulation of norepinephrine synthesis. Mol Pharmacol 3: 549-555.

Spring B, Wurtman J, Gleason R, Kessler K, Wurtman RJ. 1991. Weight gain and withdrawal symptoms after smoking cessation: A preventive intervention using d-fenfluramine. Health Psychol 10: 216-223.

Stenfors C, Ross SB. 2002. Evidence for involvement of 5-hydroxytryptamine$_{1B}$ autoreceptors in the enhancement of serotonin turnover in the mouse brain following repeated treatment with fluoxetine. Life Sci 71: 2867-2880.

Stone TW, Darlington LG. 2002. Endogenous kynurenines as targets for drug discovery and development. Nat Rev Drug Discov 1: 609-620.

Stone TW, Perkins MN. 1981. Quinolinic acid: A potent endogenous excitant at amino acid receptors in CNS. Eur J Pharmacol 72: 411-412.

Sved A, Fernstrom J. 1981. Tyrosine availability and dopamine synthesis in the striatum: Studies with gamma-butyrolactone. Life Sci 17: 743-748.

Sved AF, Fernstrom JD, Wurtman RJ. 1979a. Tyrosine administration reduces blood pressure and enhances brain norepinephrine release in spontaneously hypertensive rats. Proc Natl Acad Sci USA 76: 3511-3514.

Sved AF, Fernstrom JD, Wurtman RJ. 1979b. Tyrosine administration decreases serum prolactin levels in chronically reserpinized rats. Life Sci 25: 1293-1299.

Tam S-Y, Elsworth JD, Bradberry CW, Roth RH. 1990. Mesocortical dopamine neurons: High basal firing frequency predicts tyrosine dependence of dopamine synthesis. J Neural Transm 81: 97-110.

Tanaka T, Knox WE. 1959. The nature and mechanism of the tryptophan pyrrolase (peroxidase-oxidase) reaction of pseudomonas and of rat liver. J Biol Chem 234: 1162-1170.

Taylor MW, Feng G. 1991. Relationship between interferon-γ, indoleamine 2,3–dioxygenase, and tryptophan catabolism. FASEB J 5: 2516-2522.

Thoenen H, Mueller RA, Axelrod J. 1969. Increased tyrosine hydroxylase activity after drug-induced alteration of sympathetic transmission. Nature 221: 1264

Tourian A, Goddard J, Puck TT. 1969. Phenylalanine hydroxylase activity in mammalian cells. J Cell Physiol 73: 159-170.

Vahvelainen M-L, Oja SS. 1975. Kinetic analysis of phenylalanine-induced inhibition in the saturable influx of tyrosine, tryptophan, leucine and histidine into brain cortex slices from adult and 7-day-old rats. J Neurochem 24: 885-892.

van Praag HM. 1981. Management of depression with serotonin precursors. Biol Psychiatry 16: 291-310.

Van Woert MH, Rosenbaum D. 1979. L-5–Hydroxytryptophan therapy in myoclonus. Advances in Neurology, Vol. 26. Fahn S, Davis JN, Rowland LP, editors. New York: Raven Press; pp. 107-115.

Verrey F. 2003. System L: Heteromeric exchangers of large, neutral amino acids involved in directional transport Pflugers. Arch Eur J Physiol 445: 529-533.

Viveros OH, Arqueros L, Connett RJ, Kirshner N. 1969. Mechanism of secretion from the adrenal medulla IV. The fate of the storage vesicles following insulin and reserpine administration. Mol Pharmacol 5: 69-92.

Vogt M. 1954. The concentration of sympathin in different parts of the central nervous system under normal conditions and after the administration of drugs. J Physiol 123: 451-481.

Walther DJ, Bader M. 2003. A unique central tryptophan hydroxylase isoform. Biochem Pharmacol 66: 1673-1680.

Wang HL, Harwalkar VH, Waisman HA. 1962. Effect of dietary phenylalanine and tryptophan on brain serotonin. Arch Biochem Biophys 96: 181-184.

Wang Y, Citron BA, Ribeiro P, Kaufman S. 1991. High-level expression of rat PC12 tyrosine hydroxylase cDNA in Eschericiia coli: Purification and characterization of the cloned enzyme. Proc Natl Acad Sci USA 88: 8779-8783.

Weiner N, Rabadjija M. 1968. The effect of nerve stimulation on the synthesis and metabolism of norepinephrine in the isolated guinea-pig hypogastric nerve-vas deferens preparation. J Pharmacol Exp Ther 160: 61-71.

White FJ, Wang RY. 1984. A10 dopamine neurons: Role of autoreceptors in determining firing rate and sensitivity to dopamine agonists. Life Sci 34: 1161-1170.

Wicks WD, Kenney FT, Lee K-L. 1969. Induction of hepatic enzyme synthesis in vivo by adenosine $3',5'$-monophosphate. J Biol Chem 244: 6008-6013.

Witkovsky P, Veisenberger E, Haycock JW, Akopian A, Garcia-Espana A, et al. 2004. Activity-dependent phosphorylation of tyrosine hydroxylase in dopaminergic neurons of the rat retina. J Neurosci 24: 4242-4249.

Wretborn M, Humble E, Ragnarsson U, Engström L. 1980. Amino acid sequence at the phosphorylated site of rat liver phenylalanine hydroxylase and phosphorylation of a corresponding synthetic peptide. Biochem Biophys Res Commun 93: 403-408.

Wurtman JJ, Wurtman RJ, Growdon JH, Henry P, Lipscomb A, et al. 1981. Carbohydrate craving in obese people: Suppression by treatments affecting serotoninergic transmission. Int J Eating Dis 1: 2-15.

Wurtman JJ, Moses PL, Wurtman RJ. 1983. Prior carbohydrate consumption affects the amount of carbohydrate that rats choose to eat. J Nutr 113: 70-78.

Wurtman JJ, Wurtman RJ, Berry E, Gleason R, Goldberg H, et al. 1993. Dexfenfluramine, fluoxetine, and weight loss among female carbohydrate cravers. Neuropsychopharmacology 9: 201-210.

Wurtman RJ. 2005. Melatonin. Encyclopedia of Dietary Supplements. Coates PM, Blackman MR, Cragg GM, Levine M, Moss J, et al. editors. New York: Marcel Dekker; pp. 457-466.

Wurtman RJ, Axelrod J. 1965. The pineal gland. Sci Am 213: 50-60.

Wurtman RJ, Axelrod J. 1967. Daily rhythmic changes in tyrosine transaminase activity of the rat liver. Proc Natl Acad Sci USA 57: 1594-1598.

Wurtman JJ, Wurtman RJ. 1977. Fenfluramine and fluoxetine spare protein consumption while suppressing caloric intake by rats. Science 198: 1178-1180.

Wurtman JJ, Wurtman RJ. 1979. Drugs that enhance central serotoninergic transmission diminish elective carbohydrate consumption by rats. Life Sci 24: 895-904.

Wurtman RJ, Wurtman JJ. 1989. Carbohydrates and depression. Sci Am 68–75.

Wurtman RJ, Larin F, Axelrod J, Shein HM, Rosasco K. 1968a. Formation of melatonin and 5–hydroxyindole acetic acid from ^{14}C-tryptophan by rat pineal glands in organ culture. Nature 217: 953-954.

Wurtman RJ, Shoemaker WJ, Larin F. 1968b. Mechanism of the daily rhythm in hepatic tyrosine transaminase activity: Role of dietary tryptophan. Proc Natl Acad Sci USA 59: 800-807.

Wurtman RJ, Shein HM, Axelrod J, Larin F. 1969. Incorporation of 14C-tryptophan into 14C-protein by cultured rat pineals: Stimulation of I-norepinephrine. Proc Natl Acad Sci USA 62: 749-755.

Wurtman RJ, Larin F, Mostafapour S, Fernstrom JD. 1974. Brain catechol synthesis: Control by brain tyrosine concentration. Science 185: 183-184.

Wurtman RJ, Wurtman JJ, Regan MM, McDermott JM, Tsay RH, et al. 2003. Effects of normal meals rich in carbohydrates or proteins on the plasma tryptophan ratio. Am J Clin Nutr 77: 128-132.

Yamaoka KA, Miyasaka N, Inuo G, Saito I, Kolb J-P, et al. 1994. 1,1′-Ethylidenebis(Tryptophan) (Peak E) induces functional activation of human eosinophils and interleukin 5 production from T lymphocytes: Association of eosinophilia-myalgia syndrome with a L-tryptophan contaminant. J Clin Immun 14: 50-60.

Yamauchi T, Fujisawa H. 1979. Regulation of bovine adrenal tyrosine 3–monooxygenase by phosphorylation-dephosphorylation reaction, catalyzed by adenosine $3':5'$-monophosphate-dependent protein kinase and phosphoprotein phosphatase. J Biol Chem 254: 6408-6413.

Yamazaki F, Kuroiwa T, Takikawa O, Kido R. 1985. Human indolelamine 2,3–dioxygenase. Its tissue distribution, and characterization of the placental enzyme. Biochem J 230: 635-638.

Yokogoshi H, Wurtman RJ. 1986. Meal composition and plasma amino acid ratios: Effect of various proteins or carbohydrates, and of various protein concentrations. Metabolism 35: 837-842.

Zavisca FG, Wurtman RJ. 1978. Effects of neutral amino acids on the antihypertensive action of methyldopa in spontaneously hypertensive rats. J Pharm Pharmacol 30: 60-62.

Zhang X, Gainetdinov RR, Beaulieu J-M, Sotnikova TD, Burch LH, et al. 2005. Loss-of-function mutation in tryptophan hydroxylase-2 identified in unipolar major depression. Neuron 45: 11-16.

Zigmond RE, Schwarzschild MA, Rittenhouse AR. 1989. Acute regulation of tyrosine hydroxylase by nerve activity and by neurotransmitters via phosphorylation. Annu Rev Neurosci 12: 415-461.

5 Arginine, Citrulline, and Ornithine

H. Wiesinger

1	*Introduction*	*100*
2	*Occurrence*	*100*
2.1	Tissue and CSF	100
2.2	Cellular Localization	101
3	*Metabolism and Transport*	*102*
3.1	Synthesis	103
3.1.1	Arginine	103
3.1.2	Citrulline	104
3.1.3	Ornithine	105
3.2	Degradation	105
3.2.1	Arginine	105
3.2.2	Citrulline and Ornithine	106
3.3	Transport	107
3.3.1	Arginine	107
3.3.2	Citrulline and Ornithine	108
4	*Physiology and Pathophysiology*	*108*
4.1	Arginine	108
4.2	Citrulline and Ornithine	109
5	*Concluding Remarks*	*110*

© Springer-Verlag Berlin Heidelberg 2007

Abstract: Metabolism as well as function of the three amino acids, L-arginine, L-citrulline, and L-ornithine, are closely intertwined and related to the homeostasis of nitric oxide at the tissue and cellular levels. For the three compounds data on neural tissue and CSF levels are summarized and localization in brain mainly derived from immunohistochemical experiments is reported. A complex pattern of differential cellular localization, subcellular compartmentation and expression of isoforms holds for the enzymes involved in the synthesis and degradation of all three amino acids in the nervous system. Trafficking of arginine or citrulline between particular neural cell populations is likely, and transport mechanisms and proteins involved are discussed. Up-regulation of arginine synthesis and transport by proinflammatory agents satisfies the need of induced nitric oxide synthase for its substrate arginine especially in glial cells. Nitric oxide-related pathologies in the nervous system cannot be separated from the metabolic network established for the three amino acids; however, pharmacological intervention has to take into account the manyfold functions of arginine, citrulline, and ornithine in undisturbed neural cell physiology.

List of abbreviations: AD, Alzheimer's disease; ADC, Arginine decarboxylase; ADMA, Asymmetric N^G, N^G-dimethylarginine; AGAT, Arginine: glycine amidinotransferase; ASL, Argininosuccinate lyase; ASS, Argininosuccinate synthetase; CAT, Cationic amino acid transporter; CPS, Carbamoylphosphate synthetase; DDAH, N^G, N^G-dimethyl-L-arginine dimethylaminohydrolase; IFN-γ, Interferon-γ; LPS, Lipopolysaccharide; NMMA, N^G-monomethylarginine; NOS, Nitric oxide synthase; OAT, Ornithine aminotransferase; ODC, Ornithine decarboxylase; OTC, Ornithine carbamoyltransferase; PAD, Peptidylarginine deiminase

1 Introduction

Since the last edition of this handbook, knowledge about the metabolically interrelated amino acids, L-arginine, L-citrulline, and L-ornithine, has tremendously increased (*stereochemical denotation will be abandoned from hereon*). This is mainly due to the discovery in the late 1980s that the small radical molecule nitric oxide (NO) is ubiquitously distributed in the mammalian body and has profound physiological functions and that NO is generated in vivo exclusively from arginine with citrulline as coproduct. This perception shed new light on the role of the rudimentary urea cycle in nonhepatic tissues and in particular the various cell types of the nervous system, and recent findings on arginase added to a rethinking on established facts of arginine and ornithine in physiology and disease. Finally, it was recognized that agmatine is an important metabolite of arginine also in higher eukaryotes with functions of its own and that N-methylated arginine analogs as well as a number of guanidino compounds derived ultimately from arginine are worth being considered as (patho-)physiological mediators.

Since metabolism as well as function of the three amino acids and the arginine metabolites mentioned are closely interrelated, these compounds will be treated together under the aspects of occurrence, synthesis and degradation (including transport), and physiological role with pathophysiological implications. Emphasis will be put on findings in brain tissue and in neuronal and glial cell cultures with reference to the peripheral nervous system (PNS) whenever appropriate. It should be mentioned that all chapters of this handbook dealing with generation and functions of NO should also be consulted when information on the roles of arginine and its metabolites in normal and pathophysiology of the nervous system is desired.

2 Occurrence

2.1 Tissue and CSF

Established analytical methods have been used for quantifying amino acids in nervous tissue and CSF. Data on arginine, citrulline, and ornithine in the brain can be found in the article by Perry (1982) and on arginine in the CSF in the one by Wood (1982) in the second edition of this handbook. A compilation of data from several literature sources is given by Wiesinger (2001). A mean value of arginine concentration in human CSF can be given as about 22 μM (Wood, 1982) and does not change in several disease states such as

Alzheimer's disease (AD) or Parkinson's disease (Kuiper et al., 2000). Citrulline levels appear to be an order of magnitude lower (Kuiper et al., 2000), although a wide range of data can be taken from early reports (Wiesinger, 2001). Data on brain concentrations were gathered from common laboratory animals and the biopsied human brain; 0.1- to 0.3-µmol/g wet weight is a reliable value for arginine in whole brain, but also for frontal or cerebellar cortex (Perry, 1982; de Jonge et al., 2001; Wiesinger, 2001). Citrulline again appears to be present in somewhat lower concentrations of 0.02- to 0.1-µmol/g wet weight (Perry, 1982). With a full urea cycle lacking in the brain, the nevertheless substantial amounts of citrulline in this organ can be explained by the presence of a neural citrulline–NO cycle operating not in a stoichiometric fashion (see ▸ *Sect. 3.1*). Concentrations reported for ornithine range from 0.01- to 0.05-µmol/g wet weight (Perry, 1982; Seiler and Daune-Anglard, 1993).

An even distribution of arginine and ornithine in the brain may be assumed since data do not differ when reported for whole brain, frontal and cerebellar cortex (Perry, 1982) or brain stem, cerebellum, and medulla (Seiler and Daune-Anglard, 1993). A slightly elevated value of citrulline in cerebellar cortex (Perry, 1982) may reflect the substantial constitutive generation of NO in this region (Garthwaite and Boulton, 1995). Cerebellar concentrations of arginine were the same when measured in 3- to 5-month-old (young) and 18- to 22-month-old (aged) rats (4.4 µmol/g of soluble protein; Mistry et al., 2002).

2.2 Cellular Localization

Direct localization of amino acids by immunohistochemistry was made possible by developing protocols for the generation of antibodies directed against these small molecules. Arginine appears to be distributed unevenly among CNS cell populations. In the rat brain and spinal cord, arginine immunoreactivity was localized mainly in astrocytes whereas oligodendrocytes were not immunoreactive (Aoki et al., 1991b). Bergmann glial cells in the cerebellum stained positive for arginine, neighboring basket and Purkinje cells did not. However, some neurons in the cerebellum showed weak arginine immunoreactivity, several fibers in the brain stem and the spinal cord were prominently stained (Aoki et al., 1991b). Similarly, in the PNS arginine immunoreactivity was predominant in glial components, i.e., satellite and supporting cells surrounding neuronal structures (Aoki et al., 1991a). Arginine immunoreactivity was seen in astroglial endfeet wrapping endothelial cells of the vasculature (Aoki et al., 1991b) and a glial localization of immunoreactive arginine was reported for the ventroposterior thalamic nucleus of the rat (Kharazia et al., 1997) and the rat neurohypophysis (Pow, 1994). These immunohistochemical data correlate well with the finding of high concentrations of arginine in cultured astrocytes (Yudkoff et al., 1989).

By immunohistochemical methods, citrulline was detected in the rat brain exclusively in neurons; these neurons also contained NADPH diaphorase, an enzyme activity taken as being identical with neuronal nitric oxide synthase NOS-1 (Pasqualotto et al., 1991; for critical discussion, see Wolf, 1997). A neuronal localization of citrulline immunoreactivity was also found in the rat neurohypophysis (Pow, 1994). In addition to NOS-containing neurons, activated microglial cells in the brain stained positively for citrulline. Citrulline immunoreactivity was lacking in neurons of inducible NOS-2 knockout mice and was increased when recycling of citrulline to arginine was blocked (Keilhoff et al., 2000). With confocal laser scanning microscopy, costaining of citrulline and NOS-2 was found predominantly at postsynaptic densities (Martinelli et al., 2002). From these results it was concluded that citrulline immunohistochemistry is a reliable means for studying NOS activity in the brain at the single cell and subcellular level.

However, it should be kept in mind that citrulline can be generated posttranslationally within proteins from arginine by the action of any isoform of calcium-dependent peptidylarginine deiminase (PAD; EC 3.5.3.15; Vossenaar et al., 2003). Therefore, in addition to free citrulline generated in the brain mainly by NOS activity, antisera may also detect protein-bound citrulline. Indeed, antisera stained citrulline-containing proteins in Western blot analysis, and staining increased with increasing concentration of intracellular calcium (Keilhoff and Wolf, 2003).

Citrullinated epitopes in proteins were investigated systematically in the rat and human brains after generation of a monoclonal antibody, which was selected for reacting with human citrullinated myelin basic protein (Nicholas and Whitaker, 2002). Immunohistochemical localization of citrulline-containing proteins in the brain revealed staining of white matter myelin, but also of a subset of astrocytes associated with blood vessels and ventricular surfaces; in these cells glial fibrillary acidic protein (GFAP) may be present in multiple citrullinated isoforms (Nicholas and Whitaker, 2002; Nicholas et al., 2003). Deimination of arginine in myelin basic protein was increased in protein isolated from white matter of multiple sclerosis patients and may be related to the pathogenesis of this demyelinating disease (Kim et al., 2003). Membrane-bound PAD was elevated in a demyelinating mouse model prior to loss of myelin (Moscarello et al., 2002). Cellular localization of PAD type II revealed expression of the enzyme in immature oligodendrocytes and in astrocytes and microglial cells after kainate-induced neurodegeneration (Akiyama et al., 1999; Asaga et al., 2002).

Direct localization of ornithine has not been reported. However, localization of arginine, citrulline, and ornithine can be inferred from localization of their biosynthetic machineries. With this in mind, the reader may consult the appropriate paragraphs of this chapter (see ❯ Sect. 3.1).

3 Metabolism and Transport

The three amino acids arginine, citrulline, and ornithine are metabolically intertwined. Synthesis of one amino acid is brought about by degradation of another one, and vice versa (❯ Figure 5-1). Differential cellular localization of the enzymes involved, subcellular compartmentalization, and expression of various isoenzymes have to be considered when drawing a complete picture of the metabolic pathways involved. For proper quantitative consideration of neural homeostasis of all three amino acids, uptake from the periphery has to be taken into consideration (see ❯ Sect. 3.3). For current knowledge about the metabolism of the amino acids in peripheral tissues, about arginine being a "semi-essential" amino acid, and about interorgan trafficking of citrulline and arginine, the reader is referred to reviews (Levillain, 2003; Cederbaum et al., 2004; Morris, 2004; Wu and Morris, 2004).

◘ Figure 5-1
Overview of interrelated metabolism of arginine, citrulline, and ornithine in neural cells. For details of cosubstrates or coproducts, see ❯ Figures 5-2 and ❯ 5-3. Enzymes are written in *italics*, abbreviations as in the text. Argininosucc., argininosuccinate; GSA, glutamic semialdehyde; Glu, glutamate; PM, plasma membrane; Pro, proline. Differential cellular expression and mitochondrial localization of some of the enzymes are described in the text. Exact biochemistry of the reactions of GSA to Glu and Pro, respectively, is not shown

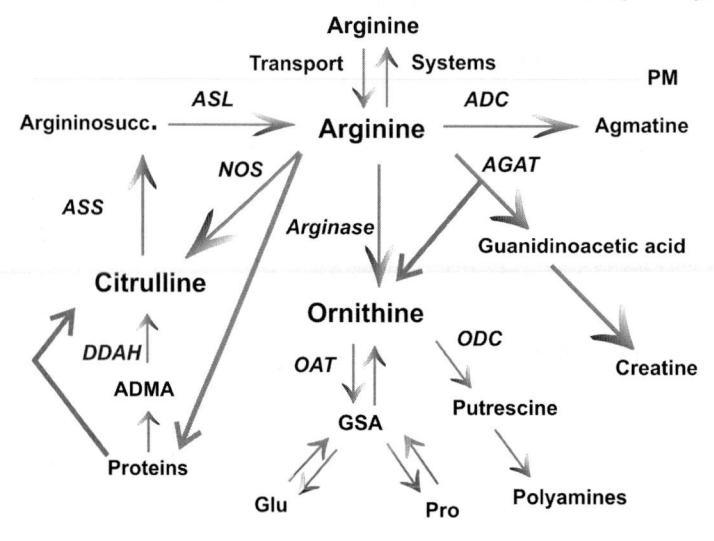

3.1 Synthesis

3.1.1 Arginine

The central pivot in the metabolism of arginine is the urea cycle as the mammalian body's means to dispose of surplus nitrogen and detoxify ammonia (Withers, 1998). The existence of urea cycle enzymes in the brain has been investigated early and is summarized in the first edition of this handbook (Buniatian, 1971; see Wiesinger, 2001). Activities of the mitochondrial enzymes carbamoylphosphate synthetase I (CPS I; EC 6.3.4.16) and ornithine carbamoyltransferase (OTC; EC 2.1.3.3) were not detected in rodent CNS, which implies that neural cells are not capable of de novo synthesis of arginine and that urea is not the molecule used for disposal of nitrogen in the brain. In contrast it was clear from the early studies that activities of the extramitochondrial enzymes of the urea cycle can be measured in brain (Buniatian, 1971; Sadasivudu and Rao, 1976). Nevertheless, a functional role of argininosuccinate synthetase (ASS; EC 6.3.4.5) and argininosuccinate lyase (ASL; EC 4.3.2.1) in neural tissue was not evident until it was conceived that a citrulline–NO cycle might explain the considerable amounts of citrulline present in the brain (see ❯ Sect. 2.1). A renaissance of investigations on arginine-synthesizing enzymes in the nervous system followed with emphasis on ASS, the rate-limiting enzyme of the pathway (❯ Figures 5-1 and ❯ 5-2).

A localization of ASS and ASL in neural cells was attempted with refined immunological and molecular biology methods. From the first reports on brain tissue, it appeared as if ASS was exclusively expressed in neurons (Nakamura et al., 1991a; Arnt-Ramos et al., 1992; summarized by Wiesinger, 2001). ASS immunoreactivity was not detected in glial cells of the brain (Arnt-Ramos et al., 1992), which also did not exhibit any signal for ASS transcript in an extensive in situ hybridization study (Braissant et al., 1999a). However, an exclusive neuronal expression of ASS in CNS tissue is in contrast to findings in cell culture. ^{15}N-labeled aspartate was incorporated into cellular arginine in astroglial cultures—a reaction, which necessitates the presence of ASS (Yudkoff et al., 1987), and ASS enzyme activity was measured in homogenates of cultured astrocytes (Jackson et al., 1996). Therefore, the presence of ASS protein in glial

◘ Figure 5-2

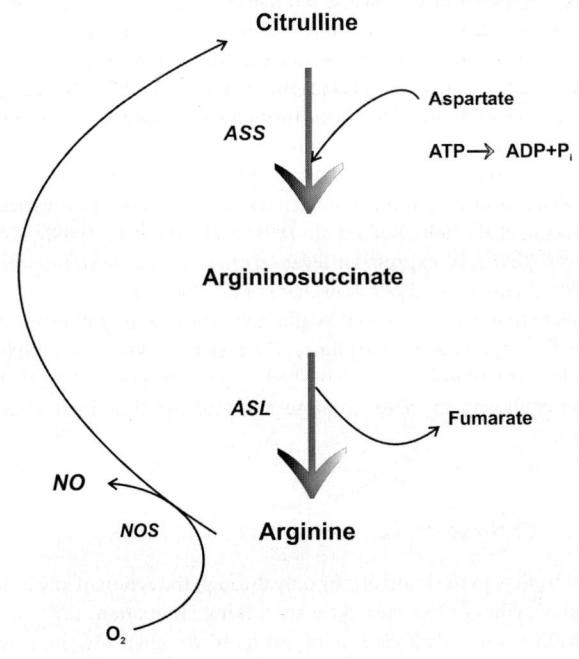

Citrulline–NO cycle. Enzymes are written in *italics*. P$_i$, inorganic phosphate; other abbreviations, see text

cells was investigated with a monospecific antiserum generated against a peptide representing a partial sequence of mouse liver ASS in homogenates of the rat glioma cell line C6-BU-1 (Schmidlin et al., 1997) or pure mouse astroglial cultures (Schmidlin and Wiesinger, 1998). In both glial paradigms, weak basal ASS immunoreactivity increased considerably when the cells had been incubated with bacterial lipopolysaccharide (LPS) and/or interferon-γ (IFN-γ), immunostimulants that also induce NO synthesis in these cells. ASS protein was reported in the C6 glioma cell line after stimulation with a mixture of immunostimulants and cytokines (Zhang et al., 1999), and ASS transcript was found in the glioma cells and in astrocytes in aggregating cell cultures from fetal rat telencephalon (Zhang et al., 1999; Braissant et al., 1999b). Coinduction of ASS and NOS-2 mRNA and protein upon incubation with LPS and IFN-γ was reported for a murine microglial cell line and rat primary microglial cells (Kawahara et al., 2001).

From the experiments described above, it became clear that ASS is expressed in unstimulated cultured glial cells at levels that are at the limits of detection with the antisera used. Glial ASS protein strongly increased after the cells had been incubated with molecules representing proinflammatory signals. Pitfalls of cell culture systems were avoided when proinflammatory conditions were generated in rat striatum by injecting a mixture of LPS and IFN-γ (Heneka et al., 1999). A detailed quantitative analysis revealed that ASS was predominantly localized in activated microglial cells and only occasionally in GFAP-positive astrocytes. Interneurons in cortex and striatum also stained positively for ASS; however, the same level of staining intensity was seen in control and injected animals (Heneka et al., 1999). Expression of ASS and NOS-2 was investigated by immunohistochemistry in hippocampus, frontal, and entorhinal cortex of brains of AD patients and nondemented age-matched controls (Heneka et al., 2001). ASS expression was detected in neurons of control brains; the number of ASS-positive neurons as well as the expression level increased in AD brains. Basal ASS expression and an increase in expression level were also observed in astrocytes without an increase in the number of ASS-positive cells. Only occasionally ASS-positive activated microglial cells were present in the surroundings of senile plaques. Colocalization of ASS and NOS-2 was evident in neurons and, although to a lesser extent, in glial cells (Heneka et al., 2001). mRNA levels of ASS and NOS-2 were increased in *postmortem* cortical tissue obtained from AD patients as compared with control brains (Haas et al., 2002). Thus, it is suggestive that a concomitant up-regulation and/or induction of ASS and NOS-2 both in neurons and glial cells may be responsible for prolonged generation of potentially deleterious NO in AD brains. Indeed, β-amyloid peptides induced transcription of NOS-2 and ASS in mixed rat neuronal–glial cultures (Haas et al., 2002).

A complex picture emerged from studies on the localization of ASL (summarized in Wiesinger, 2001). ASL-immunoreactivity was detected in the spinal cord and throughout the brain in neurons which, in most cases, were lacking ASS immunoreactivity (Nakamura et al., 1990, 1999; Nakamura, 1997). Differential neuronal distribution of ASS and ASL implies transmembrane transport of argininosuccinate which, however, is hard to conceive and is not corroborated by any experiments (see discussion in Wiesinger, 2001). ASL transcript was found in neurons as well as in astrocytes in the rat brain (Braissant et al., 1999a). ASL activity, ASL transcript, and ASL immunoreactivity were detected in a variety of glial cell cultures (Yudkoff et al., 1987; Jackson et al., 1996; Bolla et al., 1999; Braissant et al., 1999b; Kawahara et al., 2001). In contrast to ASS, however, glial ASL expression levels appear not to be influenced by proinflammatory stimuli (Bolla et al., 1999; Zhang et al., 1999; Kawahara et al., 2001).

In conclusion, it is clear that neurons as well as glial cells are able to synthesize arginine in a shortcut of the urea cycle, which was designated the citrulline–NO cycle (Hecker et al., 1990; Husson et al., 2003; ❯ *Figure 5-2*). Nevertheless, it remains to be established if neuronal generation of arginine depends on an interneuronal trafficking of the intermediate argininosuccinate and if an intercellular citrulline–NO cycle may operate in the brain.

3.1.2 Citrulline

Substantial synthesis of citrulline in the brain occurs only through the action of any isoform of NOS (❯ *Figure 5-2*). The mitochondrial enzymes of the urea cycle are lacking (Buniatian, 1971; see ❯ *Sect. 3.1.1*) and an indirect supply of citrulline from degradation of proteins in which arginine was posttranslationally deiminated by PAD (see ● Sect. 2.2) does not appear to be of quantitative importance. Citrulline can also be generated by the activity of N^G, N^G-dimethyl-L-arginine dimethylaminohydrolase (DDAH; EC 3.5.3.18; ● Figure 5-1) from N^G-methylated arginine analogs, N^G-monomethylarginine (NMMA), and asymmetric N^G,N^G-dimethylarginine (ADMA), both being inhibitors of NOS (Tran et al., 2003). However, DDAH, which is expressed in the brain (Leiper et al., 1999) and the spinal cord, is not colocalized with NOS-1 (Mishima et al., 2004). NMMA and ADMA are derived from hydrolysis of proteins in which arginine has been posttranslationally methylated. Again, generation of citrulline by DDAH after protein breakdown does not appear to be of quantitative significance given the fact that methylated proteins in the brain do not turnover rapidly under normal conditions. However, an up-regulation of protein degradation under pathological circumstances may necessitate increased hydrolysis of methylarginines as was found after motor nerve injury (Nakagomi et al., 1999), and net synthesis of citrulline from NMMA has been proposed for the striatum (Ohta et al., 1994). It may be summarized that citrulline in neural cells is almost exclusively generated as an intermediate in the citrulline–NO cycle (● Figure 5-2), which is hardly replenished with citrulline from other pathways. Since no metabolic cycle operates in a stoichiometric fashion, considerable amounts of citrulline are detected in nervous tissue (see ● Sect. 2.1).

3.1.3 Ornithine

Ornithine can be synthesized by the hydrolytic cleavage of arginine catalyzed by both isoforms of arginase and the action of arginine:glycine amidinotransferase (Seiler and Daune-Anglard, 1993). Since both enzymes are pivotal in the degradation of arginine, findings will be detailed in the next chapter (see ● Sect. 3.2.1). Ornithine synthesis from glutamate or proline is not significant in nervous tissue (Wu et al., 1997).

3.2 Degradation

3.2.1 Arginine

A major pathway of degradation of arginine in nervous tissue is the reaction catalyzed by the three isoforms of NOS (EC 1.14.13.39; ● Figure 5-2). Constitutive neuronal isoform, NOS-1, is widespread in the PNS and is established as furnishing NO as a neurotransmitter, e.g., the transmitter released from nonadrenergic, noncholinergic myenteric plexus neurons in the gastrointestinal tract (Bult et al., 1990; Grozdanovic et al., 1992). In the CNS constitutive isoform NOS-3 is particularly expressed in the endothelial cells lining the capillaries (Seidel et al., 1997). NOS-1 is found in many parts of the brain, most prominently in the cerebellum and hippocampus, but also in the cerebral cortex and the striatum. Finally, inducible isoform NOS-2 is expressed under many pathological conditions, most notably in glial cells (see ● Sect. 4). Arginine is the only physiological nitrogen-containing substrate of any NOS isoform, and neural supply of arginine by the citrulline–NO cycle or by transmembrane transport has been discussed (Wiesinger, 2001).

Arginine catabolism by the two isoforms of arginase (EC 3.5.3.1; ● Figure 5-3) yields urea and ornithine. A distinction that cytosolic arginase I is predominantly expressed in liver whereas mitochondrial arginase II is more ubiquitously distributed may become obsolete (Cederbaum et al., 2004). In the nervous system, arginase activity was detected in the rat brain in the cerebellum, cerebral cortex, and brain stem (Sadasivudu and Rao, 1976); arginase activity and protein (as probed by Western blotting) were found in the rat hippocampus and were not dependent on age or functional activity (Liu et al., 2003a, b). Arginase I immunoreactivity was localized in neurons of the olfactory bulb, the cerebellar cortex, and the facial motor nucleus (Nakamura et al., 1990). By in situ hybridization and immunohistochemistry, arginase I as well as arginase II were detected exclusively in neurons of the mouse brain with arginase I being more strongly expressed than the mitochondrial isoform (Yu et al., 2001). On the other hand, cultured microglia as well as astrocytes exhibited arginase activity (Xu et al., 2003), and the transcript for mitochondrial arginase II was present in neurons and glial cells throughout the rat brain, most notably also in oligodendrocytes of cerebellar white matter (Braissant et al., 1999a). Arginase I was strongly expressed in sympathetic ganglia of

◘ Figure 5-3
Metabolism of arginine. By hydrolysis (arginase), decarboxylation (ADC), or amidino transfer reaction (AGAT). Enzymes are written in *italics*. Abbreviations, see text

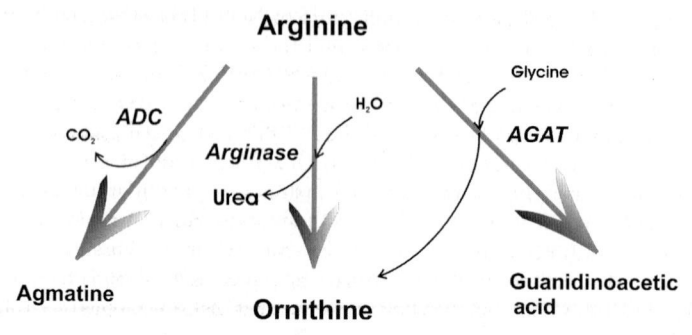

the PNS (Yu et al., 2002, 2003), which may explain the retardation symptoms in arginase I deficiency (Yu et al., 2002). A role for arginase I in autoimmune inflammation of the spinal cord was suggested since the gene for the enzyme was the most up-regulated gene in mice during experimental autoimmune encephalomyelitis and since inhibition of arginase activity ameliorated the symptoms (Xu et al., 2003). Production of polyamines downstream of arginase activity may be essential for axonal regeneration (Cai et al., 2002); however, there is no unequivocal data in the literature which isoform of arginase supplies ornithine as a precursor for polyamine synthesis in the brain.

Arginine in the brain is also catabolized by arginine:glycine amidinotransferase (AGAT; EC 2.1.4.1; ❯ *Figure 5-3*), the first enzyme in the biosynthetic pathway leading to creatine. AGAT is ubiquitously distributed in the adult rat brain in neurons and glial cells, suggesting that every cell type in the CNS is able to synthesize creatine from arginine (Braissant et al., 2001). Indeed, NMR studies hinted at creatine synthesis in cultured astroglial cells of the rat brain (Dringen et al., 1998). Deficiency in AGAT leads to retardation of speech and mental development (Item et al., 2001; Leuzzi, 2002), and the importance of creatine for proper brain functioning and in neuroprotection has been documented (Schulze, 2003; Klivenyi et al., 2004). Due to its low substrate specificity, AGAT may also be responsible for the formation of other guanidino compounds, i.e., neurotoxic guanidinopropionic acid or guanidinobutyric acid (de Deyn et al., 2001; see ❯ Sect. 4).

Finally, the irreversible decarboxylation of arginine yielding 4-(aminobutyl)guanidine (agmatine) is catalyzed by mitochondrial arginine decarboxylase (ADC; EC 4.1.1.19; ❯ *Figure 5-3*). ADC activity was detected in the brain and cultured rat glioma cells (Regunathan and Reis, 2000), and Northern blot analysis with probes for human ADC showed expression of the mRNA in the human brain (Zhu et al., 2004). Roles for agmatine at imidazoline-binding sites, α_2-receptors, and at the NMDA-type glutamate receptor, and in the regulation of cellular polyamine content and consequently cell proliferation have been described (Li et al., 1994, 2003; Olmos et al., 1999; Abe et al., 2003; Berkels et al., 2004). Agmatine (or its aldehyde metabolite) is a potent inhibitor of NOS (Satriano, 2003). Agmatine reduced NO production in cultured microglia (Abe et al., 2000); NOS-2 activity as well as the amount of NOS-2 protein were decreased when cultured astrocytes had been incubated with immunostimulants in the presence of agmatine, and synthesis of agmatine was up-regulated (Regunathan and Piletz, 2003). These latter findings may explain the neuroprotecive effects of agmatine in models of brain injury (Gilad et al., 1996; Feng et al., 2002). However, whether agmatine is indeed a key molecule in regulating the generation of potentially harmful NO by NOS-2 remains to be elucidated.

3.2.2 Citrulline and Ornithine

Recycling of citrulline to arginine in the citrulline–NO cycle is the only metabolic pathway of the NOS product in CNS. The first and rate-limiting enzyme of this transformation has been dealt with extensively in this chapter (see ❯ Sect. 3.1.1; ❯ *Figure 5-2*).

Two major catabolic pathways of ornithine in neural tissue have to be considered, i.e., decarboxylation to putrescine by ornithine decarboxylase (ODC; EC 4.1.1.17) and transamination by mitochondrial ornithine aminotransferase (OAT; EC 2.6.1.13; ❯ *Figure 5-1*). Citrulline formation from ornithine does not take place in the CNS (see ❯ *Sect. 3.1*). ODC activity in the brain is highest during pre- and early postnatal stages, whereas in the mature brain ODC activity is low not the least due to the presence of ODC-antizyme. ODC as well as antizyme transcripts and proteins have been colocalized in neurons throughout the adult rat brain (Kilpelainen et al., 2000). ODC appears to be up-regulated in some brain regions in AD (Morrison et al., 1998) and in cerebral ischemia (Babu et al., 2003). ODC is induced in glial cells upon long-lasting activation of neuronal functions (Bernstein and Müller, 1999), and in neurons and microglial cells after systemic administration of LPS (Soulet and Rivest, 2003). Early data on regulation of ODC in the brain has been summarized (Seiler and Daune-Anglard, 1993; Bernstein and Müller, 1999).

Transamination of ornithine yields glutamic semialdehyde which equilibrates nonenzymatically with L-Δ^1-pyrroline-5-carboxylate, a precursor of glutamate or proline (❯ *Figure 5-1*). Different isoforms of OAT may exist in neurons and astrocytes (Drejer and Schousboe, 1984). A possible role of ODC in the formation of GABA has been discussed (Seiler and Daune-Anglard, 1993).

3.3 Transport

3.3.1 Arginine

Several transport systems for the cationic amino acids arginine, lysine, and ornithine have been described at the physiological and molecular level (Deves and Boyd, 1998; Closs and Mann, 1999). Uptake of arginine in neural cells is mediated by systems y^+, Ly^+, $B^{0,+}$, and $b^{0,+}$, with y^+ being by far the most important transport system. A compilation of data is given by Wiesinger (2001). System y^+ is not sodium dependent, thus operating according to the laws of facilitated diffusion and consequently being unable to transport the amino acid against a concentration gradient. System y^+ is present in cultured neurons and glial cells, but also in endothelial cells of the microvessels and in the choroid plexus epithelium, thus rendering peripheral arginine available to the brain parenchyma and the CSF (Segal et al., 1990; Stoll et al., 1993; Stuhlmiller and Boje, 1995). System y^+ also seems to mediate uptake of arginine into mitochondria (Dolinska and Albrecht, 1998), where arginine catabolizing enzymes arginase II and ADC are located (see ❯ *Sect. 3.2.1*).

Upon incubation with LPS transport activity of system y^+ in cultured astroglial cells was up-regulated in parallel with the induction of NO production (Schmidlin and Wiesinger, 1994, 1995). On the other hand, arginine transport in neuronal primary cultures was insensitive against proinflammatory stimuli (Stevens and Vo, 1998). Arginine uptake into astrocytes and cells of the glioma cell line C6 was down-regulated by prolonged exposure to noradrenaline, isoproterenol, or dibutyryl cyclic AMP (Feinstein and Rozelman, 1997). Arginine transport in neural cells may be influenced by pathological conditions such as hepatic encephalopathy (Rao et al., 1997; Hazell and Norenberg, 1998).

Several transcripts related to system y^+ have been detected in brain tissue as well as in neural cell cultures. Constitutive CAT-1 has been detected in neurons of the brain, in cultured neurons and glial cells, and in microvessels and the choroid plexus, and mediates basic arginine supply to cells of the CNS (Stoll et al., 1993; Stevens et al., 1996; Stevens and Vo, 1998; Braissant et al., 1999a). CAT-2 splice variants are responsible for up-regulation of arginine transport, and a particular splice variant of CAT-2 was found in cultured astroglial cells upon incubation of the cells with immunostimulants (Stevens et al., 1996). Up-regulation of CAT-2 by LPS and IFN-γ was also reported for a murine microglial cell line (Kawahara et al., 2001), and the importance of arginine uptake by CAT-2 for astroglial NO production by NOS-2 has been stressed (Manner et al., 2003). The significance of up-regulated CAT-1 mRNA in microglial cells of mice expressing the APOE4 gene allele—a risk factor in AD—is not clear (Czapiga and Colton, 2003).

Transcript CAT-3 appears to be brain specific and was mainly localized along the midbrain-thalamus-hypothalamus axis (Hosokawa et al., 1997; Ito and Groudine, 1997). CAT-3 transcript was detected neither in capillary endothelial nor in glial cells, but prominently in neurons (Hosokawa et al., 1999; Braissant et al., 1999a). However, an antiserum prepared against a rat CAT-3 peptide stained neurons in the ventromedial part of the rat brain only weakly, whereas good colocalization of immunocytochemical and RNA hybridization signals was found in the cerebral cortex, the hippocampus, and the cerebellum. Immunoreactivity was confined to neuronal cell bodies thus confirming CAT-3 as a neuron-specific transporter. A role for human CAT-4 being transcribed in the brain remains to be established (Sperandeo et al., 1998).

3.3.2 Citrulline and Ornithine

The necessity of intercellular trafficking of citrulline in neural cells may arise if recycling of citrulline to arginine in a citrulline–NO cycle is spatially separated from the generation of NO (Wiesinger, 2001). All neural cell populations appear to have the capacity to take up citrulline, however, not by a specific transport system of its own, but with the help of the transporter classified physiologically as sodium-independent system L for large neutral amino acids (Schmidlin et al., 2000). System L can operate in both directions across the plasma membrane; thus a prerequisite of an intercellular citrulline–NO cycle is indeed given. However, effective utilization of extracellular citrulline depends on the concentration of all neutral amino acids present, which holds in particular for a possible neural uptake of citrulline from the blood. Mitochondrial transport of citrulline (Palmieri, 2004) is not relevant for brain cells that do not express the full urea cycle (see ❯ Sect. 3.1.1).

Data on transport of ornithine across the plasma membrane of neural cells are lacking. However, it is highly probable that ornithine is accepted by system y^+ since it is also a cationic amino acid and indeed inhibits uptake of arginine into neural cells (Schmidlin and Wiesinger, 1994). Since y^+ is present at the blood–brain barrier and since plasma concentrations of arginine and ornithine differ by less than an order of magnitude supply of neural ornithine by uptake from the periphery has to be considered. Mitochondrial transport of ornithine may be restricted to the liver with its fully expressed urea cycle (Begum et al., 2002).

4 Physiology and Pathophysiology

4.1 Arginine

Arginine is an amino acid with widespread roles in metabolism and the regulation of physiological functions. Nutritional and therapeutic aspects of arginine are under active investigation, and endocrine activity (Jun and Wennmalm, 1994), effects on wound healing (Witte and Barbul, 2003), modulation of the immune system (Bansal and Ochoa, 2003), and regulation of vascular tone (Böger and Bode-Böger, 2001; Ignarro and Napoli, 2004) have been reviewed. Since arginine is the only nitrogen-containing substrate of any NOS isoform, effects of arginine in vivo or in cell and tissue culture may reflect the effect of the amino acid on the generation of NO, and modulation of NO synthesis by the availability of arginine is well established (Hallemeesch et al., 2002). Indeed, NMDA receptor-stimulated generation of NO in arginine-deprived brain slices was dependent on refeeding of arginine (Garthwaite et al., 1989), and systemic as well as central administration of arginine enhanced NO production in the rat brain (Salter et al., 1996; Yamada et al., 1997). Neuroprotective effects of arginine have been reported for traumatic brain injury (Liu et al., 2002; Cherian et al., 2003) or for cerebral ischemia (Temiz et al., 2003); in contrast, arginine worsened ischemic damage when given with a delayed time frame (Zhao et al., 2003). In all cases the effect of arginine on the generation of NO has been implicated. For more information on the roles of NO in the physiology and pathophysiology of the nervous system, the reader is referred to reviews (Garthwaite and Boulton, 1995; Bredt, 1999; Prast and Philippu, 2001) and respective chapters in this handbook. Arginine as substrate for neural NO synthesis has been reviewed (Wiesinger, 2001). Some pathophysiological consequences of arginine metabolism have already been mentioned in the previous paragraphs.

A particular role for astrocytes in neural arginine metabolism has been proposed (Wiesinger, 2001). Release of arginine from astrocytes upon stimulation by glutamate or peroxynitrite (Grima et al., 1997; Vega-Agapito et al., 2002) fits well into the picture of astrocytes as storage compartments of arginine possibly for the benefit of surrounding neurons. The fact that exogenous arginine enhances NOS-mediated NO production, although the concentration of intracellular arginine is saturating for the enzyme was termed the "arginine paradox." Proximity of endothelial NOS and arginine transport system (McDonald et al., 1997) and relief of inhibition of NOS by endogenous inhibitors (Tsikas et al., 2000) have been invoked to explain the phenomenon. In astrocytes, availability of extracellular arginine regulates the expression level of inducible NOS-2 by influencing phosphorylation of eukaryotic initiation factor eIF2α; this may explain the arginine paradox in the case of NOS-2 (Lee et al., 2003). Coupling of arginine levels to synthesis of NOS-2 ensures that the enzyme is not expressed under conditions that favor its activity being directed toward generation of harmful superoxide.

Neurotoxic guanidino compounds are derived from arginine by several mechanisms and accumulate during uremia, but also hyperargininemia (De Deyn et al., 2001; De Jonge et al., 2001). AGAT may generate guanidinopropionic and guanidinobutyric acid (see ❥ Sect. 3.2.1). Formation of guanidinosuccinic acid may be a side reaction of AGAT, but also of ASS. In addition, nonenzymatic oxidation of argininosuccinate leads to guanidinosuccinic acid (Aoyagi et al., 2001), similar to a nonenzymatic formation of methylguanidine from creatinine and reactive oxygen species (Nakamura et al., 1991b). The guanidino compounds are associated with cognitive malfunction, behavioral deficits, seizures and convulsions, and coma (Hiramatsu, 2003), and molecular mechanism:s are sought in the inhibition of Na^+, K^+-ATPase or NMDA receptor activation (D'Hooge et al., 1996; da Silva et al., 1999).

Not the least arginine is a proteinogenic amino acid which is of particular interest when taking into account the high turnover of neuroactive peptides and proteins in the nervous system (see chapter in this handbook). It should also be remembered that protein-bound arginine is subject to posttranslational modification, and that citrulline and methylated arginines liberated during proteolysis can exert physiological functions and enter their appropriate metabolic routes (see ❥ Sect. 3.1.2). Arginine is involved in protein degradation since it tags unstable polypeptides that are degraded further by the ubiquitin— proteasome pathway (Bohley et al., 1991); thus one may even draw a link to the hypothesis of neurodegeneration and cell death being caused by proteasomal dysfunction (Halliwell, 2002).

As a precursor of urea, arginine has a central function in the hepatic detoxification of peripheral ammonia derived from amino acid and nucleotide catabolism. Therefore, acute or chronic liver failure as well as inborn errors of metabolism affecting the urea cycle enzymes lead to hyperammonemia with severe pathophysiological consequences, including mental retardation (Shih, 1978; Butterworth, 2003; see chapters in this handbook). Neurochemical consequences of hypoargininemia can be studied in an OTC-deficient mouse model (Qureshi and Rao, 1997). Since the OTC gene is located on the X-chromosome, affected male infants die soon after birth due to hyperammonemic coma. Chronic hyperammonemia caused by OTC-deficiency results in behavioral abnormalities and encephalopathy already in the young. Citrullinemia may be caused by a deficiency in ASS (Patejunas et al., 1994) and in this case is manifest right after birth (neonatal type I citrullinemia). That adult-onset type II citrullinemia is resulting from a defect of the gene for the mitochondrial transport protein for citrulline (Kobayashi et al., 1999; Saheki and Kobayashi, 2002) has been questioned (Sinasac et al., 2004). Nevertheless, citrullinemia patients suffer from disturbed consciousness and coma and die with cerebral edema within a few years of onset. Hyperargininemia due to arginase deficiency is associated with elevated levels of arginine in the CNS and leads to mental retardation (Shih, 1978).

4.2 Citrulline and Ornithine

Since citrulline is a precursor of arginine in the citrulline–NO cycle, biochemistry and physiology of the nonproteinogenic amino acid: are related to the generation of NO. However, extracellular citrulline cannot maintain maximal rates of NO synthesis, e.g., in cultured astrocytes (Schmidlin and Wiesinger, 1998), because it competes with other amino acids for the uptake system (see ❥ Sect. 3.3.2) or cannot enter easily the intracellular NO-generating compartment (Wiesinger, 2001).

Ornithine-derived putrescine is the precursor of the polyamines spermidine and spermine, and, therefore, activity of ODC (see ❥ Sect. 3.2.2) is pivotal for proper cell proliferation in the developing nervous system. In addition, findings on polyamines as modulators of ion channels (Williams, 1997) shed new light on neural decarboxylation of ornithine (Bernstein and Müller, 1999). Other compounds generated in pathways starting from ornithine are the amino acids proline and glutamate. A physiological significance of ornithine-derived neural glutamate, which ultimately may be decarboxylated to the inhibitory transmitter GABA, is far from being clear (Seiler and Daune-Anglard, 1993).

5 Concluding Remarks

It has become evident that also in the nervous system, arginine is "one of the most versatile amino acids" (Wu and Morris, 1998). Since arginine is metabolically tied to citrulline and ornithine, the dynamic interrelationship between the three amino acids determines the physiological roles of any of them. Interest in synthesis, degradation, and transport of the compounds has been stimulated by the discovery of arginine-dependent NO synthesis in virtually every mammalian organ. Targeted pharmacological manipulation of a complex metabolic network for treatment of disease is a challenge for future studies, not the least in the nervous system.

Acknowledgments

The author thanks J. Hullmann, Tübingen, for critical reading of the manuscript and for preparation of the figures.

References

Abe K, Abe Y, Saito H. 2000. Agmatine suppresses nitric oxide production in microglia. Brain Res 872: 141-148.

Abe K, Abe Y, Saito H. 2003. Agmatine induces glutamate release and cell death in cultured rat cerebellar granule neurons. Brain Res 990: 165-171.

Akiyama K, Sakurai Y, Asou H, Senshu T. 1999. Localization of peptidylarginine deiminase type II in a stage-specific immature oligodendrocyte from rat cerebral hemisphere. Neurosci Lett 274: 53-55.

Aoki E, Semba R, Kashiwamata S. 1991a. Evidence for the presence of L-arginine in the glial components of the peripheral nervous system. Brain Res 559: 159-162.

Aoki E, Semba R, Mikoshiba K, Kashiwamata S. 1991b. Predominant localization in glial cells of free L-arginine. Immunocytochemical evidence. Brain Res 547: 190-192.

Aoyagi K, Shahrzad S, Iida S, Tomida C, Hirayama A, et al. 2001. Role of nitric oxide in the synthesis of guanidinosuccinic acid, an activator of the N-methyl-D-aspartate receptor. Kidney Int 78 (Suppl.): S93-S96.

Arnt-Ramos LR, O'Brien WE, Vincent SR. 1992. Immunohistochemical localization of argininosuccinate synthetase in the rat brain in relation to nitric oxide synthase-containing neurons. Neuroscience 51: 773-789.

Asaga H, Akiyama K, Ohsawa T, Ishigami A. 2002. Increased and type II-specific expression of peptidylarginine deiminase in activated microglia but not hyperplastic astrocytes following kainic acid-evoked neurodegeneration in the rat brain. Neurosci Lett 326: 129-132.

Babu GN, Sailor KA, Beck J, Sun D, Dempsey RJ. 2003. Ornithine decarboxylase activity in in vivo and in vitro models of cerebral ischemia. Neurochem Res 28: 1851-1857.

Bansal V, Ochoa JB. 2003. Arginine availability, arginase, and the immune response. Curr Opin Clin Nutr Metab Care 6: 223-228.

Begum L, Jalil MA, Kobayashi K, Iijima M, Li MX, et al. 2002. Expression of three mitochondrial solute carriers, citrin, aralar1, and ornithine transporter, in relation to urea cycle in mice. Biochim Biophys Acta 1574: 283-292.

Berkels R, Taubert D, Grundemann D, Schömig E. 2004. Agmatine signaling: odds and threads. Cardiovasc Drug Rev 22: 7-16.

Bernstein H-G, Müller M. 1999. The cellular localization of the L-ornithine decarboxylase/polyamine system in normal and diseased central nervous system. Prog Neurobiol 57: 485-505.

Böger RH, Bode-Böger SM. 2001. The clinical pharmacology of L-arginine. Annu Rev Pharmacol Toxicol 41: 79-99.

Bohley P, Kopitz J, Adam G, Rist B, von Appen F, et al. 1991. Post-translational arginylation and intracellular proteolysis. Biomed Biochim Acta 50: 343-346.

Bolla T, Kalbacher H, Vogel D, Wiesinger H. 1999. Argininosuccinate lyase: generation of antisera against peptide sequences of the rat brain enzyme and immunochemical studies on glial cells. Biol Chem 380: S95.

Braissant O, Gotoh T, Loup M, Mori M, Bachmann C. 1999a. L-Arginine uptake, the citrulline-NO cycle and arginase II in the rat brain: an in situ hybridisation study. Brain Res Mol Brain Res 70: 231-241.

Braissant O, Henry H, Loup M, Eilers B, Bachmann C. 2001. Endogenous synthesis and transport of creatine in the rat brain: an in situ hybridisation study. Brain Res Mol Brain Res 86: 193-201.

Braissant O, Honegger P, Loup M, Iwase K, Takiguchi M, et al. 1999b. Hyperammonemia: regulation of argininosuccinate synthetase and argininosuccinate lyase genes in aggregating cell cultures of fetal rat brain. Neurosci Lett 266: 89-92.

Bredt DS. 1999. Endogenous nitric oxide synthesis: biological functions and pathophysiology. Free Radic Res 31: 577-596.

Bult H, Boeckxstaens GE, Pelckmans PA, Jordaens FH, Van Maercke YM, et al. 1990. Nitric oxide as an inhibitory non-adrenergic non-cholinergic neurotransmitter. Nature 345: 346-347.

Buniatian HC. 1971. The urea cycle. Handbook of Neurochemistry, Vol. V. Lajtha A, editor. New York: Plenum Press; pp. 235-247.

Butterworth RF. 2003. Molecular neurobiology of acute liver failure. Semin Liver Dis 23: 251-258.

Cai D, Deng K, Mellado W, Lee J, Ratan RR, et al. 2002. Arginase I and polyamines act downstream from cyclic AMP overcoming inhibition of axonal growth MAG and myelin in vitro. Neuron 35: 711-719.

Cederbaum SD, Yu H, Grody WW, Kern RM, Yoo P, et al. 2004. Arginases I and II: do their functions overlap? Mol Genet Metab 81 (Suppl.): 38-44.

Cherian L, Chako G, Goodman C, Robertson CS. 2003. Neuroprotective effects of L-arginine administration after cortical impact injury in rats: dose response and time window. J Pharmacol Exp Ther 304: 617-623.

Closs EI, Mann GE. 1999. Identification of carrier systems in plasma membranes of mammalian cells involved in transport of L-arginine. Methods Enzymol 301: 78-91.

Czapiga M, Colton CA. 2003. Microglial function in human APOE3 and APOE4 transgenic mice: altered arginine transport. J Neuroimmunol 134: 44-51.

da Silva CG, Parolo E, Streck EL, Wajner M, Wannmacher CMD, et al. 1999. In vitro inhibition of Na^+, K^+-ATPase activity from rat cerebral cortex by guanidine compounds accumulating in hyperargininemia. Brain Res 838: 78-84.

De Deyn PP, D'Hooge R, Van Bogaert PP, Marescau B. 2001. Endogenous guanidino compounds as uremic neurotoxins. Kidney Int 78 (Suppl.): S77-S83.

De Jonge WJ, Marescau B, D'Hooge R, De Deyn PP, Hallemeesch MM, et al. 2001. Overexpression of arginase alters circulating and tissue amino acids and guanidino compounds and affects neuromotor behavior in mice. J Nutr 131: 2732-2740.

Deves R, Boyd CAR. 1998. Transporters for cationic amino acids in animal cells: discovery, structure, and function. Physiol Rev 78: 487-545.

D'Hooge R, Raes A, Lebrun P, Diltoer M, Van Bogaert PP, et al. 1996. N-Methyl-D-aspartate receptor activation by guanidinosuccinate but not methylguanidine: behavioural and electrophysiological evidence. Neuropharmacology 35: 433-440.

Dolinska M, Albrecht J. 1998. L-Arginine uptake in rat cerebral mitochondria. Neurochem Int 33: 233-236.

Drejer J, Schousboe A. 1984. Ornithine-delta-aminotransferase exhibits different kinetic properties in astrocytes, cerebral cortex interneurons, and cerebellar granule cells in primary culture. J Neurochem 42: 1194-1197.

Dringen R, Verleysdonk S, Hamprecht B, Willker W, Leibfritz D, et al. 1998. Metabolism of glycine in primary astroglial cells: synthesis of creatine, serine, and glutathione. J Neurochem 70: 835-840.

Feinstein DL, Rozelman E. 1997. Norepinephrine suppresses L-arginine uptake in rat glial cells. Neurosci Lett 223: 37-40.

Feng Y, Piletz JE, Leblanc MH. 2002. Agmatine suppresses nitric oxide production and attenuates hypoxic-ischemic brain injury in neonatal rats. Pediatr Res 52: 606-611.

Garthwaite J, Boulton CL. 1995. Nitric oxide signaling in the central nervous system. Annu Rev Physiol 57: 683-706.

Garthwaite J, Garthwaite G, Palmer RMJ, Moncada S. 1989. NMDA receptor activation induces nitric oxide synthesis from arginine in rat brain slices. Eur J Pharmacol 172: 413-416.

Gilad GM, Salame K, Rabey JM, Gilad VH. 1996. Agmatine treatment is neuroprotective in rodent brain injury models. Life Sci 58: 41-46.

Grima G, Benz B, Do KQ. 1997. Glutamate-induced release of the nitric oxide precursor, arginine, from glial cells. Eur J Neurosci 9: 2248-2258.

Grozdanovic Z, Baumgarten HG, Brüning G. 1992. Histochemistry of NADPH-diaphorase, a marker for neuronal nitric oxide synthase, in the peripheral autonomous nervous system of the mouse. Neuroscience 48: 225-235.

Haas J, Storch-Hagenlocher B, Biessmann A, Wildemann B. 2002. Inducible nitric oxide synthase and argininosuccinate synthetase: co-induction in brain tissue of patients with Alzheimer's dementia and following stimulation with β-amyloid 1-42 in vitro. Neurosci Lett 322: 121-125.

Hallemeesch MM, Lamers WH, Deutz NE. 2002. Reduced arginine availability and nitric oxide production. Clin Nutr. 21: 273-279.

Halliwell B. 2002. Hypothesis: proteasomal dysfunction. A primary event in neurodegeneration that leads to nitrative and oxidative stress and subsequent cell death. Ann N Y Acad Sci 962: 182-194.

Hazell AS, Norenberg MD. 1998. Ammonia and manganese increase arginine uptake in cultured astrocytes. Neurochem Res 23: 869-873.

Hecker M, Sessa WC, Harris HJ, Änggard EE, Vane JR. 1990. The metabolism of L-arginine and its significance for the biosynthesis of endothelium-derived relaxing factor: cultured endothelial cells recycle L-citrulline to L-arginine. Proc Natl Acad Sci USA 87: 8612-8616.

Heneka MT, Schmidlin A, Wiesinger H. 1999. Induction of argininosuccinate synthetase in rat brain glial cells after striatal microinjection of immunostimulants. J Cereb Blood Flow Metab 19: 898-907.

Heneka MT, Wiesinger H, Dumitrescu-Ozimek L, Riederer P, Feinstein DL, et al. 2001. Neuronal and glial coexpression of argininosuccinate synthetase and inducible nitric oxide synthase in Alzheimer disease. J Neuropathol Exp Neurol 60: 906-916.

Hiramatsu M. 2003. A role for guanidino compounds in the brain. Mol Cell Biochem 244: 57-62.

Hosokawa H, Ninomiya H, Sawamura T, Sugimoto Y, Ichikawa A, et al. 1999. Neuron-specific expression of cationic amino acid transporter 3 in the adult rat brain. Brain Res 838: 158-165.

Hosokawa H, Sawamura T, Kobayashi S, Ninomiya H, Miwa S, et al. 1997. Cloning and characterization of a brain-specific amino acid transporter. J Biol Chem 272: 8717-8722.

Husson A, Brasse-Lagnel C, Fairand A, Renouf S, Lavoinne A. 2003. Argininosuccinate synthetase from the urea cycle to the citrulline-NO cycle. Eur J Biochem 270: 1887-1899.

Ignarro LJ, Napoli C. 2004. Novel features of nitric oxide, endothelial nitric oxide synthase, and atherosclerosis. Curr Atheroscler Rep 6: 281-287.

Item CB, Stockler-Ipsiroglu S, Stromberger C, Muhl A, Alessandri MG, et al. 2001. Arginine:glycine amidinotransferase deficiency: the third inborn error of creatine metabolism in humans. Am J Hum Gen 69: 1127-1133.

Ito K, Groudine M. 1997. A new member of the cationic amino acid transporter family is preferentially expressed in adult mouse brain. J Biol Chem 272: 26780-26786.

Jackson MJ, Zielke HR, Zielke CL. 1996. Induction of astrocyte argininosuccinate synthetase and argininosuccinate lyase by dibutyryl cyclic AMP and dexamethasone. Neurochem Res 21: 1161-1165.

Jun T, Wennmalm A. 1994. NO-dependent and -independent elevation of plasma levels of insulin and glucose in rats by L-arginine. Br J Pharmacol 113: 345-348.

Kawahara K, Gotoh T, Oyadomari S, Kajizono M, Kuniyasu A, et al. 2001. Co-induction of argininosuccinate synthetase, cationic amino acid transporter-2, and nitric oxide synthase in activated murine microglial cells. Mol Brain Res 90: 165-173.

Keilhoff G, Reiser M, Stanarius A, Aoki E, Wolf G. 2000. Citrulline immunohistochemistry for demonstration of NOS activity in vivo and in vitro. Nitric Oxide 4: 343-353.

Keilhoff G, Wolf G. 2003. Citrulline immunohistochemistry may not necessarily identify nitric oxide synthase activity: the pitfall of peptidylarginine deiminase. Nitric Oxide 8: 31-38.

Kharazia VN, Petrusz P, Usunoff K, Weinberg RJ, Rustioni A. 1997. Arginine and NADPH-diaphorase in the rat ventroposterior thalamic nucleus. Brain Res 744: 151-155.

Kilpelainen P, Rybnikova E, Hietala O, Pelto-Huikko M. 2000. Expression of ODC and its regulatory protein antizyme in the adult rat brain. J Neurosci Res 62: 675-685.

Kim JK, Mastronardi FG, Wood DD, Lubman DM, Zand R, et al. 2003. Multiple sclerosis: an important role for post-translational modifications of myelin basic protein in pathogenesis. Mol Cell Proteomics 2: 453-462.

Klivenyi P, Calingasan NY, Starkov A, Stavrovskaya IG, Kristal BS, et al. 2004. Neuroprotective mechanisms of creatine occur in the absence of mitochondrial creatine kinase. Neurobiol Dis 15: 610-617.

Kobayashi K, Sinasac DS, Iijima M, Boright AP, Begum L, et al. 1999. The gene mutated in adult-onset type II citrullineamia encodes a putative mitochondrial carrier protein. Nat Genet 22: 159-163.

Kuiper MA, Teerlink T, Visser JJ, Bergmans PLM, Scheltens P, et al. 2000. L-Glutamate L-arginine and L-citrulline levels in cerebrospinal fluid of Parkinson's disease, multiple system atrophy, and Alzheimer's disease patients. J Neural Transm 107: 183-189.

Lee J, Ryu H, Ferrante RJ, Morris SM Jr, Ratan RR. 2003. Translational control of inducible nitric oxide synthase expression by arginine can explain the arginine paradox. Proc Natl Acad Sci USA 100: 4843-4848.

Leiper JM, Santa Maria J, Chubb A, Mac Allister RJ, Charles IG, et al. 1999. Identification of two human dimethylarginine dimethylaminohydrolases with distinct tissue distribution and homology with microbial arginine deiminases. Biochem J 343: 209-214.

Leuzzi V. 2002. Inborn errors of creatine metabolism and epilepsy: clinical features, diagnosis, and treatment. J Child Neurol 17 (Suppl. 3): S89-S97.

Levillain O. 2003. Renal arginine synthesis in mammals. Recent Res Dev Physiol 1: 67-83.

Li G, Regunathan S, Barrow CJ, Eshragi J, Cooper R, et al. 1994. Agmatine: an endogenous clonidine-displacing substance in the brain. Science 263: 966-969.

Li YF, Gong ZH, Cao JB, Wang HL, Luo ZP, et al. 2003. Antidepressant like effects of agmatine and its possible mechanism. Eur J Pharmacol 469: 81-88.

Liu H, Goodman JC, Robertson CS. 2002. The effects of L-arginine on cerebral hemodynamics after controlled cortical impact injury in the mouse. J Neurotrauma 19: 327-334.

Liu P, Smith PF, Appleton I, Darlington CL, Bilkey DK. 2003a. Nitric oxide synthase and arginase in the rat hippocampus and the entorhinal, perirhinal, postrhinal, and temporal cortices: regional variations and age-related changes. Hippocampus 13: 859-867.

Liu P, Smith PF, Appleton I, Darlington CL, Bilkey DK. 2003b. Regional variations and age-related changes in nitric oxide synthase and arginase in the sub-regions of the hippocampus. Neuroscience 119: 679-687.

Manner CK, Nicholson B, Mac Leod CL. 2003. CAT2 arginine transporter deficiency significantly reduces iNOS-mediated NO production in astrocytes. J Neurochem 85: 476-482.

Martinelli GP, Friedrich VL Jr, Holstein GR. 2002. L-citrulline immunostaining identifies nitric oxide production sites within neurons. Neuroscience 114: 111-122.

McDonald KK, Zharikov S, Block ER, Kilberg MS. 1997. A caveolar complex between the cationic amino acid transporter 1 and endothelial nitric-oxide synthase may explain the "arginine paradox." J Biol Chem 272: 31213-31216.

Mishima T, Hamada T, Ui-Tei K, Takahashi F, Miyata Y, et al. 2004. Expression of DDAH1 in chick and rat embryos. Brain Res Dev Brain Res 148: 223-232.

Mistry SK, Greenfeld Z, Morris SM Jr, Baylis C. 2002. The "intestinal-renal" arginine biosynthetic axis in the aging rat. Mech Aging Dev 123: 1159-1165.

Morris SM Jr. 2004. Recent advances in arginine metabolism. Curr Opin Clin Nutr Metab Care 7: 45-51.

Morrison LD, Cao XC, Kish SJ. 1998. Ornithine decarboxylase in human brain: influence of aging, regional distribution, and Alzheimer's disease. J Neurochem 71: 288-294.

Moscarello MA, Pritzker L, Mastronardi FG, Wood DD. 2002. Peptidylarginine deiminase: a candidate factor in demyelinating disease. J Neurochem 81: 335-343.

Nakagomi S, Kiryu-Seo S, Kimoto M, Emson PC, Kiyama H. 1999. Dimethylarginine dimethylaminohydrolase (DDAH) as a nerve-injury associated molecule: mRNA localization in the rat brain and its coincident up-regulation with neuronal NO synthase (nNOS) in axotomized motoneurons. Eur J Neurosci 11: 2160-2166.

Nakamura H. 1997. NADPH-diaphorase and cytosolic urea cycle enzymes in the rat spinal cord. J Comp Neurol 385: 616-626.

Nakamura H, Itoh K, Kawabuchi M. 1999. NADPH-diaphorase and cytosolic urea cycle enzymes in the rat accessory olfactory bulb. J Chem Neuroanat 17: 109-117.

Nakamura H, Saheki T, Ichiki H, Nakata K, Nakagawa S. 1991a: Immunocytochemical localization of argininosuccinate synthetase in the rat brain. J Comp Neurol 312: 652-679.

Nakamura H, Saheki T, Nakagawa S. 1990. Differential cellular localization of enzymes of L-arginine metabolism in the rat brain. Brain Res 530: 108-112.

Nakamura H, Yada T, Saheki T, Noda T, Nakagawa S. 1991b. L-Argininosuccinate modulates L-glutamate response in acutely isolated cerebellar neurons of immature rats. Brain Res 539: 312-315.

Nicholas AP, King JL, Sambandam T, Echols JD, Gupta KB, et al. 2003. Immunohistochemical localization of citrullinated proteins in adult rat brain. J Comp Neurol 459: 251-266.

Nicholas AP, Whitaker JN. 2002. Preparation of a monoclonal antibody to citrullinated epitopes: its characterization and some applications to immunohistochemistry in human brain. Glia 37: 328-336.

Ohta K, Shimazu K, Komatsumoto S, Araki N, Shibata M, et al. 1994. Modification of striatal arginine and citrulline metabolism by nitric oxide synthase inhibitors. Neuroreport 5: 766-768.

Olmos G, DeGregorio-Rocasolano N, Paz-Regalado M, Gasull T, Assumpcio-Boronat M, et al. 1999. Protection by imidazol(ine) drugs and agmatine of glutamate-induced neurotoxicity in cultured cerebellar granule cells through blockade of NMDA receptor. Br J Pharmacol 127: 1317-1326.

Palmieri F. 2004. The mitochondrial transporter family (SLC25): physiological and pathological implications. Pflügers Arch 447: 689-709.

Pasqualotto BA, Hope BT, Vincent SR. 1991. Citrulline in the rat brain: immunohistochemistry and coexistence with NADPH-diaphorase. Neurosci Lett 128: 155-160.

Patejunas G, Bradley A, Beaudet AL, O'Brien WE. 1994. Generation of a mouse model for citrullinemia by targeted disruption of the argininosuccinate synthetase gene. Somat Cell Mol Genet 20: 55-60.

Perry TL. 1982. Cerebral amino acid pools. Handbook of Neurochemistry, Vol. I, 2nd edn. Lajtha A, editor. New York: Plenum Press; pp. 151-180.

Pow DV. 1994. Immunocytochemical evidence for a glial localization of arginine, and a neuronal localization of citrulline in the rat neurohypophysis: implications for nitrergic transmission. Neurosci Lett 181: 141-144.

Prast H, Philippu A. 2001. Nitric oxide as modulator of neuronal function. Prog Neurobiol 64: 51-68.

Qureshi IA, Rao KV. 1997. Sparse-fur (spf) mouse as a model of hyperammonemia: alterations in the neurotransmitter systems. Adv Exp Med Biol 420: 143-158.

Rao VLR, Audet RM, Butterworth RF. 1997. Portocaval shunting and hyperammonemia stimulate the uptake of L-[^{3}H]arginine but not of L-[^{3}H]nitroarginine into rat brain synaptosomes. J Neurochem 68: 337-343.

Regunathan S, Piletz JE. 2003. Regulation of inducible nitric oxide synthase and agmatine synthesis in macrophages and astrocytes. Ann NY Acad Sci 1009: 20-29.

Regunathan S, Reis DJ. 2000. Characterization of arginine decarboxylase in rat brain and liver: distinction from ornithine decarboxylase. J Neurochem 74: 2201-2208.

Sadasivudu B, Rao TI. 1976. Studies on functional and metabolic role of urea cycle intermediates in brain. J Neurochem 27: 785-794.

Saheki T, Kobayashi K. 2002. Mitochondrial aspartate glutamate carrier (citrin) deficiency: the cause of adult-onset type II citrullinemia (CTLN2) and idiopathic neonatal hepatitis (NICCD). J Hum Genet 47: 333-341.

Salter M, Duffy C, Garthwaite J, Strijbos PJ. 1996. Ex vivo measurement of brain tissue nitrite and nitrate accurately reflects nitric oxide synthase activity in vivo. J Neurochem 66: 1683-1690.

Satriano J. 2003. Agmatine: at the crossroads of the arginine pathways. Ann NY Acad Sci 1009: 34-43.

Schmidlin A, Fischer S, Wiesinger H. 2000. Transport of L-citrulline in neural cultures. Dev Neurosci 22: 393-398.

Schmidlin A, Kalbacher H, Wiesinger H. 1997. Presence of argininosuccinate synthetase in glial cells as revealed by peptide-specific antisera. Biol Chem 378: 47-50.

Schmidlin A, Wiesinger H. 1994. Transport of L-arginine in cultured glial cells. Glia 11: 262-268.

Schmidlin A, Wiesinger H. 1995. Stimulation of arginine transport and nitric oxide production by lipopolysaccharide is mediated by different signaling pathways in astrocytes. J Neurochem 65: 590-594.

Schmidlin A, Wiesinger H. 1998. Argininosuccinate synthetase: localization in astrocytes and role in the production of glial nitric oxide. Glia 24: 428-436.

Schulze A. 2003. Creatine deficiency syndromes. Mol Cell Biochem 244: 143-150.

Segal MB, Preston JE, Collis CS, Zlocovic BV. 1990. Kinetics and sodium independence of amino acid uptake by blood side of perfused sheep choroid plexus. Am J Physiol 258: F1288-F1294.

Seidel B, Stanarius A, Wolf G. 1997. Differential expression of neuronal and endothelial nitric oxide synthase in blood vessels of the rat brain. Neurosci Lett 239: 109-112.

Seiler N, Daune-Anglard G. 1993. Endogenous ornithine in search for CNS functions and therapeutic applications. Metab Brain Dis 8: 151-179.

Shih VE. 1978. Urea cycle disorders and other congenital hyperammonemic syndromes. Stanbury JB, Wyngaarden JB, Frederickson DS, editors. The Metabolic Basis of Inherited Disease. New York: McGraw-Hill; pp. 362-386.

Sinasac DS, Moriyama M, Jalil MA, Begum L, Li MX, et al. 2004. Slc25a13-knockout mice harbor metabolic deficits but fail to display hallmarks of adult-onset type II citrullinemia. Mol Cell Biol 24: 527-536.

Soulet D, Rivest S. 2003. Polyamines play a critical role in the control of the innate immune response in the mouse central nervous system. J Cell Biol 162: 257-268.

Sperandeo MP, Borsani G, Incerti B, Zollo M, Rossi E, et al. 1998. The gene encoding a cationic amino acid transporter (SCL7A4) maps to a region deleted in velocardiofacial syndrome. Genomics 49: 230-236.

Stevens BR, Kakuda DK, Yu K, Waters M, Vo CB, et al. 1996. Induced nitric oxide synthesis is dependent on induced alternatively spliced CAT-2 encoding L-arginine transport in brain astrocytes. J Biol Chem 271: 24017-24022.

Stevens BR, Vo CB. 1998. Membrane transport of neuronal nitric oxide synthase substrate L-arginine is constitutively expressed with CAT1 and 4F2hc, but not CAT2 or rBA T. J Neurochem 71: 564-570.

Stoll J, Wadhwani KC, Smith QR. 1993. Identification of the cationic amino acid transporter (system y^+) of the rat blood-brain barrier. J Neurochem 60: 1956-1959.

Stuhlmiller DF, Boje KM. 1995. Characterization of L-arginine and aminoguanidine uptake into isolated rat choroid plexus: differences in uptake mechanisms and inhibition by nitric oxide synthase inhibitors. J Neurochem 65: 68-74.

Temiz C, Tun K, Ugur HC, Dempsey RJ, Egemen N. 2003. L-Arginine in focal cerebral ischemia. Neurol Res 25: 465-470.

Tran CT, Leiper JM, Vallance P. 2003. The DDAH/ADMA/NOS pathway. Atherosclerosis 4 (Suppl.): 33-44.

Tsikas D, Böger RH, Sandmann J, Bode-Böger SM, Frölich JC. 2000. Endogenous nitric oxide synthase inhibitors are responsible for the L-arginine paradox. FEBS Lett 478: 1-3.

Vega-Agapito V, Almeida A, Hatzoglou M, Bolanos JP. 2002. Peroxinitrite stimulates L-arginine transport system y^+ in glial cells. A potential mechanism for replenishing neuronal L-arginine. J Biol Chem 277: 29753-29759.

Vossenaar ER, Zendman AJ, van Venrooij WJ, Pruijn GJ. 2003. PAD, a growing family of citrullinating enzymes: genes, features and involvement in disease. Bioessays 25: 1106-1118.

Wiesinger H. 2001. Arginine metabolism and the synthesis of nitric oxide in the nervous system. Prog Neurobiol 64: 365-391.

Williams K. 1997. Interaction of polyamines with ion channels. Biochem J 325: 289-297.

Withers PC. 1998. Urea: diverse functions of a "waste" product. Clin Exp Pharmacol Physiol 25: 722-727.

Witte MB, Barbul A. 2003. Arginine physiology and its implication for wound healing. Wound Repair Regen 11: 419-423.

Wolf G. 1997. Nitric oxide and nitric oxide synthase: biology, pathology, localization. Histol Histopathol 12: 251-261.

Wood JH. 1982. Physiological neurochemistry of cerebrospinal fluid. Handbook of Neurochemistry, Vol. I, 2nd edn. Lajtha A, editor. New York: Plenum Press; pp. 415-487.

Wu G, Davis PK, Flynn NE, Knabe DA, Davidson JT. 1997. Endogenous synthesis of arginine plays an important role in maintaining arginine homeostasis in postweaning growing pigs. J Nutr 127: 2342-2349.

Wu G, Morris SM Jr. 1998. Arginine metabolism: nitric oxide and beyond. Biochem J 336: 1-17.

Wu G, Morris SM Jr. 2004. Arginine metabolism in mammals. Metabolic and Therapeutic. Aspects of Amino Acids in Clinical Nutrition, 2nd edn. Cynober LA, editor. Boca Raton: CRC Press; pp. 153-167.

Xu L, Hilliard B, Carmody RJ, Tsabary G, Shin H, et al. 2003. Arginase and autoimmune inflammation in the central nervous system. Immunology 110: 141-148.

Yamada K, Nishiwaki K, Hattori K, Senzaki K, Nagata M, et al. 1997. No changes in cerebrospinal fluid levels of nitrite, nitrate and cyclic GMP with aging. J Neural Transm 104: 825-831.

Yu H, Iyer RK, Kern RM, Rodriguez WI, Grody WW, et al. 2001. Expression of arginase isozymes in mouse brain. J Neurosci Res 66: 406-422.

Yu H, Iyer RK, Yoo PK, Kern RM, Grody WW, et al. 2002. Arginase expression in mouse embryonic development. Mech Dev 115: 151-155.

Yu H, Yoo PK, Aguirre CC, Tsoa RW, Kern RM, et al. 2003. Widespread expression of arginase I in mouse tissues. Biochemical and physiological implications. J Histochem Cytochem 51: 1151-1160.

Yudkoff M, Nissim I, Nissim I, Stern J, Pleasure D. 1989. Effects of palmitate on astrocyte amino acid contents. Neurochem Res 14: 367-370.

Yudkoff M, Nissim I, Pleasure D. 1987. [^{15}N]Aspartate metabolism in cultured astrocytes. Biochem J 241: 193-201.

Zhang WY, Takiguchi M, Koshiyama Y, Gotoh T, Nagasaki A, et al. 1999. Expression of citrulline-nitric oxide cycle in lipopolysaccharide and cytokine-stimulated rat astroglioma C6 cells. Brain Res 849: 78-84.

Zhao X, Ross ME, Iadecola C. 2003. L-Arginine increases ischemic injury in wild-type mice but not in iNOS-deficient mice. Brain Res 966: 308-311.

Zhu MY, Iyo A, Piletz JE, Regunathan S. 2004. Expression of human arginine decarboxylase, the biosynthetic enzyme for agmatine. Biochim Biophys Acta 1670: 156-164.

6 Branched Chain Amino Acids (BCAAs) in Brain

S. M. Hutson · A. J. Sweatt · K. F. LaNoue

1	Introduction	118
2	Branched Chain Amino Acid Metabolism	118
3	Branched Chain Amino Acids Are Nitrogen Donors in Peripheral Tissues	121
4	Branched Chain Amino Acids as Nitrogen Donors in the Central Nervous System	121
4.1	Intercellular Nitrogen Transfer	122
5	Role of BCAA in de novo Glutamate Synthesis and the Branched Chain Aminotransferase (BCAT) Cycle Hypothesis	122
6	Localization of BCAA Catabolic Enzymes in Rodent Brain	125
6.1	The Cytosolic BCATc Isozyme	125
6.2	The Mitochondrial Enzymes BCATm and Branched Chain α-Keto Acid Dehydrogenase (BCKD)	125
6.3	The BCAT Cycle and GABA Metabolism	126
7	Conclusions, Implications for Disorders of BCAA Metabolism, and Future Directions	127

© Springer-Verlag Berlin Heidelberg 2007

Abstract: The branched chain amino acids (BCAAs) play a role in glutamate neurotransmitter synthesis by providing a ferry system to move nitrogen between astrocytes and neurons for the synthesis of glutamate. This hypothetical nitrogen cycle, the branched chain aminotransferase (BCAT) cycle, is tied to the glutamate/glutamine cycle and is an arm of the glutamate/pyruvate cycle. Immunolocalization of the BCAT isozymes in rodent brain shows that the cytosolic BCATc is localized in selected glutamatergic and GABAergic neurons whereas the mitochondrial isozyme, BCATm, is found in astroglia. These results provide support for the cycle hypothesis and suggest BCAA also play a role in γ-aminobutyric acid (GABA) metabolism. When the BCAA catabolic disposal system is genetically impaired in humans, serum BCAA and BCKA levels rise and severe neurological dysfunction occurs. In the hereditary metabolic disease that results from mutations in the branched chain α-keto acid dehydrogenase complex (BCKD), maple syrup urine disease, the most obvious symptoms are neurological. Useful animal models for the human condition are not yet available.

List of Abbreviations: BCAA, branched chain amino acid; BCAT, branched chain aminotransferase; BCATc, cytosolic branched chain aminotransferase; BCATm, mitochondrial branched chain aminotransferase; BCKA, branched chain α-keto acid; BCKD, branched chain α-keto acid dehydrogenase; E1, branched chain α-keto acid decarboxylase; E2, dihydrolipoly transacylase; E3, dihydrolipoyl dehydrogenase; GABA, γ-aminobutyric acid; GABA-T, GABA, aminotransferase; GAD, glutamate decarboxylase; αKG, α-ketoglutarate; KIC, α-ketoisocaproate; KIV, α-ketoisovalerate; KMV, α-keto-β-methylvalerate; MSUD, Maple Syrup Urine Disease; PC, pyruvate carboxylase; TCA, tricarboxylic acid

1 Introduction

The branched chain amino acids (BCAAs), leucine, isoleucine, and valine, are classified as indispensable amino acids because they cannot be synthesized de novo in mammals. Therefore, they must be obtained from dietary sources and adequate intake of BCAAs is required for normal growth and development. In addition to their role as protein building blocks, the BCAAs are involved in interorgan nitrogen transfer between peripheral tissues. In the central nervous system (CNS), they are thought to provide nitrogen for neurotransmitter glutamate synthesis. Inborn errors of BCAA metabolism that increase BCAAs and/or their metabolites are toxic to the brain. Thus, regulation of BCAA catabolism is required to limit their breakdown when dietary intake is restricted and to efficiently clear BCAAs and their metabolites when dietary intake exceeds the body's needs.

2 Branched Chain Amino Acid Metabolism

A schematic diagram of the BCAA catabolic pathways is shown in ❷ *Figure 6-1*. The first step in BCAA catabolism is reversible transamination catalyzed by the branched chain aminotransferase (BCAT) isozymes. In this step, the α-amino group of an individual BCAA is transferred to α-ketoglutarate to form glutamate and the respective branched chain α-keto acids (BCKA: α-ketoisocaproate [KIC], α-keto-β-methylvalerate [KMV] or α-ketoisovalerate [KIV]). Glutamate, in addition to its role as the major excitatory neurotransmitter in the CNS, is also a key metabolic substrate for a variety of nitrogen transfer reactions and it is the entry point for BCAA nitrogen into other metabolic pools. Given the large size of the brain glutamate pool, in short term studies BCAA nitrogen tends to accumulate in glutamate (Hutson et al., 2005). Nevertheless, BCAA nitrogen can flow from glutamate via other aminotransferase reactions to amino acids such as alanine and aspartate. In glia where glutamine synthetase is located, BCAA nitrogen flows from glutamate to the α-amino group of glutamine (Minchin and Beart, 1975; Martinez-Hernandez et al., 1977; Norenberg, 1979). In GABAergic neurons, BCAA nitrogen enters into the γ-aminobutyric acid pool (GABA) by decarboxylation of glutamate catalyzed by the vitamin B_6-dependent glutamate decarboxylase (GAD) isozymes. Free ammonia can be formed by oxidative deamination of glutamate catalyzed by mitochondrial glutamate dehydrogenase.

Figure 6-1

BCAA catabolic pathways. The first step is transamination of the BCAAs (Leu, Ile, Val) catalyzed by the branched chain aminotransferase (BCAT) isozymes (mitochondrial BCATm and cytosolic BCATc) to produce the branched chain α-keto acids (BCKAs), α-ketoisocaproate (KIC), α-keto-β-methylvalerate (KMV), and α-ketoisovalerate (KIV). α-Ketoglutarate is the α-keto acid acceptor of the BCAA α-amino group. Glutamate (Glu) is the product. Glutamate, via the action of other aminotransferases, can donate its nitrogen group to other amino acids, such as Asp and Ala, or reform α-ketoglutarate via oxidative deamination catalyzed by mitochondrial glutamate dehydrogenase forming free NH_3. Glutamate is decarboxylated by glutamate decarboxylase to form γ-aminobutyric acid (GABA) in GABAergic neurons. In astroglia, glutamate is also converted to glutamine by glutamine synthetase as part of the glutamate/glutamine cycle. The second catabolic step is oxidative decarboxylation, catalyzed by the branched chain α-keto acid dehydrogenase (BCKD) enzyme complex

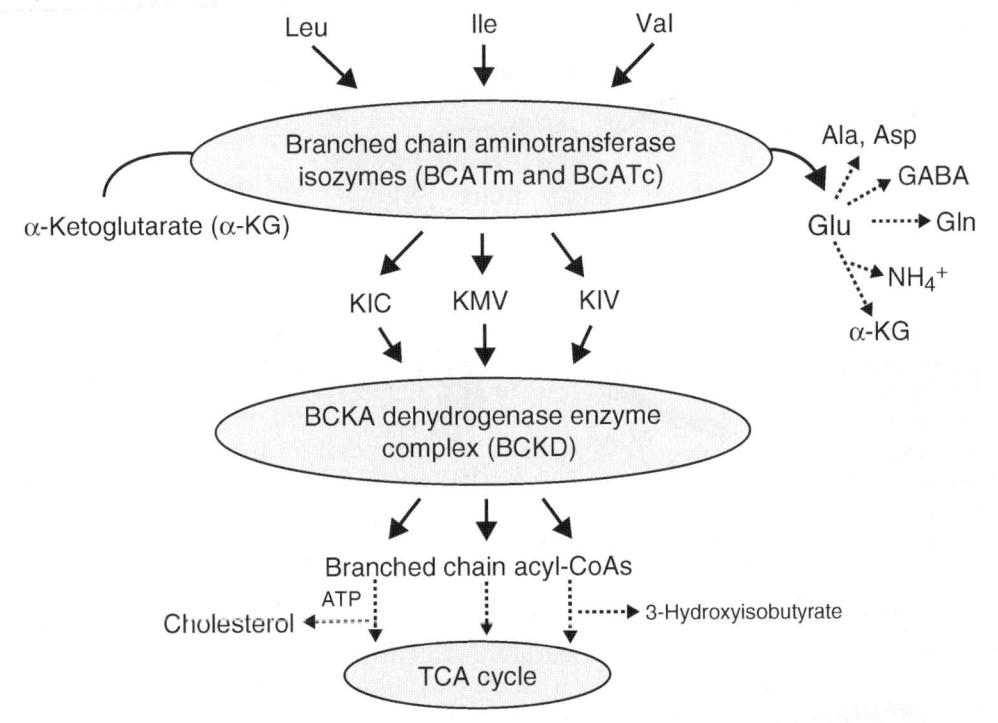

In mammals, there are two BCATs—a mitochondrial (BCATm) and a cytosolic (BCATc) isozyme that are encoded by two separate genes (Ichihara, 1985; Hutson et al., 1988; Bledsoe et al., 1997). Not only are the BCAT isozymes expressed in different cell compartments but they also show different patterns of expression in body tissues (▶ Figure 6-2). With the exception of rodent liver, the mitochondrial isozyme is expressed ubiquitously, and outside the CNS it is often localized in secretory epithelial cells (Hutson et al., 1992; Sweatt et al., 2004a). On the other hand, the cytosolic isozyme is found almost exclusively in nervous tissue (Ichihara, 1985; Hutson et al., 1988; Sweatt et al., 2004a). BCATc is the predominant BCAT activity in rat brain accounting for ∼60–70% of total brain BCAT activity (Hutson, 1988; Hall et al., 1993). In the rat, BCATc is also expressed outside the CNS, localized in peripheral nerves, but neither BCATm nor BCKD are expressed in these nerves (Sweatt et al., 2004a). Coexpression of BCATm and BCATc within the same cell type has not been observed in the CNS (Hutson et al., 2001; Sweatt et al., 2004a; Cole et al., 2005). Therefore, in the brain and spinal cord, the BCAT isozyme localization is entirely cell-specific.

The second step in BCAA catabolism is irreversible oxidative decarboxylation of the BCKAs, producing branched chain acyl-CoA derivatives. This reaction is catalyzed by the mitochondrial branched chain α-keto acid dehydrogenase (BCKD) enzyme complex (Harris et al., 1986) (▶ Figure 6-1). Because this step is

Figure 6-2

BCAA catabolic enzymes are found throughout the body. The localization of BCAA catabolic enzymes facilitates nitrogen cycling and net nitrogen transfer between and within different tissues and cell types. In the rat, BCATm is found in all tissues except liver, whereas in primates it is present in Kupffer cells of the liver but not in hepatocytes where the BCKD enzyme complex is localized (Suryawan et al., 1998; Sweatt et al., 2004a). BCATc is found primarily in the central nervous system in neurons but is also expressed in peripheral nerves (not shown) (Sweatt et al., 2004a)

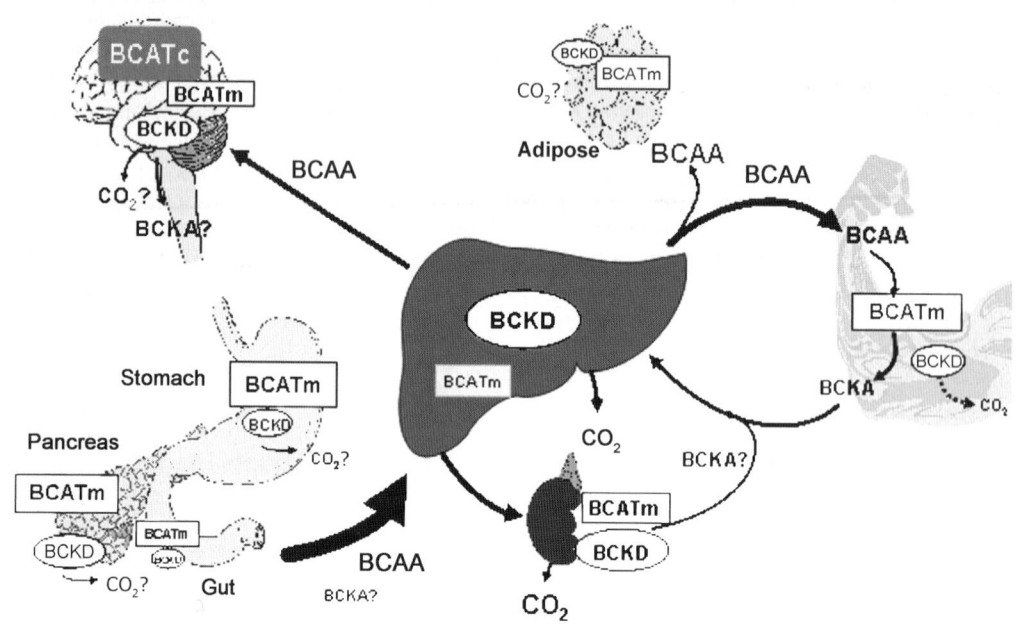

irreversible, it commits the BCAA carbon skeleton to the degradative pathway and results in irreversible transfer (net transfer) of the α-amino nitrogen from BCAA to glutamate and other dispensable amino acids. The mammalian BCKD complex contains multiple copies of three enzymes (Harris et al., 1990)—a branched chain α-keto acid decarboxylase (E1); a dihydrolipoyl transacylase (E2) and a dihydrolipoyl dehydrogenase (E3). In this respect, the BCKD complex resembles the mammalian α-ketoglutarate dehydrogenase and pyruvate dehydrogenase enzyme complexes. The activity of the BCKD complex within a tissue is regulated by its phosphorylation state. Phosphorylation by BCKD kinase inactivates the complex, and dephosphorylation by a specific phosphatase activates the complex (Harris et al., 1990, 2004). Mutations in one or more proteins in this complex result in the inborn error of BCAA metabolism known as maple syrup urine disease (MSUD) (Chuang and Shih, 2001).

The next step in the oxidation of BCAAs is dehydrogenation of the branched chain acyl-CoA BCKD reaction products catalyzed by two different dehydrogenases (Chuang and Shih, 2001). After this step, the catabolic pathways of the three BCAAs diverge. The remaining enzymatic reactions resemble β-oxidation of fatty acids. Leucine is unique in that its catabolic pathway contains an ATP- and biotin-dependent carboxylation step (β-methylcrotonyl-CoA carboxylase) and an intermediate that is a precursor of cholesterol. Leucine is ketogenic, forming acetyl-CoA and acetoacetate whereas valine is glucogenic entering the tricarboxylic acid cycle (TCA cycle) as succinyl-CoA. Both isoleucine and valine are metabolized to succinate via methylmalonyl-CoA, catalyzed by propionyl-CoA carboxylase, and subsequent rearrangement to succinyl-CoA, catalyzed by the vitamin B_{12}-dependent enzyme methylmalonyl-CoA mutase. The other product of isoleucine metabolism is acetoacetate. Neurologic consequences are observed in most inborn errors of the enzymes in BCAA metabolism (Chuang and Shih, 2001).

3 Branched Chain Amino Acids Are Nitrogen Donors in Peripheral Tissues

Early research on the metabolism of BCAAs in body organs and tissues led to the hypothesis that BCAAs play an important role in body nitrogen metabolism. They are nitrogen donors for the major nitrogen carriers, glutamine and alanine, that carry nitrogen from amino acid catabolism in peripheral tissues such as skeletal muscle to the liver and intestine (Odessey et al., 1974; Garber et al., 1976; Harper, 1989). In contrast to the degradative pathways of other indispensable amino acids, the enzymes for degradation of the BCAAs are expressed in tissues throughout the body rather than being confined to the liver (❯ *Figure 6-2*) (Ichihara, 1985; Hutson, 1988; Hutson et al., 1988; Sweatt et al., 2004a). The initial step, transamination, is reversible. BCAT activity is low in liver and high in other organs so that the initial step in BCAA degradation is located in tissues other than liver (Harper, 1989) (❯ *Figure 6-2*). BCAAs represent about 50% of skeletal muscle amino acid uptake, and most of the other plasma amino acids do not undergo catabolism in muscle. The BCAAs are used for maintenance of muscle nitrogen pools of glutamate and glutamine and provide part of the nitrogen for glutamine release from muscle and subsequent uptake by the intestines and kidney. In rat and human skeletal muscle, BCAT activity is high compared to BCKD activity (ratio of BCAT/actual BCKD activities is ≥ 90 and for total dephosphorylated BCKD activity ≥ 25) (Suryawan et al., 1998). The high ratio of BCAT/BCKD activities favors release of BCKAs rather than their oxidation (Hutson et al., 1978, 1980; Harper et al., 1984), and there is efficient transfer of BCAA nitrogen to glutamate and glutamine in skeletal muscle preparations (Hutson and Zapalowski, 1981). In addition, kinetic data indicate that in situ BCATs operate at substrate concentrations below their K_m values so tissues respond rapidly to changes in BCAA and BCKA concentrations. In rat liver, the absence of the BCATm isozyme and high activity of BCKD favor liver oxidation of circulating BCKAs. This complex metabolic scheme also permits leucine to act as an effective nutrient signal both in liver and in peripheral tissues. Leucine stimulates, via an as yet incompletely characterized signaling cascade(s), protein synthesis in skeletal muscle, adipose tissue, and liver (Anthony et al., 2000; Lynch, 2001; Lynch et al., 2002). Whether or not leucine acts specifically as a nutrient signal in the CNS is not known.

4 Branched Chain Amino Acids as Nitrogen Donors in the Central Nervous System

There are parallels between the function of BCAAs in skeletal muscle and their role in the CNS. In both organs, BCAAs are nitrogen donors to glutamate (glutamine). As described above, in the CNS the glutamate that is formed from BCAA transamination is an excitatory neurotransmitter, substrate for glutamine synthesis in the glutamate/glutamine cycle, and substrate for synthesis of the inhibitory neurotransmitter GABA. There is substantial uptake of BCAAs into the CNS in rats (Oldendorf, 1971), and in humans the cerebral arteriovenous difference for leucine exceeds that of other amino acids (Oldendorf, 1971; Grill et al., 1992). Uptake of BCAAs at the blood–brain barrier appears to be mediated by the facilitative large neutral amino acid transporter 1 system (Killian and Chikhale, 2001). Cerebral glutamine release has also been observed during BCAA uptake (Grill et al., 1992). Although release of BCKAs from peripheral tissues occurs in rodents and humans (Hutson et al., 1978; Hutson et al., 1980; Matthews et al., 1980), BCKA release from brain may be quite slow. Michaelis constants for uptake of BCKAs into rat brain (Conn and Steele, 1982) suggest there is limited transport of BCKAs across the blood–brain barrier at physiological concentrations, but it may occur under some circumstances (Matsuo et al., 1993). The low concentrations of BCKAs observed in brain (Hutson and Harper, 1981) are consistent with efficient BCKA oxidation or perhaps some release of BCKAs to the blood stream. Rat liver, which is a major site of BCKA oxidation, has BCKA concentrations that are as low as those measured in brain (Hutson and Harper, 1981).

Branched chain amino acid transport into the brain provides the major source of brain nitrogen. The exchange of α-amino nitrogen between BCAAs and other brain amino acids is very active and catalyzed by BCATc and BCATm. In rodent and human brain, maximal BCAT activity exceeds total BCKD activity (BCAT/total BCKD activity ≥ 50) (Brosnan et al., 1985; Suryawan et al., 1998). Isotope studies demonstrated the rapid transfer of nitrogen from BCAA to glutamate (glutamine) both in vitro and in vivo in animal models (Yudkoff et al., 1983; Yudkoff et al., 1993; Yudkoff et al., 1994; Yudkoff et al., 1996a; Kanamori et al., 1998; Yudkoff et al., 2003; Sakai et al., 2004). Transfer of BCAA nitrogen into glutamate and glutamine occurs in primary rat astroglia cultures and synaptosome preparations (Yudkoff et al., 1983; Yudkoff et al., 1994; Yudkoff et al., 1996a, b; Hutson et al., 1998; McKenna et al., 1998a). Incorporation of ^{15}N from leucine into glutamate, glutamine, and GABA takes place in mice in vivo (Yudkoff et al., 2003). Using NMR, Kanamori et al. (1998) estimated that leucine provides ~25% of glutamate nitrogen in rat brain, and during 9 h of continuous intragastric feeding in rats, Sakai et al. (2004) estimated that at least 50% of brain glutamate nitrogen was derived from leucine. The authors also concluded that there was significant reamination of the leucine α-keto acid, KIC (Sakai et al., 2004). Although the relative rates of brain KIC reamination (nitrogen cycling) versus oxidation or BCKA release (net nitrogen transfer) in brain in situ remain to be determined, human brain aminotransferase and actual BCKD capacities are second only to skeletal muscle. These results suggest that the brain may be a quantitatively more important site of BCAA metabolism in humans where brain represents a higher percent of total body weight than in rats (Suryawan et al., 1998).

4.1 Intercellular Nitrogen Transfer

The first evidence for metabolic compartmentalization, intercellular transfer of BCAAs and their metabolites, and transfer of BCAA nitrogen to glutamate and glutamine within the brain came from studies of BCAA metabolism in primary rat brain cell cultures and synaptosome preparations. Early studies showed that fetal and adult rat brain preparations can transaminate and oxidize BCAAs and that the rate of BCAA transamination is greater than the rate of incorporation into new protein (Chaplin et al., 1976; Shank and Aprison, 1981; Murthy and Hertz, 1987). Astrocyte cultures exhibit a high capacity for leucine uptake (Su et al., 1995), accumulate BCAAs against a concentration gradient (Su et al., 1995), readily transaminate BCAAs (Hutson et al., 1998), and transfer their nitrogen to glutamate and glutamine (Yudkoff et al., 1994). BCAAs also stimulate glutamine efflux (Hutson et al., 1998). On the other hand, cultured astrocytes oxidize BCAAs slowly (Bixel and Hamprecht, 1995; Hutson et al., 1998), releasing KIC and other leucine metabolites (Bixel et al., 1995). Although primary neuronal cultures also readily transaminate BCAAs, reamination of KIC is favored in neuronal compared to astrocyte cultures (Hutson et al., 1998). Yudkoff et al. (1996b) also concluded that in nerve endings the BCAT reaction proceeds largely in the direction of reamination.

A model that took into account the most salient features of brain BCAA metabolism was proposed by Yudkoff (Yudkoff et al., 1996a; Yudkoff, 1997) and was originally called the "leucine/glutamine cycle." In this model, leucine enters the brain across the capillary/glial interface, and then transaminates with α-ketoglutarate to form glutamate, which is then converted in the glia to glutamine. Glutamine and KIC, the α-keto acid of leucine, are released from the glia. Glutamine and KIC are then taken up by neurons. KIC is reaminated in the neuron, consuming glutamate in the process. Leucine is released into the extracellular fluid where it can again be utilized by astrocytes, completing the cycle. As originally envisioned, this cycle provides a mechanism whereby BCAA nitrogen can be used to support brain nitrogen homeostasis. In the original thesis, this cycle buffers glutamate levels in neurons, thereby preventing a buildup of glutamate in neurons. Reamination of BCKAs serves to conserve BCAA for protein synthesis. Free ammonia is produced in the glutaminase step in neurons. In this cycle, glutamate dehydrogenase uses the ammonia from the glutaminase step for resynthesis of glutamate.

5 Role of BCAA in de novo Glutamate Synthesis and the Branched Chain Aminotransferase (BCAT) Cycle Hypothesis

Hutson, LaNoue and coworkers (Hutson et al., 1998; Bixel et al., 2001; Hutson et al., 2001; LaNoue et al., 2001; Lieth et al., 2001; Xu et al., 2004) have taken the concept of the leucine/glutamine cycle a step further and proposed a BCAT cycle (❯ Figure 6-3). Because aminotransferases catalyze reversible reactions and

Figure 6-3

The BCAT cycle and de novo Glu synthesis. BCAAs donate nitrogen via BCATm in glia for de novo glutamate synthesis with reamination of BCKA in neurons by BCATc and shuttling of nitrogen to glia via BCAAs. Ammonia (NH_4^+) from Gln conversion to Glu in neurons is used to reaminate α-ketoglutarate to reform glutamate used in BCATc step. Other key enzymes of the BCAT cycle are in astrocytes (1) pyruvate carboxylase and (2) glutamine synthetase, and in neurons (3) glutaminase and (4) glutamate dehydrogenase

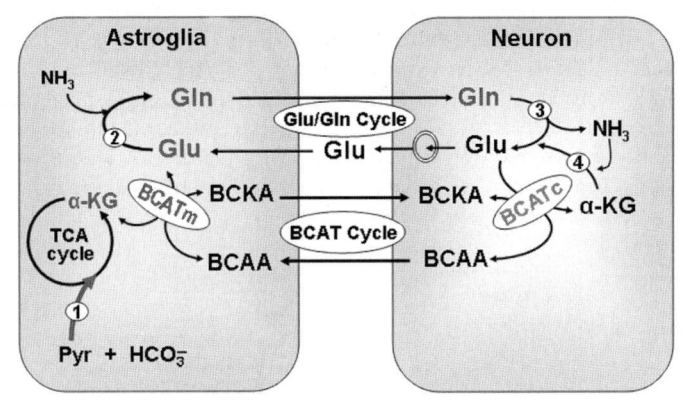

operate under near equilibrium conditions in most cells, buffering can be provided by a single isozyme in a single compartment such as the neuron. In the BCAT cycle hypothesis, emphasis is on the role of BCAAs in de novo synthesis of glutamate as opposed to a purely buffering role. De novo synthesis of glutamate is needed in the brain because some glutamate used in neurotransmission is lost in catabolic reactions. In addition, the neurotransmitter glutamate/glutamine cycle does not operate stoichiometrically. Results from studies of cultured neonatal cortical astrocytes (Sonnewald et al., 1993; Westergaard et al., 1995) have shown that a significant fraction of the glutamine returned to the neuron is not derived from the glutamate that was originally taken up by astrocytes. ◗ *Figure 6-4* highlights the pathways for glutamate oxidation to pyruvate in the glia and illustrates why some glutamine derived from glia must come from de novo synthesis of glutamate. In the degradation pathway, glutamate is first converted to α-ketoglutarate (α-KG). This step is catalyzed by transamination and/or under special circumstances by glutamate dehydrogenase (Rao and Murthy, 1993; McKenna et al., 1996a, b). As long as cycle intermediates are not removed and used to synthesize other metabolites, the citric acid cycle intermediates, including α-ketoglutarate, are not consumed. Nevertheless, studies show that a significant fraction of the glutamate taken up by astrocytes disappears because the glutamate is metabolized further to pyruvate and lactate and subsequently released by astrocytes (Sonnewald et al., 1993; Hassel and Sonnewald, 1995; Sonnewald et al., 1996) to be oxidized by neurons. This lactate is an effective substrate for maintaining energy metabolism in synaptic terminals (McKenna et al., 1998b).

In fact, Magistretti and Pellerin have proposed that lactate released by astrocytes is the principal source of energy for neurons (Pellerin and Magistretti, 1994; Magistretti et al., 1999). Lactate is the normal product of glycolysis but is also produced from the decarboxylation of the citric acid cycle intermediates, oxaloacetate (OAA), and malate, by reactions diagrammed in ◗ *Figure 6-4* (Yu et al., 1983; Shank et al., 1985). Pyruvate is reduced to lactate by lactate dehydrogenase. Alternatively, pyruvate can be transformed to oxaloacetate by pyruvate carboxylase (PC). We (Gamberino et al., 1997) and others (Schmoll and Hamprecht, 1994; Hassel and Sonnewald, 1995; McKenna et al., 1995) have identified these reactions in cultured astrocytes, and the cycle involving conversion of glutamate-derived pyruvate back to oxaloacetate has been named the glutamate/pyruvate cycle. These studies show that the cycle is active in excised retinas (Lieth et al., 2001; Xu et al., 2004). Thus, the glutamate/pyruvate cycle provides a mechanism to convert excess glutamate, which is toxic, to useful nutrients such as pyruvate and lactate.

The active conversion of glutamate to pyruvate in astrocytes would gradually lower brain glutamate concentrations unless a pathway were present in astrocytes to replenish the glutamate carbon by a process

□ Figure 6-4

Glutamate/pyruvate cycle. Glu taken up into astrocytes forms α-ketoglutarate (α-KG) in a transamination (TA) reaction. α-Ketoglutarate (α-KG) is converted to malate (Mal) and oxaloacetate (OAA) in the TCA cycle. Malic enzyme (ME) or the combined action of pyruvate kinase (PK) and phosphoenolpyruvate carboxykinase (PEPCK) produce pyruvate (Pyr). Pyruvate forms lactate, catalyzed by lactate dehydrogenase, or pyruvate carboxylase (PC) resynthesizes OAA from pyruvate

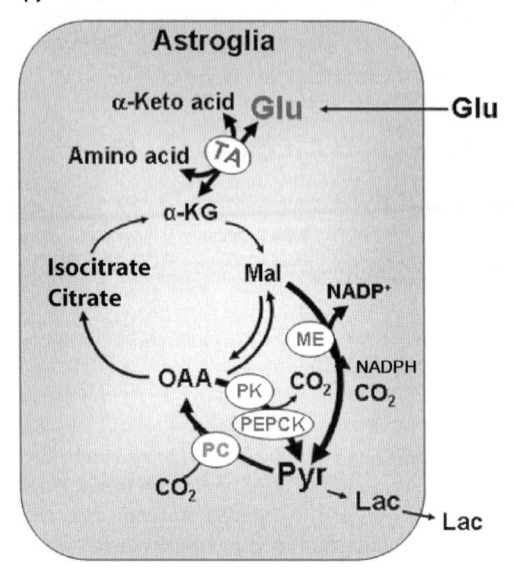

called anaplerosis. Anaplerosis is the process by which citric acid cycle intermediates are replenished by carboxylation to form oxaloacetate. The major anaplerotic enzyme in brain is PC. This biotin-dependent enzyme catalyzes an ATP-dependent reaction and uses HCO_3^- as the source of CO_2 for carboxylation of pyruvate (❯ Figure 6-4). The operation of PC produces oxaloacetate, which is rate limiting for de novo malate and α-ketoglutarate synthesis (Gamberino et al., 1997). α-Ketoglutarate provides the carbon skeleton for glutamate. There are high levels of PC in brain, and it is expressed exclusively in glia (Yu et al., 1983; Shank et al., 1985; Patel, 1989).

Conversion of α-ketoglutarate to glutamate requires nitrogen from amino acids or ammonia. In the leucine/glutamine cycle, BCAA supply the nitrogen. The need for nitrogen for de novo glutamate synthesis and the observations that in primary rat brain cell cultures, the mitochondrial isozyme BCATm is expressed in astroglia whereas the cytosolic isozyme BCATc is expressed in neurons (Bixel et al., 1997; Bixel et al., 2001; Hutson et al., 2001) led to refinement of the leucine cycle to include a role for BCAA in de novo glutamate synthesis where the different BCAT isozymes catalyze nitrogen transfer in separate cell compartments.

Hutson and coworkers (Hutson et al., 1998; Goto et al., 2005) discovered in the course of their studies on the BCAT isozymes that the antiepileptic drug gabapentin (neurontin) is a specific competitive inhibitor of cytosolic BCATc, but it does not inhibit mitochondrial BCATm. The structural basis for the specificity of gabapentin for the BCATc isozyme is now understood (Goto et al., 2005). Gabapentin was synthesized originally as a GABA analog but is actually structurally similar to the BCAA leucine (Su et al., 1995). Gabapentin is used in combination with other drugs for seizure control and is now used widely to treat neuropathic and other types of pain (Rogawski and Loscher, 2004). Gabapentin was used in the ex vivo rat retina preparation to test whether the BCAT isozymes operate as a cycle that provides nitrogen for de novo synthesis of glutamate (Hutson et al., 2001; LaNoue et al., 2001; Lieth et al., 2001). The experiments provided quantitative evidence that BCAA are necessary for optimal rates of de novo glutamate synthesis in the retina. Finally, they showed that transamination is the major source of nitrogen for de novo glutamate synthesis in the ex vivo rat retina, with ~50% of this nitrogen coming from BCAA (LaNoue et al., 2001). Rates of transfer of leucine nitrogen to glutamate were comparable to rates of de novo glutamate synthesis in whole brain (~1.3 nmol/mg protein/min) (Sakai et al., 2004; Xu et al., 2004).

This BCAT-dependent nitrogen cycle is tied to the glutamate/glutamine cycle (❷ *Figure 6-3*). In astroglia, BCAA nitrogen is used to form glutamate and BCKAs. The BCKAs and glutamine produced are transferred to neurons. In the neuron, glutamine deamination regenerates glutamate and ammonia is released (glutamate/glutamine cycle). The high levels of neuronal glutamate accelerate transamination of BCKAs to BCAAs in the neuronal compartment. The BCAAs are then free to diffuse back to the astroglia. As originally formulated, this hypothetical cycle serves to provide nitrogen for glutamate synthesis in astroglia and serves to buffer glutamate levels in neurons. Thus, the BCAT nitrogen shuttle is an accessory part of the forward arm of the glutamate/pyruvate cycle and is required as a means to regulate the size of the glutamate neurotransmitter pool.

6 Localization of BCAA Catabolic Enzymes in Rodent Brain

The BCAT cycle hypothesis predicts that in brain BCATc is localized in glutamatergic neurons and BCATm is expressed in astroglia surrounding the glutamatergic neurons. Because glutamate is the precursor of GABA, it is also likely that the BCAAs may provide nitrogen for GABA synthesis and BCATs might be expressed in GABAergic neurons. Published and unpublished results from our recent immunohistochemistry studies of the localization of BCAT isozymes and BCKD in rat brain and spinal cord show expression of the BCATc isozyme in glutamatergic and GABAergic neurons (Sweatt et al., 2004a, b).

6.1 The Cytosolic BCATc Isozyme

In rodent brain and spinal cord, BCATc is found only in neurons (Sweatt et al., 2004b; Cole et al., 2005). Immunolabeling of BCATc in adult rat cerebellum and hippocampus shows that the enzyme is expressed by both glutamatergic and GABAergic neurons (Sweatt et al., 2004b). In several brain regions, we observed striking differences in the intracellular localization of BCATc between glutamatergic and GABAergic neurons. For example, the cell bodies of the glutamatergic granule cells of the cerebellum and of the dentate gyrus (hippocampal formation) are not labeled for BCATc, but their axonal processes, including the cerebellar parallel fibers and the mossy fiber projection in the hippocampus, are labeled for BCATc (Sweatt et al., 2004b). Furthermore, in the hippocampus, BCATc is concentrated in the synaptic varicosities formed between mossy fibers and their target neurons in field CA3. From these observations, it appears that the BCATc synthesized in glutamatergic neurons is transported into the axon and toward the synaptic terminal. In contrast, in GABAergic neurons, such as the Purkinje cells of the cerebellum and the pyramidal basket cells of the dentate gyrus, immunoreactivity for BCATc is concentrated in the cell body. In the remaining brain regions and spinal cord, the pattern of BCATc is similar to what has been observed in the hippocampus and cerebellum (unpublished observations).

6.2 The Mitochondrial Enzymes BCATm and Branched Chain α-Keto Acid Dehydrogenase (BCKD)

The expression of the two mitochondrial enzymes, BCATm and BCKD is different from that of BCATc. In the adult rat brain, BCATm is found in astroglia (Hutson et al., 2001; Cole et al., 2005) whereas BCKD is found only in neurons (Cole et al., 2005). The partitioning of BCATm to astroglia is striking in the hippocampus, where a field of BCATm-expressing astroglia surrounds the BCATc-rich mossy fiber projection (Cole et al., 2005) (unpublished observations). In brain regions containing white matter tracts, BCATm is prominent in astrocytes (Cole et al., 2005). This distribution indicates that astrocytes possess

BCAA-transamination capability, as predicted by the BCAT cycle hypothesis. In preliminary studies, the second enzyme in the BCAA catabolic pathway, BCKD, has been localized to neurons and, in the hippocampus, BCKD does not appear to colocalize with BCATc, i.e., BCKD is expressed in a different population of neurons than BCATc (Cole et al., 2005). Astroglia are not immunoreactive for BCKD.

Thus, localization of BCATm in astroglia and BCATc in glutamatergic neurons positions these enzymes to catalyze the cyclic transfer of BCAA nitrogen as predicted by the BCAT cycle hypothesis. BCATm is positioned to provide nitrogen for de novo glutamate synthesis in the glia and BCATc in glutamatergic neurons reaminates BCKAs to form leucine which returns to the astrocytes. Irreversible nitrogen transfer, i.e., oxidation of branched chain α-keto acids occurs only in neurons. In areas such as the hippocampus, oxidation is confined to a separate cellular region where BCATc is not located.

6.3 The BCAT Cycle and GABA Metabolism

The expression of BCATc in GABAergic neurons suggests that BCAAs and/or the BCAT cycle play a role in GABA metabolism. GABA is the major inhibitory transmitter in the brain. The BCAT isozymes, through their involvement in the metabolism of glutamate, can influence GABA pools. Neuronal GABA is synthesized through the decarboxylation of glutamate, catalyzed by the cytosolic enzyme GAD, of which there are two isozymes. GAD67 is found in the cytosol and GAD65 is associated with synaptic vesicles (Christgau et al., 1992; Condie et al., 1997). Most of the GABA released from neurons is taken up again by the neurons, while about 20% of released GABA is taken up by astrocytes. After release at the synapse, GABA, taken up by neurons, is repackaged in synaptic vesicles. GABA catabolism occurs in the mitochondrial compartment of neurons and astrocytes. Therefore, GABA is transported into mitochondria, where it undergoes transamination with α-ketoglutarate, a TCA cycle intermediate, catalyzed by GABA aminotransferase (GABA-T). This reaction produces glutamate and succinic semialdehyde. Reduction of succinic semialdehyde produces succinate, which is also a TCA cycle intermediate. This pathway is called the GABA shunt because it shunts the mitochondrial α-ketoglutarate directly to succinate bypassing succinyl CoA thiokinase, i.e., there is no loss or gain of TCA cycle intermediates.

Depending on the cell type (astrocyte or neuron), glutamate, which is the other product of mitochondrial GABA-T activity, can follow one of several metabolic pathways. In astrocytes, if glutamate exits the mitochondria on the glutamate/hydroxyl carrier, it can be used directly for the synthesis of glutamine (glutamate/glutamine cycle). BCATm is localized in astrocyte mitochondria. So another pathway is BCATm-catalyzed transamination of glutamate with a BCKA, which would regenerate α-ketoglutarate and BCAA. This pathway essentially preserves the TCA cycle intermediate α-ketoglutarate and recycles BCAAs within the astrocyte (LaNoue and Hutson, 2005). Alternatively, glutamate can be oxidatively deaminated by mitochondrial glutamate dehydrogenase to produce α-ketoglutarate, but in the astrocyte this does not appear to be a major pathway of glutamate catabolism under physiological conditions (McKenna et al., 1996a; Gamberino et al., 1997).

In GABAergic neurons, the only BCAT isozyme that is expressed is cytosolic BCATc. If the BCAT cycle operates as proposed for glutamate, then the cycle will function to provide glutamate for subsequent GABA synthesis in GABAergic neurons. For GABA that is metabolized in neurons, glutamate is formed in the mitochondria, requiring export on the glutamate/hydroxyl carrier before it can be a substrate for BCATc. In this case BCATc, depending on the availability of BCKAs, can serve as a buffer for excess glutamate. Alternatively, deamination by glutamate dehydrogenase would generate α-ketoglutarate, thus providing substrate for BCATc-catalyzed glutamate synthesis.

BCATc is found primarily in the cell body of GABAergic neurons (Sweatt et al., 2004b), therefore, it is positioned to interact with the cytosolic pool of glutamate as opposed to the neurotransmitter pool of glutamate that is located near the synapse. Because glutamate transporters are located on GABAergic cell bodies, glutamate released from excitatory nerve terminals that synapse on GABAergic cell bodies can enter the cytoplasmic pool (Conti et al., 1998). Other studies have shown that there is a connection between glutamate and GABA metabolism. Inhibition of GABA-T resulted in an increase in GABA in the cortex and subsequent decline in glutamate and glutamine (Piérard et al., 1999).

Since glutamine is the precursor of glutamate, for the BCAT cycle to contribute directly to GABA synthesis glutaminase should be expressed in GABA neurons. To date, expression of glutaminase in GABAergic neurons has not been reported whereas published studies do support expression of glutaminase in glutamatergic neurons (Laake et al., 1999; Daikhin and Yudkoff, 2000). Thus, the BCAT cycle can only support GABA synthesis indirectly by its contribution to glutamate neurons. The role of BCATc in GABAergic neurons depends then on the supply of substrate, i.e., BCAA or BCKA. If the astrocyte arm of the cycle is active (glutamate neurotransmission), BCKAs will be available for BCATc in GABAergic neurons. Then BCATc catalyzed transamination of BCKA reduces the supply of glutamate for GABA synthesis. On the other hand, if there is active GABA neurotransmission and astrocytes are not producing BCKAs (see above and LaNoue and Hutson, 2005), then BCATc can contribute directly to GABA synthesis by transamination of BCAAs producing glutamate in the cytosol where it is available for GAD. Thus, the BCAT cycle may provide a mechanism for communication between glutamate neurotransmission and the metabolic pool of GABA.

7 Conclusions, Implications for Disorders of BCAA Metabolism, and Future Directions

Current theories of the function of BCAAs in brain and the localization of BCAA catabolic enzymes in brain and spinal cord support the involvement of the glutamatergic and/or GABAergic systems in the etiology of disorders of BCAA metabolism. The current consensus is that BCAAs donate nitrogen for synthesis of neurotransmitter glutamate and are involved in maintenance of neurotransmitter glutamate at a relatively constant level. The localization of BCAT isozymes in GABAergic neurons and surrounding astrocytes suggests a role for BCAAs in GABA metabolism, but whether or not BCAAs and BCATs operate as hypothesized for glutamate or serve a different function remains to be determined. In disorders of BCAA metabolism, such as MSUD, a disturbance and/or imbalance in the glutamatergic and GABAergic systems in brain may cause the neurological symptoms. Experiments conducted on post mortem brain tissue from MSUD calves showed changes in glutamatergic and GABAergic receptor densities compared to normal calf brains (Harper et al., 1989; Dodd et al., 1992), and the authors hypothesized that these changes might lead to an alteration in the excitation–inhibition balance and be proconvulsive (Dodd et al., 1992). Recent studies using in situ or in vitro systems incubated with concentrations of individual BCKAs or BCAAs that are equal to or exceed concentrations found in MSUD patients have generated a number of theories for the underlying cause(s) of the neurological damage observed in untreated MSUD. These theories range from metabolic imbalances to structural changes and include impaired glutamate transport, BCKA-induced apoptotic cell death, leucine-induced oxidative damage, and altered phosphorylation of intermediate filament proteins (Jouvet et al., 2000; Bridi et al., 2003; Funchal et al., 2004; Bridi et al., 2005). Cell culture systems and in vitro preparations cannot substitute for the complex system represented by the intact brain. Progress in testing the BCAT cycle hypothesis, determining the quantitative importance of BCAAs as sources of nitrogen in glutamate and GABA biosynthesis and neurotransmission, as well as understanding the physiological changes that occur in patients with inborn errors of metabolism, such as MSUD, would be facilitated by development of transgenic animal models.

Acknowledgments

The work reported here was supported by Grants NS-38641 and DK-34738 from the US National Institutes of Health.

References

Anthony JC, Anthony TG, Kimball SR, Vary TC, Jefferson LS. 2000. Orally administered leucine stimulates protein synthesis in skeletal muscle of postabsorptive rats in association with increase eIF4F formation. J Nutr 130: 139-145.

Bixel MG, Hamprecht B. 1995. Generation of ketone bodies from leucine by cultured astroglial cells. J Neurochem 65: 2450-2461.

Bixel MG, Hutson S, Hamprecht B. 1995. Cellular distribution of branched-chain aminotransferases in rat brain cells. Biol Chem Hoppe Seyler 376: S86

Bixel MG, Hutson SM, Hamprecht B. 1997. Cellular distribution of branched-chain amino acid aminotransferase isoenzymes among rat brain glial cells in culture. J Histochem Cytochem 45: 685-694.

Bixel M, Shimomura Y, Hutson S, Hamprecht B. 2001. Distribution of key enzymes of branched-chain amino acid metabolism in glial and neuronal cells in culture. J Histochem Cytochem 49: 407-418.

Bledsoe RK, Dawson PA, Hutson SM. 1997. Cloning of the rat and human mitochondrial branched chain aminotransferases (BCATm). Biochim Biophys Acta 1339: 9-13.

Bridi R, Araldi J, Sgarbi MB, Testa CG, Durigon K, et al. 2003. Induction of oxidative stress in rat brain by the metabolites accumulating in maple syrup urine disease. Int J Dev Neurosci 21: 327-332.

Bridi R, Latini A, Braum CA, Zorzi GK, Moacir W, et al. 2005. Evaluation of the mechanisms involved in leucine-induced oxidative damage in cerebral cortex of young rats. Free Radic Res 39: 71-79.

Brosnan ME, Lowry A, Wasi Y, Lowry M, Brosnan JT. 1985. Regional and subcellular distribution of enzymes of branched-chain amino acid metabolism in brains of normal and diabetic rats. Can J Physiol Pharmacol 63: 1234-1238.

Chaplin ER, Goldberg AL, Diamond I. 1976. Leucine oxidation in brain slices and nerve endings. J Neurochem 26: 701-707.

Christgau S, Aanstoot HJ, Schierbeck H, Bergley K, Tullin S, et al. 1992. Membrane anchoring of the autoantigen GAD65 to microvesicles in pancreatic beta-cells by palmitoylation in the NH2-terminal domain. J Cell Biol 118: 309-320.

Chuang DI, Shih VE. 2001. Disorders of branched chain amino acid and keto acid metabolism. The Metabolic Basis of Inherited Disease. Shriver CR, Beaudet AL, Sly S, Vaile D, editors. New York: McGraw Hill; pp. 1239-1277.

Cole J, Sweatt AJ, Wallin R, La Noue KF, Lynch CJ, et al. 2005. Branched-chain keto-acid dehydrogenase is a neuronal enzyme in brain. J Neurochem vol 94 (suppl 1): 39.

Condie BG, Bain G, Gottlieb DI. 1997. Cleft palate in mice with a targeted mutation in the gamma-aminobutyric acid-producing enzyme glutamic acid decarboxylase 67. Proc Natl Acad Sci USA 94: 11451-11455.

Conn AR, Steele RD. 1982. Transport of α-keto analogues of amino acids across blood-brain barrier in rats. Am J Physiol 243: E272-E277.

Conti F, De Biasi S, Minelli A, Rothstein JD, Melone M. 1998. EAAC1, a high-affinity glutamate transporter, is localized to astrocytes and gabaergic neurons besides pyramidal cells in the rat cerebral cortex. Cereb Cortex 8: 108-116.

Daikhin Y, Yudkoff M. 2000. Compartmentation of brain glutamate metabolism in neurons and glia. J Nutr 130: 1026S-1031S.

Dodd PR, Williams SH, Gundlach AL, Harper PA, Healy PJ, et al. 1992. Glutamate and gamma-aminobutyric acid neurotransmitter systems in the acute phase of maple syrup urine disease and citrullinemia encephalopathies in newborn calves. J Neurochem 59: 582-590.

Funchal C, Pessutto FDB, de Almeida LMV, de Lima Pelaez P, Loureiro SO, et al. 2004. α—Keto-β-methylvaleric acid increases the in vitro phosphorylation of intermediate filaments in cerebral cortex of young rats through the gabaergic system. J Neurol Sci 217: 17-24.

Gamberino WC, Berkich DA, Lynch CJ, Xu B, La Noue KF. 1997. Role of pyruvate carboxylase in facilitation of synthesis of glutamate and glutamine in cultured astrocytes. J Neurochem 69: 2312-2325.

Garber AJ, Karl IE, Kipnis DM. 1976. Alanine and glutamine synthesis and release from skeletal muscle. II. The precursor role of amino acids in alanine and glutamine synthesis. J Biol Chem 251: 836-843.

Goto M, Miyahara I, Hirotsu K, Conway M, Yennawar N, 2005. Structural determinants for branched-chain aminotransferase isozyme specific inhibition by the anticonvulsant drug gabapentin. J Biol Chem, accepted for publication Nov 4; 280(44): 37246-56.

Grill V, Bjorkhem M, Gutniak M, Lindqvist M. 1992. Brain uptake and release of amino acids in nondiabetic and insulin-dependent diabetic subjects: Important role of glutamine release for nitrogen balance. Metabolism 41: 28-32.

Hall TR, Wallin R, Reinhart GD, Hutson SM. 1993. Branched chain aminotransferase isoenzymes. Purification and characterization of the rat brain isoenzyme. J Biol Chem 268: 3092-3098.

Harper AE. 1989. Thoughts of the role of branched-chain α-keto acid dehydrogenase complex in nitrogen metabolism. Ann N Y Acad Sci 573: 267-273.

Harper AE, Miller RH, Block KP. 1984. Branched-chain amino acid metabolism. Annu Rev Nutr 4: 409-454.

Harper PA, Dennis JA, Healy PJ, Brown GK. 1989. Maple syrup urine disease in calves: A clinical, pathological and biochemical study. Aust Vet J 66: 46-49.

Harris RA, Paxton R, Powell SM, Goodwin GW, Kuntz MJ, et al. 1986. Regulation of branched-chain alpha-ketoacid dehydrogenase complex by covalent modification. Adv Enzyme Regul 25: 219-237.

Harris RA, Zhang B, Goodwin GW, Kuntz MJ, Shimomura Y, et al. 1990. Regulation of branched-chain α-ketoacid dehydrogenase and elucidation of a molecular basis for maple syrup urine disease. Adv Enzyme Regul 30: 245-263.

Harris RA, Joshi M, Jeoung NH. 2004. Mechanisms responsible for regulation of branched-chain amino acid catabolism. Biochem Biophys Res Commun 313: 391-396.

Hassel B, Sonnewald U. 1995. Glial formation of pyruvate and lactate from TCA cycle intermediates: Implications for the inactivation of transmitter amino acids? J Neurochem 65: 2227-2234.

Hutson SM. 1988. Subcellular distribution of branched-chain aminotransferase activity in rat tissues. J Nutr 118: 1475-1481.

Hutson SM, Harper AE. 1981. Blood and tissue branched-chain amino acid and α-keto acid concentrations: Effect of diet, starvation and disease. Am J Clin Nutr 34: 173-183.

Hutson SM, Zapalowski C. 1981. Relationship of branched-chain amino acids to skeletal muscle gluconeogenic amino acids. Metabolism and Clinical Implications of Branched-Chain Amino and Ketoacids. Walser M, Williamson JR, editors. New York: Elsevier/North Holland, Inc.; pp. 245-250.

Hutson SM, Cree TC, Harper AE. 1978. Regulation of leucine and α-ketoisocaproate metabolism in skeletal muscle. J Biol Chem 253: 8126-8133.

Hutson SM, Zapalowski C, Cree TC, Harper AE. 1980. Regulation of leucine and α-ketoisocaproic acid metabolism in skeletal muscle: Effects of starvation and insulin. J Biol Chem 255: 2418-2426.

Hutson SM, Fenstermacher D, Mahar C. 1988. Role of mitochondrial transamination in branched chain amino acid metabolism. J Biol Chem 263: 3618-3625.

Hutson SM, Wallin R, Hall TR. 1992. Identification of mitochondrial branched chain aminotransferase and its isoforms in rat tissues. J Biol Chem 267: 15681-15686.

Hutson SM, Berkich D, Drown P, Xu B, Aschner M, et al. 1998. Role of branched-chain aminotransferase isoenzymes and gabapentin in neurotransmitter metabolism. J Neurochem 71: 863-874.

Hutson SM, Lieth E, La Noue KF. 2001. Function of leucine in excitatory neurotransmitter metabolism in the central nervous system. J Nutr 131: 846S-850S.

Hutson SM, Sweatt AJ, Lanoue KF. 2005. Branched-chain amino acid metabolism: Implications for establishing safe intakes. J Nutr 135: 1557S–1564S.

Ichihara A. 1985. Aminotransferases of branched-chain amino acids. Transaminases. Christen P, Metzler DE, editors. New York: John Wiley and Sons; pp. 430-438.

Jouvet P, Rustin P, Taylor DL, Pocock JM, Felderhoff-Muesser U, et al. 2000. Branched chain amino acids induce apoptosis in neural cells without mitochondrial membrane depolarization or cytochrome c release: Implications for neurological impairment associated with maple syrup urine disease. Mol Biol Cell 11: 1919-1932.

Kanamori K, Ross BD, Kondrat RW. 1998. Rate of glutamate synthesis from leucine in rat brain measured in vivo by 15N NMR. J Neurochem 70: 1304-1315.

Killian DM, Chikhale PJ. 2001. Predominant functional activity of the large, neutral amino acid transporter (LAT1) isoform at the cerebrovasculature. Neurosci Lett 306: 1-4.

Laake JH, Takumi Y, Eidet J, Torgner IA, Rodberg B, et al. 1999. Postembedding immunogold labeling reveals subcellular localization and pathway-specific enrichment of phosphate activated glutaminase in rat cerebellum. Neuroscience 99: 1137-1151.

LaNoue K, Hutson SM. 2005. Interaction of mitochondria with cytosol and other organelles. In: Brain energeties: Integration of molecular and cellular processes. Handbook of neurochemistry and molecular neurobiology, 3rd ed. Berlin Heidelberg New York: Springer-Verlag; in press.

LaNoue KF, Berkich DA, Conway M, Barber AJ, Hu LY, et al. 2001. Role of specific aminotransferases in de novo glutamate synthesis and redox shuttling in the retina. J Neurosci Res 66: 914-922.

Lieth E, La Noue KF, Berkich DA, Xu B, Ratz M, et al. 2001. Nitrogen shuttling between neurons and glial cells during glutamate synthesis. J Neurochem 76: 1712-1723.

Lynch CJ. 2001. Role of leucine in the regulation of mTOR by amino acids: Revelations from structure-activity studies. J Nutr 131: 861S-865S.

Lynch CJ, Patson BJ, Anthony J, Vaval A, Jefferson LS, et al. 2002. Leucine is a direct-acting nutrient signal that regulates protein synthesis in adipose tissue. Am J Physiol Endocrinol Metab 283: E503-E513.

Magistretti PJ, Pellerin L, Rothman DL, Shulman RG. 1999. Energy on demand. Science 283: 496-497.

Martinez-Hernandez A, Bell KP, Norenberg MD. 1977. Glutamine synthetase: Glial localization in brain. Science 195: 1356-1358.

Matsuo Y, Yagi M, Walser M. 1993. Ateriovenous differences and tissue concentrations of branched-chain ketoacids. J Lab Clin Med 121: 779-784.

Matthews DE, Motil KJ, Rohrbaugh DK, Burke JF, Young VR, et al. 1980. Measurement of leucine metabolism in man from a primed continuous infusion of L-[1-^{13}C]leucine. Am J Physiol 238: E473-E479.

McKenna MC, Tildon JT, Stevenson JH, Huang X, Kingwell KG. 1995. Regulation of mitochondrial and cytosolic malic enzymes from cultured rat brain astrocytes. Neurochem Res 20: 1491-1501.

McKenna MC, Sonnewald U, Huang X, Stevenson J, Zielke HR. 1996a. Exogenous glutamate concentration regulates the metabolic fate of glutamate in astrocytes. J Neurochem 66: 386-393.

McKenna MC, Tildon JT, Stevenson JH, Huang X. 1996b. New insights into the compartmentation of glutamate and glutamine in cultured rat brain astrocytes. Dev Neurosci 18: 380-390.

McKenna MC, Sonnewald U, Huang X, Stevenson J, Johnsen SF, et al. 1998a. Alpha-ketoisocaproate alters the production of both lactate and aspartate from [U-^{13}C]glutamate in astrocytes: A ^{13}C NMR study. J Neurochem 70: 1001-1008.

McKenna MC, Tildon JT, Stevenson JH, Hopkins IB, Huang X, et al. 1998b. Lactate transport by cortical synaptosomes from adult rat brain: Characterization of kinetics and inhibitor specificity. Dev Neurosci 20: 300-309.

Minchin MCW, Beart PM. 1975. Compartmentation of amino acid metabolism in the rat dorsal root ganglion; A metabolic and autoradiographic study. Brain Res 83: 437-449.

Murthy CR, Hertz L. 1987. Acute effect of ammonia on branched-chain amino acid oxidation and incorporation into proteins in astrocytes and in neurons in primary cultures. J Neurochem 49: 735-741.

Norenberg MD. 1979. Distribution of glutamine synthetase in rat central nervous system. J Histochem Cytochem 27: 756-762.

Odessey R, Khairallah EA, Goldberg AL. 1974. Origin and possible significance of alanine production by skeletal muscle. J Biol Chem 249: 7623-7629.

Oldendorf WH. 1971. Brain uptake of radiolabeled amino acids, amines, and hexoses after arterial injection. Am J Physiol 221: 1629-1639.

Patel MS. 1989. CO_2-fixing enzymes. Neuromethods: Carbohydrate and Energy Metabolism. Boulton AA, Baker GD, editors. Clifton, NJ: Humana Press, Inc.; pp. 309-340.

Pellerin L, Magistretti PJ. 1994. Glutamate uptake into astrocytes stimulates aerobic glycolysis: A mechanism coupling neuronal activity to glucose utilization. Proc Natl Acad Sci USA 91: 10625-10629.

Piérard C, Pérès M, Satabin P, Guezennec CY, Lagarde D. 1999. Effects of GABA-transaminase inhibition on brain metabolism and amino-acid compartmentation: An in vivo study by 2D ^{1}H-NMR spectroscopy coupled with microdialysis. Exp Brain Res 127: 321-327.

Rao VL, Murthy CR. 1993. Uptake and metabolism of glutamate and aspartate by astroglial and neuronal preparations of rat cerebellum. Neurochem Res 18: 647-654.

Rogawski M, Loscher W. 2004. The neurobiology of antiepileptic drugs for the treatment of nonepileptic conditions. Nat Med 10: 685-692.

Sakai R, Cohen DM, Henry JF, Burrin DG, Reeds PJ. 2004. Leucine-nitrogen metabolism in the brain of conscious rats: Its role as a nitrogen carrier in glutamate synthesis in glial and neuronal metabolic compartments. J Neurochem 88: 612-622.

Schmoll D, Hamprecht B. 1994. Glyconeogenesis via a gluconeogenic pathway in astrogial cells. J Neurochem 63: 66c (abstract).

Shank RP, Aprison MH. 1981. Present status and significance of the glutamine cycle in neural tissues. Life Sci 28: 837-842.

Shank RP, Bennett GS, Freytag SO, Campbell GL. 1985. Pyruvate carboxylase: An astrocyte-specific enzyme implicated in the replenishment of amino acid neurotransmitter pools. Brain Res 329: 364-367.

Sonnewald U, Westergaard N, Petersen SB, Unsgård G, Schousboe A. 1993. Metabolism of [U-^{13}C]glutamate in astrocytes studied by ^{13}C NMR spectroscopy: Incorporation of more label into lactate than into glutamine demonstrates the importance of the tricarboxylic acid cycle. J Neurochem 61: 1179-1182.

Sonnewald U, Westergaard N, Jones P, Taylor A, Bachelard HS, et al. 1996. Metabolism of [U-^{13}C5] glutamine in cultured astrocytes studied by NMR spectroscopy: First evidence of astrocytic pyruvate recycling. J Neurochem 67: 2566-2572.

Su TZ, Lunney E, Campbell G, Oxender DL. 1995. Transport of gabapentin, a gamma-amino acid drug, by system l alpha-amino acid transporters: A comparative study in astrocytes, synaptosomes, and CHO cells. J Neurochem 64: 2125-2131.

Suryawan A, Hawes JW, Harris RA, Shimomura Y, Jenkins AE, et al. 1998. A molecular model of human branched-chain amino acid metabolism. Am J Clin Nutr 68: 72-81.

Sweatt A, Wood M, Suryawan A, Wallin R, Willingham MC, et al. 2004a. Branched-chain amino acid catabolism: Unique segregation of pathway enzymes in organ systems and peripheral nerves. Am J Physiol 286: E64-E76.

Sweatt AJ, Garcia-Espinosa MA, Wallin R, Hutson SM. 2004b. Branched-chain amino acids and neurotransmitter metabolism: Expression of cytosolic branched-chain aminotransferase (BCATc) in the cerebellum and hippocampus. J Comp Neurol 477: 360-370.

Westergaard N, Sonnewald U, Schousboe A. 1995. Metabolic trafficking between neurons and astrocytes: The glutamate/glutamine cycle revisited. Dev Neurosci 17: 203-211.

Xu Y, Oz G, La Noue KF, Keiger CJ, Berkich DA, et al. 2004. Whole brain glutamate metabolism evaluated by steady state kinetics using a double isotope procedure, effects of gabapentin. J Neurochem 90: 1104-1116.

Yu AC, Drejer J, Hertz L, Schousboe A. 1983. Pyruvate carboxylase activity in primary cultures of astrocytes and neurons. J Neurochem 41: 1484-1487.

Yudkoff M. 1997. Brain metabolism of branched-chain amino acids. Glia 21: 92-98.

Yudkoff M, Nissim I, Kim S, Pleasure D, Hummeler K, et al. 1983. [^{15}N]leucine as a source of [^{15}N]glutamate in organotypic cerebellar explants. Biochem Biophys Res Commun 115: 174-179.

Yudkoff M, Nissim I, Daikhin Y, Lin ZP, Nelson D, et al. 1993. Brain glutamate metabolism: Neuronal-astroglial relationships. Dev Neurosci 15: 343-350.

Yudkoff M, Daikhin Y, Lin Z-P, Nissim I, Stern J, et al. 1994. Interrelationships of leucine and glutamate metabolism in cultured astrocytes. J Neurochem 62: 1192-1202.

Yudkoff M, Daikhin Y, Grunstein L, Nissim I, Stern J, et al. 1996a. Astrocyte leucine metabolism: Significance of branched-chain amino acid transamination. J Neurochem 66: 378-385.

Yudkoff M, Daikhin Y, Nelson D, Nissim I, Erecinska M. 1996b. Neuronal metabolism of branched-chain amino acids: Flux through the aminotransferase pathway in synaptosomes. J Neurochem 66: 2136-2145.

Yudkoff M, Daikhin Y, Nissim I, Horyn O, Lazarow A, et al. 2003. Metabolism of brain amino acids following pentylenetetrazole treatment. Epilepsy Res 53: 151-162.

7 Sulfur-Containing Amino Acids

G. J. McBean

1	Introduction	134
2	Structure of the Excitatory SAAs	135
3	Biosynthesis and Metabolism of Excitatory SAAs	135
4	Cellular and Regional Localization of Excitatory SAAs in the Brain	135
5	Excitatory SAAs as Neurotransmitters	138
5.1	Stimulus-Evoked Release	139
5.2	Activation of Receptors	139
5.3	Stimulation of Neurotransmitter Release by SAAs	141
6	Transport of SAAs	141
6.1	Uptake of SAAs by the Glutamate Transporter Family	142
6.2	Transport of L-Cysteine by the ASC System	143
6.3	The L-cySS–Glutamate Exchanger	144
6.4	Interrelationship Between the Glutamate Transporters and the x_c^- Exchanger in the Uptake of L-Cystine	144
6.5	Physiological Function of Cysteine and Cystine Transport	145
7	Other Roles for Excitatory SAAs	145
8	Cytotoxicity of SAAs	146
8.1	Excitotoxicity	146
8.2	Nonexcitatory Neurotoxicity by SAAs	146
8.3	Toxicity of L-Cysteine	146
8.4	L-Homocysteine Neurotoxicity	147
8.5	S-Methylcysteine Neurotoxicity	148
8.6	Gliotoxicity	148
8.7	Protection Against SAA-Induced Cytotoxicity	149
9	Conclusions	149

© Springer-Verlag Berlin Heidelberg 2007

Abstract: This chapter reviews the structure, synthesis and biological activity of acidic sulfur-containing amino acids (SAAs) that have neuroexcitatory properties. These include the endogenous SAAs L-cysteine, L-cysteine sulfinic acid, L-cysteic acid, L-homocysteine, L-homocysteine sulfinic acid and L-homocysteic acid as well as the exogenous excitants L-serine-O-sulfate and S-sulfo-L-cysteine.

The endogenous SAAs are synthesised in situ in the brain and are derived from either L-methionine (L-homocysteine) or L-cysteine (L-cysteic acid and L-cystine sulphinate). The biosynthetic pathway leading to L-homocysteic and L-homocysteine sulfinic acids has not been identified. L-cysteine and its oxidised counterpart, L-cystine, are rate-limiting precursors for glutathione (GSH) synthesis via the γ-glutamyl cycle.

One or more subtypes of glutamate receptor mediate the neuroexcitatory activity of SAAs. However, evidence is emerging which suggests that the function of SAAs in the brain is centered more on astrocytes than is the case for 'classical' neurotransmitters such as glutamate. Transport of SAAs, as well as de novo synthesis of both SAAs and GSH, is primarily localized to astrocytes. The demonstration of L-glutamate-stimulated release of L-homocysteic acid from astrocytes is further evidence that their activity is largely glial-based.

Several of the SAAs are cytotoxic, acting either as glutamate-like toxins or as gliotoxins. An alternative mechanism of L-cysteine toxicity may be caused by the chemical reactivity of the sulfydryl group, leading to formation of hydrogen peroxide. None of the SAAs have been directly linked to neurodegenerative disease, but it is possible that certain SAAs (for example, L-homocysteine) may exacerbate glutamate-mediated toxicity in some brain pathologies.

There is still much to be discovered regarding the biological role of the SAAs, particularly as regards understanding neuronal-glial interactions and the importance of astrocytes in promoting neuronal survival. It is anticipated that further insights into this important class of neuroactive substance will emerge, leading to a better appreciation of the role of SAAs in excitatory neurotransmission.

List of Abbreviations: L-cys, L-cysteine; L-cySS, L-cystine; L-CSA, L-cysteine sulfinic acid; L-CA, L-cysteic acid; L-Hcys, L-homocysteine; L-HcySS, L-homocystine; L-HCSA, L-homocysteine sulfinic acid; L-HCA, L-homocysteic acid; L-SOS, L-serine-O-sulfate; L-SSC, S-sulfo-L-cysteine; GSH, glutathione; SAA, sulfur-containing amino acid; NMDA, N-methyl-D-aspartate; D-APV, D-2-amino-5-phosphonopentanoic acid; MPEP, 2-methyl-6-(phenylethynyl)pyridine; ASC, alanine serine cysteine; x_c^-, L-cySS–glutamate exchanger; PI, phosphatidylinositol bisphosphate

1 Introduction

A number of acidic sulfur-containing amino acids (SAAs) that have excitatory properties are known to exist in the brain or may become present in the brain under neuropathological conditions. Several of the SAAs have been classified as "putative neurotransmitters" in the sense that their existence and sphere of biological activity is consistent with a role as excitatory neurotransmitter, similar to that of L-glutamate or L-aspartate. Others, for example, L-cysteine (L-cys) and its oxidized counterpart, L-cystine (L-cySS), are precursors for synthesis of the major cellular antioxidant, glutathione (GSH). On the other hand, L-homocysteine (L-Hcys), which is normally present at extremely low concentrations, is known to accumulate in the brain in certain neuropathological situations and cause toxicity. Yet others, for example L-serine-O-sulfate (L-SOS), are not brain constituents but belong to a class of glutamate analogs that are toxic to astrocytes and are known collectively as gliotoxins.

The purpose of this chapter is to review the current understanding of the neurochemistry of excitatory SAAs in the context of their role as putative neurotransmitters or modulators, metabolic precursors, and potential toxins. Other reviews that have covered aspects of the subject include Cuénod et al. (1993), Griffiths (1993), Thompson and Kilpatrick (1996), Dringen (2000), Janaky et al. (2000), Oja et al. (2000), and Kanai and Hediger (2003).

2 Structure of the Excitatory SAAs

The endogenous SAAs include L-cys, L-cysteine sulfinic acid (L-CSA), L-cysteic acid (L-CA), L-Hcys, L-homocysteine sulfinic acid (L-HCSA), and L-homocysteic acid (L-HCA). The majority of these amino acids contain two acidic groups and are homologs of either L-aspartate (L-CSA and L-CA) or L-glutamate (L-HCSA and L-HCA). L-Cys and L-Hcys are monocarboxylic acids, but their oxidized forms (L-cysSS and L-homocystine (L-HcySS), respectively) do contain two carboxylic acid groups. L-SOS and S-sulfo-L-cysteine (SSC) are effective excitants but are not endogenously expressed. Taurine is an endogenous SAA, but since it is inhibitory, rather than excitatory, its role in neurotransmission will not be discussed in this chapter. Structures of the excitatory SAAs are shown in ❷ Figure 7-1.

3 Biosynthesis and Metabolism of Excitatory SAAs

The essential amino acid, L-methionine, is transported into the brain where it is converted into L-Hcys in a three-step pathway via S-adenosylmethionine and S-adenosylhomocysteine as intermediates (❷ Figure 7-2). L-Hcys levels are normally low, as it can be recycled back to methionine, but its concentration is known to increase in various pathological states (see ❷ Sect. 8.4). The brain is unusual, in that the transsulfuration pathway from L-Hcys to L-cys does not occur, due to the relatively low activity of cystathionine lyase in the brain compared to other tissues (Grange et al., 1992; O'Connor et al., 1995). Consequently, the brain relies on plasma L-cys derived from the liver to synthesize other SAAs and GSH (Kranich et al., 1998).

Synthesis of L-CSA and L-CA occurs by sequential oxidation of L-cys to its corresponding sulfinic (L-CSA) and sulfonic acid (L-CA) using molecular oxygen (❷ Figure 7-3). L-CSA and L-CA are metabolic precursors of taurine (Ida et al., 1985) and both are substrates for decarboxylation by cysteine sulfinate decarboxylase (Do and Tappaz, 1996). Formation of taurine and hypotaurine via L-CSA and/or L-CA is a major pathway of L-cys metabolism in astrocytes as determined by ^{13}C-NMR spectroscopy (Brand et al., 1998). An alternative route of metabolism of L-CSA does exist, since transamination of this amino acid with α-ketoglutarate followed by desulfuration yields pyruvate (❷ Figure 7-3).

L-HCSA and L-HCA have both been detected in cerebellar extracts, cultured astrocytes, and C6 glioma cells (Griffiths, 1990; Tschopp et al., 1992; Cuénod et al., 1993), yet their biosynthetic pathway has never been clarified. Cuénod et al. (1993) found no evidence of whether they appear by oxidative metabolism of L-Hcys in analogy to the route of formation of L-CSA and L-CA by oxidative metabolism of L-cys. In experiments in which brain homogenates were incubated with [^{35}S]-homocysteine and the derivatives analyzed by precolumn o-phthalaldehyde derivatization followed by HPLC separation, only a very slow, nonenzymic, nonstereospecific increase in L-HCSA could be detected. The concentration of L-HCA was below the limits of sensitivity. In vitro experiments have shown that L-HCA is an inhibitor, but not a substrate, for cysteine sulfinate decarboxylase, whereas L-HCSA is neither a substrate nor an inhibitor (Do and Tappaz, 1996). This enzyme is therefore unlikely to play any part in the metabolism of either amino acid in vivo.

L-Cys is also the rate-limiting precursor for synthesis of GSH via the γ-glutamyl cycle (❷ Figure 7-4). A unique metabolic interaction exists between neurons and astrocytes in regard to uptake of L-cys for synthesis of GSH (Kranich et al., 1996; Dringen and Hirrlinger, 2003). Astrocytes readily accumulate L-cySS (the oxidized plasma form of L-cys) for de novo GSH synthesis. GSH is then released from astrocytes and is converted to L-cys by an extracellular thiol/disulfide exchange reaction (L-cySS + GSH ↔ L-cys + L-cys-GSH disulfide) prior to its transfer into neurons (Wang and Cynader, 2000). Thus, astrocytes provide the precursors for neuronal GSH. A more detailed discussion of the astrocytic and neuronal transport systems used for L-cys and L-cySS is given in ❷ Sect. 6.

4 Cellular and Regional Localization of Excitatory SAAs in the Brain

Identification of the SAAs in mammalian CNS tissue has been undertaken by a variety of methods, including enzymatic cycling, reversed-phase, and ion-exchange HPLC (see Thompson and Kilpatrick,

Figure 7-1
Structures of sulfur-containing amino acids

Endogenous sulfur-containing amino acids

L-Cysteine (L-Cys)

L-Cystine (L-CySS)

L-Cysteine sulfinic acid (L-CSA)

L-Cysteic acid (L-CA)

L-Homocysteine (L-Hcys)

L-Homocystine (L-HcySS)

L-Homocysteine sulfinic acid (L-HCSA)

L-Homocysteic acid (L-HCA)

Exogenous sulfur-containing amino acids

L-Serine-O-sulfate (L-SOS)

S-sulfo-L-cysteine (L-SSC)

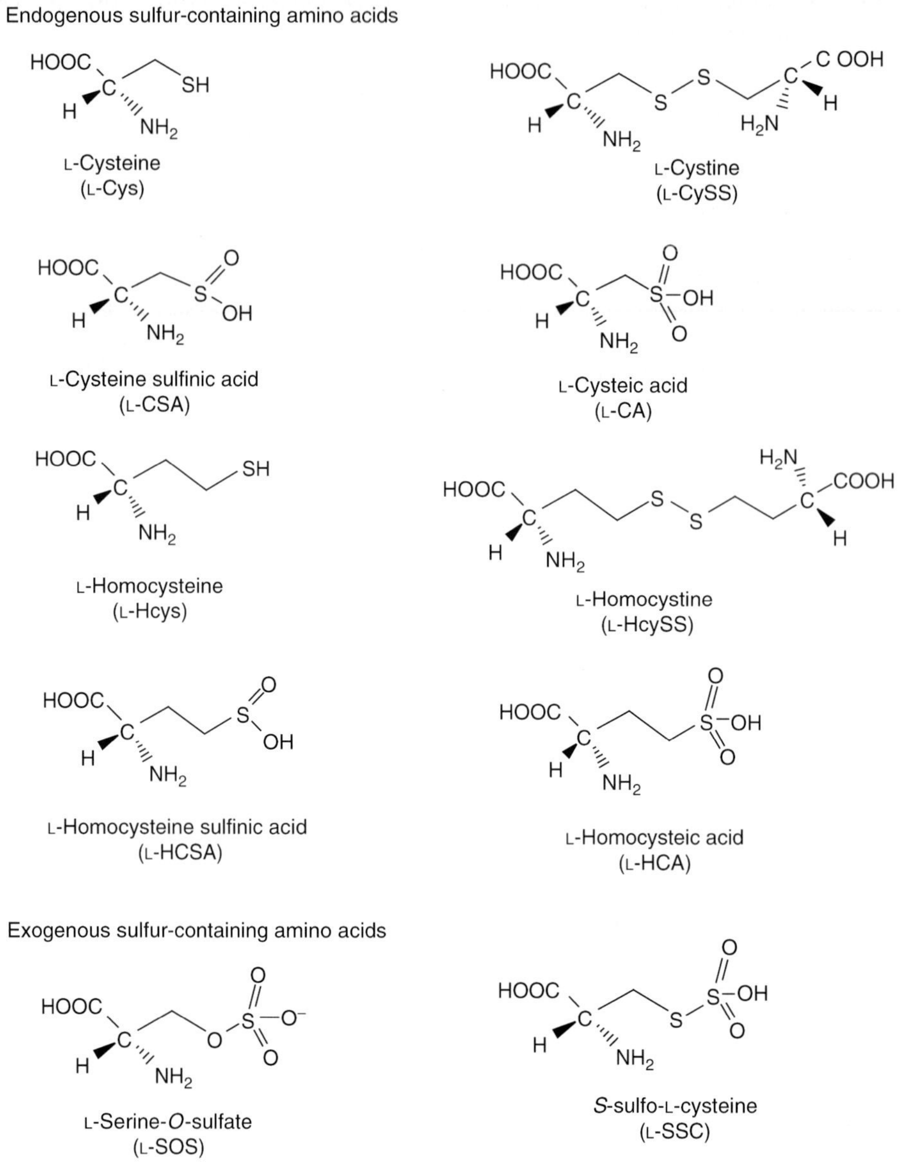

1996 and references therein). Overall, the SAAs are present in far lower concentrations (at least three orders of magnitude) than either L-glutamate or L-aspartate. Measurement of SAAs can be hampered by postmortem oxidation, which, if not prevented, may lead to inaccuracies in the estimations of their concentration in vivo. Thompson and Kilpatrick (1996) have discussed this point extensively.

One of the earliest reports on the endogenous concentration of SAAs indicates that L-HCA is the most abundant (Kilpatrick and Mozley, 1986). This view has been confirmed by Cuénod et al. (1993), who reported a concentration of L-HCA in the brain from 1.0-pmol/mg protein in the hippocampus to less than 0.2-pmol/mg protein in the spinal cord. In the case of L-HCSA, the highest concentration (0.4-pmol/mg

Figure 7-2
Synthesis of L-homocysteine. L-Hcys is formed in a three-step pathway from the essential amino acid, L-methionine. L-Hcys is recycled back to L-methionine by the action of methionine synthase, which, in mammals, is a vitamin B_{12}-dependent enzyme

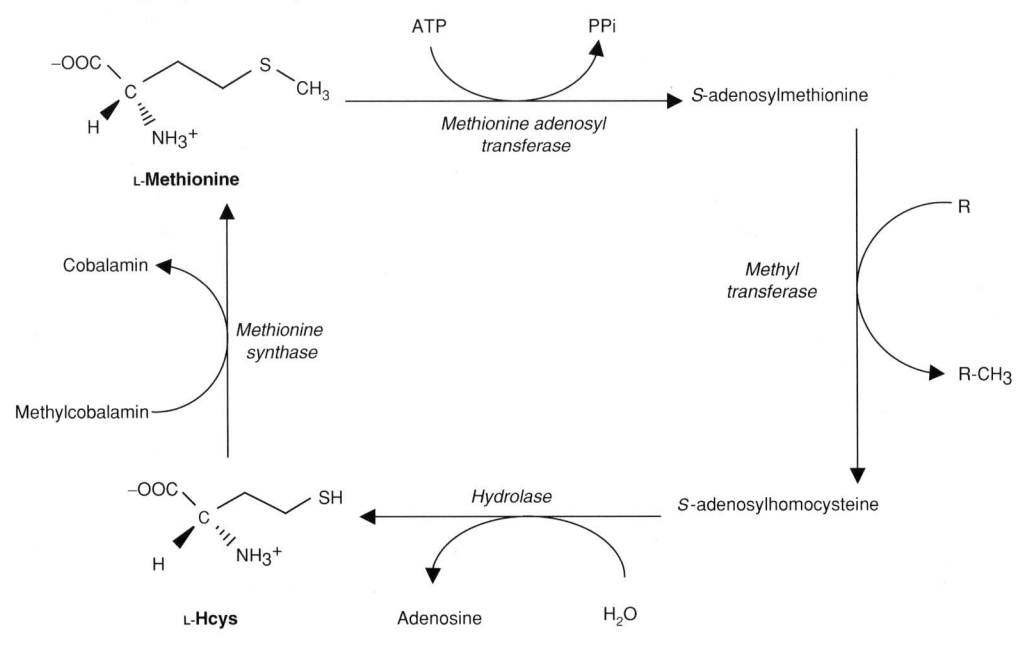

Figure 7-3
Metabolism of L-cysteine

Figure 7-4

The relationship between astrocytes and neurons in glutathione (GSH) synthesis. Transport of L-cystine (L-cySS) into astrocytes via the x_c^- cystine–glutamate exchanger (1) provides intracellular cysteine (cys) for γ-glutamylcysteine synthase (2), the rate-limiting step of the γ-glutamyl cycle. The resulting dipeptide, γ-glutamylcysteine (γ-Glu-cys) is a substrate for GSH synthase (3), which forms γ-glutamylcysteinylglycine (GSH). The tripeptide is then released from astrocytes where it is converted into cysteinylglycine (cys-Gly) in a γ-glutamyl transfer reaction with an acceptor amino acid (X), catalyzed by γ-glutamyltranspeptidase (4). Subsequent liberation of free cys occurs by the action of an ectopeptidase (5). An alternative mechanism, not shown, involves disulfide exchange between GSH and extracellular cySS, which also forms cys (Wang and Cynader, 2000). Cys is transferred into neurons by the glutamate transporters (6), where it enters the γ-glutamyl cycle (7) to form GSH

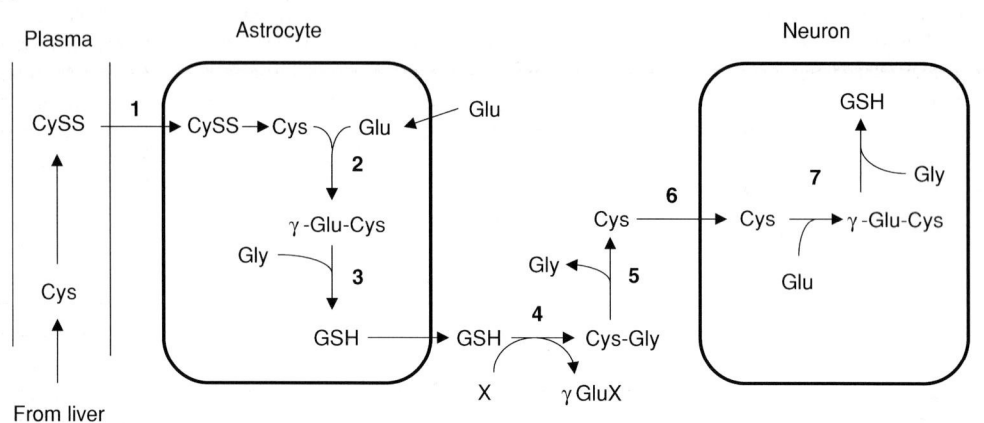

protein) was identified in both the hippocampus and cerebellum, whereas for L-CSA, the highest concentration (also 0.4 pmol/mg protein) was found in the spinal cord (Cuénod et al., 1993).

Much immunocytochemical information suggests that the SAAs are largely localized in glial cells. L-HCA-like immunoreactivity has been identified in Bergmann glial fibers as well as astrocytic processes and endfeet (Cuénod et al., 1990, 1993). Direct measurement of L-CSA, L-HCA, and L-CA by HPLC confirms that these amino acids are primarily localized in glial cells (Griffiths, 1990; Grandes et al., 1991; Grieve and Griffiths, 1992; Tschopp et al., 1992), although a presence in nerve terminals and neuronal perikarya has not been entirely excluded. Formation of ^{13}C-labeled taurine from ^{13}C-cysteine in astroglial cultures has been demonstrated using NMR spectroscopy (Brand et al., 1998). Similarly, immunocytochemical and functional analysis of cysteine sulfinate decarboxylase indicates that the enzyme is localized in glial cells (Wu et al., 1979; Almarghini et al., 1991), with the highest incidence in putamen, caudate, cerebral cortex, and hypothalamus (Wu et al., 1979). An almost complete absence of L-HCA and L-HCSA in mouse cerebellar granule cultures has been noted (Grieve and Griffiths, 1992).

5 Excitatory SAAs as Neurotransmitters

Much research effort leading up to the 1990s was devoted to establishing whether endogenous excitatory SAAs were neurotransmitters. Evidence in favor of such a role was centered on the demonstration of calcium-dependent release, activation of receptors, and initiation of a response by stimulation of second messenger pathways. There is no doubt that the endogenous SAAs, as well as the exogenous compounds, SSC and L-SOS, are potent excitants. Yet several anomalies are apparent if one compares the action of SAAs with that of L-glutamate or L-aspartate. Most obvious is the fact that the natural abundance of the SAAs is only a fraction of that of L-glutamate. They do not have their own class of receptor but activate one or more subtypes of glutamate receptor. Additionally, the nonneuronal localization of the SAAs cannot readily be explained in the context of classical neurotransmission and suggests that their sphere of activity must differ from that of the excitatory amino acids.

5.1 Stimulus-Evoked Release

Calcium-dependent evoked release is one of the hallmarks of classical neurotransmitters, particularly when efflux of endogenous, rather than preaccumulated, neurotransmitters is measured. Do and colleagues, using brain slice preparations, have been at the forefront of research into release of endogenous SAAs (Do et al., 1986; Cuénod et al., 1990, 1993). Calcium-dependent, potassium-simulated release of L-CSA, L-CA, L-HCSA, and L-HCA has been identified in the cortex, hippocampus, cerebral cortex, cerebellum, and spinal cord (Do et al., 1986; Vollenweider et al., 1990). Release of L-HCA showed the greatest level of stimulation (~sevenfold) in the cortex and hippocampus. Stimulated release of the other SAAs varied between one- and threefold. L-CA, for instance, showed only a marginal increase in release in response to stimulation by potassium that was confined to the hippocampus and mesodiencephalon. The observation that the high-frequency stimulations employed to evoke long-term potentiation in the hippocampus selectively stimulated efflux of L-CSA and L-HCA has raised the possibility that these SAAs may have a particular role in synaptic plasticity (Klancnik et al., 1992). Calcium-dependent, potassium-stimulated release of L-cys has been observed in a number of brain regions, most prominently in the neocortex, followed by the mesodiencephalon, striatum, and cerebellum (Keller et al., 1989).

Overall, release of L-HCA is the most convincing of a possible neurotransmitter role, but certain aspects of its activity still require clarification. In attempting to reconcile the fact that L-HCA is primarily stored in glial cells and that activation of a nerve pathway leads to delayed efflux of L-HCA, Cuénod et al. (1993) introduced the phrase "gliotransmitter." It was envisaged that glial receptor activation by a synaptically released neurotransmitter (such as L-glutamate) would initiate efflux of L-HCA from astrocytes. Such a scheme is not entirely hypothetical as neuronal NMDA receptor activation by L-glutamate released from astrocytes has been reported (Parpura et al., 1994). More significantly, new evidence in support of this hypothesis has recently emerged: Cuénod's group (Benz et al., 2004) have put forward a model in which glutamate-stimulated efflux of L-HCA from astrocytes is mediated by calcium-influx into the astrocyte by NMDA receptor-induced activation of the sodium–calcium exchanger (❯ *Figure 7-5*). This novel mechanism adds new insight into the role of gliotransmitter proposed more than a decade ago.

5.2 Activation of Receptors

The neuroexcitatory action of the SAAs was discovered coincidentally with the first electrophysiological experiments on the excitatory properties of the acidic amino acids in single neurons in the mammalian CNS (Curtis and Watkins, 1960). Several decades later, it is now known that the SAAs evoke activation of both NMDA and non-NMDA glutamate receptors (Griffiths, 1993), yet almost all exhibit a preference for the NMDA receptor. An extensive radioligand-binding analysis by Pullan et al. (1987), using both L- and D-enantiomers of SAAs, revealed that D-HCSA was a more selective NMDA agonist than its L-enantiomer. D-CA and D-CSA were also agonists of the NMDA receptor but were less selective than D-HCSA. Curras and Dinglediene (1992) compared L-CSA and L-HCA to L-glutamate and L-aspartate in terms of their potency, maximal activity, and selectivity as agonists of NMDA and amino-3-hydroxy-5-methyl-4-isoxazolepropionate (AMPA) receptors in *Xenopus* oocytes injected with rat brain mRNA. Their results confirmed that L-CSA and L-HCA were mixed agonists showing NMDA selectivity. In the case of NMDA receptor activation, the potency rank order was L-glutamate (EC_{50} = 2.2 µM) > L-aspartate = L-HCA (13 µM) > L-CSA (59 µM). The potency rank order for steady-state activation for AMPA receptors was L-glutamate (EC_{50} = 11 µM) > L-HCA (430 µM) > L-CSA (3,300 µM).

In their review, Thompson and Kilpatrick (1996) have cautioned against misinterpretation of data on the relative excitatory potency of SAAs in radioligand-binding experiments, because active transport

Figure 7-5

Glutamate-stimulated release of L-homocysteic acid (HCA) from astrocytes. Synaptically released Glu activates postsynaptic and astrocytic ionotropic Glu receptors (GluR), which causes an influx of Na^+. The rise in intracellular Na^+ activates the Na^+/Ca^{2+} exchanger. The rise in $[Ca^{2+}]_i$ in the astrocyte stimulates HCA release by an as yet unknown mechanism. Following its release, HCA activates postsynaptic NMDA receptors. Figure adapted from Benz et al. (2004)

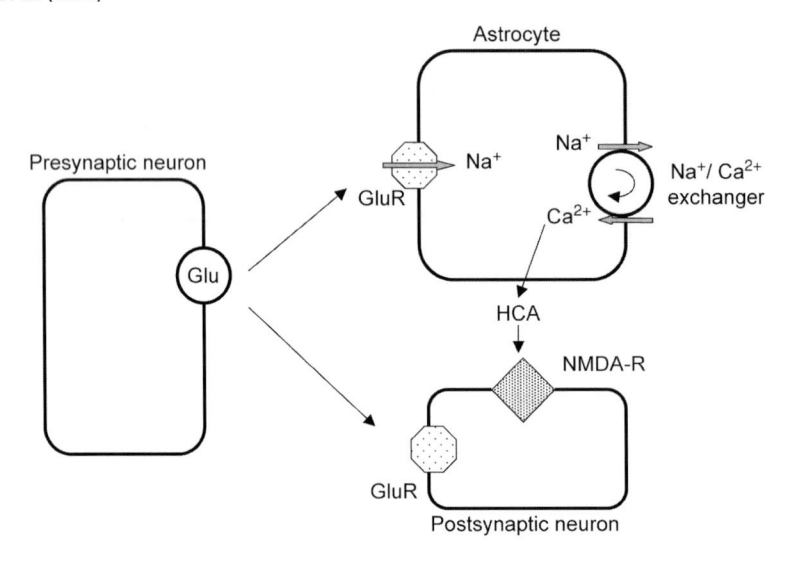

will reduce the potency of a well-transported SAA, compared to one that is not. For example, L-HCA and L-HCSA are poorly transported, compared to L-CA, L-CSA, and L-glutamate, which would make them appear as more potent receptor agonists. This interpretation is supported by the fact that displacement of L-[^{3}H]glutamate binding by L-CA and L-CSA was highly sodium dependent, making it likely that transport sites were being labeled, whereas displacement of [^{3}H]glutamate binding by L-HCA and L-HCSA is relatively sodium independent. However, regardless of their relative potency, the sensitivity of SAA-evoked depolarizations to inhibition by the competitive NMDA receptor antagonist, D-2-amino-5-phosphonopentanoic acid (D-APV) does confirm that they are all NMDA receptor agonists (Thompson and Kilpatrick, 1996).

The discovery of metabotropic glutamate receptors (mGluRs) in the late 1980s (Sugiyama et al., 1987) and the recognition that a subset (mGlu1 and mGlu5 subtypes) of these receptors are linked to inositol phospholipid metabolism initiated studies into the effect of SAAs on phosphatidylinositol bisphosphate (PI) hydrolysis. Several of the SAAs stimulate production of inositol phosphates in a number of systems including primary cultures of striatal neurons, hippocampal slices, and cerebellar granule cells (Griffiths, 1993). Porter and Roberts (1993) reported that L-CSA, DL-HCA, L-CA, and L-HCSA were all more potent mGluR agonists than glutamate, with EC_{50} values from 401 to 487 µM. Correspondingly, Croucher et al. (2001) reported that L-CSA and L-CA were potent agonists at presynaptic mGlu5 autoreceptors in rat forebrain: both compounds enhanced the electrically evoked release of preloaded D-[^{3}H]aspartate from rat forebrain slices, which was blocked by the selective mGlu5 antagonist, 2-methyl-6-(phenylethynyl)pyridine (MPEP, 10 µM). However, it is notable that the concentration response curve to L-CA was bell-shaped as had previously been observed in the experiments by Porter and Roberts (1993). In contrast, L-CSA produced a stimulation in D-[^{3}H]aspartate release that did not diminish at high concentrations of the agonist (100 µM). It was proposed that the lack of desensitization in the response to L-CSA, but not L-CA, might be due to an interaction with transporters, which would mask any loss in efficacy.

Radioligand-binding studies by Kingston et al. (1998) in a Syrian hamster tumor cell line that had been transfected with cDNA of human mGluR1α and mGluR5a receptors confirms that L-HCSA, L-SCA, L-CA, and L-SOS all bind specifically to both subtypes of mGluRs, as indicated by a stimulation of PI hydrolysis. Inhibitor constants for the SAAs in [^{3}H]glutamate receptor-binding studies in mGluR1α cells confirmed that L-HCSA had the highest affinity ($K_i = 0.44 \pm 0.2$ μM). However, L-HCSA was the only SAA tested that did not inhibit glutamate uptake in the mGluR1α cells, again raising the possibility that active transport of the other SAAs might have influenced the results. Inclusion of a nontransportable and selective uptake inhibitor would be required for absolute certainty that interaction of the SAAs at mGluRs was not affected by uptake.

5.3 Stimulation of Neurotransmitter Release by SAAs

There are a number of reports that describe stimulation of neurotransmitter release by SAAs acting at specific subtypes of excitatory amino acid receptors. Griffiths and coworkers have demonstrated SAA-stimulated release of [^{3}H]GABA from cultured mouse cerebral cortical neurons (Dunlop et al., 1991) and D-[^{3}H]aspartate release from cerebellar granule cells (Dunlop et al., 1992) that was blocked by both NMDA and non-NMDA antagonists. In both instances, blockade of the releasing action by transport inhibitors led to the proposal that the SAAs promoted a receptor-mediated reversal of neurotransmitter transporters (Griffiths, 1993). This mechanism may have relevance in pathological states, such as ischemia, where reversal of the glutamate transporters would be exacerbated by release of SAAs and would lead to a massively increased extracellular concentration of glutamate, resulting in excitotoxic cell death.

D-HCA, L-HCA, L-HCSA, L-CA, L-CSA, and L-SSC all stimulate [^{3}H]noradrenaline release from rat hippocampal slices with an order of potency that matches their affinity as NMDA receptor agonists (Selema et al., 1997). D-HCSA was the most potent of all the agonists tested, with an EC_{50} value of 750 μM. L-Glutamate showed a comparable stimulation of [^{3}H]noradrenaline release that was inhibited by NMDA receptor antagonists, from which one can conclude that, in vivo, glutamate is just as likely to mediate this effect as any of the SAAs.

In summary, while the SAAs do fulfill several of the criteria necessary for neurotransmitter function, their obvious overlap with several aspects of glutamate-mediated neurotransmission is one argument that does not favor such a role. Nevertheless, it is noteworthy that they are more potent Group 1 mGluR agonists than glutamate. This fact, coupled to their primary localization in astrocytes and the recent identification of stimulated release occurring from astrocytes (Benz et al., 2004) suggests that we may only be at the verge of understanding their function as glial-based neuromodulators. Moreover, such a role strengthens the view that glial cells actively participate in synaptic signaling and information processing.

6 Transport of SAAs

The first reports that SAAs were actively taken up by high-affinity transport systems were considered as evidence of a neurotransmitter role. As more information on transport mechanisms has emerged, it has become obvious that the function of such transporters extends beyond the confines of inactivation of synaptically released neurotransmitters. One of the most important functions of transport systems must be the supply of L-cys and L-cySS to the cytosol for use as metabolic substrates and as precursors for GSH synthesis.

It is now known that SAAs are substrates for a number of different transporter types: the family of high-affinity sodium-dependent glutamate transporters (also known as the X_{AG}^- transporters), the L-cySS – glutamate exchanger (x_c^-), and the ASC family of sodium-dependent neutral amino acid transporters (❯ *Figure 7-6*). There is a wealth of new information on the interaction of SAAs with subtypes of individual transporters, on the cellular localization of such transporters, on the response of SAA transporters to oxidative stress, and on the cytotoxic consequences of inhibition or inactivation of transporter function. The discussion in this section will be centered on the characteristics of SAA uptake in the brain. Several

Figure 7-6

L-Cysteine (cys) and L-cystine (cySS) transporters. The sodium-dependent ASC transporters take up cys in exchange for an amino acid (AA). Transport of either cys or cySS by the glutamate transporters is accompanied by the inward flow of Na$^+$ and H$^+$, with K$^+$ as the counter-ion. The L-cySS–glutamate exchanger cotransports cySS and Cl$^-$ in exchange for glutamate (Glu$^-$), which, at physiological pH, carries a net negative charge. Both the ASC and glutamate transporters are electrogenic, whereas the x_c^- exchanger is electroneutral

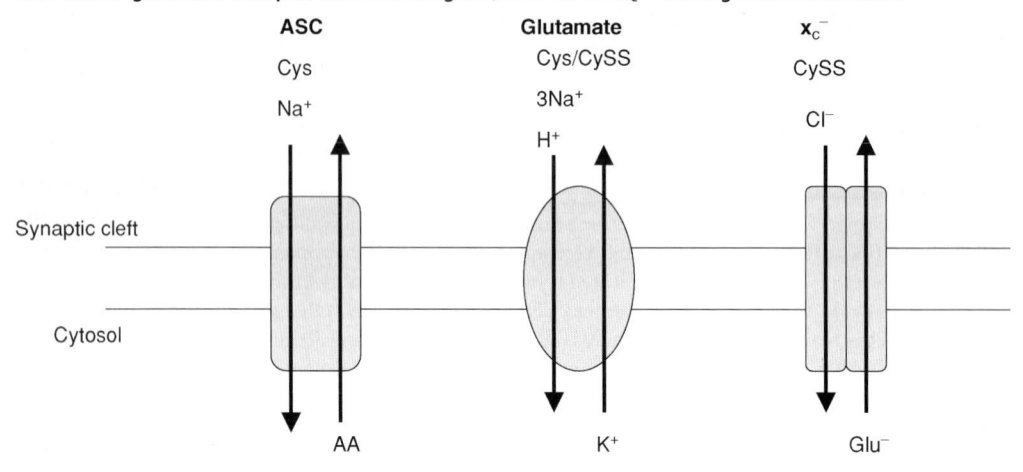

recent reviews give additional coverage of transporter structure and function (Danbolt, 2001; Shanker and Aschner, 2001; Pow, 2001b; McBean, 2002; Robinson, 2002; Kanai and Hediger, 2003).

6.1 Uptake of SAAs by the Glutamate Transporter Family

A vast quantity of information exists on the family of high-affinity glutamate transport proteins that signifies their fundamental importance as inactivators of synaptically released L-glutamate and as protectors against excitotoxic cell death (Danbolt, 2001). This family of proteins comprises five members, which in the mammalian brain are known as GLAST (EAAT1 in the human), GLT1 (EAAT2), EAAC1 (EAAT3), EAAT4, and EAAT5. The first three subtypes are widely distributed throughout the brain, whereas EAAT4 and EAAT5 have a much more limited distribution. It was originally thought that GLT1 and GLAST were expressed only on astrocytes and that EAAC1 was exclusively neuronal (Rothstein et al., 1994; Dowd et al., 1996), but all three of the proteins have since been identified in astrocytes (Miralles et al., 2001). There is also evidence that GLT1 is expressed on neurons (Chen and Swanson, 2003). Gene knockout experiments have indicated that the majority of glutamate is taken up into astrocytes (Rothstein et al., 1996).

L-CA, L-CSA, L-HCA and L-HCSA, L-cySS (Bender et al., 2000; Flynn and McBean, 2000; McBean and Flynn, 2001; Hayes et al., 2005), L-cys (Knickelbein et al., 1997; Hayes et al., 2005), and L-DOC (Vandenberg et al., 1998) are all substrates for the glutamate transporters. Calculation of the kinetic constants for transport of excitatory SAAs indicates that L-CA and L-CSA are both high-affinity substrates for glutamate uptake sites, with K_m values between 14 (L-CA uptake in the cortex) and 100 µM (L-CSA uptake in primary astrocytes), which are comparable to the K_m values for L-glutamate uptake (Griffiths, 1993). L-HCA and L-HCSA are low-affinity ($K_m > 500$ µM) substrates in each preparation, with the exception of L-HCSA uptake in astrocytes ($K_m = 225$ µM) (Griffiths, 1993). Such low affinity of L-HCSA and L-HCA for uptake sites may explain their greater potency as receptor agonists, as discussed in ● Sect. 2. Calculation of the relative clearance rates (V_{max}/K_m) of the excitatory SAAs yields values in the range of 3 ml/mg/min × 10^3 for L-HCA in primary cortical neuronal cultures to 615 ml/mg/min × 10^3 for L-CA in cerebellar granule cells. There is agreement that removal of L-CA from the extracellular medium is the most efficient among all the SAAs tested (Bouvier et al., 1991; Griffiths, 1993). In synaptosomal work, the observation that L-HCA was a potent inhibitor of cerebellar L-[^{3}H]glutamate transport but produced only a low level of inhibition of transport in cortical synaptosomes (Robinson et al., 1993) raised the question of whether L-HCA was a substrate for a specific subtype of glutamate transporter. In support of this view, recent experiments on the inhibition of D-[^{3}H]aspartate transport by L-HCA in HEK cell lines overexpressing the glutamate transporters showed that L-HCA weakly inhibited GLT1- and GLAST-mediated transport of D-[^{3}H]aspartate but had no effect on transport in HEK cells expressing EAAC1 (Hayes and McBean, unpublished observations). Analysis of electrogenic L-SOS transport by EAAT1 (GLAST) and EAAT2 (GLT1) overexpressed in *Xenopus* oocytes under voltage-clamp conditions has revealed notable differences in its mechanism of uptake. In the case of EAAT2, L-SOS transport displayed a steep voltage dependency that was not observed with the EAAT1 transporter or indeed with L-glutamate or L-aspartate uptake by either protein (Vandenberg et al., 1998). These authors proposed that the voltage-dependency of L-SOS uptake by EAAT2 might be due to the much stronger acidity of the sulfate group of L-SOS compared to the carboxyl groups of L-glutamate and L-aspartate. Nevertheless, it is notable that this difference in voltage sensitivity was only observed in the case of the EAAT2 transporter.

L-Cys is a high-affinity substrate for the glutamate transporters (Sagara et al., 1993b; Zerangue and Kavanaugh, 1996b) and is taken up into cortical neuronal cultures by both the EAAT2 (GLT1) and EAAT3 (EAAC1) proteins (Chen and Swanson, 2003). Results from measurement of L-[^{14}C]cys transport by individual glutamate transporters overexpressed in *Xenopus* oocytes and in HEK cells has shown that the EAAC1 transporter has a much greater capacity for uptake of L-cys than does either GLAST or GLT1 (Zerangue and Kavanaugh, 1996b; Hayes et al., 2005). In *Xenopus* oocytes, the affinity of L-cys for EAAC1 (K_m = 193 µM) is higher than for either GLAST (1.83 mM) or GLT1 (967 µM) (Zerangue and Kavanaugh, 1996b). Similarly, the relative clearance rate of L-cys uptake by HEK$_{EAAC1}$ cells is 48 ml/mg/min × 10^3, compared to a value of 23 ml/mg/min × 10^3 for D-[^{3}H]aspartate and 1.6 ml/mg/min × 10^3 for L-[^{14}C]cySS (Hayes et al., 2005).

Sodium-dependent uptake of L-[^{14}C]cySS by the glutamate transporters has been reported in rat cortical synaptosomes (Flynn and McBean, 2000) and primary astrocyte cultures (Bender et al., 2000; Allen et al., 2001; Shanker and Aschner, 2001). A recent investigation into L-[^{14}C]cySS transport by subtypes of the glutamate transporters individually expressed in HEK cells shows that GLT1, GLAST, and EAAC1 all take up L-[^{14}C]cySS with K_m values in the range of 20–110 µM (Hayes et al., 2005). In HEK$_{EAAC1}$ cells, the fact that the relative clearance rate of L-[^{14}C]cySS is 30 times lower than for L-[^{14}C]cys in HEK$_{EAAC1}$ cells confirms the view that L-cySS is a weak substrate for the glutamate transporter proteins (Hayes et al., 2005). Predictably, L-[^{14}C]cySS uptake by the glutamate transporters is subject to inhibition by other SAAs. L-CSA, L-HCSA, and L-SOS all potently block transport (Flynn and McBean, 2000). L-HCA is a relatively weak inhibitor (Bender et al., 2000; Flynn and McBean, 2000) that displays a preference for GLT1- and GLAST-mediated transport (IC$_{50}$ values of 66 and 67 µM, respectively), compared to EAAC1-mediated uptake (IC$_{50}$ = 200 µM) (Hayes et al., 2005).

6.2 Transport of L-Cysteine by the ASC System

The ASC (alanine, serine, cysteine) transporters are a family of sodium-dependent neutral amino acid transporters comprising two members, ASCT1 and ASCT2. They are believed to function as exchangers, as opposed to net transporters, favoring the inward translocation of L-cys in exchange for another amino acid [see reviews by Kanai (1997) and Kanai and Hediger (2003)]. These proteins have a high affinity for L-cys, alanine, serine, and threonine. L-CySS is not a substrate for either protein as indicated from experiments in cultured astrocytes (Bender et al., 2000), and L-glutamate is a low-affinity substrate for ASCT2.

ASCT1 is preferentially expressed in astrocytes, where it is thought to exchange intracellular serine for incoming substrates (Sakai et al., 2003). ASCT2 is also expressed in astrocytes and mediates efflux of glutamine in exchange for other amino acids (Broer et al., 1999). L-Hcys is an inhibitor of ASC-mediated uptake of L-[^{14}C]cys in HEK cells (Hayes et al., 2005). L-CA and L-CSA both inhibit ASCT1-mediated uptake of L-cys due to a lowering of pH. An absence of any ASC-mediated uptake of L-[^{35}S]cys in mouse cortical neuronal cultures has been noted (Chen and Swanson, 2003).

6.3 The L-cySS–Glutamate Exchanger

Experiments on L-cySS transport in the 1980s by Bannai and coworkers led to the establishment of the sodium-independent x_c^- exchanger as an important contributor to the uptake of L-cySS in nonneuronal cells (Bannai and Kitamura, 1980, 1981; Bannai, 1984; Bannai et al., 1984). In neural tissue, early investigations of the x_c^- exchanger came from studies on Cl$^-$-dependent L-[^{3}H]glutamate binding (Bridges et al., 1987a, b; Anderson et al., 1990). Since then, much work on L-cySS–glutamate exchange has been performed in CNS-derived tumor cell lines, including C6 glioma cells (Cho and Bannai, 1990; Kato et al., 1992), neuroblastoma-primary retinal hybrid cells (Murphy et al., 1989; Murphy et al., 1990), UC11 human astrocytoma cells (Johnson and Johnson, 1993), LRM55 rat spinal astrocytoma cells, and SNB-19 human astrocytoma cells (Patel et al., 2004), as well as in primary astrocytes (Allen et al., 2001; Miralles et al., 2001; Shanker and Aschner, 2001; Shanker et al., 2001), fetal brain cells (Sagara et al., 1993a), and hippocampal neurons (Allen et al., 2001).

The x_c^- exchanger is a member of the disulfide-linked heteromeric amino acid transporter family that comprise heavy and light chain subunits (Sato et al., 1999, 2000). In vivo, the exchanger operates as an electroneutral exchange system in which L-cySS is conveyed inward in a 1:1 exchange for L-glutamate traveling outward (see reviews by Shanker and Aschner, 2001 and McBean, 2002). Intracellular L-cySS is rapidly reduced to L-cys (which is not a substrate for the exchanger), whereas extracellular-oxidizing conditions favor the formation of L-cySS (but see below for a discussion of this point). The directional flow of glutamate relies on the fact that the glutamate transporters are capable of maintaining a nanomolar extracellular concentration of glutamate, compared to a millimolar concentration in the cytosol (Zerangue and Kavanaugh, 1996a). Radiolabeled L-cySS uptake by the mammalian x_c^- exchanger overexpressed in HEK cells indicates a K_m of 81 µM (Shih and Murphy, 2001).

Bridges and coworkers (Patel et al., 2004; Warren et al., 2004) used a fluorescent-based assay that detects x_c^--mediated glutamate efflux to show that L-HCA, L-SOS, and the lathryus excitotoxin, β-N-oxalylamino-L-alanine are substrates for the x_c^- exchanger (Patel et al., 2004). There is no known analog of either glutamate or L-cySS that is entirely selective for the x_c^- exchanger, although L-α-aminoadipate exhibits a higher affinity for the exchanger ($K_m = 20$ µM), compared to the glutamate transporters ($K_m = 629$ µM) (Tsai et al., 1996). Pow et al. (Pow, 2001a) raised antibodies against L-α-aminoadipate to study the regional distribution of x_c^- exchanger in the brain, from which it was concluded that the x_c^- protein was expressed on astrocytes.

In addition to its function in conveying L-cySS into the cell interior, the x_c^- exchanger plays a significant role in the regulation of extracellular levels of glutamate (Warr et al., 1999; Baker et al., 2002a, b). Bath application of L-cySS to rat brain slices has been shown to upregulate synaptic activity through activation of group II metabotropic receptors due to enhanced efflux of glutamate (Moran et al., 2003). Little is known to date about how activity of the x_c^- exchanger is controlled, but protein kinase A and calcium/calmodulin-dependent protein kinase II are known to regulate uptake of L-[^{35}S]cystine by the exchanger in primary astrocyte cultures (Tang and Kalivas, 2003).

6.4 Interrelationship Between the Glutamate Transporters and the x_c^- Exchanger in the Uptake of L-Cystine

The consensus view is that the x_c^- exchanger carries the majority of L-cySS taken up into astrocytes. The slow rate of L-cySS uptake by the glutamate transporters, coupled to the high degree of inhibition of L-cySS transport by glutamate (Hayes et al., 2005), makes it unlikely that this system is a significant supplier of L-cySS in vivo. However, in the absence of a selective nontransported inhibitor of the x_c^- exchanger, accurate predictions of the relative transport by each uptake system cannot be made. Glutamate efflux in exchange for L-cySS traveling inward by the x_c^- exchanger may lead to an overestimation of the K_m value for L-cySS by the glutamate transporters. Conversely, the functional activity of system x_c^- in certain neurochemical preparations, such as synaptosomes (Flynn and McBean, 2000), may be artificially low compared to its activity in vivo. The protein, being a dimer whose functional state is sustained by disulfide links (Sato et al., 1999, 2000), may not withstand rigorous preparation procedures.

Even if the majority of L-cySS uptake is mediated by the x_c^- exchanger, the inward flow of the substrate is indirectly dependent on the glutamate transporters for two reasons: first, to maintain the intracellular concentration of glutamate as a driving force for L-cySS-glutamate exchange (Reichelt et al., 1997) and second, to ensure that, should the activity of the exchanger increase, the extracellular concentration of glutamate is kept at a subtoxic concentration. Ye et al. (1999) have reported an increased efflux of L-glutamate by the x_c^- exchanger in C6 glioma cells in which the glutamate transporters were dysfunctional and therefore not actively taking up L-glutamate.

6.5 Physiological Function of Cysteine and Cystine Transport

There is general agreement that maintenance of intracellular GSH is the driving force for L-cySS transport into astrocytes, which accumulate L-cySS in preference to L-cys (Kranich et al., 1996, 1998). Depletion of intracellular GSH, oxidative stress, and exposure of astrocytes to dibutyryl cAMP all trigger induction of x_c^-, leading to upregulation of transport (Bannai, 1984; Li et al., 1999a; Bridges et al., 2001; Gochenauer and Robinson, 2001). Much investigation on the trafficking of GSH and its metabolites between astrocytes and neurons using NMR spectroscopy supports a model in which astrocytes actively take up L-cySS and release GSH as a means of providing precursors for neuronal GSH synthesis, either as cysteinylglycine or as L-cys (❯ *Figure 7-4*). Chief among the experimental observations to support this view is that neurons do not readily take up L-cySS (Dringen et al., 1999), although contrary evidence exists for both hippocampal (Allen et al., 2001) and cortical (Danbolt, 2001) neurons in culture. Even so, the presence of the x_c^- exchanger on astrocytes supports the view that uptake of the diamino acid is primarily directed toward astrocytes. Guebel and Torres (2004) used a mathematical approach to analyze the dynamics of sulfur–amino acid metabolism in the brain and concluded that the inward flow of L-cySS into astrocytes was more than sufficient to meet the demand for L-cys as a precursor for synthesis of GSH, taurine, and hypotaurine. Therefore, a net efflux of L-cys from astrocytes would occur, thereby maintaining a supply of extracellular L-cys for neuronal GSH synthesis. The neuronal subtypes of high-affinity glutamate transporters (Zerangue and Kavanaugh, 1996b; Chen and Swanson, 2003; Hayes et al., 2005) are well placed to mediate the transfer of L-cys into neurons. Since a rise in the extracellular concentration of GSH or cysteinylglycine will favor the reduction of L-cySS to L-cys by disulfide exchange, this will have an immediate knock-on effect in terms of the rate of delivery of L-cys to neurons.

7 Other Roles for Excitatory SAAs

Recently, it has been proposed that L-CSA may have a role in the regulation of kynurenic acid synthesis based on the observation that L-CSA selectively inhibits kynurenic acid production in rat cortical slices by inhibition of the glial-based enzyme, kynurenine aminotransferase II (Kocki et al., 2003). The consequent reduction in kynurenic acid, which is an endogenous broad-spectrum antagonist acting at the glycine site of the NMDA receptor (Stone, 2000), would presumably increase activity of the NMDA receptor. Interestingly, L-CSA was more potent ($IC_{50} \sim 2$ µM) than any of the other SAAs tested, as well as L-glutamate ($IC_{50} \sim 800$ µM) and L-aspartate ($IC_{50} \sim 70$ µM) (Kocki et al., 2003). This group has previously suggested that kynurenic acid production is mediated by the activation of mGluRs, but it is not clear whether L-CSA is acting in this way or whether it occurs subsequently to its transport into the astrocyte.

8 Cytotoxicity of SAAs

Many of the SAAs, whether normal constituents of brain tissue or as exogenous compounds, are toxic to neurons and/or glial cells. As with L-glutamate-induced cytotoxicity, various mechanisms have been proposed, some of which are better understood than others. The enduring challenge is to understand to what extent such processes occur in situ in the brain under pathological conditions and how they are initiated. Clarification of the causative factors contributing to SAA-induced cytotoxicity using in vitro systems is an invaluable step toward eventually understanding the pathophysiology of this group of compounds.

8.1 Excitotoxicity

Hyperactivation of glutamate receptors by SAAs will cause death of neurons through a "classical" excitotoxic mechanism. Examples include DL-CSA- and DL-Hcys-induced epileptic activity following direct administration to rats (Butcher et al., 1992) and L-CSA-, L-HCA-, L-CA-, and L-Hcys-induced toxicity in human neuronal cell lines (Parsons et al., 1998). Predictably, blockade of glutamate receptors prevents SAA-induced excitotoxicity. L-HCA-induced lipid peroxidation of rat brain synaptosomes was blocked by the noncompetitive NMDA receptor antagonist, dizocilpine (MK-801) (Jaro-Prado et al., 2003), indicative of a mechanism involving NMDA receptor activation, NOS formation, and associated free radical formation that results, inter alia, in lipid peroxidation. The competitive NMDA antagonist, DL-2-amino-5-phosphonovalerate (DL-APV), blocked L-CA- and L-cys-induced toxicity in energy-deprived rat hippocampal slices (Schurr et al., 1993). In accordance with the fact that L-CSA- and L-CA are good substrates for the glutamate transporters, the cytotoxicity induced in cultured cerebrocortical neurons by these SAAs is potentiated by inclusion of the glutamate uptake inhibitor, L-aspartate-β-hydroxamate (Frandsen et al., 1993).

Nevertheless, there is no evidence to date that suggests that excitotoxicity associated with endogenous SAAs is a causative factor for any human neurodegenerative disorder, with the possible exception of L-Hcys (see ❿ Sect. 8.4). On balance, the endogenous concentration of the excitatory SAAs, being much lower than for L-glutamate, makes it improbable that a level required for sustained hyperactivation of excitatory receptors could be attained. It is more likely that excitatory SAAs may exacerbate, in certain pathologies, L-glutamate-mediated excitotoxicity.

8.2 Nonexcitatory Neurotoxicity by SAAs

Murphy and coworkers (Murphy et al., 1989, 1990; Murphy and Baraban, 1990), using immature rat cortical neurons and a neuroblastoma-retinal hybrid cell line, first described a mechanism of L-glutamate toxicity that was not directly linked to activation of glutamatergic receptors. Instead, L-glutamate toxicity involved inhibition of L-cySS transport, leading to oxidative stress arising from depletion of cellular GSH. These findings highlighted, for the first time, the potential pathological consequences of x_c^- exchanger dysfunction. Several other instances of toxicity incurred by inhibition of the x_c^- exchanger have since emerged. Inhibition of L-cySS transport by lactate decreases GSH synthesis (Koyama et al., 2000) and an increase in the transport of L-cySS has been recorded as a response to ischemic stress (Koyama et al., 1995). A possible cause of methylmercury toxicity may arise from failed uptake of both L-cys and L-cySS, leading to oxidative stress associated with reduced GSH (Allen et al., 2001; Shanker et al., 2001).

8.3 Toxicity of L-Cysteine

There are several indications that the mechanism of L-cys-induced toxicity is different to that of other excitatory amino acids and cannot be explained merely by hyperactivation of glutamate receptors—not least because L-cys is a weak NMDA receptor agonist. For a comprehensive review of L-cys neurotoxicity, see

◘ **Figure 7-7**
Overview of the potential mechanisms of L-cysteine neurotoxicity associated with NMDA receptor hyperactivation. Adapted from Janáky et al. (2000)

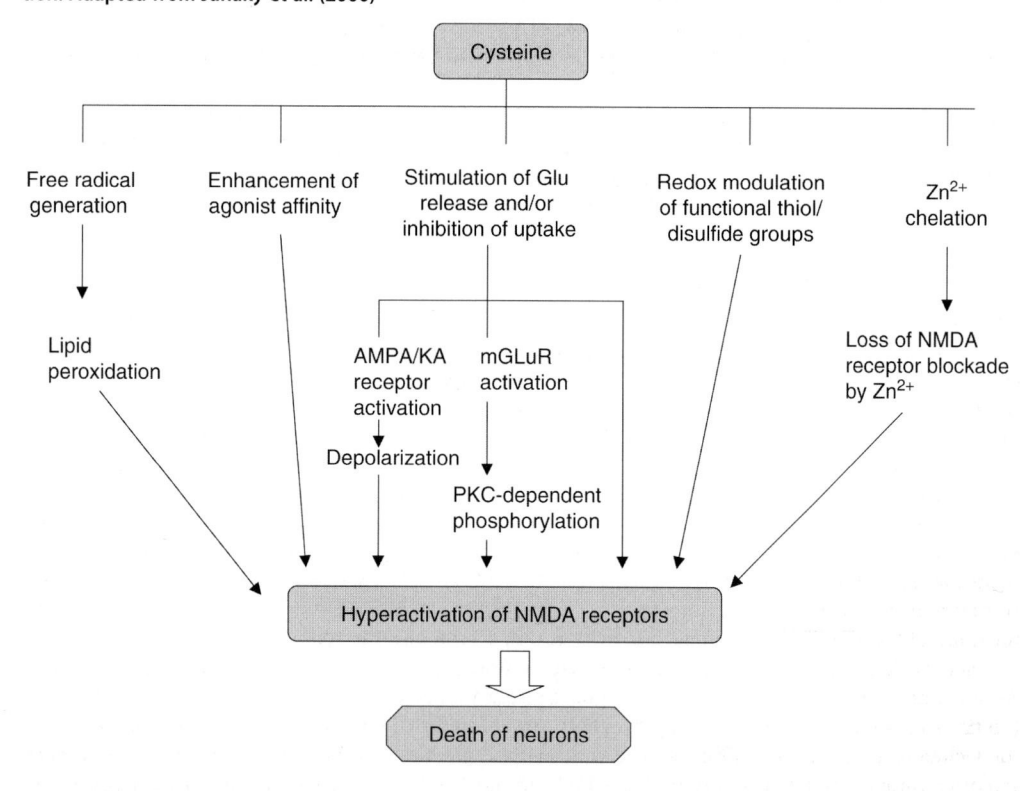

Janaky et al. (2000). Most theories are centered on the reactivity of the sulfhydryl group, either through auto-oxidation, generating hydrogen peroxide (Nath and Salahudeen, 1993), or through interaction with catecholamines, resulting in the formation of cysteinylcatechols that may inhibit mitochondrial respiration (Janaky et al., 2000). Alternatively, L-cys may increase extracellular glutamate by stimulating release or by inhibiting uptake, which would result in hyperactivation of NMDA receptors (◉ *Figure 7-7*). This scheme is thought to occur in cerebral ischemia, where the extracellular concentration of L-cys rises, possibly as a result of GSH efflux and degradation by γ-glutamyltranspeptidase (Li et al., 1999b). More recently, another mechanism of L-cys-induced neurotoxicity has been proposed by Gazit et al. (2004), who report that mice injected with L-cys (0.5- to 1.5-mg/g body weight) developed severe hypoglycemia but that resuscitation with glucose could reverse the effect.

8.4 L-Homocysteine Neurotoxicity

L-Hcys is an excitatory amino acid that is known to increase the vulnerability of neuronal cells to oxidative and excitotoxic injury in vivo and in vitro. Higher plasma L-Hcys levels in AD patients have been linked to an increased rate of temporal lobe atrophy (Clarke et al., 1998). L-Hcys has been shown to potentiate the toxicity of the amyloid β-peptide in human neuroblastoma cells (Ho et al., 2001) and in primary neuronal cultures (White et al., 2001). It has been proposed that folate deficiency contributes to excitotoxicity via increased L-Hcys generation, probably because of impaired recycling of L-Hcys back to methionine (◉ *Figure 7-2*) (Ho et al., 2003). Increased plasma L-Hcys has also been associated with an increased risk of depressive illness in humans to the extent that a recommendation has been made for dietary supplementation with vitamin B$_{12}$ and folic acid to promote L-Hcys metabolism (Coppen and Bolander-Gouaille, 2005). Hyperactivation of glutamate receptors, initiating an excitotoxic cascade that culminates in activation of caspases, DNA damage and mitochondrial dysfunction are among the proposed mechanisms of L-Hcys-induced cytotoxicity (Lipton et al., 1997; Outinen et al., 1998; Kruman et al., 2000; Ho et al., 2002; Loscalzo, 2002; Jaro-Prado et al., 2003; Mattson and Shea, 2003). Other aspects of toxicity may arise from auto-oxidation of the thiol group that contributes to increased generation of reactive oxygen species. Evidence of astrocytic cell death in vitro by L-Hcys has recently been presented (Maler et al., 2003), but the mechanism is not currently understood. The processes controlling brain L-Hcys are largely unknown, but new data has been published suggesting that activation of catechol-*O*-methyltransferase in astrocytes selectively increases synthesis of L-Hcys in these cells and its subsequent export to neurons (Huang et al., 2005).

8.5 *S*-Methylcysteine Neurotoxicity

S-methylcysteine is formed in the brain following exposure to monohalomethanes, which are known to be neurotoxic. In vitro studies, using chronic exposure of organotypic cultures to *S*-methylcysteine, revealed that, over a 24-h period, *S*-methylcysteine produced changes in membrane integrity, as determined using propidium-iodide fluorescence, that were ascribed to an interaction of the compound with GABA receptors (Pancetti et al., 2004).

8.6 Gliotoxicity

An interesting aspect of SAA-induced toxicity is that several of these amino acids have been classified as gliotoxins. This designation was first proposed by Bridges et al. (1992) to describe the fact that these compounds produced morphological changes in primary cultures of astrocytes, which culminated in lysis of the cells. Three of the five most potent gliotoxins are sulfur containing: L-SOS, L-CSA, and L-HCA (the other two being L-α-aminoadipate and L-2-amino-4-phosphonobutyrate). Neither D-HCA, L-CA nor glutamate receptor agonists, such as kainate, NMDA, and AMPA, were effective as gliotoxins, which ruled out a direct mechanism arising from receptor hyperactivation. The likelihood that the gliotoxins were targeting GSH synthesis was initially discounted on the basis that little cellular lysis was observed following depletion of GSH in mixed cortical cultures with buthionine-sulfoximine, an inhibitor of γ-glutamylcysteine synthase (Bridges et al., 1991). However, the emergence of investigations into the nonexcitatory mechanism of glutamate-induced toxicity (❯ *Sect. 8.2*), coupled with the fact that the gliotoxins are all substrates of the glutamate transporters and/or the x$_c^-$ exchanger, has led to a reappraisal of a disruption of GSH synthesis as playing a central part of the process. Certainly, the observation that L-glutamate toxicity to cultured astrocytes only occurs in the absence of GSH precursors in the medium is supportive of such a mechanism (Chen et al., 1995). An alternative proposal is that toxicity may arise from oxidative stress due to the extra demand for ATP caused by heightened activity of the glutamate transporters in the presence of the toxins (Pellerin and Magistretti, 1994; Peng et al., 2001).

More recently, metabolic fluxes measured by ^{13}C-NMR spectroscopy of cultured C6 glioma cells incubated in the presence of selected gliotoxins has highlighted important differences in their proposed mechanisms of gliotoxicity. The most dramatic effects were observed with L-SOS, which caused a profound decrease in the labeling of ^{13}C-glutamate, alanine, and lactate from [1-^{13}C]glucose (Brennan et al., 2003) and of ^{13}C-glutamate from [3-^{13}C]alanine (Brennan et al., 2004). Further experimentation revealed that L-SOS was a pseudosubstrate for alanine aminotransferase. Also noteworthy is the fact that, of the gliotoxins tested, there was a differential response on GSH levels in the cells: L-SOS (and L-α-aminoadipate) significantly increased labeling of GSH from [2-^{13}C]glycine, whereas L-CA (and D-aspartate) had the opposite effect (Brennan et al., 2004). It is concluded that, while the gliotoxins require transporter-mediated entry into glial cells, they are not acting via a common mechanism arising directly from a depletion of GSH precursors or oxidative stress initiated by heightened activity of the glutamate transporters. The current belief is that the origins of gliotoxicity may lie in the metabolic targets of these compounds inside the astrocyte.

8.7 Protection Against SAA-Induced Cytotoxicity

The ultimate goal of any research program in understanding the cytotoxicity associated with the SAAs must be to provide insight into how neurodegenerative conditions in which the SAAs may be involved could be treated, or, at best, prevented. In most cases, however, insufficient knowledge of the mechanism associated with SAA-induced toxicity means that such ambitions have not yet been realized. Nevertheless, a strategy involving elevation of GSH, through application of synthetic precursors is an option worth investigating. Thimerosal (sodium ethylmercurithiosalicylate), a one-time constituent of pediatric vaccines, is cytotoxic because of its propensity to react with thiol groups. Depletion of intracellular GSH in cultured neuroblastoma and glioblastoma cells by thimerosal was prevented by pretreatment of those cells with 100-μM GSH ethylester or N-acetylcysteine (James et al., 2005). Other studies have shown that GSH monoethylester prevents GSH depletion during focal cerebral ischemia (Anderson et al., 2004) and L-cys supplementation is effective in increasing peripheral GSH synthesis in malnourished children (Badaloo et al., 2002).

9 Conclusions

It is clear from several aspects of research into the action of SAAs that they are not, in the classical sense, neurotransmitters. While many of them are potent neuroexcitants, their sphere of activity in the brain appears to be centered much more on glial cells (astrocytes) than is the case for the excitatory neurotransmitters. This is evident from an examination of almost all aspects of their function: synthetic enzymes are localized in glia; their site of release may be from astrocytes; transport is directed into astrocytes as well as neurons; de novo synthesis of GSH occurs in astrocytes and, as gliotoxins, their pathological effects are centered on astrocytes. The role of astrocytes in promoting neuronal survival and recovery following a cerebral insult is becoming increasingly appreciated, and new instances of neuronal–glial interactions are being discovered. It is likely that additional insights into the function of this important class of neuroactive substances will emerge, leading to a better understanding of the nature of the complex interactions, at the neurochemical level, between the SAAs and excitatory neurotransmission.

References

Allen JW, Shanker G, Aschner M. 2001. Methylmercury inhibits the in vitro uptake of the glutathione precursor, cystine, in astrocytes, but not in neurons. Brain Res 894: 131-140.

Almarghini K, Remy A, Tappaz M. 1991. Immunocytochemistry of the taurine biosynthesis enzyme, cysteine sulfinate decarboxylase, in the cerebellum: Evidence for a glial localisation. Neuroscience 43: 111-119.

Anderson KJ, Monaghan DT, Bridges RJ, Tavoularis AL, Cotman CW. 1990. Autoradiographic characterization of putative excitatory amino acid transport sites. Neuroscience 38: 311-322.

Anderson MF, Nilsson M, Sims NR. 2004. Glutathione monoethylester prevents mitochondrial glutathione depletion during focal cerebral ischaemia. Neurochem Int 44: 153-159.

Badaloo A, Reid M, Forrester T, Heird WC, Jahoor F. 2002. Cysteine supplementation improves the erythrocyte glutathione synthesis rate in children with severe edematous malnutrition. Am J Clin Nutr 76: 646-652.

Baker DA, Shen H, Kalivas PW. 2002a. Cystine/glutamate exchange serves as the source for extracellular glutamate: Modifications by repeated cocaine administration. Amino Acids 23: 161-162.

Baker DA, Xi ZX, Shen H, Swanson CJ, Kalivas PW. 2002b. The origin and neuronal function of in vivo nonsynaptic glutamate. J Neurosci 22: 9134-9441.

Bannai S. 1984. Induction of cystine and glutamate transport activity in human fibroblasts by diethylmaleate and other electrophilic agents. J Biol Chem 259: 2435-2440.

Bannai S, Christensen HN, Vadgama JV, Ellory JC, Engelsberg E, et al. 1984. Amino acid transport systems. Nature 311: 308.

Bannai S, Kitamura E. 1980. Transport interaction of L-cystine and L-glutamate in human dipliod fibroblasts in culture. J Biol Chem 255: 2372-2376.

Bannai S, Kitamura E. 1981. Role of proton dissociation in the transport of cystine and glutamate in human diploid fibroblasts in culture. J Biol Chem 256: 5770-5772.

Bender AS, Reichelt W, Norenberg MD. 2000. Characterization of cystine uptake in cultured astrocytes. Neurochem Int 37: 269-276.

Benz B, Grima G, Do KQ. 2004. Glutamate-induced homocysteic acid release from astrocytes: Possible implication in glia-neuron signalling. Neuroscience 124: 377-386.

Bouvier M, Miller BA, Szatkowski M, Attwell D. 1991. Electrogenic uptake of sulphur-containing analogues of glutamate and aspartate by Muller cells from the salamander retina. J Physiol 444: 441-457.

Brand A, Leibfritz D, Hamprecht B, Dringen R. 1998. Metabolism of cysteine in astroglial cells: Synthesis of hypotaurine and taurine. J Neurochem 71: 827-832.

Brennan L, Hewage C, Malthouse JPG, McBean GJ. 2003. An NMR study of alterations in [1-^{13}C]glucose metabolism in C6 glioma cells by gliotoxic amino acids. Neurochem Int 42: 441-448.

Brennan L, Hewage C, Malthouse JPG, McBean GJ. 2004. Gliotoxins disrupt alanine metabolism and glutathione production in C6 glioma cells: a ^{13}C NMR spectroscopic study. Neurochem Int 45: 1155-1165.

Bridges CC, Kekuda R, Wang H, Prasad PD, Mehta P, et al. 2001. Structure, function, and regulation of human cystine/glutamate transporter in retinal pigment epithelial cells. Invest Ophthalmol Vis Sci 42: 47-54.

Bridges RJ, Hatalski CG, Shim SN, Cummings BJ, Vijayan V, et al. 1992. Gliotoxic actions of excitatory amino acids. Neuropharmacology 31: 899-907.

Bridges RJ, Kesslak JP, Nieto SM, Broderick JT, Yu J, et al. 1987b. A L-[^{3}H]glutamate binding site on glia: An autoradiographic study on implanted astrocytes. Brain Res 415: 163-168.

Bridges RJ, Koh J, Hatalski CG, Cotman CW. 1991. Increased excitotoxic vulnerability of cortical cultures with reduced levels of glutathione. Eur J Pharmacol 192: 199-200.

Bridges RJ, Nieto SM, Kadri M, Cotman CW. 1987a. A novel chloride-dependent L-[3H]glutamate binding site in astrocyte membranes. J Neurochem 48: 1-7.

Broer A, Brookes N, Ganapathy V, Dimmer KS, Wagner CA, et al. 1999. The astroglial ASCT2 amino acid transporter as a mediator of glutamine efflux. J Neurochem 73: 2184-2194.

Butcher SP, Cameron D, Kendall L, Griffiths R. 1992. Homocysteine-induced alterations in extracellular amino acids in rat hippocampus. Neurochem Int 20: 75-80.

Chen Q, Harris C, Brown CS, Howe A, Surmeier DJ, et al. 1995. Glutamate-mediated excitotoxic death of cultured striatal neurones is mediated by non-NMDA receptors. Exp Neurol 136: 212-224.

Chen Y, Swanson RA. 2003. The glutamate transporters EAAT2 and EAAT3 mediate cysteine uptake in cortical neuronal cultures. J Neurochem 84: 1332-1339.

Cho Y, Bannai S. 1990. Uptake of glutamate and cystine in C-6 glioma cells and in cultured astrocytes. J Neurochem 55: 2091-2097.

Clarke R, Smith AD, Jobst KA, Refsum H, Sutton L, et al. 1998. Folate, vitamin B12 and serum total homocysteine levels in confirmed Alzheimer's disease. Arch Neurol 55: 1449-1455.

Coppen A, Bolander-Gouaille C. 2005. Treatment of depression: time to consider folic acid and Vitamin B$_{12}$. J Psychopharmacol 19: 59-65.

Croucher MJ, Thomas LS, Ahmadi H, Lawrence V, Harris JR. 2001. Endogenous sulphur-containing amino acids: Potent agonists at presynaptic metabotropic glutamate autoreceptors in the rat central nervous system. Br J Pharmacol 133: 815-814.

Cuénod M, Do KQ, Grandes P, Morino P, Streit P. 1990. Localisation and release of homocysteic acid, an excitatory sulfur-containing amino acid. J Histochem Cytochem 38: 1713-1715.

Cuénod M, Grandes P, Zangerle P, Streit P, Do KQ. 1993. Sulphur-containing excitatory amino acids in intercellular communication. Biochem Soc Transact 21: 72-77.

Curras MC, Dingledine R. 1992. Selectivity of amino acid transmitters acting at N-methyl-D-aspartate and amino-3-hydroxy-5-methyl-4-isoxazolepropionate receptors. Mol Pharmacol 41: 520-526.

Curtis DR, Watkins JC. 1960. The excitation and depression of spinal neurones by structurally related amino acids. J Neurochem 6: 117-141.

Danbolt NC. 2001. Glutamate uptake. Prog Neurobiol 65: 1-105.

Do KQ, Mattenberger M, Streit P, Cuénod M. 1986. In vitro release of endogenous excitatory sulfur-containing amino acids from various rat brain regions. J Neurochem 46: 779-786.

Do KQ, Tappaz ML. 1996. Specificity of cysteine sulphinate decarboxylase (CSD) for sulphur-containing amino acids. Neurochem Int 28: 363-371.

Dowd LA, Coyle AJ, Rothstein JD, Pritchett DB, Robinson MB. 1996. Comparison of Na^{+}-dependent glutamate transport activity in synaptosomes, C6 glioma, and Xenopus oocytes expressing excitatory amino acid carrier 1 (EAAC1). Mol Pharmacol 49: 465-473.

Dringen R. 2000. Metabolism and functions of glutathione in brain. Prog Neurobiol 62: 649-671.

Dringen R, Hirrlinger J. 2003. Glutathione pathways in the brain. Biol Chem 384: 505-516.

Dringen R, Pfeiffer B, Hamprecht B. 1999. Synthesis of the antioxidant glutathione in neurons: Supply by astrocytes of CysGly as precursor for neuronal glutathione. J Neurosci 19: 562-569.

Dunlop J, Grieve A, Damgaard I, Schousboe A, Griffiths R. 1992. Sulphur-containing excitatory amino acid-evoked Ca^{2+}-independent release of D-[^{3}H]aspartate from cultured cerebellar granule cells: The role of glutamate receptor activation coupled to reversal of the acidic amino acid plasma membrane carrier. Neuroscience 50: 107-115.

Dunlop J, Grieve A, Schousboe A, Griffiths R. 1991. Stimulation of gamma-[^{3}H]aminobutyric acid release from cultured mouse cerebral cortex neurons by sulphur-containing excitatory amino acid transmitter candidates: Receptor activation mediates two distinct mechanisms of release. J Neurochem 57: 1388-1397.

Flynn J, McBean GJ. 2000. Kinetic and pharmacological analysis of L-[^{35}S]cystine transport into rat brain synaptosomes. Neurochem Int 36: 513-521.

Frandsen A, Schousboe A, Griffiths R. 1993. Cytotoxic actions and effects on intracellular Ca^{2+} and cGMP concentrations of sulphur-containing excitatory amino acids in cultured cerebrocortical neurons. J Neurosci Res 34: 331-339.

Gazit V, Ben-Abraham R, Coleman R, Weizman A, Katz Y. 2004. Cysteine-induced hypoglycaemic brain damage: An alternative mechanism to excitotoxicity. Amino Acids 26: 163-168.

Gochenauer GE, Robinson MB. 2001. Dibutyryl-cAMP (dbcAMP) up-regulates astrocytic chloride-dependent L-[^{3}H]glutamate transport and expression of both system xc(-) subunits. J Neurochem 78: 276-286.

Grandes P, Do KQ, Morino P, Cuénod M, Streit P. 1991. Homocysteate, an excitatory transmitter candidate localised in glia. Eur J Neurosci 3: 1370-1373.

Grange E, Gharib A, Lepetit P, Guillaud J, Sarda N, et al. 1992. Brain protein sythesis in the conscious rat using L-[^{35}S]methionine: relationship of methioine specific activity between plasma and precursor compartment and evaluation of methionine metabolic pathways. J Neurochem 59: 1437-1443.

Grieve A, Griffiths R. 1992. Simultaneous measurement by HPLC of the excitatory amino acid candidates homocysteate and homocysteine sulphinate supports a predominantly astrocytic localisation. Neurosci Lett 28: 1-5.

Griffiths R. 1990. Cysteine sulphinate (CSA) as an excitatory amino acid transmitter candidate in the mammalian central nervous system. Prog Neurobiol 35: 313-323.

Griffiths R. 1993. The biochemistry and pharmacology of sulphur-containing amino acids. Biochem Soc Transact 21: 63-72.

Guebel D, Torres NV. 2004. Dynamics of sulfur amino acids in mammalian brain: Assessment of the astrocytic-neuronal cysteine interaction by a mathematical hybrid model. Biochim Biophys Acta 1647: 12-28.

Hayes D, Wießner M, Rauen T, McBean GJ. 2005. Transport of L-[^{14}C]cystine and L-[^{14}C]cysteine by subtypes of high affinity glutamate transporters over-expressed in HEK cells. Neurochem Int 46: 585-594.

Ho PI, Collins SC, Dhitavat S, Ortiz D, Ashline D, et al. 2001. Homocysteine potentiates beta-amyloid neurotoxicity: Role of oxidative stress. J Neurochem 78: 249-253.

Ho PI, Ashline D, Dhitavat S, Ortiz D, Collins SC, et al. 2003. Folate deprivation induces neurodegeneration: Roles of oxidative stress and increased homocysteine. Neurobiol Dis 14: 32-42.

Ho PI, Ortiz D, Rogers E, Shea TB. 2002. Multiple aspects of homocysteine neurotoxicity: Glutamate excitotoxicity, kinase hyperactivation and DNA damage. J Neurosci Res 70: 694-702.

Huang G, Dragan M, Freeman D, Wilson JX. 2005. Activation of catechol-O-methyltransferase in astrocytes stimulates homocysteine synthesis and export to neurons. Glia 51: 47-55.

Ida S, Ohkuma S, Kimori M, Kuriyama K, Morimoto N, et al. 1985. Regulatory role of cysteine dioxygenase in cerebral biosynthesis of taurine. Analysis using cerebellum from 3-acetylpyridine-treated rats. Brain Res 344: 62-69.

James SJ, Slikker WJ, Melnyk S, New E, Pogribna M, et al. 2005. Thimerosal neruotoxicity is associated with glutathione depletion: Protection with glutathione precursors. Neurotoxicology 26: 1-8.

Janaky R, Varga V, Hermann A, Saransaari P, Oja SS. 2000. Mechanisms of L-cysteine neurotoxicity. Neurochem Res 25: 1397-1405.

Jaro-Prado A, Ortega-Vazquez A, Martinez-Ruano L, Rios C, Santamaria A. 2003. Homocysteine-induced brain lipid peroxidation: Effects of NMDA receptor blockade, antioxidant treatment and nitric oxide synthase inhibition. Neurotox Res 5: 237-243.

Johnson CL, Johnson CG. 1993. Substance P regulation of glutamate and cystine transport in human astrocytoma cells. Recept Channels 1: 53-59.

Kanai Y. 1997. Family of neutral and acidic amino acid transporters: Molecular biology, physiology and medical implications. Curr Opin Cell Biol 9: 565-572.

Kanai Y, Hediger MA. 2003. The glutamate and neutral amino acid transporter family: Physiological and pharmacological implications. Eur J Pharmacol 479: 234-247.

Kato S, Negishi K, Mawatari K, Kuo CH. 1992. A mechanism of glutamate toxicity in C6 glioma cells involving inhibition of cystine uptake leading to glutathione depletion. Neuroscience 48: 903-914.

Keller HJ, Do Q, Kollinger M, Winterhalter HK, Cuénod M. 1989. Cysteine: Depolarisation-induced release from rat brain in vitro. J Neurochem 52: 1801-1806.

Kilpatrick IC, Mozley LS. 1986. An initial analysis of the regional distribution of excitatory sulphur-containing amino acids in the rat brain. Neurosci Lett 72: 189-193.

Kingston AE, Lowndes J, Evans N, Clark B, Tomlinson R, et al. 1998. Sulphur-containing amino acids are agonists for group 1 metabotropic receptors expressed in clonal RGT cell lines. Neuropharmacology 37: 277-287.

Klancnik JM, Cuénod M, Gähwiler BH, Jiang ZP, Do KQ. 1992. Release of endogenous amino acids, including homocysteic and cysteine sulphinic acid from rat hippocampal slices evoked by electrical stimulation of Schaffer collateral-commisural fibres. Neuroscience 49: 557-570.

Knickelbein RG, Seres T, Lam G, Johnston RB, Warshaw JB. 1997. Characterisation of multiple cysteine and cystine transporters in rat alveolar type II cells. Am J Physiol 273: L1147-L1153.

Kocki T, Luchowski P, Luchowska E, Wielosz M, Turski WA, et al. 2003. L-Cysteine sulphinate, endogenous sulphur-containing amino acid, inhibits rat brain kynurenic acid production via selective interference with kynurenine aminotransferaseII. Neurosci Lett 346: 97-100.

Koyama Y, Ishibashi T, Baba A. 1995. Increase in chloride-dependent L-glutamate transport activity in synaptic membrane after in vivo ischaemic treatment. J Neurochem 65: 1798-1804.

Koyama Y, Kimura Y, Hashimoto H, Matsuda T, Baba A. 2000. L-lactate inhibits L-cystine/L-glutamate exchange transport and decreases glutathione content in rat cultured astrocytes. J Neurosci Res 59: 685-691.

Kranich O, Dringen R, Sandberg M, Hamprecht B. 1998. Utilization of cystine and cysteine precursors for the synthesis of glutathione in astroglial cultures: Preference for cystine. Glia 22: 11-18.

Kranich O, Hamprecht B, Dringen R. 1996. Different preferences in the utilisation of amino acids for glutathione synthesis in cultured neurons and astroglial cells derived from rat brain. Neurosci Lett 219: 211-214.

Kruman I, Culmsee C, Chan SL, Kruman Y, Gou Z, et al. 2000. Homocysteine elicits a DNA damage response in neurons that promotes apoptosis and hypersensitivity to excitotoxicity. J Neurosci 20: 6920-6926.

Li H, Marshall ZM, Whorton AR. 1999a. Stimulation of cystine uptake by nitric oxide: Regulation of endothelial cell glutathione levels. Am J Physiol 276: C803-C811.

Li X, Wallin C, Weber SG, Sandberg M. 1999b. Net efflux of cysteine, glutathione and related metabolites from rat hippocampal slices during oxygen/glucose deprivation: Dependence on γ-glutamyl transpeptidase. Brain Res 815: 81-88.

Lipton SA, Kim WK, Choi YB, Kumar S, D'Emilia DM, et al. 1997. Neurotoxicity associated with dual actions of homocysteine at the N-methyl-D-aspartate receptor. Proc Natl Acad Sci USA 94: 5923-5928.

Loscalzo J. 2002. Homocysteine and dementias. New Engl J Med 346: 466-468.

Maler JM, Seifert W, Hüther G, Wiltfang J, Rüther E, et al. 2003. Homocysteine induces cell death of rat astrocytes in vitro. Neurosci Lett 347: 85-88.

Mattson MP, Shea TB. 2003. Folate and homocysteine metabolism in neural plasticity and neurodegenerative disorders. Trends Neurosci 26: 137-146.

McBean GJ. 2002. Cerebral cystine uptake: a tale of two transporters. Trends Pharmacol Sci 23: 299-302.

McBean GJ, Flynn J. 2001. Molecular mechanisms of cystine transport. Biochem Soc Transact 29: 717-722.

Miralles VJ, Martínez-López I, Zaragozá R, Borrás E, García C, et al. 2001. Na^+ dependent glutamate transporters (EAAT1, EAAT2 and EAAT3) in primary astrocyte cultures: Effects of oxidative stress. Brain Res 922: 21-29.

Moran MM, Melendez R, Baker D, Kalivas PW, Seamans JK. 2003. Cystine/glutamate antiporter regulation of vesicular glutamate release. Ann N Y Acad Sci 1003: 445-447.

Murphy TH, Baraban JM. 1990. Glutamate toxicity in immature cortical neurones preceeds development of glutamate receptor currents. Brain Res Dev Brain Res 57: 146-150.

Murphy TH, Miyamoto M, Sastre A, Schnaar RL, Coyle JT. 1989. Glutamate toxicity in a neuronal cell line involves cystine transport leading to oxidative stress. Neuron 2: 1547-1558.

Murphy TH, Schnaar RL, Coyle JT. 1990. Immature cortical neurones are uniquely sensitive to glutamate toxicity by inhibition of cystine uptake. FASEB J 4: 1624-1633.

Nath KA, Salahudeen AK. 1993. Autoxidation of cysteine generates hydrogen peroxide: Cytotoxicity and attenuation of pyruvate. Am J Physiol 264: F306-F314.

O'Connor E, Devesa A, Garcia C, Puertes IR, Pellin A, et al. 1995. Biosynthesis and maintenance of GSH in primary astrocyte cultures: Role of cysteine and ascorbate. Brain Res 680: 157-163.

Oja SS, Janáky R, Varga V, Saransaari P. 2000. Modulation of glutamate receptor functions by glutathione. Neurochem Int 37: 299-306.

Outinen PA, Sood SK, Liaw PC, Sarge KD, Maeda N, et al. 1998. Characterization of the stress-inducing effects of homocysteine. Biochem J 332: 213-221.

Pancetti F, Oyarcee M, Aranda M, Parodi J, Aguayo LG, et al. 2004. S-methylcysteine may be a causative factor in monohalomethane neurotoxicity. Neurotoxicology 25: 817-823.

Parpura V, Basarsky TA, Liu F, Jeftinija K, Jeftinija S, et al. 1994. Glutamate-mediated astrocyte-neuron signalling. Nature 369: 744-747.

Parsons RB, Waring RH, Ramsden DB, Williams AC. 1998. In vitro effect of the cysteine metabolites homocysteic acid, homocysteine and cysteic acid upon human neuronal cell lines. Neurotoxicology 19: 599-603.

Patel SA, Warren BA, Rhoderick JF, Bridges RJ. 2004. Differentiation of substrate and non-substrate inhibitors of transport system x_c^-: An obligate exchanger of L-glutamate and L-cystine. Neuropharmacology 46: 273-284.

Pellerin L, Magistretti PJ. 1994. Glutamate uptake into astrocytes stimulates aerobic glycolysis: A mechanism coupling neuronal activity to glucose utilisation. Proc Natl Acad Sci USA 91: 10623-10629.

Peng L, Swanson RA, Hertz L. 2001. Effects of L-glutamate, D-aspartate and monesin on glycolytic and oxidative glucose metabolism in mouse astrocyte cultures: Further evidence that glutamate uptake is metabolically driven by oxidative metabolism. Neurochem Int 38: 437-443.

Porter RH, Roberts PJ. 1993. Glutamate metabotropic receptor activation in neonatal rat cerebral cortex by sulphur-containing excitatory amino acids. Neurosci Lett 154: 78-80.

Pow DV. 2001a. Visualising the activity of the cystine-glutamate antiporter in glial cells using antibodies to aminoadipic acid, a selectively transported substrate. Glia 34: 27-38.

Pow DV. 2001b. Amino acids and their transporters in the retina. Neurochem Int 38: 463-484.

Pullan LM, Olney JW, Price MT, Compton RP, Hood WF, et al. 1987. Excitatory amino acid receptor potency and subclass specificity of sulfur-containing amino acids. J Neurochem 49: 1301-1307.

Reichelt W, Stabel-Burow J, Pannicke T, Weichert H, Heinemann U. 1997. The glutathione level of retinal Müller glial cells is dependent on the high-affinity sodium-dependent uptake of glutamate. Neuroscience 77: 1213-1224.

Robinson MB. 2002. Regulated trafficking of neurotransmitter transporters: Common notes but different melodies. J Neurochem 80: 1-11.

Robinson MB, Sinor JD, Dowd LA, Kerwin JF. 1993. Subtypes of sodium-dependent high affinity L-[^{3}H]glutamate transport activity: Pharmacologic specificity and regulation by sodium and potassium. J Neurochem 60: 167-179.

Rothstein JD, Dykes-Hoberg M, Pardo CA, Bristol LA, Kuncl RW, et al. 1996. Knockout of glutamate transporters reveals a major role for astroglial transport in excitotoxicity and clearance of glutamate. Neuron 16: 675-686.

Rothstein JD, Martin L, Levey AI, Dykes-Hoberg M, Jin L, et al. 1994. Localisation of neuronal and glial glutamate transporters. Neuron 13: 635-643.

Sagara JI, Miura K, Bannai S. 1993a. Cystine uptake and glutathione level in foetal brain cells in primary culture and in suspension. J Neurochem 61: 1667-1671.

Sagara JI, Miura K, Bannai S. 1993b. Maintenance of neuronal glutathione by glial cells. J Neurochem 61: 1672-1676.

Sakai K, Shimizu H, Koike T, Furuya S, Watanabe M. 2003. Neutral amino acid transporter ASCT1 is preferentially expressed in L-Ser-synthetic/storing glial cells in the mouse brain with transient expression in developing capillaries. J Neurosci 23: 550-560.

Sato H, Tamba M, Ishii T, Bannai S. 1999. Cloning and expression of a plasma membrane cystine/glutamate exchange transporter composed of two distinct proteins. J Biol Chem 274: 11455-11458.

Sato H, Tamba M, Kuriyama-Matsumura K, Okuno S, Bannai S. 2000. Molecular cloning and expression of human xCT, the light chain of amino acid transport system xc. Antioxid Redox Signal 2: 665-671.

Schurr A, West CA, Heine MF, Rigor BM. 1993. The neurotoxicity of sulfur-containing amino acids in energy-deprived rat hippocampal slices. Brain Res 601: 317-320.

Selema G, Cristofol RM, Gasso S, Griffiths R, Rodriguez-Farre E. 1997. Sulphur-containing amino acids modulate noradrenaline release from hippocampal slices. J Neurochem 68: 1534-1541.

Shanker G, Allen JW, Mutkus LA, Aschner M. 2001. Methylmercury inhibits cysteine uptake in cultured primary astrocytes, but not in neurons. Brain Res 914: 159-165.

Shanker G, Aschner M. 2001. Identification and characterisation of uptake systems for cystine and cysteine in cultured astrocytes and neurones: Evidence for methylmercury-targeted disruption of astrocyte transport. J Neurosci Res 66: 998-1002.

Shih AY, Murphy TH. 2001. xCt cystine transporter expression in HEK293 cells: Pharmacology and localization. Biochem Biophys Res Commun 282: 1132-1137.

Stone TW. 2000. Development and therapeutic potential of kynurenic acid and kynurenine derivatives for neuroprotection. Trends Pharmacol Sci 21: 149-154.

Sugiyama H, Ito I, Hirono C. 1987. A new type of glutamate receptor linked to inositol phosphate metabolism. Nature 325: 531-533.

Tang XC, Kalivas PW. 2003. Bidirectional modulation of cystine/glutamate exchanger activity in cultured cortical astrocytes. Ann N Y Acad Sci 1003: 472-475.

Thompson GA, Kilpatrick IC. 1996. The neurotransmitter candidature of sulphur-containing excitatory amino acids in the mammalian central nervous system. Pharmacol Ther 72: 25-36.

Tsai MJ, Chang YF, Schwarcz R, Brookes N. 1996. Characterisation of L-α-aminoadipic acid transport in cultured rat astrocytes. Brain Res 741: 166-173.

Tschopp P, Streit P, Do KQ. 1992. Homocysteate and homocysteine sulfinate, excitatory transmitter candidates present in rat astroglial cultures. Neurosci Lett 145: 6-9.

Vandenberg RJ, Mitrovic AD, Johnston GAR. 1998. Serine-O-sulphate transport by the human glutamate transporter, EAAT2. Br J Pharmacol 123: 1593-1600.

Vollenweider FX, Cuénod M, Do KQ. 1990. Effect of climbing fibre deprivation on release of endogenous aspartate, glutamate and homocysteate in slices of rat cerebellar hemispheres and vermis. J Neurochem 54: 1533-1540.

Wang XF, Cynader MS. 2000. Astrocytes provide cysteine to neurons by releasing glutathione. J Neurochem 74: 1434-1442.

Warr O, Takahashi M, Attwell D. 1999. Modulation of extracellular glutamate concentration in rat brain slices by cystine-glutamate exchange. J Physiol 514: 783-793.

Warren BA, Patel JR, Nunn PB, Bridges RJ. 2004. The lathryus excitotoxin beta-N-oxalyl-L-alpha, beta-diaminopropinic acid is a substrate of the L-cystine/L-glutamate exchanger system xc-. Toxicol Appl Pharmacol 154: 83-92.

White AR, Huang X, Jobling MF, Barrow CJ, Beyreuther K, et al. 2001. Homocysteine potentiates copper- and amyloid beta peptide-mediated toxicity in primary neuronal cultures: possible risk factors in the Alzheimer's-type neurodegenerative pathways. J Neurochem 76: 1509-1520.

Wu JY, Moss LG, Chen MS. 1979. Tissue and regional distribution of cysteic acid decarboxylase. A new assay method. Neurochem Res 4: 201-212.

Ye ZC, Rothstein JD, Sontheimer H. 1999. Compromised glutamate transport in human glioma cells: Reduction-mislocalization of sodium-dependent glutamate transporters and enhanced activity of cystine-glutamate exchange. J Neurosci 19: 10767-10777.

Zerangue H, Kavanaugh MP. 1996a. Flux coupling in a neuronal glutamate transporter. Nature 383: 634-637.

Zerangue N, Kavanaugh MP. 1996b. Interaction of L-cysteine with a human excitatory amino acid transporter. J Physiol 493: 419-423.

8 Taurine

S. S. Oja · P. Saransaari

1	Introduction	156
2	Occurrence, Distribution, and Metabolism	156
2.1	Occurrence and Distribution	156
2.2	Metabolism	158
3	Transport	159
3.1	Uptake	159
3.2	Transporters	161
3.3	Release	162
3.3.1	General Properties	162
3.3.2	Basal Release	162
3.3.3	Release Under Cell-Damaging Conditions	163
3.3.4	Taurine Release and Cell-Volume Regulation	165
3.3.5	Potassium-Stimulated Release	167
3.3.6	Glutamate Agonist-Stimulated Release	168
3.3.7	Mechanisms of Stimulated Release	168
3.3.8	Ethanol Effects	170
3.3.9	Ammonia Stimulation	171
4	Physiological and Pharmacological Actions	172
4.1	Neuromodulatory and Possible Neurotransmitter Actions	172
4.1.1	Interactions with γ-Aminobutyrate Receptors	172
4.1.2	Interactions with Glycine Receptors	173
4.1.3	Long-Lasting Synaptic Potentiation	174
4.1.4	Putative Taurine Receptors	175
4.1.5	Effects on Other Neurotransmitter Systems	176
4.2	Other Actions	176
4.2.1	Antinociceptive Effects and Thermoregulation	176
4.2.2	Seizures and Epilepsy	177
4.2.3	Neuroprotective Actions	178
4.2.4	Retina	180
4.2.5	Ethanol Effects	182
4.2.6	Energy Drinks	182
5	Taurine Derivatives	182
6	Summary and Perspectives	184

© Springer-Verlag Berlin Heidelberg 2007

Abstract: This chapter reviews the occurrence, distribution, metabolism, transport, and actions of the simple sulphur-containing nonproteinaceous amino acid taurine. It is a ubiquitous constituent of virtually all animal cells, particularly enriched in the electrically excitable cells of the nervous system, retina, heart, and muscles. Taurine is present in both neuronal and glial cells, exhibiting moderate regional and cellular variations. It is partly synthesized in situ, but the main supply to the central nervous system (CNS) is from blood plasma. Taurine is taken up by the brain via a saturable transporter and penetrates cells requiring Na^+ and Cl^-. The release is fomented by cell swelling, depolarizing stimuli, and various cell-damaging conditions.

Taurine interferes with both $GABA_A$ and glycine receptors, depending on their subunit composition and amino acid structure. It also affects $GABA_B$ receptors, at least in specific structures. The existence of specific taurine receptors and the function of taurine as a neurotransmitter await further investigation. A number of taurine derivatives have been synthesized and tested for their efficacy in counteracting seizures, ameliorating ischemia-induced damage, and protecting neural cells from the toxic actions of xenobiotics. This is an area of research, which may produce new drugs and therapeutic strategies in the future.

List of Abbreviations: AMPA, (±)-2-amino-3-hydroxy-5-methylisoxazole-4-propionate; cGMP, cyclic GMP; CNS, central nervous system; CPP, 3-[(±)-2-carboxypiperazin-4-yl]propyl-1-phosphonate; CSF, cerebrospinal fluid; DIDS, diisothiocyanostilbene-2,2′-disulfonate; DOPAC, dihydroxyphenylacetic acid; GABA, γ-aminobutyric acid; GES, guanidinoethanesulfonate; LLP-TAU, taurine-induced long-lasting potentiation; L-NAME, N^G-nitro-L-arginine methylester hydrochloride; MPP^+, 1-methyl-4-phenylpyridinium; PKA, protein kinase A; PKC, protein kinase C; SITS, 4-acetamido-4′-isothiocyanostilbene-2,2′-disulfonate; TAG, taurine antagonist; TUG1, taurine-upregulated gene 1

1 Introduction

Taurine (2-aminoethanesulfonic acid) is a simple sulphur-containing amino acid present in virtually all cells throughout the animal kingdom. In particular, it is enriched in electrically excitable tissues such as brain, retina, heart, and skeletal muscles. Several reviews on taurine have already appeared. Jacobsen and Smith (1968) completed the first exhaustive overview of taurine. We have also a number of times reviewed the roles of taurine in the central nervous system (CNS) (Oja et al., 1977; Saransaari and Oja, 1992, 2000a; Oja and Saransaari, 1996a, 2000), one of these accounts appearing in the second edition of the *Handbook of Neurochemistry* in 1983 (Oja and Kontro, 1983a). Reviews by other authors include the excellent articles by Huxtable (1989, 1992) and Sturman (1993). The reader is referred to these texts concerning the earlier studies on taurine. In this chapter, we seek to focus on the most recent articles. Taurine has also been the subject of altogether 15 international meetings during the years 1976–2005. The proceedings of all these meetings have been published; the most recent in the series of *Advances in Experimental Medicine and Biology* (volumes 315, 359, 403, 442, 483, and 526). The proceedings of the most recent taurine meeting, held in Tampere, Finland, in 2005, will appear in the same series (Oja and Saransaari, 2006). In spite of these intensive research efforts and the abundance of taurine in animal cells, its metabolism and physiological functions are not yet entirely understood.

2 Occurrence, Distribution, and Metabolism

2.1 Occurrence and Distribution

Taurine occurs only sporadically in plants but many bacteria have the ability to synthesize it from simple building elements and to break it down (Cook and Denger, 2006). In animals, taurine is a ubiquitous constituent of tissues. In the brain, it is one of the most abundant free amino acids, glutamate alone being constantly present at higher concentrations. In the retina, the levels of taurine even exceed the concentrations in the brain (Huxtable, 1989). Taurine levels vary among different animals species, being, for instance, in the guinea pig rather low but high in the rat (Oja et al., 1968). In all animal species, however, the concentration declines to about one third in the course of postnatal development (Huxtable, 1989). In aged animals, the levels of taurine to some extent further decrease (Oja et al., 1990a; Dawson et al., 1999b).

The total tissue concentration of taurine varies from area to area within the brain but no striking differences are discernible (Oja and Kontro, 1983a; Huxtable, 1989). In the rat, high concentrations are encountered in the cerebral cortex, cerebellum, and olfactory bulb (Huxtable, 1989). Some of the highest levels also occur in the hypothalamus, pituitary gland, retina, and pineal organ (Lake et al., 1996). It is noteworthy that there is much taurine in the retina even in species in which the taurine concentration in other tissues is low. Taurine is also enriched in human brain tumors, meningiomas and gliomas (Cubillos et al., 2006).

All cell types in the brain appear to contain taurine. Taurine-specific antibodies have revealed taurine-like immunoreactivity in many structures and cell types within the brain (Huxtable, 1989). To date, in the honeybee, taurine-like immunoreactivity has been detected in neuronal perikarya, axons, and terminals, whereas glial cells were not labeled (Schafer et al., 1988). On the other hand, in the lamprey spinal cord the most intense immunolabeling was in tanocytes, a moderate level being present in ependymal cells and astrocytes (Shupliakov et al., 1994). However, in the cat perihypoglossal nuclei, only glial cells appeared to immunostain for taurine (Yingcharoen et al., 1989). Nor was there evidence in the neuronal cell bodies in the cat vestibular nuclei, whereas taurine-like immunoreactivity appeared in Purkinje cell axons and glial cells (Walberg et al., 1990). In the rat, taurine-like immunoreactivity has also been observed in Purkinje cell bodies and their dendrites, as well as within mossy fibers and Golgi axons and in glial processes (Gragera et al., 1995). Rat cerebellar Purkinje cells show a high level of taurine-like immunoreactivity (Nagelhuis et al., 1993) and in this animal species both glial cells and neurons exhibit positive staining (Torp et al., 1991). In the bovine brain, taurine is enriched in synaptic vesicles in different brain areas (Kontro et al., 1980), but vesicular taurine constitutes only a small fraction of total tissue taurine. It is not certain, however, that taurine resides in the vesicular fluid, since it has an inherent tendency to bind relatively tightly to synaptic membranes. Its complete removal from them is difficult even with the use of detergents (Kontro and Oja, 1987a; Frosini et al., 2003a).

In the olfactory bulb, the concentration of taurine is higher in primary olfactory neurons than in the postsynaptic dendrites, and taurine-like immunoreactivity is more intense in the axons than in the terminals of these neurons (Didier et al., 1994). In the guinea pig cochlea, abundant taurine-like immunoreactivity is found in all the supporting tissue, but not in the tectorial membrane and the outer pillar cells. No taurine-like immunoreactivity is discernible in the inner and outer hair cells (Horner and Aurousseau, 1997). In the rat dorsal cochlear nucleus a decreasing concentration gradient is obtained from superficial to deep layers (Godfrey et al., 2000). In this nucleus and within the rat vestibular nuclei (Li et al., 1994), a significant correlation was found between the taurine and GABA distributions, but not across brain sensory regions (Ross et al., 1995). After unilateral vestibular ganglionectomy the changes are also similar in the distributions of taurine and GABA within the vestibular nuclear complex (Li et al., 1996).

Detailed localization of taurine-like immunoreactivity has been accomplished in the rat supraoptic nucleus. The most prominent immunoreactivity was found in glial cell bodies in the ventral glial lamina and in the glial processes surrounding magnocellular neurons (Decavel and Hatton, 1995). In the posterior pituitary, which contains the axon terminals of supraoptic and paraventricular magnocellular neurons, taurine is mostly present in pituicytes, the glial cells surrounding the axonal processes, and in terminals (Pow, 1993; Miyata et al., 1997). Taurine, the most abundant amino acid in the retina (Heinämäki et al., 1986), exhibits the highest concentration in the outer nuclear layer (Ross et al., 1989). Its concentration increases from center to periphery in the teleost, but not in the mammalian retina (Lima et al., 1990). However, in the former the maximal uptake rate of taurine is higher in the central zone than in the peripheral retina (Lima et al., 1991). Taurine immunolabeling is typically associated with Müller glial and photoreceptor cells (Kalloniatis et al., 1996; Marc et al., 1998). In the pigeon (Lake et al., 1996) and mammalian retinas (Lake and Verdone-Smith, 1989), the highest levels of taurine-like immunoreactivity have been found in the cone photoreceptors, in certain cells of the inner nuclear layer, and in sublaminae of the inner plexiform layer. In the human optic nerve, immunoreactive glial cells are distributed, in particular, in the nuclei, perikaryal cytoplasm, and radiating processes (Lake, 1992; Lake et al., 1992).

The expression of taurine-like immunoreactivity is subject to substantial developmental changes in the rat retina (Lake, 1994).

The extracellular concentration of taurine has been assessed in few structures in the brain. Using the zero-net-flux approach, we have shown that under resting conditions the extracellular taurine concentration in the rat striatum amounts to 25.2 ± 5.1 μM (Molchanova et al., 2004a). Our result corresponds well to the earlier value for the rat dentate gyrus, 20.6 μM, obtained by a different method (Lerma et al., 1986). Somewhat lower but comparable values have been reported with an older methodology for the rabbit olfactory bulb (11 μM) and hippocampus (6 μM) (Jacobson and Hamberger, 1985), whereas the estimates for the striatum and cerebral cortex in fetal lambs have been markedly low, 1.23 μM and 1.53 μM, respectively (Hagberg et al., 1987). The extracellular taurine concentration estimated by us and others would appear to be quite high compared with the concentrations of other neurotransmitter amino acids. For instance, the extracellular glutamate concentration in the rat striatum estimated by the same zero-net-flux approach has been 2.6 μM (Miele et al., 1996) or 1.3 μM (Lai et al., 2000), i.e., one magnitude less than that of taurine in the same brain region. The interstitial concentrations of the nonneuroactive amino acids are much higher than that of taurine, being in the high micromolar range. The intra/extracellular concentration ratios for three groups of amino acids are as follows: more than 2000 (putative neurotransmitters), less than 100 (nonneurotransmitter amino acids), and about 400 (taurine). Taurine thus occupies an intermediate position between neuroactive and nonneuroactive compounds (Lerma et al., 1986).

2.2 Metabolism

Many animal species have limitations with respect to taurine metabolism. Mammalian tissues are only able to oxidize sulphur-containing compounds but not to reduce them. Consequently, they can only excrete taurine or first conjugate and then excrete it. Furthermore, some species possess an inadequate capability to synthesize taurine to meet the body demands. For such animals, e.g., carnivorous cats, taurine is an essential nutrient. In particular, taurine deficiency in kittens results in grave malformations and functional deficits. The retina and tapetum lucidum degenerate and pathological alterations become discernible in the electroretinogram and visual evoked potentials (Imaki et al., 1986; Sturman, 1993). In the cat cerebellum, the migration of granule cells from the external granule cell layer to the inner layers is delayed and mitotic activity persists abnormally long (Sturman et al., 1985). Taurine deficiency in dissociated mouse cerebellar cells also affects neuronal migration (Maar et al., 1995), but mice and rats are not dependent on taurine supply in nutrition. In these species a taurine deficiency state in vivo is very difficult to establish. It cannot be effected by simple dietary manipulations but only by excessive administration of guanidinoethanesulfonate (GES), an effective inhibitor of taurine transport in tissues (Lake, 1981; Lake and Cocker, 1983; Marnela and Kontro, 1984; Lehmann et al., 1987), or with β-alanine (Lu and Sturman, 1996), which also competes with taurine transport. Of these two, β-alanine is less effective, at least in the retina (Lake and De Marte, 1988). The retinal changes in GES-treated rodents resemble those observed in taurine-deficient kittens (Lake, 1981, 1983, 1986). Recently, an apparently noncoding RNA called taurine-upregulated gene 1 (TUG1) was found to be expressed in the murine brain retina. A knockdown of TUG1 resulted in a malformed or nonexistent outer segment of transfected photoreceptors, resulting from the possible dysregulation of photoreceptor gene expression (Young et al., 2005).

Primates and humans are species occupying a position between the extremes of rats and cats. During the early phases of postnatal development taurine should be considered an essential nutrient, since early taurine deficiency symptoms have manifested themselves in monkeys fed human infant formulae devoid of taurine (Neuringer et al., 1985; Imaki et al., 1987). Taurine has even been found to support the development of human embryos to the blastocyst stage (Devreker et al., 1999). Ophthalmological and electrophysiological abnormalities have also been detected in some children who have been fed totally with parenteral nutrition not containing taurine (Geggel et al., 1985; Ament et al., 1986). As a result, taurine supplementation is recommended for malnourished mothers and preterm and small-for-gestational-age babies who are not on human milk (Wasserhess et al., 1993; Aerts and Van Assche, 2002). Since the 1980s infant formulas prepared from taurine-poor cow milk are consequently nowadays fortified with taurine worldwide.

The reader might further consult the reviews of Sturman (1993) and Chesney and associates (1998) for a referred exhaustive discussion of the role of taurine in development.

In animal tissues, the sulphur-containing amino acids methionine and cysteine serve as taurine precursors. In two major synthesis pathways, cysteine is either first oxidized by cysteine dioxygenase (EC 1.13.11.20) to 3-sulfinoalanine (cysteine sulfinate), which is then decarboxylated by sulfino decarboxylase (EC 4.1.1.29) (also known as cysteine sulfinate decarboxylase) to yield hypotaurine (Huxtable, 1986), or transformed by a more devious route via coenzyme A to pantetheine and then to cysteamine, which is converted to hypotaurine by means of cysteamine dioxygenase (EC 1.13.11.19) (Coloso et al., 2006). The mechanism of the final step in taurine synthesis, oxidation of hypotaurine to taurine, has been claimed to occur only by means of nonenzymatic oxidation (Fellman and Roth, 1985), but the possibility of the involvement of enzyme(s) has also been considered (Oja and Kontro, 1981; Kontro and Oja, 1985a). This old dispute would not seem to have been unambiguously settled even now.

The brain contains the enzymatic machinery for taurine biosynthesis from its precursor amino acids. The preferred route may be through cysteine decarboxylation. However, at least in cultured rat astrocytes this metabolic pathway is a very slow process (Beetsch and Olson, 1998) in comparison with the active taurine accumulation into these cells via plasma membrane transport (Beetsch and Olson, 1996). Nor do the above enzymes participating in taurine biosynthesis respond in the rat brain to differences in the dietary supply of the amino acid precursors methionine and cysteine (Stipanuk et al., 2002). On the other hand, the activity of cysteine dioxygenase in the rat liver and hepatocytes is strongly upregulated by similar regimens (Tappaz, 2004). This would indicate that taurine levels in the brain depend mainly on the supply from other tissues, particularly from the liver. Cysteine dioxygenase and sulfino decarboxylase mRNA levels were also unchanged after acute and chronic salt loading of rats (Bitoun and Tappaz, 2000c).

Taurine is not broken down in mammalian organisms. The kidneys excrete taurine into the urine, the rate of excretion reflecting the dietary intake and plasma levels (Chesney et al., 1985). The adaptive response of the kidneys to a varying supply of taurine helps to maintain the brain taurine concentration constant. Taurine supplementation to rats increased the taurine level in the whole brain, but only moderately, and not at all in the eye (Satsu et al., 2002). However, the increase in taurine in the rat hippocampus was directly proportional to the increase in plasma after different intraperitoneal taurine doses (Lallemand and De Witte, 2004). The extracellular concentration of taurine in the rat striatum increases initially severalfold after massive intraperitoneal administration, but reverts to normal within one day (Salimäki et al., 2003). In the liver taurine is conjugated with bile acids, but in the brain the only known metabolites or derivatives of taurine are certain taurine-containing di- and tripeptides in trace amounts (Marnela et al., 1984, 1985). The synthesis routes of these peptides (Reichelt et al., 1976) and their physiological roles in the brain are not precisely known, though they have been suggested to affect aminoacidergic neurotransmission, modulating the functions of synaptic amino acid receptors (Varga et al., 1987, 1988, 1989). Of these peptides, γ-L-glutamyltaurine is the most abundant and has also been shown to exhibit a number of actions outside the central nervous system (Bittner et al., 2005).

3 Transport

3.1 Uptake

The properties of cellular taurine uptake have been studied in the past using different preparations of the brain and retina, synaptosomes, cultured cells, and slices from different regions of the brain and retina. As these studies have been extensively described and discussed in the reviews by Oja and Kontro (1983a) and Huxtable (1989), we do not cite them individually here and shall not repeat the discussion in detail. In summary of these early studies the following conclusions can be drawn.
1. Taurine uptake by brain cells is concentrative and dependent on external energy derived from the action of Na^+/K^+-ATPase. Very high concentration gradients across brain plasma membranes can be attained, ranging from 1:400 to 1:2000 in different preparations. Both neurons and glial cells actively take up taurine from the extracellular spaces.

2. The uptake is saturable, consisting of two components, low and high affinity. The estimated K_m constants for the low-affinity transport have been within the micromolar range and the K_m constants of the high-affinity transport within the millimolar range, but both greatly vary depending on the preparation studied and on the experimental approach. The affinities of transport seem to be generally higher in preparations from the developing than the adult brain. The estimated maximal velocities have likewise been somewhat variable. In addition, nonsaturable penetration has often been discernible, but its true physiological nature is ambiguous, since it may merely result from taurine fluxes though partially damaged membranes.
3. The uptake is strictly dependent on Na^+ ions, exhibiting a sigmoidal dependence on the external Na^+ concentration. The uptake is maximal at concentrations above the physiological extracellular Na^+ level. The stoichiometry is 2:1 or 3:1, i.e., two or three Na^+ ions are required for the transport of one taurine molecule. Transport is also Cl^- dependent. Other ions may have only minor effects on taurine uptake in CNS preparations.
4. The saturable uptake is structurally specific for ω-amino acids (or β-amino acids), i.e., the acidic sulfonyl group and the basic amino groups should reside at the opposite ends of the molecule. The length of the separating carbon backbone in the molecule should be two or three atoms.

In more recent studies, protein kinase C (PKC) has been shown to affect neuronal and glial taurine uptake differently, downregulating glial but not neuronal uptake in cultured cells derived from the rat cerebral cortex and in human glioma and neuroblastoma cell lines (Tchoumkeu-Nzouessa and Rebel, 1996a, b). These authors thus suggest that the neuronal and astroglial taurine transporters may be structurally different. In the rat retina, PKC stimulation and inhibition with phorbol-12-myristate-13-acetate and staurosporine, respectively, had no effect on taurine uptake, whereas chelerythrine, recognized as a potent PKC inhibitor, diminished the uptake (Militante and Lombardini, 2000b), suggesting that this effect of chelerythrine involves a PKC-independent regulatory system (Militante and Lombardini, 1999a). Taurine uptake by the retina and retinal pigment epithelium is significantly enhanced by physiological concentrations of insulin as well as by a high glucose concentration, both treatments particularly increasing the maximal velocity of low-affinity uptake, but insulin abolishing the high-affinity component (Salceda, 1999). Ischemic incubation conditions reduce taurine uptake in cerebral cortical synaptosomes, totally abolishing the high-affinity uptake in adult mice, whereas this uptake component is more resistant in developing mice, apparently reflecting the larger anaerobic metabolic capacity in the latter (Saransaari and Oja, 1996). Hypoosmotic media also obliterate the high-affinity component of the uptake in mouse cerebral cortical slices (Oja and Saransaari, 1996b), which is probably a pivotal phenomenon in cell-volume regulation, as discussed later.

Taurine is a highly lipophobic compound and its exchange with the brain and blood is slow (Oja et al., 1976). Since the de novo synthesis of taurine in the brain is also minimal, it is important to note that the saturable Na^+-dependent transporter mediates taurine transport in cultured choroid plexus cells (Ramanathan et al., 1997). In keeping with this result, the choroid plexus in vivo actively participates in taurine transport, exhibiting the same saturable and Na^+-dependent properties and structural selectivity as the plasma membrane taurine transport (Chung et al., 1994; Keep and Xiang, 1996). This transporter may play a pivotal role in the disposition of taurine in the cerebrospinal fluid (Chung et al., 1996). Taurine can also be transported from the brain into the plasma by means of a carrier-mediated efflux pump, which exhibits properties similar to the taurine transporter and is thought to regulate taurine concentrations in the brain interstitial fluid (Lee and Kang, 2004). Taurine traverses the blood–brain barrier by means of saturable transport (Benrabh et al., 1995), which is Na^+ and Cl^- dependent (Tamai et al., 1995). Systemic administration of taurine causes a logarithmic dose-dependent increase in cortical dialysates in rabbits, already reaching stable levels within 20 min (Miyamoto et al., 2006). The transport activity of taurine at the blood–brain barrier is reduced in spontaneously hypertensive rats in comparison with normotensive rats (Kang, 2000). The influx of taurine into the brain is kept constant when plasma taurine concentrations are varied in rats subjected to osmotic stress (Stummer et al., 1995). Changes in extracellular taurine may be the most important factor in the long-term loss of taurine from the brain (Keep and Xiang, 1996). Experiments in an in vitro model of the blood–brain barrier have also suggested that an increase in interstitial taurine could suppress taurine transport (Kang, 2006). Tumor necrosis factor-α and hypertonicity also affect taurine transport at the blood–brain barrier. Taurine transport activity from the brain to blood is thus regulated by both cell damage and osmolality, apparently at the transcriptional level (Kang, 2000, 2006; Lee and Kang, 2004).

3.2 Transporters

Modern molecular biology methods have made it possible to clone taurine transporters from different CNS preparations (Liu et al., 1992; Smith et al., 1992; Ramamoorthy et al., 1994). Taurine transporters in the CNS and other tissues show extremely marked homology to each other, more than 90% at the amino acid level (Miyamoto et al., 1996b; Vinnakota et al., 1997). Significant homology is also discernible with the transporters of other established neurotransmitters (Smith et al., 1992; Uchida et al., 1992). The taurine transporter also takes up β-alanine (Liu et al., 1992). mRNA for the rat taurine transporter has been localized using in situ hybridization histochemistry. The distribution is more or less universal within the CNS and among different types of cells (Lake and Orlowski, 1996). The two taurine transporters TAUT-1 and TAUT-2 cloned later are however differently localized (Pow et al., 2002). TAUT-1 is detectable in the rat pituicytes, cerebellar Purkinje cells, and in photoreceptors and bipolar cells in the retina. TAUT-2 is more widely distributed in the CNS, being however predominantly associated with Bergmann glial cells in the cerebellum and with astrocytes in other brain areas and, at lower levels, with some neurons such as CA1 pyramidal cells in the hippocampus (Pow et al., 2002). Electrophysiological experiments on rat cerebellar cells confirm the presence of taurine transporters in Bergmann glia (Barakat et al., 2002). Olson and Martinho (2006) also found strong labeling of hippocampal neurons with taurine transporter-specific antibodies but only slight staining of glial cells in this brain region. The taurine transporter is an important factor for the development and maintenance of normal retinal functions and morphology, since disruption of the taurine transporter gene in mice leads to a loss of vision due to severe retinal degeneration resembling human retinitis pigmentosa (Heller-Stilb et al., 2002). Type I astrocyte cultures (Borden et al., 1995) and brain capillary endothelial cells from the rat (Kang, 2002) also exhibit TAUT mRNA.

The regulation of the functions of taurine transporters remains a partially open matter. Cholera toxin increases taurine uptake in cultured human retinal pigment epithelial cells (Ganapathy et al., 1995; Miyamoto et al., 1996a). It is therefore suggested that either cyclic AMP (cAMP) enhances taurine uptake via an increase in the affinity of the transporter (Miyamoto et al., 1996a) or by increasing the levels of transporter mRNA (Ganapathy et al., 1995). In the same cells, nitric oxide (NO) enhances taurine transporter expression, upregulating the transporter gene and increasing the maximal velocity of taurine transport (Bridges et al., 2001). In addition to these agents, there are a variety of other potential intracellular signaling pathways which may modify the rate of taurine transport (Olson and Martinho, 2006). A consistent finding for almost every neurotransmitter transporter is the modulation of their function by PKC and protein phosphorylation states (Beckman and Quick, 1998). In line with this, TAUT in renal epithelial cell lines is activated by PKC (Jones et al., 1991) and both PKC and protein kinase A (PKA) regulate the mouse retinal TAUT by altering the number of transporters in plasma membranes (Loo et al., 1996). Manganese exposure for 6 h did not affect TAUT mRNA expression (Erikson et al., 2002) but exposure for 18 h markedly increased the expression (Erikson and Aschner, 2002) in cultured rat astrocytes. In spite of this, the short-term taurine uptake was not altered. However, astrocytes may in this manner "gear up" for increased taurine transport, which may affect the cellular resistance to manganese neurotoxicity (Erikson and Aschner, 2002). When astrocyte primary cultures are exposed to hyperosmotic media, the cells shrank and the mRNA levels of taurine transporters increase, this increase being preventable by taurine added to these hyperosmotic media (Bitoun and Tappaz, 2000a). In keeping with this, TAUT mRNA also increases in salt-loaded rats (Bitoun and Tappaz, 2000c). The gene expression of the taurine transporter appears to be under the control of two antagonistic regulations: osmolarity-induced upregulation and taurine-induced downregulation (Bitoun and Tappaz, 2000b).

The most potent inhibitors of cloned taurine transporters are hypotaurine and β-alanine, (Kulanthaivel et al., 1991; Liu et al., 1992; Smith et al., 1992; Vinnakota et al., 1997), which have also most efficiently inhibited taurine transport in brain slices (Lähdesmäki and Oja, 1973) and in cultured nervous cells (Holopainen et al., 1983; Rebel et al., 1994). The transporters also accept γ-aminobutyric acid (GABA), albeit exhibiting a much lesser affinity for it (Vinnakota et al., 1997). On the other hand, the cloned GABA transporters also transport taurine, their affinity for taurine being however much lower than for GABA (Liu et al., 1993; Melamed and Kanner, 2004). Taurine transport in the presence of varying concentrations of Na^+ shows a typical sigmoidal dependence and with varying concentrations of Cl^- a single saturation process, indicating that two Na^+ and one Cl^- are required to transport one taurine molecule via the taurine transporter (Vinnakota et al., 1997). This is the same electrogenic stoichiometry as with the GABA transporter GAT1 (Beckman and Quick, 1998). The studies on taurine transport using modern molecular biology methods thus nicely confirm the above conclusions drawn from the results previously obtained by us and others with isolated synaptosomes and brain slices (see Oja and Kontro, 1983a; Huxtable, 1989).

3.3 Release

3.3.1 General Properties

Cellular release of taurine may be mediated by different mechanisms (Saransaari and Oja, 1992): (1) Taurine molecules may simply leak through plasma membranes. We may assume that this is a purely physicochemical phenomenon influenced by the prevailing concentration and electrical gradients and the structure of membranes. In view of the zwitterionic properties and poor lipid solubility of taurine, this type of release may be assumed to represent only a minor component, but it may be increased as a result of damage to membranes. (2) The second type of release is via reversal of the functions of specific taurine transporters and other, e.g., GABA transporters, which accept taurine even though with lesser affinity. This type of release is fomented by homo- and heteroexchange and under certain ionic imbalances. (3) Ion channels in the membranes may also mediate taurine release in spite of the generally larger size of ionized taurine molecules. These channels may be governed by neurotransmitters, be voltage dependent, and respond to membrane stretch due to cell-volume alterations. (4) Finally, taurine can be extruded from the cells by exocytosis originating from the cytoplasm or synaptic vesicles. Under many circumstances the release may be mediated by means of several of the above modes. In addition, taurine molecules released from the intracellular spaces may be recaptured by taurine transporters and in this manner re-enter the cell. Particularly, in studies in vivo and in experiments with brain slices in vitro the results must always be interpreted keeping this possibility in mind, but this is equally true of any other experimental approach. The apparent release subject to estimation thus reflects a balance between release and uptake. For instance, the extracellular level of taurine has been shown to be increased in the rat nucleus accumbens in vivo by local perfusion of the taurine uptake inhibitor GES (Olive et al., 2000a).

3.3.2 Basal Release

The basal release of taurine under resting conditions may at least partially reflect leakage of taurine from the cells. In experiments in vitro with brain slices and synaptosomes the rate of taurine efflux is initially high, which certainly originates from leakage through partially damaged membranes, but it gradually diminishes with time. Na^+ omission and the absence of Ca^{2+} markedly enhance the release of taurine from mouse hippocampal (Saransaari and Oja, 1999d) and brain stem slices (Saransaari and Oja, 2006) under normoxic conditions. The level of taurine is likewise markedly increased in microdialysates from the rat striatum in vivo when the probes are perfused with Ca^{2+}-free medium (Molchanova et al., 2005). The Ca^{2+} channel blockers Ni^{2+} and Cd^{2+} further enhance release in the absence of Ca^{2+}, implying the involvement of a decreased influx of Ca^{2+}. The Cl^- channel blockers 4-acetamido-4′-isothiocyanostilbene-2,2′-disulfonate (SITS) and diisothiocyanostilbene-2,2′-disulfonate (DIDS) reduce basal taurine release from cultured cerebral cortical astrocytes (Saransaari and Oja, 1999b) and in slices from the mouse brain stem (Saransaari and Oja, 2006), indicating that these channels mediate a part of taurine release. On the other hand, protein kinases would not appear to be involved in taurine release from the brain stem, whereas the cAMP second messenger systems and phospholipases enhance the release (Saransaari and Oja, 2006).

The structural analogues hypotaurine, β-alanine, and GABA markedly enhance taurine release, which reflects the contribution of taurine transporters (Saransaari and Oja, 1999e, 2006). The inhibitor of GABA uptake, nipecotate, increases the overflow of taurine in the rat ventrolateral medulla in vivo (Kapoor et al., 1990). Madelian and associates (1988) found that adenosine stimulates cAMP-mediated taurine release from LRM55 astroglial cells and proposed that this effect was mediated by adenosine A_2 receptors. The release is also potentiated by the activation of adenosine A_1 receptors from mouse hippocampal slices (Saransaari and Oja, 2000c). In the rabbit hippocampus, rich in A_1 receptors but devoid of the A_2 receptor effector adenylate cyclase, adenosine induces taurine release in vivo, this obviously not being mediated by adenosine A_2 receptors (Miyamoto and Miyamoto, 1999). These investigators also surmised that the effect might result from activation of adenosine A_1 receptors. The activation of tachykinin receptor NK_1 by substance P and related tachykinins enhances taurine release from human astrocytoma cells (Lee et al., 1992). NO donors enhance taurine release in slices of the mouse hippocampus, the effect being more pronounced in adults than in developing mice (Saransaari and Oja, 1999a).

3.3.3 Release Under Cell-Damaging Conditions

Cell-damaging conditions, e.g., hypoxia, hypoglycemia, ischemia, and the presence of metabolic poisons or free-radical-generating compounds, markedly increase taurine release, as shown with tissue slices prepared from the mouse hippocampus (Saransaari and Oja, 1996, 1997a, 1998a, 1999a, 2000b) and brain stem (P. Saransaari and S.S. Oja, unpublished results). In these preparations the above cell-damaging conditions enhance release more in slices prepared from adult than from developing mice (Saransaari and Oja, 1996, 1997a). The release is also greatly enhanced by exposure to free radicals (Saransaari and Oja, 1997a, 2004). It was in both the adult and the developing hippocampus partly Ca^{2+}-independent, mediated by Na^+- and Cl^--dependent transporters working in reverse direction (Saransaari and Oja, 1998a, 1999e, 2004; Barakat et al., 2002), and probably resulting from disruption of cell membranes and subsequent ion imbalance. In hypoxia and ischemia, the release of endogenous glutamate, aspartate, and GABA from slices increases more than taurine release when estimated in molar amounts (Saransaari and Oja, 1998b). Both ischemia and reoxygenation have increased taurine release from slices of the rat corpus striatum (Büyükuysal, 2004), and cell-damaging conditions increased the release from both cultured neurons and astrocytes (Saransaari and Oja, 1999b, d).

In the rat striatum in vivo, transient focal cerebral ischemia increases extracellular taurine in rats (Uchiyama-Tsuyuki et al., 1994). Perinatal hypoxia–ischemia increases taurine release in the hippocampus of rat offspring (Yager et al., 2002). In fetal lambs, extreme asphyxia increases extracellular taurine in all brain areas studied, the highest enhancement being encountered in the striatum (Hagberg et al., 1987). In the spinal cord of rabbits, both ischemia and reperfusion likewise increase extracellular taurine (Simpson et al., 1990). Cyanide, which inhibits cytochrome oxidase, also augments evoked taurine release from cultured cerebellar granule cells (McCaslin and Yu, 1992) and mouse hippocampal slices (Saransaari and Oja, 1997a). The metabolic inhibition of glial cells by fluorocitrate induced a marked increase in extracellular taurine in the rat substantia nigra in vivo (García Dopico et al., 2004). D-Amphetamine administered intraperitoneally at neurotoxic dosages increases extracellular taurine severalfold in the rat striatum. Sydnocarb, a Russian-developed psychostimulant, caused a markedly smaller increase (Anderzhanova et al., 2001, 2006).

The turtle brain exhibits marked resistance to anoxic injury, and consequently even prolonged oxygen deprivation does not induce taurine release from slices from the turtle brain stem (Young et al., 1993). The massive presence of extracellular taurine in the turtle brain may prevent ischemic injury (Nilsson and Lutz, 1991). The brain of the mammalian neonate is likewise much more resistant to hypoxia than that of the adult, but its adaptation to hypoxia is different, apparently in consequence of its comparatively undifferentiated state (Pérez-Pinzón et al., 1993). The output of the phrenic nerve under acute hypoxia is biphasic, an initial hyperventilation followed by reduced ventilation. The effects on taurine release in the brain stem are similar: an initial decrease followed by an increase (Hoop et al., 1999). Tamoxifen, which is neuroprotective in several ways, inhibits the release of excitatory amino acids but also the release of taurine after middle cerebral artery occlusion in rats (Kimelberg et al., 2004). It is assumed that taurine release under cell-damaging conditions could act neuroprotectively, counteracting in several ways the effects of simultaneously released excitatory amino acids. This protection could be of great importance particularly in the developing brain (Lekieffre et al., 1992; Saransaari and Oja, 1997a, 2000b). Moreover, at this developmental stage GABA is excitatory in nature (Ben-Ari, 2002).

Taurine is lost from brain tissue during ischemia (Stummer et al., 1995). A low glucose concentration during the culture period of cerebellar and neocortical neurons is associated with reduced taurine content (Waagepetersen et al., 2005). Alterations in taurine in the brain extracellular spaces and the appearance of taurine in the cerebrospinal fluid (CSF) in vivo can be assessed in order to evaluate taurine release. Closed head injury to rats produced an immediate increase in extracellular taurine in the hippocampus in vivo (Katoh et al., 1997). After severe brain trauma in humans the increasing taurine concentration in CSF may reflect ongoing glial and neuronal impairment (Stover et al., 1999). In a global model of brain ischemia in cats, there occurred a tenfold accumulation of taurine in the extracellular space of the auditory cerebral cortex, but the initial increases in glutamate, aspartate, and GABA were even greater (Shimada et al., 1993). In the internal capsule, which consists of white matter, and in the CSF the increases were markedly smaller and delayed. In keeping with this, after short-term cerebral ischemia and reperfusion taurine increases in microdialysates from the cortical gray matter, but not in those obtained from the subcortical white matter (Dohmen et al., 2005).

Ooboshi and coworkers (1995) report that the ischemia-induced release of taurine into the extracellular space of the hippocampus was 26-fold in spontaneously hypertensive adult rats, and the releases of both aspartate and glutamate only about eightfold. In aged rats, the enhancement of taurine release diminished to 16-fold, whereas the release of aspartate and glutamate was not affected by aging. In the rat hippocampus, concomitant increases in the extracellular concentrations of both excitatory and inhibitory amino acids have been observed and the magnitude of the increase in taurine release exceeded about twofold the release of glutamate (Lekieffret et al., 1992). In the rat striatum, the relative increase in extracellular taurine exceeds the increase in glutamate, being comparable to the large relative increase in GABA, the initial basal preischemic level of which is only 1/20th of that of taurine (Molchanova et al., 2004a). In the rat parietal cortex, the increase in taurine has also much exceeded the release of glutamate after complete ischemia resulting from cardiac arrest (Scheller et al., 2000a). Transient focal ischemia temporarily increases the cerebral extracellular level of taurine in vivo in the rabbit, concomitantly with the increases in glutamate, aspartate, and GABA (Matsumoto et al., 1996). Electroacupuncture has been found to ameliorate ischemic injury, which is accompanied by a further increase in interstitial taurine levels during cerebral ischemia (Zhao and Cheng, 1997).

The onset of enhanced taurine release has often been delayed and the time course prolonged after stimuli applied to different CNS preparations (Korpi et al., 1981; Saransaari and Oja, 1992). The same phenomenon is often discernible upon exposure to ischemia (Hegstad et al., 1996; Lo et al., 1998; Büyükuysal, 2004). There probably exists more than one single release mechanism for taurine under ischemic conditions. The activation of membrane phospholipases upon damage to plasma membranes has been held responsible for a substantial part of the ischemia-induced release of excitatory amino acids (O'Regan et al., 1995). In human astrocytoma cells, the activation of PKC by phorbol-12-myristate-13-acetate increased taurine release, which was blocked by staurosporine, a potent PKC inhibitor (Lee et al., 1992). In keeping with this, PKC inhibitors significantly inhibited the ischemia-induced release in the striatum of hypertensive rats (Nakane et al., 1998). However, in our experiments phospholipase inhibitors did not reduce taurine release from hippocampal slices from either adult or developing mice. Neither did inhibitors of PKC and tyrosine phosphorylation affect the ischemia-induced release (Saransaari and Oja, 1998a, 1999a). Tyrosine phosphorylation has likewise not affected GABA release from the ischemic/reperfused rat cerebral cortex, while the efflux of other amino acid neurotransmitters was reduced (Phillis et al., 1996).

In the rat cerebral cortex in vivo, dipyramidole, an inhibitor of adenosine transport and phosphodiesterase, did not significantly affect ischemia-induced taurine release (Phillis et al., 1997). Neither did adenosine A_2 receptors and adenosine transporters affect the ischemia-induced release (Saransaari and Oja, 2000c). On the other hand, Goda et al. (1998) report that adenosine A_1 receptors are involved in the ischemia-evoked release of excitatory amino acids but not in the release of taurine in the striatum of spontaneously hypertensive rats. At variance with this, our experiments with hippocampal slices indicated that activation of adenosine A_1 receptors could potentiate the ischemia-induced taurine release in adult but not in developing mice (Saransaari and Oja, 2000c), in which the free-radical-evoked release was even reduced in the hippocampus (Saransaari and Oja, 2004). The agonists of all ionotropic glutamate receptors potentiated this release, in which the enhancement by NO-generating compounds also indicated the involvement of NMDA receptors (Saransaari and Oja, 2004). The metabotropic glutamate receptors were of only minor significance.

Ion channels would appear to be involved. Replacement of Na^+ by N-methyl-D-glucosamine reduces both basal and ischemia-induced taurine release in the rat cerebral cortex (Phillis et al., 1999). Cl^- removal also enhances taurine release in ischemia, possible because of depolarization of the cells in the absence of Cl^- (Oja and Saransaari, 1992a, b; Saransaari and Oja, 2006). The anion transport inhibitor 4,4′-dinitrostilben-2,2′-disulfonate attenuates taurine release evoked by ischemia in the rat striatum in vivo (Seki et al., 1999; Kimelberg et al., 2004). In line with this, the chloride channel blockers SITS and DIDS reduce the ischemia-induced release in mouse hippocampal slices (Saransaari and Oja, 1998a). In the rat cerebral cortex in vivo, SITS and 5-nitro-2-(3-phenylpropylamino)benzoate, a highly potent chloride channel blocker, also depress the ischemia/reperfusion-evoked release of taurine (Phillis et al., 1997). Na^+-free media reduce the ischemia-induced taurine release from slices from the developing mouse hippocampus (Saransaari and Oja, 1999a). In keeping with this, Phillis and coworkers (1999) found that replacement of Na^+ with choline enhanced taurine release from the ischemic cerebral cortex in vivo. In the absence of Ca^{2+} the release of taurine from mouse hippocampal slices is reduced during ischemia (Saransaari and Oja, 1999a). In the absence of Na^+ the function of the taurine transporter may be reversed to favor extrusion of taurine from cells. The absence of Na^+ could also inhibit the Ca^{2+}/Na^+ counter transport system with a deleterious accumulation of intracellular Ca^{2+} (Phillis et al., 1999).

3.3.4 Taurine Release and Cell-Volume Regulation

There are a great number of studies on the release of taurine during cell-volume regulation. The role of taurine in osmoregulation in marine animals has been known for decades. In the 1980s, taurine was also suggested to have an osmoregulatory function in the mammalian brain (Walz and Allen, 1987) and was soon thereafter seriously considered to play the dominant role in the regulation of cell volumes in the CNS and shown to be released from cultured glial cells and neurons by hypoosmotic media (Pasantes-Morales and Schousboe, 1988; Wade et al., 1988; Schousboe and Pasantes-Morales, 1989; Kimelberg et al., 1990; Martin et al., 1990; Olson and Goldfinger, 1990; Pasantes-Morales et al., 1990; Pazdernik et al., 1990; Schousboe et al., 1990; Oja and Saransaari, 1992a, b). Hypoosmotic media likewise enhance taurine release from retinal Müller cells (Faff-Michalak et al., 1994). Taurine release from synaptosomes is affected by hypoosmolality in more or less the same manner (Sánchez Olea and Pasantes-Morales, 1990). An altered pattern of taurine release and cell-volume regulation is seen in cerebral cortical slices from chronically hypernatremic rats (Law, 1995). Taurine enhances volume regulation in hippocampal slices swollen osmotically (Kreisman and Olson, 2003). When rat cerebral astrocytes grown in hyperosmotic media are transferred to medium of normal osmolality, the cells rapidly release taurine (Olson and Goldfinger, 1990). These effects are considered to represent the so-called regulatory volume decrease in cells (Vitarella et al., 1994; Aschner et al., 1998). The threshold for volume-induced release of taurine is lower than that for the release of other amino acids participating in the regulatory volume decrease in neural cells (Franco et al., 2000).

Hypoosmotic solutions likewise markedly increase extracellular taurine in vivo (Solís et al., 1988a; Nagelhuis et al., 1994). For example, such release has been demonstrated in the rat hippocampus (Lehmann, 1989) and pituitary cells (Miyata et al., 1997). Chronic hyponatremia (Lien et al., 1991), the hyponatremia of pregnancy (Law, 1989), and the hyponatremia induced by 1-desamino-8-D-arginine vasopressin (Cordoba et al., 1998) lower the level of taurine in the rat brain. Taurine-depleted cats are particularly vulnerable to hypernatremic dehydration (Trachtman et al., 1988a), and taurine deficiency protects against cerebral edema during acute hyponatremia (Trachtman et al., 1990). In the photochemical stroke model in rats extracellular taurine increases in the peri-infarct zone, possibly compensating for tissue swelling in this area (Scheller et al., 1997). Taurine is also released into the CSF in patients with severe brain trauma as an apparent sign of astrocytic cell swelling (Seki et al., 2005). In the rat substantia nigra, taurine is assumed to act as a volume transmitter (García Dopico et al., 2004). On the other hand, when distilled water is injected intraperitoneally into rats, the taurine content in the brain gray matter does not acutely change (Olson et al., 1994, 1997). Nevertheless, a great number of investigations thus corroborate the assumption of a role of taurine as an osmolyte and unanimously witness the pivotal role of taurine in cell-volume regulation in the CNS.

The taurine transporter mRNA level is increased by exposure of primary rat astrocytes in culture to hyperosmotic medium, which is fully prevented by increasing the cell taurine content (Bitoun and Tappaz, 2000a). The significance of this finding remains open. For instance, the release of taurine induced from rat supraoptic glial cells by hypotonic stimulus is not affected by the taurine transport inhibitor GES (Brès et al., 2000). This and other findings would appear to exclude the involvement of taurine transporters (Pasantes-Morales and Schousboe, 1997), even though taurine release from cortical synaptosomes has been thought to result possibly from blockage of the carrier reversal operation (Tuz et al., 2004). The phospholipase A_2 inhibitors had no effect on the hypoosmotically induced release from the rat cerebral cortex in vivo (Estevez et al., 1999a). Furthermore, the release of taurine from hippocampal slices evoked by hypoosmolality was not affected by the PKC blocker chelerythrine or cytochalasin E (Franco et al., 2001). At variance with this, however, the same laboratory found that the hypoosmolarity-evoked release from rat cerebral cortical synaptosomes is accelerated by the PKC activator phorbol-12-myristate-13-acetate and reduced by chelerythrine (Tuz et al., 2004). On the other hand, again, Estevez and associates (1999a) report that hypoosmotically induced taurine release from the rat cerebral cortex in vivo is reduced by chelerythrine but not affected by phorbol-12-myristate-13-acetate. Taurine release associated with cell swelling is thus likely to be reduced by tyrosine kinase blockers and increased by tyrosine phosphatase inhibitors, but the specific tyrosine kinases involved are as yet unknown (Pasantes-Morales and Franco, 2002). In swollen glial cells the efflux channel activation is mediated by intracellular kinase pathways, including ERK1/2 and PI3 (Pasantes-Morales et al., 2000a, b).

The release of taurine depends on pH. Acidosis reduces the extracellular concentration of taurine, as shown by perfusion of the rat hippocampus in vivo with acidic bicarbonate buffers (Puka and Lehmann, 1994). The Cl^-/HCO_3^- antiporter probably mediates the effects of pH (Law, 1994; Puka and Lehmann, 1994). In keeping with these findings, in the bullfrog sympathetic ganglia the hypoosmolarity-induced taurine release is suppressed by the antagonists of the Cl^-/HCO_3^- exchange system, but also by the anion channel blockers SITS, DIDS, and niflumic acid (Sakai and Tosaka, 1999). The existence of volume-sensitive taurine- and Cl^--permeable channels has been demonstrated (Galietta et al., 1997). All anion channel blockers tested have reduced the hypoosmotically induced taurine release in vivo from the rat cerebral cortex, indicating that a part of taurine release occurs via volume-sensitive chloride channels (Estevez et al., 1999b). In keeping with these results again, furosemide, an inhibitor of Cl^- transport, greatly reduces the increase in extracellular taurine evoked by K^+ and permeant anions (Solís et al., 1988b), and the anion channel blockers SITS and DIDS lower the basal extracellular level of taurine in the rat striatum (Molchanova et al., 2004b). On the basis of these findings, we estimated that about one-half of taurine exits from brain cells through these volume-sensitive anion channels under resting conditions (Molchanova et al., 2006). Tamoxifen, an antiestrogen and chloride channel blocker, may attenuate chloride-related osmotic cell swelling, and hence reduce the associated volume-regulatory release of taurine in the ischemic rat cerebral cortex (Phillis et al., 1998). Vasopressin induces taurine release from rat pituicytes, which may corroborate the conception of glial control of neurohormone output (Rosso et al., 2004). On the other hand, although a number of anion channel inhibitors, including 5-nitro-2-(3-phenylpropylamino)benzoate, N-phenylanthranylate, niflumic acid, and DIDS, have reduced the osmodependent release of taurine from the rat supraoptic glial cells, other inhibitors such as dideoxyforskolin, 4-bromophenacylbromide, mibefradil, and tamoxifen have inhibited the release only marginally or not at all (Brès et al., 2000). The last mentioned authors thus suggest that the osmodependent taurine-permeable channels are formed by yet unidentified proteins.

The calcium/calmodulin-dependent processes may affect the release of taurine in anionic form from cultured rat astrocytes (Liu et al., 2002). Sulfhydryl groups seem to be essential for the volume-sensitive release of taurine from astrocytes, since the sulfhydryl group-modifying reagents N-ethylmaleimide, mersalyl, and p-chloromercuribenzenesulfonate reduce hypoosmolarity-evoked taurine release from cultured astrocytes, whereas the membrane-impermeant 5-eosin maleinimide is not effective (Martínez et al., 1994). In keeping with this, methylmercury inhibits the regulatory volume decrease of astrocytes (Aschner et al., 1998). This effect is partially attenuated by the Na^+/H^+ antiporter blocker amiloride but not by SITS, DIDS, furosemide, or bumetadine. Metallothioneins also reverse the impairment of regulatory volume decrease by methylmercury and concomitantly inhibit taurine release (Aschner, 1997). A decrease in intracellular ionic strength seems to be involved in the activation of swelling-elicited taurine release from cultured cortical astrocytes (Cardin et al., 1999).

The mechanisms by which taurine participates in cell-volume regulation have not yet been thoroughly characterized. They are apparently complex in nature (Pasantes-Morales and Franco, 2002). Taurine is not the only organic compound able to act as an osmolyte in the brain (Pasantes-Morales et al., 2002a, b). Its significance is also likely to vary in different CNS preparations and at different stages of development. For example, it may be quantitatively more important in the developing brain than the adult (Miller et al., 2000), since the taurine content in the brain decreases markedly during ontogenic development, as discussed above. The participation of taurine and other compounds, certain amino acids in particular, in cell-volume regulation is reviewed in greater detail by Pasantes-Morales in *Chapter 10*. She and her coauthors also deal thoroughly with the general mechanisms involved in cell-volume regulation (Pasantes-Morales et al., 2002a, b).

Taurine is sometimes claimed to constitute a particularly suitable osmolyte in that it is metabolically inert and has no other specific physiological functions. The latter assumption does not hold, however, since taurine is also a neuromodulator (Oja and Saransaari, 1996a). Nor is the enhanced release of taurine from neural cells invariably associated with cell swelling. The release from brain slices is under certain conditions accompanied by intracellular shrinking (Oja and Saransaari, 1992a). For example, the release was enhanced from cerebral cortical slices which shrank intracellularly when incubated in media in which chloride ions were replaced by the impermeant anion gluconate (Oja and Saransaari, 1992b). Cell shrinkage is a distinctive feature of apoptosis, and taurine release from cerebellar granule neurons cultured in conditions resulting in apoptotic death is markedly increased (Morán et al., 2000). Concordant with these results is the observation that taurine release into extracellular spaces is dramatically increased when the rat cerebral cortex is exposed to artificial CSF rendered hyperosmotic with impermeant sucrose (Phillis et al., 1999). Mannitol and gluconate exhibit the same effect in ischemia. The properties of taurine release in different brain structures may differ essentially. For example, taurine release in the substantia nigra, probably from the terminals of striatonigral neurons following stimulation of their cell bodies in the neostriatum, resembles the similarly evoked release of GABA, but taurine release in the neostriatum may not be related to neuronal activity (Bianchi et al., 1998).

3.3.5 Potassium-Stimulated Release

Electrical stimulation evokes release of endogenous taurine from hippocampal, cerebral, cortical, and cerebellar slices from the rat (Kubo et al., 1992). The release of preloaded labeled taurine evoked by depolarizing concentrations of K^+ has been demonstrated in many neuronal preparations of the CNS in vitro, e.g., synaptosomes (Kontro, 1979), cerebral cortical slices (Kontro and Oja, 1987b), cultured cerebellar granule cells (Holopainen and Kontro, 1989; Rogers et al., 1991), striatal, hippocampal, and brain stem slices (Oja et al., 1990a; Saransaari and Oja, 2006), molecular, fusiform, and deep layers of dorsal cochlear slices (Zheng et al., 2000), and Müller cells cultured from the rabbit retina (Faff-Michalak et al., 1994). K^+ stimulation likewise evokes release of endogenous taurine in vitro from cerebral cortical slices (Oja and Saransaari, 1995) and in vivo from the spinal cord (Sundström et al., 1995), substantia nigra

(Biggs et al., 1995), nucleus tractus solitarius (Sved and Curtis, 1993), striatum (Böckelmann et al., 1998; Colivicchi et al., 1998; Molchanova et al., 2004a), and cerebral cortex (Estevez et al., 2000). On the other hand, K^+ stimulation induced only insignificant release or was totally ineffective in enhancing taurine release from glial cell preparations (Holopainen and Kontro, 1989; Puka et al., 1992). In particular, when cell swelling is prevented, K^+ stimulation fails to evoke any release from astrocytes (Oja and Saransaari, 1992a). Cotransport of taurine and Na^+ also depolarizes neuronal membranes (Holopainen et al., 1990).

3.3.6 Glutamate Agonist-Stimulated Release

Other types of stimuli also induce taurine release from neural cells. The agonists of all ionotropic glutamate receptor classes evoke taurine release from cerebral cortical slices (Saransaari and Oja, 1991), cultured cerebellar granule neurons (McCaslin et al., 1992), and hippocampal slices (Magnusson et al., 1991; Saransaari and Oja, 1994, 1997b, c). N-Methyl-D-aspartate (NMDA) was effective with hippocampal slices (Menéndez et al., 1993), the rat spinal cord (Sundström et al., 1995) and parietal cortex (Scheller et al., 2000a, b), and kainate in the rabbit olfactory bulb (Jacobson and Hamberger, 1985) in vivo. Endogenous glutamate apparently increases the extracellular levels of taurine through both NMDA and ($\pm$)-2-amino-3-hydroxy-5-methylisoxazole-4-propionate (AMPA)/kainate receptors in the striatum of freely moving rats (Segovia et al., 1997). The involvement of AMPA receptors has also been confirmed by showing that AMPA markedly enhances extracellular taurine in the primary motor cortex of freely moving rats (La Bella and Piccoli, 2000). The NMDA and non-NMDA agonists evoke taurine release through a mechanism independent of cell swelling (Menéndez et al., 1990; Shibanoki et al., 1993). The NMDA receptor antagonist D-2-amino-5-phosphonovalerate blocks the NMDA-evoked taurine release from cerebral cortical slices (Saransaari and Oja, 1991), and pretreatment of rats with dizocilpine (MK-801), a noncompetitive NMDA antagonist, inhibits the NMDA-evoked release of taurine from the rat hippocampus in vivo (Katoh et al., 1997). The NMDA-evoked taurine release from the hippocampus is receptor mediated in both the adult and the developing mouse (Saransaari and Oja, 2003b). The release from the rat spinal cord is also inhibited by the NMDA antagonist 3-[($\pm$)-2-carboxypiperazin-4-yl]propyl-1-phosphonate (CPP) and, surprisingly, by the agonist at the glycine site of the NMDA receptor 3-amino-1-hydoxypyrrolidine-2-one (HA-966) (Sundström et al., 1995).

Agonists of all ionotropic glutamate receptors also potentiate the free-radical-evoked release in the hippocampus of developing mice in a receptor-mediated manner (Saransaari and Oja, 2004). On the other hand, kainate and AMPA, but not NMDA, increased extracellular taurine in the rat substantia nigra (García Dopico et al., 2004). The evoked release in the striatum after local application of kainate may be extraneuronal and may not involve non-NMDA receptors, whereas the simultaneous release in the globus pallidus results from propagation of action potentials and activation of non-NMDA receptors (Bianchi et al., 2006). To our best knowledge we have made the only investigation of the effects of metabotropic glutamate receptors on taurine release (Saransaari and Oja, 1999c). S-3,5-Dihydroxyphenylglycine, among the agonists of group I metabotropic glutamate receptors, and, (2S,2'R,3'R)-2-(2',3'-dicarboxycyclopropyl) glycine, among the group II agonists, potentiate the basal release of taurine in a receptor-mediated manner, and group II agonists are only slightly effective. In slices from the brain stem, agonists of groups I and II attenuate the K^+-evoked release in adult mice, whereas agonists of group I and III cause the same effect in developing mice (Saransaari and Oja, 2006).

3.3.7 Mechanisms of Stimulated Release

As also discussed above, there is ample evidence that taurine release is dependent on cAMP (Shain et al., 1992). The regulation of the NMDA-evoked release from the immature hippocampus appears to be somewhat complex, being affected by both PKC and adenosine (Saransaari and Oja, 2003b). The NMDA-induced taurine release is thought to be mediated via the NO cascade (Scheller et al., 2000b).

NO stimulates the formation of cyclic GMP, and hence enhances the release of neurotransmitters and taurine (Guevara-Guzman et al., 1994; Saransaari and Oja, 2002). The unstimulated basal release of taurine is generally enhanced by addition of the NO donors to the incubation medium of slices from the mouse hippocampus (Saransaari and Oja, 1999a). In keeping with this, the extracellular concentrations of taurine have increased in vivo in the rat striatum upon exposure to the NO release-inducing substances (Guevara-Guzman et al., 1994) and in the rat parietal cortex by administration of the NO donor 3-morpholinyl-sydnoneimine (SIN-1) (Scheller et al., 2000b). The NO synthase inhibitor N^G-nitro-L-arginine methylester hydrochloride (L-NAME) inhibits the NMDA-mediated taurine release (Scheller et al., 2000b). The same NO synthase inhibitor also attenuates the increase in extracellular taurine evoked in the rat striatum by K^+ stimulation (Böckelmann et al., 1998). The NO generators S-nitroso-N-acetylpenicillamine and sodium nitroprusside also dose dependently increase taurine release from cerebral cortical neurons and these enhancements are completely abolished by hemoglobin, a NO scavenger (Chen et al., 1996). On the other hand, the same compounds have attenuated the K^+-evoked release, as well as the NMDA-enhanced release, from hippocampal slices (Saransari and Oja, 1999a). The NO-generating compounds also enhance the release evoked by free radicals in the mouse hippocampus (Saransaari and Oja, 2004). An increase in cerebral cyclic GMP (cGMP) levels induces taurine release from the hippocampus, as witnessed by the stimulatory effects of 8-bromo-cyclic GMP, the phosphodiesterase inhibitor 2-(2-propyloxyphenyl)-8-azapurin-6-one (zaprinast), the soluble guanylyl cyclase inhibitor 1H-(1,2,4)oxadiazolo-4,3a)quinoxalin-1-one (ODQ), and PKC, particularly in adult mice (Saransaari and Oja, 2002). The effects are discernible in both normoxia and ischemia. The free-radical-stimulated release, potentiated by activation of ionotropic glutamate receptors and subsequent NO production, could constitute a part of the neuroprotective properties of taurine in the hippocampus, preventing excitotoxicity (Saransaari and Oja, 2004). In the rat hypothalamic supraoptic nucleus, extracellular taurine is increased by both direct administration of NO and forced swimming (Engelmann et al., 2002). The effects of NO agents depend markedly on the duration and timing of the exposure and are not identical when different NO donors are used. We have extensively reviewed the roles of glutamate receptors and NO in taurine release not long ago, and the reader is referred to Oja and Saransaari (2000) for further discussion.

Veratridine, which opens the voltage-dependent Na^+ channels, also stimulates taurine release from superfused slices from the rat substantia nigra (Della Corte et al., 1990). The selective Na^+/K^+-ATPase inhibitor ouabain evokes a marked taurine release in vitro in slices from the adult (Saransaari and Oja, 1998a) and developing mouse hippocampus (Saransaari and Oja, 1999e), from human neuroblastoma cells (Basavappa et al., 1998), and in vivo in the rat cerebral cortex (Estevez et al., 2000). Noradrenaline produces in rat astrocytes a biphasic effect: stimulation, preceded by inhibition (Puka et al., 1992). Aluminium causes changes in neurotransmission and exposure of primary astrocyte cultures from the neonatal rat cerebral cortex to ammonium chloride markedly stimulates release of endogenous taurine (Albrecht et al., 1991).

Furthermore, the adenosine transport inhibitors dipyradimole and nitrobenzylthioinosine significantly enhance taurine release from the rat hippocampus in vivo (Hada et al., 1996). Dipyramidole and the adenosine receptor A_1 agonist N^6-cyclohexyladenosine have also enhanced the K^+-stimulated release in slices from the adult hippocampus, but only in ischemia (Saransaari and Oja, 2003a). All experimental conditions which cause membrane depolarization would thus appear to enhance taurine release. On the other hand, membrane hyperpolarization shifts taurine transport from release toward uptake, resulting in seemingly diminished release (Lewin et al., 1994). Correspondingly, the anticonvulsant taurine derivatives 2-phthalimidoethanesulfon-N-isopropylamide (MY-117, taltrimide) and 2-phthalimidoethanesulfonamide (MY-103) reduce the K^+-stimulated taurine release from mouse cerebral cortical slices (Kontro and Oja, 1987c) and hippocampal slices (Saransaari and Oja, 1999f).

The mechanisms of the release of taurine may not be simple and it is often difficult to demarcate different mechanisms from each other and to analyze them separately. For instance, it is known that excitation of the CNS is accompanied by ionic movements across plasma membranes and by an increase in cell volumes (Walz and Hinks, 1985). Furthermore, stimulation by means of equimolar replacements of Na^+ by depolarizing concentrations of K^+ causes neural cells to swell and taurine is then released (Pasantes-Morales and Schousboe, 1989). The agonists of ionotropic glutamate receptors also induce cell swelling

(Menéndez et al., 1990), in particular in preparations from the developing brain (Saransaari and Oja, 1991), and hence the evoked taurine release must stem partially from this phenomenon. The release of glutamate, and possibly other compounds as well, may also secondarily affect taurine release, as suggested by experiments with the inhibitor of glutamate transport DL-threo-β-benzyloxyaspartate (Phillis et al., 2000).

Calcium dependence is generally assumed for the exocytotic release of a neurotransmitter originating from the emptying of synaptic vesicles into the synaptic cleft (Saransaari and Oja, 1992). The question of the Ca^{2+} dependency of taurine release is not yet settled. The K^+-stimulated release of taurine from cultured cerebellar neurons has been reported to be independent of external Ca^{2+} (Rogers et al., 1991). Likewise, the release from goldfish cerebellar slices (Lucchi et al., 1994; Rosati et al., 1995) and the P_2 fraction of the rat retina (Lombardini, 1992a, b; 1993a) has not evinced Ca^{2+} dependence. Furthermore, substitution of Ca^{2+} by Co^{2+} in the perfusion medium does not affect the K^+-evoked taurine release from the rat spinal cord in vivo (Sundström et al., 1995). Neither does the evoked release of taurine show any Ca^{2+} dependency in astrocytes (Holopainen et al., 1985) and in the hippocampus in vivo (Solís et al., 1986). On the other hand, the K^+-stimulated release has been found to be Ca^{2+} dependent in synaptosomes from the rat cerebral cortex (Kamisaki et al., 1996b) and the rat substantia nigra in vivo (García Dopico et al., 2004) and the release evoked by NMDA receptor activation as well (Menéndez et al., 1993). In a computer analysis of K^+-evoked release of taurine from cerebral cortical synaptosomes, a release component disappeared in the absence of Ca^{2+} (Kontro, 1979). Outside the CNS, Ca^{2+} modulates taurine release in the rat vas deferens (Diniz et al., 1999). Ca^{2+} dependence may thus differ among tissues, between the methods used to elicit the release, and with the intensity of stimulus. In the absence of external Ca^{2+} the release of this ion from intracellular stores may also support taurine release, if any part of it requires Ca^{2+} (Menéndez et al., 1993). One part of the evoked taurine release from cortical synaptosomes was also recently estimated to be exocytotic in nature (Tuz et al., 2004).

In a few studies, release of taurine has been associated with learning. The learning process of imprinting enhances taurine release from slices from the intermediate and medial hyperstriatum ventrale in the forebrain region in the domestic chick (McCabe et al., 2001). In particular, in the fast-learning chicks the release tends to increase. The enhanced taurine release upon learning may stem from the learning-induced increase in the number of NMDA receptors in this brain structure (McCabe and Horn, 1988). In both trained and untrained chicks, the evoked releases of taurine and GABA have been seen to be significantly positively correlated (Meredith et al., 2004). This correlation may be associated with the mutual modulation of their releases (Kontro and Oja, 1989). Furthermore, antagonists of all three $GABA_A$, $GABA_B$, and $GABA_C$ receptors have reduced the K^+-evoked taurine release potentiated by GABA in hippocampal slices from adult and developing mice (Saransaari and Oja, 2000a). Rats exposed to inescapable electric footshock show enhanced releases of taurine and several other neuroactive amino acids in the locus coeruleus, but conditioned fear selectively enhances the release of aspartate and taurine, the release of taurine being strikingly long lasting (Kaehler et al., 2000). Noxious tail pinch also induces a tetrodotoxin-sensitive release of taurine in the rat locus coeruleus (Singewald et al., 1995).

3.3.8 Ethanol Effects

The effects of alcohol (ethanol) administration on the brain levels of taurine have been shown in many investigations. Administration of ethanol elicited increases in extracellular taurine in the rat frontal cortex and hippocampus (Dahchour and De Witte, 2000a), amygdala (Quartemont et al., 1998a, 1999), and nucleus accumbens (Dahchour et al., 1994, 1996), and in the nucleus accumbens in PKCε-deficient mice (Olive et al., 2000b). The effect in the nucleus accumbens is due to ethanol itself, not to the ethanol metabolite acetaldehyde (Kashkin and De Witte, 2004). Isoosmotic exposure of primary astrocytes in culture to ethanol has led to both cell swelling and taurine release (Kimelberg et al., 1993). Chronic exposure (96 h) of cultured cerebral cortical astrocytes to alcohol has also significantly increased taurine uptake and the release was likewise significantly increased following ethanol withdrawal (Allen et al., 2002).

At variance with most other results, in ethanol-dependent rats the taurine concentration in the brain has been observed to be reduced by ethanol but to return to normal after its withdrawal (Iwata et al., 1980). Even low doses of ethanol elicit significant temporal increase in extracellular taurine in the rat ventral striatum, whereas the changes in the dorsal striatum are less pronounced (Smith et al., 2004). The effects of higher doses of ethanol are more long lasting. The cell bodies of immunoreactively taurine-positive astrocytes are hypertrophied and their processes elongated, and some of them are colocalized with glial fibrillary acidic protein-immunoreactive cells in the mouse hippocampus owing to short-time ethanol exposure (Sakurai et al., 2003). The precise mechanisms of ethanol-induced taurine release are not fully known, but taurine has been surmised to compensate for the osmotic changes caused by ethanol in the extracellular fluid in the brain (Dahchour et al., 1996; Quartemont et al., 1999; Allen et al., 2002). Indeed, hypoosmotic states enhance and hyperosmotic states prevent the ethanol-induced taurine release in the nucleus accumbens (Quartemont et al., 2003). It is, however, possible that osmotic manipulation simply adds one extra dimension to taurine release, being only additive to the effect of ethanol. The ethanol-induced increases in extracellular taurine may thus not be related solely to osmoregulatory processes. The interactions of ethanol and taurine were extensively reviewed not long ago by Olive (2002) and the interactions of ethanol, amino acids, and acamprosate, an anti-craving derivative of taurine, by Dahchour and De Witte (2000a). The reader is referred to these articles for further discussion.

3.3.9 Ammonia Stimulation

Exposure to ammonium chloride increases taurine release from neonatal rat primary astrocytes in culture (Albrecht et al., 1994), cultured cerebellar astrocytes and granule neurons (Wysmyk et al., 1994), and cerebrocortical minislices (Zielińska et al., 1999). Ammonia also enhances taurine release from Müller cells cultured from the rabbit retina (Faff-Michalak et al., 1994; Faff et al., 1996, 1997). Administration of ammonium chloride in vivo leads to a significant increase in extracellular taurine in the rat frontal cortex, but not in portacaval-shunted animals (Rao et al., 1995). Alterations in the release of taurine are also encountered in fulminant liver failure, which is accompanied by an increase in the ammonia concentration in the brain. Both hepatic devascularization and treatment with hepatotoxic thioacetamide increased taurine levels in the rat cerebrospinal fluid collected by means of an indwelling cisterna magna catheter (Swain et al., 1992). Thioacetamide-induced hepatic encephalopathy and experimental hyperammonemia have likewise increased spontaneous release of preloaded labeled taurine from slices from the rat basal ganglia and cerebral cortex (Hilgier et al., 1996). On the other hand, at the asymptomatic stage of hepatic failure 21 days after thioacetamide administration, the basal microdialysate content of taurine in the rat striatum was significantly lower than in the controls, but K^+ stimulation still evoked markedly more taurine release in the treated animals (Hilgier et al., 1999).

Changes in cell volumes offer the most straightforward explanation for the enhancement of taurine release evoked by hyperammonemia and hepatic encephalopathy. Taurine is extruded from the cells upon swelling, regulating the intracellular osmolality (Swain et al., 1992) and improving cell-volume adjustments (Hilgier et al., 1999). However, the underlying mechanism may not be simply a cell-volume regulatory response (Zielińska et al., 1999). Taurine may also be released via an osmoresistant, cAMP-controlled channel or a cAMP-activated taurine transporter (Faff et al., 1996). Moreover, acute ammonia treatments may produce an extra taurine release which is cAMP-independent (Faff et al., 1997). Ammonia neurotoxicity is a complex phenomenon, involving alterations in the functions of inhibitory GABA and excitatory glutamate receptors, together with possible alterations in other transmitter systems (Albrecht, 1998). As discussed above, particularly the activation of ionotropic glutamate receptors has been shown to enhance taurine release from brain cells. Indeed, ammonia administration induces extracellular accumulation of taurine in vivo in the rat striatum, in which the ionotropic glutamate receptors are involved (Zielińska et al., 2002). Taurine as an inhibitory amino acid and cell membrane protectant may serve to counteract the deleterious effects of the increased excitatory transmission accompanying acute hyperammonemic insults (Zielińska et al., 1999).

4 Physiological and Pharmacological Actions

4.1 Neuromodulatory and Possible Neurotransmitter Actions

Taurine has been thought to act either as a neurotransmitter or as a neuromodulator in the CNS (Oja and Kontro, 1983a, 1992a, 1996a; Frosini et al., 2003a, b; El Idrissi and Trenkner, 2004). The first function implies the existence of specific taurine receptors, and the latter, interference of taurine with the functions of other transmitter systems. There is but scant evidence to corroborate the first assumption, but ample for the latter, as discussed below.

4.1.1 Interactions with γ-Aminobutyrate Receptors

Taurine has been shown to displace the binding of labeled GABA to the GABA–benzodiazepine receptor complex in brain membranes (Malminen and Kontro, 1986). Similar effects have been shown with the solubilized GABA–benzodiazepine receptor complex (Malminen and Kontro, 1987). Taurine also interferes with the binding of ligands to the benzodiazepine site in this receptor (Iwata et al., 1984; Medina and De Robertis, 1984; Malminen and Kontro, 1986, 1987), modulating allosterically flunitrazepam binding to synaptic membranes (Quinn and Miller, 1992). Taurine has also been shown to act at GABA receptors in *Xenopus* oocytes injected with mouse brain mRNA (Horikoshi et al., 1988). Both taurine and GABA can activate GABA$_A$ receptors (Zhu and Vicini, 1997; del Olmo et al., 2000a; Barakat et al., 2002), but the affinity of these receptors for taurine is less marked. Taurine and GABA thus act at the same receptor-governed Cl^- channel (El Idrissi, 2006). Treatment of cultured cerebellar granule cells with taurine leads to the formation of low-affinity GABA receptors (Abraham and Schousboe, 1989).

At low concentrations taurine inhibits GABA-induced Cl^- fluxes into synaptic membrane microsacs prepared from different brain areas, but at higher concentrations, apparently owing to inhibitory properties of its own, it enhances the fluxes (Oja et al., 1990b). In keeping with such observations, taurine increases Cl^- conductance in cerebellar Purkinje cell dendrites in vitro (Okamoto et al., 1983a), in rat substantia nigra pars reticulate neurons (Ye et al., 1997), and in mitral and tufted cells in slices from the rat main olfactory bulb (Puopolo et al., 1998; Belluzzi et al., 2004). Chronic supplementation of taurine in drinking water for mice induces a downregulation of expression of GABA$_A$ receptor β subunit (a key subunit present in virtually all GABA$_A$ receptors), apparently because of the sustained interaction of taurine with GABA$_A$ receptors (El Idrissi, 2006). On the other hand, glutamate decarboxylase-positive neurons increase in number in the cerebral cortex of taurine-fed mice (Levinskaya et al., 2006).

At the GABA$_A$ receptor complex taurine appears to act as a full agonist at the GABA binding site (Quinn and Harris, 1995), but apparently only at a subclass of GABA$_A$ receptors. These taurine-sensitive GABA sites seem to be preferentially enriched only in certain brain areas, e.g., the dentate gyrus, substantia nigra, cerebellar molecular layer, median thalamic nuclei, and hippocampal CA3 field (Bureau and Olsen, 1991). The different subunit composition of GABA$_A$ receptors apparently underlies differences in taurine sensitivity. In the nucleus accumbens of young rats taurine could be a partial agonist of GABA$_A$ receptors (Jiang et al., 2004). In cultured cerebellar granule cells, inhibition of K^+-evoked aspartate release by taurine is partially blocked by the GABA$_A$ receptor antagonist bicuculline (Wahl et al., 1994). The GABA$_A$ antagonists bicuculline and picrotoxin also block the actions of taurine on solubilized GABA$_A$ receptors (Malminen and Kontro, 1987). Likewise, in mitral and tufted cells in the rat main olfactory bulb, in which taurine has been seen to reduce the input resistance and shift the membrane potential toward the chloride equilibrium potential, these antagonists were effective blockers, whereas the GABA$_B$ and glycine antagonists were not (Belluzzi et al., 2004). In this preparation, taurine did not influence granule and periglomerular cells, and the kinetics of GABA and taurine effects were dissimilar. The effective concentrations of taurine were higher, and the GABA-evoked current was fast and the current induced by taurine slow. Taurine may thus sustain a prolonged and modest inhibition of the mitral and tufted cells in the olfactory bulb. The taurine responses were nevertheless enhanced by flurazepam and pentobarbital in the same manner as the GABA responses (Belluzzi et al., 2004).

In addition to GABA$_A$ receptors, taurine also affects GABA$_B$ receptors (Kontro et al., 1990). It was an even more potent displacer of GABA at GABA$_B$ than at GABA$_A$ receptors in rat brain membranes when comparing its IC$_{50}$ values (Malminen and Kontro, 1986). The IC$_{50}$ values of taurine in both the [^{3}H] baclofen binding and the isoguvacine-insensitive [^{3}H]GABA binding are within the physiological concentration range of extracellular taurine in the mouse brain (Kontro and Oja, 1990). In the frog olfactory bulb in vivo, taurine suppresses the spontaneous firing of mitral cells and the effect is mediated mainly by GABA$_B$ receptors, as witnessed by the sensitivity of the effects to the GABA$_B$ receptor antagonist saclofen and by the relative insensitivity to picrotoxin, the antagonist of GABA$_A$ and GABA$_C$ receptors (Chaput et al., 2004). Since the odor responses were not affected, these authors think that taurine may favor the processing of primary sensory information in these structures by increasing the signal-to-noise ratio. In the rat, taurine suppressed olfactory nerve-evoked monosynaptic responses of mitral and tufted cells and blocked chloride conductance. These effects were mimicked by baclofen and abolished by a GABA$_B$ receptor blocker, probably because of the GABA$_B$ receptor-mediated inhibition of glutamate release (Belluzzi et al., 2004). There is also some evidence that taurine regulates GABA release from rat brain olfactory bulb synaptosomes through the activation of presynaptic GABA$_B$ receptors (Kamisaki et al., 1993, 1996a). However, the paucity of reports on taurine effects on GABA$_B$ receptors may allow an interpretation whereby it acts at these receptors only in limited cases.

4.1.2 Interactions with Glycine Receptors

There are a number of reports on the interactions of taurine with glycinergic transmitter systems in the CNS. Taurine displaces glycine from its strychnine-sensitive binding sites in the mouse brain stem (Kontro and Oja, 1987d). It also acts at glycine receptors in *Xenopus* oocytes injected with mouse brain mRNA (Horikoshi et al., 1988). In the rat supraoptic magnocellular neurons, taurine acts as the agonist on the glycine receptors (Hussy et al., 1997). Also in dopamine neurons in the rat substantia nigra (Häuser et al., 1992) and in freshly isolated neurons from the medulla oblongata and hippocampus (Krishtal et al., 1988) taurine and glycine activate the same Cl$^-$ conductance and apparently act at the same recognition site. Taurine and glycine also act on a strychnine-sensitive site to open the same Cl$^-$ channels in the rat sacral dorsal commissural neurons (Wang et al., 1998), and endogenous taurine, together with β-alanine, activates glycine receptors in cultured rat hippocampal slices (Mori et al., 2002). In acutely dissociated rat hippocampal neurons, taurine may act predominantly upon the glycine receptor-associated Cl$^-$ channels as a functional neurotransmitter or modulator (Yu et al., 2003). Isolated rat basolateral amygdala neurons respond to extracellular application of taurine with pronounced changes in the resting membrane current (McCool and Botting, 2000).

Taurine also in some instances acts through both GABA and glycine receptors simultaneously. For instance, taurine controls hormone release from the rat neurohypophysis by acting via both GABA$_A$ receptors and one or more glycine receptors (Song and Hatton, 2003) and similarly prevents the ammonia-induced accumulation of cGMP in the rat striatum by interaction with both GABA$_A$ and glycine receptors (Hilgier et al., 2005). Taurine induces inward currents in cultured rat cerebellar granule cells. In the generation of these currents a metabotropic receptor-coupled second-messenger system seems to be involved in addition to the ionotropic receptor-governed ion channels (Linne et al., 1996).

Intracellular cAMP suppresses responses to high concentrations of taurine in acutely dissociated substantia nigra neurons. This inhibition might be partly mediated by an activation of PKA (Inomata et al., 1993). On the other hand, α$_1$ adrenoceptors potentiate glycine receptor-mediated taurine responses through PKC (Nabekura et al., 1996b) and α$_2$ adrenoceptors do so through PKA in rat substantia nigra neurons (Nabekura et al., 1996a). Taurine increases the amplitude of spontaneous inhibitory postsynaptic currents in dopaminergic neurons in the ventral tegmental area in young rats, in which GABA is excitatory due to variations in the extracellular content of Cl$^-$, with a potency of one-tenth of that of glycine (Ye et al., 2004). By reason of developmental changes in the extracellular concentration of Cl$^-$, the activation of Cl$^-$ channels by taurine (Yoshida et al., 2004) during cortical circuit formation in the postnatally developing mouse alters its physiological effects from excitatory to inhibitory. Within the same time frame the target of taurine shifts from glycine receptors to GABA receptors (Yoshida et al., 2004). At low concentrations, taurine activates glycine receptors in CA1 neurons in the immature rat hippocampus while at high concentrations it activates both glycine and GABA receptors (Wu and Xu, 2003). This may signify that taurine is prone to act as a native ligand of glycine receptors. Since taurine is also a glycine agonist in the ventral tegmental area in adult animals, it may there be an important endogenous ligand for presynaptic glycine receptors.

Taurine acts mainly on nonsynaptic glycine receptors in the ventral tegmental area and the dopaminergic neurons studied could be exposed tonically to taurine (Wang et al., 2005). Taurine may thus act as an excitatory extrasynaptic neurotransmitter in the ventral tegmental area during early development. Intracellular Ca^{2+} potentiates the taurine-activated Cl^- currents in immature hippocampal CA1 neurons (Wu and Xu, 2003). In line with this, influx of extracellular Ca^{2+} through the L-type Ca^{2+} channels increases anion conductance in cultured rat astrocytes via their calmodulin-dependent activation, and a similar process affects the osmotic release of taurine (Li et al., 2002). The activation by taurine of glycine receptors in neurohypophysial nerve terminals is also thought to participate in the osmoregulation of vasopressin secretion (Hussy et al., 2001). A striking anatomical association is obtained between the glycine receptor clusters and the glial fibrillary acidic protein-positive astroglial processes, which contain high levels of taurine in the rat supraoptic nucleus (Deleuze et al., 2005). Taurine may have an important role as a regulator of the hormonal stress responses in the hypothalamic-neurohypophysial system (Engelmann et al., 2003). The very specialized localization of taurine in this system has given grounds for considering taurine also to have a general role in the regulation of the whole-body fluid balance, as discussed by Hussy and his coauthors in a review of the pertinent investigations of this intriguing topic (Hussy et al., 2000).

Taurine has been shown to inhibit the glycine responses of oocytes injected with rat α1 subunits of the glycine receptor or spinal cord poly(A)+ RNA. The effect was due to mutually exclusive binding to a common agonist site (Schmieden et al., 1989). Taurine has been found to be a full agonist of the glycine receptors in the ventral tegmental area (Wang et al., 2005). The amino acid residue 167 in the α2 subunit in the glycine receptor seems to be implicated in the inhibition of glycine responses by taurine, but the human and rat α2 and α3 glycine receptors display rather low responses to taurine (Kuhse et al., 1990a, b). On the other hand, taurine efficiently gates the α1 receptors, eliciting about fourfold greater inward currents when compared with the responses of the α2 receptors (Schmieden et al., 1992). Nevertheless, glycine is more effective than taurine as the agonist in both subtypes. In the mammalian α1 subunits, the amino acids in positions 111 and 212 were identified as important determinants of taurine activation (Schmieden et al., 1992).

Two separate point mutations at the same base pair in the gene encoding the human glycine receptor α1 subunit, resulting in arginine at position 271 being substituted by leucine (R271L) or glutamine (R271Q), have converted taurine from a full agonist into a competitive antagonist (Laube et al., 1995; Rajendra et al., 1995). The same changes occur with the Y279C and K276E mutations (tyrosine to cysteine and lysine to glutamate, respectively) (Lynch et al., 1997). It is assumed that these mutations either simply disconnect the binding site of taurine from the Cl^- channel activation site or selectively disrupt a common agonist recognition subsite and thereby unmask the antagonist subsite for taurine (Laube et al., 1995; Rajendra et al., 1995; Schmieden and Betz, 1995). The glycine receptor α_1 subunit of the zebrafish is very similar to the rat α1 receptor in both length and amino acid residue composition (David-Watine et al., 1999). However, the homomeric α1 receptor in the zebrafish exhibits exceptionally high sensitivity for taurine, almost comparable to glycine sensitivity. It contains valine instead of isoleucine in position 111, and sequence differences in the immediate vicinity of Val^{111} may thus underlie this high efficacy of taurine.

Very recently, Albrecht and Schousboe (2005) wrote a short review of the interactions of taurine with glycine, $GABA_A$, and $GABA_B$ receptors. In this article this aspect of taurine functions is informatively overviewed.

4.1.3 Long-Lasting Synaptic Potentiation

Prolonged exposure of the hippocampus to millimolar concentrations of taurine produces an increase in synaptic efficacy and axon excitability (Galarreta et al., 1996a, b). This phenomenon is called taurine-induced long-lasting potentiation (LLP-TAU). It is similar in nature to the long-lasting potentiation induced by GABA and is assumed to have the same underlying mechanism. LLP-TAU needs extra- and intracellular sources of Ca^{2+} for its induction (del Olmo et al., 2000b, c). After taurine withdrawal, LLP-TAU lasts several hours (del Olmo et al., 2003). This long-lasting synaptic potentiation is not caused by the activation of hippocampal dopamine receptors and is not affected by the Ca^{2+}/calmodulin-dependent protein kinase, but requires the cyclic AMP-dependent protein kinase A and protein synthesis (del Olmo et al., 2003). Taurine and GABA transport inhibitors have failed to abolish LLP-TAU in transverse slices from the rat hippocampus, and the development of LLP-TAU appears to be associated with the intracellular accumulation of taurine rather than the activation of its known transporters (Dominy et al., 2004).

Taurine also seems to trigger the sequence of some of the molecular events involved in the induction of the late phase of long-term potentiation. Moreover, this synaptic phenomenon is mutually occlusive with LLP-TAU (del Olmo et al., 2003). However, the induction of long-term potentiation is not related to the intracellular accumulation of taurine and not impaired by the taurine uptake inhibitor GES in the hippocampus (del Olmo et al., 2004). Taurine uptake may thus be required for the induction of synaptic plasticity phenomena such as long-term potentiation. Exposure to lead produces exactly opposite effects of taurine- and high-frequency-stimulated long-term potentiations (Yu et al., 2005). Lead exposure inhibited the high-frequency-stimulated potential whereas the amplitude of taurine-induced long-term potentiation was significantly larger than in controls.

The mechanisms of LLP-TAU are apparently different in different brain regions, e.g., in the hippocampus and striatum. In corticostriatal synaptic transmission, LLP-TAU is not prevented by the antagonists of NMDA and GABA receptors but is sensitive to the glycine receptor antagonist strychnine and blocked by 1 mM GES, the taurine transport inhibitor (Chepkova et al., 2002). On the other hand, a higher GES concentration (10 mM) evoked an enhancement of field potentials similar to LLP-TAU. In the striatum, LLP-TAU is PKC dependent, unaffected by calmodulin and PKA (Chepkova et al., 2006), and significantly smaller in juvenile than in mature rodents (Chepkova et al., 2002). Moreover, LLP-TAU has been significantly reduced in the striatum by dopamine D_1 and D_2, muscarine, and nicotine receptor antagonists, indicating the involvement of dopaminergic and cholinergic mechanisms (Chepkova et al., 2005, 2006). Taurine transporter-deficient mice have displayed a low ability to develop LLP-TAU in the striatum, but not in the hippocampus, even though taurine levels were very low in both structures (Sergeeva et al., 2003). The different mechanisms of taurine-induced synaptic plasticity apparently underlie the different vulnerabilities of the striatum and hippocampus under cell-damaging conditions.

4.1.4 Putative Taurine Receptors

It is difficult to determine whether all physiological functions of taurine are mediated by GABA and glycine receptors or whether specific taurine receptors are involved in certain brain areas, at developmental stages and in different animal species. No agonists or antagonists sufficiently specific for definite identification of possible taurine receptors are known. Only 6-aminomethyl-3-methyl-4H-1,2,4-benzothiadiazine-1,1-dioxide has been promoted as a taurine antagonist (TAG) (Girard et al., 1982). Indeed, TAG antagonizes the taurine-induced hyperpolarization in Purkinje cell dendrites in the guinea pig cerebellum (Okamoto et al., 1983b, c). Several authors have in the past failed to demonstrate any specific binding of taurine to synaptic membranes. We were the first to report some success (Kontro and Oja, 1983). In those early experiments it emerged that taurine binding to mouse brain synaptic membranes was a consistent finding only if the membranes were subjected to freezing–thawing cycles and detergent treatments to extract as much as possible of the endogenous taurine attached to the preparations tested (Kontro and Oja, 1987a). The binding proved to be Na^+ independent and exhibited sigmoidal dependence on the ligand concentration (Kontro and Oja, 1985b, 1987e). The taurine antagonist TAG was an effective displacer of this binding, but it was also affected by the glycine receptor antagonist strychnine and to a lesser extent by GABA antagonists. It was therefore not possible to demarcate the binding to glycine and GABA receptors from the possible binding to taurine receptors.

A few other reports have since appeared claiming detection of more or less specific taurine binding to mammalian brain membranes (Wu et al., 1990, 1992). In the frog spinal cord, two taurine receptor subtypes were surmised to exist on the basis of their pharmacological properties, differing from the responses to GABA and glycine (Kudo et al., 1988). In the olfactory organ of the lobster, preliminary identification and partial characterization of the putative taurine receptor proteins have been made (Sung et al., 1996). On the other hand, the demonstration of physiologically relevant taurine binding to crude retinal preparations involves a great number of problems (Militante and Lombardini, 2003b). Recently, Frosini and coworkers (2003a) made a very thorough characterization of [^{3}H]taurine binding to washed and detergent-treated synaptic membranes from the rabbit brain. They could show that taurine binding was displaced by TAG, 2-aminoethylarsonate, 2-hydroxyethanesulfonate and ($\pm$)cis-2-aminocyclohexanesulfonate. These effectors did not interact with GABA$_A$ and GABA$_B$ receptors, taurine and GABA uptake systems, or GABA transaminase. The taurine binding studied thus does not represent binding to GABA receptors, but the effects of glycine agonists and antagonists on this taurine binding were not tested. As also discussed above, only minor changes in the amino acid chains may alter the relative affinities of receptors for glycine and taurine, and there exist marked species differences and alterations during development. The matter of independent identity of taurine receptors is thus very much in need of further investigation.

4.1.5 Effects on Other Neurotransmitter Systems

There is only somewhat fragmentary information on the effects of taurine on the function of other neurotransmitters in the CNS. Taurine modulates noradrenaline uptake and release in rat cerebral cortical slices (Kontro et al., 1984), interferes with the binding of labeled spiperone (serotonin 5-HT$_{2A}$ and dopamine D$_2$ receptor antagonist) to cerebral cortical membranes (Kontro and Oja, 1986), and interacts with dopaminergic neurotransmission in the striatum (Kontro, 1987). The interactions of taurine, GABA, and dopamine are subject to developmental changes in the rat striatum (Kontro and Oja, 1988a). The age-related decline in striatal taurine is correlated with a loss of dopaminergic markers (Dawson et al., 1999a). The intracerebroventricular administration of taurine into mice (Ahtee and Vahala, 1985), intraperitoneal administration into rats (Salimäki et al., 2003), and direct administration into the striatum (Ruotsalainen et al., 1998) have significantly increased extracellular dopamine in a tetrodotoxin-dependent manner in the striatum (Ruotsalainen and Ahtee, 1996). At variance with these results, in other studies direct taurine administration into the rat striatum did not markedly affect extracellular dopamine, but the basal levels of the dopamine metabolite dihydroxyphenylacetic acid (DOPAC) were reduced and the NMDA-induced decrease in DOPAC accentuated (Anderzhanova et al., 2001, 2006). Taurine injected intranigrally modulates striatal dopaminergic neurotransmission (O'Neill, 1986) and reduces extracellular dopamine in the rat striatum (Leviel et al., 1979; Ruotsalainen et al., 1996). The effects of intrastriatal injections of taurine, GABA, and homotaurine are different, and these amino acids seem to affect striatal dopaminergic neurons via more than one mechanism (Ruotsalainen et al., 1998).

Endogenous taurine has been shown to inhibit the synthesis and release of serotonin in the rat rostral, but not in caudal, rhombencephalic raphe cells (Becquet et al., 1993). Taurine also affects the binding of phencyclidine (which blocks NMDA receptor activation) in the cerebral cortex of developing and adult mice (Saransaari and Oja, 1993). Finally, taurine reduces the K$^+$-stimulated adenosine release in slices from the mouse immature hippocampus without affecting either the unstimulated release in them or the unstimulated and K$^+$-stimulated release in the adult hippocampus (Saransaari and Oja, 2003a). However, the information obtained on the effects of taurine on these other neurotransmitter systems is not yet sufficiently systematic to enable any definite inferences to be drawn, and it should also be borne in mind that most preparations used contain partially intact neural networks in which the effects may be complex.

4.2 Other Actions

4.2.1 Antinociceptive Effects and Thermoregulation

Taurine has been shown to be antinociceptive. It is effective in the acetic acid test, but not in the hot-plate test (Serrano et al., 1990), and blocks the biting and scratching behaviors elicited by intrathecal administration of substance P (Smullin et al., 1990). Taurine inhibits strychnine-induced scratching and self-biting in rats, possibly by displacing strychnine from its binding sites (Beyer et al., 1988), and intrathecal NMDA- and kainate-induced nociception in mice (Hornfeldt et al., 1992). The analgesic effects manifest themselves at the spinal cord level, where taurine modulates nociceptive impulses generated by the action of substance P (Beyer et al., 1989). To date, taurine has also been observed to reduce nociceptive behavior in the mouse formalin test (Silva et al., 1993) and in the rat tail-flick test (Serrano et al., 1994) and autonomy after neurectomy in rats (Belfer et al., 1998). The analgesic effect is dose dependent and at least opioid and GABAergic mechanisms are involved (Serrano et al., 1994). No differences were found in the analgesic effect of taurine between prepubertal and adult mice, but taurine did not induce antinociception in aged mice (Serrano et al., 2002). Taurine administration to stressed mice offspring from birth until postnatal day 21 modifies behavioral programming in male mice in adult age. The elevated pain threshold in stressed animals in the tail-flick test, which mostly measures the spinal mechanisms of pain, was not affected by neonatal taurine administration, whereas in the hot-plate test, which reflects mainly the supraspinal mechanisms of pain, it was completely prevented (Franconi et al., 2004). It was assumed that this effect resulted from modification of hippocampal CA1 neurotransmission.

Data are also available on the participation of taurine in thermoregulation. Intracerebroventricularly administered taurine induced dose-related hypothermia in rabbits, accompanied by depression of gross motor behavior (Sgaragli et al., 1981), whereas the taurine antagonist TAG increases the core temperature (Sgaragli et al., 1994). Furthermore, heat stress increases the cerebrospinal fluid content of taurine in conscious rabbits. This effect is likely to counteract the hyperthermia resulting from exposure to heat, since the rectal temperature starts to decline immediately when the taurine content starts to rise in the cerebrospinal fluid (Frosini et al., 2000a). It is assumed that a specific taurine recognition site, i.e., taurine receptor, in the rabbit brain is responsible for the taurine effects on thermoregulation (Frosini et al., 2003b). The involvement of brain calcium metabolism in the action of taurine in mammalian thermoregulation has also been suggested (Palmi et al., 1987).

4.2.2 Seizures and Epilepsy

The inhibitory nature of taurine has led to the conception that this molecule functions as an endogenous anticonvulsive agent and has prompted evaluation of its possible applicability as an antiepileptic drug. The early experiments in the 1960s and 1970s with various animal seizure models proved the efficacy of taurine, and clinical trials with human patients in general showed that about one-third of epileptic patients benefit by administration of taurine; the symptoms are alleviated at least temporarily (Airaksinen et al., 1980; Oja and Kontro, 1983b). Ten years later, Anyanwu and Harding (1993) also reviewed the involvement of taurine in the treatment of epilepsy and drew similar inferences. Recently, El Idrissi and coworkers (2003) have also turned attention to the same subject. They showed that acute injections of taurine to mice increased the onset latency and reduced the occurrence of tonic seizures, but that long-term taurine supplementation increased susceptibility to kainate-induced seizures.

Taurine has had a significant antiepileptic effect in a rat model of penicillin-induced seizures. The behavior and frequency and amplitude of EEG improved markedly (Yang et al., 2006). Diminishing of the brain taurine levels by administration of GES markedly lowered the threshold for seizure activity elicited by excitatory 4-aminopyridine (Pasantes-Morales et al., 1987). GES also reduced the antiepileptic actions of phenytoin and phenobarbital (Izumi et al., 1985). There is an inherent problem, however, when GES is used to lower brain taurine levels. It is a weak $GABA_A$ agonist (Mellor et al., 2000) and a competitive antagonist of strychnine-sensitive glycine receptors (Sergeeva et al., 2002), and by means of these effects, in particular with interference with GABA receptors (Herranz et al., 1990), can increase seizure susceptibility. This reservation does not invalidate inferences of the attenuation of seizures by taurine. For instance, taurine inhibited wet-dog shakes produced by [D-Ala^2, Met^3]enkephalinamide (Yoshida et al., 1986), seizures in amygdaloid-kindled rats (Uemura et al., 1991), and convulsions induced by hypoxia (Malcangio et al., 1989). Furthermore, acute administration of taurine was also recently shown to suppress seizure-like events induced by removal of Mg^{2+} in combined rat entorhinal cortex-hippocampal slices in vitro

(Kirchner et al., 2003), and to attenuate the pentylenetetrazol-induced increases in intracellular Ca^{2+}, NO synthase, and lactate dehydrogenase, and decreases in superoxide dismutase and reduced glutathione in the mouse cerebral cortex (El-Abhar and Abd El Gawad, 2003). Electroacupuncture significantly increased the taurine level in the hippocampus in both kainate- and penicillin-induced epilepsies (Yang et al., 2006). Acupuncture and taurine may interact in a synergistic manner.

At variance with the investigations cited above, Lehmann (1987) saw no change in the threshold for pentylenetetrazol seizures in kittens, and Eppler and coworkers (1999) saw no attenuation of kainate-induced seizures in rats after dietary taurine supplementation. In contrast, a taurine-deficient diet decreased the number of clonic–tonic and partial seizures (Eppler et al., 1999). However, kainate has been shown to increase extracellular taurine markedly in the rat piriform cortex after systemic administration of taurine (Wade et al., 1987). It is often difficult, however, to induce any long-term increase in taurine levels in the brain with dietary supply, as Eppler and coworkers (1999) note. Taurine is generally effective when the administration route bypasses the blood–brain barrier.

Seizures and epilepsy induce changes in brain amino acids, including taurine, but the results of earlier studies have been variable, apparently depending on the timing of analyses and seizures, the brain areas studied, and the causative agent underlying the seizure activity. The situation is the same in more recent investigations. For example, in mice exposed to hyperbaric oxygen, the taurine level increased in the cerebral cortex and brain stem in the exposed animals, which did not convulse, whereas in mice, which did convulse, no differences were discernible. No changes were seen in the striatum (Mialon et al., 1995). In the frontal cortex of rats, the convulsive dose of pentylenetetrazol reduced the taurine level, while no significant changes were seen in pentylenetetrazol-kindled rats (Li et al., 2000). On the other hand, pentylenetetrazol administration increased the total tissue taurine concentration in the hippocampus and parietotemporal cortex in rats (Bikjdaouene et al., 2003). In penicillin-induced seizures in rats, the extracellular level of taurine increased significantly in the hippocampus (Shen and Lai, 2002). In the pentylenetetrazol-kindled rat hippocampus, the release of taurine was markedly reduced during the interictal period, whereas it significantly increased during the seizure period (Li et al., 2004). Melatonin, which exhibits anticonvulsant properties, increases taurine concentrations in most rat brain areas studied (Bikjdaouene et al., 2003). In newly diagnosed, untreated human patients with epilepsy, the cerebrospinal fluid taurine is reduced after tonic–clonic seizures (Rainesalo et al., 2004).

Administration of the antiepileptic drug levetiracetam to rats significantly reduces extracellular taurine in the hippocampus and frontal cortex (Tong and Patsalos, 2001). Acute administration of the antiepileptic lamotrigine does not affect the basal or high K^+-enhanced extracellular levels of taurine in the hippocampus of freely moving rats, whereas it markedly reduces the veratridine-evoked release (Ahmad et al., 2004). On the other hand, chronic administration of lamotrigine to rats increases the taurine content in the hippocampus (Hassel et al., 2001). Valproate has not been found to affect the basal extracellular level of taurine, but together with lamotrigine it increases taurine (Ahmad et al., 2005a). Carbamazepine markedly reduces the veratridine-stimulated release of taurine in the hippocampus in freely moving rats, whereas phenytoin does not affect this stimulated release (Ahmad et al., 2005b).

Blood platelets, known to be rich in taurine, are thought to reflect changes in amino acid concentrations in the brain. The mean platelet taurine content has been found to be significantly lower in patients with epilepsy than in controls. This may have resulted from medication, since there obtained a significant negative correlation between serum valproate and platelet taurine contents (Goodman et al., 1989).

4.2.3 Neuroprotective Actions

Taurine has been proved to be neuroprotective in a number of situations, for example, in a rat hypoxic model, possibly because of its antiacidotic and membrane-stabilizing actions (Mankovskaya et al., 2000). Exogenously applied taurine alleviates neuronal damage evoked by a variety of pathological impacts, including ischemia (Kingston et al., 2004). For example, taurine improves the recovery of neural functions in brain slices after hypoxic conditions (Schurr et al., 1987), and neurons in the rat hippocampus containing the taurine-synthesizing enzyme sulfino decarboxylase are more resistant to ischemic insult (Wu et al., 1994). Cells which contain GABA in addition to taurine are even less vulnerable (Wu et al., 1987). The increased release of taurine from neural cells gave the impetus to the hypothesis that taurine could attenuate the excessive neuronal accumulation of Ca^{2+}, which predisposes cells to damage (Schurr and Rigor, 1987). Indeed, taurine has been shown to inhibit NMDA-evoked Ca^{2+} accumulation in brain slices (Lehmann et al., 1984), to attenuate Ca^{2+} influx into slices from developing mice (Kontro and Oja, 1988b), and to protect neurons from glutamate-induced excitotoxicity (Tang et al., 1996), this by preventing or reducing the glutamate-induced elevation of intracellular Ca^{2+} (Chen et al., 2001). Taurine also inhibits the reverse mode of the Na^+/Ca^{2+} exchangers (Wu et al., 2000) and protects cultured rat astrocytes against reperfusion injury (Matsuda et al., 1996). Taurine may not directly affect the rate of Ca^{2+} uptake but rather the duration of the maximal response to glutamate (El Idrissi and Trenkner, 1999). It also inhibits the glutamate-induced release of Ca^{2+} from the internal pools (Wu et al., 2000) and regulates cytoplasmic and mitochondrial calcium homeostasis (El Idrissi and Trenkner, 2003). In addition to this, taurine inhibits the glutamate-induced Ca^{2+} influx through L-, P/Q-, and N-types of voltage-gated calcium channels and the NMDA receptor-governed calcium channel in whole-brain primary neuronal cultures (Wu et al., 2005). On the other hand, it has been shown to increase the accumulation of $^{45}Ca^{2+}$ in mitochondria of the rat cerebral cortex, thus attenuating the cytosolic free Ca^{2+} concentration, which, in turn, inhibits specific protein phosphorylation and phosphoinositide turnover (Li and Lombardini, 1991). On the whole, taurine may have an essential role in the modulation of intracellular calcium homeostasis (Chen et al., 2001; Foos and Wu, 2002).

As discussed above, glutamate and glutamate receptor agonists evoke taurine release from neural cells. This evoked taurine release might be neuroprotective in balancing glutamate activity under cell-damaging conditions (Trenkner et al., 1996; Katoh et al., 1997; Saransaari and Oja, 1997a, 1999e). Taurine efficiently counteracts the glutamate-induced Ca^{2+} uptake in cerebellar granule cells (El Idrissi et al., 1998; Trenkner et al., 1998). In brain slices taurine forestalls cell damage evoked by overactivation of ionotropic glutamate receptors (Zielińska et al., 2003). Overactivation of NMDA receptors leads to mitochondrial damage associated with Ca^{2+} influx, which results in the generation of free radicals, including superoxide. It is thought that a part of the neuroprotective action of taurine depends on its antioxidant properties (Schaffer et al., 2003). Taurine has been shown to antagonize the Ca^{2+} overload induced by glutamate and hypoxia in cultured rat hippocampal neurons (Zhao et al., 1999). Infusion of ammonium chloride or NMDA into the striata of rats via microdialysis probes increases the content of cGMP, an indicator of NO production, and hydroxyl radicals in the microdialysates (Hilgier et al., 2003). Co-infusion of taurine virtually abolishes both the ammonia- and NMDA-induced accumulation of cGMP and markedly attenuates accumulation of hydroxyl radicals (Hilgier et al., 2003; Anderzhanova et al., 2006). The effects of taurine on the accumulation of cGMP are mediated by interactions with $GABA_A$ and glycine receptors, since these could be blocked by co-infusion of the $GABA_A$ receptor antagonist bicuculline and the glycine receptor antagonist strychnine (Hilgier et al., 2005). Taurine has also been shown to enhance excitatory amino acid-induced phosphatidylinositol turnover in the neonatal rat cerebellum by activating $GABA_B$ receptors (Smith and Li, 1991).

The nootropic drug aniracetam aids in maintaining high extracellular taurine levels in conscious gerbils under transient cerebral ischemia (Yu and Cai, 2003). The anticonvulsant valproate doubles the anoxic survival time of developing mice, which may be attributable at least in part to increases in cerebral levels of taurine (Thurston and Hauhart, 1989). In keeping with this, taurine prolongs rat survival and reduces striatal damage caused by 3-nitropropionic acid (Rivas-Arancibia et al., 2001). On the other hand, dietary depletion of brain taurine by β-alanine results in cerebellar (Lu et al., 1996) and retinal (Sturman et al., 1996) damage in adult cats. Systemic intravenous administration of taurine protects the rabbit spinal cord against 60-min ischemia in both normothermia and hypothermia (Miyamoto et al., 2006). Administration of taurine to rats treated with 3-nitropropionic acid, which increases reactive oxygen species in tissues, partially protected the hippocampus from oxidative damage, and also exhibited morphological protection in surviving cells (Rodríguez-Martínez et al., 2004). Taurine protects rat cerebellar granule cells from free-radical damage (Boldyrev et al., 1999). However, taurine has been proved to be only a weak scavenger of peroxynitrite in cultured PC12 cells (Mehta and Dawson, 2001). Nevertheless, it restores the ethanol-induced depletion of antioxidants and attenuates oxidative stress in the rat (Pushpakiran et al., 2004). The activity of the superoxide dismutase is increased in the brain of lead-intoxicated rats treated with taurine, which may damp the toxic effects of lead (Flora et al., 2004). Taurine protects brain slices against acute GES neurotoxicity and blocks GES-induced cell swelling (Law, 2006).

Taurine protects cultured neurons in vitro and mice in vivo against carbon tetrachloride toxicity (Vohra and Hui, 2001). It also attenuates 1-methyl-4-phenylpyridinium (MPP^+) neurotoxicity in coronal slices from the rat brain (O'Byrne and Tipton, 2000). It is likely that this effect is mediated by extracellular mechanism(s), possibly by means of the activation of $GABA_A$ receptors, as witnessed by tests with phosphonotaurine, GES, and trimethyltaurine. Electroacupuncture has been shown to further augment taurine protection in transient focal cerebral ischemia in rats, whereas in taurine-depleted animals the efficacy of electroacupuncture has been lowered (Guo et al., 2006). Taurine also blocks the neurotoxicity of β-amyloid and glutamate receptor agonists, possibly by activation of GABA receptors. This may imply that taurine could have beneficial effects in Alzheimer's disease and other neurological disorders (Louzada et al., 2004). Indeed, taurine increases the levels of acetylcholine in the brain of animals, and decreased levels of taurine have been found in patients with Alzheimer's disease (Birdsall, 1998). Moreover, taurine improves memory in rats exposed to ozone (Rivas-Arancibia et al., 2000a), but fails to do so in healthy untreated mice (Vohra and Hui, 2000). However, it is effective in attenuating the amnesia produced by ethanol, pentobarbital, cycloheximide, and sodium nitrite (Vohra and Hui, 2000). Furthermore, taurine is able to facilitate the survival of hNT neurons transplanted in the adult rat striatum and attenuate the immune response generated by this xenograft (Rivas-Arancibia et al., 2000b).

At variance with the results from most other studies, continuous intracerebral infusion of taurine did not cause any postischemic glutamate surge suppression after transient global forebrain ischemia in gerbils and provided no histologically demonstrable neuroprotection (Khan et al., 2000). However, in the experiments in question taurine infusion did not significantly elevate extracellular taurine levels. Administration of taurine to rats treated with 6-hydroxydopamine significantly reduced the loss of tyrosine hydroxylase-positive cells and the levels of dopamine decreased less (Ward et al., 2006). In the case of the complex neurotoxic actions of 6-hydroxydopamine taurine afforded only partial neuroprotection (Hayes et al., 2001). The neurotoxic mechanisms of 6-hydroxydopamine are thus sufficiently different from the cytoprotective function of taurine to render taurine relatively ineffective. In keeping with these results, taurine also failed to ameliorate behavioral effects in rats in which dopamine nerve endings had been destroyed with 6-hydroxydopamine (Hashimoto-Kitsukawa et al., 1988b). However, taurine enhances the caudate spindle and reduces the suppressive effects of apomorphine and methamphetamine on the spindle in pargyline-pretreated rats (Hashimoto-Kitsukawa et al., 1988a). There is also a possibility that taurine could slightly potentiate the oxidation of catecholamines (Dawson et al., 2000). One new conception is that taurine could serve as a universal carrier of lipid-soluble vitamins, turning them into water-soluble complexes (Petrosian and Haroutounian, 2000).

4.2.4 Retina

As already briefly outlined above and as discussed by Lombardini (1991) and Militante and Lombardini (2002), taurine is essential for the normal development and functions of the retina in kittens and primates. It has a trophic effect in the retina (Lima et al., 1992; Lima, 1999). Taurine depletion also leads to a loss of optic nerve axons in the rat (Lake et al., 1988). In the goldfish, the level of taurine increases significantly at two days after crushing of the optic nerve (Lima et al., 1998). Taurine addition increases the length and density of neurite outgrowth from retina explants (Lima et al., 1988). This process is due in part to a calcium-mediated mechanism (Lima et al., 1993). In the goldfish, taurine increases the number of cells presenting neurites, whereas in developing rats cell proliferation is not affected (Matus et al., 1997). As a trophic factor in the goldfish retina taurine modulates phosphorylation of cellular proteins (Lima and Cubillos, 1998a), and a certain level of protein phosphorylation is critical for its trophic effects (Lima and

Cubillos, 1998b). Furthermore, PKC activation inhibits the high-affinity transport of taurine in both the goldfish and the rat retina (Lima et al., 2000). The activation of PKC reduces the outgrowth of neurites from the retinal explant and impairs the stimulatory effect of taurine. The mechanisms by which taurine exerts its trophic potential in the regenerating retina are related to modifications of protein phosphorylation and increases in calcium fluxes (Lima et al., 2001). The trophic effects of taurine are greatly influenced by the medium composition (Cubillos et al., 2002), by the substrates available and by their concentrations (Lima et al., 1989, 2003), and by the presence or absence of pieces of the optic tectum (Cubillos et al., 2000, 2002, 2006). Zinc has been found to be necessary for normal outgrowth of retinal fibers and taurine could counteract the effects of zinc chelators (Nusetti et al., 2005, 2006). However, no positive correlation has been obtained between the zinc and taurine levels in the rat retina (Lima et al., 2004). Taurine also promotes differentiation of retinal cells in vitro (Altshuler et al., 1993). By analogy, it also promotes proliferation and differentiation of human fetal neuronal cells (Chen et al., 1998).

The cytoprotective effects of taurine have been attributed to its roles as osmoregulator, antioxidant, regulator of Ca^{2+} fluxes, membrane stabilizer, and GABA agonist (Hayes et al., 2001). It has been thought to participate in rhodopsin regeneration and to protect the rod outer segments against osmotic, mechanical, and light-induced damage (Petrosian and Haroutounian, 1998, 2000). Taurine stimulates the ATP-dependent Ca^{2+} uptake in synaptosomal and mitochondrial fractions of the rat retina (Lombardini, 1988). In the rat retina, it also stimulates the ATP-dependent Ca^{2+} uptake in the rod outer segments, the effects being concentration dependent (Militante and Lombardini, 1998a), but independent of taurine uptake (Militante and Lombardini, 1999b). Taurine potentiates the effect of ATP (Militante and Lombardini, 2003a), but taurine and ATP limit the capacity of rat rod outer segments to take up Ca^{2+} up to a stable level, protecting in this manner against calcium toxicity (Militante and Lombardini, 2000a). Rats treated with the taurine transport inhibitor GES display an altered retinogram (Cocker and Lake, 1987), whereby pathophysiological processes are accelerated by light exposure (Cocker and Lake, 1989). Taurine may have a role particularly in the light adaptation processes (Barabás et al., 2003). As already mentioned, the taurine transporter has been characterized and localized in the retina (Liu et al., 1992; Vinnakota et al., 1997). In fact, there may be two immunologically heterogeneous taurine transporters in the retina, differentiating neuronal and glial transport systems (Pow et al., 2002). Disruption of the taurine transporter gene results in retinal degeneration in mice (Heller-Stilb et al., 2002). The taurine effects in the retina are also ATP dependent and antagonized by the agents which block cyclic nucleotide-gated cation channels (Militante and Lombardini, 1998b).

In the frog retina, taurine has been shown to depress the b-wave signal, which is reversed by strychnine (Haroutounian and Petrosian, 1998). Taurine and its precursor hypotaurine inhibit light-induced lipid peroxidation and protect the rod outer segment structure in the retina (Pasantes-Morales and Cruz, 1985). Taurine itself is a rather weak antioxidant but in combination with retinol a synergistic effect is obtained. Effective protection of lipids from oxidation in bovine photoreceptor cell membranes is then seen, which supports findings that taurine protects rod outer segment lipids during exposure to cyclic light (Keys and Zimmerman, 1999). Conversion of taurine to N-chlorotaurine and sulfoacetaldehyde may also be a mechanism underlying its protective effects in oxidative stress, though there is as yet no convincing experimental evidence to corroborate this stipulation (Cunningham et al., 1998). Taurine inhibits the phosphorylation of a 20-kDa protein (Lombardini, 1992a, 1993b). In rats, the phosphorylation of this mitochondrial protein is increased when the retinal taurine content had been depleted with GES administration, and the effect is reversed when the animals are subsequently given taurine (Lombardini, 1998). Taurine inhibits the stimulatory effect of chelerythrine on the phosphorylation of this protein (Lombardini and Props, 1996). The protein affected is probably histone H2B (Lombardini, 1998). At the functional level, taurine supplementation has been reported to alleviate visual fatigue induced by stressing work with visual display terminals (Zhang et al., 2004).

In the case of other eye structures, taurine has also been shown to protect cultured lens against oxidative stress (Devamanoharan et al., 1998). This protective effect of the ocular surface has likewise been attributed to the antioxidant properties of taurine (Nakamori et al., 1993). Millimolar concentrations of taurine also exhibit scavenging activity against both superoxide and peroxide, which may underlie the ability of taurine to reduce damage in a model of diabetic cataract in rats (Kilic et al., 1999).

4.2.5 Ethanol Effects

Several studies have demonstrated significant interactions between taurine and the effects of ethanol in rodents. Low doses of ethanol stimulate and high doses reduce motor activity in mice. Taurine modulates this ethanol-stimulated locomotor activity. Low taurine doses are inhibitory, whereas higher doses enhance the activity (Aragón et al., 1992). These effects are specific for ethanol. On the other hand, taurine prolongs ethanol-induced sedation when given intracerebroventricularly (Ferko, 1987; Ferko and Bobyock, 1988), but attenuates the sedating effects of ethanol when given intraperitoneally only shortly before ethanol administration (McBroom et al., 1986). The reason for these opposite effects is not known, but they may be due to differential pharmacokinetics/pharmacodynamics after intraventricular and peripheral administrations (Olive, 2002). However, intraperitoneal taurine lengthens pentobarbitone-induced sleep in mice (Ahtee et al., 1985).

Ethanol exhibits both rewarding and aversive motivational properties, these being strongly concentration dependent. The ability of taurine to modify these properties also depends on the dose of ethanol administered (Aragón and Amit, 1993; Quartemont et al., 1998b). It is likely that these effects are mediated by pharmacological interactions in the CNS (Olive, 2002), even though taurine can modulate alcohol dehydrogenase activity and oxidation to acetaldehyde (Theofanopoulos and Lau-Cam, 1998; Ward et al., 2001). Acute administration of taurine moderately reduces voluntary consumption of ethanol in rats (Olive et al., 2002). These effects are probably mediated by the CNS and might result from synergistic effects of taurine and ethanol on ion channels and neurotransmitter receptors. For example, taurine inhibits the ethanol withdrawal-induced increase in extracellular glutamate in the nucleus accumbens of ethanol-dependent rats (Dahchour and De Witte, 2000b).

4.2.6 Energy Drinks

Recently, taurine-containing so-called "energy drinks" have become particularly popular and have been claimed to augment many different seemingly unrelated functions, e.g., from increased energy supply to sexual performance. Scientific evidence supporting such promises is scanty. A taurine-containing drink has been reported to improve exercise performance in endurance athletes (Geiss et al., 1994), a taurine- and caffeine-containing drink to stimulate cognitive performance and well-being (Seidl et al., 2000), and a taurine-, caffeine-, and vitamin-containing drink to improve aerobic endurance and anaerobic performance (Alford et al., 2001). In addition to these positive effects the drink improved mental performance, concentration, and memory (Alford et al., 2001). In keeping with these observations, a caffeinated taurine beverage has been found to improve attention and verbal reasoning, but not to affect memory functions (Warburton et al., 2001). It is not clear, however, whether caffeine alone underlies the improvements or whether they are also enhanced by taurine. Sometimes youngsters admix ethanol with these energy drinks. Taurine could mediate some of the neurochemical effects of ethanol (Quertemont and Grant, 2004) but the pharmacological effects of these taurine-containing energy drinks fortified with ethanol have not yet been explored.

5 Taurine Derivatives

A large number of taurine derivatives have already been reported in the literature and their efficacies tested in different biological models, with partial or marked success. Among them, acamprosate, the calcium salt of N-acetylhomotaurine, is already in clinical use as an anticraving drug in ethanolism (Ward et al., 2000), and tauromustine, 1-(2-chloroethyl)-3-(2-dimethylaminosulfonyl)ethyl-1-nitrosourea, as an anticancer drug (Molineux et al., 1987). Both homotaurine and acamprosate dose dependently reduce ethanol intake and preference in rats (Olive et al., 2002). At the beginning of the 1980s, in a search for potent anticonvulsants, we synthesized a series of 2-acylamidoethanesulfonamides, derivatives of taurine more lipophilic than the parent compound (Andersen et al., 1984). Several of them proved to be effective against seizures in several animal models (Lindén et al., 1983; Oja et al., 1983). These derivatives interfered with the binding of GABA and taurine to mouse cerebral synaptic membranes (Kontro and Oja, 1987f) and modified GABA and taurine release in the mouse hippocampus (Saransaari and Oja, 1999f). 2-Phthalimidoethanesulfon-N-isopropylamide (taltrimide, MY-117) also raises the threshold for audiogenic seizures and reduces seizure severity (Huxtable and Nakagawa, 1985). This compound was subjected to phase I clinical trials on patients with intractable epilepsy. The good antiepileptic activity observed in animal models could not however be confirmed in humans (Koivisto et al., 1986; Airaksinen et al., 1987). N, N-Phthaloyltaurinamides also effectively antagonize electrically induced seizures upon oral administration to rats (Usifoh et al., 2001). Furthermore, homotauryl-homotaurine, carbobenzoxyhomotauryl-homotaurine, and γ-aminobutyltaurine have exhibited anticonvulsant actions, possibly via interaction with the GABA$_A$ binding site in the brain (Sgaragli et al., 1994).

Taltrimide and 2-phtalimidoethanesulfonamide (MY-103) were further evaluated as to their potential for cerebral osmoprotection during chronic hypernatremic dehydration (Trachtman et al., 1988b). Taurine shows antiaggressive effects and several taurine analogues with a free amino group or with a phthalimido moiety have been synthesized and studied in the context of aggressive behavior (Mandel et al., 1985). Taltrimide and taurinamide were more effective than taurine itself. Several sulfonamide derivatives of taurine were similarly tested when administered into the olfactory bulb, intraperitoneally or orally. All derivatives were more potent than taurine itself in inhibiting the muricidal activity in rats. A dipeptide, γ-L-glutamyltaurine, exhibits a variety of effects on the CNS. For example, it reduces aversion, phobia, and anxiety synergistically with the anxiolytic diazepam, probably modulating excitatory aminoacidergic neurotransmission, in addition to a number of other effects on brain functions (Bittner et al., 2005).

Trimethyltaurine has been shown to bind to GABA$_A$ receptors and to exert neuroprotective properties similar to those of taurine (Marangolo et al., 1997). Intraperitoneally administered N-pivaloyltaurine causes significant hypothermia and elevates the striatal dopamine concentration in mice (Ahtee et al., 1985). It penetrates better into the brain than taurine and temporarily elevates taurine tissue concentration, since it is a lipophilic derivative. A series of other taurine analogues have also been tested, including *cis-* and *trans-*isomers of 2-aminocyclohexanesulfonate (Frosini et al., 2000b). *cis-*2-Aminocyclohexane sulfonate, but not the *trans-*isomer, induces hypothermia, this not being mediated by the GABA receptor but due to interaction with the putative taurine receptor.

The sulfone derivatives of taurine, 2-aminoethylmethylsulfone, thiomorpholine-1,1-dioxide, N-methylthiomorpholine-1,1-dioxide, and *cis-*2-aminocyclohexanesulfonate have proved more potent than taurine in stimulating ATP-dependent, but not ATP-independent, Ca^{2+} uptake in the retina. (Liebowitz et al., 1986), whereas *trans-*2-aminocyclohexanesulfonate, *cis-* and *trans-*2-aminocyclopentanesulfonate, 3-aminotetrahydrothiophene-1,1,-dioxide, aminomethanesulfonate, and 1,2,3,4-tetrahydroquinoline-8-sulfonate are inhibitory in the presence of ATP (Liebowitz et al., 1987, 1988; Lombardini et al., 1989; Lombardini and Liebowitz, 1990a). In the absence of ATP, *trans-*2-aminocyclopentanesulfonate and *trans-*2-aminocyclohexanesulfonate are stimulatory (Lombardini and Liebowitz, 1990b). Similar effects have been seen in the incorporation of phosphate into retinal proteins, 2-aminoethylmethylsulfone, 3-aminotetrahydrothiophene-1,1,-dioxide, 3-aminotetrahydrothiopyran-1,1-dioxide, and N-methylthiomorpholine-1,1-dioxide being equipotent inhibitors (Liebowitz et al., 1989). Tauropyrone [2-(2,6-dimethyl-3,5-diethoxycarbonyl-1,4-dihydropyridine-4-carboxamido)ethansulfo acid] added to a culture of cerebellar granule cells was capable of protecting cells against damage induced by oxygen and glucose deprivation (Kluša et al., 2006). It also protects cells from glutamate-induced cell death, but does not affect MPP$^+$-induced free-radical formation. Tauropyrone also attenuates 6-hydroxydopamine-induced damage to dopaminergic cells in vivo in rats and reduces the inflammatory agent-evoked release of NO from glial cells in culture (Ward et al., 2006). Intraperitoneally administered taurepar [2-(1-phenylethylamino)ethanesulfonic acid isopropylamide hydrochloride] in rats produced positive effects on mnestic functions and exhibited antiamnestic activity after electroconvulsive shock or intraperitoneal introduction of pentylenetetrazol and scopolamine (Sapronov and Gavrovskaya, 2006). Taurepar as well as taurhythman [2-(1-methyl-2-phenylethylamino)ethanesulfonic acid isopropylamide hydrochloride] and IEM-1702

[2-(benzylamino)ethanesulfonic acid isopropylamide hydrochloride] have prolonged survival time of mice and rats after acute hypoxia (Gavrovskaya et al., 2006). Taurepar did likewise after bilateral occlusion of carotic arteries and ameliorated neurological defects after spinal cord compression trauma in rats (Krylova et al., 2006).

Retinyledin taurine (tauret) is a water-soluble taurine derivative that is spontaneously hydrolyzed with a half-time of 24 min (Petrosian and Haroutounian, 1990). It has been suggested that tauret is an endogenous substance in the retina, involved in the processes of removal of all-*trans* and 11-*cis* retinals and rhodopsin regeneration in darkness (Petrosian et al., 1996, 2000b). Tauret is synthesized in frogs in the pigment epithelium rather than in the retina in the eye adapted to dark (Petrosian et al., 2000a).

Della Corte and coauthors (2002) have informatively discussed the use and potential therapeutic applications of taurine derivatives. In addition to their article, two recent reviews have been published on biologically active taurine derivatives (Gupta et al., 2005; Gupta, 2006). The reader is referred to these texts for a more detailed discussion of derivatives which affect CNS functions. In these reviews a number of other taurine derivatives not affecting the CNS are also described.

6 Summary and Perspectives

Although the present knowledge of the functions and role of taurine in the CNS is by no means complete, the following inferences can be drawn. Taurine appears to be an essential amino acid during the early developmental phases in many mammals, including primates and humans. However, the ubiquitous occurrence of taurine in the animal kingdom very soon guarantees a sufficient dietary supply, unless consumers are very strict vegans. The taurine contents in the CNS of various species are highly variable, characteristic of the species in question. The distribution of taurine in the CNS has been fairly thoroughly characterized. Particularly high concentrations are encountered in the retina. The only matter still open in taurine biosynthesis is the conversion of hypotaurine to taurine: an enzymatic or nonenzymatic mechanism. The uptake of taurine by neural cells is concentration dependent, saturable, and dependent on Na^+ and Cl^-. The properties of taurine uptake have already been characterized in early investigations with crude CNS preparations, and recent experiments with cloned taurine transporters have yielded only marginally new information. Taurine release has likewise been studied extensively but the mechanisms involved have not yet been completely explored.

Taurine is an essential component in volume adjustments in neural cells. The dispute over whether taurine is only an osmolyte or whether it also functions as a modulator of neural activity seems to be settled. Both functions are widely documented. Taurine has been demonstrated to interfere with both GABA and glycine receptors, the efficacy depending on the amino acid chain in the receptors. A systematic investigation of the structures of GABA and glycine receptors would determine which requirements of receptor subunits render them taurine sensitive. The question of the existence of a possible functional taurine receptor is not yet settled. It necessitates the discovery of agonists and antagonists specific for the possible taurine receptor and characterization of the receptor by molecular biology methods.

Taurine has been shown to be neuroprotective under a number of different conditions, including hypoxia, ischemia, energy shortage, toxicity of xenobiotics, etc. The effects stem partially from the effects of taurine on the glycinergic and GABAergic neurotransmitter systems. Taurine is also an anticonvulsive agent, but its applicability in the treatment of human epilepsy is hampered by its low lipid solubility and difficulties in penetrating into brain tissue in sufficient amounts. The inherent homeostatic regulation of taurine contents in the brain is another obstacle to its prolonged efficacy. The development of further novel taurine derivatives would be most welcome; they may prove invaluable in the treatment of epilepsies, brain insults, and neurodegenerative diseases.

Acknowledgments

The financial support of the Medical Fund of Tampere University Hospital is gratefully acknowledged.

References

Abraham JH, Schousboe A. 1989. Effects of taurine on cell morphology and expression of low-affinity GABA receptors in cultured cerebellar granule cells. Neurochem Res 14: 1031-1038.

Aerts L, Van Assche FA. 2002. Taurine and taurine-deficiency in the perinatal period. J Perinat Med 30: 281-286.

Ahmad S, Fowler LJ, Whitton PS. 2004. Effects of acute and chronic lamotrigine treatment on basal and stimulated extracellular amino acids in the hippocampus of freely moving rats. Brain Res 1029: 41-47.

Ahmad S, Fowler LJ, Whitton PS. 2005a. Effects of combined lamotrigine and valproate on basal and stimulated extracellular amino acids and monoamines in the hippocampus of freely moving rats. Naunyn Schmiedebergs Arch Pharmacol 371: 1-8.

Ahmad S, Fowler LJ, Whitton PS. 2005b. Lamotrigine, carbamazepine and phenytoin differentially alter extracellular levels of 5-hydroxytryptamine, dopamine and amino acids. Epilepsy Res 63: 141-149.

Ahtee L, Vahala, ML. 1985. Taurine and its derivatives alter brain dopamine metabolism similarly to GABA in mice and rats. Prog Clin Biol Res 179: 331-341.

Ahtee L, Auvinen H, Mäenpää AR, Vahala ML, Lehtinen M, et al. 1985. Comparison of central nervous system actions of taurine and N-pivaloyltaurine. Acta Pharmacol Toxicol 57: 96-105.

Airaksinen EM, Koivisto K, Keränen T, Pitkänen A, Riekkinen PJ, et al. 1987. Biochemical and clinical studies on epileptic patients during two phase I trials with the novel anticonvulsant taltrimide. Epilepsy Res 1: 308-311.

Airaksinen EM, Oja SS, Marnela KM, Leino E, Pääkkönen L. 1980. Effects of taurine treatment on epileptic patients. Prog Clin Biol Res 3: 157-166.

Albrecht J. 1998. Roles of neuroactive amino acids in ammonia neurotoxicity. J Neurosci Res 51: 133-138.

Albrecht J, Schousboe, 2005. Taurine interaction with neurotransmitter receptors in the CNS: An update. Neurochem Res 30: 1615-1621.

Albrecht J, Bender AS, Norenberg MD. 1994. Ammonia stimulates the release of taurine from cultured astrocytes. Brain Res 660: 288-292.

Albrecht J, Simmons M, Dutton GR, Norenberg MD. 1991. Aluminium chloride stimulates the release of endogenous glutamate, taurine and adenosine from cultured rat cortical astrocytes. Neurosci Lett 127: 105-107.

Alford C, Cox H, Wescott R. 2001. The effects of Red Bull energy drink on human performance and mood. Amino Acids 21: 139-150.

Allen JW, Mutkus LA, Aschner M. 2002. Chronic ethanol produces increased taurine transport and efflux in cultured astrocytes. Neurotoxicology 23: 693-700.

Altshuler D, Lo Turco JJ, Rush J, Cepko C. 1993. Taurine promotes the differentiation of a vertebrate retinal cell type in vitro. Development 119: 1317-1328.

Ament ME, Geggel HS, Heckenlively JR, Martin DA, Kopple JD. 1986. Taurine supplementation in infants receiving long-term total parenteral nutrition. J Am Coll Nutr 5: 127-135.

Andersen L, Sundman LO, Lindén IB, Kontro P, Oja SS. 1984. Synthesis and anticonvulsant properties of some 2-aminoethanesulfonic acid (taurine) derivatives. J Pharm Sci 73: 106-108.

Anderzhanova E, Rayevsky KS, Saransaari P, Riitamaa E, Oja SS. 2001. Effects of sydnocarb and D-amphetamine on the extracellular levels of amino acids in the rat caudate-putamen. Eur J Pharmacol 428: 87-95.

Anderzhanova E, Saransaari P, Oja SS. 2006. Neuroprotective mechanisms of taurine in vivo. Taurine 6. Oja SS, Saransaari P, editors. New York: Springer/Kluwer Academic; pp. 377-387.

Anyanwu E, Harding GF. 1993. The involvement of taurine in the action mechanism of sodium valproate (VPA) in the treatment of epilepsy. Acta Physiol Pharmacol Ther Latinoamer 43: 20-27.

Aragón CM, Amit Z. 1993. Taurine and ethanol-induced conditioned taste aversion. Pharmacol Biochem Behav 44: 236-263.

Aragón CM, Trudeau LE, Amit Z. 1992. Effect of taurine on ethanol-induced changes in open-field locomotor activity. Psychopharmacology 107: 337-340.

Aschner, M. 1997. Astrocyte metallothioneins (MTs) and their neuroprotective role. Ann N Y Acad Sci 825: 334-347.

Aschner M, Vitarella D, Allen JW, Conklin DR, Cowan KS. 1998. Methylmercury-induced inhibition of regulatory volume decrease in astrocytes: Characterization of osmoregulatory efflux and its reversal by amiloride. Brain Res 811: 133-142.

Barabás P, Kovács I, Kardos J, Schousboe A. 2003. Exogenous glutamate and taurine exert differential actions on light-induced release of two endogenous amino acids in isolated rat retina. J Neurosci Res 73: 731-736.

Barakat L, Wang D, Bordey A. 2002. Carrier-mediated uptake and release from Bergman glia in rat cerebellar slices. J Physiol 541: 753-767.

Basavappa S, Mobasheri A, Errington R, Huang CC, Al-Adawi S, et al. 1998. Inhibition of Na^+, K^+-ATPase activates swelling-induced taurine efflux in a human neuroblastoma cell line. J Cell Physiol 174: 145-153.

Beckman ML, Quick MW. 1998. Neurotransmitter transporters: Regulators of function and functional regulation. J Membr Biol 164: 1-10.

Becquet D, Hery M, Francois-Bellan AM, Giraud P, Deprez P, et al. 1993. Glutamate, GABA, glycine and taurine modulate serotonin synthesis and release in rostral and caudal rhombencephalic raphe cells in primary cell cultures. Neurochem Int 23: 269-283.

Beetsch JW, Olson JE. 1996. Hyperosmotic exposure alters total taurine quantity and cellular transport in rat astrocyte cultures. Biochim Biophys Acta 1290: 141-148.

Beetsch JW, Olson JE. 1998. Taurine synthesis and cysteine metabolism in cultured rat astrocytes: Effects of hyperosmotic exposure. Am J Physiol 274: C866-C874.

Belfer I, Davidson E, Ratner A, Beery E, Shir Y, et al. 1998. Dietary supplementation with the inhibitory amino acid taurine suppresses autonomy in HA rats. Neuroreport 13: 3103-3107.

Belluzzi D, Puopolo M, Benedusi M, Kratskin I. 2004. Selective neuroinhibitory effects of taurine in slices of rat main olfactory bulb. Neuroscience 124: 929-944.

Ben-Ari Y. 2002. Excitatory actions of GABA during development: The nature of the nurture. Nature Rev 3: 728-739.

Benrabh H, Bourre JM, Lefauconnier JM. 1995. Taurine transport at blood-brain barrier: An in vivo brain perfusion study. Brain Res 692: 57-65.

Beyer C, Banas C, Gomora P, Komisurak BR. 1988. Prevention of the convulsant and hyperalgesic action of strychnine by intrathecal glycine and related amino acids. Pharmacol Biochem Behav 29: 73-78.

Beyer C, Banas C, Gonzalez-Flores O, Komisaruk BR. 1989. Blockage of substance P-induced scratching behavior in rats by the intrathecal administration of inhibitory amino acid agonists. Pharmacol Biochem Behav 34: 491-495.

Bianchi L, Colivicchi MA, Bolami JP, Della Corte L. 1998. The release of amino acids from rat neostriatum and substantia nigra in vivo: A dual microdialysis probe analysis. Neuroscience 87: 171-180.

Bianchi L, Colivicchi MA, Ballini C, Fattori M, Venturi C, et al. 2006. Taurine 6. Oja SS, Saransaari P, editors. New York: Springer/Kluwer Academic; pp. 443-448.

Biggs CS, Fowler LJ, Whitton PS, Starr MS. 1995. Impulse-dependent and tetrodotoxin-sensitive release of GABA in the rat's substantia nigra measured by microdialysis. Brain Res 684: 172-178.

Bikjdaouene L, Escames G, León J, Ferrer JMR, Khaldy H, et al. 2003. Changes in brain amino acids and nitric oxide after melatonin administration in rats with pentylenetetrazol-induced seizures. J Pineal Res 35: 54-60.

Birdsall TC. 1998. Therapeutic applications of taurine. Altern Med Rev 3: 128-136.

Bitoun M, Tappaz M. 2000a. Gene expression of the transporters and biosynthetic enzymes of the osmolytes in astrocyte primary cultures exposed to hyperosmotic conditions. Glia 32: 165-176.

Bitoun M, Tappaz M. 2000b. Taurine down-regulates basal and osmolarity-induced gene expression of its transporter, but not the gene expression of its biosynthetic enzymes, in astrocyte primary cultures. J Neurochem 75: 919-924.

Bitoun M, Tappaz M. 2000c. Gene expression of taurine transporter and taurine biosynthetic enzymes in brain of rats with acute or chronic hyperosmotic plasma. A comparative study with gene expression of myo-inositol transporter, betaine transporter and sorbitol. Brain Res Mol Brain Res 77: 10-18.

Bittner S, Win T, Gupta R. 2005. γ-L-Glutamyltaurine. Amino Acids 28: 343-356.

Böckelmann R, Reiser M, Wolf G. 1998. Potassium-stimulated taurine release and nitric oxide synthase activity during quinolinic acid lesion of the rat striatum. Neurochem Res 23: 469-475.

Boldyrev AA, Johnson P, Wei Y, Tan Y, Carpenter DO. 1999. Carnosine and taurine protect rat cerebellar granular cels from free radical damage. Neurosci Lett 263: 169-172.

Borden LA, Smith KE, Vaysse PJ, Gustafson EL, Weinshank RL, et al. 1995. Re-evaluation of GABA transport in neuronal and glial cell cultures: Correlation of pharmacology and mRNA localization. Receptors Channels 3: 129-146.

Brès V, Hurbin A, Duvoid A, Orcel H, Moos FC, et al. 2000. Pharmacological characterization of volume-sensitive, taurine permeable anion channel in rat supraoptic glial cells. Br J Pharmacol 130: 1976-1982.

Bridges CC, Ola MS, Prasad PD, EL-Sherbeny A, Ganapathy V, Smith SB. 2001. Regulation of transporter expression by NO in cultured human retinal pigment epithetial cells. Am J phusiol Cell Physiol 287: C1825-1836.

Bureau MH, Olsen RW. 1991. Taurine acts on a subclass of $GABA_A$ receptors in mammalian brain in vitro. Eur J Pharmacol 207: 9-16.

Büyükuysal RL. 2004. Ischemia and reoxygenation-induced amino acid release and tissue damage in the slices of rat corpus striatum. Amino Acids 27: 57-67.

Cardin V, Peña-Segura C, Pasantes-Morales H. 1999. Activation and inactivation of taurine efflux in hypoosmotic and isosmotic swelling in cortical astrocytes: Role of ionic strength and cell volume decrease. J Neurosci Res 56: 659-667.

Chaput MA, Palouzier-Paulignan B, Delaleu JC, Duchamp-Viret P. 2004. Taurine action on mitral cell activity in the frog olfactory bulb in vivo. Chem Senses 29: 83-91.

Chen D-Z, Oheuma, Kurizama K. 1996. Characteristics of nitric oxide-evoked [^{3}H]taurine release from cerebral cortial neurons. Neurochem Int 28: 110-117.

Chen WQ, Jin H, Nguyen M, Carr J, Lee YJ, et al. 2001. Role of taurine in regulation of intracellular calcium level and neuroprotective function in cultured neurons. J Neurosci Res 66: 612-619.

Chen XC, Pan ZL, Liu DS, Han X. 1998. Effect of taurine on human fetal neuron cells: Proliferation and differentiation. Adv Med Exp Biol 442: 397-403.

Chepkova AN, Doreulee N, Yanovsky Y, Mukhopadhyay D, Haas HL, et al. 2002. Long-lasting enhancement of corticostriatal neurotransmission by taurine. Eur J Neurosci 16: 1523-1530.

Chepkova AN, Sergeeva OA, Haas HL. 2005. Long-lasting enhancement of corticostriatal transmission by taurine: Role of dopamine and acetylcholine. Cell Mol Neurobiol 25: 767-776.

Chepkova AN, Sergeeva OA, Haas HL. 2006. Mechanisms of long-lasting enhancement of corticostriatal neurotransmission by taurine. Taurine 6. Oja SS, Saransaari P, editors. New York: Springer/Kluwer Academic; pp. 401-410.

Chesney RW, Gusowski N, Dabbaugh S. 1985. Renal cortex taurine regulates the adaptive response to altered dietary intake of sulfur amino acids. J Clin Invest 76: 2213-2221.

Chesney RW, Helms RA, Christensen M, Budreau AM, Han X, et al. 1998. The role of taurine in infant nutrition. Adv Exp Med Biol 442: 463-476.

Chung SJ, Ramanathan V, Giacomini KM, Brett CM. 1994. Characterization of a sodium-dependent taurine transporter in rabbit choroid plexus. Biochim Biophys Acta 1193: 10-16.

Chung SJ, Ramanathan VK, Brett CM, Giacomini KM. 1996. Saturable disposition of taurine in the rat cerebrospinal fluid. J Pharmacol Exp Ther 276: 676-682.

Cocker SE, Lake N. 1987. Electroretinographic alterations and their reversal in rats treated with guanidinoethyl sulfonate, a taurine depletory. Exp Eye Res 45: 977-987.

Cocker SE, Lake N. 1989. Effects of dark maintenance on retinal biochemistry and function during taurine depletion in the adult rats. Vis Neurosci 3: 33-38.

Colivicchi MA, Bianchi L, Boloam JP, Galeffi F, Frosini M, et al. 1998. The in vivo release of taurine in the striatonigral pathway. Adv Exp Med Biol 442: 363-370.

Coloso RM, Hirschberger LL, Dominy JE, Lee JI, Stipanuk MH. 2006. Cysteamine dioxygenase: Evidence for the physiological conversion of cysteamine to hypotaurine in rat and mouse tissues. Taurine 6. Oja SS, Saransaari P, editors. New York: Springer/Kluwer Academic; pp. 25-35.

Cook AM, Denger K. 2006. Metabolism of taurine in microorganisms. A primer in molecular biodiversity. Taurine 6. Oja SS, Saransaari P, editors. New York: Springer/Kluwer Academic; pp. 3-13.

Cordoba J, Gottstein J, Blei AT. 1998. Chronic hyponatremia exacerbates ammonia-induced brain edema in rats after portacaval anastomosis. J Hepatol 29: 589-594.

Cubillos S, Fazzino F, Lima L. 2002. Medium requirements for neuritic outgrowth from goldfish retinal explants and the trophic effect of taurine. Int J Dev Neurosci 20: 607-617.

Cubillos S, Obregón F, Fernanda Vargas M, Salazar LA, Lima L. 2006. Taurine concentration in human gliomas and meningiomas: Tumoral, peritumoral, and extratumoral tissue. Taurine 6. Oja SS, Saransaari P, editors. New York: Springer/Kluwer Academic; pp. 419-422.

Cubillos S, Urbina M, Lima L. 2000. Differential taurine effect on outgrowth from goldfish retinal ganglion cells after optic crush or axotomoy. Influence of the optic tectum. Int J Dev Neurosci 18: 843-853.

Cunningham C, Tipton KF, Dixon HBF. 1998. Conversion of taurine into N-chlorotaurine (taurine chloramines) and sulphoacetaldehyde in response to oxidative stress. Biochem J 330: 939-945.

Dahchour A, De Witte P. 2000a. Ethanol and amino acids in the central nervous system: Assessment of the pharmacological actions of acamprosate. Prog Neurobiol 60: 343-362.

Dahchour A, De Witte P. 2000b. Taurine blocks the glutamate increase in the nucleus accumbens microdialysate of ethanol-dependent rats. Pharmacol Biochem Behav 65: 345-350.

Dahchour A, Quartemont E, De Witte P. 1994. Acute ethanol increases taurine but neither glutamate nor GABA in the nucleus accumbens of male rats: A microdialysis study. Alcohol Alcohol 29: 485-487.

Dahchour A, Quartemont E, De Witte P. 1996. Taurine increases in the nucleus accumbens microdialysate after acute ethanol administration to naive and chronically alcoholised rats. Brain Res 735: 9-19.

David-Watine B, Goblet C, de Saint Jan D, Fucile S, Devignot V, et al. 1999. Cloning, expression and electrophysiological characterization of glycine receptor α subunit from zebrafish. Neuroscience 90: 303-317.

Dawson R Jr, Baker D, Eppler B, Tang E, Shih D, et al. 2000. Taurine inhibition of metal-stimulated catecholamine oxidation. Neurotox Res 2: 1-15.

Dawson R Jr, Pelleymounter MA, Cullen MJ, Gollub M, Liu S. 1999a. An age-related decline in striatal taurine is correlated with a loss of dopaminergic markers. Brain Res Bull 48: 319-324.

Dawson R Jr, Liu S, Eppler B, Patterson T. 1999b. Effects of dietary taurine supplementation or deprivation in aged male Fischer 344-rats. Mech Ageing Dev 107: 73-91.

Decavel C, Hatton GI. 1995. Taurine immunoreactivity in the rat supraoptic nucleus: Prominent localization in glial cells. J Comp Neurol 354: 13-26.

Deleuze C, Alonso G, Lefevre IA, Duvoid-Guillou A, Hussy N. 2005. Extrasynaptic localization of glycine receptors in the rat supraoptic nucleus: Further evidence for their involvement in glia-to-neuron communication. Neuroscience 133: 175-183.

Della Corte L, Bolam JP, Clarke DJ, Parry DM, Smith AD. 1990. Sites of [^{3}H]taurine uptake in the rat substantia nigra in relation to the release of taurine from the nigrostriatal pathway. Eur J Neurosci 2: 50-61.

Della Corte L, Crichton RR, Duburs G, Nolan K, Tipton KF, et al. 2002. The use of taurine analogues to investigate taurine functions and their potential therapeutic applications. Amino Acids 23: 367-379.

del Olmo N, Bustamante J, Martín del Río R, Solís JM. 2000a. Taurine activates GABA$_A$ but not GABA$_B$ receptors in rat hippocampal CA1 area. Brain Res 864: 298-307.

del Olmo N, Galarreta M, Bustamante J, Martín del Río R, Solís JM. 2000b. Taurine-induced synaptic potentiation: Role of calcium and interaction with LTP. Neuropharmacology 39: 40-54.

del Olmo N, Galarreta M, Bustamante J, Martín del Río R, Solís JM. 2000c. Taurine-induced synaptic potentiation: Dependence on extra- and intracellular taurine sources. Adv Exp Med Biol 483: 283-292.

del Olmo N, Handler A, Alvarez L, Bustamante J, Martín del Río R, et al. 2003. Taurine-induced synaptic potentiation and the late phase of long-term potentiation are related mechanistically. Neuropharmacology 44: 26-39.

del Olmo N, Suárez LM, Orensanz LM, Suárez F, Bustamante J, et al. 2004. Role of taurine uptake on the induction of long-term synaptic potentiation. Eur J Neurosci 19: 1875-1886.

Devamanoharan PS, Ali AH, Varma SD. 1998. Oxidative stress to rat lens in vitro: Protection by taurine. Free Radic Res 29: 189-195.

Devreker F, Bergh Van den M, Biramane J, Winston RML, Englert Y, et al. 1999. Effects of taurine on human embryo development in vitro. Hum Reprod 14: 2350-2356.

Didier A, Ottersen OP, Storm-Mathisen J. 1994. Differential subcellular distribution of glutamate and taurine in primary olfactory neurons. Neuroreport 6: 145-148.

Diniz C, da Cruz Fresco P, Gonçalves J. 1999. Taurine release in the rat vas deferens is modulated by Ca^{2+} and is independent of concentrations. Eur J Pharmacol 376: 273-278.

Dohmen C, Kumura E, Rosner G, Heiss WD, Graf R. 2005. Extracellular correlates of glutamate toxicity in short-term cerebral ischemia and reperfusion: A direct in vivo comparison between white and gray matter. Brain Res 1017: 43-51.

Dominy J Jr, Thinschmidt JS, Peris J, Dawson R Jr, Papke RL. 2004. Taurine-induced long-lasting potentiation in the rat hippocampus shows a partial dissociation from total hippocampal taurine content and independence from activation of known taurine transporters. J Neurochem 89: 1195-1205.

El-Abhar HS, Abd El Gawad HM. 2003. Modulation of cortical nitric oxide synthase, glutamate, and redox state by nifedipine and taurine in PTZ-kindled mice. Epilepsia 44: 276-281.

El Idrissi A. 2006. Taurine and brain excitability. Taurine 6. Oja SS, Saransaari P, editors. New York: Springer/Kluwer Academic; pp. 315-322.

El Idrissi A, Trenkner E. 1999. Growth factors and taurine protect against excitotoxicity by stabilizing calcium homeostasis and energy metabolism. J Neurosci 19: 9459-9468.

El Idrissi A, Trenkner E. 2003. Taurine regulates mitochondrial calcium homeostasis. Adv Exp Med Biol 526: 527-536.

El Idrissi A, Trenkner E. 2004. Taurine as a modulator of excitatory and inhibitory neurotransmission. Neurochem Res 29: 189-197.

El Idrissi A, Harris C, Trenkner E. 1998. Taurine modulates glutamate and growth factors-mediated signaling mechanisms. Adv Exp Med Biol 442: 385-396.

El Idrissi A, Messing H, Scalia J, Trenkner E. 2003. Prevention of epileptic seizures by taurine. Adv Exp Med Biol 526: 515-525.

Engelmann M, Landgraf R, Wotjak CT. 2003. Taurine regulates corticotropin secretion at the level of the supraoptic nucleus during stress in rats. Neurosci Lett 348: 120-122.

Engelmann M, Wolf G, Horn TFW. 2002. Release patterns of excitatory and inhibitory amino acids within the hypothalamic supraoptic nucleus in response to direct nitric oxide administration during forced swimming in rats. Neurosci Lett 324: 252-254.

Eppler B, Patterson TA, Zhou W, Millard WJ, Dawson R Jr. 1999. Kainic acid (KA)-induced seizures in Sprague-Dawley rats and the effect of dietary taurine (TAU) supplementation or deficiency. Amino Acids 16: 133-147.

Erikson K, Aschner M. 2002. Manganese causes differential regulation of glutamate transporter (GLAST), taurine transporter and metallothionein in cultured rat astrocytes. Neurotoxicology 23: 595-602.

Erikson KM, Suber RL, Aschner M. 2002. Glutamate/aspartate transporter (GLAST), taurine transporter and metallothionein mRNA levels are differentially altered in astrocytes exposed to manganese chloride, manganese phosphate or manganese sulfate. Neurotoxicology 23: 281-288.

Estevez AY, O'Regan MH, Song D, Phillis JW. 1999a. Hyposmotically induced amino acid release from the rat cerebral cortex: Role of phospholipases and protein kinases. Brain Res 844: 1-9.

Estevez AY, O'Regan MH, Song D, Phillis JW. 1999b. Effects of anion channel blockers on hyposmotically induced amino acid release from the in vivo rat cerebral cortex. Neurochem Res 24: 447-452.

Estevez AY, Song D, Phillis JW, O'Regan MH. 2000. Effects of the anion channel blocker DIDS on ouabain- and high K^+-induced release of amino acids from the rat cerebral cortex. Brain Res Bull 52: 45-50.

Faff L, Reichenbach A, Albrecht J. 1996. Ammonia-induced taurine release from cultured rabbit Müller cells is an osmoresistant process mediated by intracellular accumulation of cyclic AMP. J Neurosci Res 46: 231-238.

Faff L, Reichenbach A, Albrecht J. 1997. Two models of stimulation by ammonia of taurine release from cultured rabbit Müller cells. Neurochem Int 31: 301-305.

Faff-Michalak L, Reichenbach A, Dettmer D, Kellner K, Albrecht J. 1994. K^+-, hypoosmolarity-, and NH_4^+-induced taurine release from cultured rabbit Müller cells: Role of Na^+ and Cl^- ions and relation to cell volume changes. Glia 10: 114-120.

Fellman JH, Roth ES. 1985. The biological oxidation of hypotaurine to taurine: Hypotaurine as an antioxidant. Prog Clin Biol Res 179: 71-82.

Ferko AP. 1987. Ethanol-induced sleep time: Interaction with taurine and a taurine antagonist. Pharmacol Biochem Behav 27: 235-238.

Ferko AP, Bobyock E. 1988. Effect of taurine on ethanol-induced sleep time in mice genetically bred for differences in ethanol sensitivity. Pharmacol Biochem Behav 31: 667-673.

Flora SJ, Pande M, Bhadauria S, Kannan GM. 2004. Combined administration of taurine and *meso*-2,3-dimercaptosuccinic acid in the treatment of chronic lead intoxication in rats. Human Exp Toxicol 23: 157-166.

Foos TM, Wu JY. 2002. The role of taurine in the central nervous system and the modulation of intracellular calcium homeostasis. Neurochem Res 27: 21-26.

Franco R, Quesada O, Pasantes-Morales H. 2000. Efflux of osmolyte amino acids during isovolumic regulation in hippocampal slices. J Neurosci Res 61: 701-711.

Franco R, Torres-Márquez ME, Pasantes-Morales H. 2001. Evidence for two mechanisms of amino acid osmolyte release from hippocampal slices. Pflügers Arch Eur J Physiol 442: 791-800.

Franconi F, Diana G, Fortuna A, Galietta G, Trombetta G, et al. 2004. Taurine administration during lactation modifies hippocampal CA1 neurotransmission and behavioural programming in adult male mice. Brain Res Bull 63: 491-497.

Frosini M, Sesti C, Dragoni S, Valoti M, Palmi M, et al. 2003a. Interactions of taurine and structurally related analogues with the GABAergic system and taurine binding sites of rabbit brain. Br J Pharmacol 138: 1163-1171.

Frosini M, Sesti C, Saponara S, Ricci L, Valoti M, et al. 2003b. A specific taurine recognition site in the rabbit brain is responsible for taurine effects on thermoregulation. Br J Pharmacol 139: 487-494.

Frosini M, Sesti C, Palmi M, Valoti M, Fusi F, et al. 2000a. Heat-stress-induced hyperthermia alters CSF osmolality and composition in conscious rabbits. Am J Physiol Regul Integr Comp Physiol 279: 2095-2103.

Frosini M, Sesti C, Saponara S, Donati A, Palmi M, et al. 2000b. Effects of taurine and some structurally related analogues on the central mechanism of thermoregulation. A structure-activity relationship study. Adv Exp Med Biol 483: 273-282.

Galarreta M, Bustamante J, Martín del Río R, Solís JM. 1996a. A new neuromodulatory action of taurine: Long-lasting increase of synaptic potentials. Adv Exp Med Biol 403: 463-471.

Galarreta M, Bustamante J, Martín del Río R, Solís JM. 1996b. Taurine induces a long-lasting increase of synaptic efficacy and axon excitability in the hippocampus. J Neurosci 16: 92-102.

Galietta LJ, Falzoni S, Di Virgilio F, Romeo G, Zegarra-Moran O. 1997. Characterization of volume-sensitive taurine- and Cl^--permeable channels. Am J Physiol 273: C57-C66.

Ganapathy V, Ramamoorthy JD, Del Monte MA, Leibach FH, Ramamoorthy S. 1995. Cyclic AMP-dependent up-regulation of the taurine transporter in a human retinal pigment epithelial cell line. Curr Eye Res 14: 843-850.

García Dopico J, Perdomo Díaz J, Alonso TJ, González Hernández T, Castro Fuentes R, et al. 2004. Extracellular taurine in the substantia nigra: Taurine–glutamate interaction. J Neurosci Res 76: 528-538.

Gavrovskaya LK, Krylova IB, Selina EN, Safonova AF, Petrova NN, et al. 2006. Antihypoxic properties of taurinamide derivatives. Taurine 6. Oja SS, Saransaari P, editors. New York: Springer/Kluwer Academic; pp. 523-528.

Geggel HS, Ament ME, Heckenlively JR, Martin DA, Kopple JD. 1985. Nutritional requirement for taurine in patients receiving long-term parenteral nutrition. New Eng J Med 312: 142-146.

Geiss KR, Jester I, Falke W, Hamm M, Waag KL. 1994. The effect of a taurine-containing drink on performance in 10 endurance-athletes. Amino Acids 7: 45-56.

Girard Y, Atkinson JG, Haubrich DR, Williams M, Yarbrough GG. 1982. Aminomethyl-1,2,4-benzothiadiazines as potential analogues of γ-aminobutyric acid. Unexpected discovery of a taurine antagonist. J Med Chem 25: 113-116.

Goda H, Ooboshi H, Nakane H, Ibayashi S, Sadoshima S, et al. 1998. Modulation of ischemia-evoked release of excitatory and inhibitory amino acids by adenosine A_1 receptor agonist. Eur J Pharmacol 357: 149-155.

Godfrey DA, Farms WB, Godfrey TG, Mikesell NL, Liu J. 2000. Amino acid concentrations in rat cochlear nucleus and superior olive. Hear Res 150: 189-205.

Goodman HO, Shihabi Z, Oles KS. 1989. Antiepileptic drugs and plasma and platelet taurine in epilepsy. Epilepsia 30: 201-207.

Gragera RR, Muniz E, De Esteban G, Alonso MJ, Martínez-Rodríguez R. 1995. Immunochemical demonstration of taurine in the rat cerebellar cortex. Evidence for its location within mossy fibers and Golgi axons. J Hirnforsch 36: 269-276.

Guevara-Guzman R, Emson PC, Kendrick KM. 1994. Modulation of in vivo striatal transmitter release by nitric oxide and cyclic GMP. J Neurochem 62: 807-810.

Guo J, Zhao P, Xia Y, Zhou F, Yang R, et al. 2006. Involvement of taurine in cerebral ischemia and electroacupuncture anti-ischemia. Taurine 6. Oja SS, Saransaari P, editors. New York: Springer/Kluwer Academic; pp. 395-400.

Gupta RC. 2006. Taurine analogues and taurine transport: Therapeutic advantages. Taurine 6. Oja SS, Saransaari P, editors. New York: Springer/Kluwer Academic; pp. 449-467.

Gupta RC, Win T, Bittner S. 2005. Taurine analogues; a new class of therapeutics: Retrospect and prospects. Curr Med Biol 12: 2021-2039.

Hada J, Kaku T, Morimoto K, Hayashi Y, Nagai K. 1996. Adenosine transport inhibitors enhance high K^+-evoked taurine release from rat hippocampus. Eur J Pharmacol 305: 101-107.

Hagberg H, Andersson P, Kjellmer I, Thiringer K, Thordstein M. 1987. Extracellular overflow of glutamate, aspartate, GABA and taurine in the cortex and basal ganglia of fetal lambs during hypoxia-ischemia. Neurosci Lett 78: 311-317.

Haroutounian JE, Petrosian AM. 1998. Effects of taurine and light on retinal GABA content and the efflux of ^{14}C-GABA and ^{14}C-aspartate from frog retina. Adv Exp Med Biol 442: 415-421.

Hashimoto-Kitsukawa S, Okuyama S, Aihara H. 1988a. Enhancing effect of taurine in the rat caudate spindle. I. Interaction of taurine with the nigro-striatal dopamine system. Pharmacol Biochem Behav 31: 411-416.

Hashimoto-Kitsukawa S, Okuyama S, Aihara H. 1988b. Enhancing effect of taurine in the rat caudate spindle. II. Effect of bilateral 6-hydroxydopamine lesions of the nigro-striatal dopamine system. Pharmacol Biochem Behav 31: 417-423.

Hassel B, Taubøll E, Gjerstad L. 2001. Chronic lamotrigine treatment increases rat hippocampal GABA shunt activity and elevates cerebral taurine levels. Epilepsy Res 43: 153-163.

Häuser MA, Yung WH, Lacey MG. 1992. Taurine and glycine activate the same Cl^- conductance in substantia nigra dopamine neurons. Brain Res 571: 103-108.

Hayes J, Tipton KF, Bianchi L, Della Corte L. 2001. Complexities in the neurotoxic actions of 6-hydroxydopamine in relation to the cytoprotective properties of taurine. Brain Res Bull 55: 239-245.

Hegstad E, Berg-Johnsen J, Haugstad TS, Hauglie-Hanssen E, Langmoen IA. 1996. Amino-acid release from human cerebral cortex during simulated ischemia in vitro. Acta Neurochir (Wien) 138: 234-241.

Heinämäki AA, Muhonen AS, Piha RS. 1986. Taurine and other free amino acids in the retina, vitreous, lens, iris-ciliary body, and cornea of the rat eye. Neurochem Res 11: 535-542.

Heller-Stilb B, van Roeyen C, Rascher K, Hartwig HG, Huth A, et al. 2002. Disruption of the taurine transporter gene (*taut*) leads to retinal degeneration in mice. FASEB J 16: 231-233.

Herranz AS, Solís JM, Herreras O, Menéndez N, Ambrosio E, et al. 1990. The epileptogenic action of the taurine analogue guanidinoethane sulphonate may be caused by a blockade of GABA receptors. J Neurosci Res 26: 98-104.

Hilgier W, Anderzhanova E, Oja SS, Saransaari P, Albrecht J. 2003. Taurine reduces ammonia- and N-methyl-D-aspartate-induced accumulation of cyclic GMP and hydroxyl radicals in microdialysates of the rat striatum. Eur J Pharmacol 468: 21-25.

Hilgier W, Oja SS, Saransaari P, Albrecht J. 2005. Taurine prevents ammonia-induced accumulation of cyclic AMP in rat striatum by interaction with $GABA_A$ and glycine receptors. Brain Res 1043: 242-246.

Hilgier W, Olson JE, Albrecht J. 1996. Relation of taurine transport and brain edema in rats with simple hyperammonemia or liver failure. J Neurosci Res 45: 69-74.

Hilgier W, Zielińska M, Borkowska HD, Gadamski R, Walski M, et al. 1999. Changes in the extracellular profiles of neuroactive amino acids in the rat striatum at the asymptomatic stage of hepatic failure. J Neurosci Res 56: 76-84.

Holopainen I, Kontro P. 1989. Uptake and release of glycine in cerebellar granule cells and astrocytes in primary culture: The potassium-stimulated release from granule cells is Ca-dependent. J Neurosci Res 24: 374-383.

Holopainen I, Kontro P, Frey HJ, Oja SS. 1983. Taurine, hypotaurine, and GABA uptake by cultured neuroblastoma cells. J Neurosci Res 10: 83-92.

Holopainen I, Kontro P, Oja SS. 1985. Release of preloaded taurine and hypotaurine from astrocytes in primary culture: Stimulation by calcium-free media. Neurochem Res 10: 123-131.

Holopainen I, Lidén E, Nilsson A, Sellström Å. 1990. Depolarization of the neuronal membrane caused by cotransport of taurine and sodium. Neurochem Res 15: 89-94.

Hoop B, Beagle JL, Maher TJ, Kazemi H. 1999. Brainstem amino acid transmitters and hypoxic ventilatory response. Respir Physiol 118: 117-129.

Horikoshi T, Asanuma A, Yanagisawa K, Anzai K, Goto S. 1988. Taurine and β-alanine act on both GABA and glycine receptors in *Xenopus* oocyte injected with mouse brain messenger RNA. Brain Res 464: 97-105.

Horner KC, Aurousseau C. 1997. Immunoreactivity for taurine in the cochlea: Its abundance in supporting cells. Hearing Res 109: 135-142.

Hornfeldt CS, Smullin DH, Schamber CD, Sun X, Larson AA. 1992. Antinociceptive effects of intrathecal taurine and calcium in the mouse. Life Sci 50: 1925-1934.

Hussy N, Brès V, Rochette M, Duvoid A, Alonso G, et al. 2001. Osmoregulation of vasopressin secretion via activation of neurohypophysial nerve terminals glycine receptors by glial taurine. J Neurosci 21: 7110-7116.

Hussy N, Deleuze C, Desarménien MG, Moos FC. 2000. Osmotic regulation of neuronal activity: A new role for taurine and glial cells in a hypothalamic neuroendocrine structure. Prog Neurobiol 62: 113-134.

Hussy N, Deleuze C, Pantaloni A, Desarménien MG, Moos F. 1997. Agonist action of taurine on glycine receptors in rat supraoptic magnocellular neurons: Possible role in osmoregulation. J Physiol 502: 609-621.

Huxtable RJ. 1986. Taurine and the oxidative metabolism of cysteine. Biochemistry of Sulfur. New York: Plenum Press; pp. 121-198.

Huxtable RJ. 1989. Taurine in the central nervous system and the mammalian actions of taurine. Prog Neurobiol 32: 471-533.

Huxtable RJ. 1992. The physiological actions of taurine. Physiol Rev 72: 101-163.

Huxtable RJ, Nakagawa K. 1985. The anticonvulsant actions of two taurine derivatives in genetic and chemically induced seizures. Prog clin Biop Res 179: 435-448.

Imaki H, Moretz RC, Wisniewski HM, Sturman JA. 1986. Feline maternal taurine deficiency: Effects on retina and tapetum of the offspring. Dev Neurosci 8: 160-181.

Imaki H, Moretz RC, Wisniewski HN, Neuringer M, Sturman JA. 1987. Retinal degeneration in three-month-old rhesus monkey infants fed a taurine-free human infant formula. J Neurosci Res 18: 602-614.

Inomata H, Nabekura J, Akaike N. 1993. Suppression of taurine response in acutely dissociated substantia nigra neurons by intracellular cyclic AMP. Brain Res 615: 347-350.

Iwata H, Matsuda T, Lee E, Yamagami S, Baba A. 1980. Effect of ethanol on taurine concentration in the brain. Experientia 36: 332-333.

Iwata H, Nakayama K, Matsuda T, Baba A. 1984. Effect of taurine on a benzodiazepine-GABA-chloride ionophore receptor complex in rat brain membranes. Neurochem Res 9: 535-544.

Izumi H, Kishita C, Nakagawa K, Huxtable RJ, Shimizu T, et al. 1985. Modification of the antiepileptic actions of phenobarbital and phenytoin by the taurine transport inhibitor, guanidinoethane sulfonate. Eur J Pharmacol 110: 219-224.

Jacobsen JG, Smith LH Jr. 1968. Biochemistry and physiology of taurine and taurine derivatives. Physiol Rev 48: 424-511.

Jacobson I, Hamberger A. 1985. Kainic acid-induced changes of extracellular amino acid levels, evoked potentials and EEG activity in the rabbit olfactory bulb. Brain Res 348: 289-296.

Jiang Z, Krnjević K, Wang F, Ye JH. 2004. Taurine activates strychnine-sensitive glycine receptors in neurons freshly isolated from nucleus accumbens of young rats. J Neurophysiol 91: 248-257.

Jones DP, Miller LA, Dowling C, Chesney RW. 1991. Regulation of taurine transporter activity in LLC-PK$_1$ cells: Role of protein synthesis and protein kinase C activation. J Am Soc Nephrol 2: 1021-1029.

Kaehler ST, Sinner C, Kouvelas D, Philippu A. 2000. Effects of inescapable shock and conditioned fear on the release of excitatory and inhibitory amino acids in the locus coeruleus. Naunyn Schmiedebergs Arch Pharmacol 361: 193-199.

Kalloniatis M, Marc RE, Murry FR. 1996. Amino acid signatures in the primate retina. J Neurosci 16: 6807-6829.

Kamisaki Y, Maeda K, Ishimura M, Omura H, Itoh T. 1993. Effects of taurine on depolarization-evoked release of amino acids from rat cortical synaptosomes. Brain Res 627: 181-185.

Kamisaki Y, Wada K, Nakamoto K, Itoh T. 1996a. Effect of taurine on GABA release from synaptosomes of rat olfactory bulb. Amino Acids 10: 49-57.

Kamisaki Y, Wada K, Nakamoto K, Itoh T. 1996b. Release of taurine and its effects on release of neurotransmitter amino acids in rat cerebral cortex. Adv Exp Med Biol 403: 445-454.

Kang YS. 2000. Taurine transport mechanism through the blood-brain barrier in spontaneously hypertensive rats. Adv Exp Med Biol 483: 321-324.

Kang YS. 2006. The effect of oxidative stress on the transport of taurine in an in vitro model of the blood-brain barrier. Taurine 6. Oja SS, Saransaari P, editors. New York: Springer/Kluwer Academic; pp. 291-298.

Kang YS, Ohtsuki S, Takanaga H, Tomi M, Hosoya K, et al. 2002. Regulation of taurine transport at the blood-brain barrier by tumor necrosis factor-α, taurine and hypertonicity. J Neurochem 83: 1188-1195.

Kapoor V, Nakahara D, Blood RJ, Chalmers JP. 1990. Preferential release of neuroactive amino acids from the ventrolateral medulla of the rat in vivo as measured by microdialysis. Neuroscience 37: 187-191.

Kashkin VA, De Witte P. 2004. Ethanol but not acetaldehyde induced changes in brain taurine: A microdialysis study. Amino Acids 26: 117-124.

Katoh H, Sima K, Nawashiro H, Wada K, Chigasaki H. 1997. The effect of MK-801 on extracellular neuroactive amino acids in hippocampus after closed head injury followed by hypoxia in rats. Brain Res 758: 153-162.

Keep RF, Xiang J. 1996. Choroid plexus taurine transport. Brain Res 715: 17-24.

Keys SA, Zimmerman WF. 1999. Antioxidant activity of retinol, glutathione, and taurine in bovine photoreceptor cell membranes. Exp Eye Res 68: 693-702.

Khan SNH, Banigesh A, Baziani A, Todd KG, Miyashita H, et al. 2000. The role of taurine in neuronal protection following transient global forebrain ischemia. Neurochem Res 25: 217-223.

Kilic F, Bhardwaj R, Caulfield J, Trevithick JR. 1999. Modelling cortical cataractogenesis 22: Is in vitro reduction of damage in model diabetic rat cataract by taurine due to its antioxidant activity? Exp Eye Res 69: 291-300.

Kimelberg HK, Cheema M, O'Connor ER, Tong H, Goderie SK, et al. 1993. Ethanol-induced aspartate and taurine release from primary astrocyte cultures. J Neurochem 60: 1682-1689.

Kimelberg HK, Goderie SK, Higman S, Pang S, Waniewski RA. 1990. Swelling-induced release of glutamate, aspartate, and taurine from astrocyte cultures. J Neurosci 10: 1583-1591.

Kimelberg HK, Nestor NB, Feustel PJ. 2004. Inhibition of release of taurine and excitatory amino acids in ischemia and neuroprotection. Neurochem Res 29: 267-274.

Kingston R, Kelly CJ, Murray P. 2004. The therapeutic role of taurine in ischaemia-reperfusion injury. Curr Pharm Des 10: 2401-2410.

Kirchner A, Breustedt J, Rosche B, Heinemann UF, Schmieden V. 2003. Effects of taurine and glycine on epileptiform activity induced by removal of Mg^{2+} in combined rat entorhinal cortex-hippocampal slices. Epilepsia 44: 1145-1152.

Kluša V, Klimaviciusa L, Duburs G, Poikans J, Zharkovsky A. 2006. Anti-neurotoxic effect of tauropyrone, a taurine analogue. Taurine 6. Oja SS, Saransaari P, editors. New York: Springer/Kluwer Academic; pp. 499-508.

Koivisto K, Sivenius J, Keränen T, Partanen J, Riekkinen P, et al. 1986. Clinical trial with an experimental taurine derivative, taltrimide, in epileptic patients. Epilepsia 27: 87-90.

Kontro P. 1979. Components of taurine efflux in rat brain synaptosomes. Neuroscience 4: 1745-1749.

Kontro P. 1987. Interactions of taurine and dopamine in the striatum. Adv Exp Med Biol 217: 347-355.

Kontro P, Oja SS. 1983. Sodium-independent taurine binding to brain synaptic membranes. Cell Mol Neurobiol 3: 183-187.

Kontro P, Oja SS. 1985a. Hypotaurine oxidation by mouse liver tissue. Prog Clin Biol Res 179: 83-90.

Kontro P, Oja SS. 1985b. Properties of sodium-independent taurine binding to brain synaptic membranes. Prog Clin Biol Res 179: 249-259.

Kontro P, Oja SS. 1986. Taurine interferes with spiperone binding in the striatum. Neuroscience 19: 1007-1010.

Kontro P, Oja SS. 1987a. Taurine and GABA binding in mouse brain: Effects of freezing, washing and Triton X-100 treatment on membranes. Int J Neurosci 32: 881-889.

Kontro P, Oja SS. 1987b. Taurine and GABA release from mouse cerebral cortex slices: Potassium stimulation releases more taurine than GABA from developing brain. Dev Brain Res 37: 277-291.

Kontro P, Oja SS. 1987c. Effects of the anticonvulsant taurine derivative, taltrimide, on membrane transport and binding of GABA and taurine in the mouse cerebrum. Neuropharmacology 26: 19-23.

Kontro P, Oja SS. 1987d. Glycinergic systems in the brain stem of developing and adult mice: Effects of taurine. Int J Dev Neurosci 5: 461-470.

Kontro P, Oja SS. 1987e. Co-operativity in sodium-independent taurine binding to brain membranes in the mouse. Neuroscience 23: 567-570.

Kontro P, Oja SS. 1987f. Effects of the new anticonvulsant taurine derivative, taltrimide, on membrane transport and binding of GABA and taurine in the mouse cerebrum. Neuropharmacology 26: 19-23.

Kontro P, Oja SS. 1988a. Release of taurine, GABA and dopamine from rat striatal slices: Mutual interactions and developmental aspects. Neuroscience 24: 49-58.

Kontro P, Oja SS. 1988b. Effects of taurine on the influx and efflux of calcium in brain slices of adult and developing mice. Int J Neurosci 38: 103-109.

Kontro P, Oja SS. 1989. Release of taurine and GABA from cerebellar slices from developing and adult mice. Neuroscience 29: 413-423.

Kontro P, Oja SS. 1990. Interactions of taurine with $GABA_B$ binding sites in mouse brain. Neuropharmacology 29: 243-247.

Kontro P, Korpi ER, Oja OS, Oja SS. 1984. Modulation of noradrenaline uptake and release by taurine in rat cerebral slices. Neuroscience 13: 663-666.

Kontro P, Korpi ER, Oja SS. 1990. Taurine interacts with $GABA_A$ and $GABA_B$ receptors in the brain. Prog Clin Biol Res 351: 83-94.

Kontro P, Marnela K-M, Oja SS. 1980. Free amino acids in the synaptosome and synaptic vesicle fractions of different bovine brain areas. Brain Res 184: 129-141.

Korpi ER, Kontro P, Nieminen K, Marnela K-M, Oja SS. 1981. Spontaneous and depolarization-induced efflux of hypotaurine from mouse cerebral cortex slices: Comparison with taurine and GABA. Life Sci 29: 811-816.

Kreisman NR, Olson JE. 2003. Taurine enhances volume regulation in hippocampal slices swollen osmotically. Neuroscience 120: 635-642.

Krishtal OA, Osipchuk YV, Vrublevsky SV. 1988. Properties of glycine-activated conductances in rat brain neurons. Neurosci Lett 84: 271-276.

Krylova IB, Bulion VV, Gavrovskaya LK, Selina EN, Kuznetzova NN, et al. 2006. Neuroprotective effect of a new taurinamide derivative–taurepar. Taurine 6. Oja SS, Saransaari P, editors. New York: Springer/Kluwer Academic; pp. 543-550.

Kubo T, Takano A, Tokushige N, Miyata N, Sato M, et al. 1992. Electrical stimulation-evoked release of endogenous taurine from slices of the hippocampus, cerebral cortex and cerebellum of the rat. J Pharmacobiodyn 15: 519-525.

Kudo Y, Akiyoshi E, Akagi H. 1988. Identification of two taurine receptor subtypes on the primary afferent terminal of frog spinal cord. Br J Pharmacol 94: 1051-1056.

Kuhse J, Schmieden V, Betz H. 1990a. Identification and functional expression of a novel ligand binding subunit of the inhibitory glycine receptor. J Biol Chem 265: 22317-22320.

Kuhse J, Schmieden V, Betz H. 1990b. A single amino acid exchange alters the pharmacology of neonatal rat glycine receptor subunit. Neuron 5: 867-873.

Kulanthaivel P, Cool DR, Ramamoorthy S, Mahesh VB, Leibach FH, et al. 1991. Transport of taurine and its regulation by protein kinase C in the JAR human placental choriocarcinoma cell line. Biochem J 277: 53-58.

La Bella V, Piccoli F. 2000. Differential effect of β-N-oxalylamino-L-alanine, the *Lathyrus sativus* neurotoxin, and (±)-α-amino-3-hydroxy-5-methylisoxazole-4-propionate on the excitatory amino acid and taurine levels in the brain of freely moving rat. Neurochem Int 36: 523-530.

Lähdesmäki P, Oja SS. 1973. On the mechanism of taurine transport at brain cell membranes. J Neurochem 20: 1411-1417.

Lai YJ, Shen EY, Pan WHT. 2000. Effects of ascorbate in microdialysis perfusion medium on the extracellular basal concentration of glutamate in rat's striatum. Neurosci Lett 279: 145-148.

Lake N. 1981. Depletion of retinal taurine by treatment with guanidinoethyl sulfonate. Life Sci 29: 445-448.

Lake N. 1983. Taurine depletion of lactating rats: Effects on developing pups. Neurochem Res 8: 881-887.

Lake N. 1986. Electroretinographic deficits in rats treated with guanidinoethyl sulfonate, a depletor of taurine. Exp Eye Res 42: 87-91.

Lake N. 1992. Localization of taurine and glial fibrillary acidic protein in human optic nerve using immunocytochemical techniques. Adv Exp Med Biol 315: 303-307.

Lake N. 1994. Taurine and GABA in the rat retina during postnatal development. Vis Neurosci 11: 253-260.

Lake N, Cocker SE. 1983. In vitro studies of guanidinoethyl sulfonate and taurine transport in the rat retina. Neurochem Res 8: 1557-1563.

Lake N, De Marte L. 1988. Effects of β-alanine treatment on the taurine and DNA content of the rat heart and retina. Neurochem Res 13: 1003-1006.

Lake N, Orlowski J. 1996. Cellular studies of the taurine transporter. Adv Exp Med Biol 403: 371-376.

Lake N, Verdone-Smith C. 1989. Immunocytochemical localization of taurine in the mammalian retina. Curr Eye Res 8: 163-173.

Lake N, Malik N, De Marte L. 1988. Taurine depletion leads to loss of rat optic nerve axons. Vision Res 28: 1071-1076.

Lake N, Ueck M, Hach A, Verdone-Smith C. 1996. Ultrastructural localization of taurine immunoreactivity in the pineal organ and retina of the pigeon. Adv Exp Med Biol 403: 329-331.

Lake N, Verdone-Smith C, Brownstein S. 1992. Immunocytochemical localization of taurine and glial acidic fibrillary protein in human optic nerve. Vis Neurosci 8: 251-255.

Lallemand F, De Witte P. 2004. Taurine concentration in the brain and in the plasma following intraperitoneal injections. Amino Acids 26: 111-116.

Laube B, Langosch D, Betz H, Schmieden V. 1995. Hyperekplexia mutations of the glycine receptor unmask the inhibitory subsite for β-amino acids. Neuroreport 6: 897-900.

Law RO. 1989. The effect of pregnancy on the content of water, taurine, and total amino nitrogen in rat cerebral cortex. J Neurochem 53: 300-302.

Law RO. 1994. Effects of extracellular bicarbonate ions and pH on volume-regulatory taurine efflux from rat cerebral cortical slices in vitro: Evidence for separate neutral and anionic transport mechanisms. Biochim Biophys Acta 1224: 377-383.

Law RO. 1995. Taurine efflux and cell volume regulation in cerebral cortical slices during chronic hypernatraemia. Neurosci Lett 185: 56-59.

Law RO. 2006. The mechanisms of taurine's protective action against acute guanidino neurotoxicity. Taurine 6. Oja SS, Saransaari P, editors. New York: Springer/Kluwer Academic; pp. 359-364.

Lee CM, Tung WL, Young JD. 1992. Tachykinin-stimulated inositol phospholipid hydrolysis and taurine release from human astrocytoma cells. J Neurochem 59: 406-414.

Lee NY, Kang YS. 2004. The brain-to-blood efflux transport of taurine and changes in the blood-brain barrier transport system by tumor necrosis factor-α. Brain Res 1023: 141-147.

Lehmann A. 1987. Pentylenetetrazol seizure threshold and extracellular levels of cortical amino acids in taurine-deficient kittens. Acta Physiol Scand 131: 453-458.

Lehmann A. 1989. Effects of microdialysis-perfusion with anisosmotic media on extracellular amino acids in the rat hippocampus and skeletal muscle. J Neurochem 53: 525-535.

Lehmann A, Hagberg. H, Hamberger A. 1984. A role for taurine in the maintenance of homeostasis in the central nervous system during hyperexcitation? Neurosci Lett 52: 341-346.

Lehmann A, Huxtable RJ, Hamberger A. 1987. Taurine deficiency in the rat and cat: Effects on neurotoxic and biochemical actions of kainate. Adv Exp Med Biol 213: 331-339.

Lekieffre D, Callebert J, Plotkine M, Boulu RG. 1992. Concomitant increases in the extracellular concentrations of excitatory and inhibitory amino acids in the rat hippocampus during forebrain ischemia. Neurosci Lett 137: 78-82.

Lerma J, Herranz AS, Herreras O, Abraira V, Martín del Río R. 1986. In vivo determination of extracellular concentration of amino acids in the rat hippocampus. A method based on brain dialysis and computerized analysis. Brain Res 384: 145-155.

Leviel V, Chéramy A, Nieoullon A, Glowinski J. 1979. Symmetric bilateral changes in dopamine release from the caudate nuclei of the cat induced by unilateral nigral application of glycine and GABA-related compounds. Brain Res 175: 259-270.

Levinskaya N, Trenkner E, El Idrissi A. 2006. Increased GAD-positive neurons in the cortex of taurine-fed mice. Taurine 6. Oja SS, Saransaari P, editors. New York: Springer/Kluwer Academic; pp. 411-417.

Lewin L, Rassin DK, Sellström Å. 1994. Net taurine transport and its inhibition by a taurine antagonist. Neurochem Res 19: 347-352.

Li G, Liu Y, Olson JE. 2002. Calcium/calmodulin-modulated chloride and taurine conductances in cultured rat astrocytes. Brain Res 925: 1-8.

Li H, Godfrey DA, Rubin AM. 1994. Quantitative distribution of amino acids in the rat vestibular nuclei. J Vest Res 4: 437-452.

Li H, Godfrey TG, Godfrey DA, Rubin AM. 1996. Quantitative changes of amino acid distributions in the rat vestibular nuclear complex after unilateral vestibular ganglionectomy. J Neurochem 66: 1550-1564.

Li YP, Lombardini JB. 1991. Inhibition by taurine of the phosphorylation of specific synaptosomal proteins in the rat cortex: Effects of taurine on the stimulation of calcium uptake in mitochondria and inhibition of phosphoinositide turnover. Brain Res 553: 89-96.

Li ZP, Zhang XY, Lu X, Zhong MK, Ji YH. 2004. Dynamic release of amino acid transmitters induced by valproate in PTZ-kindled epileptic rat hippocampus. Neurochem Int 44: 264-270.

Li ZQ, Yamamoto Y, Morimoto T, Ono J, Okada S, et al. 2000. The effect of pentylenetetrazole-kindling on the extracellular glutamate and taurine levels in the frontal cortex of rats. Neurosci Lett 282: 117-119.

Liebowitz SM, Lombardini JB, Allen IA. 1986. 2-Amino-ethymethy-sulfone (AEMS): A potent stimulator of ATP-dependent calcium uptake. Eur J Pharmacol 120: 111-113.

Liebowitz SM, Lombardini JB, Allen CI. 1988. Effects of aminocycloalkanesulfonic acid analogs of taurine on ATP-dependent calcium ion uptake and protein phosphorylation. Biochem Pharmacol 37: 1303-1309.

Liebowitz SM, Lombardini JB, Allen CI. 1989. Sulfone analogues of taurine as modifiers of calcium uptake and protein phosphorylation in rat retina. Biochem Pharmacol 38: 399-406.

Liebowitz SM, Lombardini JB, Salva PS. 1987. Cyclic taurine analogs. Synthesis and effects on ATP-dependent Ca^{2+} uptake in rat retina. Biochem Pharmacol 36: 2109-2114.

Lien YH, Shapiro JI, Chan L. 1991. Study of brain electrolytes and organic osmolytes during correction of chronic hyponatremia. Implications for the pathogenesis of central pontine myelinolysis. J Clin Invest 88: 303-309.

Lima L. 1999. Taurine and its trophic effects in the retina. Neurochem Res 24: 1333-1338.

Lima L, Cubillos S. 1998a. Taurine might be acting as a trophic factor in the retina by modulating phosphorylation of cellular proteins. J Neurosci Res 53: 377-384.

Lima L, Cubillos S. 1998b. Taurine-stimulated outgrowth from the retina is impaired by protein kinase C activators and phosphatase inhibitors. Adv Exp Med Biol 442: 423-430.

Lima L, Cubillos S, Fazzino F. 2003. Taurine effect on neuritic outgrowth from goldfish retinal explants in the absence and presence of fetal calf serum. Adv Exp Med Biol 526: 507-514.

Lima L, Cubillos S, Guerra A. 2000. Regulation of high affinity taurine transport in goldfish and rat retinal cells. Adv Exp Med Biol 483: 431-440.

Lima L, Drujan B, Matus P. 1990. Spatial distribution of taurine in the teleost retina and its role in retinal tissue regeneration. Prog Clin Biol Res 351: 103-112.

Lima L, Matus P, Drujan B. 1988. Taurine effect on neuritic growth from goldfish retinal explants. Int J Dev Neurosci 6: 417-424.

Lima L, Matus P, Drujan B. 1989. The interaction of substrate and taurine modulates the outgrowth from regenerating goldfish retinal explants. Int J Dev Neurosci 7: 375-382.

Lima L, Matus P, Drujan B. 1991. Differential taurine uptake in central and peripheral regions of goldfish retina. J Neurosci Res 28: 422-427.

Lima L, Matus P, Drujan B. 1992. The trophic role of taurine in the retina. A possible mechanism of action. Adv Exp Med Biol 315: 287-294.

Lima L, Matus P, Drujan B. 1993. Taurine-induced regeneration of goldfish retina in culture may involve a calcium-mediated mechanism. J Neurochem 60: 2153-2158.

Lima L, Obregón F, Cubillos S, Fazzino F, Jaimes I. 2001. Taurine as micronutrient in development and regeneration of the central nervous system. Nutr Neurosci 4: 439-443.

Lima L, Obregón F, Matus P. 1998. Taurine, glutamate and GABA modulate the outgrowth from goldfish retinal explants and its concentrations are affected by the crush of the optic nerve. Amino Acids 15: 195-209.

Lima L, Roussó T, Quintal M, Benzo Z, Auladell C. 2004. Content and concentration of taurine, hypotaurine, and zinc in the retina, the hippocampus, and the dentate gyrus of the rat at various postnatal days. Neurochem Res 29: 247-255.

Lindén IB, Gothóni G, Kontro P, Oja SS. 1983. Anticonvulsant activity of 2-phthalimidoethanesulfonamides: New derivatives of taurine. Neurochem Int 5: 319-324.

Linne ML, Jalonen TO, Saransaari P, Oja SS. 1996. Taurine-induced single-channel currents in cultured rat cerebellar granule cells. Adv Exp Med Biol 403: 455-462.

Liu G, Liu Y, Olson JE. 2002. Calcium/calmodulin-modulated chloride and taurine conductances in cultured rat astrocytes. Brain Res 925: 1-8.

Liu QR, López-Corcuera B, Mandiyan S, Nelson H, Nelson N. 1993. Molecular characterization of four pharmacologically distinct γ-aminobutyric acid transporters in mouse brain. J Biol Chem 268: 2106-2112.

Liu QR, López-Corcuera B, Nelson H, Mandiyan S, Nelson N. 1992. Cloning and expression of cDNA encoding the transporter of taurine and β-alanine in mouse brain. Proc Natl Acad Sci USA 89: 12145-12149.

Lo EH, Pierce AR, Matsumoto K, Kano T, Evans CJ, et al. 1998. Alterations in K^+ evoked profiles of neurotransmitter and neuromodulator amino acids after focal ischemia-reperfusion. Neuroscience 83: 449-458.

Lombardini JB. 1988. Effects of taurine and mitochondrial metabolic inhibitors on ATP-dependent Ca^{2+} uptake in synaptosomal and mitochondrial subcellular fractions of rat retina. J Neurochem 51: 200-205.

Lombardini JB. 1991. Taurine: Retinal function. Brain Res Rev 16: 151-169.

Lombardini JB. 1992a. Effects of taurine on the phosphorylation of specific proteins in subcellular fractions of the rat retina. Neurochem Res 17: 821-824.

Lombardini JB. 1992b. Potassium-stimulated release of taurine in a crude retinal preparation obtained from the rat is calcium independent. Adv Med Exp Biol 315: 399-404.

Lombardini JB. 1993a. Spontaneous and evoked release of [^{3}H]taurine from a P_2 subcellular fraction of the rat retina. Neurochem Res 18: 193-202.

Lombardini JB. 1993b. Partial characterization of an approximately 20 K M_r retinal protein whose phosphorylation is inhibited by taurine. Biochem Pharmacol 46: 1445-1451.

Lombardini JB. 1998. Increased phosphorylation of specific rat cardiac and retinal proteins in taurine-depleted animals: Isolation and identification of the phosphoproteins. Adv Exp Med Biol 442: 441-447.

Lombardini JB, Liebowitz SM. 1990a. Inhibitory and stimulatory effects of structural and conformational analogues of taurine on ATP-dependent calcium ion uptake in the rat retina: Deductions concerning the conformation of taurine. Prog Clin Biol Res 351: 197-206.

Lombardini JB, Liebowitz SM. 1990b. Taurine analogues as modifiers of the accumulation of ^{45}calcium ions in a rat retinal membrane preparation. Curr Eye Res 9: 1147-1156.

Lombardini JB, Props C. 1996. Effects of kinase inhibitors and taurine analogues on the phosphorylation of specific proteins in mitochondrial fractions of rat heart and retina. Adv Exp Med Biol 403: 343-350.

Lombardini JB, Liebowitz SM, Chou TC. 1989. Analogues of taurine as stimulators and inhibitors of ATP-dependent calcium ion uptake in rat retina: Combination kinetics. Mol Pharmacol 36: 256-264.

Loo DDF, Hirsch JR, Sarkar HK, Wright EM. 1996. Regulation of the mouse retinal taurine transporter (TAUT) by protein kinases in Xenopus oocytes. FEBS Lett 392: 250-254.

Louzada PR, Lima ACP, Mendonça-Silva DL, Noël F, De Mello FG, et al. 2004. Taurine prevents the neurotoxicity of β-amyloid and glutamate receptor agonists: Activation of GABA receptors and possible implications for Alzheimer's disease and other neurological disorders. FASEB J 18: 511-518.

Lu P, Xu W, Sturman JA. 1996. Dietary β-alanine results in taurine depletion and cerebellar damage in adult cats. J Neurosci Res 43: 112-119.

Lucchi R, Poli A, Traversa U, Barnabei O. 1994. Functional adenosine A_1 receptors in goldfish brain: Regional distribution and inhibition of K^+-evoked glutamate release from cerebellar slices. Neuroscience 58: 237-243.

Lynch JW, Rajendra S, Pierce KD, Handford CA, Barry PH, et al. 1997. Identification of intracellular and extracellular domains mediating signal transduction in the inhibitory glycine receptor chloride channel. EMBO J 16: 110-120.

Maar T, Morán J, Schousboe A, Pasantes-Morales H. 1995. Taurine deficiency in dissociated mouse cerebellar neurons affects neuronal migration. Int J Dev Neurosci 13: 491-502.

Madelian V, Silliman S, Shain W. 1988. Adenosine stimulates cAMP-mediated taurine release from LRM55 glial cells. J Neurosci Res 20: 176-181.

Magnusson KR, Koerner JF, Larson AA, Smullin DH, Skilling SR, et al. 1991. NMDA-, kainate- and quisqualate-stimulated release of taurine from electrophysiologically monitored rat hippocampal slices. Brain Res 549: 1-8.

Malcangio M, Bartolini A, Ghelardini C, Bennardini F, Malmberg-Aiello P, et al. 1989. Effects of ICV taurine on the impairment of learning, convulsions and death caused by hypoxia. Psychopharmacology 98: 316-320.

Malminen O, Kontro P. 1986. Modulation of the GABA-benzodiazepine receptor complex by taurine in rat brain membranes. Neurochem Res 11: 85-94.

Malminen O, Kontro P. 1987. Actions of taurine on the GABA-benzodiazepine receptor complex solubilized from rat brain. Neurochem Int 11: 113-117.

Mandel P, Gupta RC, Bourguignon JJ, Wermuth CG, Molina V, et al. 1985. Effects of taurine and taurine analogues on aggressive behavior. Prog Clin Biol Res 179: 449-458.

Mankovskaya IN, Serebrovskaya TV, Swanson RJ, Vavilova GL, Kharlamova ON. 2000. Mechanisms of taurine antihypoxic and antioxidant action. High Alt Med Biol 1: 105-110.

Marangolo M, Zisterer D, Williams DC, Tipton KF, Della Corte L. 1997. Different specificities for taurine analogues and their target sites in brain. Neurochemistry: Cellular, Molecular and Clinical Aspects. Teelken A, Korf J, editors. New York: Plenum Press; pp. 59-962.

Marc RE, Murry RF, Fisher SK, Linberg KA, Lewis GP, et al. 1998. Amino acid signatures in the normal cat retina. Invest Ophthalmol Vis Sci 39: 1685-1693.

Marnela K-M, Kontro P. 1984. Free amino acids and the uptake and binding of taurine in the central nervous system of rats treated with guanidinoethanesulphonate. Neuroscience 12: 323-328.

Marnela K-M, Morris HR, Panico M, Timonen M, Lähdesmäki P. 1985. Glutamyl-taurine is the predominant synaptic peptide. J Neurochem 44: 752-754.

Marnela K-M, Timonen M, Lähdesmäki P. 1984. Mass spectrometric analyses of brain synaptic peptides containing taurine. J Neurochem 43: 1650-1653.

Martin DL, Madelian V, Seligman B, Shain W. 1990. The role of osmotic pressure and membrane potential in K^+-stimulated taurine release from cultured astrocytes and LRM55 cells. J Neurosci 10: 571-577.

Martínez A, Muñoz-Clares RA, Guerra G, Morán J, Pasantes-Morales H. 1994. Sulfhydryl groups essential for the volume-sensitive release of taurine from astrocytes. Neurosci Lett 176: 239-242.

Matsuda T, Takuma K, Kishida Y, Azuma J, Baba A. 1996. Protective effect of taurine against reperfusion injury in cultured rat astrocytes. Adv Exp Med Biol 403: 491-497.

Matsumoto K, Lo EH, Pierce AR, Halpern EF, Newcomb R. 1996. Secondary elevation of extracellular neurotransmitter amino acids in the reperfusion phase following focal cerebral ischemia. J Cereb Blood Flow Metab 16: 114-124.

Matus P, Cubillos S, Lima L. 1997. Differential effect of taurine and serotonin on the outgrowth from explants or isolated cells of the retina. Int J Dev Neurosci 15: 785-793.

McBroom MJ, Elkhawad AO, Dlouha H. 1986. Taurine and ethanol-induced sleeping time in mice: Route and time course effects. Gen Pharmacol 17: 97-100.

McCabe BJ, Horn G. 1988. Learning and memory: Regional changes in N-methyl-D-aspartate receptors in the chick brain after imprinting. Proc Natl Acad Sci USA 85: 2849-2853.

McCabe BJ, Horn G, Kendrick KM. 2001. GABA, taurine and learning: Release of amino acids from slices of chick brain following filial imprinting. Neuroscience 105: 317-324.

McCaslin PP, Yu XZ. 1992. Cyanide selectively augments kainate—but not NMDA—induced release of glutamate and taurine. Eur J Pharmacol 228: 73-75.

McCaslin PP, Yu XZ, Ho IK, Smith TG. 1992. Amitriptyline prevents N-methyl-D-aspartate (NMDA)-induced toxicity, does not prevent NMDA-induced elevations of extracellular glutamate, but augments kainate-induced elevations of glutamate. J Neurochem 59: 401-405.

McCool BA, Botting SK. 2000. Characterization of strychnine-sensitive glycine receptors in acutely isolated adult rat basolateral amygdala neurons. Brain Res 859: 341-351.

Medina JH, De Robertis E. 1984. Taurine modulation of the benzodiazepine-γ-aminobutyric acid receptor complex in brain membranes. J Neurochem 42: 1212-1217.

Mehta TR, Dawson R Jr. 2001. Taurine is a weak scavenger of peroxynitrite and does not attenuate sodium nitroprusside toxicity to cells in culture. Amino Acids 20: 419-433.

Melamed N, Kanner BI. 2004. Transmembrane domains I and II of the γ-aminobutyric acid transporter GAT-4 contain molecular determinants of substrate specificity. Mol Pharmacol 65: 1452-1461.

Mellor RJ, Gunthorpe MJ, Randall AD. 2000. The taurine uptake inhibitor guanidinoethyl sulphonate is an agonist at γ-aminobutyric acid$_A$ receptors in cultured murine cerebellar granule cells. Neurosci Lett 286: 25-28.

Menéndez N, Solís JM, Herreras O, Galarreta M, Conejero C, et al. 1993. Taurine release evoked by NMDA receptor activation is largely dependent on calcium mobilization from intracellular stores. Eur J Neurosci 5: 1273-1279.

Menéndez N, Solís JM, Herreras O, Herranz AS, Martín del Río R. 1990. Role of endogenous taurine on the glutamate analogue-induced neurotoxicity in the rat hippocampus in vivo. J Neurochem 55: 714-717.

Meredith RM, McCabe BJ, Kendrick KM, Horn G. 2004. Amino acid neurotransmitter release and learning: A study of visual imprinting. Neuroscience 126: 249-256.

Mialon P, Joanny P, Gibey R, Cann-Moisan C, Caroff J, et al. 1995. Amino acids and ammonia in the cerebral cortex, the corpus striatum and the brain stem of the mouse prior to the onset and after a seizure induced by hyperbaric oxygen. Brain Res 676: 352-357.

Miele M, Berners M, Boutelle MG, Kusakabe H, Fillenz M. 1996. The determination of the extracellular concentration of brain glutamate using quantitative microdialysis. Brain Res 707: 131-133.

Militante JD, Lombardini JB. 1998a. Effect of taurine on chelerythrine inhibition of calcium uptake and ATPase activity in the rat retina. Biochem Pharmacol 55: 557-565.

Militante JD, Lombardini JB. 1998b. Pharmacological characterization of the effects of taurine on calcium uptake in the rat retina. Amino Acids 15: 99-108.

Militante JD, Lombardini JB. 1999a. Taurine uptake activity in the rat retina: Protein kinase C-independent inhibition by chelerythrine. Brain Res 818: 368-374.

Militante JD, Lombardini JB. 1999b. Stimulatory effect of taurine on calcium ion uptake in rod outer segments of the rat retina is independent of taurine uptake. J Pharmacol Exp Ther 291: 383-389.

Militante JD, Lombardini JB. 2000a. Stabilization of calcium uptake in rat outer rod segments by taurine and ATP. Amino Acids 19: 561-570.

Militante JD, Lombardini JB. 2000b. Characterization of taurine uptake in the rat retina. Adv Exp Med Biol 483: 461-467.

Militante JD, Lombardini JB. 2002. Taurine: Evidence of physiological function in the retina. Nutr Neurosci 5: 75-90.

Militante JD, Lombardini JB. 2003a. Taurine stimulation of calcium uptake in the retina. Adv Exp Med Biol 526: 547-554.

Militante JD, Lombardini JB. 2003b. The nature of taurine binding in the retina. Adv Exp Med Biol 526: 555-560.

Miller TJ, Hanson RD, Yancey PH. 2000. Developmental changes in organic osmolytes in prenatal and postnatal rat tissues. Comp Biochem Physiol A Mol Integr Physiol 125: 45-56.

Miyamoto KJ, Miyamoto MR, Miyamoto TA. 2006. Systemically administered taurine. Taurine 6. Oja SS, Saransaari P, editors. New York: Springer/Kluwer Academic; pp. 323-351.

Miyamoto TA, Miyamoto KJ. 1999. Does adenosine release taurine in the A1-receptor-rich hippocampus? J Anesth 13: 94-98.

Miyamoto Y, Marczin N, Catravas JD, Del Monte MA. 1996a. Cholera toxin enhances taurine uptake in cultures of human retinal pigment epithelial cells. Curr Eye Res 15: 229-236.

Miyamoto Y, Liou GI, Sprinkle TJ. 1996b. Isolation of a cDNA encoding a taurine transporter in the human retinal pigment epithelium. Curr Eye Res 15: 345-349.

Miyata S, Matsushima O, Hatton GI. 1997. Taurine in rat posterior pituitary: Localization in astrocytes and selective release by hypoosmotic stimulation. J Comp Neurol 381: 513-523.

Molchanova S, Kööbi P, Oja SS, Saransaari P. 2004a. Interstitial concentrations of amino acids during global forebrain ischemia and potassium-evoked spreading depression. Neurochem Res 29: 1519-1527.

Molchanova S, Oja SS, Saransaari P. 2004b. Characteristics of basal taurine release in the rat striatum measured by microdialysis. Amino Acids 27: 261-268.

Molchanova SM, Oja SS, Saransaari P. 2005. Mechanisms of enhanced taurine release under Ca^{2+} depletion. Neurochem Int 47: 343-349.

Molchanova SM, Oja SS, Saransaari P. 2006. Properties of basal taurine release in the rat striatum in vivo. Taurine 6. Oja SS, Saransaari P, editors. New York: Springer/Kluwer Academic; pp. 365-375.

Molineux H, Schofield R, Testa NG. 1987. Hematopoietic effects of TCNU in mice. Cancer Treat Rep 71: 837-841.

Morán J, Hernández-Pech X, Merchant-Larios H, Pasantes-Morales H. 2000. Release of taurine in apoptotic cerebellar granule neurons in culture. Pflügers Arch 439: 271-277.

Mori M, Gähwiler BH, Gerber U. 2002. β-Alanine and taurine as endogenous agonists at glycine receptors in rat hippocampus in vitro. J Physiol 539: 191-200.

Nabekura J, Omura T, Akaike N. 1996a. α2 adrenoceptor potentiates glycine receptor-mediated taurine response through protein kinase A in rat substantia nigra neurons. J Neurophysiol 76: 2447-2454.

Nabekura J, Omura T, Horimoto N, Ogawa T, Akaike N. 1996b. α1 adrenoceptor activation potentiates taurine response through protein kinase C in rat substantia nigra neurons. J Neurophysiol 76: 2455-2460.

Nagelhuis EA, Amiry-Moghaddam M, Lehmann A, Ottersen OP. 1994. Taurine as an organic osmolyte in the intact brain: Immunocytochemical and biochemical studies. Adv Exp Med Biol 359: 325-334.

Nagelhuis EA, Lehmann A, Ottersen OP. 1993. Neuronal-glial exchange of taurine during hypo-osmotic stress: A combined immunocytochemical and biochemical analysis in rat cerebellar cortex. Neuroscience 54: 615-631.

Nakamori K, Koyama I, Nakamura T, Nemoto M, Yoshida T, et al. 1993. Quantitative evaluation of the effectiveness of taurine in protecting the ocular surface against oxidant. Chem Pharm Bull 41: 335-338.

Nakane H, Yao H, Ibayashi S, Kitazono T, Ooboshi H, et al. 1998. Protein kinase C modulates ischemia-induced amino acids release in the striatum of hypertensive rats. Brain Res 782: 290-296.

Neuringer M, Sturman JA, Wen GY, Wisniewski HM. 1985. Dietary taurine is necessary for normal retinal development in monkeys. Prog Clin Biol Res 179: 53-62.

Nilsson GE, Lutz PL. 1991. Release of inhibitory neurotransmitters in response to anoxia in turtle brain. Am J Physiol 261: R32-R37.

Nusetti S, Obregón F, Lima L. 2006. Neuritic outgrowth from goldfish retinal explants, interaction of taurine and zinc. Taurine 6. Oja SS, Saransaari P, editors. New York: Springer/Kluwer Academic; pp. 435-440.

Nusetti S, Obregón F, Quintal M, Benzo Z, Lima L. 2005. Taurine and zinc modulate outgrowth from goldfish retinal explants. Neurochem Res 30: 1483-1492.

O'Byrne MB, Tipton KF. 2000. Taurine-induced attenuation of MPP^+ neurotoxicity in vitro: A possible role for the $GABA_A$ subclass of GABA receptors. J Neurochem 74: 2087-2093.

Oja SS, Kontro P. 1981. Oxidation of hypotaurine in vitro by mouse liver and brain tissues. Biochim Biophys Acta 677: 350-357.

Oja SS, Kontro P. 1983a. Taurine. Handbook of Neurochemistry, Vol. 3. Lajtha A, editor. 2nd edn, New York: Plenum Press; pp. 501-533.

Oja SS, Kontro P. 1983b. Free amino acids in epilepsy: Possible role of taurine. Acta Neurol Scand 67 (Suppl. 93): 5-20.

Oja SS, Saransaari P. 1992a. Cell volume changes and taurine release in cerebral cortical slices. Adv Exp Med Biol 315: 369-374.

Oja SS, Saransaari P. 1992b. Taurine release and swelling of cerebral cortex slices from adult and developing mice in media of different ionic compositions. J Neurosci Res 32: 551-561.

Oja SS, Saransaari P. 1995. Chloride ions, potassium stimulation and release of endogenous taurine from cerebral cortical slices from 3-day-old and 3-month-old mice. Neurochem Int 27: 313-318.

Oja SS, Saransaari P. 1996a. Taurine as osmoregulator and neuromodulator in the brain. Metab Brain Dis 11: 153-164.

Oja SS, Saransaari P. 1996b. Kinetic analysis of taurine influx into cerebral cortical slices from adult and developing mice in different incubation conditions. Neurochem Res 21: 161-166.

Oja SS, Saransaari P. 2000. Modulation of taurine release by glutamate receptors and nitric oxide. Prog Neurobiol 62: 407-425.

Oja SS, Saransaari P, editors. 2006. Taurine 6. New York: Springer/Kluwer Academic.

Oja SS, Holopainen I, Kontro P. 1990a. Stimulated taurine release from different brain preparations: Changes during development and aging. Prog Clin Biol Res 351: 277-287.

Oja SS, Korpi ER, Saransaari P. 1990b. Modification of chloride flux across brain membranes by inhibitory amino acids in developing and adult mice. Neurochem Res 15: 797-804.

Oja SS, Kontro P, Lähdesmäki P. 1977. Amino acids as inhibitory neurotransmitters. Prog Pharmacol 1/3: 1-119.

Oja SS, Kontro P, Lindén IB, Gothóni G. 1983. Anticonvulsant action of some 2-aminoethanesulfonic acid (taurine) derivatives. Eur J Pharmacol 87: 191-198.

Oja SS, Lehtinen I, Lähdesmäki P. 1976. Taurine transport rates between plasma and tissues in adult and 7-day-old mice. Q J Exp Physiol 61: 133-143.

Oja SS, Uusitalo AJ, Vahvelainen ML, Piha RS. 1968. Changes in cerebral and hepatic amino acids in the rat and guinea pig during development. Brain Res 11: 655-661.

Okamoto K, Kimura H, Sakai Y. 1983a. Taurine-induced increase of the Cl^- conductance of cerebellar Purkinje cell dendrites in vitro. Brain Res 259: 319-323.

Okamoto K, Kimura H, Sakai Y. 1983b. Evidence for taurine as an inhibitory neurotransmitter in cerebellar stellate interneurons: Selective antagonism by TAG (6-aminomethyl-4H1,2,4-benzothiadiazine-1,1-dioxide). Brain Res 265: 163-168.

Okamoto K, Kimura H, Sakai Y. 1983c. Antagonistic action of 6-aminomethyl-4H1,2,4-benzothiadiazine-1,1-dioxide (TAG), and evidence for a transmitter role of taurine in stellate interneurons in the cerebellum. Prog Clin Biol Res 125: 151-160.

Olive MF. 2002. Interactions between taurine and ethanol in the central nervous system. Amino Acids 23: 345-357.

Olive MF, Mehmert KK, Hodge CW. 2000a. Modulation of extracellular neurotransmitter levels in the nucleus accumbens by a taurine uptake inhibitor. Eur J Pharmacol 409: 291-294.

Olive MF, Mehmert KM, Messing RO, Hodge CW. 2000b. Reduced operant ethanol self-administration and in vivo mesolimbic dopamine responses to ethanol in PKCε deficient mice. Eur J Neurosci 12: 4131-4140.

Olive MF, Nannini MA, Ou CJ, Koenig HN, Hodge CW. 2002. Effects of acute acamprosate and homotaurine on ethanol intake and ethanol-stimulated mesolimbic dopamine release. Eur J Pharmacol 437: 55-61.

Olson JE, Goldfinger MD. 1990. Amino acid content of rat cerebral astrocytes adapted to hyperosmotic medium in vitro. J Neurosci Res 27: 241-246.

Olson JE, Martinho E Jr. 2006. Taurine transporter regulation in hippocampal neurons. Taurine 6. Oja SS, Saransaari P, editors. New York: Springer/Kluwer Academic; pp. 307-314.

Olson JE, Banks M, Dimlich RV, Evers J. 1997. Blood-brain barrier water permeability and brain osmolyte content during edema development. Acad Emerg Med 4: 662-673.

Olson JE, Evers JA, Banks M. 1994. Brain osmolyte content and blood-brain barrier water permeability surface area product in osmotic edema. Acta Neurochir (Suppl.) 60: 571-573.

O'Neill R. 1986. Effects of intranigral injection of taurine and GABA on striatal dopamine release monitored voltammetrically in the unanaesthetized rat. Brain Res 382: 28-32.

Ooboshi H, Sadoshima S, Yao H, Ibayashi S, Matsumoto T, et al. 1995. Ischemia-induced release of amino acids in the hippocampus of aged hypertensive rats. J Cereb Blood Flow Metab 15: 227-234.

O'Regan MH, Smith-Barbour M, Perkins LM, Phillis JW. 1995. A possible role of phospholipases in the release of transmitter amino acids from ischemic rat cerebral cortex. Neurosci Lett 185: 191-194.

Palmi M, Frosini M, Sgaragli GP. 1987. Possible involvement of brain calcium metabolism in the action of taurine in mammalian thermoregulation. Adv Exp Med Biol 217: 237-244.

Pasantes-Morales H, Cruz C. 1985. Taurine and hypotaurine inhibit light-induced lipid peroxidation and protect rod outer segment structure. Brain Res 330: 154-157.

Pasantes-Morales H, Franco R. 2002. Influence of protein tyrosine kinases and cell volume change-induced taurine release. Cerebellum 1: 103-109.

Pasantes-Morales H, Schousboe A. 1988. Volume regulation in astrocytes: A role of taurine as an osmoeffector. J Neurosci Res 20: 505-509.

Pasantes-Morales H, Schousboe A. 1989. Release of taurine from astrocytes during potassium-evoked swelling. Glia 2: 45-50.

Pasantes-Morales H, Schousboe A. 1997. Role of taurine in osmoregulation in brain cells: Mechanisms and functional implications. Amino Acids 12: 281-292.

Pasantes-Morales H, Arzate ME, Quesada O, Huxtable RJ. 1987. Higher susceptibility of taurine-deficient rats to seizures induced by 4-aminopyridine. Neuropharmacology, 26: 1721-1725.

Pasantes-Morales H, Cardin V, Tuz K. 2000a. Signaling events during swelling and regulatory volume decrease. Neurochem Res 25: 1301-1314.

Pasantes-Morales H, Franco R, Torres-Marquez ME, Hernandez-Fonseca K, Ortega A. 2000b. Amino acid osmolytes in regulatory volume decrease and isovolumetric regulation in brain cells: Contribution and mechanisms. Cell Physiol Biochem 10: 361-370.

Pasantes-Morales H, Franco R, Ochoa L, Ordaz B. 2002a. Osmosensitive release of neurotransmitter amino acids: Relevance and mechanisms. Neurochem Res 27: 59-65.

Pasantes-Morales H, Franco R, Ordaz B, Ochoa LD. 2002b. Mechanisms counteracting swelling in brain cells during hyponatremia. Arch Med Res 33: 237-244.

Pasantes-Morales H, Moran J, Schousboe A. 1990. Volume-sensitive release of taurine from cultured astrocytes: Properties and mechanism. Glia 4: 427-432.

Pazdernik TL, Wade JV, Nelson SR, Samson FE. 1990. Is taurine involved in cerebral osmoregulation? Adv Exp Med Biol 403: 117-124.

Pérez-Pinzón MA, Nilsson GE, Lutz PL. 1993. Relationship between ion gradients and neurotransmitter release in the newborn rat striatum during anoxia. Brain Res 602: 228-233.

Petrosian AM, Haroutounian JE. 1990. Tauret: Further studies on the role of taurine in retina. Prog Clin Biol Res 351: 471-475.

Petrosian AM, Haroutounian JE. 1998. The role of taurine in osmotic, mechanical, and chemical protection of the retinal outer segments. Adv Exp Med Biol 442: 407-413.

Petrosian AM, Haroutounian JE. 2000. Taurine as a universal carrier of lipid soluble vitamins: A hypothesis. Amino Acids 19: 409-421.

Petrosian AM, Haroutounian JE, Fugelli K, Kanli H. 2000a. Effects of osmotic and light stimulation on ^{3}H-taurine efflux from isolated rod outer segments and synthesis of tauret in the frog retina. Adv Exp Med Biol 483: 441-451.

Petrosian AM, Haroutounian JE, Gundersen TE, Blomhoff R, Fugelli K, et al. 2000b. New HPLC evidence on endogenous tauret in retina and pigment epithelium. Adv Exp Med Biol 483: 453-460.

Petrosian AM, Haroutounian JE, Zueva LV. 1996. Tauret: A taurine-related endogenous substance in the retina and its role in vision. Adv Exp Med Biol 403: 333-342.

Phillis JW, Ren J, O'Regan MH. 2000. Transporter reversal as a mechanism of glutamate release from the ischemic rat cerebral cortex: Studies with DL-threo-β-benzyloxyaspartate. Brain Res 868: 105-112.

Phillis JW, Song D, O'Regan MH. 1996. Inhibition of tyrosine phosphorylation attenuates amino acid neurotransmitter release from the ischemic/reperfused rat cerebral cortex. Neurosci Lett 207: 151-154.

Phillis JW, Song D, O'Regan MH. 1997. Inhibition of anion channel blockers of ischemia-evoked release of excitotoxic and other amino acids from rat cerebral cortex. Brain Res 758: 9-16.

Phillis JW, Song D, O'Regan MH. 1998. Tamoxifen, a chloride channel blocker, reduces glutamate and aspartate release from the ischemic cerebral cortex. Brain Res 780: 352-355.

Phillis JW, Song D, O'Regan MH. 1999. Effects of hyperosmolarity and ion substitutions on amino acid efflux from the ischemic rat cerebral cortex. Brain Res 828: 1-11.

Pow DV. 1993. Immunocytochemistry of amino-acids in the rodent pituitary using extremely specific, very high titre antisera. J Neuroendoc 5: 349-356.

Pow DV, Sullivan R, Reye P, Hermanussen S. 2002. Localization of taurine transporters, taurine and ^{3}H taurine accumulation in the rat retina, pituitary, and brain. Glia 37: 153-168.

Puka M, Lehmann A. 1994. In vivo acidosis reduces extracellular concentrations of taurine and glutamate in the rat hippocampus. J Neurosci Res 37: 641-646.

Puka M, Albrecht J, Lazarewicz JW. 1992. Noradrenaline- and glutamate-induced taurine release from bulk isolated adult rat astrocytes. Acta Neurobiol Exp 52: 31-35.

Puopolo M, Kratskin I, Belluzzi O. 1998. Direct inhibitory effect of taurine on relay neurons of the rat olfactory bulb. Neuroreport 9: 2319-2322.

Pushpakiran G, Mahalakshmi K, Anuradha CV. 2004. Taurine restores ethanol-induced depletion of antioxidants and attenuates oxidative stress in rat tissues. Amino Acids 27: 91-96.

Quertemont E, Grant KA. 2004. Discriminative stimulus effects of ethanol: Lack of interaction with taurine. Behav Pharmacol 15: 495-501.

Quartemont E, Dahchour A, Ward RJ, De Witte P. 1999. Ethanol induces taurine release in the amygdala: An in vivo microdialysis study. Addict Biol 4: 47-54.

Quartemont E, de Neuville J, De Witte O. 1998a. Changes in the amygdala amino acid microdialysate after conditioning with a cue associated with ethanol. Psychopharmacology 139: 71-78.

Quartemont E, Goffaux V, Vlaminck AM, Wolff C, De Witte P. 1998b. Oral taurine supplementation modulates ethanol-conditioned stimulus preference. Alcohol 16: 201-206.

Quartemont E, Devitgh A, De Witte P. 2003. Systemic osmotic manipulations modulate ethanol-induced taurine release: A brain microdialysis study. Alcohol 29: 11-19.

Quinn MR, Harris CL. 1995. Taurine allosterically inhibits binding of [^{35}S]-*t*-butyl-bicyclophosphonothionate (TBPS) to rat brain synaptic membranes. Neuropharmacology 34: 1607-1613.

Quinn MR, Miller CI. 1992. Taurine allosterically modulates flunitrazepam binding to synaptic membranes. J Neurosci Res 33: 136-141.

Rainesalo S, Keränen T, Palmio J, Peltola J, Oja SS, et al. 2004. Plasma and cerebrospinal fluid amino acids in epileptic patients. Neurochem Res 29: 319-324.

Rajendra S, Lynch JW, Pierce KD, French CR, Barry PH, et al. 1995. Mutation of an arginine residue in the human glycine receptor transforms β-alanine and taurine from agonists into competitive antagonists. Neuron 14: 169-175.

Ramamoorthy D, Del Monte MA, Leibach FH, Ganapathy V. 1994. Molecular identity and calmodulin-mediated regulation of the taurine transporter in a human retinal pigment epithelial cell line. Curr Eye Res 13: 523-529.

Ramanathan VK, Chung SJ, Giacomini KM, Brett CM. 1997. Taurine transport in cultured choroid plexus. Pharm Res 14: 406-409.

Rao VLR, Audet RM, Butterworth RF. 1995. Selective alterations of extracellular brain amino acids in relation to function in experimental portal-systemic encephalopathy: Results of an in vivo microdialysis study. J Neurochem 65: 1221-1228.

Rebel G, Petegnief V, Lleu PL, Gupta RC, Guérin P, et al. 1994. New data on the regulation of taurine uptake in cultured nervous cells. Adv Exp Med Biol 359: 225-233.

Reichelt KL, Edminson PD, Kvamme E. 1976. The formation of peptide-amines from constituent amino acids and histamine in hypothalamic tissue. J Neurochem 26: 811-815.

Rivas-Arancibia S, Dorado-Martínez C, Borgonio-Pérez G, Hiriart-Urdanivia M, Verdugo-Díaz L, et al. 2000a. Effects of taurine on ozone-induced memory deficits and lipid peroxidation levels in brains of young, mature, and old rats. Environ Res 82: 7-17.

Rivas-Arancibia S, Willing AE, Zigova T, Rodríguez AI, Cahill DW, et al. 2000b. The effects of taurine on hNT neurons transplanted in adult rat striatum. Cell Transplant 9: 751-758.

Rivas-Arancibia S, Rodríguez AI, Zigova T, Willing AE, Brown WD, et al. 2001. Taurine increases rat survival and reduces striatal damage caused by 3-nitropropionic acid. Int J Neurosci 108: 55-67.

Rodríguez-Martínez E, Rugerio-Vargas C, Rodríguez AI, Borgonio-Pérez G, Rivas-Arancibia S. 2004. Antioxidant effects of taurine, vitamin C, and vitamin E on oxidative damage in hippocampus caused by the administration of 3-nitropropionic acid in rats. Int J Neurosci 114: 1133-1145.

Rogers KL, Philibert RA, Dutton GR. 1991. K$^+$-Stimulated amino acid release from cultured cerebellar neurons: Comparison of static and dynamic stimulation paradigms. Neurochem Res 16: 899-904.

Rosati AM, Traversa U, Lucchi R, Poli A. 1995. Biochemical and pharmacological evidence for the presence of A$_1$ but not A$_{2a}$ adenosine receptors in the brain of the low vertebrate teleost *Cerassium auratus* (goldfish). Neurochem Int 26: 411-423.

Ross CD, Godfrey DA, Parli JA. 1995. Amino acid concentrations and selected enzyme activities in rat auditory, olfactory, and visual systems. Neurochem Res 20: 1483-1490.

Ross CD, Parli JA, Godfrey DA. 1989. Quantitative distribution of six amino acids in rat retinal layers. Vision Res 29: 1079-1084.

Rosso L, Peteri-Brunback B, Poujeol P, Hussy N, Mienville JM. 2004. Vasopressin-induced taurine efflux from rat pituicytes: A potential negative feedback for hormone secretion. J Physiol 554: 731-742.

Ruotsalainen M, Ahtee L. 1996. Intrastriatal taurine increases striatal extracellular dopamine in a tetrodotoxin-sensitive manner in rats. Neurosci Lett 212: 175-178.

Ruotsalainen M, Heikkilä M, Lillsunde P, Seppälä T, Ahtee L. 1996. Taurine infused intrastriatally elevates, but intranigrally decreases striatal extracellular dopamine concentration in anaesthetized rats. J Neural Transm 103: 935-946.

Ruotsalainen M, Majasaari M, Salimäki J, Ahtee L. 1998. Locally infused taurine, GABA and homotaurine alter differentially the striatal extracellular concentration of dopamine and its metabolites in rats. Amino Acids 15: 117-134.

Sakai S, Tosaka T. 1999. Analysis of hyposmolarity-induced taurine efflux pathways in the bullfrog sympathetic ganglia. Neurochem Int 34: 203-212.

Sakurai T, Miki T, Li HP, Miyatake A, Satriotomo I, et al. 2003. Colocalization of taurine and glial fibrillary acidic protein immunoreactivity in mouse hippocampus induced by short-term ethanol exposure. Brain Res 959: 160-164.

Salceda R. 1999. Insulin-stimulated taurine uptake in rat retina and retinal pigment epithelium. Neurochem Int 35: 301-306.

Salimäki J, Scriba G, Piepponen TP, Rautolahti N, Ahtee L. 2003. The effects of systemically administered taurine and N-pivaloyltaurine on striatal extracellular dopamine and taurine in freely moving rats. Naunyn Schmiedebergs Arch Pharmacol 368: 134-141.

Sánchez Olea R, Pasantes-Morales H. 1990. Chloride dependence of the K^+-stimulated release of taurine from synaptosomes. Neurochem Res 15: 535-540.

Sapronov NS, Gavrovskaya LK. 2006. Taurinamide derivatives—drugs with the metabolic type of action. Taurine 6. Oja SS, Saransaari P, editors. New York: Springer/Kluwer Academic; pp. 509-514.

Saransaari P, Oja SS. 1991. Excitatory amino acids evoke taurine release from cerebral cortex slices from adult and developing mice. Neuroscience 45: 451-459.

Saransaari P, Oja SS. 1992. Release of GABA and taurine from brain slices. Prog Neurobiol 38: 455-482.

Saransaari P, Oja SS. 1993. Phencyclidine binding sites in mouse cerebral cortex during development and ageing: Effects of inhibitory amino acids. Mech Ageing Dev 68: 125-136.

Saransaari P, Oja SS. 1994. Taurine release form mouse hippocampal slices: Effects of glutamatergic substances and hypoxia. Adv Med Exp Biol 359: 279-287.

Saransaari P, Oja SS. 1996. Taurine and neural cell damage. Transport of taurine in adult and developing mice. Adv Exp Med Biol 403: 481-489.

Saransaari P, Oja SS. 1997a. Enhanced taurine release in cell-damaging conditions in the developing and ageing mouse hippocampus. Neuroscience 79: 847-854.

Saransaari P, Oja SS. 1997b. Taurine release from the developing and ageing hippocampus: Stimulation by agonists of ionotropic glutamate receptors. Mech Ageing Dev 99: 219-232.

Saransaari P, Oja SS. 1997c. Glutamate-agonist-evoked taurine release from the adult and developing mouse hippocampus in cell-damaging conditions. Amino Acids 9: 323-335.

Saransaari P, Oja SS. 1998a. Mechanisms of ischemia-induced taurine release in mouse hippocampal slices. Brain Res 807: 118-124.

Saransaari P, Oja SS. 1998b. Release of endogenous glutamate, aspartate, GABA and taurine from hippocampal slices from adult developing mice in cell-damaging conditions. Neurochem Res 23: 563-570.

Saransaari P, Oja SS. 1999a. Taurine release modified by nitric oxide-generating compounds in the developing and adult mouse hippocampus. Neuroscience 89: 1103-1111.

Saransaari P, Oja SS. 1999b. Taurine release is enhanced in cell-damaging conditions in cultured cerebral cortical astrocytes. Neurochem Res 24: 1523-1529.

Saransaari P, Oja SS. 1999c. Involvement of metabotropic glutamate receptors in taurine release in the adult and developing hippocampus. Amino Acids 16: 165-179.

Saransaari P, Oja SS. 1999d. Enhanced taurine release in cultured cerebellar granule cells in cell-damaging conditions. Amino Acids 17: 323-334.

Saransaari P, Oja SS. 1999e. Characteristics of ischemia-induced taurine release in the developing mouse hippocampus. Neuroscience 94: 949-954.

Saransaari P, Oja SS. 1999f. Anticonvulsant taurine derivatives modify taurine and GABA release in the mouse hippocampus. Proc West Pharmacol Soc 42: 27-29.

Saransaari P, Oja SS. 2000a. Taurine release modified by GABAergic agents in hippocampal slices from adult and developing mice. Amino Acids 18: 17-30.

Saransaari P, Oja SS. 2000b. Taurine and neural cell damage. Amino Acids 19: 509-526.

Saransaari P, Oja SS. 2000c. Modulation of the ischemia-induced taurine release by adenosine receptors in the developing and adult mouse hippocampus. Neuroscience 97: 425-430.

Saransaari P, Oja SS. 2002. Taurine release in the developing and adult mouse hippocampus: Involvement of cyclic guanosine monophosphate. Neurochem Res 27: 15-20.

Saransaari P, Oja SS. 2003a. Interactions of taurine and adenosine in the mouse hippocampus in normoxia and ischemia. Adv Exp Med Biol 526: 445-451.

Saransaari P, Oja SS. 2003b. Characterization of N-methyl-D-aspartate-evoked taurine release in the developing and adult mouse hippocampus. Amino Acids 24: 213-221.

Saransaari P, Oja SS. 2004. Characteristics of taurine release induced by free radicals in mouse hippocampal slices. Amino Acids 26: 91-98.

Saransaari P, Oja SS. 2006. Characteristics of taurine release in slices from adult and developing mouse brain stem. Amino Acids, in press.

Satsu H, Kobayashi Y, Yokohama T, Terasawa E, Shimizu M. 2002. Effect of dietary sulfur amino acids on the taurine content of rat tissues. Amino Acids 23: 447-452.

Schafer S, Bicker G, Ottersen OP, Storm-Mathisen J. 1988. Taurine-like immunoreactivity in the brain of the honey-bee. J Comp Neurol 368: 60-70.

Schaffer SW, Azuma J, Takahashi K, Mozaffari M. 2003. Why is taurine cytoprotective? Adv Exp Med Biol 526: 307-321.

Scheller D, de Ryck M, Clincke G, Tegtmeier F. 1997. Extracellular changes of taurine in the peri-infarct zone: Effect of lubeluzole. Acta Neurochir (Suppl.) 70: 185-187.

Scheller D, Szathmary S, Kolb J, Tegtmeier F. 2000a. Observations on the relationship between the extracellular changes of taurine and glutamate during cortical spreading depression, during ischemia, and within the area surrounding a thrombotic infarct. Amino Acids 19: 571-583.

Scheller D, Korte M, Szathmary S, Tegtmeier F. 2000b. Cerebral taurine release mechanisms in vivo: Pharmacological investigation in rats using microdialysis for proof of principle. Neurochem Res 25: 801-807.

Schmieden V, Betz H. 1995. Pharmacology of the inhibitory glycine receptor: Agonist and antagonist actions of amino acids and piperidine carboxylic acid compounds. Mol Pharmacol 48: 919-927.

Schmieden V, Grenningloh G, Schofield PR, Betz H. 1989. Functional expression in Xenopus oocytes of the strychnine binding 48 kd subunit of the glycine receptor. EMBO J 8: 695-700.

Schmieden V, Kuhse J, Betz H. 1992. Agonist pharmacology of neonatal and adult glycine receptor α subunits: Identification of amino acid residues involved in taurine activation. EMBO J 11: 2025-2032.

Schousboe A, Pasantes-Morales H. 1989. Potassium-stimulated release of [^{3}H]taurine from cultured GABAergic and glutamatergic neurons. J Neurochem 53: 1309-1315.

Schousboe A, Morán J, Pasantes-Morales H. 1990. Potassium-stimulated release of taurine from cultured cerebellar granule neurons is associated with cell swelling. J Neurosci Res 27: 71-77.

Schurr A, Rigor BM. 1987. The mechanism of neuronal resistance and adaptation to hypoxia. FEBS Lett 224: 4-8.

Schurr A, Tseng MT, West CA, Rigor BM. 1987. Taurine improves the recovery of neural function following cerebral hypoxia: An in vitro study. Life Sci 40: 2059-2066.

Segovia G, Del Arco A, Mora F. 1997. Endogenous glutamate increases extracellular concentrations of dopamine, GABA, and taurine through NMDA and AMPA/kainate receptors in striatum of the freely moving rat: A microdialysis study. J Neurochem 69: 1476-1483.

Seidl R, Peyrl A, Nicham R, Hauser E. 2000. A taurine and caffeine-containing drink stimulates cognitive performance and well-being. Amino Acids 19: 635-642.

Seki Y, Feustel PJ, Keller RW Jr, Tranmer BI, Kimelberg HK. 1999. Inhibition of ischemia-induced glutamate release in rat striatum by dihydrokainate and an anion channel blocker. Stroke 30: 433-440.

Seki Y, Kimura M, Mizutani N, Fujita M, Aimi Y, et al. 2005. Cerebrospinal fluid taurine after traumatic brain injury. Neurochem Res 30: 123-128.

Sergeeva OA, Chepkova AN, Doreulee N, Eriksson KS, Poelchen W, et al. 2003. Taurine-induced long-lasting enhancement of synaptic transmission in mice: Role of transporters. J Physiol 550: 911-919.

Sergeeva OA, Chepkova AN, Haas HL. 2002. Guanidinoethyl sulphonate is a glycine receptor antagonist in striatum. Br J Pharmacol 137: 855-860.

Serrano MI, Goicoechea C, Serrano JS, Serrano-Martino MC, Sánchez E, et al. 2002. Age-related changes in the antinociception induced by taurine in mice. Pharmacol Biochem Behav 73: 863-867.

Serrano MI, Serrano JS, Guerrero MR, Fernández A. 1994. Role of $GABA_A$ and $GABA_B$ receptors and peripheral cholinergic mechanisms in the antinociceptive action of taurine. Gen Pharmacol 24: 1123-1129.

Serrano JS, Serrano MI, Guerrero MR, Ruiz R, Polo J. 1990. Antinociceptive effect of taurine and its inhibition by naloxone. Gen Pharmacol 21: 333-336.

Sgaragli GP, Carlà V, Magnani M, Galli A. 1981. Hypothermia induced in rabbits by intracerebroventricular taurine: Specificity and relationship with central serotonin (5-HT) systems. J Pharmacol Exp Ther 219: 778-785.

Sgaragli GP, Frosini M, Palmi M, Bianchi L, Della Corte L. 1994. Calcium and taurine interaction in mammalian brain metabolism. Adv Exp Med Biol 359: 299-308.

Shain W, Bausback D, Fiero A, Madelian V, Turner JN. 1992. Regulation of receptor-mediated shape change in astroglial cells. Glia 5: 223-238.

Shen EY, Lai YJ. 2002. In vivo microdialysis study of excitatory and inhibitory amino acid levels in the hippocampus following penicillin-induced seizures in mature rats. Acta Paediatr Taiwan 43: 313-318.

Shibanoki S, Kogure M, Sugahara M, Ishikawa K. 1993. Effect of systemic administration of N-methyl-D-aspartic acid on extracellular taurine level measured by microdialysis in the hippocampal CA1 field and striatum of rats. J Neurochem 61: 1689-1704.

Shimada N, Graf R, Rosner G, Keiss WD. 1993. Ischemia-induced accumulation of extracellular amino acids in cerebral cortex, white matter, and cerebrospinal fluid. J Neurochem 60: 66-71.

Shupliakov O, Brodin L, Srinivasan M, Grillner S, Cullheim S, et al. 1994. Extrasynaptic localization of taurine-like immunoreactivity in the lamprey spinal cord. J Comp Neurol 347: 301-311.

Silva MA, Cunha GM, Viana GS, Rao VS. 1993. Taurine modulates chemical nociception in mice. Braz J Med Biol Res 26: 1319-1324.

Simpson RK Jr, Robertson CS, Goodman JC. 1990. Spinal cord ischemia-induced elevation of amino acids. Extracellular measurement with microdialysis. Neurochem Res 15: 635-639.

Singewald N, Zhou GY, Schneider C. 1995. Release of excitatory and inhibitory amino acids from the locus coeruleus of conscious rats by cardiovascular stimuli and various forms of acute stress. Brain Res 704: 42-50.

Smith A, Watson CJ, Frantz KJ, Eppler B, Kennedy RT, et al. 2004. Differential increase in taurine levels by low-dose ethanol in the dorsal and ventral striatum revealed by microdialysis with on-line capillary electrophoresis. Alcohol Clin Exp Res 28: 1028-1038.

Smith KE, Borden LA, Wang CH, Hartig PR, Branchek TA, et al. 1992. Cloning and expression of a high affinity taurine transporter from rat brain. Mol Pharmacol 42: 563-569.

Smith SS, Li J. 1991. GABA$_B$ receptor stimulation by baclofen and taurine enhances excitatory amino acid induced phosphatidylinositol turnover in neonatal rat cerebellum. Neurosci Lett 132: 59-64.

Smullin DH, Schamber CD, Skilling SR, Larson AA. 1990. A possible role for taurine in analgesia. Prog Clin Biol Res 351: 129-132.

Solís JM, Herranz AS, Herreras O, Lerma J, Martín del Río R. 1988a. Does taurine act as an osmoregulatory substance in the rat brain? Neurosci Lett 91: 53-58.

Solís JM, Herranz AS, Herreras O, Lerma J, Martín del Río R. 1988b. Low chloride-dependent release of taurine by a furosemide-sensitive process in the in vivo rat hippocampus. Neuroscience 24: 885-891.

Solís JM, Herranz AS, Herreras O, Muñoz MD, Martín del Río R, et al. 1986. Variation of potassium ion concentrations in the rat hippocampus specifically affects extracellular taurine level. Neurosci Lett 66: 263-268.

Song Z, Hatton GI. 2003. Taurine and the control of basal hormone release from rat neurohypophysis. Exp Neurol 183: 330-337.

Stipanuk MH, Londono M, Lee JI, Hu M, Yu AF. 2002. Enzymes and metabolites of cysteine metabolism in nonhepatic tissues of rats show little response to changes in dietary protein or sulfur amino acid levels. J Nutr 132: 3369-3378.

Stover JF, Morganti-Kosmann MC, Lenzlinger PM, Stocker R, Kempski OS. 1999. Glutamate and taurine are increased in ventricular cerebrospinal fluid of severely brain-injured patients. J Neurotrauma 16: 135-142.

Stummer W, Betz AL, Shakui P, Keep RF. 1995. Blood-brain barrier taurine transport during osmotic stress and in focal cerebral ischemia. J Cereb Blood Flow Metab 15: 852-859.

Sturman JA. 1993. Taurine in development. Physiol Rev 73: 119-147.

Sturman JA, Lu P, Messing JM, Imaki H. 1996. Depletion of feline taurine levels by β-alanine and dietary taurine restriction. Adv Exp Med Biol 403: 19-36.

Sturman JA, Moretz RC, French JH, Wisniewski HM. 1985. Taurine deficiency in the developing cat: Persistence of the cerebellar external granule cell layer. J Neurosci Res 13: 403-416.

Sundström E, Mo LL, Seiger Å. 1995. In vivo studies on NMDA-evoked release of amino acids in the rat spinal cord. Neurochem Int 27: 185-193.

Sung DY, Walthall WW, Derby CD. 1996. Identification and partial purification of putative taurine receptor proteins from the olfactory organ of the spiny lobster. Comp Biochem Physiol B Biochem Mol Biol 115: 19-26.

Sved AF, Curtis JT. 1993. Amino acid neurotransmitters in nucleus tractus solitarius: An in vivo microdialysis study. J Neurochem 61: 2089-2098.

Swain MS, Bergeron M, Audet R, Blei AT, Butterworth RF. 1992. Monitoring of neurotransmitter amino acids by means of an indwelling cisterna magna catheter: A comparison of two rodent models of fulminant liver failure. Hepatology 16: 1028-1035.

Tamai I, Senmaru M, Terasaki T, Tsuji A. 1995. Na$^+$- and Cl$^-$-dependent transport of taurine at the blood-brain barrier. Biochem Pharmacol 50: 1783-1793.

Tang XW, Deupree DL, Sun Y, Wu JY. 1996. Biphasic effect of taurine on excitatory amino acid-induced neurotoxicity. Adv Exp Med Biol 403: 499-505.

Tappaz ML. 2004. Taurine biosynthetic enzymes and taurine transporter: Molecular identification and regulations. Neurochem Res 29: 83-96.

Tchoumkeu-Nzouessa GC, Rebel G. 1996a. Characterization of taurine transport in human glioma GL15 cell line: Regulation by protein kinase C. Neuropharmacology 35: 37-44.

Tchoumkeu-Nzouessa GC, Rebel G. 1996b. Activation of protein kinase C down-regulates glial but not neuronal taurine uptake. Neurosci Lett 206: 61-64.

Theofanopoulos V, Lau-Cam C. 1998. Modification by taurine of the metabolism and hypothermic effect of ethanol in the rat. Adv Exp Med Biol 442: 309-318.

Thurston JH, Hauhart RE. 1989. Valproate doubles the anoxic survival time of normal developing mice: Possible relevance to valproate-induced decreases in cerebral levels of glutamate and aspartate, and increases in taurine. Life Sci 45: 59-62.

Tong X, Patsalos PN. 2001. A microdialysis study of the novel antiepileptic drug levetiracetam: Extracellular pharmacokinetics and effect on taurine in rat brain. Br J Pharmacol 133: 867-874.

Torp R, Andiné P, Hagberg H, Karagülle T, Blackstad TW, et al. 1991. Cellular and subcellular redistribution of glutamate-, glutamine- and taurine-like immunoreactivities during forebrain ischemia: A semiquantitative electron microscopic study in rat hippocampus. Neuroscience 41: 433-447.

Trachtman H, Barbour R, Sturman JA, Finberg L. 1988a. Taurine and osmoregulation: Taurine is a cerebral osmoprotective molecule in chronic hypernatremic dehydration. Pediatr Res 23: 35-39.

Trachtman H, Del Pizzo R, Sturman JA, Huxtable RJ, Finberg LH. 1988b. Taurine and osmoregulation. II Administration of taurine analogues affords cerebral osmoprotection during chronic hypernatremic dehydration. Am J Dis Child 142: 1194-1198.

Trachtman H, del Pizzo R, Sturman JA. 1990. Taurine and osmoregulation. III. Taurine deficiency protects against cerebral edema during acute hyponatremia. Pediatr Res 27: 85-88.

Trenkner E, El Idrissi A, Dumas R, Rabe A. 1998. Functional consequences of calcium uptake modulation by taurine in vivo and in vitro. Adv Exp Med Biol 442: 277-284.

Trenkner E, El Idrissi A, Harris C. 1996. Balanced interaction of growth factors and taurine regulate energy metabolism, neuronal survival, and function of mouse cultured cerebellar cells under depolarizing conditions. Adv Exp Med Biol 403: 507-517.

Tuz K, Peña-Segura C, Franco R, Pasantes-Morales H. 2004. Depolarization, exocytosis and amino acid release evoked by hyposmolarity from cortical synaptosomes. Eur J Neurosci 19: 916-924.

Uchida S, Kwon HM, Yamauchi A, Preston AS, Marumo F, et al. 1992. Molecular cloning of the cDNA for an MDCK cell Na^+- and Cl^--dependent taurine transporter that is regulated by hypertonicity. Proc Natl Acad Sci USA 89: 8230-8234.

Uchiyama-Tsuyuki Y, Araki H, Yae T, Otomo S. 1994. Changes in the extracellular concentrations of amino acids in the rat striatum during transient focal cerebral ischemia. J Neurochem 62: 1074-1078.

Uemura S, Ienaga K, Higashiura K, Kimura H. 1991. Effects of intraamygdaloid injection of taurine and valyltaurine on amygdaloid kindled seizure in rats. Jap J Psychiatry Neurol 45: 383-385.

Usifoh CO, Lambert DM, Woulters J, Scriba GKE. 2001. Synthesis and anticonvulsant activity of N,N-phthaloyl derivatives of central nervous system inhibitory amino acids. Arch Pharm Pharmacol Med Chem 234: 323-331.

Varga V, Janáky R, Marnela KM, Gulyás J, Kontro P, et al. 1989. Displacement of excitatory aminoacidergic receptor ligands by acidic oligopeptides. Neurochem Res 14: 1223-1227.

Varga V, Kontro P, Oja SS. 1988. Modulation of GABA-ergic neurotransmission in the brain by dipeptides. Neurochem Res 13: 1027-1034.

Varga V, Marnela KM, Kontro P, Gulyás J, Vadász Z, et al. 1987. Effects of acidic dipeptides on aminoacidergic neurotransmission in the brain. Adv Exp Med Biol 217: 357-368.

Vinnakota S, Qian X, Egal H, Sarthy V, Sarkar HK. 1997. Molecular characterization and in situ localization of a mouse retinal taurine transporter. J Neurochem 69: 2238-2250.

Vitarella D, Di Risio DJ, Kimelberg HK, Aschner M. 1994. Potassium and taurine release are highly correlated with regulatory volume decrease in neonatal primary rat astrocyte cultures. J Neurochem 63: 1143-1149.

Vohra BP, Hui X. 2000. Improvement of impaired memory in mice by taurine. Neural Plast 7: 245-259.

Vohra BP, Hui X. 2001. Taurine protects against carbon tetrachloride toxicity in the cultured neurons and in vivo. Arch Physiol Biochem 109: 90-94.

Waagepetersen H, Melo T, Schousboe A, Sonnewald U. 2005. Homeostasis of neuroactive amino acids in cultured cerebellar and neocortical neurons is influenced by environmental cues. J Neurosci Res 79: 97-105.

Wade JV, Olsson JP, Samson FE, Nelson SR, Pazdernik TL. 1988. A possible role for taurine in osmoregulation within the brain. J Neurochem 51: 740-745.

Wade JV, Samson FE, Nelson SR, Pazdernik TL. 1987. Changes in intracellular amino acids during soman- and kainic acid-induced seizures. J Neurochem 49: 645-649.

Wahl P, Elster L, Schousboe A. 1994. Identification and function of glycine receptors in cultured cerebellar granule cells. J Neurochem 62: 2457-2463.

Walberg F, Ottersen OP, Rinvik E. 1990. GABA, glycine, aspartate, glutamate and taurine in the vestibular nuclei: An immunocytochemical investigation in the cat. Exp Brain Res 79: 547-563.

Walz W, Allen AF. 1987. Evaluation of the osmoregulatory function of taurine in brain cells. Exp Brain Res 68: 290-298.

Walz W, Hinks EC. 1985. Carrier-mediated KCl accumulation accompanied by water movements is involved in the control of physiological K^+ levels by astrocytes. Brain Res 343: 44-51.

Wang DS, Xu TL, Pang ZP, Li JS, Akaike N. 1998. Taurine-activated chloride currents in the rat sacral dorsal commissural neurons. Brain Res 792: 41-47.

Wang F, Xiao C, Ye JH. 2005. Taurine activates excitatory non-synaptic glycine receptors on dopamine neurons in ventral tegmental area of young rats. J Physiol 565: 503-516.

Warburton DM, Bersellini E, Sweeney E. 2001. An evaluation of a caffeinated taurine drink on mood, memory and information processing in healthy volunteers without caffeine abstinence. Psychopharmacology 158: 322-328.

Ward R, Cirkovic-Vellichovia T, Ledeque F, Tirizitis G, Dubars G, et al. 2006. Neuroprotection by taurine and taurine analogues. Taurine 6. Oja SS, Saransaari P, editors. New York: Springer/Kluwer Academic; pp. 299-306.

Ward RJ, Kest W, Bruyeer P, Lallemand P, De Witte, 2001. Taurine modulates catalase, aldehyde dehydrogenase, and ethanol elimination rates in rat brain. Alcohol Alcohol 36: 39-43.

Ward RJ, Martinez J, Ball D, Marshall EJ, De Witte P. 2000. Investigation of the therapeutic efficacy of a taurine analogue during the initial stages of ethanol detoxification: Preliminary studies in chronic alcohol abusers. Adv Exp Med Biol 483: 375-381.

Wasserhess P, Becker M, Staab D. 1993. Effect of taurine on synthesis of neutral and acidic sterols and fat absorption in preterm and full-term infants. Am J Clin Nutr 58: 349-353.

Wu H, Jin Y, Wei J, Jin H, Sha D, et al. 2005. Mode of action of taurine as a neuroprotector. Brain Res 1038: 123-131.

Wu JY, Chen W, Tang XW, Jin H, Foos T, et al. 2000. Mode of action of taurine and regulation dynamics of its synthesis in the CNS. Adv Exp Med Biol 483: 35-44.

Wu JY, Johansen FF, Lin CT, Liu JW. 1987. Taurine system in the normal and ischemic rat hippocampus. Adv Exp Med Biol 217: 265-274.

Wu JY, Liao C, Lin CJ, Lee YH, Ho JY, et al. 1990. Taurine receptor in the mammalian brain. Prog Clin Biol Res 351: 147-156.

Wu JY, Lin CT, Johansen FE, Liu JW. 1994. Taurine neurons in rat hippocampal formation are relatively inert to cerebral ischemia. Adv Exp Med Biol 359: 289-298.

Wu JY, Tang XW, Tsai WH. 1992. Taurine receptor: Kinetic analysis and pharmacological studies. Adv Exp Med Biol 315: 263-268.

Wu ZY, Xu TL. 2003. Taurine-evoked chloride current and its potentiation by intracellular Ca^{2+} in immature rat hippocampal CA1 neurons. Amino Acids 24: 155-162.

Wysmyk U, Oja SS, Saransaari P, Albrecht J. 1994. Long-term treatment with ammonia affects the content and release of taurine in cultured cerebellar astrocytes and granule neurons. Neurochem Int 24: 317-322.

Yager JY, Armstrong EA, Miyashita H, Wirrell EC. 2002. Prolonged neonatal seizures exacerbate hypoxic-ischemic damage: Correlation with cerebral energy metabolism and excitatory amino acid release. Dev Neurosci 24: 367-381.

Yang R, Li Q, Guo JC, Jin HB, Liu J, et al. 2006. Taurine participates in the anticonvulsive effect of electroacupuncture. Taurine 6. Oja SS, Saransaari P, editors. New York: Springer/Kluwer Academic; pp. 389-394.

Ye GI, Tse ACO, Yung WH. 1997. Taurine inhibits rat substantia nigra pars reticulate neurons by activation of GABA- and glycine-linked chloride conductance. Brain Res 749: 175-179.

Ye JH, Wang F, Krnjević K, Wang W, Xiong ZG, et al. 2004. Presynaptic glycine receptors on GABAergic terminals facilitate discharge of dopaminergic neurons in ventral tegmental area. J Neurosci 24: 8961-8974.

Yingcharoen K, Rinvik E, Storm-Mathisen J, Ottersen OP. 1989. GABA, glycine, glutamate, aspartate and taurine in the perihypoglossal nuclei: An immunocytochemical investigation in the cat with particular reference to the issue of amino acid localization. Exp Brain Res 78: 345-357.

Yoshida M, Fukuda S, Tozuka Y, Miyamoto Y, Hisatsune T. 2004. Developmental shift in bidirectional functions of taurine-sensitive chloride channels during cortical circuit formation in postnatal mouse brain. J Neurobiol 60: 166-175.

Yoshida M, Izumi K, Koja T, Fukuda T, Munekata E, et al. 1986. Inhibitory effect of taurine on wet-dog shakes produced by [D-Ala^2, Met^3]enkephalinamide with reference to effects on hippocampal epileptic discharges. Neuropharmacology 25: 1373-1378.

Young RSK, During MJ, Donnelly DF, Aquila WJ, Perry VL, et al. 1993. Effect of anoxia on excitatory amino acids in brain slices of rats and turtles: In vitro microdialysis. Am J Physiol 264: R716-R719.

Young TL, Matsuda T, Cepko CL. 2005. The noncoding RNA taurine upregulated gene 1 is required for differentiation of the murine retina. Curr Biol 15: 501-512.

Yu K, Ge SY, Ruan DY. 2003. Fe^{2+} decreases the taurine-induced Cl^- current in acutely dissociated rat hippocampal neurons. Brain Res 960: 25-35.

Yu K, Yu SS, Ruan DY. 2005. Opposite effects of lead on taurine- and HFS-induced LTP in rat hippocampus. Brain Res Bull 64: 525-531.

Yu S, Cai J. 2003. Effects of aniracetam on extracellular levels of transmitter amino acids in the hippocampus of the conscious gerbils: An intracranial microdialysis study. Neurosci Lett 339: 187-190.

Zhang M, Bi LF, Ai YD, Yang LP, Wang HB, et al. 2004. Effects of taurine supplementation on VDT work induced visual stress. Amino Acids 26: 59-63.

Zhao P, Cheng J. 1997. Effects of electroacupuncture on extracellular contents of amino acid neurotransmitters in rat striatum following transient focal cerebral ischemia. Acupunct Electrother Res 22: 119-126.

Zhao P, Huang YL, Cheng JS. 1999. Taurine antagonizes calcium overload induced by glutamate or chemical hypoxia in cultured rat hippocampal neurons. Neurosci Lett 268: 25-28.

Zheng L, Godfrey DA, Waller HJ, Godfrey TG, Chen K, et al. 2000. Effects of high-potassium-induced depolarization on amino acid chemistry of the dorsal cochlear nucleus in rat brain slices. Neurochem Res 25: 823-835.

Zhu WJ, Vicini S. 1997. Neurosteroid prolongs $GABA_A$ channel deactivation by altering kinetics of desensitized states. J Neurosci 17: 4022-4031.

Zielińska M, Hilgier W, Borkowska HD, Oja SS, Saransaari P, et al. 2002. Ammonia-induced extracellular accumulation of taurine in the rat striatum in vivo: Role of ionotropic glutamate receptors. Neurochem Res 27: 37-42.

Zielińska M, Hilgier W, Law RO, Goryński P, Albrecht J. 1999. Effects of ammonia in vitro on endogenous taurine efflux and cell volume in rat cerebrocortical minislices: Influence of inhibitors of volume-sensitive amino acid transport. Neuroscience 91: 631-638.

Zielińska M, Law RO, Albrecht J. 2003. Excitotoxic mechanism of cell swelling in rat cortical slices treated acutely with ammonia. Neurochem Int 43: 299-303.

9 Neurobiology of D-Amino Acids

E. Dumin · H. Wolosker

1	Introduction	208
2	*Neurobiology of D-Serine*	*208*
2.1	D-Serine: A Physiological Regulator of NMDA Receptors	209
2.2	Serine Racemase: Biosynthesis of D-Serine	211
2.3	Regulation of D-Serine Production	212
2.4	D-Serine Metabolism and Transport	212
2.5	D-Serine Release	214
2.6	New Roles for D-Serine	215
2.7	D-Serine and NMDA Receptor Dysfunction	215
2.8	Neuromodulator or Neurotransmitter?	216
3	*Neurobiology of D-Aspartate*	*217*
3.1	D-Aspartate in the Nervous and Endocrine System	217
3.2	Role of Endogenous D-Aspartate: Still Mysterious After Two Decades	218
3.3	D-Aspartate Transport and Release	218
3.4	Biosynthesis of D-Aspartate	218
3.5	D-Aspartate Metabolism	219
3.6	D-Aspartate as Precursor for Endogenous NMDA	219
4	*Conclusions and perspectives*	*219*

© Springer-Verlag Berlin Heidelberg 2007

Abstract: D-Amino acids are generally perceived as unnatural isomers and are thought not to play a role in mammals. In the recent years, data obtained from several laboratories established the occurrence of large amounts of D-serine and D-aspartate in the mammalian brain, where they seem to play unique roles. Endogenous D-serine is a high-affinity ligand of the glycine site of NMDA receptors and is required for NMDA receptor activation. Synthesis of endogenous D-serine is catalyzed by serine racemase enzyme, which directly converts L- into D-serine in the brain. D-Aspartate is enriched in neuroendocrine tissues and has been implicated in the regulation of hormone synthesis and release. D-Serine and D-aspartate are substrates for plasma membrane amino acid transporters and metabolized by specific peroxisomal oxidases. We now review several aspects of the neurobiology of these D-amino acids, including their actions, target receptors, metabolic pathways, and roles in pathological conditions related to NMDA receptor dysfunction.

List of Abbreviations: AMPA, α-Amino-3-Hydroxy-5-Methylisoxasole-4-Propionic Acid; NMDA, N-Methyl D-Aspartate; PLP, Pyridoxal 5′-Phosphate

1 Introduction

The origin of life on Earth is intimately related to the selection of stereochemical favorable compounds. Except for glycine, all common amino acids exhibit a chiral center resulting in the occurrence of two different enantiomers (L and D). Although the chemical properties of L- and D-amino acids are identical, L-amino acids were selected as the protein-building blocks as proteins are synthesized exclusively from L- and not D-amino acids. The sporadic presence of D-amino acid residues in proteins is due to posttranslational modifications in aged proteins rather than direct incorporation during translation (Fujii et al., 1999; Young et al., 2005).

The presence of D-amino acids in bacteria has been recognized a long time ago, where they constitute the bacterial cell wall, making it more resistant to proteases (Izaki et al., 1968). Later studies identified free D-amino acids in the cytosol of invertebrate cells, but their role remained elusive (Srinivasan et al., 1962; Corrigan, 1969; D'Aniello and Giuditta, 1978). Till recently, D-amino acids were thought not to exist in substantial quantities in higher organisms, and little or no attention has been directed toward their study.

In recent years, remarkable quantities of D-amino acids were discovered in the mammalian brain, especially D-serine and D-aspartate (Dunlop et al., 1986; Fisher et al., 1991; Hashimoto et al., 1993a). They are not incorporated into proteins or peptides, constituting a free amino acid pool in the brain. In the last few years, there has been considerable progress in the understanding of the role of D-serine and D-aspartate in the brain. There is experimental evidence that these enantiomers play important neurobiological roles, with implications for the pathophysiology of human diseases. The biosynthetic enzyme for D-serine in the brain has been discovered (Wolosker et al., 1999a, b; De Miranda et al., 2000), and D-serine seems to be a major modulator of NMDA receptor transmission (Mothet et al., 2000). D-Aspartate role is less clear and it has been linked to the regulation of hormone release (D'Aniello et al., 1996, 2000b; Takigawa et al., 1998; Wang et al., 2000). In the following sessions, we discuss several aspects of D-amino acid distribution and role in the nervous system, as well as their recently discovered biosynthetic and metabolic pathways. Although several D-amino acids exist at trace levels in the brain, we now review only D-serine and D-aspartate since these are the only D-amino acids present at significant levels (Wolosker et al., 2002). The study of D-amino acids in the brain can now be considered an exciting new field in molecular neuroscience.

2 Neurobiology of D-Serine

D-Serine has been shown to occur at high levels in the brain, where it is an endogenous coagonist of NMDA receptors. ❯ Table 9-1 summarizes the properties and disposition of D-serine, which suggest a major role for this D-amino acid in the nervous system.

Table 9-1
Disposition and properties of D-serine

	D-Serine	References
Occurrence	Enriched in the mammalian brain	Hashimoto et al. (1993a, 1992a)
Cellular distribution	Astroglia and some neurons	Schell et al. (1995, 1997a), Kartvelishvily (2006)
Action	Endogenous coagonist of NMDA receptor	Mothet et al. (2000)
Receptor	Coagonist site of NMDA receptors	Kleckner and Dingledine (1988)
Biosynthesis	Serine racemase	Wolosker et al. (1999a, b), De Miranda et al. (2000)
Metabolism	Peroxisomal D-amino acid oxidase, α,β-elimination catalyzed by serine racemase	Nagata (1992), Foltyn et al. (2005)
Transport	Neutral amino acid transporters	Hayashi et al. (1997), Ribeiro et al. (2002)
Release	Elicited by glutamate through reversal of ASCT-like transporter and vesicular release from cultured astrocytes	Schell et al. (1995), Ribeiro et al. (2002), Mothet et al. (2005)
Potential therapeutic use	Adjuvant for neuroleptic treatment in schizophrenia	Tsai et al. (1998), Heresco-Levy et al. (2005)
Biosynthesis inhibitors	PLP inhibitors, L-serine O-sulfate, cysteine, phenazines	Wolosker et al. (1999b), Panizzutti et al. (2001), Kim et al. (2005)

2.1 D-Serine: A Physiological Regulator of NMDA Receptors

NMDA receptors are one of the most important neurotransmitter receptors in the central nervous system. Broadly distributed throughout the brain, they play a major role in excitatory neurotransmission. NMDA receptor mediates Ca^{2+} influx that is required for neuronal plasticity, development, leaning, and memory (Kemp and McKernan, 2002). Knockout of NMDA receptor subunit NR1 leads to embryonic lethality, demonstrating its prominent role in CNS (Forrest et al., 1994).

A unique feature of NMDA receptors relates to the requirement of more than one agonist for the channel to operate. In addition to glutamate, binding of glycine to the receptor is essential for NMDA receptor function (Johnson and Ascher, 1987; McBain et al., 1989). Glycine binding to NMDA receptor is insensitive to strychnine, and this coagonist site is generally referred to as the "glycine site." The coagonist site is located at the NR1 subunit, while the glutamate/NMDA site is found at the NR2 subunit.

The affinity of NMDA receptors to glycine is in the low micromolar range (Matsui et al., 1995; Danysz and Parsons, 1998). Due to the high values of extracellular glycine concentration, it has been assumed that the glycine site of NMDA receptors will always be saturated and that this site would not play a dynamic role in NMDA neurotransmission (Danysz and Parsons, 1998). While there are several experimental examples in which the glycine site is saturated in vitro (Kemp et al., 1988; Fletcher et al., 1989; Mothet et al., 2000), this seems to be an exception to the rule. A large number of studies clearly show that the glycine site is not saturated under most experimental conditions both in vitro and in vivo (Wood et al., 1989; Danysz and Parsons, 1998). This implies that the glycine site of NMDA receptors may play an important role in regulating glutamatergic neurotransmission.

Another aspect of the regulatory role of the glycine site relates to the identity of the endogenous coagonist. Early studies have shown that, similar to glycine, D-serine binds to the coagonist site of NMDA receptors with high affinity (Kleckner and Dingledine, 1988; McBain et al., 1989). Since this D-amino acid was considered to be "unnatural," D-serine was not thought to physiologically regulate NMDA receptors. This notion was overturned when Hashimoto et al. (1992a, 1993a) discovered large quantities of D-serine in the mammalian brain. Levels of D-serine in the brain are about one-third those of L-serine, while peripheral tissues contain only trace amounts of D-serine. Moreover, brain levels of D-serine are comparable or higher than many common L-amino acids such as asparagine, isoleucine, valine, and tryptophan (Hashimoto et al., 1992b). This led Hashimoto et al. (1992a) to suggest that D-serine could be an endogenous coagonist of NMDA receptors.

The possibility that D-serine was an endogenous coagonist of NMDA receptors was soon strengthened by detailed immunohistochemical studies carried out by Snyder group (Schell et al., 1995, 1997a). Using a selective antibody against D-serine, Schell et al. (1995) found that D-serine is enriched in areas containing highest levels of NMDA receptors, including the rostral cerebral cortex, hippocampus, and striatum. D-Serine densities are much less in the caudal part of the brain, including the adult cerebellum. By contrast, glycine immunoreactivity is higher in caudal areas of the brain, where densities of NMDA receptors are lower (Schell et al., 1997a). The inverse localizations of D-serine and glycine led Schell and coworkers to propose that endogenous D-serine was closer to NMDA receptors than glycine. Moreover, the authors found that D-serine densities were localized in astrocytes in the gray matter, raising the possibility that D-serine would be released from astrocytes ensheathing the synapse to activate neuronal NMDA receptors (Schell et al., 1995; Snyder and Ferris, 2000).

Additional data support the notion that D-serine is a relevant endogenous coagonist of NMDA receptors. Microdialysis experiments demonstrated that extracellular D-serine concentration is similar or higher than glycine in some brain areas such as cerebral cortex and striatum (Hashimoto et al., 1995, 2000; Ciriacks and Bowser, 2004).

The three-dimensional structure of D-serine is similar to that of glycine, and its affinity for NMDA receptors is comparable or even higher than glycine (Matsui et al., 1995). A structural explanation for this finding comes from the analysis of X-ray structures of the binding core of NR1 subunit of NMDA receptors (Furukawa and Gouaux, 2003). The crystal structures revealed that agonist binding is critically dependent on a series of hydrogen bonds to side-chain and main-chain atoms, as well as to water molecules. Furukawa and Gouaux (2003) proposed that D-serine binds more tightly to the receptor in comparison with glycine because it makes three additional hydrogen bonds and displaces a water molecule. Interestingly, L-serine binds 300-fold less tightly than D-serine on NMDA receptors. Modeling of L-serine binding using D-serine as a guide showed that the hydroxyl group of L-serine unfavorably interacts in the binding pocket, providing astonishing D-isomer specificity (Furukawa and Gouaux, 2003).

More direct evidence that NMDA receptor is the target for endogenous D-serine came from experiments utilizing D-amino acid oxidase treatment to selectively deplete endogenous D-serine in hippocampal cell cultures (Mothet et al., 2000). D-Amino acid oxidase specifically degrades D- but not L-amino acids or glycine (Mattevi et al., 1996). Thus, the application of purified D-amino acid oxidase preparation led to depletion of endogenous D-serine without affecting glycine levels in hippocampal primary cultures. Depletion of D-serine led to a 60%–70% decrease in spontaneous activity attributed to postsynaptic NMDA receptor, while AMPA responses were unaffected (Mothet et al., 2000). Although the demonstration that D-serine is an endogenous coagonist of NMDA receptors in cell cultures was clear, its role in a more physiological preparation was uncertain. D-Amino acid oxidase treatment was much less effective in hippocampal slices, in which it only depleted about 19% of total D-serine (Mothet et al., 2000), precluding any conclusion about the role of D-serine in this preparation.

Subsequent studies confirmed that addition of D-amino acid oxidase led to a decrease in NMDA receptor responses in the retina and in long-term potentiation of synaptic activity in the hippocampus (Stevens et al., 2003; Yang et al., 2003, 2005; Miller, 2004). Despite the reported ineffectiveness of D-amino acid oxidase in hippocampal slices, these studies did not check if D-serine levels were actually depleted after D-amino acid oxidase treatment. One concern with experiments employing D-amino acid oxidase is that it displays very low affinity for D-serine (about 50 mM at pH 7.4) (D'Aniello et al., 1993b). In addition, commercial preparations of D-amino acid oxidase may contain many impurities, including D-aspartate oxidase, which quickly degrades NMDA (Shleper et al., 2005). Thus, unless levels of D-serine are not carefully monitored, it is not possible to evaluate the specificity of the treatment with D-amino acid oxidase.

Recently, a new developed technique overcame many caveats encountered with the use of D-amino acid oxidase as a tool to remove D-serine. The bacterial D-serine deaminase enzyme degrades D-serine to pyruvate and ammonia (Marceau et al., 1988), being three orders of magnitude more efficient than D-amino acid oxidase as judged by its kinetic parameters. By using this enzyme, it was possible to completely remove endogenous D-serine from hippocampal organotypic slices preparation (Shleper et al., 2005). Depletion of endogenous D-serine in organotypic slices virtually abolished NMDA receptor-elicited neurotoxicity. By contrast, depletion of D-serine did not affect kainate neurotoxicity. This indicates that endogenous D-serine is the dominant and necessary coagonist for NMDA receptor neurotoxicity.

Endogenous D-serine may also regulate important developmental processes. During cerebellar development, granule cells migrate from the external to the internal granule cell layer (Komuro and Rakic, 1993). Bergman glia, which contains high levels of endogenous D-serine, serves as scaffold for granule cell migration. Blockage of NMDA receptor at granule cells decreases the rate of migration (Komuro and Rakic, 1993). Recent observation suggests that endogenous D-serine presumably released by Bergman glia mediates the NMDA receptor-dependent neuronal migration in the cerebellum (Kim et al., 2005). Since migrating granule cells do not make conventional synaptic connections, the modulatory action of glial-released D-serine reflects a novel mechanism for neuromodulation (Kim et al., 2005).

2.2 Serine Racemase: Biosynthesis of D-Serine

Because amino acid racemases were thought not to exist in mammals, the origin of brain D-serine has been a mystery. The most compelling evidence for an important role for D-serine came from the identification of a specific biosynthetic enzyme. The initial studies of the origin of D-serine were contradictory. Patients lacking the ability to degrade glycine by mutations in the glycine cleavage system (GCS) had decreased levels of brain D-serine, suggesting the possible involvement of GCS in D-serine synthesis (Iwama et al., 1997). On the other hand, administration of L-serine to rats increases brain D-serine, implicating L-serine as a precursor for D-serine synthesis (Dunlop and Neidle, 1997; Takahashi et al., 1997).

The biosynthetic pathway from L-serine was established by the purification and cloning of serine racemase, an enzyme that converts L- to D-serine in rat (Wolosker et al., 1999a, b; Panizzutti et al., 2001) and human brain (De Miranda et al., 2000). Serine racemase comprises 339 amino acids with a predicted molecular weight of 36.3 kDa (Wolosker et al., 1999b). The enzyme utilizes pyridoxal 5′-phosphate (PLP) as cofactor and displays high-sequence homology with the fold-type II group of PLP enzymes such as serine/threonine dehydratase (De Miranda et al., 2000). Serine racemase does not bear significant homology to bacterial racemases. The enzyme occurs in postnatal glia but was also present at lower levels in prenatal neurons (Wolosker et al., 1999b). Transfection of serine racemase gene conferred to mammalian-cultured cells the ability to synthesize large amounts of D-serine. The distribution of serine racemase is closely similar to that of endogenous D-serine, with highest concentrations in the forebrain.

The racemization of L- into D-serine catalyzed by serine racemase was initially thought to be solely dependent on PLP (Wolosker et al., 1999b). Accordingly, PLP forms a Schiff's base with the lysine56 (Lys^{56}) and mutations of this amino acid abolished serine racemase activity (Wolosker et al., 1999a). Subsequent studies showed that magnesium and ATP are physiological ligands of the enzyme and stimulate the racemization of L-serine several fold (De Miranda et al., 2002; Foltyn et al., 2005). Calcium also binds and activates serine racemase (Cook et al., 2002), but the affinity for calcium is much lower than for magnesium, indicating that magnesium rather than calcium is the physiological cofactor (De Miranda et al., 2002). Magnesium stimulates serine racemase by two distinct mechanisms: the first is related to a direct binding of magnesium to the enzyme and the second mechanism involves an increase in the affinity to ATP due to the formation of a Mg–ATP complex (De Miranda et al., 2002).

The cofactors magnesium and ATP also disclosed new reactions catalyzed by serine racemase as they strongly stimulate the α,β-elimination of water from L-serine to form pyruvate and ammonia (De Miranda et al., 2002; Neidle and Dunlop, 2002). For each molecule of D-serine synthesized, about three to four molecules of pyruvate are formed in the presence of magnesium and ATP (De Miranda et al., 2002; Foltyn et al., 2005).

It has been proposed that generation of pyruvate by the elimination activity of serine racemase within astrocytes may play a metabolic role by providing extra-energy in situations of increased energy demands

(De Miranda et al., 2002). This is the first indication that serine racemase is linked to energy metabolism. However, further studies on the role of serine racemase-derived pyruvate will be required to clarify its role in cellular metabolism.

L-Threonine is also a substrate for serine racemase both in vitro and in vivo. Elimination of water from L-threonine is strictly dependent on ATP, generating 2-oxobutyrate and ammonia (Foltyn et al., 2005). Different from L-serine, threonine is not epimerized by serine racemase. The physiological role of L-threonine utilization by serine racemase is not clear.

The α,β-elimination catalyzed by serine racemase can be explained by the known reactions catalyzed by PLP (Morino and Snell, 1967; Schnackerz et al., 1979; Faraci and Walsh, 1988). ◉ *Figure 9-1* describes the chemical mechanism for racemization and α,β-elimination of L-serine (Foltyn et al., 2005). PLP is bound to the enzyme through an internal aldimine with Lys^{56}. Reaction with the substrate L-serine generates an external aldimine intermediate. Subsequent α-proton abstraction forms a resonance-stabilized carbanion, which has a planar configuration. Reprotonation of this intermediate on the opposite face of the planar carbanion generates the D-serine external aldimine intermediate. D-Serine is released via transimination with Lys^{56}, regenerating the free enzyme. The resonance-stabilized carbanion is also an intermediate for the α,β-elimination reaction. Elimination of the β-hydroxyl group from the carbanionic intermediate leads to the formation of the aminoacrylate-PLP intermediate. Subsequent transimination releases the initial aminoacrylate product and regenerates free enzyme. The aminoacrylate released undergoes rapid nonenzymatic hydrolysis to give pyruvate and ammonia (◉ *Figure 9-1*).

2.3 Regulation of D-Serine Production

Modulation of serine racemase activity may be crucial for regulating brain D-serine levels and consequently NMDA receptor neurotransmission. Serine racemase protein does not contain any obvious regulatory region, though it contains potential sites for phosphorylation by protein kinase C and casein kinase. Recently, Snyder and coworkers showed that the glutamate receptor-interacting protein (GRIP-1) binds to and activates mouse serine racemase (Kim et al., 2005). The interaction is mediated by the PDZ domain 6 of GRIP-1 and the last four C-terminal amino acid of serine racemase. GRIP-1 was found in astrocytes, and its overexpression strongly stimulates D-serine synthesis and enhanced granule cell migration in the cerebellum in vivo (Kim et al., 2005). Additionally, AMPA receptor stimulation increases D-serine synthesis and mediates granule cell migration in the developing cerebellum. After AMPA receptor activation, GRIP-1 seems to dissociate from the receptor and binds to mouse serine racemase, activating the enzyme. The molecular mechanism by which GRIP-1 activates mouse serine racemase is not clear, and it may result either from a direct effect of GRIP-1 on the enzyme or due to recruitment of additional interacting proteins. Nevertheless, this study provides the first mechanism by which a transmitter signal may alter D-serine levels, with implications for important developmental processes such as neuronal migration (Kim et al., 2005). Curiously, the cloned rat serine racemase gene lacks the C-terminal PDZ-binding motif found in the mouse enzyme (Konno, 2003). This indicates that regulation of serine racemase enzyme may significantly differ among rodent species.

2.4 D-Serine Metabolism and Transport

Brain D-serine exhibits a half-life of about 16 h (Dunlop and Neidle, 1997). The only enzyme known to degrade D-serine in mammals is D-amino acid oxidase, which occurs in the cerebellum and brainstem. Metabolism of D-serine by D-amino acid oxidase generates hydroxypyruvate, H_2O_2, and NH_3. In the brain, the enzyme is mainly expressed in astrocytes of the cerebellum and brain stem (Horiike et al., 1994; Urai et al., 2002). Mice carrying a mutation in D-amino acid oxidase gene exhibit augmented levels of D-serine in the cerebellum, brain stem, and spinal cord, demonstrating that D-amino acid oxidase physiologically metabolizes D-serine (Nagata, 1992; Hashimoto et al., 1993a). These mice display higher sensitivity to

◘ Figure 9-1
Reaction mechanism of L-serine racemization and α,β-elimination. For details, see text and Foltyn et al. (2005). Reproduced with permission

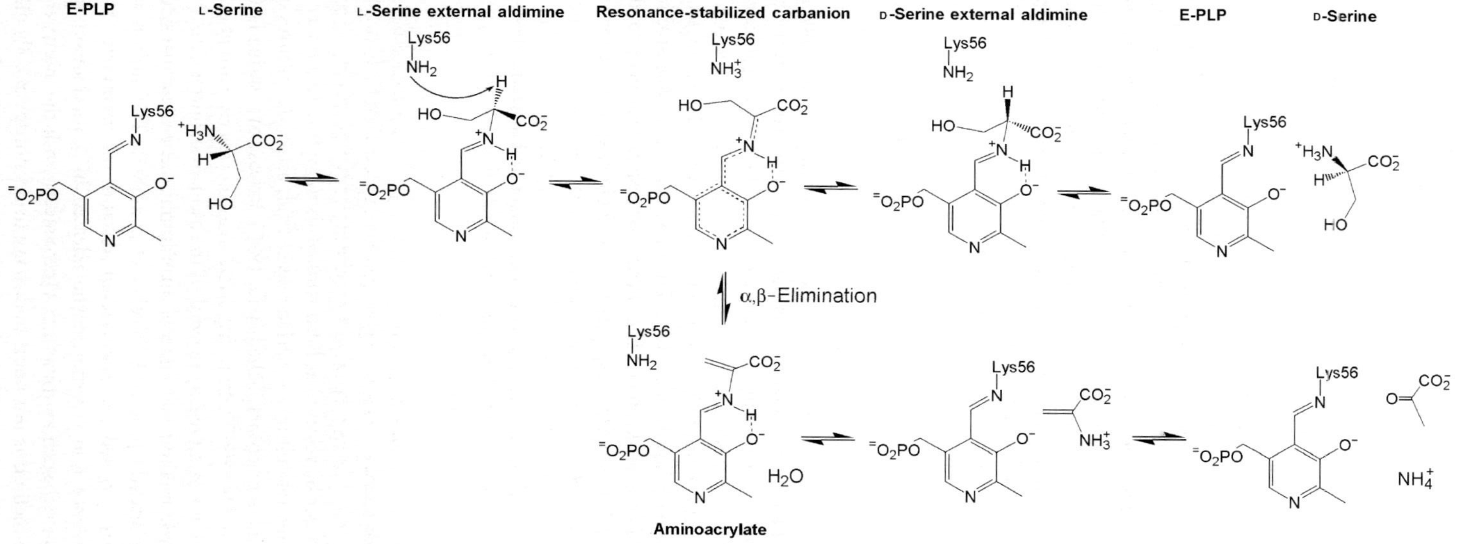

pain caused by exacerbation of NMDA receptor activity in the spinal cord dorsal horn neurons (Wake et al., 2001). D-Amino acid oxidase, however, occurs only at very low levels in the forebrain areas, such as the cerebral cortex and hippocampus, where high levels of D-serine are quite constant from 3 to 86 weeks old rats. Moreover, mutant mice possessing inactive D-amino acid oxidase enzyme exhibit large increases in D-serine in the cerebellum and brain stem but do not display increased D-serine in forebrain areas (Hashimoto et al., 1993b). This strongly suggests that D-serine is not significantly degraded by D-amino acid oxidase enzyme in the forebrain.

Recently, it has been shown that serine racemase catalyzes the degradation of cellular D-serine itself, through the α,β-elimination of water and subsequent production of pyruvate and ammonia. α,β-Elimination with D-serine is observed both in vitro and in cells (Foltyn et al., 2005). Thus, addition of D-serine to cells overexpressing serine racemase or astrocytes infected with lentivirus containing serine racemase leads to increased metabolism of D-serine (Foltyn et al., 2005). Several serine racemase mutants at a flexible loop of serine racemase display impairment of α,β-elimination activity and increased racemization activity. This suggests a novel role for the α,β-elimination reaction, which is to limit the achievable D-serine concentration. α,β-Elimination also competes with the reverse serine racemase reaction (conversion of D- into L-serine). No significant reversal is observed with physiological levels of D-serine due to the transformation of D-serine into pyruvate (Foltyn et al., 2005).

It has been proposed that elimination of water from D-serine may be important to limit D-serine levels in brain areas where D-amino acid oxidase is poorly expressed (Foltyn et al., 2005). On its synthesis and release from the cells, D-serine stimulates NMDA receptors at the coagonist site. D-Serine action could be terminated by D-serine reuptake into the cells. It is possible that serine racemase further degrades excess D-serine taken up into the cells through its α,β-elimination activity. This might constitute a mechanism for regulating intracellular D-serine concentration upon D-serine reuptake. One caveat of such model is that D-serine reuptake, rather than metabolism, is the key for termination of D-serine actions.

Selective transporters for D-serine have not been identified yet, but at least two different neutral amino acid transport systems are candidates to mediate D-serine transport. The ASCT and asc-1 types of neutral amino acid transporters can not only transport D-serine but also display high affinity for L-serine, L-alanine, L-cysteine, among others (Hayashi et al., 1997; Fukasawa et al., 2000; Nakauchi et al., 2000; Javitt et al., 2002; Ribeiro et al., 2002; Helboe et al., 2003). The affinity for D-serine has been reported to be 0.65 and 0.05 mM for ASCT-like and asc-1 transporter, respectively (Nakauchi et al., 2000; Ribeiro et al., 2002). Despite its low affinity, the ASCT-like transporter present in astrocytes mediates D-serine release by hetero-exchange mechanism (Ribeiro et al., 2002). While ASCT-type transporters are present in glia (Hayashi et al., 1997; Ribeiro et al., 2002), asc-1 transporter is mainly neuronal (Helboe et al., 2003; Matsuo et al., 2004). Further studies on the regulation of D-serine transport and metabolism will shed light on the regulation of NMDA receptor coagonist levels in the brain.

2.5 D-Serine Release

The mechanism of D-serine release has been studied using radioactive tracer or indirect determination of D-serine released in neural cultures. D-Serine release from astrocyte cultures can be induced by glutamate, acting on AMPA/kainate receptors (Schell et al., 1995). Release of D-serine is also induced by hetero-amino acid exchange catalyzed by ASCT-like transporters in astrocytes (Ribeiro et al., 2002). To date, no evidence exists on D-serine release catalyzed by reversal of the asc-1 transporter.

By using an indirect method with luminol chemiluminescence, vesicular release of D-serine from cultured astrocytes has been recently proposed (Mothet et al., 2005). The authors observed that AMPA-elicited D-serine release was blocked by removal of calcium or treatment with botulinum toxin. This provides strong evidence that D-serine maybe released by a mechanism similar to that employed for conventional transmitters such as glutamate. One concern with the interpretation of luminol chemiluminescence is that the method is not linear and many drugs nonspecifically affect luminescence. Moreover, one caveat of all the studies on D-serine release is the use of astrocyte cultures, which may not reflect in vivo situation.

A very promising approach to study the mechanism of D-serine release has been recently developed by Bowser group (Ciriacks and Bowser, 2004). Using capillary electrophoresis, the authors were able to measure endogenous D-serine release elicited by glutamate in brain slices. Further analysis of D-serine release in brain slices as well as direct demonstration of vesicular D-serine by immunoelectron microscopy will shed light on the mechanism of D-serine release.

2.6 New Roles for D-Serine

New possible roles for serine racemase and D-serine come from recent localization studies. Serine racemase and D-serine occurs in microglia, and synthesis of D-serine can be induced by inflammatory stimuli (Wu et al., 2004b). It has been proposed that D-serine released from microglia may mediate the neuronal cell death through overactivation of NMDA receptors that occurs after challenge with β-amyloid (Wu and Barger, 2004; Wu et al., 2004b). Nevertheless, direct demonstration of D-serine and serine racemase in intact brain microglia is still missing. Serine racemase immunoreactivity was also demonstrated in Schwann cells of sciatic nerve preparation (Wu et al., 2004a, 2005). Since NMDA receptor does not occur along the sciatic nerve fibers, the role of D-serine in the peripheral nervous system is different from that in the central nervous system.

Serine racemase expression may be even wider than previously thought. Using a new antibody, we are able to detect expression of serine racemase also in neurons of the cerebral cortex (Kartvelishvily et al., 2006). This matches earlier observation that D-serine is detectable in some neurons of the cerebral cortex (Yasuda et al., 2001). This raises new questions on the role of D-serine and indicates a possible important role for the neuronal D-serine pool. For instance, Neuron-Derived D-serine can be released by KCl depolarization and ionotropic glutamate receptor activation. Moreover, Neuronal D-serine release mediates a significant fraction of the NMDA receptor-elicited neurotoxicity in virtually pure neuronal cultures (Kartvelishvily et al., 2006).

2.7 D-Serine and NMDA Receptor Dysfunction

As an endogenous coagonist of NMDA receptors, D-serine synthesized by serine racemase may play a role in several pathological conditions related to NMDA receptor dysfunction. NMDA receptor function is thought to be decreased in schizophrenia. The so-called NMDA receptor hypofunction hypothesis relies on clinical and animal observations. NMDA antagonists, such as phencyclidine, induce schizophrenic-like symptoms in healthy volunteers and precipitate thought disorder and delusions in schizophrenia patients (Javitt, 2001; Coyle et al., 2003). In mice, D-serine antagonizes phencyclidine- and MK-801-induced stereotyped behavior and ataxia (Contreras, 1990). Mice expressing only 5% of normal levels of NMDAR1 (NR1) subunit display behavioral abnormalities, including increased motor activity and stereotypy and deficits in social and sexual interactions (Mohn et al., 1999). This condition is ameliorated by conventional antipsychotics.

On the basis of the NMDA receptor hypofunction hypothesis, several clinical trials evaluated the effectiveness of NMDA coagonists in schizophrenia. Administration of D-serine greatly improved the negative, cognitive, and positive symptoms of the disease even when associated with conventional and new atypical antipsychotics (Tsai et al., 1998; Heresco-Levy et al., 2005). Since the negative symptoms and cognitive impairment is normally resistant to conventional antipsychotic therapy, D-serine is a promising new adjuvant for schizophrenia treatment.

Converging data indicate that endogenous D-serine metabolism may be altered in the disease as well. Using linkage analysis, Cohen et al. (2002) found association of the disease with single nucleotide polymorphisms (SNP) in the G72 gene. This gene interacts and activates D-amino acid oxidase activity in vitro (Chumakov et al., 2002). Additionally, Cohen and coworkers found an association of schizophrenia with SNPs in D-amino acid oxidase gene in a French–Canadian population. Association of schizophrenia with D-amino acid oxidase polymorphisms was subsequently confirmed in German and Chinese population studies (Liu et al., 2004; Schumacher et al., 2004). In support of a role of D-serine in schizophrenia, a statistically significant decrease in serum D-serine was described in patients in schizophrenic subjects (Hashimoto et al., 2003).

Despite the significant association of schizophrenia with SNPs in D-amino acid oxidase gene, the genetic data should be interpreted with caution. Schizophrenia is a complex and multifactorial disease, in which several genes are likely to be involved. The contribution of a single gene may be very difficult to ascertain. Additional studies with larger populations will be required to establish a role of D-amino acid oxidase in schizophrenia. Individual contribution of each gene may be small and difficult to quantify. Also, biochemical data on the activity of D-amino acid oxidase in schizophrenic tissue will be crucial to determine its role in the disease.

The overproduction of glutamate has been widely implicated in a large number of acute and chronic degenerative diseases. The harmful effects of excessive glutamate occur mainly through activation of the NMDA receptors and consequently by massive calcium influx into the cell (Danysz and Parsons, 1998). Since D-serine is an endogenous agonist of NMDA receptors, selective inhibitors of serine racemase will be valuable tools for investigating the regulation of NMDA transmission. Overstimulation of NMDA receptors is implicated in neural damage following stroke (Choi and Rothman, 1990). Elevated extracellular concentrations of D-serine are observed after transient cerebral ischemia in rats (Lo et al., 1998), and drugs that block the "glycine site" of NMDA receptors prevent stroke damage in animal models (Wenk et al., 1998). It has been proposed that inhibitors of serine racemase provide a new strategy to decrease NMDA receptor coactivation and may be useful in conditions, such as stroke and neurodegenerative diseases, where overstimulation of NMDA receptors play a pathological role (Panizzutti et al., 2001). On the other hand, a possible caveat in using serine racemase inhibitors to prevent stroke damage is the long half-life of D-serine in the brain. Conceivably, inhibitors of both racemization and α,β-elimination activities of serine racemase will block D-serine synthesis, but removal of preformed D-serine will be slow due to the absence of D-amino acid oxidase activity in the forebrain. In this context, ligands that stimulate selectively D-serine α,β-elimination while inhibiting racemization will be more effective in decreasing brain D-serine.

2.8 Neuromodulator or Neurotransmitter?

The coagonist site of NMDA receptors clearly modulates NMDA receptor responses. The affinity of NMDA receptor for glutamate increases by preexposure to coagonists D-serine or glycine (Benveniste et al., 1990; Lerma et al., 1990; Krasteniakov et al., 2005). Glycine and D-serine binding to NMDA receptors in the absence of glutamate also primes the receptors for internalization (Nong et al., 2003). Despite the ample evidence for the modulatory activity of the coagonist site, D-serine may be more than a neuromodulator. The operation of NMDA receptors requires the obligatory binding of the coagonist, otherwise the channel does not open (Johnson and Ascher, 1987). The term neuromodulator does not incorporate the obligatory presence of D-serine for NMDA receptor activity.

Snyder and Ferris (2000) elegantly build the case for a transmitter role for D-serine by adopting a more liberal conceptualization of a neurotransmitter. The existence of metabolic pathways, a specific receptor, and mechanisms of transport and release point to a possible transmitter role for D-serine (❯ *Table 9-1*). Different from classical transmitters, D-serine has been thought to be exclusively made in glia.

In support for a "gliotransmitter" role, it has been recently shown that D-serine release from cultured glial cells can occur through exocytosis (Mothet et al., 2005). Rapid release of D-serine was calcium dependent and induced by AMPA receptor stimulation. This provides a strong evidence that D-serine can function as a gliotransmitter (Mothet et al., 2005). The detection of D-serine and serine racemase in neurons of the cerebral cortex, however, indicates that D-serine also has a neuronal origin (Kartvelishvily et al., 2006). It has been suggested that neuron-derived D-serine stimulates NMDA receptors by an autocrine or Paracrine manner (Kartvelishvily et al., 2006).

A common aspect of all transmitters is that their levels are dynamic and rapidly change at the synapse. To date, there is no experimental evidence on whether local D-serine levels are dynamic. Identification of specific signals that stimulate D-serine synthesis and/or release together with their effect on NMDA receptor neurotransmission will help to establish D-serine as a transmitter. Future studies that address this important issue will require the technical capability to directly measure tiny amounts of D-serine in a very fast timescale.

3 Neurobiology of D-Aspartate

D-Aspartate is the only other D-amino acid reported to occur at significant levels in the central nervous system. Lajtha and associates first detected free D-aspartate in mammalian brain and other tissues (Dunlop et al., 1986). Though aged proteins may exhibit a limited posttranslational racemization of L- into D-aspartate, brain D-aspartate is not directly incorporated into proteins (Dunlop et al., 1986). ◗ *Table 9-2* summarizes the current knowledge on D-aspartate disposition and possible roles in neuroendocrine tissues.

◘ Table 9-2
Disposition and properties of D-aspartate

	D-Aspartate	References
Occurrence	Neuroendocrine tissues and developing neurons	Dunlop et al. (1986), Hashimoto et al. (1993a), Schell et al. (1997b), Wolosker et al. (2000)
Action	Not known, endocrine modulator?	D'Aniello et al. (1996), D'Aniello et al. (2000b)
Receptor	Unclear. Glutamate site of NMDA receptors?	Olverman et al. (1988)
Biosynthesis	From L-aspartate in cultured cells. Biosynthetic enzyme not known	Long et al. (1998), Wolosker et al. (2000)
Metabolism	Peroxisomal D-aspartate oxidase	Dixon and Kenworthy (1967)
Transport	High-affinity excitatory amino acid transporters	Karlsen and Fonnum (1978), Taxt and Storm-Mathisen (1984)
Release	Elicited by acetylcholine, nicotine and KCl depolarization from adrenal slices	Wolosker et al. (2000)
Biosynthesis inhibitors	PLP inhibitor	Wolosker et al. (2000)

3.1 D-Aspartate in the Nervous and Endocrine System

D-Aspartate occurs early during brain development, its levels dramatically fall after birth (Dunlop et al., 1986; Hashimoto et al., 1993a). Immunohistochemistry with a selective antibody against D-aspartate revealed that it occurs exclusively in neurons with no evidence of staining in glia (Schell et al., 1997b). D-Aspartate localizations are associated to regions that play a major role in regulating development. Particularly intense immunoreactivity occurs in the cortical plate and the subventricular zone at P0 (Wolosker et al., 2000). The hippocampus displays pronounced staining in pyramidal cells of CA1-CA3 and the dentate gyrus. In the cerebellum, the most intense immunoreactivity occurs in the external granular layer associated with granule cells that have not yet undergone migration to the adult granule cell layer. Very high levels of D-aspartate are also observed in embryonic primary neuronal cultures, where it accounts for almost half of total aspartate levels (Wolosker et al., 2000).

In adult animals, little D-aspartate is left in forebrain areas (Wolosker et al., 2000). Highest concentrations of D-aspartate persist in the pineal, pituitary, adrenal glands, and testis (Dunlop et al., 1986; Lee et al., 1997, 1999; Imai et al., 1995; D'Aniello et al., 2000a). In the adrenal gland, D-aspartate is confined to the medulla where it is concentrated specifically in epinephrine cells (Schell et al., 1997b).

3.2 Role of Endogenous D-Aspartate: Still Mysterious After Two Decades

Despite the abundance of D-aspartate in neuroendocrine tissue, its role remains mysterious, as a specific receptor for endogenous D-aspartate has not been identified yet.

NMDA receptors bind D-aspartate, but it is not clear whether endogenous D-aspartate will activate NMDA receptors in vivo, as the affinity for D-aspartate is several fold lower than for glutamate (Olverman et al., 1988). Recently, it has been shown that AMPA receptors are blocked by D-aspartate, but not by the L-enantiomer (Gong et al., 2005). D-Aspartate competitively blocks L-glutamate and kainate effects in rat hippocampal neurons and AMPA receptors expressed in *Xenopus* oocytes, but the affinity of AMPA receptors for D-aspartate was much lower than for glutamate (Gong et al., 2005).

Several studies suggested possible roles for D-aspartate by employing exogenous administration of D-aspartate. D-Aspartate greatly stimulates testosterone synthesis in rat testes and in isolated Leydig cells (D'Aniello et al., 1996, 1998, 2000b; Di Fiore et al., 1998; Sakai et al., 1998) by apparently increasing the messenger RNA levels of a steroidogenic acute regulatory protein (Nagata et al., 1999a, b). D-Aspartate administration to rats also enhances oxytocin (Wang et al., 2000), prolactin (D'Aniello et al., 2000b), and luteinizing hormone release (D'Aniello et al., 2000a). D-Aspartate taken up by pinealocytes depresses melatonin synthesis (Takigawa et al., 1998). Taken together, these studies suggest that D-aspartate is an important endocrine modulator, but they do not provide any evidence that the endogenous D-aspartate actually affects hormone synthesis and release. Further studies are needed to address the role of endogenous D-aspartate. This will require the identification of its molecular targets and biosynthetic pathway. Techniques to deplete D-aspartate from the tissues, similar to those previously employed for D-serine, will be valuable to access the role of endogenous D-aspartate.

3.3 D-Aspartate Transport and Release

D-Aspartate is taken by excitatory amino acid transporters, and its tritiated form has been extensively used in transport studies (Karlsen and Fonnum, 1978; Taxt and Storm-Mathisen, 1984; Gundersen et al., 1995). Immunohistochemical studies showed that D-aspartate is not taken up in synaptic vesicles (Gundersen et al., 1995). This seems to be due to the very high specificity of the vesicular glutamate transporters (Wolosker et al., 1996).

D-Aspartate can also be mobilized from endogenous stores in vivo, implying an important role. Epinephrine is released from the adrenal in response to activation of nicotinic cholinergic receptors following firing of the splanchnic innervation of the adrenal (Seidler and Slotkin, 1976; Nussdorfer, 1996). In vivo treatment of rats with high doses of nicotine depletes catecholamine stocks in adrenals (Seidler and Slotkin, 1976). Nicotine treatment is also very effective in releasing D-aspartate, which are reduced by 70% in the adrenal medulla after chronic nicotine treatment, with no significant change in L-aspartate or L-glutamate concentrations (Wolosker et al., 2000). Mobilization of endogenous D-aspartate is also observed in adrenal slices after depolarization with KCl or acetylcholine treatment (Wolosker et al., 2000).

3.4 Biosynthesis of D-Aspartate

Endogenous origin of D-aspartate was first demonstrated by a developmental increase in D-aspartate concentration in chicken egg (Neidle and Dunlop, 1990). Since the egg is a closed system, any increase in D-aspartate should be due to a biosynthetic process. Evidence that D-aspartate is synthesized in mammalian tissue comes from an ingenious experiment carried out by D'Aniello et al. (1998). The author carefully measured the concentration of D-aspartate in the testicular venous blood and compared it to the peripheral blood. Values for D-aspartate were 20 higher in the testicular venous blood, indicating that testis synthesize and secrete D-aspartate into the blood in vivo (D'Aniello et al., 1998).

Further evidence supporting the presence of a specific biosynthetic route for D-aspartate comes from experiments in cell cultures. Cultured pheochromocytoma cells (PC12) release D-aspartate to the culture medium over time, suggesting a biosynthetic pathway in cancer cells (Long et al., 1998). In neuronal primary cultures, D-aspartate synthesis seems to be mediated by L-aspartate racemization. Biochemical evidence for racemization comes from the detection of D-aspartate synthesis from L-[^{3}H]aspartate in primary neuronal cultures of rat cerebral cortex (Wolosker et al., 2000). Addition of amino-oxyacetic acid, an inhibitor of PLP-dependent enzymes, inhibited D-aspartate synthesis (Wolosker et al., 2000). This indicates that, like D-serine, a specific racemase toward L-aspartate might exist. Nevertheless, attempts to biochemically purify an aspartate racemase have failed. Identification of aspartate racemase will be crucial to understand the biological role of this D-amino acid.

Despite the general believe that most of D-aspartate is endogenous, one cannot discard that a significant fraction of D-aspartate has an exogenous origin or is carried away from its synthesis site to target tissues that do not possess a biosynthetic enzyme. Of notice is the occurrence of D-aspartate in tissues that are blood–brain barrier deficient such as pituitary and pineal glands. Blood levels of D-aspartate may be derived from D-aspartate absorbed from the gut that escaped liver metabolism or from D-aspartate synthesized elsewhere in the body (e.g., testis). Conceivably, D-aspartate could be carried out by the blood to the pituitary and pineal gland, which possess a high-affinity transporter that takes up extracellular D-aspartate (Takigawa et al., 1998). Further studies will be required to evaluate if tissues like the pituitary and pineal gland synthesize their own D-aspartate.

3.5 D-Aspartate Metabolism

Degradation of D-aspartate is mediated by D-aspartate oxidase enzyme, which is able to oxidize D-aspartate, D-glutamate, and N-methyl D-aspartate (Dixon and Kenworthy, 1967; D'Aniello et al., 1993a). Like D-amino acid oxidase, D-aspartate oxidase is a peroxisomal enzyme, with highest expression in the kidney and liver (D'Aniello et al., 1993b). However, D-aspartate oxidase does not use D-serine as substrate. Highest levels of this enzyme occur in areas with lowest D-aspartate levels, suggesting that D-aspartate oxidase degrades endogenous D-aspartate (Schell et al., 1997b).

3.6 D-Aspartate as Precursor for Endogenous NMDA

In invertebrates, occurrence of NMDA has been documented, but till recently NMDA was thought not to exist in mammals (Sato et al., 1987; Todoroki et al., 1999). Interestingly, D'Aniello et al. (2000a) have shown that D-aspartate administration to rats elicits NMDA accumulation in the pituitary gland (D'Aniello et al., 2000a). The authors demonstrated the existence of endogenous NMDA in rats by an enzymatic method using D-aspartate oxidase followed by HPLC analysis. Most importantly, the authors reconstituted the methylation of D-aspartate into NMDA with rat brain homogenates utilizing s-adenosyl-methionine as a methyl group donor (D'Aniello et al., 2000a). If confirmed by additional analytical methods, the presence of endogenous NMDA in the brain will be of great importance. The enzyme responsible for its biosynthesis has not been identified yet. Identification of the biosynthetic route for NMDA will be important to definitively establish the presence of endogenous biosynthesis of NMDA in the mammalian brain.

4 Conclusions and perspectives

The field of D-amino acids research experienced significant advances in the last few years. Several studies demonstrated an important physiological role of D-serine in excitatory neurotransmission. The physiological role of endogenous D-aspartate and its derivative NMDA is still unclear, but major breakthrough is expected when their biosynthetic enzymes will be isolated. D-Serine can now be considered a novel "transmitter-like" molecule in the brain, and the study of its neurobiology will shed light on the regulation of NMDA receptor neurotransmission. The isolation of serine racemase opened a new era in D-amino acid research and will allow the study of the mechanisms regulating NMDA receptor coagonist levels in the brain. Although much remains to be understood, the demonstration of a physiological role for D-serine and the presence of a biosynthetic apparatus in mammals overturn classic concepts in biology, which usually consider D-amino acids as "unnatural" products. In light of the NMDA receptor dysfunction in several human diseases, D-serine is likely may be relevant in several pathological states and lead to new therapies. Blockers of D-serine synthesis may be useful for conditions, such as stroke and neurodegenerative diseases, in which NMDA receptor stimulation plays a pathological role. In sum, D-amino acid research now constitutes an exciting new field in neurobiology and biochemistry, which is likely to attract increasing numbers of investigators.

References

Benveniste M, Clements J, Vyklicky L Jr, Mayer ML. 1990. A kinetic analysis of the modulation of N-methyl-D-aspartic acid receptors by glycine in mouse cultured hippocampal neurones. J Physiol 428: 333-357.

Choi DW, Rothman SM. 1990. The role of glutamate neurotoxicity in hypoxic-ischemic neuronal death. Annu Rev Neurosci 13: 171-182.

Chumakov I, Blumenfeld M, Guerassimenkom O, Cavarec L, Palicio M, et al. 2002. Genetic and physiological data implicating the new human gene G72 and the gene for D-amino acid oxidase in schizophrenia. Proc Natl Acad Sci USA 99: 13675-13680.

Ciriacks CM, Bowser MT. 2004. Monitoring D-serine dynamics in the rat brain using online microdialysis-capillary electrophoresis. Anal Chem 76: 6582-6587.

Contreras PC. 1990. D-Serine antagonized phencyclidine- and MK-801-induced stereotyped behavior and ataxia. Neuropharmacology 29: 291-293.

Cook SP, Galve-Roperh I, Martinez del Pozo A, Rodriguez-Crespo I. 2002. Direct calcium binding results in activation of brain serine racemase. J Biol Chem 277: 27782-27792.

Corrigan JJ. 1969. D-Amino acids in animals. Science 164: 142-149.

Coyle JT, Tsai G, Goff D. 2003. Converging evidence of NMDA receptor hypofunction in the pathophysiology of schizophrenia. Ann NY Acad Sci 1003: 318-327.

D'Aniello A, D'Onofrio G, Pischetola M, D'Aniello G, Vetere A, et al. 1993a. Biological role of D-amino acid oxidase and D-aspartate oxidase. Effects of D-amino acids. J Biol Chem 268: 26941-26949.

D'Aniello A, De Simone A, Spinelli P, D'Aniello S, Branno M, et al. 2002. A specific enzymatic high-performance liquid chromatography method to determine N-methyl-D-aspartic acid in biological tissues. Anal Biochem 308: 42-51.

D'Aniello A, Di Cosmo A, Di Cristo C, Annunziato L, Petrucelli L, et al. 1996. Involvement of D-aspartic acid in the synthesis of testosterone in rat testes. Life Sci 59: 97-104.

D'Aniello A, Di Fiore MM, D'Aniello G, Colin FE, Lewis G, et al. 1998. Secretion of D-aspartic acid by the rat testis and its role in endocrinology of the testis and spermatogenesis. FEBS Lett 436: 23-27.

D'Aniello A, Di Fiore MM, Fisher GH, Milone A, Seleni A, et al. 2000a. Occurrence of D-aspartic acid and N-methyl-D-aspartic acid in rat neuroendocrine tissues and their role in the modulation of luteinizing hormone and growth hormone release. FASEB J 14: 699-714.

D'Aniello A, Giuditta A. 1978. Presence of D-aspartate in squid axoplasm and in other regions of the cephalopod nervous system. J Neurochem 31: 1107-1108.

D'Aniello A, Vetere A, Petrucelli L. 1993b. Further study on the specificity of D-amino acid oxidase and D-aspartate oxidase and time course for complete oxidation of D-amino acids. Comp Biochem Physiol B 105: 731-734.

D'Aniello G, Tolino A, D'Aniello A, Errico F, Fisher GH, et al. 2000b. The role of D-aspartic acid and N-methyl-D-aspartic acid in the regulation of prolactin release. Endocrinology 141: 3862-3870.

Danysz W, Parsons AC. 1998. Glycine and N-methyl-D-aspartate receptors: Physiological significance and possible therapeutic applications. Pharmacol Rev 50: 597-664.

De Miranda J, Panizzutti R, Foltyn VN, Wolosker H. 2002. Cofactors of serine racemase that physiologically stimulate the synthesis of the N-methyl-D-aspartate (NMDA) receptor coagonist D-serine. Proc Natl Acad Sci USA 99: 14542-14547.

De Miranda J, Santoro A, Engelender S, Wolosker H. 2000. Human serine racemase: Molecular cloning genomic organization and functional analysis. Gene 256: 183-188.

Di Fiore MM, Assisi L, Botte V, D'Aniello A. 1998. D-Aspartic acid is implicated in the control of testosterone production by the vertebrate gonad. Studies on the female green frog Rana esculenta. J Endocrinol 157: 199-207.

Dixon M, Kenworthy P. 1967. D-Aspartate oxidase of kidney. Biochim Biophys Acta 146: 54-76.

Dunlop DS, Neidle A. 1997. The origin and turnover of D-serine in brain. Biochem Biophys Res Commun 235: 26-30.

Dunlop DS, Neidle A, McHale D, Dunlop DM, Lajtha A. 1986. The presence of free D-aspartic acid in rodents and man. Biochem Biophys Res Commun 141: 27-32.

Faraci WS, Walsh CT. 1988. Racemization of alanine by the alanine racemases from *Salmonella typhimurium* and *Bacillus stearothermophilus*: Energetic reaction profiles. Biochemistry 27: 3267-3276.

Fisher GH, D'Aniello A, Vetere A, Padula L, Cusano GP, et al. 1991. Free D-aspartate and D-alanine in normal and Alzheimer brain. Brain Res Bull 26: 983-985.

Fletcher EJ, Millar JD, Zeman S, Lodge D. 1989. Non-competitive antagonism of N-methyl-D-aspartate by displacement of an ndogenous glycine-like substance. Eur J Neurosci 1: 196-203.

Foltyn VN, Bendikov I, De Miranda J, Panizzutti R, Dumin E, et al. 2005. Serine racemase modulates intracellular D-serine levels through an alpha beta-elimination activity. J Biol Chem 280: 1754-1763.

Forrest D, Yuzaki M, Soares HD, Ng L, Luk DC, et al. 1994. Targeted disruption of NMDA receptor 1 gene abolishes NMDA response and results in neonatal death. Neuron 13: 325-338.

Fujii N, Momose Y, Ishii N, Takita M, Akaboshi M, et al. 1999. The mechanisms of simultaneous stereoinversion racemization and isomerization at specific aspartyl residues of aged lens proteins. Mech Ageing Dev 107: 347-358.

Fukasawa Y, Segawa H, Kim JY, Chairoungdua A, Kim DK, et al. 2000. Identification and characterization of a Na$^+$-independent neutral amino acid transporter that associates with the 4F2 heavy chain and exhibits substrate selectivity for small neutral D- and L-amino acids. J Biol Chem 275: 9690-9698.

Furukawa H, Gouaux E. 2003. Mechanisms of activation inhibition and specificity: Crystal structures of the NMDA receptor NR1 ligand-binding core. EMBO J 22: 2873-2885.

Gong XG, Frandsen A, Lu WY, Wan YD, Zabek RL, et al. 2005. D-Aspartate and NMDA but not L-aspartate block AMPA receptors in rat hippocampal neurons. Br J Pharmacol 145: 449-459.

Gundersen V, Shupliakov O, Brodin L, Ottersen OP, Storm-Mathisen J. 1995. Quantification of excitatory amino acid uptake at intact glutamatergic synapses by immunocytochemistry of exogenous D-aspartate. J Neurosci 15: 4417-4428.

Hashimoto K, Fukushima T, Shimizu E, Komatsu N, Watanabe H, et al. 2003. Decreased serum levels of D-serine in patients with schizophrenia: Evidence in support of the N-methyl-D-aspartate receptor hypofunction hypothesis of schizophrenia. Arch Gen Psychiatry 60: 572-576.

Hashimoto A, Kanda J, Oka T. 2000. Effects of N-methyl-D-aspartate kainate or veratridine on extracellular concentrations of free D-serine and L-glutamate in rat striatum: An in vivo microdialysis study. Brain Res Bull 53: 347-351.

Hashimoto A, Kumashiro S, Nishikawa T, Oka T, Takahashi K, et al. 1993a. Embryonic development and postnatal changes in free D-aspartate and D-serine in the human prefrontal cortex. J Neurochem 61: 348-351.

Hashimoto A, Nishikawa T, Hayashi T, Fujii N, Harada K, et al. 1992a. The presence of free D-serine in rat brain. FEBS Lett 296: 33-36.

Hashimoto A, Nishikawa T, Konno R, Niwa A, Yasumura Y, et al. 1993b. Free D-serine D-aspartate and D-alanine in central nervous system and serum in mutant mice lacking D-amino acid oxidase. Neurosci Lett 152: 33-36.

Hashimoto A, Nishikawa T, Oka T, Takahashi K, Hayashi T. 1992b. Determination of free amino acid enantiomers in rat brain and serum by high-performance liquid chromatography after derivatization with N-tert.-butyloxycarbonyl-L-cysteine and o-phthaldialdehyde. J Chromatogr 582: 41-48.

Hashimoto A, Oka T, Nishikawa T. 1995. Extracellular concentration of endogenous free D-serine in the rat brain as revealed by in vivo microdialysis. Neuroscience 66: 635-643.

Hayashi F, Takahashi K, Nishikawa T. 1997. Uptake of D- and L-serine in C6 glioma cells. Neurosci Lett 239: 85-88.

Helboe L, Egebjerg J, Moller M, Thomsen C. 2003. Distribution and pharmacology of alanine-serine-cysteine transporter 1 (asc-1) in rodent brain. Eur J Neurosci 18: 2223-2227.

Heresco-Levy U, Javitt DC, Ebstein R, Vass A, Lichtenberg P, et al. 2005. D-Serine efficacy as add-on pharmacotherapy to risperidone and olanzapine for treatment-refractory schizophrenia. Biol Psychiatry 57: 577-585.

Horiike K, Tojo H, Arai R, Nozaki M, Maeda T. 1994. D-Amino-acid oxidase is confined to the lower brain stem and cerebellum in rat brain: Regional differentiation of astrocytes. Brain Res 652: 297-303.

Imai K, Fukushima T, Hagiwara K, Santa T. 1995. Occurrence of D-aspartic acid in rat brain pineal gland. Biomed Chromatogr 9: 106-109.

Iwama H, Takahashi K, Kure S, Hayashi F, Narisawa K, et al. 1997. Depletion of cerebral D-serine in non-ketotic hyperglycinemia: Possible involvement of glycine cleavage system in control of endogenous D-serine. Biochem Biophys Res Commun 231: 793-796.

Izaki K, Matsuhashi M, Strominger JL. 1968. Biosynthesis of the peptidoglycan of bacterial cell walls 8 Peptidoglycan transpeptidase and D-alanine carboxypeptidase: Penicillin-sensitive enzymatic reaction in strains of *Escherichia coli*. J Biol Chem 243: 3180-3192.

Javitt DC. 2001. Management of negative symptoms of schizophrenia. Curr Psychiatry Rep 3: 413-417.

Javitt DC, Balla A, Sershen H. 2002. A novel alanine-insensitive D-serine transporter in rat brain synaptosomal membranes. Brain Res 941: 146-149.

Johnson JW, Ascher P. 1987. Glycine potentiates the NMDA response in cultured mouse brain neurons. Nature 325: 529-531.

Karlsen RL, Fonnum F. 1978. Evidence for glutamate as a neurotransmitter in the corticofugal fibres to the dorsal lateral geniculate body and the superior colliculus in rats. Brain Res 151: 457-467.

Kartvelishily E, Shleper M, Balan L, Dumin E, Wolosker H. 2006. Neuron-Derived D-serine release provides a novel means to activate N-methyl D-Aspartate receptors. J Biol Chem 281: 14151-14162.

Kemp JA, Foster AC, Leeson PD, Priestley T, Tridgett R, et al. 1988. 7-Chlorokynurenic acid is a selective antagonist at the glycine modulatory site of the N-methyl-D-aspartate receptor complex. Proc Natl Acad Sci USA 85: 6547-6550.

Kemp JA, McKernan RM. 2002. NMDA receptor pathways as drug targets. Nat Neurosci 5 (Suppl.): 1039-1042.

Kim PM, Aizawa H, Kim PS, Huang AS, Wickramasinghe SR, et al. 2005. Serine racemase: Activation by glutamate neurotransmission via glutamate receptor interacting protein and mediation of neuronal migration. Proc Natl Acad Sci USA 102: 2105-2110.

Kleckner NW, Dingledine R. 1988. Requirement for glycine in activation of NMDA-receptors expressed in *Xenopus* oocytes. Science 241: 835-837.

Komuro H, Rakic P. 1993. Modulation of neuronal migration by NMDA receptors. Science 260: 95-97.

Konno R. 2003. Rat cerebral serine racemase: Amino acid deletion and truncation at carboxy terminus. Neurosci Lett 349: 111-114.

Krasteniakov NV, Martina M, Bergeron R. 2005. Role of the glycine site of the N-methyl-D-aspartate receptor in synaptic plasticity induced by pairing. Eur J Neurosci 21: 2782-2792.

Lee JA, Homma H, Sakai K, Fukushima T, Santa T, et al. 1997. Immunohistochemical localization of D-aspartate in the rat pineal gland. Biochem Biophys Res Commun 231: 505-508.

Lee JA, Homma H, Tashiro K, Iwatsubo T, Imai K. 1999. D-Aspartate localization in the rat pituitary gland and retina. Brain Res 838: 193-199.

Lerma J, Zukin RS, Bennett MV. 1990. Glycine decreases desensitization of N-methyl-D-aspartate (NMDA) receptors expressed in *Xenopus oocytes* and is required for NMDA responses. Proc Natl Acad Sci USA 87: 2354-2358.

Liu X, He G, Wang X, Chen Q, Qian X, et al. 2004. Association of DAAO with schizophrenia in the Chinese population. Neurosci Lett 369: 228-233.

Lo EH, Pierce AR, Matsumoto K, Kano T, Evans CJ, et al. 1998. Alterations in K^+-evoked profiles of neurotransmitter and neuromodulator amino acids after focal ischemia-reperfusion. Neuroscience 83: 449-458.

Long Z, Homma H, Lee JA, Fukushima T, Santa T, et al. 1998. Biosynthesis of D-aspartate in mammalian cells. FEBS Lett 434: 231-235.

Marceau M, McFall E, Lewis SD, Shafer JA. 1988. D-Serine dehydratase from Escherichia coli DNA sequence and identification of catalytically inactive glycine to aspartic acid variants. J Biol Chem 263: 16926-16933.

Matsui T, Sekiguchi M, Hashimoto A, Tomita U, Nishikawa T, et al. 1995. Functional comparison of D-serine and glycine in rodents: The effect on cloned NMDA receptors and the extracellular concentration. J Neurochem 65: 454-458.

Matsuo H, Kanai Y, Tokunaga M, Nakata T, Chairoungdua A, et al. 2004. High affinity D- and L-serine transporter Asc-1: cloning and dendritic localization in the rat cerebral and cerebellar cortices. Neurosci Lett 358: 123-126.

Mattevi A, Vanoni MA, Todone F, Rizzi M, Teplyakov A, et al. 1996. Crystal structure of D-amino acid oxidase: A case of active site mirror-image convergent evolution with flavocytochrome b2. Proc Natl Acad Sci USA 93: 7496-7501.

McBain CJ, Kleckner NW, Wyrick S, Dingledine R. 1989. Structural requirements for activation of the glycine coagonist site of N-methyl-D-aspartate receptors expressed in *Xenopus oocytes*. Mol Pharmacol 36: 556-565.

Miller RF. 2004. D-Serine as a glial modulator of nerve cells. Glia 47: 275-283.

Mohn AR, Gainetdinov RR, Caron MG, Koller BH. 1999. Mice with reduced NMDA receptor expression display behaviors related to schizophrenia. Cell 98: 427-436.

Morino Y, Snell EE. 1967. The subunit structure of tryptophanase I. The effect of pyridoxal phosphate on the subunit structure and physical properties of tryptophanase. J Biol Chem 242: 5591-5601.

Mothet JP, Parent AT, Wolosker H, Brady RO Jr, Linden DJ, et al. 2000. D-Serine is an endogenous ligand for the glycine site of the N-methyl-D-aspartate receptor. Proc Natl Acad Sci USA 97: 4926-4931.

Mothet JP, Pollegioni L, Ouanounou G, Martineau M, Fossier P, et al. 2005. Glutamate receptor activation triggers a calcium-dependent and SNARE protein-dependent release of the gliotransmitter D-serine. Proc Natl Acad Sci USA 102: 5606-5611.

Nagata Y. 1992. Involvement of D-amino acid oxidase in elimination of D-serine in mouse brain. Experientia 48: 753-755.

Nagata Y, Homma H, Lee JA, Imai K. 1999a. D-Aspartate stimulation of testosterone synthesis in rat Leydig cells. FEBS Lett 444: 160-164.

Nagata Y, Homma H, Matsumoto M, Imai K. 1999b. Stimulation of steroidogenic acute regulatory protein (StAR) gene expression by D-aspartate in rat Leydig cells. FEBS Lett 454: 317-320.

Nakauchi J, Matsuo H, Kim DK, Goto A, Chairoungdua A, et al. 2000. Cloning and characterization of a human brain Na^+-independent transporter for small neutral amino acids that transports D-serine with high affinity. Neurosci Lett 287: 231-235.

Neidle A, Dunlop DS. 1990. Developmental changes in free D-aspartic acid in the chicken embryo and in the neonatal rat. Life Sci 46: 1517-1522.

Neidle A, Dunlop DS. 2002. Allosteric regulation of mouse brain serine racemase. Neurochem Res 27: 1719-1724.

Nong Y, Huang YQ, Ju W, Kalia LV, Ahmadian G, et al. 2003. Glycine binding primes NMDA receptor internalization. Nature 422: 302-307.

Nussdorfer GG. 1996. Paracrine control of adrenal cortical function by medullary chromaffin cells. Pharmacol Rev 48: 495-530.

Olverman HJ, Jones AW, Mewett KN, Watkins JC. 1988. Structure/activity relations of N-methyl-D-aspartate receptor ligands as studied by their inhibition of [^{3}H]D-2-amino-5-phosphonopentanoic acid binding in rat brain membranes. Neuroscience 26: 17-31.

Panizzutti R, De Miranda J, Ribeiro CS, Engelender S, Wolosker H. 2001. A new strategy to decrease N-methyl-D-aspartate (NMDA) receptor coactivation: Inhibition of D-serine synthesis by converting serine racemase into an eliminase. Proc Natl Acad Sci USA 98: 5294-5299.

Ribeiro CS, Reis M, Panizzutti R, de Miranda J, Wolosker H. 2002. Glial transport of the neuromodulator D-serine. Brain Res 929: 202-209.

Sakai K, Homma H, Lee JA, Fukushima T, Santa T, et al. 1998. Localization of D-aspartic acid in elongate spermatids in rat testis. Arch Biochem Biophys 351: 96-105.

Sato M, Inoue F, Kanno N, Sato Y. 1987. The occurrence of N-methyl-D-aspartic acid in muscle extracts of the blood shell Scapharca broughtonii. Biochem J 241: 309-311.

Schell MJ, Brady RO Jr, Molliver ME, Snyder SH. 1997a. D-Serine as a neuromodulator: Regional and developmental localizations in rat brain glia resemble NMDA receptors. J Neurosci 17: 1604-1615.

Schell MJ, Cooper OB, Snyder SH. 1997b. D-Aspartate localizations imply neuronal and neuroendocrine roles. Proc Natl Acad Sci USA 94: 2013-2018.

Schell MJ, Molliver ME, Snyder SH. 1995. D-Serine an endogenous synaptic modulator: Localization to astrocytes and glutamate-stimulated release. Proc Natl Acad Sci USA 92: 3948-3952.

Schnackerz KD, Ehrlich JH, Giesemann W, Reed TA. 1979. Mechanism of action of D-serine dehydratase. Identification of a transient intermediate. Biochemistry 18: 3557-3563.

Schumacher J, Jamra RA, Freudenberg J, Becker T, Ohlraun S, et al. 2004. Examination of G72 and D-amino-acid oxidase as genetic risk factors for schizophrenia and bipolar affective disorder. Mol Psychiatry 9: 203-207.

Seidler FJ, Slotkin TA. 1976. Effects of chronic nicotine administration on the denervated rat adrenal medulla. Br J Pharmacol 56: 201-207.

Shleper M, Kartvelishvily E, Anowolosker H. 2005. D-serine is the dominant endogenous coagonist for NMDA receptor neurotoxicity in organotypic hippocampal slices. J Neurosci 25: 9413-9417.

Snyder SH, Ferris CD. 2000. Novel neurotransmitters and their neuropsychiatric relevance. Am J Psychiatry 157: 1738-1751.

Srinivasan NG, Corrigan JJ, Meister A. 1962. D-Serine in the blood of the silkworm Bombyx mori and other lepidoptera. J Biol Chem 237: 3844-3845.

Stevens ER, Esguerra M, Kim PM, Newman EA, Snyder SH, et al. 2003. D-Serine and serine racemase are present in the vertebrate retina and contribute to the physiological activation of NMDA receptors. Proc Natl Acad Sci USA 100: 6789-6794.

Takahashi K, Hayashi F, Nishikawa T. 1997. In vivo evidence for the link between L- and D-serine metabolism in rat cerebral cortex. J Neurochem 69: 1286-1290.

Takigawa Y, Homma H, Lee JA, Fukushima T, Santa T, et al. 1998. D-Aspartate uptake into cultured rat pinealocytes and the concomitant effect on L-aspartate levels and melatonin secretion. Biochem Biophys Res Commun 248: 641-647.

Taxt T, Storm-Mathisen J. 1984. Uptake of D-aspartate and L-glutamate in excitatory axon terminals in hippocampus: Autoradiographic and biochemical comparison with gamma-aminobutyrate and other amino acids in normal rats and in rats with lesions. Neuroscience 11: 79-100.

Todoroki N, Shibata K, Yamada T, Kera Y, Yamada R. 1999. Determination of N-methyl-D-aspartate in tissues of bivalves by high-performance liquid chromatography. J Chromatogr B Biomed Sci Appl 728: 41-47.

Tsai G, Yang P, Chung LC, Lange N, Coyle JT. 1998. D-Serine added to antipsychotics for the treatment of schizophrenia. Biol Psychiatry 44: 1081-1089.

Urai Y, Jinnouchi O, Kwak KT, Suzue A, Nagahiro S, et al. 2002. Gene expression of D-amino acid oxidase in cultured rat astrocytes: Regional and cell type specific expression. Neurosci Lett 324: 101-104.

Wake K, Yamazaki H, Hanzawa S, Konno R, Sakio H, et al. 2001. Exaggerated responses to chronic nociceptive stimuli and enhancement of N-methyl-D-aspartate receptor-mediated synaptic transmission in mutant mice lacking D-amino-acid oxidase. Neurosci Lett 297: 25-28.

Wang H, Wolosker H, Pevsner J, Snyder SH, Selkoe DJ. 2000. Regulation of rat magnocellular neurosecretory system by D-aspartate: Evidence for biological role(s) of a naturally occurring free D-amino acid in mammals. J Endocrinol 167: 247-252.

Wenk GL, Baker LM, Stoehr JD, Hauss-Wegrzyniak B, Danysz W. 1998. Neuroprotection by novel antagonists at the NMDA receptor channel and glycineB sites. Eur J Pharmacol 347: 183-187.

Wolosker H, Blackshaw S, Snyder SH. 1999a. Serine racemase: A glial enzyme synthesizing D-serine to regulate glutamate-N-methyl-D-aspartate neurotransmission. Proc Natl Acad Sci USA 96: 13409-13414.

Wolosker H, D'Aniello A, Snyder SH. 2000. D-Aspartate disposition in neuronal and endocrine tissues: Ontogeny biosynthesis and release. Neuroscience 100: 183-189.

Wolosker H, Panizzutti R, De Miranda J. 2002. Neurobiology through the looking-glass: D-serine as a new glial-derived transmitter. Neurochem Int 41: 327-331.

Wolosker H, Reis M, Assreuy J, de Meis L. 1996. Inhibition of glutamate uptake and proton pumping in synaptic vesicles by S-nitrosylation. J Neurochem 66: 1943-1948.

Wolosker H, Sheth KN, Takahashi M, Mothet JP, Brady RO Jr, et al. 1999b. Purification of serine racemase: Biosynthesis of the neuromodulator D-serine. Proc Natl Acad Sci USA 96: 721-725.

Wood PL, Emmett MR, Rao TS, Mick S, Cler J, et al. 1989. In vivo modulation of the N-methyl-D-aspartate receptor complex by D-serine: Potentiation of ongoing neuronal activity as evidenced by increased cerebellar cyclic GMP. J Neurochem 53: 979-981.

Wu S, Barger SW. 2004. Induction of serine racemase by inflammatory stimuli is dependent on AP-1. Ann N Y Acad Sci 1035: 133-146.

Wu S, Barger SW, Sims TJ. 2004a. Schwann cell and epineural fibroblast expression of serine racemase. Brain Res 1020: 161-166.

Wu SZ, Bodles AM, Porter MM, Griffin WS, Basile AS, et al. 2004b. Induction of serine racemase expression and D-serine release from microglia by amyloid beta-peptide. J Neuroinfl 1: 2.

Wu SZ, Jiang S, Sims TJ, Barger SW. 2005. Schwann cells exhibit excitotoxicity consistent with release of NMDA receptor agonists. J Neurosci Res 79: 638-643.

Yang S, Qiao H, Wen L, Zhou W, Zhang Y. 2005. D-Serine enhances impaired long-term potentiation in CA1 subfield of hippocampal slices from aged senescence-accelerated mouse prone/8. Neurosci Lett 379: 7-12.

Yang Y, Ge W, Chen Y, Zhang Z, Shen W, et al. 2003. Contribution of astrocytes to hippocampal long-term potentiation through release of D-serine. Proc Natl Acad Sci USA 100: 15194-15199.

Yasuda E, Ma N, Semba R. 2001. Immunohistochemical evidences for localization and production of D-serine in some neurons in the rat brain. Neurosci Lett 299: 162-164.

Young GW, Hoofring SA, Mamula MJ, Doyle HA, Bunick GJ, et al. 2005. Protein L-isoaspartyl methyltransferase catalyzes in vivo racemization of asparate-25 in mammalian histone H2B. J Biol Chem 280 (28): 26094-26098.

10 Amino Acids and Brain Volume Regulation: Contribution and Mechanisms

H. Pasantes-Morales

1	Introduction	226
2	*Hyposmotic Swelling and Regulatory Volume Decrease: Amino Acid Contribution*	227
2.1	Amino Acid Levels in Brain During Volume Adjustment to Hyposmolarity	227
2.2	The Mechanism of Regulatory Volume Decrease	228
2.2.1	The Amino Acid Efflux Pathway	229
2.2.2	The Osmotransduction Signaling	230
2.3	Hyposmolarity-Induced Release of Amino Acids from Nerve Endings	231
2.4	Volume Regulation After Gradual Decreases in Osmolarity: The Relevance of Amino Acids	233
2.5	Taurine as an Osmotransmitter of Neurohormone Output	235
3	*Isosmotic Swelling*	235
3.1	Amino Acids and Isosmotic Swelling	236
4	*Brain Cell Volume Decrease*	239
4.1	Mechanism of the Compensatory Increase in Amino Acid Levels During Hypernatremia	239
4.1.1	Hypertonicity and Osmolyte Transporters	239
4.1.2	Hypertonicity and Osmolyte Biosynthesis	241
5	*Proliferation, Apoptosis and Cell Volume*	242
5.1	Proliferation	242
5.2	Apoptosis	242

© Springer-Verlag Berlin Heidelberg 2007

Abstract: Cell volume is continuously compromised by the generation of local and transient osmotic microgradients associated with uptake of nutrients, secretion, cytoskeletal remodeling and transynaptic ionic gradients. It is also disturbed in pathological conditions including those leading to hyponatremia or those associated with ion redistribution such as ischemia, trauma and epilepsy. Changes in cell volume, swelling or shrinkage, appear as key signals in directing the cell death type to necrosis or apoptosis and as signals for proliferation. Cell swelling in brain is critical since the limited expansion imposed by the rigid cranium results in vascular rupture and the consequent ischemic episodes and neuronal death. Besides, disturbing the extracellular/intracellular ionic equilibrium in the brain, as occurs in isosmotic swelling or during volume recovery after hyposmotic swelling results in hyperexcitability and hypersynchrony of neuronal activity. Therefore, the role for amino acids as osmolytes in volume regulation, particularly those being synaptically inhibitory or inert is of particular importance. However, others such as glutamate exacerbate neuronal excitability and lead ultimately to excitotoxicity and neuronal death. Understanding the implication of amino acids in cell volume control, and elucidating the signals and mechanisms underlying their participation is crucial to the design of strategies to prevent swelling and to protect brain cells, neurons particularly, from the deleterious effects of ionic disequilibrium and excitotoxicity. This is important also for avoiding the dangers of a rapid correction of the osmolarity of external fluids in hyponatremic conditions. This review presents an overview of the available information about the amino acid contribution to volume regulation after swelling in hyposmotic and isosmotic conditions, their role in volume recovery after cell shrinkage and their implication in cell volume changes during apoptosis and proliferation.

List of Abbreviations: CaMKII, calcium-calmodulin kinase II; CGN, cerebellar granule neurons; EAAC1, excitatory amino acid transporter-1; EAAT, excitatory amino acid transporter; EGTA-AM, EGTA Acetoxymethyl ester; ERK, extracellular signal-regulated kinase; FAK, focal adhesion kinase; JNK, Jun N-terminal kinase; MAP, mitogen-activated protein; PI3K, phosphatidyl inositol 3'-kinase; PKA, protein kinase A; PKC, protein kinase C; RVD, regulatory volume decrease; TAUT, taurine transporter; TBOA, DL-threo-β-benzyloxyaspartate; TonEBP, tonicity-responsive enhancer binding protein; TonE, tonicity-responsive enhancer element

1 Introduction

The ability to regulate cell volume is a trait preserved in essentially all species throughout evolution. The maintenance of a constant cell volume is a homeostatic imperative in animal cells, since even small changes in cell water content may modify the concentration of messenger molecules and disturb the complex signaling network, crucial for cell functioning and intercellular communication. Although in physiological conditions the extracellular fluids have a highly controlled osmolarity, numerous diseases lead to alterations of systemic osmolarity. The intracellular volume constancy is also continuously compromised by the generation of local and transient osmotic microgradients associated with uptake of nutrients, secretion, cytoskeletal remodeling, and transynaptic ionic gradients (Lang et al., 1998a).

Cell volume perturbation is a challenge for cells in all animal organs but has particularly dramatic consequences in brain, since the rigid skull gives narrow margins for buffering intracranial volume changes. As expansion occurs, the constraining and eventual rupture of small vessels, generate episodes of ischemia, infarct, excitotoxicity, and neuronal death. In extreme conditions, caudal herniation of the brain parenchyma affects brain stem nuclei, resulting in death by respiratory and cardiac arrest. Cellular edema in brain occurs by a decrease in external osmolarity (anisosmotic swelling) or in isosmotic conditions, by changes in ion redistribution or by accumulation of ammonia or lactate. This is named cytotoxic or cellular edema. These two types of swelling are characteristically different from the vasogenic edema in which the hallmark is the brain–blood barrier disruption. However, in most pathologies, one type of edema gradually results in the development of the other type.

2 Hyposmotic Swelling and Regulatory Volume Decrease: Amino Acid Contribution

The role of amino acids in cell volume control in brain was evident since the early studies in chronic hyponatremia, the most common cause of hyposmotic swelling in brain cells. Hyponatremia occurs in pathologies such as renal or hepatic failure, inappropriate secretion of antidiuretic hormone, glucocorticoid deficiency, hypothyroidism, excessive use of thiazide diuretics, and psychotic polydipsia. Conditions such as head trauma, brain tumor, and cerebrovascular accidents also result in hyponatremia associated with inappropriate secretion of antidiuretic hormone or the cerebral salt wasting syndromes. Hyponatremia may also be caused by rapid correction of uremia by excessive hemodialysis or by infusion of hypotonic solutions in the perioperative period. It is a common condition in the elderly and during pregnancy. Fatal hyponatremia-induced cerebral edema has been recently associated with "ecstasy" use.

2.1 Amino Acid Levels in Brain During Volume Adjustment to Hyposmolarity

Early studies in experimental hyponatremia showed that brain is not behaving as a perfect osmometer but that the initial swelling is followed by progressive water loss until almost complete normalization, despite the persistence of hyponatremia. A compensatory displacement of fluids from the interstitial to the ventricular spaces first occurs, followed by the decrease of intracellular electrolytes. This however, was found not sufficient to compensate for the loss of water observed, particularly in the long term, and the contribution of other osmotically active solutes had to be considered. These molecules initially referred as "idiogenic osmolytes" were further identified as organic molecules, including myoinositol, phosphocreatine/creatine, glycerophosphorylcholine, betaine, N-acetylaspartate, and the amino acids, glutamate, glutamine, taurine, GABA, and glycine. Creatine, myoinositol, glutamate, and taurine are those more importantly contributing to volume control (❯ *Figure 10-1*) (Verbalis and Gullans,

◘ Figure 10-1
Decrease of osmolyte levels in brain of hyponatremic rats. (a) Change in the concentration of K^+, Cl^-, and of the main organic osmolytes in the brain of rats during moderate chronic hyponatremia. (b) Brain content of osmolytes in the normotremic condition. Abbreviations: Glu, glutamate; Cr, creatine; Gln, glutamine; myo-I, myoinositol. Other organic molecules with marginal contribution to the hyposmolarity corrections in brain are N-acetylaspartate, aspartate, GABA, glycerophosphorylcholine, and betaine. Data from Verbalis and Gullans (1991)

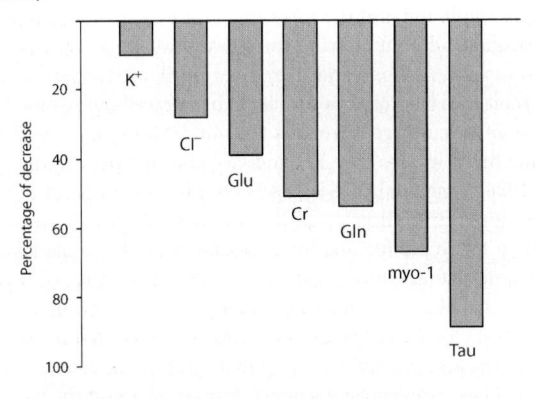

Brain content (mmol/kg dry weight)	
K^+	472.0 ± 5.0
Cl^-	153.0 ± 2.0
Glutamate	48.5 ± 1.4
Creatine	36.3 ± 2.5
Glutamine	15.3 ± 1.2
myoinositol	14.8 ± 5.3
Taurine	18.8 ± 1.1

1991). Interestingly, while decreases of electrolyte reverse with time, those of organic osmolytes, particularly that of taurine, are sustained as long as the hyponatremic conditions persists, suggesting that the electrolyte loss is an emergency mechanism to rapidly counteract brain swelling, but may be potentially harmful on long term basis, in contrast to the relatively innocuous organic osmolytes. Taurine in particular, may be a perfect osmolyte because it is metabolically inert and exhibits only weak synaptic interactions.

In experimental chronic hyponatremia in vivo, the contribution of electrolytes and organic osmolytes to the total brain osmolarity change has been estimated as 62–70% and 23–29%, respectively. Among organic osmolytes, creatine and myoinositol are the major contributors to volume adjustment, together with the amino acids glutamate, glutamine, and taurine, which show a level reduction of 39%, 54%, and 89%, respectively, after 7 days in chronic moderate hyponatremia (❷ *Figure 10-1*) (Verbalis and Gullans, 1991). This pattern of amino acid change is found, in general, in all brain regions (Massieu et al., 2004) as well as in a variety of brain preparations, including slices, cultured cells, and in vivo superfusion. Taurine is not only the amino acid showing the largest reduction in response to the osmolarity decrease, but it also exhibits the highest sensitivity to the stimulus (Solis et al., 1998; Estevez et al., 1999; Olson, 1999). This may be due to a higher efficiency of the taurine efflux pathway or to the fact that taurine pools are more available for release than those of amino acids involved in other functions such as neurotransmitters or metabolic elements.

This marked change in the brain content of amino acids and other organic osmolytes resulting from the adaptation to external osmolarity should be considered during the correction of hyponatremia to avoid brain injury (Berl, 1990). In a chronic hyponatremic condition, the intracellular osmolarity is in equilibrium with the external hyposmotic environment and therefore, when the correction in plasma tonicity restores the normal, isosmotic condition, this condition is now sensed as hyperosmotic by brain cells, which dehydrates until activation of new adaptive mechanisms. The main risk of this situation is a neurological sequel of demyelinating lesions, a pathology known as osmotic demyelination syndrome, or pontine demyelination (because of its preferential location at the basis pontis), whose salient clinical features are motor abnormalities, progressing to flaccid quadriplegia, occasional respiratory paralysis, mental state disturbances, lethargy, and coma. Although not yet fully understood, it seems that dehydration is affecting the tight junctions of the blood–brain barrier exposing oligodendrocytes to substances normally excluded from the brain, such as complement, that could be the precipitous factor of demyelination (Baker et al., 2000).

2.2 The Mechanism of Regulatory Volume Decrease

Insights into the role of amino acids in cell volume regulation and the mechanisms involved, have derived from in vitro studies showing the activation of amino acid efflux during the process known as regulatory volume decrease (RVD). Following the pattern in vivo, neurons and astrocytes swell when exposed to hyposmotic external solutions, and then recover their initial volume by extrusion of intracellular osmolytes followed by water. This is illustrated in ❷ *Figure 10-2a* for cultured astrocytes (Pasantes-Morales, 1996). The osmolytes involved in RVD are the same as in the brain in vivo, i.e., K^+ and Cl^- and organic osmolytes, including amino acids (Pasantes-Morales et al., 1996; Lang et al., 1998a). RVD has been studied in detail in cultured astrocytes and neurons, in neuroblastoma and glioma cell lines, and in the snail neurons (Pasantes-Morales, 1996). RVD is a complex chain of events formed by a volume sensor(s), a signaling cascade for transducing the information about volume changes into the activation of pathways for osmolyte extrusion, leading to volume correction. The sensor machinery has memory of the original cell volume and sets the timing for inactivation of the regulatory process. In the last years, most efforts have been directed to identify and characterize the osmolyte efflux pathways, and it is only recently that interest has emerged for understanding the osmotransduction mechanisms. There is so far only scarce information about the nature of the volume sensing mechanisms.

Figure 10-2

Regulatory volume decrease (RVD) and osmolyte efflux pathways in cultured glial cells. (a) When exposed to hyposmotic solutions, cells first swell and then activate a process of volume recovery, which occurs despite the persistence of the hyposmotic condition. (b) RVD occurs by the extrusion via leak pathways, of intracellular osmolytes including amino acids, here illustrated by the release of labeled taurine, glutamate, alanine, and glycine from cultured astrocytes. Cells loaded with the labeled amino acids are superfused with isosmotic medium and then with a medium of reduced osmolarity. Samples are collected every minute and the points represent the radioactivity released per minute as percent of total incorporated radioactivity. (c, d) The hyposmolarity-evoked efflux of K^+ and Cl^- in glial cells occurs through volume-sensitive channels. Cl^- and K^+ currents activated by hyposmolarity reduction were measured in the whole-cell recording mode of patch-clamp technique with a holding potential of -70 mV. For Cl^- currents (c) the pipette was filled with 140 mM CsCl and for K^+ currents (d) with K^+-aspartate. Data from Pasantes-Morales et al. (1993) and Ordaz et al. (2004a, b)

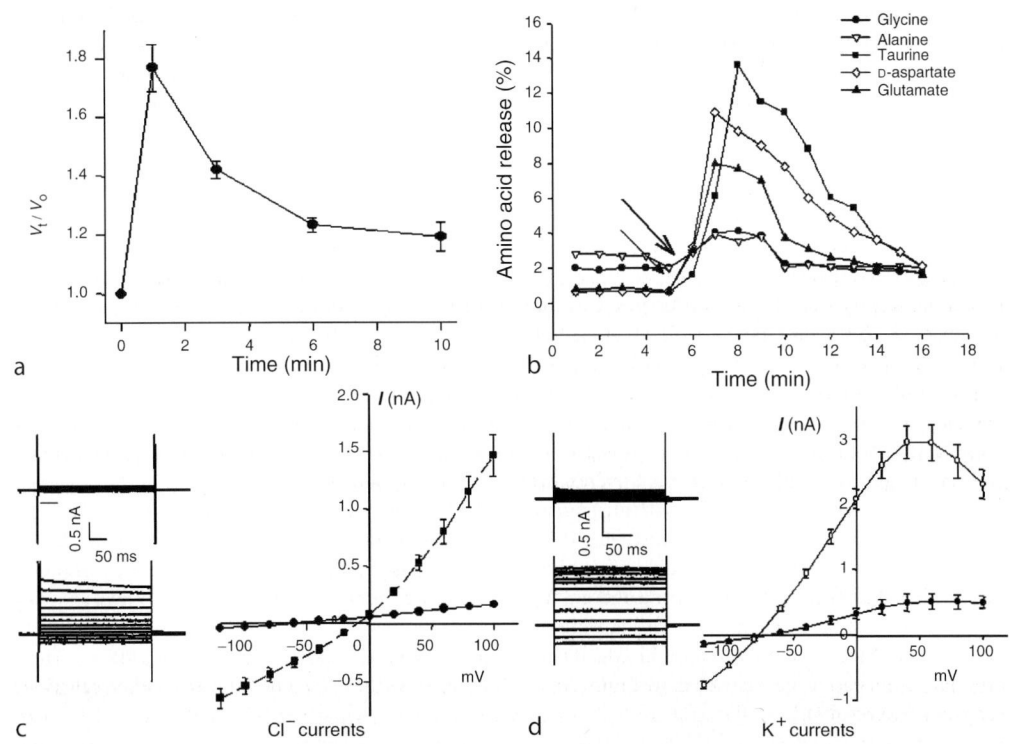

2.2.1 The Amino Acid Efflux Pathway

In most cells so far examined, the corrective osmolyte fluxes occur via diffusive pathways. K^+ and Cl^- permeate through separate channels, with marginal participation of electroneutral cotransporters (❯ *Figure 10-2c* and ❯ *d*) (Nilius et al., 1997; Niemeyer et al., 2001), and organic osmolytes through leak pathways, with essentially no contribution of energy-dependent carriers (Kirk, 1997). Amino acid efflux during RVD has been reported in a large number of brain cells and preparations, including glial cells and neurons in culture (❯ *Figure 10-2b*) (Pasantes-Morales and Schousboe, 1988; Kimelberg et al., 1990; Pasantes-Morales et al., 1993; Pasantes-Morales and Schousboe, 1997), slices from brain regions (Lehmann,

1989; Law, 1994; Deleuze et al., 2000) and in the brain in vivo using paradigms of microdialysis or superfusion (Solis et al., 1998; Estevez et al., 1999; Massieu et al., 2004). In accordance with the decrease in brain content in vivo, the amino acids preferentially released by hyposmolarity are taurine and glutamate (Pasantes-Morales et al., 1993; Massieu et al., 2004). N-acetylaspartate seems to be also released, particularly from neurons (Davies et al., 1998). In most preparations, taurine is the most sensitive to the osmotic perturbation, with the lowest release threshold and the largest amount released. For these reasons, together with some particular features such as its metabolic inertness and high intracellular concentrations, taurine seems a suitable molecule for the study of RVD mechanisms and has been often considered as representative of organic osmolytes. However, this has to be handled with caution since particularly in brain, the features of hyposmotic taurine efflux do not always match those of other organic osmolytes, including amino acids (Mongin et al., 1999; Franco et al., 2001).

The osmosensitive taurine release occurs via a leak pathway, and taurine translocation is driven by the concentration gradient (Pasantes-Morales and Schousboe, 1997; Hoffmann et al., 1988). In most cell types, including cultured astrocytes and neurons, the swelling activated taurine efflux is sensitive to Cl^- channel blockers, a finding which raised the proposal of an anion channel-like molecule as a common pathway for the corrective fluxes of Cl^- and amino acids during RVD (Strange and Jackson, 1995). However, recent evidence favors the notion of two separate pathways for taurine and Cl^- (Stutzin et al., 1999). In this case, the effect of Cl^- channel blockers on taurine efflux may reflect a close interconnection between the two pathways or the need of a specific change in intracellular Cl^- for taurine efflux activation. Other candidates to act as taurine translocating molecules are the anion exchanger (band 3) and the phospholemman. The anion exchanger seems involved mainly in fish erythrocytes (Perlman and Goldstein, 2004), and its possible role in brain cells has not been examined. The phospholemman is a member of a superfamily of proteins with single transmembrane domains, forming homomeric channels. Phospholemman, first found in heart, is present in a variety of cells and tissues, including neurons and astrocytes (Moorman and Jones, 1998; Moran et al., 2001). A characteristic of phospholemman is its markedly high permeability to taurine, a feature due possibly to the presence within the pore of binding sites for cations and anions, facilitating the transport of zwitterionic molecules. Besides the high taurine permeability, other findings are in support of phospholemman as an osmosensitive taurine efflux pathway such as taurine efflux inhibition by antisense oligonucleotide blockade of phospholemman function in cultured astrocytes and increased RVD efficiency and taurine release by phospholemman overexpression (Moran et al., 2001).

Besides taurine, hyposmolarity activates fluxes of GABA, glutamate, and glycine in cultured astrocytes and neurons (Pasantes-Morales et al., 1993). It has been often assumed that a pathway similar to that of taurine serves for other amino acid efflux, but recent results challenge this hypothesis by showing remarkable differences between taurine and glutamate. These differences include the time course, showing slower activation and inactivation for taurine efflux as compared to glutamate and the sensitivity to Cl^- channel blockers and tyrosine kinase blockers (Mongin et al., 1999; Franco et al., 2001). These differences appear to be restricted to brain cells or tissue since they are not found in other cell types (unpublished results).

2.2.2 The Osmotransduction Signaling

Identification of the signaling chains connecting cell swelling and the activation of osmolyte efflux pathways for volume adjustment has been complicated by the fact that during the volume increase and recovery, a number of reactions occur which are unrelated to osmotransduction itself. The change in cell volume is a complex process, with numerous concurrent phenomena such as stress, reorganization of the cytoskeleton, and adhesion or retraction mechanisms, among others. All of them activate their own signals, which may or may not be implicated in the activation of corrective osmolyte fluxes. In brain cells, in addition, swelling leads to depolarization and $[Ca^{2+}]_i$ rise, two conditions involved in neurotransmitter release. Since the main neurotransmitter amino acids, i.e., GABA and glutamate, are playing also a role as osmolytes, these may be the signals for their release. Besides the $[Ca^{2+}]$ increase, hyposmotic swelling leads to changes in the concentration of second messengers, such as cAMP, IP$_3$, and arachidonic acid, as well as activation of enzymes, mainly protein tyrosine kinases and phospholipases (Pasantes-Morales and Morales-Mulia, 2000; Wehner et al., 2003; Lambert, 2004). The role of each one of them in osmotransduction and their possible interplay are now actively investigated (Lambert, 2004). Despite the swelling-evoked [Ca^{2+}]$_i$ rise, the osmosensitive amino acid fluxes are often Ca^{2+}-independent (Pasantes-Morales and Morales-Mulia, 2000), although a modulatory role for Ca^{2+} on the taurine and glutamate efflux pathway has now been found. An increase of [Ca^{2+}]$_i$ over that elicited by hyposmolarity as that elicited by ATP or ionomycin potentiates taurine and glutamate efflux once the pathway has been activated (Cardin et al., 2003; Franco et al., 2004a; Mongin and Kimelberg, 2005).

Protein tyrosine kinases play a role in the hyposmotic taurine efflux, as documented by the marked inhibitory effect of tyrosine kinase blockade on taurine release, as well as by the corresponding potentiation by tyrosine phosphatase inhibition (Pasantes-Morales and Franco, 2002). A number of protein tyrosine kinases activate by hyposmolarity including p125^{FAK}, p38, JNK, p56^{lck}, p72syk, and ERK1/ERK2 (Hubert et al., 2000; van der Wijk et al., 2000). This, however, does not necessarily imply a link with osmolyte fluxes. In fact, hyposmotic activation of ERK1/ERK2, p38, and JNK appears unrelated to amino acid fluxes in cultured astrocytes and neurons (Pasantes-Morales and Franco, 2002). In contrast, phosphorylation of the tyrosine kinase membrane receptors appears closely involved in the taurine efflux pathway. Hyposmolarity activates the epidermal growth factor receptor in fibroblasts and the ErbB4 receptor in cerebellar granule neurons (● Figure 10-3a), and blockade of this reaction reduces the osmosensitive taurine efflux (● Figure 10-3b) (Franco et al., 2004b; Lezama et al., 2005). In erythrocytes, the swelling-induced tyrosine phosphorylation of band 3 (anion exchanger) is linked to p72syk and p56lyn (Hubert et al., 2000), this being the first report showing direct tyrosine phosphorylation of the osmolyte translocation molecule. The activity of the phosphoinositide-3-kinase (PI3K), a target of protein tyrosine kinases, has also an influence on the taurine efflux pathway. In cultured neurons and astrocytes, in hippocampal brain slices, and in the retina, PI3K activates by hyposmolarity, and its blockade with wortmannin impairs volume regulation and inhibits the hyposmotic taurine efflux (● Figure 10-3a and ● b) (Franco et al., 2004b). In cerebellar granule neurons, phosphorylation by hyposmolarity of ErbB4 and PI3K are interconnected reactions (Lezama et al., 2005).

Other possible elements of the osmotransduction signaling are the small GTPase p21Rho, and its downstream kinase Rho kinase connecting with the light myosin chain. The role of phospholipases on amino acid fluxes related to hyposmolarity is still unclear, with conflicting results about the effect of blockers of several phospholipases (Lambert, 2004). The connection between all these enzymes and their hierarchy remain to be established. An association between PI3K, Rho GTPases, and phospholipases has been shown in a variety of pathways, some of them regulating the dynamics of the cortical and cytoplasmic actin cytoskeleton, which may be modulatory of the amino acid efflux pathway (van der Wijk et al., 2000).

2.3 Hyposmolarity-Induced Release of Amino Acids from Nerve Endings

The effect of volume changes on subcellular compartments, such as dendrites or nerve endings, has not been studied in detail. This is of interest, since dendrites are particularly sensitive to swelling and nerve endings are continuously exposed to osmotic gradients due to constant ion redistribution and neurotransmitter uptake and release. Nerve ending swelling occurs in conditions of damage and hyperexcitability such as in head trauma and seizures. The question is raised on whether nerve ending swelling may result in the release of amino acids, this being a most important subject since amino acids acting as osmolytes and neurotransmitters, such as glutamate and GABA, have a prominent role in the control of brain excitability.

A study in isolated nerve endings (synaptosomes) showed a hyposmolarity-evoked release of taurine, glutamate, and GABA (Tuz et al., 2004) (● Figure 10-4a). The release mechanism differs for the three amino acids. Most of the hyposmotic taurine efflux shows the typical features of the osmolyte translocation,

Figure 10-3

The volume-sensitive taurine efflux pathway is influenced by the activity of tyrosine-kinase membrane receptors and the associated tyrosine-kinase target phosphoinositide-3 kinase (PI3K). (a) Illustrates the effect of hyposmolarity in cerebellar granule neurons activating phosphorylation of the tyrosine-kinase membrane receptor ErbB4 and of PI3K. The *upper panel* shows ErbB4 phosphorylation in the following conditions: Isos, isosmotic medium; H30%, hyposmotic medium −30%; Herβ, effect of the receptor ligand heregulin β (200 ng/ml). (b) Blockade of these reactions as shown in (a), with 50 μM AG 213 (ErbB4) or 100 nM wortmannin (PI3K), markedly reduces the volume-sensitive taurine efflux. Taurine release was measured as in ▶ *Figure 10.2b*. ErbB4 phosphorylation was assayed by immunoblotting with the appropriate antibodies, and PI3K activity was assayed by measuring phosphorylation of its main target AKT. Data are from Lezama et al. (2005)

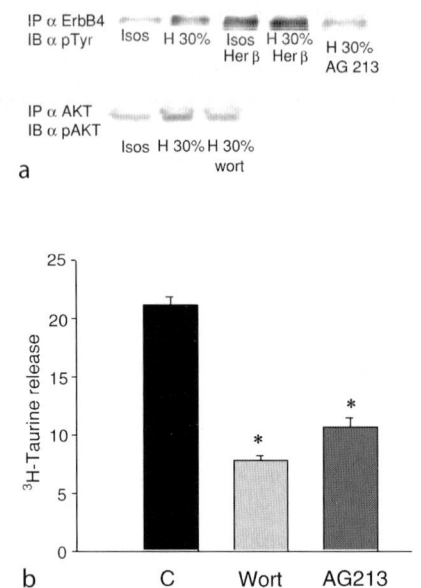

such as inhibition by Cl^- channel blockers, tyrosine kinase blockers, and PI3K blockers, while glutamate efflux is essentially insensitive to these blockers but is markedly Na^+-dependent, blocked by toxins interfering with the exocytotic release and modulated by protein kinase C (PKC). GABA efflux shows the two types of release (▶ *Figure 10-4b*). In nerve endings, hyposmolarity evokes depolarization and $[Ca^{2+}]$ increase and thus, glutamate, but not taurine, efflux may be a response to these events and mediated by exocytosis or carrier reversal operation (Tuz et al., 2004).

The nerve ending release of amino acids may explain the increased susceptibility to seizures associated with hyponatremia. Also the hyposmotic release of GABA and glutamate from nerve endings may be responsible for the increase in duration and amplitude of excitatory and inhibitory postsynaptic potentials in hyposmotic conditions (Chebabo et al., 1995; Baraban and Schwartzkroin, 1998). Thus, swelling may be affecting brain excitability and synaptic transmission. The hyposmotic taurine release from nerve endings, and likely also from synaptic vesicles, is an interesting finding since taurine even when it is highly concentrated in nerve endings (Kontro et al., 1980) has a marginal function as neurotransmitter. The presence of this swelling-responsive synaptic taurine pool may reflect a need for mechanisms to correct volume in the nerve terminal and/or in synaptic vesicles, disturbed by ion and neurotransmitter redistribution during synaptic activity. In pathological conditions, the release of taurine may have a dual benefit, relieving swelling first and once translocated into the extracellular space, acting as a neuroprotectant, a widely documented action of this amino acid.

◻ Figure 10-4
Hyposmolarity induced release of amino acids from rat brain synaptosomes. (a) Reducing osmolarity by 20% elicits a rapid efflux of glutamate, taurine, and GABA (followed by ^{3}H-tracers) from isolated nerve endings. Synaptosomes loaded with the labeled amino acids were superfused with isosmotic medium, and at the *arrow* the superfusion medium was changed to a 20% hyposmotic medium. Results are expressed as radioactivity released per min as percentage of the total radioactivity incorporated. (b) Contribution of depolarization-exocytosis [estimated by the inhibitory effect of La^{3+}, ethyleneglycol-bis-(β-aminoethylether)-N,N′-tetraacetic acid-AM (EGTA-AM) and tetanus toxin], carrier reversal {efflux sensitive to the carrier blockers for GABA and DL-threo-β-benzyloxyaspartate (TBOA) for glutamate} or leak pathway (efflux reduced by Cl$^-$ channel- and tyrosine kinase blockers). Data in (b) correspond to the amino acid efflux during the time of stimulation with the hyposmotic medium (10 min). For details see Tuz et al. (2004)

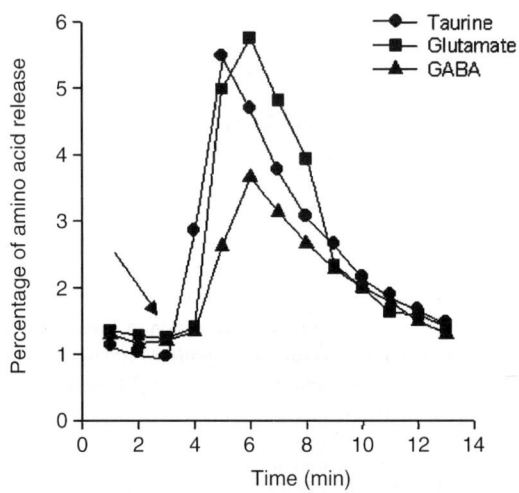

a

Contribution of depolarization-exocytosis, leak pathway, and carrier reversal to the hyposmolarity evoked release of glutamate, GABA, and taurine

	Fraction release (%)		
	Glutamate	GABA	Taurine
Depolarization-exocytosis	34–44	21–29	18–22
Leak-pathway	–	10–18	41–55
Carrier reversal	37	28	ND

b

2.4 Volume Regulation After Gradual Decreases in Osmolarity: The Relevance of Amino Acids

In most studies about volume regulation, cells are exposed to sudden and large osmolarity reductions, of a magnitude probably never occurring in brain under physiological conditions or even in pathological situations. An approach closer to normal was devised by Lohr and Grantham (1986) in renal cells, which consisted in exposing the cells to small and gradual changes in external osmolarity. Cells were then able to maintain a constant volume over a wide range of tonicities if the rate of change in osmolarity was lower than 2.2 mOsm/min. This response was named "isovolumetric regulation," a term reflecting the active nature of this process, as the constant volume was not due to the absence of swelling but to a continuous volume adjustment accomplished by the extrusion of intracellular osmolytes. This paradigm has been applied to other cell types and marked differences have been found with respect to the efficiency of the process. Interestingly, these differences appear to be related to the contribution of amino acids.

According to their behavior facing gradual osmolarity decreases, three types of cell response have been so far observed. Renal cells and cerebellar granule neurons show a volume constancy, even when the external osmolarity has dropped by 50% (Lohr and Grantham, 1986; Van Driessche et al., 1997; Tuz et al., 2001) (❯ *Figure 10-5a*). Glioma C6 cells, cultured myocytes, and astrocytes (Souza et al., 2000; Ordaz et al., 2004a, b) respond to gradual osmolarity changes by a volume increase, which nevertheless is lower than when an

◻ **Figure 10-5**
Changes in cell volume and taurine efflux in cerebellar granule neurons or astrocytes, exposed to conditions of gradual osmolarity reduction. Cells were exposed to media in which the osmolarity decrease rate was 1.8 mOsm/min. At the end of the experiment, the osmolarity of the medium has decreased by 50% (150 mOsm). Cell volume change was measured by large-angle light scattering as described in Ordaz et al. (2005) (astrocytes) or by dilution of calcein-AM followed by spectrofluorometry (granule neurons) as in Tuz et al. (2001). Volume is expressed as V_t/V_0. (a) Cerebellar granule neurons (♦) maintained a constant volume facing the osmolarity reduction while in astrocytes (●) volume increased but still it is lower than in cells exposed to a sudden decrease in osmolarity of the same magnitude, as shown in the upper curve (○), suggesting a more efficient volume control. (b, c) Taurine efflux in response to gradual reductions in osmolarity. Taurine efflux was measured as described in ❯ *Figure 10-2*, following the release of [^{3}H]taurine. The taurine efflux threshold is notably lower in neurons (b) than in astrocytes (c)

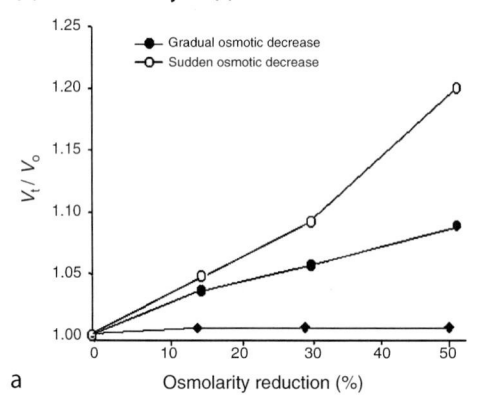

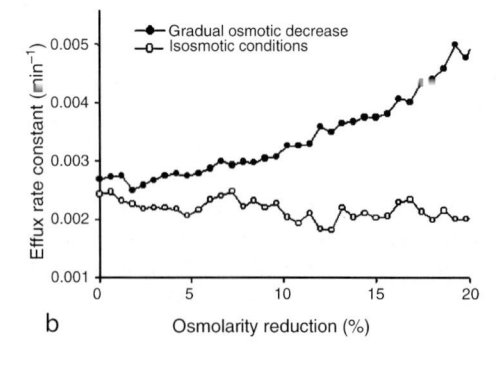

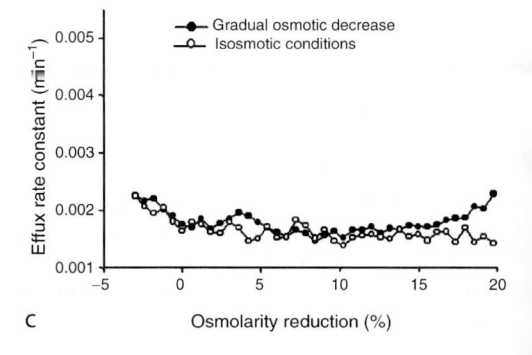

osmotic stimulus of the same magnitude is suddenly imposed (❯ *Figure 10-5a*). Finally, trout erythrocytes (Godart et al., 1999) respond with similar swelling to gradual or sudden exposure to hyposmotic solutions.

The osmolytes involved in volume corrective mechanisms during gradual osmolarity reductions are K^+, Cl^-, and organic molecules, but the activation threshold and the efficacy of the osmolyte translocation pathways show differences in the various cell types, which appears to correspond with the efficiency to counteract changes in cell volume. Thus, a correlation is observed between the taurine and glutamate efflux threshold and the extent of swelling in conditions of gradual osmolarity changes. In cerebellar granule neurons, which show the typical isovolumetric regulation, the efflux of taurine and glutamate activates very early after the osmotic stimulus, as early as a 2% osmolarity reduction for taurine (Tuz et al., 2001) (❯ *Figure 10-5b*), whereas in astrocytes, taurine efflux occurs later (❯ *Figure 10-5c*) (Ordaz et al., 2004a), and in trout erythrocytes (Godart et al., 1999), the efflux of taurine is markedly delayed and seems insufficient to contribute to volume regulation. The higher ability of neurons as compared to astrocytes to resist changes in external osmolarity, which seems based primarily on the contribution of organic osmolytes, may represent a protective mechanism to spare neurons from the deleterious consequences of swelling.

2.5 Taurine as an Osmotransmitter of Neurohormone Output

Besides the role of taurine as a ubiquitous osmolyte, in some specific cells the osmosensitive taurine efflux may have additional functions. In the hypothalamic-neurohypophysial system, a role for taurine has been proposed in the regulation of the whole body fluid balance through the control of plasmatic levels of vasopressin and oxytocin, two hormones critically involved in the body water balance. This effect of taurine is suggested by experiments showing a regulatory role of the glial hyposmotic release of taurine on the neuroendocrine cells at the supraoptic nucleus, resulting in vasopressin release by a mechanism likely involving activation of glycine receptors by the high extracellular taurine levels, reached after its hyposmolarity-induced release from glial cells (Hussy et al., 2000). Moreover, recent work has shown an osmotic regulation of the depolarization-evoked vasopressin secretion in the neurohypophysis, similarly mediated by the osmosensitive release of taurine from pituicytes, an effect involving also glycine receptors (Hussy et al., 2001).

3 Isosmotic Swelling

Isosmotic (cytotoxic) swelling occurs in association with various brain pathologies including epilepsy, ischemia, and head trauma. A number of causal factors of brain cell swelling are common to these three conditions, such as energy failure, elevated extracellular K^+ levels, depolarization, extracellular glutamate increase, lactacidosis, and in the late phases, reactive oxygen species generation, membrane lipid peroxidation, and excitotoxicity (Arundine and Tymianski, 2004). In hepatic encephalopathy, another pathology in which brain edema is a fatal outcome, other mechanisms are involved initially, but the energy failure and oxidative stress also concur in the late phases of edema (Felipo and Butterworth, 2002; Butterworth, 2003).

In epilepsy, the abnormally high-firing rate of neurons elevates K^+ levels, and the resulting depolarization activates the release of glutamate from the widespread glutamatergic synapses. The astrocyte capacity of K^+ clearance by spatial buffering is exceeded and K^+ accumulates intracellularly, followed by Cl^- and water (Walz, 2000). During the early phases of ischemia, the interruption of glucose and oxygen supply leads to energy failure, ATP depletion, and the progressive collapse of transmembrane ionic gradients, resulting in glial and neuronal depolarization, the increase in extracellular K^+, and glutamate rise (Hansson et al., 2000; Walz, 2000; Nishizawa, 2001). Additional elements for cell swelling in brain, particularly in astrocytes, result from the activation of anaerobic glycolysis, a pathway which remains operating in astrocytes under hypoxic conditions as long as there is a remnant delivery of glucose. This alternate route contributes to swelling by generation of protons and lactate. In the late ischemic phases, the mechanism of glutamate removal is inefficiently operating, as the disturbed ionic gradients make the carriers to work in reverse, further increasing the concentration of extracellular glutamate (Rossi et al., 2000; Nelson et al., 2003; Phillis and O'Regan, 2003). At this moment, the cascade of excitotoxicity is triggered, with its sequel of intracellular $[Ca^{2+}]$ rise, phospholipid degradation, free fatty acid release, and formation of reactive oxygen species (Arundine and Tymiansky, 2004). All these factors substantially contribute, by different mechanisms, to further aggravate cell swelling. The situation may be different at the ischemic penumbra, where energy depletion and ionic disruption are less severely affected. The normal rather than reversed operation of the glutamate transporter, particularly the astrocyte type, may then lead to swelling (Feustel et al., 2004).

Traumatic brain injury is a most common situation leading to cytotoxic swelling, resulting from a number of mechanical and biochemical events, with different time course and strong interplaying. The most directly related to swelling is the posttraumatic widespread depolarization, with its associated increase in intracellular Na^+, Cl^-, and water influx. Additional swelling factors derive from the transient sheer forces that mechanically deform membranes, resulting in ion permeability changes. When, as often happens, restriction of cerebral blood flow concurs with trauma, the cascade of reactions characteristic of ischemia is also triggered. An indirect mechanism of swelling after brain trauma is the hyponatremia, evolving as a consequence of a disturbed secretion of the antidiuretic hormone.

In hepatic encephalopathy, brain edema is a most characteristic neuropathological feature resulting from the acute liver failure, and it is the major cause of death. Brain cellular edema, in hepatic encephalopathy is essentially restricted to astrocytes (Haussinger et al., 2000). The initial and causal factor is an increase in brain ammonia levels subsequent to its blood rise (Felipo and Butterworth, 2002; Butterworth, 2003). Ammonia detoxification in brain, in both normal or hyperammonemic conditions is essentially carried out by astrocytes. Since the brain lacks the key enzymes to remove ammonia in the form of urea, ammonia metabolism occurs basically via the synthesis of glutamine, through the amidation of glutamate by the glutamine synthetase, an enzyme localized almost exclusively in astrocytes (Martinez-Hernandez et al., 1997). Consequently, brain glutamine production in astrocytes is dramatically increased following hyperammonemia. Besides the accumulation of glutamine, which may be per se a swelling inductor, several mechanisms are put in motion in association with the ammonium/glutamine rise in astrocytes, including blockade of key enzymes in the oxidative metabolism and lactate production, free radical generation, and mitochondrial dysfunction by induction of the permeability transition pore, all of them leading to cell swelling (Norenberg et al., 2004).

3.1 Amino Acids and Isosmotic Swelling

In contrast to chronic hyponatremia, cytotoxic swelling in brain seems not being followed by efficient mechanisms of volume adjustment. This is not unexpected, since as discussed in the preceding section, volume regulation relies on water extrusion carried by the efflux of K^+, Cl^-, and organic molecules. This occurs via diffusion pathways, with the osmolytes moving in their concentration gradient direction. In ischemic conditions, the imbalance of K^+, Na^+, and Cl^- gradients due to the energy failure excludes ion efflux as an effective mechanism for volume correction, since gradients may favor influx instead of efflux. Besides, the swelling-sensitive Cl^- channel, which is crucial for an adequate process of volume regulation, requires ATP for activation and will thus be inoperative in conditions of ATP depletion. Furthermore, a recent study (Mori et al., 2002) has shown impaired activity of the volume-sensitive Cl^- channel during lactacidosis, a condition associated with various situations generating cellular astrocyte swelling.

The release of amino acids and other organic osmolytes may occur in isosmotic swelling if the extrusion pathways are not impaired. This may reduce the extent of cell swelling but seems insufficient to permit cell volume recovery. A marked increase of amino acid efflux, of manyfold over basal release, has been observed in essentially all models of ischemia, either in vivo, in models of total or focal ischemia, or in vitro, in a variety of brain preparations, in experimental models of chemical ischemia (Benveniste et al., 1984; Saransaari and Oja, 1999; Phillis and O'Reagan, 2003). Aspartate, glutamate, GABA and taurine are the amino acids preferentially released, with manyfold increases in the ischemic condition. Efflux of glycine, alanine, serine, and phosphoethanolamine occurs mainly upon reperfusion (Phillis and O'Regan, 2003). Some differences are observed between brain regions but the general pattern is essentially similar.

For amino acids, such as GABA, aspartate, glutamate, and glycine, which besides serving as osmolytes, have a widespread function as neurotransmitters, the key point is to identify whether the efflux responds to swelling or to other events concurrent with the ischemic situation such as depolarization, Ca^{2+} influx, and ion redistribution. Glutamate release is of special interest, since the rise in glutamate is largely responsible for neuronal death by excitotoxicity during ischemia. Moreover, the survival chances of cells around the ischemic focus, at the penumbra area, appear to depend also on the control of extracellular glutamate levels (Feustel et al., 2004). As mentioned before, the ischemic condition is a multifactorial situation, in which a sequential series of responses are evoked as the ischemic episode proceeds, with events characteristic for each step, which may put in motion different mechanisms for amino acid release and/or their persistence at the extracellular space (❯ *Figure 10-6*). Thus, amino acid efflux may occur by one or several events, one of which may be swelling. Most studies on the mechanisms leading to amino acid overflow in ischemia have focused on glutamate. Available evidence supports the involvement of at least three mechanisms: (1) Ca^{2+}-dependent exocytotic release, (2) the reverse operation of the energy-dependent glutamate transporters, due to intracellular Na^+ accumulation as a consequence of the energy failure, and (3) swelling activated efflux (Benveniste et al., 1984; Seki et al., 1999; Rossi et al., 2000; Nelson et al., 2003). The second option is considered at present, as the main contributor to the elevated extracellular glutamate, with less participation of the swelling-related mechanisms. The same considerations are valid for other amino acids with the dual role of osmolyte and neurotransmitters (Allen et al., 2004).

The contribution of swelling and the reverse carriers to the extracellular glutamate increase has been examined in brain regions less severely affected by ischemia. In a model of middle cerebral artery occlusion in rats, the efflux of glutamate collected by microdialysis from regions of incomplete ischemia was found to be reduced by tamoxifen, a blocker of the volume-sensitive Cl^- channel, but not by dihydrokainate, a glutamate carrier blocker, leading to the conclusion that during incomplete ischemia, glutamate overflow occurs predominantly by swelling and not by the reversal carrier operation (Feustel et al., 2004). This is an important notion for a rational design of strategies to improve survival of brain cells in the ischemic penumbra. In the same line, a reduction of ischemia-evoked amino acid fluxes by anion channel inhibitors has been reported in cortical superfusates of a four vessel cerebral ischemia. The blockers markedly reduced the efflux of aspartate, glutamate, taurine, and phosphoethanolamine, with less effect on GABA, and no effect on serine, alanine, or glutamine (Phillis and O'Regan, 2003). It is worthy to notice that the glutamate efflux decrease by Cl^- channel blockers may reflect the inhibition of the glutamate efflux pathway or an effect of the blockers preventing Cl^- influx and swelling. The efflux of inhibitory amino acids, such as taurine, GABA, or glycine, which in contrast to glutamate, do not generate per se a secondary volume increase, nor excitotoxicity, may contribute not only to attenuate swelling but also to counteract the hyperexcitability generated during ischemia. However, recent results suggest an effect of GABA released during ischemia as a factor of neuronal swelling (Allen et al., 2004).

Amino acid efflux including excitotoxic amino acids is also observed in brain during cytotoxic swelling by hyperamonemia or head trauma (Stover and Unterberg, 2000). The mechanism of this release is still unclear, but at least for taurine in hepatic encephalopathy models, it seems not directly caused by swelling (Zielińska et al., 1999), and in vivo, no clear correlation between taurine efflux has been established neither in experimental models of hepatic encephalopathy nor in patients (Butterworth, 2003). Extracellular brain glutamate levels are enhanced possibly due to the ammonia effect decreasing the expression of EAAT1, the glial glutamate transporter (Filipo and Butterworth, 2002). Then, hyperactivation of glutamate receptors by both glutamate and depolarization triggers the cascade of reactive oxygen species generation and mitochondrial dysfunction, all this contributing to brain edema. A microdialysis study on NH_4^+-induced acidosis, reports an increase in extracellular N-acetyl-aspartate, an amino acid present in large amounts in neurons, which may have an important role as an osmolyte. (Davies et al., 1998).

◻ Figure 10-6
Isosmotic swelling and glutamate efflux from astrocytes and neurons during hypoxia and ischemia. At early ischemic phases or during hypoxia, the energy failure, ATP depletion and progressive collapse of ionic gradients, leads to glutamate efflux by the reversal operation of the neuronal carrier. The glial transporter may be still working for glutamate influx, due to ATP generation by glycolysis, which operates if a remnant glucose supply persists. This glutamate influx may cause astrocyte swelling by the Na$^+$ influx and water cotransport intrinsic to the carrier. At late ischemic phases, also the astrocyte transporter operates for glutamate efflux. The increased extracellular glutamate activates ionotropic glutamate receptors, primarily in neurons, precipitating a chain of uncontrolled swelling by generation of reactive oxygen species, membrane lipid peroxidation, ion overload, and further swelling. At this point, the volume-sensitive glutamate pathway in both neurons and astrocytes contributes to glutamate efflux. EAAT, excitatory amino acid transporter; ROS, reactive oxygen species

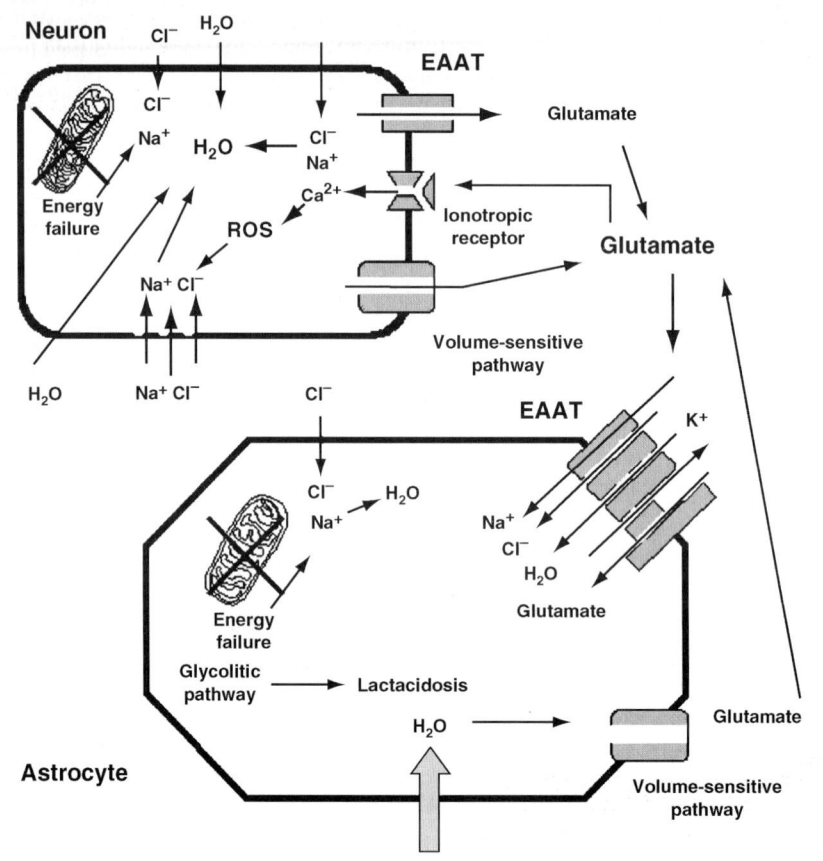

It has been consistently observed that cytotoxic edema in vivo is more prominent in astrocytes than in neurons, being so far unclear whether this difference is due to a selective localization of the swelling-generating mechanisms in astrocytes or to the presence of more efficient mechanisms of cell volume control in neurons (Pasantes-Morales and Franco, 2005). In this respect, a most interesting observation is the transfer of taurine and glutamate from neurons to astrocytes during experimental ischemia (Torp et al., 1991). By this mechanism, neurons are spared and protected from the deleterious effects of swelling.

4 Brain Cell Volume Decrease

Hypernatremia is the main clinical condition resulting in brain cell shrinkage. It is most prevalent in the geriatric population. Hypernatremia is caused by a loss of water or a gain in Na$^+$, the two conditions being related in most cases. The mechanisms for defective water intake involve mainly impaired thirst mechanisms, due to diseases or defects in the hypothalamic thirst center or in the frontal lobe thirst perception. Deficient water intake may occur also during impaired mental function by coma or confusion, during anesthesia, or by unavailability of water. Other situations associated with or causing hypernatremia include nephrogenic diabetes insipidus when the kidney fails to respond to vasopressin, in diuretic therapy, or by salt poisoning. Central diabetes insipidus due to failure of the hypothalamus to make vasopressin may occur as a consequence of cerebrovascular diseases, ischemia, and head trauma, resulting in hypernatremia. In all cases, the brain cell water content decreases as a consequence of the osmotic gradient between blood and brain. This is then followed by an adaptive response of brain, consisting of the accumulation of osmotically active intracellular solutes, decreasing the osmotic gradient and tending to restore brain water. The brain osmolytes involved in the volume regulatory increase are K$^+$, Cl$^-$, and Na$^+$, as well as a number of organic molecules (Heilig et al., 1989).

The contribution of organic osmolytes is particularly important in chronic hypernatremia. The same three major groups of organic molecules involved in the adjustment of brain water content in hyponatremia, i.e., polyols, trimethylamines, amino acids and their derivatives, serve this function of osmolytes in hypernatremia. Myoinositol, betaine, and glycerophosphoryl choline rise in brain in moderate or severe hypernatremia, while betaine only accumulates in severe hypernatremia (Lien et al., 1990). The change in the brain amino acid content in hypernatremia has been examined in chronic or acute hypernatremic condition. A study in rats showed that amino acid levels are not substantially modified in acute hypernatremia, but in the chronic moderate condition, there is a substantial increase in glutamate, glutamine, and taurine of 45%, 70%, and 39%, respectively (❯ *Figure 10-7a*). In severe chronic hypernatremia, the amino acid increase is of 84%, 143%, and 78%, respectively (Lien et al., 1990). In some animal models, the brain taurine increase accounts for as much as 50% of the osmolytes required for brain volume adjustment.

4.1 Mechanism of the Compensatory Increase in Amino Acid Levels During Hypernatremia

4.1.1 Hypertonicity and Osmolyte Transporters

The increase in brain content of organic osmolytes, including amino acids, may occur by an osmoregulation of the transport or/and by increasing biosynthesis. The first studies on the mechanisms of hyperosmotic regulation of organic osmolyte transporters were carried out in renal cells (García-Pérez and Bourg, 1991), but other studies soon followed, showing similar mechanisms in brain cells, particularly in astrocytes. Hyperosmolarity-induced increased levels and transporter activity in astrocytes have been described mainly for myoinositol (Strange et al., 1991; Paredes et al., 1992; Isaacks et al., 1994), betaine (García-Pérez and Bourg, 1991), and taurine (Olson and Goldfinger, 1990; Beetsch and Olson, 1996) (❯ *Figure 10-7b*). In general, hyperosmolarity increases V_{max} without significant changes in K_m, suggesting an increase in the number of transporters. The effect of hyperosmolarity on the specific Na$^+$-dependent transporter for taurine, TAUT, has been studied in detail. TAUT is specific for the β-amino acids taurine, hypotaurine, and β-alanine. It is a 620 amino acid protein of a molecular weight of 70 kDa, 12 transmembrane domains, and more than 90% homology in mammalian cells. The molecule requires two Na$^+$ ions and one Cl$^-$ to transport one taurine molecule (Tappaz, 2004). The transporter has several putative consensus sites for phosphorylation by PKC and protein kinase A (PKA). Increased expression of TAUT by hyperosmotic conditions, and a resultant increase in taurine levels by a more efficient accumulation, has been found with a rather similar pattern in numerous mammalian cell types, including hepatoma cells, endothelial cells from liver, aorta, and brain capillaries, corneal and lens epithelium, cultured astrocytes as well as in retinal cells

◻ **Figure 10-7**
Increase of osmolyte levels in brain of hypernatremic rats. (a) Change in the concentration of the main organic osmolytes in the brain of rats during moderate chronic hypernatremia. Abbreviations: Glu, glutamate; Cr, creatine; Gln, glutamine; myo-I, myoinositol. (b) Increased taurine content, taurine transport, and TAUT gene expression induced by hyperosmolarity in rat brain cultured astrocytes. (a) Results are expressed as percentage increase in each case of the hyperosmotic over the normosmotic condition. Data are from Lien et al. (1990), Olson et al. (1996), and Bitoun and Tappaz (2000)

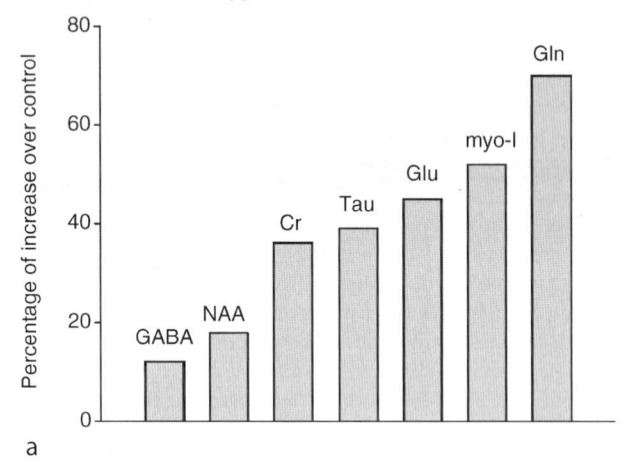

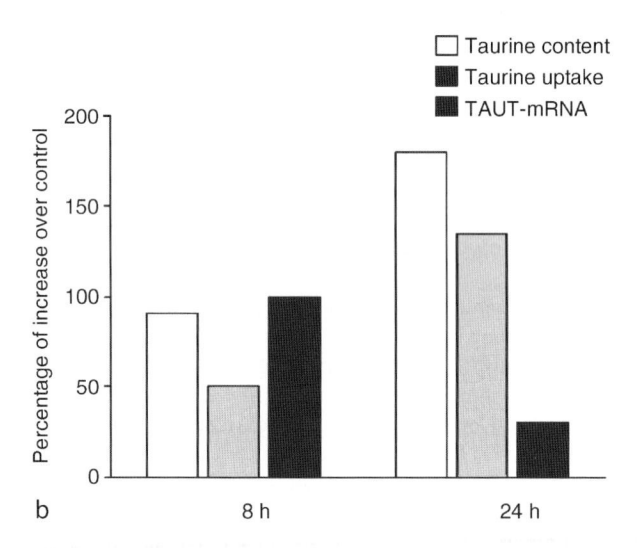

(Tappaz, 2004). The hypertonicity-induced expression of TAUT seems to result from the transcription activation of osmosensitive genes. Genes responsive to hyperosmolarity have been identified for the myoinositol and the betaine/GABA transporters (Uchida et al., 1992; Burg et al., 1997). The 5′-flanking region of these genes contains a short 11 base pair consensus sequence considered as a tonicity-responsive enhancer element (TonE) whose mutation or deletion prevents the hypertonicity-induced transcription of the osmosensitive genes (Takenaka et al., 1994; Ferraris et al., 1999). Furthermore, a protein named TonEBP, which is induced by hypertonicity, is acting as the main transcriptional activator of the osmosensitive genes by binding to TonE (Ferraris et al., 1999; Miyakawa et al., 1999). The osmosensitivity of the TAUT gene has not yet been conclusively demonstrated, but a sequence in the 5′-flanking region showing a sequence that fits the functional consensus sequence for TonE (Han et al., 2000) suggest a mechanism of gene osmoregulation similar to that found for other organic osmolyte transporters. TonEBP is rapidly and strongly overexpressed in the nucleus of neurons (Loyher et al., 2004) in hippocampus and cerebral cortex following acute systemic hypertonicity. Interestingly, the tonicity expression of TonEBP in neurons is not bidirectionaly regulated since hyposmolarity does not result in a reduction in its expression. This study showed that TonEBP is very weakly expressed in nonneuronal cells, including astrocytes. Since a robust hypertonicity-induced expression of organic osmolyte transporters, including TAUT, has been found in astrocytes, this observation is puzzling and requires further studies on the time course and specific features of the osmosensitivity of the transporter genes in astrocytes. Osmoregulation of the EAAC1 glutamate transporter has been found in a line of renal cells NBL-1, with features similar to those described for TAUT and the other organic osmolyte transporters, i. e., a large increase in EAAC1 mRNA levels and in the immunoreactive EAAC1 protein (McGivan and Nicholson, 1999). However, these adaptative changes may be restricted to renal cells, not being observed in several other cell types, including possibly the brain cells.

The hyperosmotic regulation of TAUT seems to be modulated by Ca^{2+} and the Ca^{2+}/calmodulin chain. A study in the human intestinal cell line Caco-2 shows a significant inhibition of TAUT induction by Ca^{2+} chelators or by blockers of Ca^{2+}/calmodulin, and its total suppression by Ca^{2+} blockers (Tappaz, 2004). Protein tyrosine kinases may also participate. Hypertonicity has been shown to induce at least three MAP kinase pathways: (1) the ERK pathway which is rapidly activated by hypertonicity and could extend the signal cascade to the nucleus, (2) the JNK pathway which has c-jun and AP-1 transcription factors as pathway elements, and (3) the p38 pathway. A specific role for these signaling pathways in the regulation of gene expression for organic osmolytes/TAUT transporters has not yet been ascribed. Heat shock proteins may also be involved since gene expression for those proteins is activated by hypertonicity but link with TAUT has not yet been established (Tappaz, 2004). All these enzymes may be involved in the activation of events associated with the cell volume change, which are numerous and complex, involving cell responses such as cytoskeleton organization, adhesion reactions, and stress.

4.1.2 Hypertonicity and Osmolyte Biosynthesis

The intracellular increase of organic osmolytes as a mechanism to counteract cell shrinkage may be achieved also by adaptive changes in the rate of synthesis. This is known to occur in renal cells, where the increase in sorbitol during hypernatremia results essentially from the enhanced activity of aldose reductase, the enzyme forming sorbitol from glucose (García-Perez and Bourg, 1991), following an effect of hypertonicity on the aldose reductase gene transcription (Ferraris et al., 1994). The question is whether a change in biosynthesis also contributes to the increase of other organic osmolytes, in particular that of taurine. Taurine biosynthesis in the nervous tissue occurs from the precursor cysteine to cysteinesulfinate and hypotaurine. The key enzyme in this biosynthetic route is the cysteinesulfinate decarboxylase. A study in cultured astrocytes from rat brain has shown that taurine synthesis is enhanced by hypertonicity (Beetsch and Olson, 1998). However, hypertonicity did not increase the mRNA levels of either cysteine dehydrogenase or cysteinesulfinate decarboxylase, suggesting that the enzyme genes may not be tonicity-sensitive (Tappaz, 2004). Moreover, the hypertonicity-evoked increase in cell taurine content is essentially dependent on the presence of the amino acid in the extracellular milieu. Furthermore, the activity of cysteinesulfinate decarboxylase, the rate limiting enzyme for taurine biosynthesis is very small in the human brain, thus making marginal if any, the contribution of this mechanism for taurine adaptation to hypertonicity in humans.

5 Proliferation, Apoptosis and Cell Volume

5.1 Proliferation

Cells have to reach a certain size before initiating division. Cell size is attained after an interplay of events during the various phases of the cell cycle. Mitogens and growth factors stimulate the activity or expression of carriers and channels involved in nutrient transport and in numerous other aspects of the dynamics of the growing cell. This finally results in accumulation of a variety of molecules, some of them acting as osmolytes and driving water influx, changing the concentration of intracellular elements involved in the cell cycle progression, and affecting the rate of cell proliferation. The increase in cell volume may, in turn, act as trigger for growth factor receptor activation (Franco et al., 2004b; Lezama et al., 2005), thus recycling the chain of volume-related proliferative events The cellular ion content resulting from these events seems to play a key role in proliferation as pointed out by the potent effect of Cl^- channel blockers modulating proliferation in different cell types (Voets et al., 1995; Lang et al., 2000; Wondergem et al., 2001). This effect may be related either to a decrease in Cl^- influx reducing swelling or to a specific requirement of Cl^- channel functional expression at some point of the proliferation cycle. As mentioned in the above sections, Cl^- channel blockers are potent inhibitors of the volume-sensitive efflux of organic osmolytes, notably amino acids, thus raising the question of whether these osmolytes are participating in the proliferative process. Studies on the effect of taurine addition show either inhibition of cell proliferation as in hepatic stellate cells and aortic vascular smooth muscle cells (Imada et al., 2003; Chen et al., 2004) or increasing or restoring proliferation as in human fetal neurons (Chen et al., 1998) or fetal pancreatic cells (Boujendar et al., 2002). The effect of decreasing intracellular taurine levels on cell volume and proliferation rate has not been examined.

In physiological conditions, taurine may play a role in the homeostasis of the growing and proliferating cells. This is suggested by the markedly higher taurine concentration found in the developing brain as compared to the adult brain, the difference being up to fivefold higher taurine levels in the immature brain (Sturman and Gaull, 1975). The reason for this difference has not been fully explained. Acting as an osmolyte, taurine may contribute to maintain the cell size required for the progress of proliferation. As development progresses, there is an increase in protein expression and a parallel decrease in taurine levels, giving some support to this possible role of taurine in regulating the cell size in face of the continuous changes in metabolite cell content occurring during development. Taurine deficiency is known to impair the maturation and subsequent migration of neurons in cerebellum and visual cortex by a still unknown mechanism (Sturman et al., 1985; Neuringer et al., 1990). The sequencial steps of brain ontogeny include neuronal proliferation, maturation, migration, differentiation, and synaptogenesis. In the taurine-deficient cats, the presence of numerous mitotic figures at a time in which they are not found anymore in the normal cats suggests a delay in the process of proliferation. This delayed proliferation occurring when taurine levels are lower than normal, may fit the hypothesis discussed above of its requirement to attain the cell volume necessary for a normal cell proliferation rate.

5.2 Apoptosis

A decrease in cell volume has been always recognized as a hallmark of apoptosis, but it is only recently that this volume reduction as been considered as an element in the signaling mechanisms of the program for cell death. Cell shrinkage normally precedes apoptotic events such as cytochrome c release, caspase-3 activation, nuclear condensation, and DNA fragmentation (Maeno et al., 2000). The mechanisms of cell shrinkage during apoptosis include K^+ and Cl^- loss through activated K^+ and Cl^- channels. It is then likely that experimental manipulation of the intracellular concentration of these ions, including channel blockade, exert a profound influence on apoptosis (Maeno et al., 2000; Bortner and Cidlowski, 2004). The question is raised of whether organic osmolytes including amino acids may have a similar role in modulating apoptosis. The release of taurine in connection with apoptosis has been reported in cerebellar granule neurons grown in the absence of depolarizing K^+ and in Jurkat lymphocytes after stimulation of the

CD95-receptors (Lang et al., 1998b; Moran et al., 2000) (▶ *Figure 10-8*). Taurine efflux in these cells is not reduced but rather notably increased by Cl⁻ channel blockers (▶ *Figure 10-8*) and is insensitive to tyrosine kinase blockers, in contrast to the diffusive pathway activated by swelling. The Na^+ and temperature

▣ **Figure 10-8**
Taurine efflux from apoptotic cerebellar granule neurons (CGN) and Jurkat lymphocytes. Apoptotic death was induced in CGN by growing cultures in K^+ 5 mM instead of the depolarizing K^+ concentrations (K^+ 25 mM), required for cell survival, and in Jurkat lymphocytes by activation of the Fas (CD95) receptor. (a) Evolution of apoptosis followed by activation of caspase-3 in CGN. (b) Taurine efflux from CGN at different days in culture in apoptosis-inducing conditions. Taurine efflux was followed by the labeled tracer [³H]taurine. The efflux was measured at the indicated time in culture, and the efflux rate was obtained from samples collected every 30 min during 5 h. (c) Effect of Na^+ or Cl⁻ omission and of Cl⁻ channel blockers on the efflux of taurine from CGN and Jurkat cells. In Na^+-free or Cl⁻-free media, NaCl was replaced by choline chloride or by Na^+-gluconate, respectively. Bars represent the percentage change in taurine efflux with respect to efflux control cells not exposed to the apoptotic conditions. Data are from Lang et al. (1998a, b) and Moran et al. (2000)

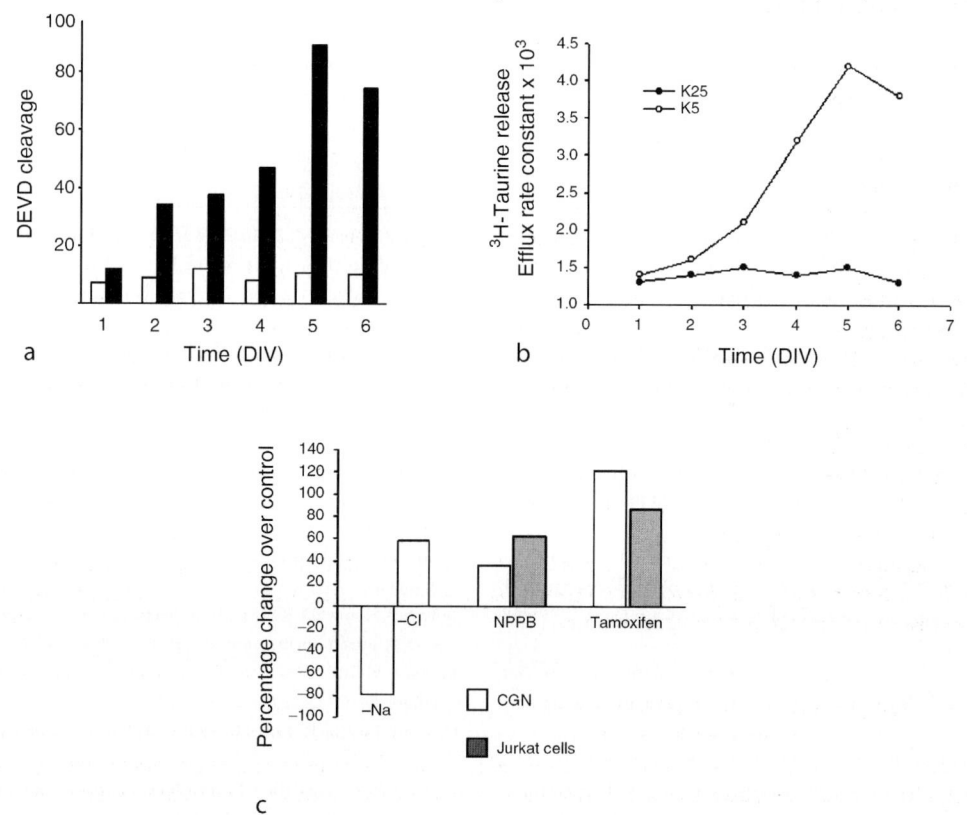

dependence of apoptotic taurine release suggests the involvement of a carrier-mediated transport. In cerebellar granule neurons, apoptosis increases also the efflux of glutamate and GABA, which may then also contribute to cell shrinkage.

Effects of manipulating K^+ and Cl⁻ cell content on apoptosis appear mediated by two main events of the process, i.e., the caspase pathway and the mitochondrial cytochrome *c* release (Maeno et al., 2000;

Bortner and Cidlowski, 2004). Taurine has also some effects attenuating apoptosis induced by different factors in several cell types (Verzola et al., 2002). In the ischemia-induced apoptosis in cardiomyocytes, taurine affects the caspase pathway, and inhibits the assembly of the Apaf-1/caspase-9 apoptosome (Takatani et al., 2004), but has no influence on the mitochondrial potential and cytochrome c release. As yet, there is not sufficient information to clarify whether the effect of taurine attenuating apoptosis is linked to a reduction of the apoptotic-activated taurine efflux or to an effect preventing cell shrinkage. The concentration of external taurine used in most studies, of about 20–40 mM, would presumably block the efflux of intracellular taurine and possibly also the apoptotic Cl^- efflux and might thus prevent or reduce the cell shrinkage.

Taurine efflux, cell shrinkage, and apoptosis may be associated with the photoreceptor death known to occur during taurine deficiency. This link is suggested by the photoreceptor appearance observed in early studies of taurine-deficiency–induced retinal degenerations, showing the typical shrinkage preceding apoptotic cell death (Lake and Malik, 1987). A recent report, which seems to confirm this possibility, shows a severe and progressive photoreceptor apoptotic death in TAUT knockout mice (Heller-Stilb et al., 2002).

Acknowledgments

Work in the author's laboratory has been supported by grants No. 46465 from CONACYT and IN209507 from DGAPA, UNAM.

References

Allen NJ, Rossi DJ, Attwell D. 2004. Sequential release of GABA by exocytosis and reversed uptake leads to neuronal swelling in simulated ischemia of hippocampal slices. J Neurosci 24: 3837-3849.

Arundine M, Tymianski M. 2004. Molecular mechanisms of glutamate-dependent neurodegeneration in ischemia and traumatic brain injury. Cell Mol Life Sci 61: 657-668.

Baker EA, Tian Y, Adler S, Verbalis JG. 2000. Blood-brain barrier disruption and complement activation in the brain following rapid correction of chronic hyponatremia. Exp Neurol 165: 221-230.

Baraban SC, Schwartzkroin PA. 1998. Effects of hyposmolar solutions on membrane currents of hippocampal interneurons and mossy cells in vitro. J Neurophysiol 79: 1108-1112.

Beetsch JW, Olson JE. 1996. Hyperosmotic exposure alters total taurine quantity and cellular transport in rat astrocyte cultures. Biochim Biophys Acta 1290: 141-148.

Beetsch JW, Olson JE. 1998. Taurine synthesis and cysteine metabolism in cultured astrocytes: Effects of hyperosmotic exposure. Am J Physiol Cell Physiol 274: C866-C874.

Benveniste H, Drejer J, Schousboe A, Diemer NH. 1984. Elevation of the extracellular concentrations of glutamate and aspartate in rat hippocampus during transient cerebral ischemia monitored by intracerebral microdialysis. J Neurochem 43: 1369-1374.

Berl T. 1990. Treating hyponatremia: Damned if we do and damned if we don't. Kidney Int 37: 1006-1018.

Bitoun M, Tappaz M. 2000. Gene expression of the transporters and biosynthetic enzymes of the osmolytes in astrocyte primary cultures exposed to hyperosmotic conditions. Glia 32: 165-176.

Bortner CD, Cidlowski JA. 2004. The role of apoptotic volume decrease and ionic homeostasis in the activation and repression of apoptosis. Eur J Physiol Pflugers Arch 448: 313-318.

Boujendar S, Reusens B, Merezak S, Ahn MT, Arany E, et al. 2002. Taurine supplementation to a low protein diet during foetal and early postnatal life restores a normal proliferation and apoptosis of rat pancreatic islets. Diabetologia 45: 856-866.

Burg MB, Know ED, Kultz D. 1997. Regulation of gene expression by hypertonicity. Annu Rev Physiol 59: 437-455.

Butterworth RF. 2003. Molecular neurobiology of acute liver failure. Semin Liver Dis 23: 251-258.

Cardin V, Lezama R, Torres-Márquez ME, Pasantes-Morales H. 2003. Potentiation of the osmosensitive taurine release and cell volume regulation by cytosolic Ca rise in cultured cerebellar astrocytes. Glia 44: 119-128.

Chebabo SR, Hester MA, Aitken PG, Somjen GG. 1995. Hypotonic exposure enhances synaptic transmission and triggers spreading depression in rat hippocampal tissue slices. Brain Res 695: 203-216.

Chen XC, Pan ZL, Liu DS, Han X. 1998. Effect of taurine on human fetal neuron cells: Proliferation and differentiation. Adv Exp Med Biol 442: 397-403.

Chen YX, Zhang XR, Xie WF, Li S. 2004. Effects of taurine on proliferation and apoptosis of hepatic stellate cells in vitro. Hepatobiliary Pancreat Dis Int 3: 106-109.

Davies SE, Gotoh M, Richards DA, Obrenovitch TP. 1998. Hypoosmolarity induces an increase of extracellular N-acetylaspartate concentration in the rat striatum. Neurochem Res 23: 1021-1025.

Deleuze C, Duvoid A, Moos FC, Hussy N. 2000. Tyrosine phosphorylation modulates the osmosensitivity of volume-dependent taurine efflux from glial cells in the rat supraoptic nucleus. J Physiol 523: 291-299.

Estevez AY, O'Regan MH, Song D, Phillis JW. 1999. Effects of anion channel blockers on hyposmolitally induced amino acid release from the in vivo rat cerebral cortex. Neurochem Res 24: 447-452.

Felipo V, Butterworth RF. 2002. Neurobiology of ammonia. Prog Neurobiol 67: 259-279.

Ferraris JD, Williams CK, Martin BM, Bourg MB, García-Pérez A. 1994. Cloning, genomic organization and osmotic response of the aldose reductase gene. Proc Natl Acad Sci USA 91: 10742-10746.

Ferraris JD, Williams CK, Ohtaka A, Garcia-Perez A. 1999. Functional consensus for mammalian osmotic response elements. Am J Physiol Cell Physiol 276: C667-C673.

Feustel PJ, Jin Y, Kimelberg HK. 2004. Volume-regulated anion channels are the predominant contributors to release of excitatory amino acids in the ischemic cortical penumbra. Stroke 35: 1164-1168.

Franco R, Torres-Márquez ME, Pasantes-Morales H. 2001. Evidence of two mechanisms for amino acid osmolyte release from hippocampal slices. Eur J Physiol Pflugers Arch 442: 791-800.

Franco R, Rodríguez R, Pasantes-Morales H. 2004a. Mechanisms of the ATP potentiation of hyposmotic taurine release in Swiss 3T3 fibroblasts. Eur J Physiol Pflugers Arch 449: 159-169.

Franco R, Lezama R, Ordaz B, Pasantes-Morales H. 2004b. Epidermal growth factor receptor is activated by hyposmolarity and is an early signal modulating osmolyte efflux pathways in Swiss 3T3 fibroblasts. Eur J Physiol Pflugers Arch 447: 830-839.

García-Perez A, Bourg MB. 1991. Renal medullary organic osmolytes. Physiol Rev 71: 1081-1115.

Godart H, Ellory JC, Motais R. 1999. Regulatory volume response of erythrocytes exposed to a gradual and slow decrease in medium osmolality. Eur J Physiol Pflugers Arch 437: 776-779.

Han X, Budreau AM, Chesney RW. 2000. Cloning and characterization of the promoter region of the rat taurine transporter (TauT) gene. Adv Exp Med Biol 483: 97-108.

Hansson E, Muyderman H, Isonova J, Allanson L, Sinclair J, et al. 2000. Astroglia and glutamate in physiology and pathology: Aspects on glutamate transport, glutamate-induced cell swelling and gap-junction communication. Neurochem Int 37: 317-329.

Haussinger D, Kircheis G, Fischer R, Schliess F, vom Dahl S. 2000. Hepatic encephalopathy in chronic liver disease: A clinical manifestation of astrocyte swelling and low-grade cerebral edema? J Hepatol 32: 1035-1038.

Heilig CW, Stromski ME, Blumenfeld JB, Lee JP, Gullans SR. 1989. Characterization of the major brain osmolytes that accumulate in salt-loaded rats. Am J Physiol Renal Physiol 26: F1108-F1116.

Heller-Stilb B, van Roeyen C, Rascher K, Hartwig HG, Huth A, et al. 2002. Disruption of the taurine transporter gene (taut) leads to retinal degeneration in mice. FASEB J 16: 231-233.

Hoffmann EK, Lambert IH, Simonsen LO. 1988. Mechanisms in volume regulation in Ehrlich ascites tumor cells. Renal Physiol Biochem 11: 221-247.

Hubert EM, Musch MW, Goldstein L. 2000. Inhibition of volume-stimulated taurine efflux and tyrosine kinase activity in the skate red blood cell. Eur J Physiol Pflugers Arch 440: 132-139.

Hussy N, Deleuze C, Bres V, Moos FC. 2000. New role of taurine as an osmomediator between glial cells and neurons in the rat supraoptic nucleus. Adv Exp Med Biol 483: 227-237.

Hussy N, Bres V, Rochette M, Duvoid A, Alonso G, et al. 2001. Osmoregulation of vasopressin secretion via activation of neurohypophysial nerve terminals glycine receptors by glial taurine. J Neurosci 21: 7110-7116.

Imada K, Hosokawa Y, Terashima M, Mitani T, Tanigawa Y, et al. 2003. Inhibitory mechanism of taurine on the platelet-derived growth factor BB-mediated proliferation in aortic vascular smooth muscle cells. Adv Exp Med Biol 526: 5-15.

Isaacks RE, Bender AS, Kim CY, Prieto NM, Norenberg MD. 1994. Osmotic regulation of myo-inositol uptake in primary astrocyte cultures. Neurochem Res 19: 331-338.

Kimelberg HK, Goderie SK, Higman S, Pang S, Waniewsky RA. 1990. Swelling-induced release of glutamate, aspartate, and taurine from astrocyte cultures. J Neurosci 10: 1583-1591.

Kirk K. 1997. Swelling-activated organic osmolyte channels. J Membr Biol 158: 1-6.

Kontro P, Marnela K-M, Oja SS. 1980. Free amino acids in the synaptosome and synaptic vesicle fractions of different bovine brain areas. Brain Res 184: 129-141.

Lake N, Malik N. 1987. Retinal morphology in rats treated with a taurine transport antagonist. Exp Eye Res 44: 331-346.

Lambert IH. 2004. Regulation of the cellular content of the organic osmolyte taurine in mammalian cells. Neurochem Res 29: 27-63.

Lang F, Busch GL, Ritter M, Völki H, Waldegger S, et al. 1998a Functional significance of cell volume regulatory mechanisms. Physiol Rev 78: 247-306.

Lang F, Madlung J, Uhlemann AC, Risler T, Gulbins E. 1998b. Cellular taurine release triggered by stimulation of the Fas (CD95) receptor in Jurkat lymphocytes. Eur J Physiol Pflugers Arch 436: 377-383.

Lang F, Ritter M, Gamper N, Huber S, Fillon S. 2000. Cell volume in the regulation of cell proliferation and apoptotic cell death. Cell Physiol Biochem 10: 417-428.

Law RO. 1994. Taurine efflux and the regulation of cell volume in incubated slices of rat cerebral cortex. Biochim Biophys Acta 1221: 21-28.

Lehmann A. 1989. Effects of microdialysis-perfusion with anisoosmotic media on extracellular amino acids in the rat hippocampus and skeletal muscle. J Neurochem 53: 525-535.

Lezama R, Ortega A, Ordaz B, Pasantes-Morales H. 2005. Hyposmolarity-induced ErbB4 phosphorylation and its influence on the non-receptor tyrosine kinase network response in cultured cerebellar granule neurons. J Neurochem 93: 1189-1198.

Lien YH, Shapiro JI, Chan L. 1990. Effects of hypernatremia on organic brain osmoles. J Clin Invest 85: 1427-1435.

Lohr JW, Grantham JJ. 1986. Isovolumetric regulation of isolated S2 proximal tubules in anisotonic media. J Clin Invest 78: 1165-1672.

Loyher ML, Mutin M, Woo SK, Kwon HM, Tappaz ML. 2004. Transcription factor tonicity-responsive enhancer-binding protein (TonEBP) which transactivates osmoprotective genes is expressed and upregulated following acute systemic hypertonicity in neurons in brain. Neuroscience 124: 89-104.

Maeno E, Ishizaki Y, Kanaseki T, Hazama A, Okada Y. 2000. Normotonic cell shrinkage because of disordered volume regulation is an early prerequisite to apoptosis. Proc Natl Acad Sci USA 97: 9487-9492.

Martinez-Hernandez A, Bell KP, Norenberg MD. 1997. Glutamine synthetase: Glial localization in brain. Science 195: 1356-1358.

Massieu L, Montiel T, Robles G, Quesada O. 2004. Brain amino acids during hyponatremia in vivo: Clinical observations and experimental studies. Neurochem Res 29: 73-82.

McGivan JD, Nicholson B. 1999. Regulation of high affinity glutamate transport by amino acid deprivation and hyperosmotic stress. Am J Physiol Renal Physiol 277: F498-F500.

Miyakawa H, Woo SK, Dahl SC, Handler JS, Kwon HM. 1999. Tonicity-responsive enhancer binding protein, a rel-like protein that stimulates transcription in response to hypertonicity. Proc Natl Acad Sci USA 96: 2538-2542.

Mongin AA, Kimelberg HK. 2005. ATP regulates anion channel-mediated organic osmolyte release from cultured rat astrocytes via multiple Ca^{2+}-sensitive mechanisms. Am J Physiol Cell Physiol 288: C204-C213.

Mongin AA, Reddi JM, Charniga C, Kimelberg HK. 1999. [^{3}H]Taurine and D-[^{3}H]aspartate release from astrocyte cultures are differently regulated by tyrosine kinases. Am J Physiol Cell Physiol 276: C1226-C1230.

Moorman JR, Jones L. 1998. Phospholemman: A cardiac taurine channel involved in regulation of cell volume. Adv Exp Med Biol 442: 219-228.

Moran J, Hernandez-Pech X, Merchant-Larios H, Pasantes-Morales H. 2000. Release of taurine in apoptotic cerebellar granule neurons in culture. Eur J Physiol Pflugers Arch 439: 271-277.

Moran J, Morales-Mulia M, Pasantes-Morales H. 2001. Reduction of phospholemman expression decreases osmosensitive taurine efflux in astrocytes. Biochim Biophys Acta 1538: 313-320.

Mori SI, Morishima S, Takasaki M, Okada Y. 2002. Impaired activity of volume-sensitive anion channel during lactacidosis-induced swelling in neuronally differentiated NI08–15 cells. Brain Res 957: 1-11.

Nelson RM, Lambert DG, Green RA, Hainsworth AH. 2003. Pharmacology of ischemia-induced glutamate efflux from rat cerebral cortex in vitro. Brain Res 964: 1-8.

Neuringer M, Palackal T, Kujawa M, Moretz RC, Sturman JA. 1990. Visual cortex development in rhesus monkeys deprived of dietary taurine. Prog Clin Biol Res 351: 415-422.

Niemeyer MI, Cid LP, Barros LF, Sepulveda FV. 2001. Modulation of the two pore domain acid-sensitive K^+ channel TASK-2 (KCNK5) by changes in cell volume. J Biol Chem 16: 43166-43174.

Nilius B, Eggermont J, Voets T, Buyse G, Manolopoulos V, et al. 1997. Properties of volume-regulated anion channels in mammalian cells. Prog Biophys Mol Biol 68: 69-119.

Nishizawa Y. 2001. Glutamate release and neuronal damage in ischemia. Life Sci 69: 369-381.

Norenberg MD, Jayakumar AR, Rama Rao KV. 2004. Oxidative stress in the pathogenesis of hepatic encephalopathy. Metab Brain Dis 19: 313-329.

Olson JE. 1999. Osmolyte contents of cultured astrocytes grown in hypoosmotic medium. Biochim Biophys Acta 1453: 175-179.

Olson JE, Goldfinger MD. 1990. Amino-acid content of rat cerebral astrocytes adapted to hyperosmotic medium in vitro. J Neurosci Res 27: 241-246.

Ordaz B, Tuz K, Ochoa LD, Lezama R, Peña-Segura C, et al. 2004a. Osmolytes and mechanisms involved in regulatory volume decrease under conditions of sudden or gradual osmolarity decrease. Neurochem Res 29: 65-72.

Ordaz B, Vaca L, Franco R, Pasantes-Morales H. 2004b. Volume changes and whole cell membrane currents activated during gradual osmolarity decrease in C6 glioma cells: Contribution of two types of K^+ channels. Am J Physiol Cell Physiol 286: C1399-C1409.

Paredes A, McManus M, Know HM, Strange K. 1992. Osmoregulation of Na^+-inositol cotransporter activity and messenger RNA levels in brain glial cells. Am J Physiol Cell Physiol 263: C1282-C1288.

Pasantes-Morales H. 1996. Volume regulation in brain cells: Cellular and molecular mechanisms. Metab Brain Dis 11: 187-204.

Pasantes-Morales H, Franco R. 2002. Influence of tyrosine kinases on cell volume change-induced taurine release. Cerebellum 1: 103-109.

Pasantes-Morales H, Franco R. 2005. Astrocyte cellular swelling: Mechanisms and relevance to brain edema. The Role of Glia in Neurotoxicity. Aschner M, Costa L editors. Boca Raton: CRC-Press; pp. 173-190.

Pasantes-Morales H, Morales Mulia S. 2000. Influence of calcium on regulatory volume decrease: Role of potassium channels. Nephron 86: 414-427.

Pasantes-Morales H, Schousboe A. 1988. Volume regulation in astrocytes: A role for taurine as osmoeffector. J Neurosci Res 20: 505-509.

Pasantes-Morales H, Schousboe A. 1997. Role of taurine in osmoregulation: Mechanisms and functional implications. Amino Acids 12: 281-293.

Pasantes-Morales H, Alavez S, Sanchez-Olea R, Moran J. 1993. Contribution of organic and inorganic osmolytes to volume regulation in rat brain cells in culture. Neurochem Res 18: 445-452.

Perlman DF, Goldstein L. 2004. The anion exchanger as an osmolyte channel in the skate erythrocyte. Neurochem Res 29: 9-15.

Phillis JW, O'Regan MH. 2003. Characterization of modes of release of amino acids in the ischemic/reperfused rat cerebral cortex. Neurochem Int 43: 461-467.

Rossi DJ, Oshima T, Attwell D. 2000. Glutamate release in severe brain ischaemia is mainly by reversed uptake. Nature 403: 316-321.

Saransaari P, Oja SS. 1999. Taurine release is enhanced in cell-damaging conditions in cultured cerebral cortical astrocytes. Neurochem Res 24: 1523-1529.

Seki Y, Feustel PJ, Keller RW, Trammer BI, Kimelberg HK. 1999. Inhibition of ischemia-induced glutamate release in rat striatum by dihydrokainate and an anion channel blocker. Stroke 30: 433-440.

Solis JM, Herranz AS, Herreras O, Lerma J, Martín del Río R. 1998. Does taurine act as an osmoregulatory substance in the rat brain? Neurosci Lett 91: 53-58.

Souza MM, Boyle RT, Lieberman M. 2000. Different physiological mechanisms control isovolumetric regulation and regulatory volume decrease in chick embryo cardiomyocytes. Cell Biol Int 24: 713-721.

Stover JF, Unterberg AW. 2000. Increased cerebrospinal fluid glutamate and taurine concentrations are associated with traumatic brain edema formation in rats. Brain Res 875: 51-55.

Strange K, Jackson PS. 1995. Swelling-activated organic osmolyte efflux: A new role for anion channel. Kidney Int 48: 994-1003.

Strange K, Morrison R, Heilig C, Dipietro S, Gullans M. 1991. Up-regulation of inositol transport mediates inositol accumulation in hyperosmolar brain cells. Am J Physiol 260: C780-C790.

Sturman JA, Gaull GE. 1975. Taurine in the brain and liver of the developing human and monkey. J Neurochem 25: 831-835.

Sturman JA, Moretz RM, French JH, Wisniewsky HM. 1985. Taurine deficiency in the developing cat: Persistence of the cerebellar external granule cell layer. J Neurosci Res 13: 405-416.

Stutzin A, Torres R, Oporto M, Pacheco P, Eguiguren AL, et al. 1999. Separate taurine and chloride efflux pathways activated during regulatory volume decrease. Am J Physiol Cell Physiol 277: C392-C402.

Takatani T, Takahashi K, Uozumi Y, Shikata E, Yamamoto Y, et al. 2004. Taurine inhibits apoptosis by preventing formation of the apaf-1/caspase-9 apoptosome. Am J Physiol Cell Physiol 287: C949-C953.

Takenaka M, Preston AS, Kwon HM, Handler JS. 1994. The tonicity-sensitive element that mediates increased transcription of the betaine transporter gene in response to hypertonic stress. J Biol Chem 269: 29379-29381.

Tappaz ML. 2004. Taurine biosynthetic enzymes and taurine transporter: Molecular identification and regulations. Neurochem Res 29: 83-96.

Torp R, Andine P, Hayberg H, Karagulle T, Blackstad TW, et al. 1991. Cellular and subcellular redistribution of glutamate-, glutamine- and taurine-like immunoreactivities during forebrain ischemia: A semiquantitative electron microscopic study in rat hippocampus. Neuroscience 41: 433-447.

Tuz K, Ordaz B, Vaca L, Quesada O, Pasantes-Morales H. 2001. Isovolumetric regulation mechanisms in cultured cerebellar granule neurons. J Neurochem 79: 143-151.

Tuz K, Peña-Segura C, Franco R, Pasantes-Morales H. 2004. Depolarization, exocytosis and amino acid release evoked by hyposmolarity from cortical synaptosomes. Eur J Neurosci 19: 916-924.

Uchida S, Kwon HM, Yamauchi A, Preston AS, Marumo F, et al. 1992. Molecular cloning of the cDNA for an MDCK cell Na$^+$ and Cl$^-$-dependent taurine transporter that is regulated by hypertonicity. Proc Natl Acad Sci USA 89: 8230-8234.

van der Wijk T, Tomassen SF, de Jonge HR, Tilly BC. 2000. Signalling mechanisms involved in volume regulation of intestinal epithelial cells. Cell Physiol Biochem 10: 289-296.

Van Driessche W, De Smet P, Li J, Allen S, Zizi M, et al. 1997. Isovolumetric regulation in a distal nephron cell line (A6). Am J Physiol Cell Physiol 272: C1890-C1898.

Verbalis JG, Gullans SR. 1991. Hyponatremia causes large sustained reductions in brain content of multiple organic osmolytes in rats. Brain Res 567: 274-282.

Verzola D, Bertolotto MB, Villaggio B, Ottonello L, Dallegri F, et al. 2002. Taurine prevents apoptosis induced by high ambient glucose in human tubule renal cells. J Invest Med 50: 443-451.

Voets T, Szucs G, Droogmans G, Nilius B. 1995. Blockers of volume-activated Cl$^-$ currents inhibit endothelial cell proliferation. Eur J Physiol Pflugers Arch 431: 132-134.

Walz W. 2000. Role of astrocytes in the clearance of excess extracellular potassium. Neurochem Int 36: 291.

Wehner F, Olsen H, Tinel H, Kinne-Saffran E, Kinne RKH. 2003. Cell volume regulation: Osmolytes, osmolyte transport and signal transduction. Rev Physiol Biochem Pharmacol 148: 1-80.

Wondergem R, Gong W, Monen SH, Dooley SN, Gonce JL, et al. 2001. Blocking swelling-activated chloride current inhibits mouse liver cell proliferation. J Physiol 532: 661-672.

Zielińska M, Hilgier W, Law RO, Gorynski P, Albrecht J. 1999. Effects of ammonia in vitro on endogenous taurine efflux and cell volume in rat cerebrocortical minislices: Influence of inhibitors of volume-sensitive amino acid transport. Neuroscience 91: 631-638.

11 Urea Cycle Enzymopathies

R. Butterworth

1	*Experimental Urea Cycle Enzymopathies*	251
2	*Neurotransmitter Changes in OTC Deficiency*	252
2.1	Amino Acids	252
2.2	Peripheral-Type (Mitochondrial) Benzodiazepine Receptors	253
2.3	The Cholinergic System	253
3	*Serotonin*	254
4	*Pathogenesis of Neuronal Cell Death in Congenital OTC Deficiency*	255
4.1	Brain Energy Deficit	255
4.2	Nitric Oxide	255
4.3	NMDA Receptor-Mediated Excitotoxicity	255
5	*Current Therapy in Congenital OTC Deficiency*	255
5.1	Sodium Benzoate, Phenylacetate	255
5.2	L-Carnitine	256
5.3	Acetyl-L-Carnitine	256
5.4	Liver Transplantation	256
5.5	Gene Therapy	257

© Springer-Verlag Berlin Heidelberg 2007

11 Urea cycle enzymopathies

Abstract: Inborn errors of urea cycle enzymes and transporters related to the urea cycle have been described. Ornithine transcarbamylase (OTC) deficiency, an X-linked genetic disorder is well characterised from the molecular genetic standpoint and over 100 mutations have been described. Clinical symptomatology varies according to the residual enzyme activity and may include seizures, mental retardation and cerebral palsy. Neuropathologic evaluation reveals brain atrophy, ventricular dilatation and delayed myelination. A well characterised animal model of OTC deficiency, the "sparse fur (*spf*)" mouse has been employed to study the neurochemistry of the disorder; abnormalities of the glutamatergic, cholinergic and serotoninergic neurotransmitter systems have been identified. Studies on the pathogenesis of the neuronal cell death in congenital OTC deficiency suggest that cerebral energy compromise and NMDA receptor-mediated excitotoxicity are implicated. Current therapy in congenital urea cycle disorders involves the reduction of circulating ammonia (using agents such as sodium benzoate and phenylacetate), L-carnitine administration to improve cellular energy metabolism and liver transplantation. Clinical trials using gene therapy are currently under evaluation.

List of Abbreviations: ATP, adenosine triphosphate; αKGOH, α-ketoglutarate dehydrogenase; CP, candate-puramen; CPSase, carbamyl phosphate synthetase; CSF, cerebrospinal fluid; CAT, choline actyltransferase; CoA, coenzyme A; CT, computed tomographic; FC, frontal cortex; GABA, γ-aminobutyric acid; GP, globus pallidus; 5HIAA, 5-hydroxyindoleacetic acid; MAO, monoamine oxidase; NMDA, N-methyl-D-aspartate; NOS, nitric oxide synthase; OTC, ornithine transcarbamylase; OLT, orthotopic liver transplantation; PC, parietal cortex; 5HT, serotonin; *spf, sparse fur*

The major pathway for ammonia detoxification in mammalian systems depends on its conversion to urea in the liver by a series of enzymatic reactions shown schematically in ❯ *Figure 11-1*. Although a complete urea cycle is only expressed in the liver, other tissues including the brain may express some of the constituent enzymes (❯ *Table 11-1*). Inborn errors of six constituent enzymes and two transporters related to the urea cycle have been described (Brusilow and Horwich, 1995). The prevalence of these disorders in the USA is approximately 1 in 8,000. All congenital disorders of the urea cycle are inherited as autosomal

◘ **Figure 11-1**
Schematic representation of the urea cycle showing intra- and extramitochondrial localization of constituent enzymes, their substrate cofactors, and the ornithine transporter

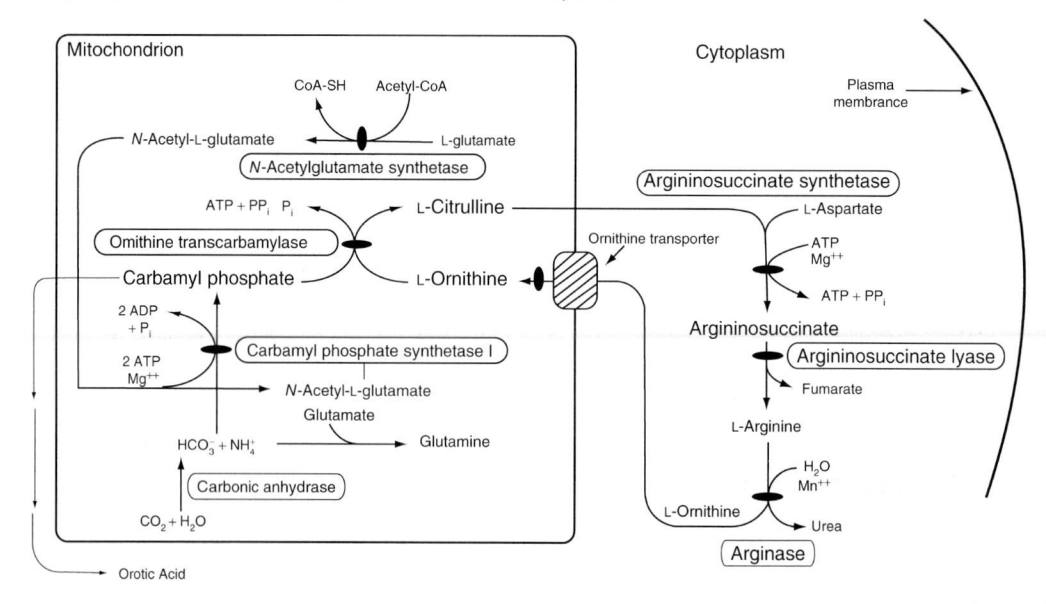

◘ Table 11-1
Constituent enzymes of the urea cycle[a]

Enzyme	Compartment	Activity (μmol/h/g)	Tissue distribution
Argininosuccinate synthetase EC 6.3.4.5	Cytoplasm	90	Liver, kidney, brain (trace)
Arginosuccinate lyase EC 4.3.2.1	Cytoplasm	220	Liver, kidney
Arginase EC 3.5.3.1	Cytoplasm	86,600	Liver, kidney, brain (trace)
N-acetylglutamate synthetase EC 2.3.11	Mitochondrial matrix	0.30–1.49	Liver, intestine
Carbamyl phosphate synthetase EC 6.3.4.16	Mitochondrial matrix	279	Liver, intestine
Ornithine transcarbamylase EC 2.1.3.3	Mitochondrial matrix	6,600	Liver, intestine

[a]Enzyme activity is expressed as micromoles per hour per gram wet weight for human liver

recessive traits with the exception of ornithine transcarbamylase (OTC) deficiency, which is X-linked. OTC deficiency is the best studied of the urea cycle disorders from the point of view of molecular genetics, pathophysiology, and treatment. OTC is a mitochondrial enzyme that catalyzes the conversion of ornithine and carbamyl phosphate to citrulline (❯ *Figure 11-1*). More than 100 mutations have been described in families with OTC deficiency where large deletions are encountered in 8% of patients, and small deletions/insertions in a further 10%. The remaining mutations are single-base substitutions (Tuchman et al., 1996). Many hemizygous males die (presumably) from severe ammonia intoxication during the neonatal period.

Heterozygous females manifest a variable clinical course (Brusilow and Horwich, 1995). A recent study in a 2-year-old female patient revealed OTC deficiency presenting with strokelike episodes (Christodoulou et al., 1993). Depending on the residual enzyme activity and the degree of hyperammonemia, infants and children with congenital OTC deficiency present with a spectrum of clinical symptoms, including failure to thrive, lethargy, and hypotonia. Seizures are not uncommon and a large percentage of survivors go on to manifest severe developmental disabilities, including mental retardation and cerebral palsy (Msall et al., 1984). Neuropathology in congenital OTC deficiency consists of atrophy, cystic degeneration, ventricular dilatation, delayed myelination, and Alzheimer type II astrocytosis (Harding et al., 1984; Dolman et al., 1988; Harper and Butterworth, 1997). Cerebral edema sufficient to lead to intracranial hypertension and brain herniation has been described in congenital OTC deficiency (Kendall et al., 1983). Computed tomographic (CT) abnormalities were reported in the brain of a 2-year-old patient with a residual liver OTC activity of 7% (Takayanagi et al., 1984). Such abnormalities included diffuse hypointensities in frontal and parietal lobes. Follow-up CT scanning showed bilateral ventricular dilatation and cerebral atrophy. It has repeatedly been suggested that such neuropathologic damage in congenital OTC deficiency may be acquired in utero (Harding et al., 1984; Filloux et al., 1986). Deficiency of carbamyl phosphate synthetase-1 (CPSase-1) results (in addition to hyperammonemia) in increases of plasma glutamine and alanine with low to absent citrulline and arginine. CPSase-1 deficiency results in high mortality; survivors have a high prevalence of mental retardation and suffer frequent hyperammonemic crises during intercurrent illness or catabolic stress. Citrullinemia results from arginosuccinate synthase deficiency.

1 Experimental Urea Cycle Enzymopathies

Experimental animal models of congenital urea cycle disorders have been described; some are spontaneous mutants while others have appeared with the advent of gene targeting paradigms. By far the most widely used and best characterized model of a congenital urea cycle disorder is the sparse-fur (*spf*) mouse with a defect in OTC. The X-linked *spf* mutation arose spontaneously in the progeny of an irradiated male mouse, at Oakridge Laboratories, USA, maintained on various genetic backgrounds (Demars et al., 1976; Qureshi, 1992). Young *spf/Y* males are characterized by their small size, relatively sparse fur, and wrinkled skin—traits that are visible as early as 7 days postnatally. The *spf* mouse harbors a structural mutation in the OTC molecule consisting of a single point substitution mutation of a cytosine to adenine within the coding region of OTC cDNA, which results in a change of amino acid 117 from a histidine to an arginine residue (Veres et al., 1987). This mutation affects the pH dependence of the enzyme and limits the enzyme's capacity to form citrulline from carbamyl phosphate and ornithine. Residual OTC activities in the livers of *spf/Y* mice are in the 10–15% range (Qureshi, 1992). Brain ammonia concentrations are increased in the *spf* mutant compared to littermate controls (● *Table 11-2*) (Batshaw et al., 1988; Ratnakumari et al., 1992) and,

◘ Table 11-2
Hepatic activities of OTC, brain ammonia, and amino acid concentrations in *spf* (OTC-deficient) mice

	Control	OTC deficient
Hepatic OTC activity (μmol/min/mg)	64.2 ± 4.0	6.3 ± 0.8*
Brain Ammonia (μmol/g)	1.41 ± 0.05	2.85 ± 0.16*
Brain Amino acids		
Ornithine (nmol/g)	11.7 ± 3.1	6.9 ± 0.7*
Citrulline (nmol/g)	12.0 ± 0.3	8.0 ± 0.1*
Arginine (nmol/g)	128.0 ± 3.1	78.0 ± 7.5*
Glutamine (μmol/g)	3.93 ± 0.7	10.11 ± 1.3*
Glutamate (μmol/g)	8.78 ± 1.0	5.91 ± 0.8*

*$p < 0.01$ compared to control values

since glutamine synthesis is the major route for ammonia removal by brain, glutamine concentrations are also increased (Batshaw et al., 1988; Ratnakumari et al., 1994a). Increased brain glutamine concentrations have also been reported using magnetic resonance spectroscopy in children with OTC deficiency (Connelly et al., 1993).

2 Neurotransmitter Changes in OTC Deficiency

2.1 Amino Acids

Glutamate concentrations in cerebral cortex of *spf* mice are significantly reduced (● *Table 11-2*) (Ratnakumari et al., 1994a), whereas concentrations of aspartate are increased in several brain regions of these animals. GABA concentrations, although unchanged in cerebral cortex of *spf* mice, are significantly increased in striatum, hippocampus, and midbrain, and it was suggested that these increases reflect increased flux through the GABA shunt pathway (Ratnakumari et al., 1994a). However, activities of the GABA synthetic enzyme, glutamic acid decarboxylase, are unchanged (Ratnakumari et al., 1993a). Glutamate (NMDA)-binding sites were evaluated using the selective ligand [^{3}H]-MK801 in the brains of *spf* mice. [^{3}H]-MK801 sites were reduced in density by up to 70% in several brain regions (Ratnakumari et al., 1995b) (● *Figure 11-2*). Whether these changes result from a loss of neurons expressing NMDA receptors or are the consequence of "downregulation" as a result of increased exposure to endogenous ligands (such as glutamate or quinolinate) is unclear, but it was suggested that the loss of NMDA receptor sites in congenital OTC deficiency could represent an adaptive mechanism of protection against further excitotoxic injury (Ratnakumari et al., 1995b). More recently it has been suggested that NMDA sites may also be expressed by

◘ Figure 11-2
Regional distribution of [^{3}H]-MK801-binding sites in the brains of control (□) and OTC-deficient (■) mice. Values represent mean ± S.E. of data from five animals per group. Values significantly differ from control indicated by *$p < 0.01$ by Student's t-test. Abbreviations: FC, frontal cortex; PC, parietal cortex; CP, caudate-putamen; GP, globus pallidus; CA1, CA2, CA3: regions of hippocampus

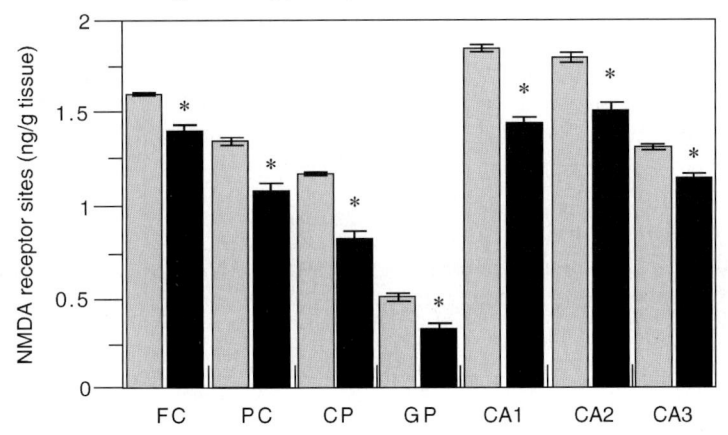

astrocytes suggesting that the loss of [^{3}H]-MK801 sites could also be the consequence of astrocytic changes in OTC deficiency.

2.2 Peripheral-Type (Mitochondrial) Benzodiazepine Receptors

"Peripheral-type" benzodiazepine receptors (PTBRs) are localized on the outer mitochondrial membrane of peripheral tissues and brain where they are predominantly astrocytic in localization. Using [^{3}H]-PK11195, a radioligand with high selectivity for the PTBR, these sites were assessed in brains and peripheral tissues of *spf* mice. Densities of [^{3}H]-PK11195 binding sites were significantly increased in brain, kidney, liver, and testes of OTC-deficient animals (Raghavendra Rao et al., 1993), and it was suggested that these increases were the result of exposure to increased ammonia concentrations. Increased expression of PTBRs in brain could also be the result of reactive astrocytosis following neuronal cell loss in congenital OTC deficiency since reactive astrocytes are known to express increased quantities of PTBRs. However, this explanation is unlikely since PTBR increases were particularly apparent in the brainstem, a brain region which shows relatively little loss of neurons in *spf* mice brains. More likely, increased densities of PTBRs reflect astrocytic changes resulting from exposure to ammonia. Putative roles for the PTBR in brain include oxidative metabolism (Anholt, 1986) and in the synthesis of neurosteroids, some of which have positive allosteric modulatory properties at GABA$_A$ receptors (Simmonds, 1991). Activation of PTBRs in the brain in OTC deficiency, therefore, could indirectly result in increased GABAergic neurotransmission.

2.3 The Cholinergic System

Ratnakumari et al. (1994b, 1995a, 1996a) performed a series of studies of the cholinergic system in OTC deficiency. Activities of choline acetyltransferase (CAT) and acetylcholinesterase (AChE) were measured in the brains of *spf/Y* mice and CD-l/Y controls. CAT activities were found to be reduced in frontal cortex, striatum, hippocampus, thalamus, and brainstem of *spf* mice (◉ *Figure 11-3*), and CAT-positive neurons were reduced in number in cerebral cortex, septal area, and diagonal band of these animals (Ratnakumari

Figure 11-3
Choline acetyltransferase activities in the brain regions of control (□) and congenitally hyperammonemic (■) *spf/Y* mice. FC, frontal cortex; CS, striatum; HC, hippocampus. Each value represents mean ± S.E. of triplicate determinations in five animals per group. Significant differences between data from *spf/Y* and control are indicated by *$p < 0.05$, **$p < 0.01$ by unpaired Student's *t*-test

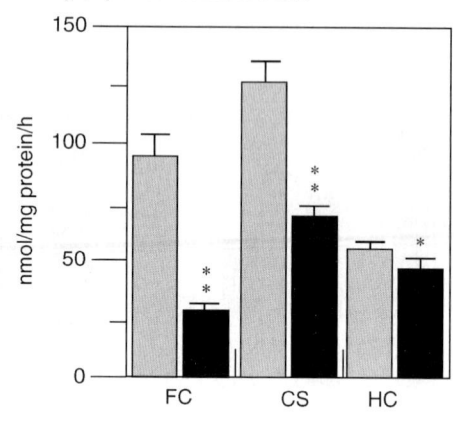

et al., 1994b). These findings are consistent with a loss of forebrain cholinergic neurons in OTC deficiency. In a follow-up study conducted at various times during postnatal development, CAT activities were found to be reduced as early as 35 days postnatally in the brainstem of *spf* mice and high-affinity [^{3}H]-choline uptake by cerebral cortical synaptosomes from *spf* mice was significantly reduced as early as 21 days postnatally (Ratnakumari et al., 1995a). Subsequent studies of muscarinic cholinergic M_1- and M_2-binding sites in the brains of *spf* mice using quantitative receptor autoradiography revealed up to 54% increased binding sites for the M_1 receptor ligand [^{3}H]-pirenzepine and a concomitant loss of binding sites for the M_2 receptor ligand [^{3}H]-AFDX384 (Ratnakumari et al., 1996a). Since the M_1 sites are predominantly postsynaptic and the M_2 sites presynaptic in localization, significant reductions of M_2 sites and concomitant increases of M_1 sites are, in the light of the previous finding of a loss of the cholinergic nerve terminal enzyme CAT, confirmative of a selective loss of cholinergic neurons in this model of congenital OTC deficiency. The increase in binding-site densities for the postsynaptic M_1 ligand [^{3}H]-pirenzepine in brain regions of *spf* mice could be the consequence of receptor upregulation following loss of presynaptic neurons, which is similar to findings in autopsied brain tissues from patients with Alzheimer's disease (Arauja et al., 1988). Basal forebrain cholinergic neuronal loss could readily explain the severe cognitive dysfunction characteristic of congenital OTC deficiency (Gropman and Batshaw, 2004).

3 Serotonin

Increased brain concentrations of the serotonin metabolite 5-hydroxyindoleacetic acid (5HIAA) have been reported in the brains of *spf* mice (Bachmann and Colombo, 1984) and in the cerebrospinal fluid (CSF) of children with congenital hyperammonemia (Hyman et al., 1988). Concentrations of tryptophan, the amino acid precursor of serotonin (5HT), are increased in the blood and brain tissues of *spf* mice (Chaouloff et al., 1985; Inoue et al., 1989), and increases in the 5HIAA/5HT ratio, indicative of increased flux through the 5HT pathway, were also reported. One possible explanation for the increased 5HT metabolism in the brains of *spf* mice is provided by the subsequent reporting of increased activity of the 5HT-metabolizing enzyme monoamine oxidase-A (MAO-A) (Raghavendra Rao et al., 1994). Receptor-binding studies revealed a significant loss of binding sites for the $5HT_2$ receptor ligand [^{3}H]-ketanserin and a concomitant increase in binding sites for the $5HT_{1A}$ site ligand [^{3}H]-8-OHDPAT (Robinson et al., 1992). Neurobehavioral assessment of *spf* mice revealed abnormalities in 5HT-related behaviors in these animals.

4 Pathogenesis of Neuronal Cell Death in Congenital OTC Deficiency

4.1 Brain Energy Deficit

In vitro studies demonstrate that millimolar concentrations of ammonia inhibit α-ketoglutarate dehydrogenase (αKGDH) in both brain and heart preparations (Lai and Cooper, 1986). Brain tissue from *spf* mice contains significantly increased concentrations of alanine and lactate (Ratnakumari et al., 1994a) consistent with diminished pyruvate oxidation in the brains of these animals, a phenomenon which could result from an inhibitory effect of ammonia on αKGDH. If αKGDH inhibition is sufficiently prolonged and severe, a cerebral energy deficit could result. Consistent with such a possibility, direct measurements reveal significant losses of ATP in the brains of *spf* mice (Ratnakumari et al., 1992).

4.2 Nitric Oxide

Both plasma and brain concentrations of L-arginine are reduced in human and experimental OTC deficiency (Ratnakumari et al., 1996b). In the *spf* mouse, brain concentrations of several guanidino compounds including L-arginine were found to be significantly decreased (Ratnakumari et al., 1996b). Nitric oxide is synthesized by nitric oxide synthase (NOS) from L-arginine via oxidation of the guanidino nitrogen. Not surprisingly, L-arginine deficiency in the *spf* mouse resulted in decreased brain activities of NOS, and it was suggested that decreased synthesis of nitric oxide could be of pathophysiologic importance in relation to the central nervous system dysfunction characteristic of congenital OTC deficiency. Supplementary L-arginine forms part of the treatment regimen currently used in congenital OTC deficiency; its beneficial effects could conceivably result not only from priming of the urea cycle but also from normalization of nitric oxide synthesis.

4.3 NMDA Receptor-Mediated Excitotoxicity

Tryptophan may be oxidized to quinolinic acid, an excitotoxic compound which causes neuronal cell death via its action on NMDA receptors (Schwarcz et al., 1983). Quinolinic acid concentrations are increased by up to tenfold in CSF of severely affected children with congenital OTC deficiency (Batshaw et al., 1993) and studies in the *spf* mouse reveal twofold increases of quinolinic acid in several brain regions (Robinson et al., 1992, 1995a) including striatum where neuropathologic evaluation showed a loss of medium spiny neurons and increases in microglia (Robinson et al., 1995a); similar neuropathologic patterns of cellular changes were previously reported following direct infusions of quinolinic acid into the brains of normal mice (Bakker and Foster, 1991). The loss of NMDA sites reported in the brains of *spf* mice (Ratnakumari et al., 1995b) could reflect downregulation of these sites following chronic exposure to quinolinic acid and in this way represent a protective mechanism against further NMDA receptor-mediated excitotoxicity and neuronal cell death. These findings suggest the potential for the eventual use of NMDA receptor antagonists for the prevention of neuronal loss in congenital OTC deficiency.

5 Current Therapy in Congenital OTC Deficiency

5.1 Sodium Benzoate, Phenylacetate

In 1979, Brusilow and coworkers proposed that the conjugation of sodium benzoate with glycine to form hippurate could represent an alternative pathway for the excretion of waste nitrogen in children with congenital urea cycle deficiencies. Subsequent studies in these disorders demonstrated a lowering effect on ammonia with concomitantly improved neurological outcome and survival. Sodium benzoate prescribed in combination with L-arginine and sodium phenylacetate remains the treatment of choice for congenital OTC deficiency (Brusilow and Horwich, 1995). With this therapy, survival rates for patients with OTC deficiency have increased to >50%. However, more than 75% of children with neonatal-onset OTC deficiency still suffer severe brain damage resulting in seizures, mental retardation, and developmental disabilities (Msall et al., 1984). In studies in *spf* mice, sodium benzoate was found to be effective in lowering brain ammonia and glutamine concentrations (Ratnakumari et al., 1993b). However, high doses of sodium benzoate led to a rebound in brain ammonia and to a significant reduction in acetyl-CoA and ATP concentrations (Ratnakumari et al., 1993b). Similar changes had previously been reported in the livers of rats following sodium benzoate treatment (Kalbag and Palekar, 1988; Michalak and Qureshi, 1995a). In an analogous fashion to sodium benzoate, sodium phenylacetate removes ammonia by conjugating with glutamine to form phenylacetylglutamine (Brusilow et al., 1979).

5.2 L-Carnitine

L-Carnitine administration significantly attenuates ammonia neurotoxicity (Matsuoka et al., 1991), an effect which is mediated by maintenance of cerebral energy metabolites (Matsuoka and Igisu, 1993). It was suggested that the potentiation of ammonia toxicity at higher doses of sodium benzoate (see above) could be prevented by the administration of L-carnitine (O'Connor et al., 1986). A subsequent study demonstrated a dose-related beneficial effect of L-carnitine on liver (Michalak and Qureshi, 1995b) and brain (Ratnakumari et al., 1993c) acetyl-CoA and on ATP deficits caused by sodium benzoate in *spf* mice. On the basis of these findings, it was suggested that L-carnitine supplementation could be beneficial during benzoate therapy in congenital OTC deficiency in humans.

5.3 Acetyl-L-Carnitine

Acetyl-L-carnitine is a physiological substance found in many tissues including brain where it appears to act as a carrier for acetyl groups across the inner mitochondrial membrane (Tucek, 1985) and, by virtue of its proposed role in cholinergic neurotransmission, has been suggested to be beneficial in the treatment of cognitive disorders (Bassi et al., 1988). In view of the cholinergic deficit described in OTC deficiency (Ratnakumari et al., 1994b, 1995a), acetyl-L-carnitine was administered to pregnant *spf* rats and its effects on cholinergic parameters in the brains of offspring were assessed (Ratnakumari et al., 1995a). Acetyl-L-carnitine treatment resulted in a partial correction of the developmental profile for the cholinergic nerve terminal enzyme CAT, suggesting a potential therapeutic role for this compound in the prevention of cholinergic neuronal loss in OTC deficiency.

5.4 Liver Transplantation

Liver transplantation continues to gain acceptance as an effective treatment of urea cycle enzymopathies. To date, it has been used to treat cases of OTC deficiency, arginosuccinic aciduria, and citrullinemia. Recent improvements in organ procurement, surgical techniques, and immunosuppression have significantly improved morbidity and mortality. Several cases of OTC deficiency have been successfully treated by orthotopic liver transplantation (OLT). In all cases, OLT led to correction of hyperammonemia and plasma amino acid profiles. Survival rates are currently >80% and most survivors have a normal mental status posttransplantation (Todo et al., 1992; Hasegawa et al., 1995; McBride et al., 2004). Living donor (partial) liver transplants have successfully been used in cases of citrullinemia (Takenaka et al., 2000) and OTC deficiency (Nagasaka et al., 2001). Initial studies using isolated hepatocyte transplantation in a patient with OTC deficiency resulted in metabolic improvement and temporary relief of hyperammonemia prior to full liver transplantation (Horslen et al., 2003).

5.5 Gene Therapy

Clinical trials of gene therapy have been considered for patients with congenital urea cycle disorders. On the basis of studies in the *spf* mouse, a two-step process was suggested (Robinson et al., 1995b) consisting of both in vivo and ex vivo gene therapy. Step one would address the treatment of neonatal hyperammonemic coma using in vivo gene therapy and once stabilized, long-term correction would be attempted by ex vivo therapy using recombinant retroviruses. Modifications of adenoviral vectors to decrease immunogenicity could then potentially permit long-term correction (Morsy and Caskey, 1994; Robinson et al., 1995b). A series of adenoviral vectors containing mouse OTC cDNA used in a study in the *spf* mouse revealed that recombinant adenoviruses deleted in E_1 and with a temperature-sensitive mutation in E_2 successfully corrected the OTC deficiency and subsequent metabolic abnormalities for several months (Ye et al., 1996). Adenoviral vectors have been explored as a treatment for citrullinemia in two models of arginosuccinate synthase deficiency namely a naturally occurring bovine model and a murine model created by molecular mutagenesis (Patejunas et al., 1998). Mice treated with adenoviral vectors expressing arginosuccinate synthase lived longer than the untreated animals, but the treatment was suboptimal. In the bovine model, reports of either no improvement (Patejunas et al., 1998) or significant clinical improvement and normalization of plasma glutamine (Lee et al., 1999) have been reported using a different adenoviral vector. Studies so far in human OTC deficiency have not been encouraging with both lethal complications and adverse effects including flulike episodes and thrombocytopenia being reported (Raper et al., 2002).

Acknowledgments

Studies from the authors' research unit were funded by The Canadian Institute of Health Research and the Canadian Liver Foundation.

References

Anholt RRH. 1986. Mitochondrial benzodiazepine receptors as potential modulators of intermediary metabolism. TiPS 7: 506-511.

Arauja DM, Lapchak PA, Robitaille Y, Gauthier S, Quirion R. 1988. Differential alterations of various cholinergic markers in cortical and subcortical regions of human brain in Alzheimer's disease. J Neurochem 50: 1914-1923.

Bachmann C, Colombo JP. 1984. Increase of tryptophan and 5-hydroxyindoleacetic acid in the brain of ornithine carbamoyl transferase deficient sparse-fur mice. Pediatr Res 18: 372-375.

Bakker MHM, Foster AC. 1991. An investigation of the mechanisms of delayed neuron degeneration caused by direct injection of quinolinic acid into the rat striatum in vivo. Neuroscience 42: 387-395.

Bassi S, Ferrarese C, Finola MG, Frattola L, Lannucelli M, et al. 1988. L-Acetyl-carnitine in Alzheimer disease (AD) and Senile Dementia of the Alzheimer type (SDAT). Senile Dementias (Second International Symposium). Agnoli A, Calin J, Larsen N, Mayeux R, editors. Paris: John Libbey Eurotext; pp. 461-466.

Batshaw ML, Hyman SL, Coyle JT, Robinson MB, Qureshi IA, et al. 1988. Effect of sodium benzoate and sodium phenylacetate on brain serotonin turnover in the ornithine transcarbamylase-deficient sparse-fur mouse. Pediatr Res 23: 368-374.

Batshaw ML, Robinson MB, Hyland K, Djali S, Heyes MP. 1993. Quinolinic acid in children with congenital hyperammonemia. Ann Neurol 34: 676-681.

Brusilow SW, Horwich AL. 1995. Urea cycle enzymes. The Metabolic and Molecular Basis of Inherited Disease, 7th edn. Scriver CR, Beaudet AL, Sly WS, et al. editors. New York: McGraw-Hill; pp.1187-1232.

Brusilow SW, Valle DL, Batshaw M. 1979. New pathways of nitrogen excretion in inborn errors of urea synthesis. Lancet 2: 452-454.

Chaouloff F, Laude D, Mignot E, Kamoun P, Elghozi JL. 1985. Tryptophan and serotonin turnover rate in the brain of genetically hyperammonemic mice. Neurochem Int 7: 143-153.

Christodoulou J, Qureshi IA, McInnes RF, Clarke JTR. 1993. Ornithine transcarbamylase deficiency presenting with stroke-like episodes. J Pediatr 122: 423-425.

Connelly A, Cross JH, Gadian DG, Hunter JV, Kirkham FJ, et al. 1993. Magnetic resonance spectroscopy shows increased brain glutamine in ornithine carbamoyl transferase deficiency. Pediatr Res 33: 77-81.

Demars R, Levan SL, Trend BL, Russel LB. 1976. Abnormal ornithine carbamyl transferase in mice having the sparse-fur mutation. Proc Natl Acad Sci USA 23: 1693-1698.

Dolman CL, Clasen RA, Dorovini-Zis K. 1988. Severe cerebral damage in ornithine transcarbamylase deficiency. Clin Neuropathol 7: 10-15.

Filloux F, Townsend JJ, Leonard C. 1986. Ornithine transcarbamylase deficiency: Neuropathologic changes acquired in utero. J Pediat 108: 942-945.

Gropman AL, Batshaw ML. 2004. Cognitive outcome in urea cycle disorders. Mol Genet Metab 81 (Suppl. 1): S58-S62.

Harding BN, Leonard JV, Erdohazi M. 1984. Ornithine transcarbamylase deficiency: Neuropathological study. Eur J Pediatr 141: 215.

Harper CG, Butterworth RF. 1997. Nutritional and metabolic disorders. Greenfield's Neuropathology. Graham DI, Lantos PL, editors. London: Arnold; pp. 601-655.

Hasegawa T, Tzakis AG, Todo S, Reyes J, Nour B, et al. 1995. Orthotopic liver transplantation for ornithine transcarbamylase deficiency with hyperammonemic encephalopathy. J Pediat Surg 30: 863-865.

Horslen SP, McGowan TC, Goertzen TC, Warkentin PI, Cai HB, et al. 2003. Isolated hepatocyte transplantation in an infant with severe urea cycle disorder. Pediatrics 111 (6 Part 1): 1262-1267.

Hyman SL, Porter CA, Page JJ, Iwata BA, Kissel R, et al. 1988. Behavioral management of feeding disturbances in urea cycle and organic acid disorders. J Pediatr 111: 558-562.

Inoue I, Shimizu T, Saheki T, Noda T, Fukuda T. 1989. Serotonin- and catecholamine-related substances in the brain of ornithine transcarbamylase-deficient sparse-fur mice in the hyperammonemic state: comparison of two procedures for obtaining brain extract, decapitation and microwave irradiation. Biochem Med Metab Biol 42: 232-239.

Kalbag SS, Palekar AG. 1988. Sodium benzoate inhibits fatty acid oxidation in rat liver. Effect on ammonia levels. Biochem Med Metab Biol 40: 133-142.

Kendall BE, Kingsley DPE, Leonard JV, Lingam S, Oberholzer VG. 1983. Neurological features and computed tomography of the brain in children with ornithine carbamoyl transferase deficiency. J Neurol Neurosurg Psychiatry 46: 28-34.

Lai JCK, Cooper AJL. 1986. Brain a-ketoglutarate dehydrogenase complex: Kinetic properties, regional distribution and effects of inhibitors. J Neurochem 47: 1376-1386.

Lee B, Dennis JA, Healy PJ, Mull B, Pastore L, et al. 1999. Hepatocyte gene therapy in a large animal: A neonatal bovine model of citrullinemia. Proc Natl Acad Sci USA 96(7): 3981-3986.

Matsuoka M, Igisu H. 1993. Effects of L- and D-carnitine on brain energy metabolites in mice given sublethal doses of ammonium acetate. Pharmacol Toxicol 72: 145-147.

Matsuoka M, Igisu H, Kohriyama K, Inoue N. 1991. Suppression of neurotoxicity of ammonia by L-carnitine. Brain Res 567: 328-331.

McBride KL, Miller G, Carter S, Karpen S, Gross J, et al. 2004. Developmental outcomes with early orthotopic liver transplantation for infants with neonatal-onset urea cycle defects and a female patient with late-onset ornithine transcarbamylase deficiency. Pediatrics 114(4): e523-e526.

Michalak A, Qureshi IA. 1995a. Free and esterified coenzyme A in the liver and muscles of chronically hyperammonemic mice treated with sodium benzoate. Biochem Mol Med 54: 96-104.

Michalak A, Qureshi IA. 1995b. Tissue acylcarnitine and acylcoenzyme A profiles in chronically hyperammonemic mice treated with sodium benzoate and supplementary L-carnitine. Biomed Pharmacother 49: 350-357.

Morsy MA, Caskey CT. 1994. Ornithine transcarbamylase deficiency: A model for gene therapy. editors. Hepatic Encephalopathy, Hyperammonemia and Ammonia Toxicity. Felipo V, Grisolia S, editors. New York: Plenum Press; pp. 145-154.

Msall M, Batshaw ML, Suss R, Brusilow SW, Mellits ED. 1984. Neurologic outcome in children with inborn errors of urea synthesis. New Engl J Med 310: 1500-1505.

Nagasaka H, Yorifuji T, Egawa H, Kikuta H, Tanaka K, et al. 2001. Successful living-donor liver transplantation from an asymptomatic carrier mother in ornithine transcarbamylase deficiency. J Pediatr 138(3): 432-434.

O'Connor JE, Costell M, Miguez MP, Grisolia S. 1986. Influence of the route of administration on the protective effect of L-carnitine on acute hyperammonemia. Biochem Pharmacol 35: 3173-3176.

Patejunas G, Lee B, Dennis JA, Healy PJ, Reeds PJ, et al. 1998. Evaluation of gene therapy for citrullinaemia using murine and bovine models. J Inherit Metab Dis 21 (Suppl. 1): 138-150.

Qureshi IA. 1992. Animal models of hereditary hyperammonemias. Neuromethods, Vol. 22: Animal Models of Neurological Disease, II. Boulton A, Baker G, Butterworth R, editers. Clifton, NJ: Humana Press; pp. 329-356.

Raghavendra Rao VL, Qureshi IA, Butterworth RF. 1993. Increased densities of binding sites for the peripheral-type benzodiazepine receptor ligand PH]PK11195 in congenital ornithine transcarbamylase-deficient sparse fur mouse. Pediatr Res 34: 777-780.

Raghavendra Rao VL, Qureshi IA, Butterworth RF. 1994. Activities of monoamine oxidase-A and -B are altered in the brains of congenitally hyperammonemic sparse-fur (*spf*) mice. Neurosci Lett 170: 27-30.

Raper SE, Yudkoff M, Chirmule N, Gao GP, Nunes F, et al. 2002. A pilot study of in vivo liver-directed transfer with an adenoviral vector in partial ornithine transcarbamylase deficiency. Hum Gene Ther 13(1): 163-175.

Ratnakumari L, Qureshi LA, Butterworth RF. 1992. Effects of congenital hyperammonemia on the cerebral and hepatic levels of the intermediates of energy metabolism in *spf* mice. Biochem Biophys Res Commun 184: 746-751.

Ratnakumari L, Qureshi IA, Butterworth RF. 1993a. Evidence for a severe cholinergic neuronal deficit in brain in congenital ornithine transcarbamylase (OTC) deficiency. Soc Neurosci Abstr 19: 122.

Ratnakumari L, Qureshi IA, Butterworth RF. 1993b. Effect of sodium benzoate on cerebral and hepatic energy metabolites in *spf* mice with congenital hyperammonemia. Biochem Pharmacol 45: 137-146.

Ratnakumari L, Qureshi IA, Butterworth RF. 1993c. Effect of L-carnitine on cerebral and hepatic energy metabolites in congenitally hyperammonemic sparse-fuce mice and its role during benzoate therapy. Metabolism 42: 1039-1046.

Ratnakumari L, Qureshi IA, Butterworth RF. 1994a. Regional amino acid neurotransmitter changes in brains of *spf Pi* mice with congenital ornithine transcarbamylase deficiency. Metab Brain Dis 9: 43-51.

Ratnakumari L, Qureshi IA, Butterworth RF. 1994b. Evidence for cholinergic neuronal loss in brain in congenital ornithine transcarbamylase deficiency. Neurosci Lett 178: 63-65.

Ratnakumari L, Qureshi IA, Maysinger D, Butterworth RF. 1995a. Developmental deficiency of the cholinergic system in congenitally hyperammonemic *spf* mice: Effect of acetyl-L-carnitine. J Pharmacol Exp Ther 274: 437-443.

Ratnakumari L, Qureshi IA, Butterworth RF. 1995b. Loss of [^{3}H]MK801 binding sites in brain in congenital ornithine transcarbamylase deficiency. Metab Brain Dis 10: 249-255.

Ratnakumari L, Qureshi IA, Butterworth RF. 1996a. Central muscarinic cholinergic M, and M2 receptor changes in congenital ornithine transcarbamylase deficiency. Pediatr Res 40: 25-28.

Ratnakumari L, Qureshi IA, Butterworth RF, Marescau B, De Deyn PP. 1996b. Arginine-related guanidino compounds and nitric oxide synthase in the brain of ornithine transcarbamylase deficient *spf* mutant mouse: Effect of metabolic arginine deficiency. Guanidino Compounds. De Deyn PP, et al. editors. John Libby and Co.; pp. 17-20.

Robinson MB, Anegawa NJ, Gorry E, et al. 1992. Brain serotonin2 and serotonin, receptors are altered in the congenitally hyperammonemic sparse fur mouse. J Neurochem 58: 1016-1022.

Robinson MB, Hopkins K, Batshaw ML, McLaughlin BA, Heyes MP, et al. 1995a. Evidence of excitotoxicity in the brain of the ornithine carbamoyltransferase deficient *sparse fur* mouse. Dev Brain Res 90: 35-44.

Robinson MB, Batshaw ML, Ye X, Wilson JM. 1995b. Prospects for gene therapy in ornithine carbamoyltransferase deficiency and other urea cycle disorders. MRDS Res Rev 1: 62-70.

Schwarcz R, Whetsell WO Jr, Mangano RM. 1983. Quinolinic acid: An endogenous metabolite that produces axon-sparing lesions in rat brain. Science 219: 316-318.

Simmonds MA. 1991. Modulation of the GABAA receptor by steroids. Semin Neurosci 3: 231-239.

Takayanagi M, Ohtake A, Ogura N, Nakajima H, Hoshino M. 1984. A female case of ornithine transcarbamylase deficiency with marked computed tomographic abnormalities of the brain. Brain Dev 6: 58.

Takenaka K, Yasuda I, Araki H, Naito T, Fukutomi Y, et al. 2000. Type II citrullinemia in an elderly patient treated with living related partial liver transplantation. Intern Med 39(7): 553-558.

Todo S, Starzl IE, Tzakis A, Benkov KJ, Kalousek P, et al. 1992. Orthotopic liver transplantation for urea cycle enzyme deficiency. Hepatology 15: 419-422.

Tucek S. 1985. Regulation of acetylcholine synthesis in the brain. J Neurochem 44: 11-24.

Tuchman M, Plante RJ, Garcia-Perez MA, Rubio V. 1996. Relative frequency of mutation causing ornithine transcarbamylase deficiency in 78 families. Hum Genet 97: 274-276.

Veres G, Gibbs RA, Scherer SE, Caskey CT. 1987. The molecular basis of the sparse fur mutation. Science 237: 415-417.

Ye X, Robinson MB, Batshaw ML, Furth EE, Smith I, Wilson JM. 1996. Prolonged metabolic correction in adult ornithine transcarbamylase-deficient mice with adenoviral vectors. J Biol Chem 271: 3639-3646.

12 Ammonia Toxicity in the Central Nervous System

J. Albrecht

1	Introduction: Issues Under Consideration	262
2	Hyperammonemic Encephalopathies: Clinical Conditions and Model Systems	263
3	Penetration of Ammonia from Blood to Brain and Changes in the Blood–Brain Barrier Function	264
4	Effects of Ammonia on Astrocytes, Neurons and Astrocyte–Neuron Interactions	265
4.1	Morphological Changes, Disturbances of Energy Metabolism, and Oxidative Stress in Astrocytes	265
4.2	Changes in Astrocyte–Neuron Interaction Resulting in the Imbalance Between the Excitatory (Glutamatergic) and Inhibitory (GABAergic) Neurotransmission	267
4.3	Changes in the NMDA Receptor/Nitric Oxide/cGMP Pathway and Their Pathophysiological Effects	268
4.3.1	NMDA Receptor-Mediated Nitrosative and Oxidative Stress	268
4.3.2	Inhibition of cGMP Production in Chronic Hyperammonemia: Pathophysiological Implications	269
4.4	Factors Involved in Hyperammonemic Brain Edema	269
4.4.1	Accumulation of Glutamine and Other Osmolytes: Osmotic Versus Metabolic Effects and Critical Coupling to Bioenergetic Failure	269
4.4.2	Changes in the Cerebral Blood Flow	270
5	Derangements Resulting from Increased Accumulation in Brain of Aromatic Amino Acids: Catecholaminergic Transmission	270
6	Protection Against Neurotoxic Effects of Ammonia	271
7	General Conclusions: Missing Links in the Knowledge	272

© Springer-Verlag Berlin Heidelberg 2007

Abstract: Ammonia is a regular metabolite in the central nervous system (CNS). However, when it enters the brain from blood in excessive quantities it becomes toxic to CNS cells. Therefore, ammonia is a causative factor in neurological disorders associated with increased blood ammonia, among which hepatic encephalopathy (HE) is a public health problem. Astrocytes are the cells that in the CNS where ammonia is metabolized in a reaction of glutamine synthesis from glutamate and ammonia, and are the primary victim of ammonia toxicity. Bioenergetic failure, oxidative or nitrosative stress and excessive accumulation of glutamine are the interrelated aspects of ammonia-induced astrocytic impairment that contribute to cerebral edema – a major cause of death associated with acute HE. Effects of ammonia on astrocytes radiate to neurons and affect the astrocytic-neuronal interactions. Interference of ammonia with the various steps of the glutamine-glutamate cycle in astrocytes lead to alterations in amino acidergic (gluta-minergic and GABAergic) neurotransmission. Ammonia increases GABA-ergic tone by stimulating peripheral benzodiazepine receptors on astrocytes, in this way enhancing the synthesis of neurosteroids that are positive modulators of the GABAA receptors. Direct effects of ammonia on neurons are highlighted by the changes in the NMDA receptor/nitric oxide/cGMP pathway. Overactivation of NMDA receptors in the acute phase of ammonia toxicity is responsible for oxidative and nitrosative stress in neurons (and perhaps in astrocytes), whereas their downregulation upon prolonged exposure to ammonia.leads to impairment of cGMP synthesis held responsible for intellectual and memory deficits of chronic HE patients. Acute hyperammonemia is often associated with increased cerebral blood flow that by a complex mechanism contributes to hyperammonemic brain edema. Ammonia increases the transport across the blood-brain barrier of aromatic amino acids that are precursors of catecholamines: serotonin and dopamine. Ensuing derangements of catecholaminergic transmission are held responsible for sedative effects anfd motor impairment, respectively. Pharmacological interventions to attenuate individual neurotoxic effects of ammonia have in no case reached the stage of clinical trial: slowing the general metabolism with hypothermia is the most recently introduced life-saving paradigm in patients with advanced HE.

List of Abbreviations: AAT, aspartate aminotransferase; AE, astrocytic end-feet; ATP, adenosine tripjhosphate; CAMP, cyclic adenosine–3′-monophosphate; CBF, cerebral blood flow; cGMP, cyclic guanosine –3′-monophosphate; CNS, central nervous system; CoA, coenzyme A; CsA, cyclosporin A; EAAT, excitatatory amino acid transporter; GABA, gamma-amino butyric acid; GC, guanylate cyclase; GGT, gamma-glutamyl transpeptidase; GS, glutamine synthetase; HE, hepatic encephalopathy; 5-HT, 5-hydroxytryptamine; α-KGDHC, alpha-ketoglutarate dehydrogenase complex; LNAA, large neutral amino acids; MAO-A, monoamino oxidase A; MAO-B, monoamino oxidase B; MDH, malate dehydrogenase; MPT, mitochondrial permeability transition; MSO, methionyl-D,L-sulfoximine; Na/K ATPase, sodium/potassium-dependent adenosine triphosphatase; NMDA, N-methyl-D-aspartate; NMR, nuclear magnetic resonance spectroscopy; NO, nitric oxide; NOS, nitric oxide synthase; nNOS, neuronal nitric oxide synthase; PBR, peripheral benzodiazepine receptors; PET, positron emission tomography; PKC, protein kinase C; PTN, protein tyrosine nitration; PTP, permeability transition pore; TAA, thioacetamide; $\Delta\psi_m$, mitochondrial membrane potential

1 Introduction: Issues Under Consideration

Ammonia is a metabolite in all the mammalian tissues and organs, including the brain. Several aspects of cerebral ammonia metabolism under conditions when its supply from the blood to brain is within physiological range have been discussed in the chapters written by Drs. Butterworth and Schousboe. The present chapter describes how the metabolism of ammonia is changed when it enters the brain from the blood in excessive quantities. As discussed in ❷ *Sect. 2*, hyperammonemia is a causative factor in several neurological diseases, which can be modeled in experimental animals or in in vitro studies on isolated or cultured CNS cells. ❷ *Section 3* analyzes the mechanism of penetration of ammonia from the blood to the brain and how ammonia modifies the functioning of the blood–brain barrier. ❷ *Section 4* discusses metabolic alterations taking place in astrocytes, cells that in the CNS are the primary site of ammonia metabolism, and the victim of ammonia-induced bioenergetic failure and oxidative stress, and how these astrocytic changes radiate to neurons and affect the astrocytic-neuronal interactions. Focus will be on the effects of ammonia that are responsible for the pathophysiological manifestations of hyperammonemic encephalopathy: alterations in aminoacidergic (glutaminergic and GABAergic) transmission and cerebral edema, a major cause of death associated with acute hyperammonemia. The contribution of excessive accumulation of glutamine—the astrocytic end-product of ammonia detoxification—will be highlighted, and so will be the changes in the activity of the N-methyl-D-aspartate (NMDA) receptor/nitric oxide (NO)/cGMP pathway and their neurophysiological roles. Alterations of other neurotransmitter systems, mostly the serotoninergic and dopaminergic system, which are associated with aromatic amino acid imbalance will be briefly discussed in ◑ Sect. 5. Attempts of attenuating the consequences of ammonia neurotoxicity in experimental and clinical settings and also the potential autoprotective mechanisms will be discussed in ◑ Sect. 6. ◑ Section 7 briefly outlines missing links in the knowledge and further perspectives.

It must be born in mind that in neurological disorders associated with hyperammonemia, ammonia is usually not the only pathogen interfering with brain metabolism and function. Discussion in this chapter is confined to events that clearly reflect the toxic effects of ammonia.

2   Hyperammonemic Encephalopathies: Clinical Conditions and Model Systems

Ammonia is a strong endogenous neurotoxin, a major pathogenic factor in a rather heterogeneous group of diseases of different etiologies collectively referred to as hyperammonemic encephalopathies (◑ Table 12-1). The most common form of hyperammonemic encephalopathy is hepatic encephalopathy

◻ Table 12-1
Encephalopathies associated with hyperammonemia

Metabolic encephalopathies
Acute hepatic encephalopathy (viral or toxic liver injury)
Chronic hepatic encephalopathy (cirrhosis, portacaval shunt)
Uremic encephalopathy
Diabetic encephalopathy
Hypoglycemic coma
Idiopathic hyperammonemic encephalopathies (associated with chemotherapy or bone marrow transplantation)
Encephalopathies associated with genetic changes
Reye's syndrome (medium chain acyl-CoA dehydrogenase deficiency)
Ornithine transcarbamylase deficiency
Organic acidemias (propionyl CoA carboxylase deficiency)

(HE), where accumulation of ammonia results from liver damage. Acute and chronic forms of HE are distinguished depending on the type, duration, and intensity of action of the liver-damaging factors (Mullen, 1999). Acute HE is often associated with overexposure to hepatotoxic chemicals (anesthetics, paracetamol, etc.) or with infections by hepatotropic viruses (HBV, HDV, HAV). Death rate in acute HE patients reaches 80%, and cerebral edema is the major precipitating factor (Cordoba and Blei, 1996). Chronic HE is a complex neurological disorder associated with liver cirrhosis. HE symptoms include a spectrum of motor and intellectual impairments, which according to their severity can be classified into four stages (or grades), the most severe (grade IV) being characterized by loss of consciousness and referred to as hepatic coma (Conn, 1994; Albrecht and Jones, 1999; Rahman and Hodgson, 2001). HE is often characterized by disturbances in the sleep–wake cycle (Corbalan et al., 2002a).

Epidemiological data concerning HE are only available for liver cirrhosis patients without specific consideration of HE symptoms. According to a WHO report for EU countries from June 2000 (Regional Office, Copenhagen), in this region liver cirrhosis has been diagnosed in 14 cases/100,000 with the death rate of ~60,000/year. Although the number of deaths directly associated with hyperammonemia has not been estimated, at least 50% of cases of liver cirrhosis are periodically accompanied by hyperammonemia and symptoms typical of hyperammonemic encephalopathy. Other disease units listed in (❯ *Table 12-1*), although epidemiologically insignificant, are severe in their course and often lethal. Of note, cerebral edema is a precipitating factor in most of them.

Mechanistic considerations regarding ammonia neurotoxicity at the whole brain level have been derived from studies on animal models of hyperammonemia or HE. Acute ammonia toxicity in mice or rats is analyzed following i.p. injection into these animals of ammonium salts (chloride or acetate) (reviewed by Felipo and Butterworth, 2002). Acute toxic liver damage is routinely modeled using hepatotoxins: galactosamine (Ede et al., 1987) or more recently thioacetamide (TAA). Subsequent i.p. administrations of TAA at daily intervals allow to follow stages of HE, from acute, symptomatic to chronic, subclinical (for a review and references see Saran et al., 2004). Cellular and subcellular events underlying ammonia neurotoxicity have been delineated in studies using ammonia-treated brain slices, astrocytes and neurons in culture, or subcellular fractions including cerebral microvessels or mitochondria. Studies using these different approaches will be referred to in the following sections.

3 Penetration of Ammonia from Blood to Brain and Changes in the Blood–Brain Barrier Function

The term "ammonia" applies to the sum of ammonia base (NH_3) and ammonium ions (NH_4^+). However, as the pKa for ammonia in the blood is estimated to be around 9.1–9.2, not less than 98% of circulating ammonia is in the form of NH_3. It has long been assumed that ammonia base which freely penetrates the lipid bilayer of the blood–brain barrier accounts for most of ammonia that enters the brain, whereas the contribution of ammonium ions may be neglected. However, tracer studies in rhesus monkey performed in the late 1970s have indicated that around 20%–25% of ammonia actually may enter the brain in an ionized form (for a recent review see Ott and Larsen, 2004). Two complimentary mechanisms have been invoked to account for the cerebral capillary transport of NH_4^+: (1) substitution of other cations normally taken up by specific membrane ion transporters or exchangers present in the capillary endothelium and (2) the use of specific ammonium ion transporters. Accumulation of potassium ions in the extracellular space of the brain in rats with impaired intracerebral ammonium clearance has been interpreted to favor possibility (1) (Sugimoto et al., 1997). Possibility (2) is hypothetical, based indirectly on a single demonstration of the presence in brain of mRNA coding for one of these transporters (RhCG; Huang and Liu, 2001).

Ammonia transport from the blood to brain is substantially increased during hyperammonemia. The blood/brain ammonia concentration ratio, which in control rats is ∼5, may under severe hyperammonemic conditions go up to ∼8, the brain ammonia reaching the 5-mM concentration (reviewed in Butterworth, 2002; Felipo and Butterworth, 2002). The factors responsible for the increased extraction of ammonia by the brain have not been delineated.

Ammonia per se does not disintegrate the blood–brain barrier: for instance, hyperammonemia does not change the permeability-to-surface-area product for acetate (Jessy et al., 1990). However, ammonia affects the blood–brain barrier passage of a variety of nutrients by different mechanisms. A change consistently observed in hyperammonemic animals is an increased blood to brain transport of large neutral amino acids (LNAA), mostly tyrosine, tryptophan, and leucine (Jonung et al., 1985). There is evidence that the increased uptake of LNAA across the blood–brain barrier involves increased exchange with glutamine mediated by the amino acid transport system L (Hilgier et al., 1992). The latter effect is subsequent to the excessive glutamine synthesis in astrocytes (see ❯ *Sect. 4*). The tryptophan–glutamine exchange appears to be partly mediated by gamma glutamyl-transpeptidase (GGT) (Bachmann, 2002), whose activity in the brain microvessels is increased in hyperammonemia (Stastny et al., 1988). Increased entry to the brain of aromatic amino acids, tyrosine and tryptophan, which are precursors of the catecholamines dopamine and serotonin, respectively, contributes to disturbances of dopaminergic and serotoninergic transmission (see ❯ *Sect. 5*). Hyperammonemia also affects the blood–brain barrier transport of basic amino acids: it increases the brain uptake index for ornithine but decreases that of arginine (Albrecht et al., 1994, 1996). Pathophysiological implications of the basic amino acids imbalance have not been studied in detail.

4 Effects of Ammonia on Astrocytes, Neurons and Astrocyte–Neuron Interactions

4.1 Morphological Changes, Disturbances of Energy Metabolism, and Oxidative Stress in Astrocytes

Excess of ammonia that crosses the blood–brain barrier primarily affects the astrocytes. This is due to the topographic positioning of the cells between the vascular bed and neurons, which makes them absorb the ammonia wave (◗ *Figure 12-1*). As discussed somewhere in the volume, astrocytes are

◘ Figure 12-1
Entry and metabolism of ammonia in the CNS in a topographic perspective. Astrocytic end-feet (AE) wrap the blood–brain barrier-forming cerebral capillaries (Capillary) at one end and neuronal bodies (Neuron) and synapses (Syn) at the other end, forming a transmission belt for metabolites, and a buffering zone, between the cerebral vascular bed and neurons. Ammonia entering the brain from the periphery passes the astrocytic end-feet and is neutralized in the cell body (Astrocyte) by reacting with glutamate (Glu) to form glutamine (Gln), in a glutamine-synthetase (GS)-mediated reaction. Gln is then transferred to neurons and degraded by phosphate-activated glutaminase (PAG) to Glu and ammonia. A portion of Glu feeds the neurotransmitter pool, is released to the synaptic cleft, and interacts with Glu receptors (filled quadrangles) located on synapses and astrocytes

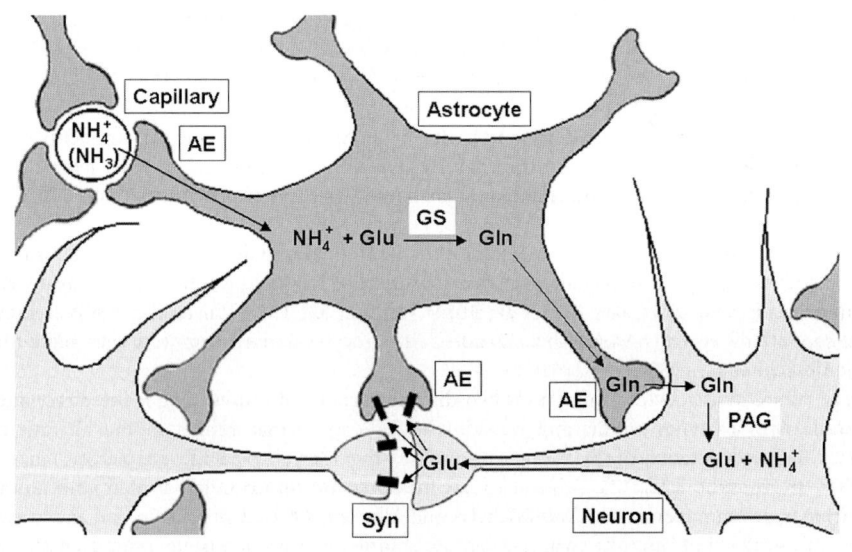

equipped to metabolize ammonia: they are the locus of glutamine synthetase, the key ammonia-metabolizing enzyme (Norenberg and Martinez-Hernandez, 1979) (◗ *Figure 12-1*). As outlined below, ammonia-induced astrocytic dysfunction may be viewed as a cooperative effect of bioenergetic failure, reduction of the ammonia-metabolizing capacity of the cells and of the metabolic stress associated with excessive glutamine synthesis. Astrocytes are the cells whose morphology is primarily affected by ammonia. Their transformation to so-called Alzheimer II astrocytes (◗ *Figure 12-2*), coincident with their

◘ Figure 12-2
A section of cerebral cortex from the autopsy material from a patient who died with symptoms of hyperammonemic coma. Alzheimer II astrocytes (*arrows*) are characterized by a large pale nucleus with margination of chromatin and weakly discernible cytoplasm. H + E staining, original magnification × 1,000. Courtesy of Dr. Ewa Matyja from the Department of Neuropathology, Medical Research Centre, Polish Academy of Sciences, Warsaw

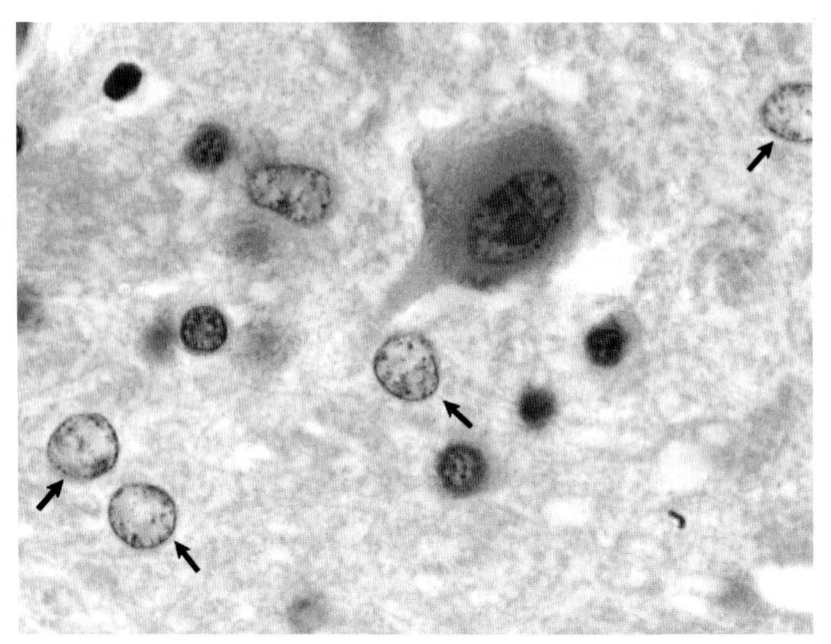

increase in size and/or diffuse proliferation, is the hallmark of chronic hyperammonemia. Neuronal morphology is affected only rarely, and the changes, if present, are confined to focal neuronal necrosis (Pilbeam et al., 1983). The link of morphological changes in astrocytes to their dysfunction has not been elucidated.

In vitro studies have provided strong evidence that ammonia impairs oxidative metabolism in astrocytes. In cultured astrocytes, ammonia decreased oxidation of key energy substrates: pyruvate, branched chain amino acids, and glutamate (Yu et al., 1984; Hertz et al., 1987; Lai et al., 1989). Depression of pyruvate-supported oxygen consumption was noted in astrocytes derived from rats with acute toxic liver damage (Albrecht et al., 1987).

Energy failure in astrocytes results from two different groups of causes. One is the direct interaction of ammonia with enzymes within and outside the TCA cycle that results in the alteration of the cycle rate. Reduced activity of pyruvate carboxylase, the astroglia-specific anaplerotic enzyme that replenishes tricarboxylic acid cycle substrates, was measured in mitochondria treated with ammonia or derived from hyperammonemic rats (Faff-Michalak and Albrecht, 1991). Ammonia in vivo and hyperammonemia in vitro inhibit not only the malate–aspartate shuttle enzymes: aspartate aminotransferase (AAT) (Faff-Michalak and Albrecht, 1991; Ratnakumari and Murthy, 1992) and malate dehydrogenase (MDH) (Faff-Michalak and Albrecht, 1991) but also the E1 and E3 components of the α-ketoglutarate dehydrogenase complex (αKGDHC) (Faff-Michalak and Albrecht, 1993). The inhibitory effects have been specifically noted in the nonsynaptic mitochondria in a large degree derived from astrocytes but not in synaptic mitochondria.

The other cause of bioenergetic failure is increased utilization of energy by ammonia-affected cells. Increased ATP consumption in astrocytes exposed to excess ammonia has two major causes: increased glutamine synthesis and enhanced activity of the Na$^+$/K$^+$-ATPase. Both events represent the early response to acute ammonia overload in vivo and in vitro (Albrecht et al., 1985; Farinelli and Nicklas, 1992; Waniewski, 1992). The resulting energy depletion leads to astrocytic mitochondrial damage, which at the microscopic level is manifested by swelling of mitochondrial matrix (Gregorios et al., 1985). In functional terms, this leads to a cyclosporin A (CsA)-inhibitable collapse of the inner mitochondrial membrane potential ($\Delta\psi_m$) and MPT (Bai et al., 2001; Rama Rao et al., 2003) and to intracellular accumulation of free radicals (Murthy et al., 2001). Interestingly, these manifestations of mitochondrial dysfunction appear to depend in a large degree on glutamine accumulation: inhibition of glutamine synthesis attenuates or abolishes the effects (Bai et al., 2001; Murthy et al., 2001; Rama Rao et al., 2003). Glutamine added to astrocytes in culture (Jayakumar et al., 2004) or mitochondria mimics the ammonia-induced mitochondrial swelling (Ziemińska et al., 2000), and the effect depends on mitochondrial glutamine uptake (Ziemińska et al., 2000). Interestingly, ammonia increases the accumulation of exogenously added glutamine in mitochondria (Dolińska et al., 1996). The pathogenic role of glutamine will be discussed in more detail in ❯ Sect. 4.4.1, in the context of the mechanism underlying ammonia-induced astrocytic swelling.

One other aspect of astrocytic energy metabolism pertains to the breakdown of glycogen. This process, which is stimulated by noradrenaline and associated with generation of cAMP, is involved in meeting the glucose demand of adjacent neurons (reviewed by Fillenz et al., 1999). Treatment of cultured astrocytes with ammonia reduces both noradrenaline-dependent generation of cAMP (Liskowsky et al., 1986) and cAMP-mediated glycogen breakdown in these cells (Dombro et al., 1993). This may limit the neuron-supporting capability of astrocytes.

4.2 Changes in Astrocyte–Neuron Interaction Resulting in the Imbalance Between the Excitatory (Glutamatergic) and Inhibitory (GABAergic) Neurotransmission

As discussed elsewhere in the book, one of the key neuromodulatory functions of astrocytes is the clearance from the synaptic cleft of the excitatory amino acid glutamate. Ammonia impairs astrocytic glutamate reuptake, and there is unambiguous evidence from in vivo and in vitro studies that this is due to downregulation of the astrocytic glutamate transporters EAAT-1 and EAAT-2 (Knecht et al., 1997; Norenberg et al., 1997). Impaired glutamate reuptake at first results in its increased accumulation in the extrasynaptic space, which leads to an increase of the glutaminergic tone. One other factor that cannot be ignored as a trigger of excitation is instant depolarization of nerve cells leading to an increase of the calcium-dependent release of the neurotransmitter pool of glutamate (reviewed by Rose, 2002). Acute neurotoxic effects associated with overactivation of the NMDA class of glutamate receptors will be discussed later. Prolonged exposure to excess glutamate associated with chronic treatment of astrocytes with ammonia downregulates NMDA receptors and shifts the neurotransmission balance from excitation to inhibition (Peterson et al., 1990; Saransaari et al., 1997). Changes in glutaminergic neurotransmission related to long-term ammonia exposure appear to be related to alterations in the protein kinase C (PKC)-mediated phosphorylation of NMDA receptors and, less directly, Na$^+$/K$^+$-ATPase (reviewed by Monfort et al., 2002). A recent study revealed that chronic ammonia exposure of cerebellar granule neurons in culture induces redistribution of different PKC isoforms between membranes and cytosol (Giordano et al., 2005). Most interestingly in the context of glutaminergic transmission, ammonia treatment alters the regulation of PKC translocation by NMDA receptor activation (Giordano et al., 2005). Hence, ammonia appears to trigger a vicious circle of events involving PKC and NMDA receptors.

Indirect evidence suggests that ammonia, especially following long-term exposure, impairs the synthesis of the neurotransmitter pool of glutamate. As discussed in ❯ Sect. 4.1.1, ammonia depresses the activity of the malate–aspartate shuttle enzymes, which in astrocytes produce metabolic intermediates of neuronal glutamate synthesis (Palaiologos et al., 1989). Ammonia also inhibits the αKGDHC activity, which is directly involved in glutamate syntheses (Peng et al., 1991). More recently, Sonnewald et al. (1996) have used ^{13}C-NMR spectroscopy to demonstrate that the flux of astrocytic precursors of glutamate to neurons is markedly decreased in rats with portal systemic encephalopathy, a model of chronic hyperammonemia. However, the degree in which the impairment of synthesis affects the glutamate pool that is released from the synaptic endings remains to be directly demonstrated.

Ammonia also promotes changes in the inhibitory GABAergic neurotransmission. The contribution of astrocytes is associated with the location on the outer membrane of their mitochondria, of peripheral benzodiazepine receptors (PBR), a hydrophobic transmembrane protein (Anholt et al., 1986). Naturally occurring PBR ligands have been identified in the CNS (reviewed by Costa and Guidotti, 1991). PBRs are thought to exert an indirect control upon the GABAergic system: their activation elicits the astrocytic synthesis of prognenelone-derived neurosteroids, some of which are positive modulators of the $GABA_A$ receptor complex. PBR concentrations have been found increased in the autopsy brain samples of hyperammonemic patients (Lavoie et al., 1990) and in the rat brain in experimentally induced hyperammonemia (Itzhak et al., 1995). PBRs are likely to contribute to the ammonia-induced oxidative and nitrosative stress. This possibility will be discussed in some detail in ❷ *Sect. 4.3.1*. PBR is a component of the permeability transition pore (PTP) whose opening leads to MPT, and benzodiazepines acting at PBR in astrocytes induce the formation of free radicals (Jayakumar et al., 2002).

Direct interference of ammonia with GABAergic neurons has long been a matter of controversy, but more recent data tend to favor this concept. In cultured neurons, ammonia added at concentrations that fall within the range of those that occur in severe hyperammonemic encephalopathy induced a concentration-dependent increase in GABA-induced chloride current. Radioligand-binding assays provided evidence for direct interaction of ammonia on the $GABA_A$/PBR receptor complex: ammonia increased selectively maximal binding of agonist ligands (e.g., muscimol, flunitrazepam) to the $GABA_A$ receptor complex. Higher ammonia concentrations (up to 2.0 mM) returned the ligand binding to the control level. In addition, ammonia and PBR agonists were found to synergistically enhance the binding of the GABA agonist muscimol (Ha and Basile, 1996). Thus, ammonia not only directly enhances the ability of GABA to depress neuronal activity but can also further inhibit CNS function by its synergistic interactions with natural benzodiazepine receptor ligands.

4.3 Changes in the NMDA Receptor/Nitric Oxide/cGMP Pathway and Their Pathophysiological Effects

Experimental studies using hyperammonemia models in vivo and in vitro and recent observations on brain autopsy material derived from patients with different forms of hyperammonemic encephalopathy point to the disturbances in NO synthesis and the ensuing changes in the guanylate cyclase (GC) activity and cGMP as key factors contributing in variable ways to acute or chronic ammonia neurotoxicity. The prevailing view is that ammonia induces changes in the NO/cGMP pathway by altering the activity of NMDA receptors located on the nerve cell membranes. Their overactivation occurs in the acute phase and downregulation upon prolonged exposure to ammonia.

4.3.1 NMDA Receptor-Mediated Nitrosative and Oxidative Stress

The acute phase response was documented by the reduction of the death rate of mice-administered lethal doses of ammonia by coadministration of NMDA receptor antagonists and nitric oxide synthase (NOS) inhibitors (Marcaida et al., 1992; Hermenegildo et al., 1996). Metabolic disturbances and neurological symptoms resulting from NMDA receptor activation by ammonia are related to the degree of oxidative stress and the rate of formation of free radicals, and enhancement of NO synthesis in acute hyperammonemic models is reflected by extracellular accumulation of cGMP as measured in brain microdialysates (Hermenegildo et al., 2000).

For a long time, overactivation of the NMDA receptor/NO pathway has been interpreted to be a domain of neurons. More recent studies counter this view. Ammonia was found to induce NOS in cultured astrocytes, and the process appeared sensitive to NMDA receptor blockade (Schliess et al., 2002).

NO and superoxide anion O_2^- conjugate to form peroxinitrite, which exerts its damaging effects by nitration of functional proteins. Indeed, activation of the NMDA/NO pathway in astrocytes promoted protein tyrosine nitration (PTN) (Schliess et al., 2002) and inhibited glutamine synthetase activity (Miñana et al., 1997), supporting the involvement of PTN in tonic regulation of ammonia detoxication in astrocytes. Collectively, the results point to the concerted contribution of astrocytes and neurons to the oxidative and nitrosative stress evoked by ammonia.

Of note, activation of an NMDA receptor-independent component of NO synthesis has been suspected on the basis of the observation that the ammonia neurotoxicity attenuating effect of the NOS blocker, nitroarginine was qualitatively different from NMDA receptor antagonists (Kosenko et al., 1995). This NMDA receptor-independent NO synthesis takes place in the nerve endings and ensues in the increased synaptic uptake of the NO precursor, arginine, and/or increased expression of nNOS mRNA and protein (Rao et al., 1997; Rao, 2002).

4.3.2 Inhibition of cGMP Production in Chronic Hyperammonemia: Pathophysiological Implications

Long-term exposure to ammonia decreases the activity of the NMDA/NO/cGMP pathway. This decrease is noticeable at three distinct steps of the process, i.e., (1) desensitization of NMDA receptors (Peterson et al., 1990; Saransaari et al., 1997; see also ❯ *Sect. 4.2*), (2) decreased ability of NO to activate GC (Hermenegildo et al., 1998), and (3) decreased expression and spontaneous activity of GC (Monfort et al., 2002). Decreased cGMP accumulation in the brains of rats with chronic hyperammonemia is correlated in time with the occurrence with changes in pathophysiological parameters that are causally related to the glutamate-NO-cGMP pathway such as diurnal rhythm or learning ability (Corbalan et al., 2002b). Brain region-dependent changes in the expression of the α subunit of GC and in the susceptibility of the enzyme to activation by NO have been recorded in autopsy material derived from liver cirrhotic patients (Corbalan et al., 2002a). Decreased accumulation of cGMP is held responsible for intellectual and memory deficits of chronic hyperammonemic patients. Accordingly, enrichment of cGMP in the brain by pharmacological manipulation, i.e., application of inhibitors of a phosphodiesterase that degrades cGMP (zaprinast, sildenafil), restores some of the deficits in hyperammonemic rats (Erceg et al., 2005a, b).

4.4 Factors Involved in Hyperammonemic Brain Edema

4.4.1 Accumulation of Glutamine and Other Osmolytes: Osmotic Versus Metabolic Effects and Critical Coupling to Bioenergetic Failure

Hyperammonemic brain edema is a direct consequence of astrocytic swelling. As mentioned in ❯ *Sect. 1* and elsewhere in this volume, glutamine synthesis from glutamate and ammonia, the major route of ammonia metabolism in the brain, occurs almost exclusively in astrocytes (see also ❯ *Figure 12-2*). Hence, excessive glutamine accumulation in the brain is the major consequence of overloading astrocytes with ammonia under hyperammonemic conditions. For many years, glutamine accumulating in excess has been considered as neutral to the brain function. As briefly discussed before (❯ *Sect. 4.1.1*), this view has been changed on the basis of observations made in different laboratories that treatment with the glutamine synthetase inhibitor, methionyl-D,L-sulfoximine (MSO) attenuates or abolishes the pathophysiological and metabolic brain abnormalities in hyperammonemic animals (Takahashi et al., 1991; Hawkins et al., 1993; Willard-Mack et al., 1996; for a review see Albrecht and Dolińska, 2001; Albrecht, 2003). In particular, the treatment resulted in attenuation of cerebral edema, astrocytic swelling, and a spectrum of edema-related changes in the cerebral blood flow (CBF) and metabolism. The original interpretation ascribed the edema-inducing effect of glutamine to its intraastrocytic osmotic action. This simple view has been seriously challenged in studies in which hypothermia was effectively employed to attenuate hyperammonemic brain edema. The beneficial effect occurred despite the absence of reduction of cerebral glutamine content

(Chatauret et al., 2003). Reduction of edema correlated well with the reduced accumulation of other osmotically active metabolites, lactate and alanine, whose overproduction in hyperammonemia manifests excessive anaerobic glycolysis. Ammonia-induced brain cell swelling in cerebral cortical slices was reduced by antioxidative and antiexcitotoxic manipulations. This was likewise not accompanied by reduced glutamine content (Zielińska et al., 2003). Reduction of hyperammonemic brain edema was also achieved with the use of an NMDA receptor antagonist memantine (Vogels et al., 1997). The two above studies highlight the contribution of the excitotoxic component and oxidative (or nitrosative) stress to the mechanism by which ammonia brings about cerebral edema.

With reference to glutamine, the results collectively indicate that the nonosmotic (or not directly osmotic) aspect of glutamine-mediated ammonia neurotoxicity is associated with its ability to accumulate in astrocytic mitochondria, to cause mitochondrial permeability transition, and to mediate generation of free radicals. Hence, glutamine specifically interferes with astrocytic energy metabolism: this effect is likely to be less significant under hypothermia, when the brain energy demand is decreased. Of note, as opposed to astrocytes, neurons appear resistant to the deteriorating effects of treatment with excess glutamine (Jayakumar et al., 2004). In summary, ammonia-induced astrocytic swelling, which is the primary cause of hyperammonemic brain edema, reflects a vicious circle of oxidative stress, bioenergetic failure, and increased accumulation of glutamine and other osmolytes.

4.4.2 Changes in the Cerebral Blood Flow

Increased CBF often accompanies acute hyperammonemia. The cause/consequence relationship between cerebral hyperemia and hyperammonemic brain edema has not been unequivocally established. Nonetheless, pharmacological increase or decrease of CBF in hyperammonemic rats promoted or attenuated cerebral edema, respectively, bespeaking the increased CBF as a causative factor (Chung et al., 2001, 2003). Data on the mechanism underlying the cerebral hyperemic effect of ammonia have been controversial. Earlier data have implicated activation of NOS, coupled by a not clearly defined mechanism to increased glutamine synthesis (Master et al., 1999). This finding is difficult to reconcile with the lack of effect of NOS inhibitors on cerebral hyperemia in another hyperammonemic model (Larsen et al., 2001). Recent preliminary data point to the possible involvement of activation of heme oxygenase, providing a logical link to oxidative stress as a primary event in ammonia neurotoxicity (discussed by Vaquero and Blei, 2004).

5 Derangements Resulting from Increased Accumulation in Brain of Aromatic Amino Acids: Catecholaminergic Transmission

As mentioned earlier, increased accumulation in the hyperammonemic brain of glutamine promotes, by exchange, increased inward transport of the aromatic amino acids. Of these, tryptophan and tyrosine serve as precursors of the catecholamine neurotransmitters, serotonin [5-hydroxytryptamine (5-HT)] and dopamine. However, evidence in favor of the serotoninergic or dopaminergic dysfunction has been mostly derived from studies on animal models of hepatic failure or from analysis of autopsy material derived from cirrhotic patients. In neither case, it was possible to make a clear distinction between the effects of ammonia and other toxic factors. Nonetheless, considering that blood or brain ammonia have been found increased in all the conditions examined the available evidence deserves a mention.

Increased serotoninergic inhibitory tone has been implicated in hyperammonemic encephalopathy on the basis of brain region- and ligand-dependent differences in 5-HT receptor binding in HE patients (Rao and Butterworth, 1994) and in mice with congenital hyperammonemia (Robinson et al., 1992). Derangement of motor activity in rats with hyperammonemia associated with acute liver failure was partly corrected following the administration of the wide spectrum serotonin receptor antagonist methysergide (Yurdaydin et al., 1996). One other interesting tryptophan metabolite is oxindole, which produces sedative and hypotensive effects in the rat. High-oxindole levels were recorded in the blood of cirrhotic patients with hyperammonemia (Moroni et al., 1998) and in the blood and brain of rats with experimentally induced acute liver failure (Carpenedo et al., 1998).

The fact that hyperammonemia is associated with motor deficits has prompted interest in the role of dopamine. A loss of striatal dopamine D_2 receptors has been documented both in the brain tissue of cirrhotic patients (Mousseau et al., 1993) and in rats with TAA-induced acute liver failure (Waśkiewicz et al., 2001). The above findings plus the increased activities of dopamine-metabolizing enzymes, monoamineoxidases A (MAO-A) and B (MAO-B) observed in the brain tissue of HE patients (Rao et al., 1993), suggest that a decrease in dopaminergic tone may indeed contribute to motor disturbances associated with HE. An impairment of a glutamate-dependent aspect of dopamine function appears to be impaired in a model of HE; in particular, the NMDA-dependent dopamine release in striatum in vitro (Borkowska et al., 1997), and accumulation of dopamine metabolites in vivo (Borkowska et al., 1999), is decreased in rats with TAA-induced acute liver failure. Recently, ammonia directly administered to the rat striatum by microdialysis has been shown to evoke a prompt release of dopamine and its metabolite, 3,4-dihydroxyphenylacetate, which occurred partly by an NMDA receptor-dependent mechanism (Anderzhanova et al., 2003).

6   Protection Against Neurotoxic Effects of Ammonia

An overwhelming majority of attempts at counteracting neurotoxic effects of ammonia overdose have so far been undertaken in experimental animals. The studies have served well the purpose of elucidating steps in the mechanism, and only recently they have begun to set stage for clinical trials. NMDA receptor antagonists and NOS inhibitors ameliorated various aspects of acute ammonia neurotoxicity, adding credence to the role of overactivation of the NMDA receptor/NO/cGMP pathway (❷ Sect. 4.3.1). Also memory deficits have been corrected in chronic hyperammonemic rats by counteracting the decrease of brain cGMP content using phosphodiesterase inhibitors (❷ Sect. 4.3.2). Substantial information as to the underlying mechanisms of toxicity, and the potential neuroprotection targets, has been gained in studies employing taurine, a nonproteinaceous sulfur amino acid which exerts cytoprotection by a plethora of mechanisms (for a review see Saransaari and Oja, 2000). Cell volume increase in rat cerebral cortical slices treated with ammonia is thought to mimic hyperammonemic brain edema in vivo (Zielińska et al., 1999, 2003). Incubation with taurine ameliorated this effect with potency similar to NMDA receptor antagonists and NOS blockers (Zielińska et al., 2003). In vivo, taurine reduced to zero value ammonia-induced accumulation of cGMP and substantially attenuated the accumulation of free radicals in striatal microdialysates (Hilgier et al., 2003). The beneficial effects of taurine in vivo and in vitro were counteracted by $GABA_A$ and glycine receptor antagonists (Zielińska et al., 2003; Hilgier et al., 2005). This result supports $GABA_A$ and glycine receptor agonistic properties of taurine inferred from electrophysiological tests (Chepkova et al., 2002). Taurine also prevents derangements in ammonia-induced dopamine release and metabolism in vivo (Anderzhanova et al., 2003). Ammonia is a trigger of taurine release from CNS cells in vivo and in vitro, and it has been speculated that the release renders taurine available for neuroprotective action (Zielińska, et al., 2002; for discussion see Albrecht and Zielińska, 2002; Albrecht and Węgrzynowicz, 2005). However, poor penetration of taurine across the blood–brain barrier prevents its future application in the therapy of hyperammonemic encephalopathy.

As mentioned earlier, hypothermia ameliorated many of the toxic effects of ammonia in hyperammonemic animals. Most recently, Jalan et al. (2004) successfully employed moderate hypothermia in a clinical setting. The authors were able to control the elevated intracranial pressure, a condition related to cerebral edema, in a group of patients with acute liver failure. This was a life-saving treatment allowing bridging all the patients to liver transplantation. Interestingly, this simple "patient cooling" procedure normalized the key aspects of ammonia metabolism in these patients: it decreased blood ammonia, reduced the extraction of brain ammonia, and brought to normal level the brain ammonia flux. Moreover, it normalized CBF. Emphasizing the similarity of metabolic and pathophysiological responses noted in hypothermia-treated hyperammonemic animals and now in ALF patients, the authors of the pertinent and exhaustive editorial make a fully justified comment that "Interactions between bench and bedside have been keenly successful in unraveling the pathogenesis of brain edema during 2 decades of work ..." (Vaquero and Blei, 2004).

7 General Conclusions: Missing Links in the Knowledge

Excessive entry of ammonia to the brain primarily impairs the metabolism of astrocytes, which leads to changes in their morphology and impairment of their neuromodulatory function. Ammonia-induced changes in neural transmission can thus be viewed as mostly secondary to astrocytic dysfunction or damage. On the other hand, studies on cultured neurons and some in vivo observations favor the direct neurotoxic action of ammonia mediated by activation of the NMDA class of glutamate receptors, thus excitotoxic in nature. Clearly, further studies will have to assess in quantitative terms the relative contribution of the different cell types of the CNS: astrocytes, neurons, and cerebrovascular endothelial cells to the toxic damage evoked by ammonia in the intact brain. Combination of modern imaging techniques in situ (PET, NMR) with selective interference with the metabolism of these different cell types should render the desired answers.

References

Albrecht J. 2003. Glucose-derived osmolytes and energy impairment in brain edema accompanying liver failure: The role of glutamine reevaluated. Gastroenterology 125: 976-978.

Albrecht J, Dolińska M. 2001. Glutamine as a pathogenic factor in hepatic encephalopathy. J Neurosci Res 65: 1-6.

Albrecht J, Jones EA. 1999. Hepatic encephalopathy: Molecular mechanisms underlying the clinical syndrome. J Neurol Sci 170: 138-146.

Albrecht J, Hilgier W, Januszewski S, Kapuściński A, Quack G. 1994. Increase of the brain uptake index for L-ornithine in rats with hepatic encephalopathy. Neuroreport 5: 671-673.

Albrecht J, Hilgier W, Januszewski S, Quack G. 1996. Contrasting effects of toxic liver damage on the brain uptake indices of ornithine, arginine and lysine: Modulation by treatment with ornithine aspartate. Metab Brain Dis 11: 229-236.

Albrecht J, Węgrzynowicz M. 2005. Endogenous neuroprotectants in ammonia toxicity in the CNS: Facts and hypotheses. Metab Brain Dis, 20: 253-263.

Albrecht J, Wysmyk-Cybula U, Rafałowska U. 1985. Na^+/K^+ ATPase activity and GABA uptake in astroglial cell-enriched fractions and synaptosomes derived from rats in the early stage of experimental hepatogenic encephalopathy. Acta Neurol Scand 72: 317-320.

Albrecht J, Wysmyk-Cybula U, Rafałowska U. 1987. Cerebral oxygen consumption in experimental hepatic encephalopathy: Different responses in astrocytes, neurons and synaptosomes. Exp Neurol 97: 418-422.

Albrecht J, Zielińska M. 2002. The role of inhibitory amino acidergic neurotransmission in hepatic encephalopathy: A critical overview. Metab Brain Dis 17: 283-294.

Anderzhanova E, Oja SS, Saransaari P, Albrecht J. 2003. Changes in the striatal extracellular levels of dopamine and dihydroxyphenylacetic acid evoked by ammonia and N-methyl-D-aspartate: Modulation by taurine. Brain Res 977: 290-293.

Anholt RRH, Pedersen PL, De Souza EB, Snyder S. 1986. The peripheral benzodiazepine receptor: Localization to the mitochondrial membrane. J Biol Chem 261: 576-583.

Bachmann C. 2002. Mechanisms of hyperammonemia. Clin Chem Lab Med 40: 653-662.

Bai G, Rama Rao KV, Murthy CR, Panickar KS, Jayakumar AR, et al. 2001. Ammonia induces the mitochondrial permeability transition in primary cultures of rat astrocytes. J Neurosci Res 66: 981-991.

Borkowska HD, Oja SS, Saransaari P, Albrecht J. 1997. Release of [^{3}H]dopamine from striatal and cerebral cortical slices from rats with thioacetamide-induced hepatic encephalopathy: Different responses to stimulation by potassium ions and agonists of ionotropic glutamate receptors. Neurochem Res 22: 101-106.

Borkowska HD, Oja SS, Saransaari P, Hilgier W, Albrecht J. 1999. N-Methyl-D-aspartate-evoked changes in the striatal extracellular levels of dopamine and its metabolites in vivo in rats with acute hepatic encephalopathy. Neurosci Lett 268: 151-154.

Butterworth RF. 2002. Pathophysiology of hepatic encephalopathy: A new look at ammonia. Metab Brain Dis 17: 221-227.

Carpenedo R, Mannaioni G, Moroni F. 1998. Oxindole, a sedative tryptophan metabolite, accumulates in blood and brain of rats with acute hepatic failure. J Neurochem 70: 1998-2003.

Chatauret N, Zwingmann C, Rose C, Leibfritz D, Butterworth RF. 2003. Effects of hypothermia on brain glucose metabolism in acute liver failure: A H/C-nuclear magnetic resonance study. Gastroenterology 125: 815-824.

Chepkova AN, Doreulee N, Yanovsky Y, Mukhopadhyay D, Haas HL, et al. 2002. Long-lasting enhancement of corticostriatal neurotransmission by taurine. Eur J Neurosci 16: 1523-1530.

Chung C, Gottstein J, Blei AT. 2001. Indomethacin prevents the development of experimental ammonia-induced brain edema in rats after portacaval anastomosis. Hepatology 34: 249-254.

Chung C, Vaquero J, Gottstein J, Blei AT. 2003. Vasopressin accelerates experimental ammonia-induced brain edema in rats after portacaval anastomosis. J Hepatol 39: 193-199.

Conn HO. 1994. The hepatic encephalopathies. Hepatic Encephalopathy: Syndromes and Therapies. Conn HO, Bircher J, editors. Bloomington: Medi-Ed Press; pp. 1-12.

Corbalan R, Chatauret N, Behrends S, Butterworth RF, Felipo V. 2002a. Region selective alterations of soluble guanylate cyclase content and modulation in brain of cirrhotic patients. Hepatology 36: 1155-1162.

Corbalan R, Hernandez-Viadel M, Llansola M, Montoliu C, Felipo V. 2002b. Chronic hyperammonemia alters protein phosphorylation and glutamate receptor-associated signal transduction in brain. Neurochem Int 41: 103-108.

Cordoba J, Blei AT. 1996. Brain edema and hepatic encephalopathy. Semin Liver Dis 16: 271-280.

Costa E, Guidotti A. 1991. Diazepam binding inhibitor (DBI): A peptide with multiple biological actions. Life Sci 49: 325-344.

Dolińska M, Hilgier W, Albrecht J. 1996. Ammonia stimulates glutamine uptake to nonsynaptic cerebral mitochondria of the rat. Neurosci Lett 11: 175-184.

Dombro RS, Hutson DG, Norenberg MD. 1993. The action of ammonia on astrocyte glycogen and glycogenolysis. Mol Chem Neuropathol 19: 259-268.

Ede RJ, Gove CD, Hughes RD, Marshall W, Williams R. 1987. Reduced brain Na^+, K^+-ATPase activity in rats with galactosamine-induced hepatic failure: Relationship to encephalopathy and cerebral oedema. Clin Sci 72: 365-371.

Erceg S, Monfort P, Hernandez-Viadel M, Llansola M, Montoliu C, et al. 2005a. Restoration of learning ability in hyperammonemic rats by increasing extracellular cGMP in brain. Brain Res 1036: 115-121.

Erceg S, Monfort P, Hernandez-Viadel M, Rodrigo R, Montoliu C, et al. 2005b. Oral administration of sildenafil restores learning ability in rats with hyperammonemia and with portacaval shunts. Hepatology 41: 299-306.

Faff-Michalak L, Albrecht J. 1991. Aspartate aminotransferase, malate dehydrogenase, and pyruvate-carboxylase activities in rat cerebral synaptic and nonsynaptic mitochondria: Effects of in vitro treatment with ammonia, hyperammonemia and hepatic encephalopathy. Metab Brain Dis 6: 187-197.

Faff-Michalak L, Albrecht J. 1993. The two catalytic components of the 2-oxoglutarate dehydrogenase complex in rat cerebral synaptic and nonsynaptic mitochondria. Comparison of the response to in vitro treatment with ammonia, hyperammonemia and hepatic encephalopathy. Neurochem Res 18: 119-123.

Farinelli SE, Nicklas WJ. 1992. Glutamate metabolism in rat cortical astrocyte cultures. J Neurochem 58: 1905-1915.

Felipo V, Butterworth RF. 2002. Neurobiology of ammonia. Prog Neurobiol 67: 259-279.

Fillenz M, Lowry JP, Boutelle MG, Fray AE. 1999. The role of astrocytes and noradrenaline in neuronal glucose metabolism. Acta Physiol Scand 167: 275-284.

Giordano G, Sanchez-Perez AM, Burgal M, Montoliu C, Costa LG, et al. 2005. Chronic exposure to ammonia induces isoform-selective alterations in the intracellular distribution and NMDA receptor-mediated translocation of protein kinase C in cerebellar neurons in culture. J Neurochem 92: 143-157.

Gregorios JB, Mozes LW, Norenberg MD. 1985. Morphologic effects of ammonia on primary astrocyte cultures. II. Electron microscopic studies. J Neuropathol Exp Neurol 44: 404-414.

Ha JH, Basile AS. 1996. Modulation of ligand binding to components of the $GABA_A$ receptor complex by ammonia: Implications for the pathogenesis of hyperammonemic syndromes. Brain Res 720: 35-44.

Hawkins RA, Jessy J, Mans AM, De Joseph MR. 1993. Effect of reducing brain glutamine synthesis on metabolic symptoms of hepatic encephalopathy. J Neurochem 60: 1000-1006.

Hermenegildo C, Marcaida G, Montoliu C, Grisolia S, Minana M, et al. 1996. NMDA receptor antagonists prevent acute ammonia toxicity in mice. Neurochem Res 21: 1237-1244.

Hermenegildo C, Monfort P, Felipo V. 2000. Activation of N-methyl-D-aspartate receptors in rat brain in vivo following acute ammonia intoxication: Characterization by in vivo brain microdialysis. Hepatology 31: 709-715.

Hermenegildo C, Montoliu C, Llansola M. 1998. Chronic hyperammonemia impairs the glutamate-nitric oxide-cyclic GMP pthway in cerebellar neurons in culture and in the rat in vivo. Eur J Neurosci 10: 3201-3209.

Hertz L, Murthy CR, Lai JC, Fitzpatrick SM, Cooper AJ. 1987. Some metabolic effects of ammonia on astrocytes and neurons in primary cultures. Neurochem Pathol 6: 97-129.

Hilgier W, Anderzhanova E, Oja SS, Saransaari P, Albrecht J. 2003. Taurine reduces ammonia- and N-methyl-D-aspartate-induced accumulation of cyclic GMP and hydroxyl radicals in microdialysates of the rat striatum. Eur J Pharmacol 468: 21-25.

Hilgier W, Oja SS, Saransaari P, Albrecht J. 2005. Taurine prevents ammonia-induced accumulation of cyclic GMP in rat striatum by interaction with GABA$_A$ and glycine receptors. Brain Res, 1043: 242-246.

Hilgier W, Puka M, Albrecht J. 1992. Characteristics of large neutral amino acid- induced release of preloaded glutamine from rat cerebral capillaries in vitro: Effects of ammonia, hepatic encephalopathy and γ-glutamyltranspeptidase inhibitors. J Neurosci Res 32: 221-226.

Huang CH, Liu PZ. 2001. New insights into the Rh superfamily of genes and proteins in erythroid cells and nonerythroid tissues. Blood Cells Mol Dis 27: 90-101.

Itzhak Y, Roig-Cantisano A, Dombro RS, Norenberg MD. 1995. Acute liver failure and hyperammonemia increase peripheral-type benzodiazepine receptor binding and pregnenolone synthesis in mouse brain. Brain Res 705: 345-348.

Jalan R, Olde Damink SW, Deutz NE, Hayes PC, Lee A. 2004. Moderate hypothermia in patients with acute liver failure and uncontrolled intracranial hypertension. Gastroenterology 127: 1338-1346.

Jayakumar AR, Panickar KS, Norenberg MD. 2002. Effects on free radical generation by ligands of the peripheral benzodiazepine receptor in cultured neural cells. J Neurochem 83: 1226-1234.

Jayakumar AR, Rao KVR, Schousboe A, Norenberg MD. 2004. Glutamine-induced free radical production in cultured astrocytes. Glia 46: 296-301.

Jessy J, Mans AM, De Joseph MR, Hawkins RA. 1990. Hyperammonaemia causes many of the changes found after portacaval shunting. Biochem J 272: 311-317.

Jonung T, Rigotti P, James JH, Brackett K, Fischer JE. 1985. Effect of hyperammonemia and methionine sulfoximine on the kinetic parameters of blood-brain transport of leucine and phenylalanine. J Neurochem 45: 308-318.

Knecht K, Michalak A, Rose C, Rothstein JD, Butterworth RF. 1997. Decreased glutamate transporter (GLT-1) expression in frontal cortex of rats with acute liver failure. Neurosci Lett 229: 201-203.

Kosenko E, Kaminsky Y, Grau E, Miñana MD, Grisolia S, et al. 1995. Nitroarginine, an inhibitor of nitric oxide synthetase, attenuates ammonia toxicity and ammonia-induced alterations in brain metabolism. Neurochem Res 20: 451-456.

Lai JC, Murthy CR, Cooper AJ, Hertz E, Hertz L. 1989. Differential effects of ammonia and beta-methylene-DL-aspartate on metabolism of glutamate and related amino acids by astrocytes and neurons in primary culture. Neurochem Res 14: 377-389.

Larsen FS, Gottstein J, Blei AT. 2001. Cerebral hyperemia and nitric oxide synthase in rats with ammonia-induced brain edema. J Hepatol 34: 548-554.

Lavoie J, Pomier-Layrargues G, Butterworth RF. 1990. Increased densities of "peripheral-type" benzodiazepine receptors in autopsied brain tissue from cirrhotic patients with hepatic encephalopathy. Hepatology 11: 874-878.

Liskowsky DR, Norenberg LO, Norenberg MD. 1986. Effect of ammonia on cyclic AMP production in primary astrocyte cultures. Brain Res 386: 386-388.

Marcaida G, Felipo V, Hermenegildo C, Miñana M, Grisolia S. 1992. Acute ammonia toxicity is mediated by the NMDA type of glutamate receptors. FEBS Lett 296: 67-68.

Master S, Gottstein J, Blei AT. 1999. Cerebral blood flow and the development of ammonia-induced brain edema in rats after portacaval anastomosis. Hepatology 30: 876-880.

Miñana MD, Kosenko E, Marcaida G, Hermenegildo C, Montoliu C, et al. 1997. Modulation of glutamine synthesis in cultured astrocytes by nitric oxide. Cell Mol Neurobiol 17: 433-445.

Monfort P, Muñoz MD, El Ayadi A, Kosenko E, Felipo V. 2002. Effects of hyperammonemia and liver failure on glutamatergic neurotransmission. Metab Brain Dis 17: 237-250.

Moroni F, Carpenedo R, Venturini I, Baraldi M, Zeneroli ML. 1998. Oxindole in pathogenesis of hepatic encephalopathy. Lancet 351: 1861.

Mousseau DD, Perney P, Pomier Layrargues G, Butterworth RF. 1993. Selective loss of palladial dopamine D2 receptor density in hepatic encephalopathy. Neurosci Lett 162: 192-196.

Mullen KD. 1999. Terminology of hepatic encephalopathy work in evolution. Advances in Hepatic Encephalopathy & Metabolism in Liver Disease. Yourdaydin C, Bozkaya H, editors. Istanbul; ISHE pp. 139-150.

Murthy CR, Rama Rao KV, Bai G, Norenberg MD. 2001. Ammonia-induced production of free radicals in primary cultures of rat astrocytes. J Neurosci Res 66: 282-288.

Norenberg MD, Hugo Z, Neary JT, Roig-Cantesano A. 1997. The glial glutamate transporter in hyperammonemia and hepatic encephalopathy: Relation to energy metabolism and glutamatergic neurotransmission. Glia 21: 124-133.

Norenberg MD, Martinez-Hernandez A. 1979. Fine structural localization of glutamine synthetase in astrocytes of rat brain. Brain Res 161: 303-310.

Ott P, Larsen FS. 2004. Blood-brain barrier permeability to ammonia in liver failure: A critical reappraisal. Neurochem Int 44: 185-198.

Palaiologos G, Hertz L, Schousboe A. 1989. Role of aspartate aminotransferase and mitochondrial dicarboxylate transport for release of endogenously and exogenously supplied neurotransmitter in glutamatergic neurons. Neurochem Res 14: 359-366.

Peng L, Schousboe A, Hertz L. 1991. Utilization of alpha-ketoglutarate as a precursor for transmitter glutamate in cultured cerebellar granule cells. Neurochem Res 16: 29-34.

Peterson C, Giguere JF, Cotman CW, Butterworth RF. 1990. Selective loss of N-methyl-D-aspartate binding sites in rat brain following portacaval anastomosis. J Neurochem 55: 386-390.

Pilbeam CM, Anderson RM, Bhathal PS. 1983. The brain in experimental portal-systemic encephalopathy. I. Morphological changes in three animal models. J Pathol 140: 331-345.

Rahman T, Hodgson H. 2001. Clinical management of acute hepatic failure. Intensive Care Med 27: 467-476.

Rama Rao KV, Chen M, Simard JM, Norenberg MD. 2003. Suppression of ammonia-induced astrocyte swelling by cyclosporin A. J Neurosci Res 74: 891-897.

Rao VL. 2002. Nitric oxide in hepatic encephalopathy and hyperammonemia. Neurochem Int 41: 161-170.

Rao VL, Audet RM, Butterworth RF. 1997. Portacaval shunting and hyperammonemia stimulate the uptake of L-[^{3}H] arginine but not of L-[^{3}H]nitroarginine into rat brain synaptosomes. J Neurochem 68: 337-443.

Rao VL, Butterworth RF. 1994. Alterations of [^{3}H]8-OH-DPAT and [^{3}H]ketanserin binding sites in autopsied brain tissue from cirrhotic patients with hepatic encephalopathy. Neurosci Lett 182: 69-72.

Rao VL, Giguere JF, Pomier Layrargues G, Butterworth RF. 1993. Increased activites of MAO$_A$ and MAO$_B$ in autopsied brain tissue from cirrhotic patients with hepatic encephalopathy. Brain Res 621: 349-352.

Ratnakumari L, Murthy CR. 1992. In vitro and in vivo effects of ammonia on glucose metabolism in the astrocytes of rat cerebral cortex. Neurosci Lett 148: 85-88.

Robinson MB, Anegawa NJ, Gorry E, Qureshi IA, Coyle JT, et al. 1992. Brain serotonin 2 and serotonin 1A receptors are altered in the congenitally hyperammonemic sparse fur mouse. J Neurochem 58: 1016-1022.

Rose C. 2002. Increased extracellular brain glutamate in acute liver failure: Decreased uptake or increased release? Metab Brain Dis 17: 251-261.

Saran T, Hilgier W, Urbańska EM, Turski WA, Albrecht J. 2004. Kynurenic acid synthesis in cerebral cortical slices of rats with progressing symptoms of thioacetamide-induced hepatic encephalopathy. J Neurosci Res 75: 436-440.

Saransaari P, Oja SS. 2000. Taurine and neural cell damage. Amino Acids 19: 509-526.

Saransaari P, Oja SS, Borkowska HD, Kostinaho J, Hilgier W, et al. 1997. Effects of thioacetamide-induced hepatic failure on the N-methyl-D-aspartate receptor complex in the rat cerebral cortex, striatum and hippocampus: Binding of different ligands and expression of receptor subunit mRNA's. Mol Chem Neuropathol 32: 179-194.

Schliess F, Gorg B, Fischer R, Desjardins P, Bidmon HJ, et al. 2002. Ammonia induces MK-801-sensitive nitration and phosphorylation of protein tyrosine residues in rat astrocytes. FASEB J 16: 739-741.

Sonnewald U, Therrien G, Butterworth RF. 1996. Portacaval anastomosis results in altered neuron-astrocytic metabolic trafficking of amino acids: Evidence from ^{13}C-NMR studies. J Neurochem 46: 1711-1717.

Stastny F, Hilgier W, Albrecht J, Lisy V. 1988. Changes in the activity of γ-glutamyl transpeptidase in brain microvessels, astroglial cells and synaptosomes derived from rats with hepatic encephalopathy. Neurosci. Lett. 84: 323-328.

Sugimoto H, et al. 1998. Changes in the activity of γ-glutamyl transpeptidase in brain microvessels, astroglial cells and synaptosomes derived from rats with hepatic encephalopathy. Neurosci. Lett. 84: 323-328.

Sugimoto H, Koehler RC, Wilson DA, Brusilow SW, Traystman RJ. 1997. Methionine sulfoximine, a glutamine synthetase inhibitor, attenuates increased extracellular potassium activity during acute hyperammonemia. J Cereb Blood Flow Metab 17: 44-49.

Takahashi H, Koehler RC, Brusilow SW, Traystman RJ. 1991. Inhibition of brain glutamine accumulation prevents cerebral edema in hyperammonemic rats. Am J Physiol 261: H825-H829.

Vaquero J, Blei AT. 2004. Cooling the patient with acute liver failure. Gastroenterology 127: 1626-1629.

Vogels BAPM, Maas MAW, Daalhuisen J, Quack G, Chamuleau RAFM. 1997. Memantine, a noncompetitive NMDA receptor antagonist improves hyperammonemia-induced encephalopathy and acute hepatic encephalopathy in rats. Hepatology 25: 820-827.

Waniewski RA. 1992. Physiological levels of ammonia regulate glutamine synthesis from extracellular glutamate in astrocyte cultures. J Neurochem 58: 167-174.

Waśkiewicz J, Fręśko I, Lenkiewicz A, Albrecht J. 2001. Reversible decrease of dopamine D2 receptor density in the striatum of rats with acute hepatic failure. Brain Res 900: 143-145.

Willard-Mack CL, Koehler RC, Hirata T, Cork LC, Takahashi H, et al. 1996. Inhibition of glutamine synthetase reduces ammonia-induced astrocyte swelling in the rat. Neuroscience 71: 589-599.

Yu AC, Schousboe A, Hertz L. 1984. Influence of pathological concentrations of ammonia on metabolic fate of ^{14}C-labeled glutamate in astrocytes in primary cultures. J Neurochem 42: 594-597.

Yurdaydin C, Herneth AM, Püspök A, Steindl P, Singer E, et al. 1996. Modulation of hepatic encephalopathy in rats with thioacetamide-induced acute liver failure by serotonin antagonists. Eur J Hepatol Gastroenterol 8: 667-671.

Ziemińska E, Dolińska M, Lazarewicz JW, Albrecht J. 2000. Induction of permeability transition and swelling of rat brain mitochondria by glutamine. Neurotoxicology 21: 295-300.

Zielińska M, Hilgier W, Borkowska HD, Oja SS, Saransaari P, et al. 2002. Ammonia-induced extracellular accumulation of taurine in the rat striatum in vivo: Role of ionotropic glutamate receptors. Neurochem Res 27: 37-42.

Zielińska M, Hilgier W, Law RO, Gorynski P, Albrecht J. 1999. Effects of ammonia in vitro on endogenous taurine efflux and cell volume in rat cerebrocortical minislices: Influence of inhibitors of volume-sensitive amino acid transport. Neuroscience 91: 631-638.

Zielińska M, Law RO, Albrecht J. 2003. Excitotoxic mechanism of cell swelling in rat cerebral cortical slices treated acutely with ammonia. Neurochem Int 43: 299-303.

13 Disorders of Amino Acid Metabolism

M. Yudkoff

1	***Introduction: General Features of the Aminoacidopathies***	**279**
1.1	History	279
1.2	Enzymatic Deficiency	280
1.3	Metabolite Accumulation and Neurotoxicity	280
1.4	Nutrient Deficiency	280
1.5	Autosomal Recessive Inheritance	280
1.6	Treatability	281
1.7	Diagnosis	281
1.8	Clinical Heterogeneity	282
2	***Brain Injury in Aminoacidopathies***	**282**
2.1	Overview	282
2.2	Acute Versus Chronic Encephalopathies	282
2.3	Aminoacidopathies and Brain Energy Metabolism	283
2.4	Effects on Brain Amino Acid Uptake	283
2.5	Impact of Aminoacidopathies on Myelin Synthesis and Turnover	284
2.6	Aminoacidopathies and Neurotransmitters	284
2.7	Effects on Brain Water Metabolism	285
3	***Disorders of Phenylalanine Metabolism***	**285**
3.1	Phenylketonuria	285
3.1.1	Phenylalanine Hydroxylase Deficiency	285
3.1.2	Defects of Biopterin Synthesis and Metabolism	287
4	***Disorders of Branched-Chain Amino Acids***	**287**
4.1	Maple Syrup Urine Disease	287
5	***Disorders of Glycine Metabolism: Nonketotic Hyperglycinemia***	**289**
6	***Disorders of Sulfur Amino Acid Metabolism***	**291**
6.1	Homocystinuria	291
6.1.1	Cystathionine Synthase Deficiency	291
6.1.2	Remethylation Deficiency Homocystinuria	294
6.1.3	Methylenetetrahydrofolate Reductase Deficiency	294
6.1.4	Methionine Synthase Deficiency	295
6.1.5	Cobalamin-C Disease	295
6.1.6	Hereditary Folate Malabsorption	295
7	***The Urea Cycle Defects***	**295**
7.1	Ureagenesis	295
7.1.1	Carbamyl Phosphate Synthetase Deficiency	297
7.1.2	*N*-Acetylglutamate Synthetase Deficiency	297

© Springer-Verlag Berlin Heidelberg 2007

7.1.3	Ornithine Transcarbamylase Deficiency	297
7.1.4	Citrullinemia	298
7.1.5	Argininosuccinic Aciduria	298
7.1.6	Arginase Deficiency	298
7.1.7	Hyperornithinemia, Hyperammonemia, and Homocitrullinuria	298
7.1.8	Lysinuric Protein Intolerance	298
7.2	Management of Urea Cycle Defects	299

8	***Disorders of Glutathione Metabolism***	**299**
8.1	Glutathione Metabolism	299
8.1.1	5-Oxoprolinuria	299
8.1.2	γ-Glutamylcysteine Synthetase Deficiency	300
8.1.3	γ-Glutamyltranspeptidase Deficiency	300
8.1.4	5-Oxoprolinase Deficiency	300

9	***Disorders of GABA Metabolism***	**300**
9.1	GABA Metabolism	300
9.1.1	Pyridoxine Dependency	300
9.1.2	GABA Transaminase Deficiency	301
9.1.3	Succinic Semialdehyde Dehydrogenase Deficiency	301

| **10** | ***Disorders of N-Acetyl Aspartate Metabolism: Canavan's Disease*** | **301** |

1 Introduction: General Features of the Aminoacidopathies

1.1 History

Disorders of amino acid metabolism were among the first of the metabolic defects to be described by Archibald Garrod, the English physician who originated the notion of an inherited enzymatic deficiency that would give rise to a clinical syndrome of varying symptomatology, depending upon the nature of the underlying defect (Garrod, 1923). Garrod himself was aware of a relatively limited number of disorders—alcaptonuria, cystinuria, congenital porphyria, and pentosuria. In the years since his seminal insight, a large number of inherited diseases of amino acid metabolism have been described. These are outlined in ❯ Table 13-1.

Table 13-1
Disorders of Amino Acid Metabolism

Disorders	Examples
Disorders of phenylalanine metabolism	Phenylketonuria
	Phenylalanine hydroxylase deficiency
	Defects of biopterin metabolism
Disorders of the branched-chain amino acids	Maple syrup urine disease
	Classical disease
	Intermittent form
	Thiamine-responsive form
Disorders of glycine metabolism	Nonketotic hyperglycinemia
Disorders of sulfur amino acid metabolism	Homocystinuria
	Cystathionine synthetase deficiency
	Remethylation deficiency homocystinuria
	Methylenetetrahydrofolate reductase deficiency
	Methionine synthase deficiency (Cobalamin-E disease)
	Cobalamin-C disease
	Hereditary folate malabsorption
The urea cycle defects	Carbamyl phosphate synthetase deficiency
	N-Acetylglutamate synthetase deficiency
	Ornithine transcarbamylase deficiency
	Citrullinemia
	Argininosuccinic aciduria
	Arginase deficiency
	Hyperammonemia, hyperornithinemia, and homocitrullinuria syndromes
	Lysinuric protein intolerance
Disorders of glutathione metabolism	5-Oxoprolinuria (glutathione synthetase deficiency)
	γ-Glutamyl synthetase deficiency
	γ-Glutamyl transpeptidase deficiency
	5-Oxoprolinase deficiency
Disorders of GABA metabolism	Pyridoxine dependency
	GABA transaminase deficiency
	Succinic semialdehyde dehydrogenase deficiency
Disorders of N-acetylaspartate metabolism	Canavan's disease

Each disorder presents a series of biochemical, clinical, and therapeutic findings that are unique to that particular disturbance. Indeed, it is remarkable that a single category of disease should involve so pleiotropic a mode of clinical expression. However, there are certain features that typify most of these defects.

1.2 Enzymatic Deficiency

In almost all instances the cause is a mutation in a gene encoding an enzyme that is important for the oxidation of an amino acid or a group of amino acids. Rarely the mutation involves a transport system. A common result is the production of an enzyme protein that is lacking in catalytic efficiency or that weakly binds the amino acid or another coreactant. In some instances no enzyme protein at all is formed or an unstable species is made that is quickly degraded.

1.3 Metabolite Accumulation and Neurotoxicity

Normal degradative metabolism of amino acids entails complete conversion to CO_2, H_2O, and NH_3. For most amino acids this is a relatively rapid process, with a plasma half-life that typically is less than 30 min. A consequence of this rapid turnover is that a deficiency of an enzyme in a degradative pathway quickly leads to the accumulation of the parent compound. If there is a block in the conversion of compound A to compound B, not only will compound A accumulate but it also will be converted in excess amount to other species. Thus, the failure in phenylketonuria (PKU) to convert phenylalanine to tyrosine because of a congenital deficiency of phenylalanine hydroxylase will be associated with extremely high blood, urine, and cerebrospinal fluid (CSF) levels of phenylalanine itself as well as excess formation of phenylpyruvic and phenyllactic acids. The latter organic acids are normally produced from phenylalanine, but only in relatively modest amount because much phenylalanine is converted to tyrosine.

The most important clinical consequence of metabolite accumulation is that high concentrations of certain amino acids and metabolites of these compounds have very deleterious effects on brain function and development. Indeed, the developing brain is exquisitely sensitive to such syndromes of neurotoxicity. The severity of the syndrome and the age of onset will depend upon two factors: the nature of the accumulating metabolites and the completeness of the underlying enzyme deficiency. A complete or near-complete deficiency tends to cause symptoms early in life, even in the first days after birth. In contrast, if the defect is partial, the onset of symptoms may be delayed, sometimes even to adult life (Scriver et al., 2001; Kahler and Fahey, 2003).

1.4 Nutrient Deficiency

As emphasized in the preceding section, metabolite accumulation is a common feature of the disorders of amino acid metabolism. On occasion, the pathology of these disorders may be attributable not only to such toxicity but also to the failure to produce a compound essential to normal brain function. The deficient compound characteristically is a neurotransmitter produced from amino acid precursors. Examples of this pathophysiology would be the deficiencies of serotonin, dopamine, and other neurotransmitters that are the consequence of the defects of biopterin metabolism, which occur in PKU as well as in aromatic amino acid decarboxylase deficiency.

1.5 Autosomal Recessive Inheritance

Nearly all of the disorders of amino acid metabolism are inherited according to an autosomal recessive pattern. The only exceptions are those rare instances of sex-linked inheritance, the most important example of which probably is ornithine transcarbamylase (OTC) deficiency, one of the urea cycle defects. Dominant inheritance patterns are extremely unusual, and heterozygotes for most of these disorders almost never manifest clinical symptomatology.

1.6 Treatability

The clinical syndrome that attaches to many aminoacidopathies may be very severe. Cognitive impairment and mental retardation often are the rule in untreated cases, and death may be the unfortunate outcome. Most disorders are amenable to treatment with a special diet purposefully low in the "offending" amino acid. Many decades of experience in the management of affected patients indicates that scrupulous dietotherapy usually will allow a relatively happy outcome, often with normal cognitive development. Effective treatment presupposes an understanding of human nutrition, since the goal of therapy is not only to avoid undue excursions in the blood and tissue concentration of a particular amino acid but also to provide sufficient amino acid to enable normal or, at least, near-normal rates of protein synthesis. Indeed, overly stringent dietary control may prove to be as deleterious to the patient as lax control.

In addition to dietotherapy, many disorders are treatable by the administration of a vitamin or a group of vitamins. Unfortunately, this is not helpful in most cases, but there are instances in which the inherited gene mutation is of such nature that treatment with high doses of a particular water-soluble vitamin will "coax" the abnormal enzyme to function at a level that is compatible with normal or near-normal metabolic function. Thus, on occasion a patient with homocystinuria secondary to a deficiency of cystathionine synthetase, which requires pyridoxal phosphate as a cofactor in the reaction, will respond to the administration of high doses (25–100 mg/day) of vitamin B_6 with a sharp diminution of the blood homocystine concentration.

There also are individual syndromes of nutritional neurotoxicity for which an "antidote" to the toxic state is available. The best example of this therapeutic "motif" is the use of acylation therapy in the urea cycle defects. Affected patients are unable to convert ammonia to urea because of a deficiency of one of the five constituent enzymes of the urea cycle. The result is a severe hyperammonemia that often results in profound neurologic disability. It is possible to treat these patients by administration of an acylation agent, usually phenylacetate or benzoate, that forms a complex with either glutamine (phenylacetate) or glycine (benzoate). This complex is rapidly excreted in the urine, in the process removing waste nitrogen not as urea, but as either phenylacetylglutamine or hippurate (benzoylglycine). Such treatment is remarkably effective: the amount of nitrogen that can be so mobilized approximates that which would have been excreted via the urea cycle.

1.7 Diagnosis

The availability of reasonably effective treatments has prompted significant advances in the diagnosis of these disturbances. For most aminoacidopathies, the long-term outcome is generally favorable as long as treatment can be instituted before frank encephalopathy ensues. The more profound the encephalopathy and the more chronic its duration, the more guarded is the prognosis. Diagnosis usually presupposes the detection of an elevation of the concentration of a relevant amino acid in either blood or urine. This should be feasible with modern chromatographic methods for the separation and quantitation of the naturally occurring amino acids.

A notable conceptual and technologic advance has been the advent of mass newborn screening for the aminoacidurias. This approach is predicated on the use of triple-stage quadrupole mass spectrometry for the semiquantitative detection of amino acids in dried blood spots taken from newborn infants (usually by pricking the skin of the heel) at 2–3 days of life. After deproteinization of the blood specimen, aliquots of stable isotopic species of the amino acids to be measured are added. The amino acids are derivatized to yield the butyl esters and the latter are analyzed according to their characteristic mass spectra. The amino acids of clinical interest then can be quantitated with isotope dilution techniques.

The great advantage of this approach is that in only a few minutes of analytic time it is feasible to obtain semiquantitative measures of many different amino acids. Such information allows for the daily screening of a large number of infants with a "turnaround time" of minutes. Once a presumptive diagnosis of an aminoacidopathy can be made in this manner, treatment can be instituted promptly and steps can be undertaken to perform more definitive tests—including enzymatic assay and molecular DNA analysis in order to confirm or refute the screening result. If the final result is positive for a disturbance of amino acid metabolism, the infant will benefit enormously from rapid intervention. In many cases the institution of prompt treatment may mean the difference between incurring marked cognitive disability and a relatively normal outcome. Indeed, in extreme cases rapid intervention will be lifesaving.

1.8 Clinical Heterogeneity

The disorders of amino acid metabolism share in common many clinical features, the most salient of which is a tendency to sustain brain damage because of the toxic accumulation of an amino acid or an amino acid metabolite. However, it should be stressed that nearly each of these diseases is characterized by clinical features that are peculiar to the enzyme deficiency in question. The range of clinical presentation may vary from a relatively mild disorder to a syndrome of profound toxicity that may be fatal. Some patients manifest little more than biochemical findings, which appear to be wholly detached from any clinical counterpart. This, for example, is the rule in the patient with histidinemia secondary to histidase deficiency or the patient with prolinemia secondary to proline oxidase deficiency. In contrast, the patient with the classical form of nonketotic hyperglycinemia (NKH) secondary to a congenital lesion in the glycine cleavage system (GCS) typically manifests a very serious—and therapeutically resistant—seizure disorder that commonly begins in utero. Thus, generalizations about the clinical findings in the disturbances of amino acid metabolism are usually erroneous. Each syndrome must be considered as an individual entity.

An exhaustive discussion of each of the disorders listed in ❷ *Table 13-1* would oblige a book-length tract that would be beyond the scope of this chapter. We restrict ourselves to a exploration of three disorders that typify most of the features of the aminoacidopathies and for which recent researches have disclosed important new information with respect to the etiology of the neurologic damage. These three disorders are phenylketonuria (PKU), maple syrup urine disease (MSUD), and nonketotic hyperglycinemia (NKH).

2 Brain Injury in Aminoacidopathies

2.1 Overview

We still do not know the precise mechanism of brain damage in any of the aminoacidopathies, even though this subject has been studied intensively for decades. Our interpretation of these mechanisms has paralleled our more general understanding of brain development, neurophysiology, and neurochemistry. Advances in each of these fields have been applied with success toward an understanding of the pathophysiology of the aminoacidurias. The remarkable progress in recent years in terms of in vivo imaging of the human brain has proved to be of special value in terms of characterizing and interpreting pathologic events during a period of active intoxication.

2.2 Acute Versus Chronic Encephalopathies

In thinking about the pathophysiology of the aminoacidopathies, it is important to discriminate between acute and chronic intoxication. Acute encephalopathies, which are typified by diseases such as MSUD and the urea cycle defects, entail a sudden onset of obtundation, coma, intractable vomiting and dehydration, abnormal posturing, temperature instability, and seizures. In the most severe cases, respiratory and cardiovascular compromise will ensue within a matter of days or even hours. The price of survival will then be assisted ventilation. Many affected infants and youngsters will succumb. Pathologic examination of the brain usually discloses evidence of frank neuronal degeneration coupled with gliosis and dysmyelination. If the toxic state persists for a sufficiently long period the patient will sustain cortical atrophy. The lamentable clinical counterpart of the brain pathology will be cognitive impairment, including frank mental retardation, severe spasticity, and a chronic epileptic disorder (Wagner et al., 2000; Felipo and Butterworth, 2002).

Acute encephalopathies occur secondary to the relatively abrupt accumulation of a neurotoxin, which affects critical aspects of brain function. Among the best known examples would be ammonia, the branched-chain amino acids, and the organic acids (methylmalonic acid, propionic acid, 3-methylglutaconic acid, etc.) that are the intermediate products of the metabolism of the branched-chain amino acids. A sudden rise of the blood ammonia level usually occurs in the urea cycle defects, which are discussed in more detail below. Branched-chain amino acid (leucine, isoleucine, and valine) accumulation is the hallmark of MSUD. These perturbations are the result of sudden metabolic decompensation brought about by a stress, usually the stress of an acute infection. The unwanted consequence of such stress is a sudden increase in the rate of body protein catabolism. This leads to the presentation of a largeload of amino acid (or ammonia or an organic acid) to an enzyme system that already is compromised. The defective system cannot cope with the load and the offending metabolite accumulates within the brain to a toxic concentration.

The chronic encephalopathies correspond not to the sudden rise of a particular amino acid or amino acid derivative but to the persistence over time of a metabolite that adversely impacts brain development. This process is typified by PKU in which the affected child does not sustain the sudden onset of coma, but a more gradual encephalopathy that manifests primarily as a failure to achieve developmental milestones according to the usual schedule. A similar process may occur in homocystinuria, although in this disorder stroke may occur.

2.3 Aminoacidopathies and Brain Energy Metabolism

It is thought that most encephalopathies, whether acute or chronic, involve an adverse impact on brain energetics, a conclusion suggested by a wealth of studies in in vitro systems and in animal models. Recent in vivo investigations involving magnetic resonance imaging confirm that energy metabolism is disturbed, often seriously so, in patients with aminoacidopathies. The use of ^{31}P magnetic resonance spectroscopy shows increased brain ADP in individuals with PKU in the basal state, and an even greater rise when affected adults receive a phenylalanine load (Pietz et al., 2003). Increased brain lactate is seen during metabolic decompensation in patients with MSUD (Jan et al., 2003), an indication of compromised oxidative metabolism and a consequent rise of the brain redox state.

The factors that lead from amino acid or ammonia accumulation to a compromise of brain energy metabolism are not certain. The skein of reactions and transport processes that constitute brain energy metabolism is enormously complex, and any toxin likely exerts its effects through more than one mechanism. Among the factors that have been implicated in the aminoacidopathies are direct inhibition of the tricarboxylic acid cycle, the uncoupling of oxidative phosphorylation, impairments of glucose homeostasis, alterations of the intracellular redox potential, and excessive stimulation of excitatory amino acid receptors (usually NMDA receptors) that ultimately leads to heightened consumption of ATP in response to untoward activation of Na^+/K^+ATPase (Felipo and Butterworth, 2002; Pilla et al., 2003).

2.4 Effects on Brain Amino Acid Uptake

Amino acid metabolism directs an individual amino acid toward one of three different fates: (1) A small portion becomes a messenger or informational molecule. Examples are neurotransmitters and hormones derived from tyrosine or tryptophan or both. (2) A much larger fraction is utilized for the synthesis of nucleic acids and proteins, with the latter constituting a much more prominent destiny. (3) Amino acids are oxidized to completion, forming CO_2, H_2O, and NH_3. Almost all aminoacidopathies involve congenital defects of the latter process, which quantitatively subsumes a fairly significant portion of amino acid flux.

As a result, in the untreated state the concentration of the amino acid in question will increase in body fluids to extravagantly high levels, even as much as 20–30 times the usual value. At such an extremely high concentration of phenylalanine, for example, the importation into brain of tyrosine and tryptophan, each of which must compete with phenylalanine for a binding locus on a transport protein (the so-called L system in the case of phenylalanine) of the blood–brain barrier, may become compromised. Such competitive inhibition may be a powerful factor in giving rise to clinical symptomatology (Wagner et al., 2000), since the rate of protein synthesis will be reduced if the brain does not have available sufficient stores of all amino acids. In addition, if the intracerebral level of tyrosine and tryptophan are low enough, the brain may no longer be able to synthesize adequate amounts of either dopamine or serotonin, both of which are critical for normal neurologic function. Recent in vivo investigations with magnetic resonance spectroscopy are consistent with this formulation. If adults with PKU are treated with dietary supplements of neutral amino acids, the serum level of the administered compounds increases and the level of phenylalanine does not change. In contrast, the concentration of phenylalanine in brain declines toward the level that is regularly seen in heterozygotes for PKU (Koch et al., 2003). It is not entirely certain whether there is any clinical counterpart to this novel approach to treatment, although it is worth noting that some patients reported a relief from the depression that had bothered them before the study, a phenomenon that could denote increased brain levels of dopamine and/or serotonin (Smith and Kang, 2000).

The interplay of brain amino acid uptake with genetics and dietary status is extremely complex. On occasion large changes of amino acid transport may even result from the low-protein diets that commonly are employed to treat the aminoacidopathies. Patients sense that eating protein makes them feel lethargic, confused, or nauseated. As a result they tend to avoid protein and, in its place, eat relatively large amounts of carbohydrate, which evokes the secretion of insulin from the endocrine pancreas. Insulin favors an increase in the ratio of tryptophan to other amino acids, particularly the branched-chain amino acids. More tryptophan consequently enters the central nervous system, where it can be converted to serotonin. Excess serotonin may have manifold effects on brain function, including the suppression of appetite. It has been speculated, for example, that the anorexia which regularly accompanies patients with urea cycle defects may be attributable to such pathophysiology.

2.5 Impact of Aminoacidopathies on Myelin Synthesis and Turnover

A common feature of most aminoacidopathies is a disturbance in brain lipid metabolism. Thus, in MSUD intramyelinic edema is quite common, particularly during the acute phase of metabolic decompensation (Jan et al., 2003). In most syndromes a decrease of brain lipids, proteolipids, and cerebrosides is a prominent pathologic finding. One etiologic factor may be a derangement of brain protein synthesis, a normal rate of which is essential to the maintenance of proteins that constitute myelin. In addition, the changes of brain energy metabolism that accompany most of the aminoacidopathies (see above) will give rise to a disturbance of the synthesis of brain lipids, a process extremely intense during brain development and very dependent on the availability of ATP. Finally, an important element in the hypomyelination and dysmyelination of these syndromes appears to be the fact that certain amino acids or their metabolites can directly inhibit lipid synthesis, including the synthesis of arachidonic and docosahexaenoic acids (Infante et al., 2001). A recent finding of significance in this regard is that elevated levels of phenylalanine may impair brain cholesterol synthesis by inhibiting 3-hydroxy-3-methyl-glutaryl-CoA reductase, the rate-limiting step of cholesterol synthesis (Shefer et al., 2000).

2.6 Aminoacidopathies and Neurotransmitters

The brain injury that accompanies many of the aminoacidopathies may be a consequence of aberrant function of neurotransmitter receptor systems. In NKH, which involves a mutation of the GCS and is associated with extreme increases of glycine in the blood and brain, there may be excess excitation of the N-methyl-D-aspartate (NMDA) receptor, which has a glycine-binding site. The fact that this disorder gives rise to a severe seizure disorder seems consistent with the formulation that there is a massive increase of excitatory tone. Indeed, this conceptualization provided a rationale for treating this disorder with NMDA receptor blockers, but the effort was not of obvious therapeutic benefit (Deutsch et al., 1998). In a similar vein, marked elevations of the brain ammonia concentration impairs high-affinity uptake of glutamate into astrocytes and neurons, a perturbation that also would be expected to augment excitatory tone. The response to the failure of glutamate uptake appears to be a downregulation of NMDA receptors (Chan and Butterworth, 2003).

2.7 Effects on Brain Water Metabolism

Clinicians who care for patients with aminoacidopathies, particularly during a period of acute crisis, note the tendency to develop brain edema and increased intracranial pressure of alarming degree. We have noted (above) a tendency to develop intramyelinic edema during the acute phase of MSUD. Similarly, brain edema is a common occurrence during metabolic decompensation in patients with urea cycle defects, in whom the severe increase of brain ammonia favors the synthesis of glutamine from glutamate and ammonia. In the brain, the glutamine synthetase system is restricted to the astrocytes, and the abrupt rise of internal glutamine is focused primarily on these glial cells. Glutamine is a relatively small molecule (M, 146 Da) that then draws water into the astrocytes. Furthermore, an increase of brain blood flow, a phenomenon that often accompanies hyperammonemia, would favor the entry of water into that compartment (Felipo and Butterworth, 2002).

3 Disorders of Phenylalanine Metabolism

3.1 Phenylketonuria

3.1.1 Phenylalanine Hydroxylase Deficiency

Phenylketonuria is a relatively common disturbance of amino acid metabolism. It occurs with a frequency (in European populations) of approximately 1:20,000 live births. The cause in almost all cases is a mutation in the gene that encodes phenylalanine hydroxylase, the enzyme system that mediates the conversion of phenylalanine to tyrosine (❯ *Figure 13-1*). Most children born with such a mutation will manifest little or no activity of the hydroxylase. In the untreated state, the blood phenylalanine regularly exceeds 20 mg/dL (1.2 mmol/L; normal 0.05 mmol/L), but some patients have enough residual activity that they display only hyperphenylalaninemia (usually 3–15 mg/dL), even when they receive an unrestricted diet. There is some evidence to suggest that even this relatively mild biochemical perturbation can cause neurologic impairment and, therefore, obliges therapy (White et al., 2001; Cedarbaum, 2002).

Phenylalanine hydroxylase normally is constituted as a trimer of three identical subunits with a combined molecular weight of ~150,000. Nearly all activity is confined to the liver, with virtually no activity being detectable in the central nervous system, which, therefore, must import tyrosine from the blood in order to synthesize dopamine, epinephrine, norepinephrine, and related metabolites. In humans, the hydroxylase is coded on chromosome 12q22–24.1. The coding region includes 13 exons that extend over 90 kb of genomic DNA.

Human PKU usually results from a mutation within the coding sequence rather than frank deletions of the gene. Among northern European populations a common mutation is a G → A transition at the 5′-donor splice site of intron 12. As a consequence, the C terminus is absent. Another common mutation among Europeans (about 20% of cases) is a C → G transversion in exon 12 that results in the substitution of a tryptophan for an arginine (Eisensmith and Woo, 1991). The number of individual mutations is relatively large, and compound heterozygosity is the rule with about 75% of all cases so affected. In the American population, more than 70 discrete mutations have been identified to date (Guldberg et al., 1996). Genetic analysis, as reflected in the study of restriction length polymorphisms, is relevant to clinical outcome.

Figure 13-1

Phenylalanine metabolism and phenylketonuria (PKU). Classical PKU is caused by a congenital deficiency of phenylalanine hydroxylase (reaction 1). In rare instances the cause is a defect in the metabolism of tetrahydrobiopterin (BH$_4$), a cofactor for the hydroxylase. Enzymes: (1) phenylalanine hydroxylase; (2) dihydropteridine reductase; (3) GTP cyclohydrolase; (4) 6-pyruvoyltetrahydrobiopterin synthase. Abbreviations: QH$_2$, dihydrobiopterin; BH$_4$, tetrahydrobiopterin; GTP, guanosine-5'-triphosphate; DEDT, D-*erythro*-dihydroneopterin triphosphate

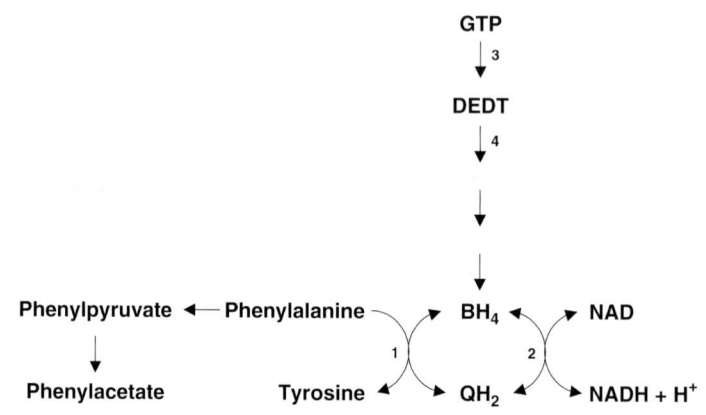

```
                          GTP
                           ↓ 3
                          DEDT
                           ↓ 4
                           ↓
                           ↓
Phenylpyruvate ← Phenylalanine    BH₄ ←     NAD
      ↓                      ⟩  ⟨      ⟩  ⟨
Phenylacetate       Tyrosine      QH₂        NADH + H⁺
                          1          2
```

The clinical presentation usually is one of an infant who seems quite normal at birth. There are no pathognomonic physical findings, although some youngsters seem very fair, perhaps because of a defect of melanin synthesis (Farishian and Whittaker, 1980). On occasion these patients may display a musty odor because of excess excretion in their urine of phenylacetic acid. Indeed, it is the presence of this odor that prompts some parents to seek medical help for their children (Centerwall and Centerwall, 2000). Patients also can display a severe eczema that will remit with careful therapy. Patients with long-standing disease may manifest autistic features, including stereotypic movements, self-injurious behavior, and profound mental retardation. Motor dysfunction, including Parkinsonism, can occur with long-standing disease, and there even are reports of this clinical finding in patients with hyperphenylalaninemia (Kikuchi et al., 2004).

The mainstay of management for PKU is a diet restricted in its content of phenylalanine. In most instances, this will mandate a daily phenylalanine intake of 200–500 mg. Rigorous monitoring of blood phenylalanine concentrations is essential in order to keep this parameter within the therapeutic range, usually construed as 2–6 mg/dL (~ 0.1–0.3 mmol/L) (Cedarbaum, 2002). It has been speculated that there may be a role for tyrosine supplementation, even without a phenylalanine-restricted diet, in PKU. The rationale was that the failure of normal conversion of phenylalanine to tyrosine makes the latter an essential amino acid. However, there is no evidence for this approach, although there may be a rationale for the administration of tyrosine supplementation in addition to dietary phenylalanine restriction (van Spronsen et al., 2001).

A persistent controversy has involved the duration of therapy. It once had been thought—and hoped— that patients could dispense with the dietary restriction once the brain was "fully formed," usually assumed to occur by later childhood. However, the result of more recent research suggests that it is prudent to maintain dietary restriction of phenylalanine throughout adolescence, and, conceivably, indefinitely. Discontinuation of the diet has been associated with retrogressive changes and even with frank deterioration of the IQ score. With scrupulous dietary control almost all patients can enjoy normal intelligence and development, but there probably is an increased incidence of perceptual-learning disabilities, attention deficit/hyperactivity disorder, emotional problems, and "minor" motor difficulties (Diamond and Herzberg, 1996; Huijbregts et al., 2002).

The success in managing female patients with PKU over the past several decades has led to the fortunate outcome that these women are able to bear normal children, although all of their progeny will be at least carriers for one mutation of phenylalanine hydroxylase. It is imperative to carefully monitor phenylalanine levels in such women during their pregnancy, and to administration dietotherapy in order to keep levels in a therapeutic range. It is clear that the failure to restrict phenylalanine in such women can have severe teratogenic effects in their offspring, including microcephaly and cardiac defects (Levy and Ghavami, 1996).

3.1.2 Defects of Biopterin Synthesis and Metabolism

The hydroxylation of phenylalanine to form tyrosine requires a cofactor that serves as an electron donor. Tetrahydrobiopterin (BH_4) occupies this pivotal role by transferring electrons to molecular oxygen, in the process generating tyrosine and dihydrobiopterin (❯ *Figure 13-1*). In order to maintain functional capacity, the system then must regenerate BH_4, a reaction catalyzed by dihydropteridine reductase (DHPR), an NADH-dependent enzyme present in many cells. In the brain, this reductase is particularly important because it enables the hydroxylation of tyrosine in the dopamine synthesis pathway as well as the hydroxylation of serotonin in the pathway leading to the synthesis of serotonin. The enzyme, which is coded on chromosome 4p15.1–p16.1, bears little homology to other reductases such as dihydrofolate reductase.

The classical form of PKU almost always (>95% of cases) occurs secondary to a mutation in the phenylalanine hydroxylase gene (above). However, there are instances when this enzyme is perfectly normal and the disturbance of phenylalanine metabolism results from a mutation in the synthesis of BH_4, which is produced from GTP with sepiapterin as an intermediate (❯ *Figure 13-1*). The unique clinical feature of this syndrome is that neurologic deterioration may be progressive even with careful dietary control of the phenylalanine concentration. The presumed reason for the neurologic disarray is that phenylalanine restriction fails to rectify the deficit in the synthesis of dopamine, serotonin, and related biogenic amines. This phenomenon is reflected in the fact that the urine of affected patients shows very low levels of metabolites of serotonin, norepinephrine, and dopamine. Furthermore, since the reductase appears to play a role in maintaining tetrahydrofolate in the brain, the blood folate level may be low.

Treatment of this disorder with dietotherapy is unavailing, as indicated above. However, the therapeutic picture is not always bleak because the administration of carbidopa and tryptophan may serve to replete depleted CNS stores of these neurotransmitters. The administration of folinic acid will serve to replete brain stores of tetrahydrofolate.

The diagnosis requires the demonstration of DHPR deficiency in either skin fibroblasts or amniotic cells. It is important to demonstrate the presence of a normal level of activity of phenylalanine hydroxylase.

There are even rare forms of defects of biopterin metabolism than can lead to PKU. Examples would include GTP cyclohydrolase deficiency and 6-pyruvoyltetrahydrobiopterin synthase deficiency (see ❯ *Figure 13-1*). Affected patients characteristically manifest truncal hypotonia, psychomotor retardation that may be severe, seizures, and limb hypertonia. Seizures are frequent. There have been efforts to treat such disorders with intravenous BH_4, but this compound only slowly crosses the blood–brain barrier. An alternative may be the administration of synthetic pterin derivatives.

4 Disorders of Branched-Chain Amino Acids

4.1 Maple Syrup Urine Disease

Maple syrup urine disease acquired its piquant name because of the unusual urine odor that the branched-chain ketoacids confer. (In Europe, which lacks maple syrup, the odor has been likened to that of burned sugar). The cause is a congenital deficiency of a mitochondrial enzyme, branched-chain ketoacid decarboxylase (❯ *Figure 13-2*), that mediates the conversion of the ketoacids to coenzyme A derivatives. The ketoacids consequently accumulate and are freely reaminated to the parent branched-chain amino acids, which attain a very high concentration in body fluids. The latter biochemical finding is of diagnostic importance.

◼ **Figure 13-2**
Defects of branched-chain amino acid metabolism. The initial metabolic transformation (reaction 1) is transamination to a branched-chain ketoacid. Maple syrup urine disease (MSUD) is caused by a congenital deficiency of the next step (reaction 2), or branched-chain ketoacid decarboxylase. The accumulated ketoacids are freely reaminated to the parent amino acid, thereby accounting for the extreme elevations of the latter in body fluids. Enzymes: (1) branched-chain amino acid transaminase; (2) branched-chain amino acid decarboxylase. Abbreviations: T-PP, thiamine pyrophosphate; LipA, lipoic acid; NAD, nicotinamide adenine dinucleotide

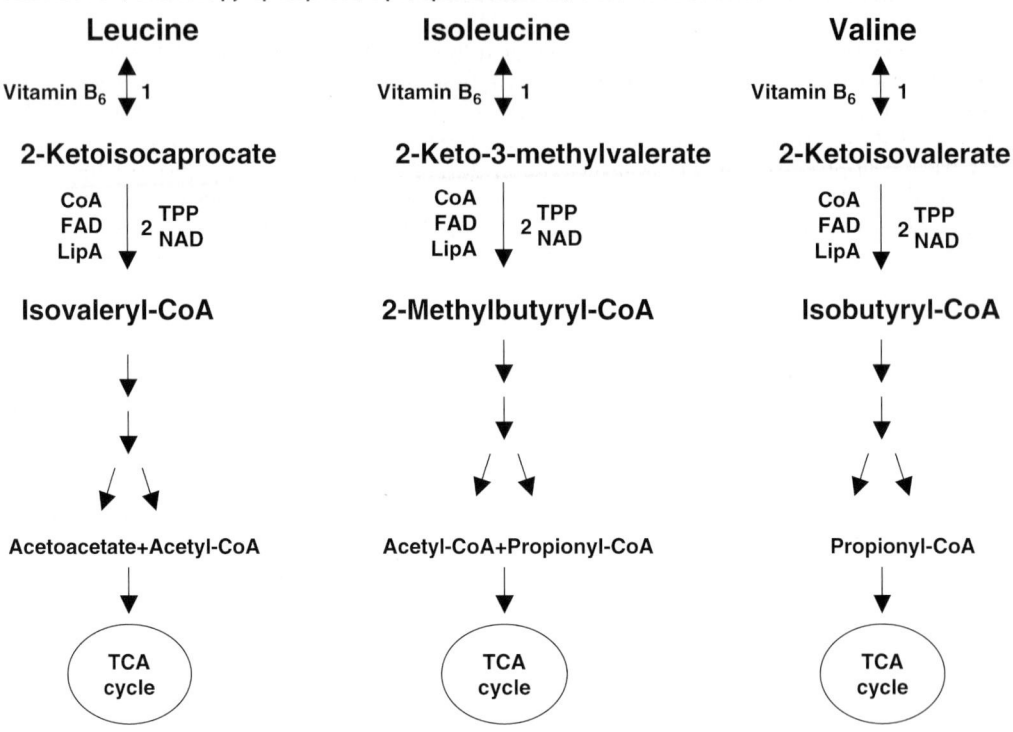

Leucine	Isoleucine	Valine
Vitamin B_6 ↕ 1	Vitamin B_6 ↕ 1	Vitamin B_6 ↕ 1
2-Ketoisocaprocate	2-Keto-3-methylvalerate	2-Ketoisovalerate
CoA, FAD, LipA ↓ 2 TPP, NAD	CoA, FAD, LipA ↓ 2 TPP, NAD	CoA, FAD, LipA ↓ 2 TPP, NAD
Isovaleryl-CoA	2-Methylbutyryl-CoA	Isobutyryl-CoA
↓↓	↓↓	↓↓
Acetoacetate+Acetyl-CoA	Acetyl-CoA+Propionyl-CoA	Propionyl-CoA
↓	↓	↓
TCA cycle	TCA cycle	TCA cycle

The decarboxylase is a very large complex composed of four subunits: E1-α, E1-β, E2, and E3. Activity of the complex is regulated by a specific kinase and phosphatase. Patients with the classical form of MSUD typically will manifest congenital lesions owing to the deficiency of the E1-α subunit, although other defects have been described (Chuang et al., 1995). A large number of different mutations in the E1-α subunit are known, but most involve faulty assembly of the heterotetrameric ($\gamma_2\beta_2$) E1 protein. The E3 subunit is common not only to the branched-chain complex but also to the other decarboxylases, including pyruvate dehydrogenase and 2-oxo-glutarate dehydrogenase. A mutation of this protein, therefore, is associated not only with the accumulation of the branched-chain amino acids but also with lactic acidosis and dysfunction of the tricarboxylic acid cycle.

The classical form of MSUD becomes clinically manifest within 1–3 days of birth. During gestation, the placenta is able to maintain the infant's blood amino acid levels within a nontoxic range, but soon after ingesting the first protein-containing feeds the child manifests alternating periods of lethargy and irritability. Within a day or two this syndrome will progress to one of vomiting, hypotonia, and coma. By the end of the first week the untreated child sustains progressive respiratory embarrassment, possible seizures, increased intracranial pressure and, in the most extreme cases, a need for mechanical ventilation. If the infant survives, the infant is at high risk of suffering brain damage and permanent impairment, often including mental retardation.

The primary threat to the older youngsters with MSUD is a metabolic relapse. This usually occurs because an intercurrent infection causes the catabolism of endogenous protein stores, and the consequent release of branched-chain amino acids overwhelms the child's already compromised ability to metabolize branched-chain ketoacids. In the most severe cases the result is neurologic dysfunction and even increased intracranial pressure. Vigorous therapy must be instituted in order to rapidly diminish the body burden of the branched-chain amino acids.

In rare instances the enzymatic defect is incomplete and the clinical presentation is correspondingly delayed until later childhood or, very rarely, until adult life.

Thiamine (vitamin B_1) is a cofactor for the branched-chain decarboxylase reaction, and an occasional patient harbors a mutation that is responsive to the administration of thiamine in very large doses (10–30 mg/day). It is thought that these cases involve an increased K_m of the enzyme for thiamine so that binding of the cofactor is diminished.

The overall incidence of MSUD in the general population is approximately 1/250,000 live births. There are genetic isolates that have a much greater incidence, an example being the Mennonite community, in whom the carrier state probably is as great as 1/7 individuals. A major advance in combating this disease has been the advent of mass screening programs (see above). These allow a presumptive diagnosis to be made relatively early in the course of the illness, perhaps even before the onset of frank symptomatology. Definitive diagnosis requires either demonstration of the underlying enzymatic defect or molecular analysis of DNA. Antenatal testing is feasible.

For the patient with classical disease, the best available therapy is dietary restriction of the branched-chain amino acids to allow only the minimum amount necessary to support a normal rate of growth. Fortunately, special formulae now are available from which these compounds have been eliminated. A relatively new approach to the affected child is orthotopic liver transplantation, which seems to almost completely rectify the underlying biochemical abnormality, at least when the patient is not in a state of decompensation. Of course, liver transplantation entails real risks to the patient in terms of the surgical procedure itself and the subsequent risks of infection and malignant transformation. Gene therapy for MSUD is not yet clinically available, but studies in vitro indicate that it is possible to use a retroviral vector to transfer both the E1-α and E2 subunits of the decarboxylase complex (Chuang et al., 1995; Mueller et al., 1995).

Normal cognitive development is possible for these youngsters (Kaplan et al., 1991). The advent of mass newborn screening (see above) should mean that the long-term prognosis is even more favorable.

5 Disorders of Glycine Metabolism: Nonketotic Hyperglycinemia

As shown in ❶ *Figure 13-3*, the main route for glycine breakdown is the GCS, a mitochondrial complex found primarily in the liver and kidney. Essential cofactors in the reaction are pyridoxal phosphate and tetrahydrofolic acid. In addition to catalyzing the breakdown of glycine, the GCS furnishes precursors to the "one carbon pool" of folate intermediates, which is essential to many synthetic reactions, including the formation of purine and pyrimidine bases as well as the provision of methyl groups needed for the synthesis of neurotransmitters and hormones (Kikuchi, 1973).

The term "nonketotic hyperglycinemia" originally was intended to differentiate this disorder from "ketotic hyperglycinemia," which formerly referred to propionic acidemia and methylmalonic acidemia, both of which can cause extreme elevations of blood glycine and ketosis during periods of metabolic decompensation. With the discovery of the causes of the enzyme deficiencies that cause "ketotic hyperglycinemia" this term fell into disuse. The term "nonketotic hyperglycinemia" has been retained to describe the disorder caused by a congenital mutation of the GCS, usually in the initial step that involves the decarboxylation of glycine (Toone et al., 2000).

The clinical syndrome is among the most dramatic of all the aminoacidopathies. An intractable seizure disorder dominates the clinical picture in classical cases. The onset probably occurs in utero in many cases and is manifest soon after birth, when hiccuping, myoclonic jerks, and a profound hypotonia became evident. When the infant is not suffering convulsions, he or she appears calmly asleep. The electroencephalogram usually shows an admixture of burst–suppression pattern and hypsarrhythmia.

Figure 13-3

Glycine cleavage system (GCS) and related reactions. Glycine is decarboxylated in a B_6-dependent reaction (reaction 1) that involves transfer of the methylene carbon to tetrahydrofolate to yield 5,10-methylenetetrahydrofolate. The latter reacts with glycine to form serine (reaction 2), or it is reduced to methyltetrahydrofolate (reaction 3), a key methyl donor. Once the methyl group is surrendered, tetrahydrofolate is regenerated and the cycle can begin again. Enzymes: (1) glycine cleavage system; (2) serine hydroxymethyltransferase; (3) $N^{5,10}$-methylenetetrahydrolate reductase. Abbreviations: $N_{5,10}$-CH_2-FH_4, $N_{5,10}$-methylenetetrahydrolate; FH_4 - Tetrahydrofolic acid

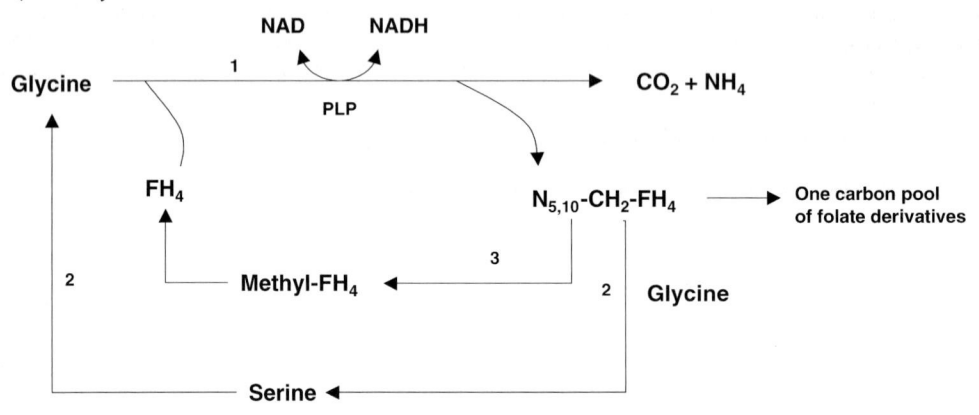

The long-term outlook is very poor. Most babies succumb during the first weeks of life, but the few who survive typically sustain profound mental retardation and neurologic disability. Magnetic resonance imaging of the brain confirms a pervasive loss of myelin and frank atrophy of the cortex.

A rare child will have only a partial defect of the GCS and will not present symptoms until later in life when the cause of growth failure and psychomotor retardation is found to be a variant form (Flusser et al., 2005). An even rarer presentation is that of the older child with no evidence of major mental impairment but with spinocerebellar degeneration and a progressive motor disorder (Steiner et al., 1996).

The biochemical hallmark of this disorder is the extravagantly high elevation of the blood glycine concentration, which commonly exceeds 1 mmol/L (normal: 150–350 mmol/L). Measurement of CSF amino acids always shows a glycine level in excess of 100 μmol/L (normal: 10 μmol/L). The ratio of glycine in the CSF compared with blood (CSF/blood ratio) typically is 5–10 times the control value (0.02).

It is important to emphasize that, in addition to the classical disease, there is a transient form that occurs in the newborns who have seizures but an otherwise normal neurologic examination. The glycine level in both the blood and the CSF is quite high. Urine organic acids are within normal limits. Patients usually do not have a burst–suppression pattern in their electroencephalogram. In most instances, the seizures remit by the time the child is 2 months of age. The infants do not usually suffer any permanent neurologic sequelae. The presumed cause is developmental immaturity of the GCS.

The treatment of NKH is quite frustrating. There is no specific therapy, although heroic efforts have been made to employ exchange transfusion or hemodialysis or both in an effort promptly to reduce the glycine concentration. These typically have no beneficial effects.

A pharmacologic approach is the administration of sodium benzoate (250–500 mg/kg/day), which the liver complexes with glycine to yield hippuric acid. The latter is rapidly excreted in the urine, thereby removing glycine from the body. Unfortunately, the clinical response has been minimal. A combination of benzoate and carnitine may be more useful (Van Hove et al., 1995), but it is clear that such treatment fails to improve outcome in many, and perhaps in most, youngsters. Dietotherapy with protein restriction is not helpful because glycine is a nonessential amino acid that the body freely synthesizes.

Our understanding the relevant pathophysiology of NKH has been greatly enabled by recent research into the importance of glycine to normal neurotransmission. Glycine serves as a postsynaptic inhibitor in the spinal cord and in selected central neurons. Early attempts at treatment, therefore, included strychnine, a glycine receptor blocker. Unfortunately, this approach was unavailing. An alternative is the administration of diazepam, which is known to displace strychnine from its binding sites. It may be that selected patients will respond to a combination of treatment with benzoate, which may lower glycine levels in the brain, and diazepam, which theoretically should block the glycine effect.

Yet another pharmacologic alternative is based on the observation that glycine potentiates the action of the NMDA receptor. This fact has prompted efforts to manage affected infants with ketamine and/or dextromethorphan, both of which are NMDA blockers (Alemzadeh et al., 1996). Although this approach seems theoretically sound, the results to date have been inconclusive, with only a fraction of patients manifesting any sort of robust clinical response. In a few instances, there appears to have been some improvement of the electroencephalogram and irritability. Treatment with dextromethorphan (5 mg/kg/day) is usually well tolerated.

6 Disorders of Sulfur Amino Acid Metabolism

6.1 Homocystinuria

As shown in ❯ Figure 13-4, the transsulfuration pathway mediates the transfer of the sulfur atom of methionine to serine to yield cysteine. The first step in the reaction is the activation of methionine to produce S-adenosylmethionine (SAM), which is the major methyl donor and an essential element in the synthesis of many neurotransmitters, hormones, and creatine. Decarboxylation of SAM also is the initial step in the production of the polyamines spermidine and spermine.

Once SAM donates its methyl group for synthetic reactions, S-adenosylhomocysteine is produced. This compound is a potent inhibitor of methyltransferases, a fact that probably accounts for at least a portion of the pathology of homocystinuria (below). Ordinarily, tissue concentrations of S-adenosylhomocysteine are kept extremely low because of the efficient action of a specific hydrolase that cleaves the compound to homocysteine and adenosine (❯ Figure 13-4).

Approximately half of the homocysteine generated in this manner becomes remethylated to yield methionine. In this reaction, catalyzed by 5-methyltetrahydrofolate-betaine methyltransferase, the immediate source of the methyl group is methylcobalamin, but the "original" source of –CH$_3$ is either betaine or 5-methyltetrahydrofolate. Reaction kinetics favor remethylation. A form of homocystinuria—so-called remethylation homocystinuria—will occur if remethylation is compromised. This may occur secondary to: (1) nutritional factors such as a deficiency of vitamin B$_{12}$; (2) a primary deficiency of the apoenzyme; (3) an inability to methylate either folate or vitamin B$_{12}$; or (4) treatment with antifolate agents as part of cancer chemotherapy.

Homocystinuria occurs most commonly as an aminoacidopathy because of a mutation in cystathionine-β-synthase, a vitamin B$_6$-dependent enzyme that mediates the condensation of homocysteine and serine to yield cystathionine. The gene is found on human chromosome 21. The formation of cystathionine is kinetically favored and the reaction rate is enhanced by SAM (Kluijtmans et al., 1996). In this manner, homocysteine levels are kept quite low since there are two discrete routes for the disposal of the amino acid: the "forward" reaction toward cystathionine as well as the "reverse" reaction to remethylate homocysteine and regenerate methionine. Cleavage of cystathionine is accomplished by the action of cystathionase, another vitamin B$_6$- dependent enzyme that is coded on human chromosome 16. This enzyme functions in the direction of cysteine synthesis, there being essentially no reversal of the reaction.

6.1.1 Cystathionine Synthase Deficiency

Various mutations give rise to classical homocystinuria, including the production of an unstable enzyme protein, an enzyme that loosely binds one of the reactants, or a truncated enzyme that differs in size from the normal variety (Kraus, 1994). The enzyme is found in most tissues, including the brain. Thus, most

Figure 13-4

Homocystinuria and the transsulfuration pathway. The pathway involves transfer of the sulfur of methionine to serine to form cysteine. Classical homocystinuria results from a congenital deficiency of cystathionine-β-synthase (reaction 5). A rarer form of homocystinuria is caused by a failure to remethylate homocysteine in a reaction for which the methyl group is donated by either betaine (reaction 4a) or methyltetrahydrofolate (reaction 4b). The enzyme for the latter pathway (reaction 4b) utilizes vitamin B_{12} as a prosthetic group. Remethylation deficiency homocystinuria can occur because of a failure to generate methylfolate or methylcobalamin. With a generalized failure of cobalamin activation or absorption, methylmalonic aciduria as well as homocystinuria results because cobalamin derivatives are essential to both pathways. Enzymes: (1) methionine-activating enzyme; (2) generic depiction of methyl group transfer from S-adenosylmethionine (SAM); (3) S-adenosylhomocysteine hydrolase; (4a) betaine/homocysteine methyltransferase; (4b) 5-methyltetrahydrofolate:homocysteine methyltransferase; (5) cystathionine-β-synthase; (6) cystathionase; (7) sulfite oxidase; (8) propionyl-CoA carboxylase; (9) methylmalonyl-CoA mutase; (10) $N^{5,10}$-methylenetetrahydrolate reductase; (11) and (12) glycine cleavage system (GCS); (13) and (14) hydroxycobalamin reductases; (15) cobalamin adenosyltransferase. Abbreviations: OH-B_{12}, hydroxycobalamin; Ado-B_{12}, adenosylcobalamin; methyl-B_{12}, methylcobalamin

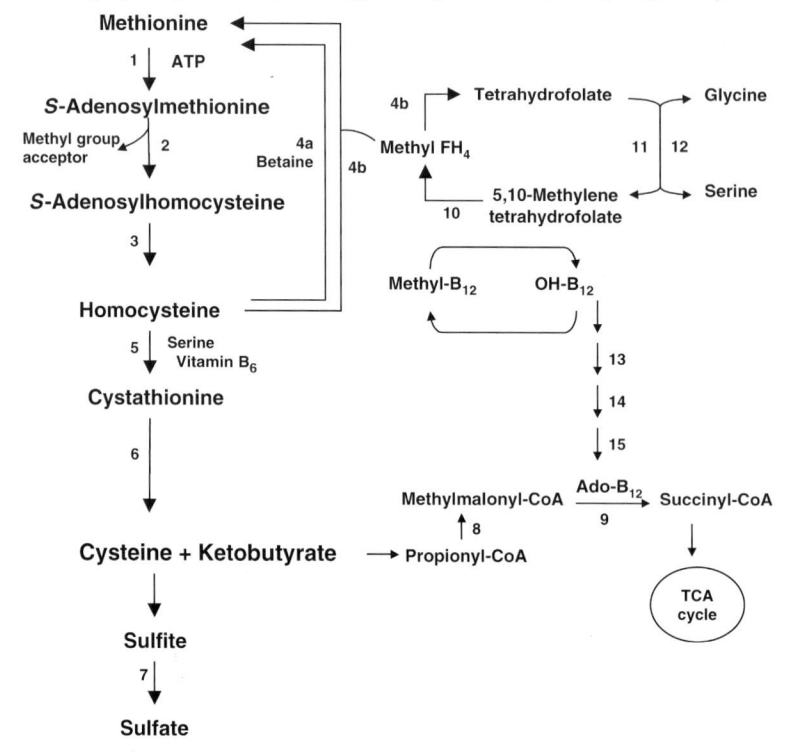

mutations result in a deficiency that becomes expressed in a variety of cell types. The biochemical counterpart of the enzyme deficiency is an elevation of blood homocystine to 50–200 μmol/L (normal: under 10 μmol/L). The blood cysteine concentration tends to be low, although this is not a diagnostic finding. As homocysteine no longer is converted to cystathionine, more homocysteine becomes available for remethylation to methionine (◆ Figure 13-4), and the blood concentration of methionine consequently is elevated (normal: 20–40 μmol/L).

A minority of patients respond favorably to treatment with pharmacologic doses of pyridoxine (25–100 mg daily) by manifesting a reduction of the blood homocysteine and methionine concentrations. Siblings tend to be concordant for this characteristic. In addition, pyridoxine responders usually will display a milder clinical course than nonresponders. Studies of enzymatic activity in skin fibroblasts of responders also show an augmentation of residual activity. It is of interest that kinetic studies have refuted the notion that responders have a mutation in which there is loose binding of pyridoxal phosphate to the enzyme. Indeed, the precise biochemical mechanism that would explain B_6 responsiveness is not clear, and even exposure to a high pyridoxine level does not increase enzyme activity to a normal level. It may be that in some individuals the administration of vitamin B_6 helps to stabilize an otherwise unstable species.

An important diagnostic feature is the presence of a downward displacement of the ocular lens, or *ectopia lentis*. This is present in at least half of patients by 5–10 years of age. Many cases are diagnosed by an ophthalmologist.

Mental retardation is the rule. The median IQ score for B_6 nonresponders is 56 and for responders it is 78. Mental retardation may be the presenting symptom. Convulsions occur in about 20% of cases. Clinicians who care for these youngsters often have the impression of an increased incidence of psychiatric problems, including depression, personality disorders, and difficulties with impulsiveness. Approximately half of cases are so affected.

The most arresting clinical feature is a thromboembolic diathesis that can affect any blood vessel. Thrombi are common in peripheral vessels as well as those of brain, heart, and kidney. Among pyridoxine non-responders the incidence of a major vessel occlusion during childhood is about 25%. Even pyridoxine responders can suffer thromboses, with about 25% of such patients being so affected by age 20 years. Dehydration secondary to vomiting and diarrhea is a major risk factor, as is the stress of surgery (including, at times, even "minor" surgery) and the administration of anesthesia. Patients with both homocystinuria and the relatively common Leiden mutation of clotting factor V appear to be at particularly high risk (Mandel et al., 1996).

Physical findings in cystathionine-β-synthase deficiency include a Marfanoid habitus of arachnodactyly, a high-arched palate, tall stature, and pes cavus. Most patients display abnormalities in bones, including osteoporosis and scoliosis. Surgical correction should be considered only in severe cases because of the thromboembolic risk attached to any surgical procedure.

Magnetic resonance imaging often shows evidence of multiinfarcts in almost any region of the brain. Pathologic examination discloses thickening of the intima of the endothelium and splitting of the smooth musculature of the media. There also may be evidence of frank demyelination and spongy degeneration of the white matter.

It is not certain how the biochemical derangements are translated into clinical symptomatology, but it is probable that hyperhomocystinemia rather than hypermethioninemia is primarily responsible. Homocysteine increases platelet adhesiveness, perhaps by enhancing the synthesis of thromboxanes. When homocysteine is infused into animals, there is a tendency for endothelial injury and dehiscence. The vascular endothelium shows evidence of becoming denuded to the point of providing an atherogenic nidus. Platelet survival time is diminished. Homocysteine also affects the clotting cascade by promoting activation of factor V, thereby favoring the conversion of prothrombin to thrombin.

There is evidence that homocysteine causes the deposition of increased copper in the vascular endothelium. This leads to the oxidation of ceruloplasmin and the subsequent release of enough H_2O_2 to injure endothelial cells. These studies have been performed with an in vitro system. When the culture medium is supplemented with catalase, protection against such oxidant injury is achieved (Starkebaum and Harlan, 1986).

In addition to toxic effects on the brain vasculature, homocysteine may directly injure the brain. Thus, administration of homocysteine to rats can cause grand mal convulsions, perhaps through blockade of the GABA receptor. In addition, homocysteine is oxidized in the brain to form homocysteic acid, which has glutamatergic activity (Folbergrova et al., 2005). As noted above, the presence of high levels of S-adenosylhomocysteine would inhibit methylation reactions in the brain. This would adversely affect the methylation of proteins, a key element in metabolic control, and of phosphatidylethanolamine. A failure of methylation also would affect the activity of catechol-O-methyltransferase and the biosynthesis of compounds such as norepinephrine.

Patients who are vitamin B$_6$ responders enjoy the most favorable prognosis in terms of the magnitude of reduction of blood homocysteine and the cognitive outcome. In many such instances, it will be feasible to lower the homocysteine concentration to a normal or near-normal level. Folate supplementation (2–5 mg/day) should be given at the same time because of the risk of incurring folic acid deficiency. In rare instances patients who take pyridoxine will sustain peripheral neuropathy and hepatotoxicity. These potential complications should be carefully monitored.

Treating a nonvitamin responder is challenging. It is possible to reduce blood homocysteine through a diet that is purposefully low in methionine, but this may be unpalatable, especially to a newly diagnosed older patient who is accustomed to an unrestricted diet. The advent of mass newborn screening programs for homocystinuria may be very helpful in this regard, since treatment may become feasible very early in life, even in the newborn period, before the patient has become acclimated to any particular diet.

A useful therapeutic adjunct is treatment with betaine (6–12 g/day). This will reduce the blood homocysteine concentration by favoring the remethylation of homocysteine to methionine (Dudman et al., 1996). There are no known major side effects attached to this treatment. It may impart a "fishy" odor to the urine, but this usually is tolerable. Betaine supplementation also will cause the blood methionine concentration to increase, often to a major extent (~1 mmol/L), but this perturbation has not yet been associated with ill effects.

6.1.2 Remethylation Deficiency Homocystinuria

The form of homocystinuria we have reviewed until now involves a failure of converting homocysteine to cystathionine (❯ *Figure 13-4*). Another form of the disease involves aberrations in the metabolism of methylfolate or methylcobalamin, the major cofactors for the remethylation of homocysteine to methionine (❯ *Figure 13-4*). These patients tend to become ill relatively early in life with a syndrome of lethargy, poor feeding, psychomotor retardation, and growth failure. There may be hematologic abnormalities such as megaloblastosis, macrocytosis, thrombocytopenia, and hypersegmentation of the leukocytes. On occasion a patient will present in later life with seizures, dementia, hypotonia, mental retardation, spasticity, or myelopathy.

Usually the blood levels of folate and vitamin B$_{12}$ are within normal limits. An important diagnostic finding is the presence of hyperhomocysteinemia without hypermethioninemia. Indeed, the blood methionine concentration tends to be low in these individuals, a finding that helps to discriminate this group from patients with homocystinuria secondary to cystathionine-β-synthase deficiency (see above). If there is a disruption in the metabolism of vitamin B$_{12}$ there may be increased excretion of methylmalonic acid because vitamin B$_{12}$ is needed for the conversion of methylmalonic acid to succinate in the methylmalonyl-CoA mutase reaction.

6.1.3 Methylenetetrahydrofolate Reductase Deficiency

This enzyme reduces 5,10-methylenetetrahydrofolate to methyltetrahydrofolate in a NADPH-dependent reaction, which is inhibited by SAM. The same enzyme is important in brain for the reduction of pteridines to yield BH$_4$, an essential cofactor in the phenylalanine hydroxylase system (see ❯ *Sect. 3*).

The initial clinical presentation often occurs at 6–12 months with a syndrome of profound psychomotor retardation, convulsions, microcephaly, and homocysteinemia (about 50 µmol/L) coupled with hypomethioninemia (<20 µmol/L). The level of vitamin B$_{12}$ in blood is normal and these individuals do not display either anemia or methylmalonic acidemia. Blood folate tends to be somewhat low.

Patients may sustain a thromboembolic diathesis as well as microgyri, demyelination, gliosis, and brain atrophy. On occasion lipid-laden macrophages have been reported.

Patients should receive supplements of folinic acid (5-formyltetrahydrofolic acid) and folate. Other agents have been attempted, including betaine, methionine, pyridoxine, cobalamin, and carnitine.

6.1.4 Methionine Synthase Deficiency

Methionine synthase catalyzes transfer of a methyl group from methyltetrahydrofolate to homocysteine to yield methionine (❯ *Figure 13-4*). There is a cobalamin group bound to the enzyme that must be converted to a methylcobalamin group before the actual transfer and formation of methionine.

In cobalamin-E disease (methionine synthase deficiency) methyl-B_{12} does not bind to the methionine synthase. It is uncertain whether this biochemical lesion represents a primary lesion in the synthase or the lack of yet another enzyme activity. Affected patients can display megaloblastic changes coupled with pancytopenia, homocystinuria, and hypomethioninemia. Methylmalonic aciduria is absent because these individuals continue to form adenosyl-B_{12}, which is cofactor for the methylmalonyl-CoA mutase reaction. Onset of symptoms usually is in infancy with vomiting, developmental retardation, and lethargy. Most patients respond nicely to injections of hydroxocobalamin.

6.1.5 Cobalamin-C Disease

The precise nature of the lesion in vitamin B_{12} metabolism is not certain. Complementation analysis has assigned patients to three different groups: cblC (the most common variant), cblD, and cblF. Affected babies become ill early in life with hypotonia, lethargy, growth failure, optic atrophy, and retinal changes. There tends to be excessive excretion of methylmalonic acid, but much less than in classical methylmalonic aciduria. Rarely is there ketoaciduria and/or metabolic acidosis.

Fibroblasts from patients with this disease do not convert cyanocobalamin or hydroxocobalamin to methylcobalamin or adenosylcobalamin. Thus, there is diminished flux through both methyltetrahydrofolate/homocysteine and methyltransferase/methylmalonyl-CoA mutase. These patients require supplementation with hydroxocobalamin to correct the underlying metabolic defect.

6.1.6 Hereditary Folate Malabsorption

Most patients have presented with megaloblastic anemia, seizures, and a progressive syndrome of neurologic deterioration. Levels of folate in both the blood and the CSF have been very low. The anemia is correctable with injections of folate, or with the administration of large oral doses, but the concentration of folate in the CSF is still low, suggesting that a distinct carrier system mediates folate uptake into the brain and that this system is the same as that facilitating intestinal transport.

7 The Urea Cycle Defects

7.1 Ureagenesis

Urea cycle defects (❯ *Figure 13-5*) usually derive from mutations in one of the five enzymes of the urea cycle, which converts ammonia to urea. Affected patients may sustain hyperammonemic encephalopathy and irreversible brain injury.

The first reaction is mitochondrial and involves condensation of NH_3, HCO_3^-, and ATP to form carbamyl phosphate via carbamyl phosphate synthetase (CPS). *N*-acetylglutamate (NAG), formed from glutamate and acetyl-CoA via NAG synthetase (❯ *Figure 13-5*, reaction 10), is an obligatory effector of CPS and an important regulator of ureagenesis. A variety of influences, including dietary protein, arginine, and corticosteroids, augment the concentration of NAG in mitochondria.

In the ornithine transcarbamylase (OTC) reaction (❯ *Figure 13-5*, reaction 2), carbamyl phosphate condenses with ornithine to yield citrulline. OTC is coded for on band p21.1 of the X chromosome, where the gene contains 8 exons and spans 85 kb. The activity of this enzyme is directly related to dietary

◘ **Figure 13-5**
The urea cycle and related reactions. Urea cycle defects usually are caused by a deficiency of a constituent enzyme (reactions 1–5). Ammonia also can be metabolized to glutamine (reaction 6) and glycine (reaction 7). Treatment with phenylacetate or benzoate leads to formation of phenylacetylglutamine (reaction 8) or hippurate (reaction 9), thereby providing an effective "antidote" to ammonia toxicity. Formation of N-acetylglutamate (reaction 10) provides the system with obligatory effector of carbamyl phosphate synthetase (CPS) (reaction 1). Enzymes: (1) carbamyl phosphate synthetase; (2) ornithine transcarbamylase (OTC); (3) argininosuccinate synthetase (AS); (4) argininosuccinate lyase (AL); (5) arginase; (6) glutamine synthetase; (7) glycine cleavage system; (8) glutamine/glycine-N-acylase; (9) cytosolic pathway of orotic acid synthesis; (10) N-acetylglutamate synthetase. Abbreviation: NAG, N-acetylglutamate. The + symbols denote that arginine and NAG are positive effectors for reactions 12 and 1, respectively

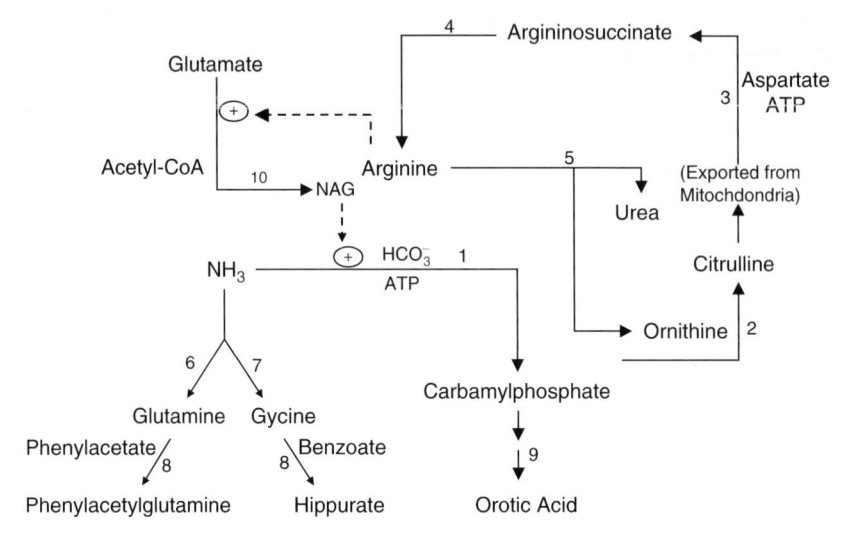

protein. There may be "tunneling" of ornithine transported from the cytosol to OTC, with the availability of intramitochondrial ornithine regulating the reaction.

In the cytosol, citrulline condenses with aspartate to form argininosuccinate via argininosuccinate synthetase (AS). The mRNA for this enzyme is increased by starvation, treatment with corticosteroids, or cAMP. Citrulline itself induces the mRNA.

Argininosuccinate lyase (AL) cleaves argininosuccinate to form fumarate, which is oxidized in the tricarboxylic acid cycle, and arginine, which is hydrolyzed to urea and ornithine via arginase. Both AL and arginase are induced by starvation, dibutyryl cAMP, and corticosteroids.

Profound defects become manifest in newborns, who display coma, convulsions, and vomiting. Sometimes these babies are thought to be septic and receive futile therapy with antibiotics. Hyperammonemia may exceed 1 mmol/L (normal: <100 μmol/L).

The aminogram is important to diagnosis. Plasma concentrations of glutamine and alanine, the major nitrogen-carrying amino acids, usually are high and that of arginine is low. Patients with citrullinemia (deficiency of AS) or argininosuccinic aciduria (deficiency of AL) will manifest marked increases of the blood citrulline and argininosuccinate, respectively.

An important sign is increased urinary orotic acid in babies with OTC deficiency and normal or even low orotic acid in the infant with CPS deficiency. Orotic acid is high because it is synthesized from carbamyl phosphate that spills into the cytoplasm, where it enters the pyrimidine synthesis pathway (● Figure 13-5, reaction 9).

Diagnosis of CPS or OTC deficiency may not always be apparent from the aminogram. Ornithine levels typically are normal in the latter disorder. Hyperammonemia, hyperglutaminemia, hyperalaninemia, and orotic aciduria suggests OTC deficiency. Without orotic aciduria this pattern suggests CPS deficiency.

Older patients may present with psychomotor retardation, growth failure, vomiting, behavioral abnormalities, perceptual difficulties, recurrent cerebellar ataxia, and headache. Whenever a patient shows unexplained neurologic symptoms, it is worth considering a urea cycle defect. In rare cases blood NH_3 may be normal, so quantitation of blood amino acids and of urinary orotate may be indicated.

Hyperammonemia can occur in other inborn errors of metabolism, especially organic acidurias. Thus, urine organic acids should be measured in all patients with significant hyperammonemia.

The brain of patients shows abnormal myelination with cystic degeneration of neurons. Swelling of the astrocytes is common. Frank cortical atrophy may occur in long-standing disease.

In argininosuccinic aciduria, there may be varying degrees of hepatic fibrosis, but urea cycle defects show few pathologic changes outside of the central nervous system.

7.1.1 Carbamyl Phosphate Synthetase Deficiency

Neonates with this rare disorder develop lethargy, hypothermia, vomiting, irritability, and profound hyperammonemia that may exceed 1 mmol/L. Occasional patients with partial enzyme deficiency have had a relapsing syndrome of lethargy and irritability upon exposure to protein. Brain damage can occur in both neonatal and late-onset groups.

7.1.2 N-Acetylglutamate Synthetase Deficiency

NAG synthetase, an obligatory effector of carbamyl phosphate synthase (❯ Figure 13-5, reaction 10), catalyzes the formation of NAG from glutamate and acetyl-CoA. Both neonatal and late-onset forms of the deficiency have been described. A promising new treatment for this disorder may be the use of carbamylglutamate (Caldovic et al., 2004), which corrects a similar form of hyperammonemia in propionic acidemia (Gebhardt et al., 2005).

7.1.3 Ornithine Transcarbamylase Deficiency

The most common urea cycle defect, OTC deficiency presents in diverse forms, including a fulminant, fatal disorder of neonates to a schizophrenic-like illness in an otherwise healthy adult. In this sex-linked disease, males fare more poorly than females because of random inactivation (lyonization) of the X chromosome. If inactivation affects the chromosome bearing the mutant gene, a more favorable outcome can be anticipated. The converse also holds true.

Diagnosis can be made with analysis of restriction fragment length polymorphisms. More than 80% of heterozygotes can be detected, and ante-natal diagnosis often is possible. Approximately one-third of the mothers of males and two-thirds of the mothers of females have been found to be noncarriers, reflecting the greater propensity for mutation in the male gamete.

Carrier diagnosis (85%–90%) can be made with protein-loading tests using urinary orotic acid as a marker. Administration of allopurinol also will favor orotic acid excretion. Loading studies with $^{15}NH_4Cl$ as metabolic tracer indicate that symptomatic female carriers produce less [^{15}N]urea compared with controls. Asymptomatic heterozygotes form urea at a normal rate, but they over-produce [5-^{15}N]glutamine. Thus, whole-body nitrogen metabolism is abnormal even in this group (Yudkoff et al., 1996).

There are animal models, including the sparse fur (spf) mouse (15% control enzyme activity) and the sparse fur-abnormal skin and hair (spf-ash) mouse (5% of control). Both kinds of mice manifest hyperammonemia, orotic aciduria, growth failure, and sparse fur.

OTC deficiency should be suspected in any patient with unexplained neurologic symptoms. The absence of hyperammonemia should not rule out diagnosis, especially with a history of protein intolerance or an untoward reaction to infections. The family history also may be suggestive. The blood amino acids and urinary orotic acid should be quantitated in such individuals.

7.1.4 Citrullinemia

Complete AS deficiency often is fatal. Survivors will have major brain injury. Patients with a partial deficiency may have a milder course, and a few individuals with citrullinemia have been normal.

The diagnosis usually is apparent from the hyperammonemia and the extreme hypercitrullinemia. The activity of AS can be determined in both fibroblasts and chorionic villi samples, thus simplifying the problem of ante-natal diagnosis.

7.1.5 Argininosuccinic Aciduria

Patients manifest extremely high levels of argininosuccinate in urine, blood, and CSF. Neonates have a stormy clinical course that often is fatal or leads to severe brain injury. A peculiar finding is trichorrhexis nodosa, or dry brittle hair with nodular protrusions that are best visible with light microscopy. The precise cause is unknown.

7.1.6 Arginase Deficiency

Many patients show psychomotor retardation during the first year of life. The dominant presentation is a progressive, spastic tetraplegia, especially in the lower extremities. Seizures and growth failure may occur, although some patients are of normal size. The motor dysfunction usually comes to clinical attention by age 2–3 years. Leukodystrophic changes are seen. Hyperammonemia usually is less severe than that seen in neonatal-onset disorders. Plasma arginine is 2–5 times normal. Urine orotic acid excretion is extremely high, perhaps because arginine stimulates flux through the CPS reaction by favoring the synthesis of NAG.

7.1.7 Hyperornithinemia, Hyperammonemia, and Homocitrullinuria

Affected neonates have growth failure and varying degrees of mental retardation. Sometimes symptoms are deferred until adulthood. Vomiting, lethargy, and hypotonia are noted after protein ingestion. Recurrent hospitalizations for hyperammonemia are the rule. Some patients have manifested a bleeding diathesis and hepatomegaly. Electron microscopy of liver samples has shown irregularities of mitochondrial shape.

The underlying defect is diminished mitochondrial uptake of ornithine that leads to a failure of citrulline synthesis and consequent hyperammonemia. Urinary orotic acid is high, presumably because of underutilization of carbamyl phosphate. In contrast, excretion of creatine is low, reflecting the inhibition of glycine transamidinase by excessive levels of ornithine.

7.1.8 Lysinuric Protein Intolerance

Symptomatology starts after weaning, when infants manifest growth failure, hepatomegaly, splenomegaly, vomiting, hypotonia, recurrent lethargy, coma, abdominal pain and, in rare instances, psychosis. Rarefaction of the bones is common, with both fractures and vertebral compression having been reported. Not all patients are mentally retarded. Some patients sustain a potentially fatal interstitial pneumonia that may respond to corticosteroids.

The cause is a failure of reabsorption of lysine, ornithine, and arginine by the proximal tubule and dibasic aminoaciduria. There also is intestinal failure of dibasic amino acid uptake. It is the basolateral rather than the luminal transporter that is involved. Hyperammonemia is caused by a deficiency of intramitochondrial ornithine. Treatment involves oral citrulline supplementation, which corrects the hyperammonemia by allowing replenishment of the mitochondrial pool of ornithine.

7.2 Management of Urea Cycle Defects

It is important to restrict protein to minimize ammonia formation. Unfortunately, the tolerance for dietary protein may be so limited that it is not possible to support normal growth.

An important therapeutic advance is treatment with sodium benzoate and sodium phenylacetate (❯ *Figure 13-5*, reaction 8). The liver rapidly condenses benzoyl-CoA with glycine to form hippurate. Similarly the liver acylates phenylacetyl-CoA with glutamine to yield phenylacetylglutamine. Both hippurate and phenylacetylglutamine are efficiently excreted in the urine, thereby allowing waste nitrogen elimination not as urea but as conjugates of benzoate and phenylacetate (Brusilow et al., 1980; Maestri et al., 1991, 1996). Excretion of nitrogen as phenylacetylglutamine eliminates two moles of nitrogen with each mole of phenylacetylglutamine. Unfortunately, the clinical utility of phenylacetate is limited by its objectionable odor. An alternative is sodium phenylbutyrate, which is less malodorous and which is converted in liver to phenylacetate. Acylation therapy has greatly improved survival and morbidity. Thus, the outlook for female heterozygotes with OTC deficiency is favorable for those who are treated from an early age (Ye et al., 1996).

Surviving neonates can be maintained on a combination of low-protein diet and sodium benzoate. Supplementation with arginine is useful for therapy of citrullinemia and argininosuccinic aciduria, which enhances the ability to eliminate nitrogen as either citrulline or argininosuccinate. In addition, the maintenance of arginine levels in the normal range facilitates protein synthesis.

Liver transplantation is a novel approach that allows metabolic correction, although relatively minor deviations of amino acid concentration may persist postoperatively. Obviously, this approach incurs all the risks and morbidity of organ transplantation.

Either hemodialysis or peritoneal dialysis relieves acute toxicity during fulminant hyperammonemia. Exchange transfusion has not been useful.

Gene therapy remains a theoretical possibility (Ye et al., 1996). An adenoviral vector containing a cDNA for the OTC gene has corrected hyperammonemia in mice with a congenital deficiency of OTC.

8 Disorders of Glutathione Metabolism

8.1 Glutathione Metabolism

Glutathione (γ-glutamyl-cysteinyl-glycine) is a tripeptide that is the major intracellular antioxidant. It is synthesized via these reactions:
1. Glutamate + cysteine + ATP → γ-glutamylcysteine + ADP + P_i
2. γ-Glutamylcysteine + glycine + ATP → glutathione + ADP + P_i
   Glutathione subsequently is metabolized in the γ-glutamyl cycle:
3. Glutathione + amino acid → γ-glutamyl amino acid + cysteinylglycine
4. γ-Glutamyl amino acid → 5-oxoproline + amino acid
5. 5-Oxoproline + ATP + $2H_2O$ → glutamate + ADP + P_i
6. Cysteinylglycine → cysteine + glycine

The cycle is renewed after the cysteine formed in reaction 6 and the glutamate derived from reaction 5 are converted to γ-glutamylcysteine via γ-glutamylcysteine synthetase (reaction 1). Various defects of glutathione metabolism have been described (Ristoff and Larsson, 1998).

8.1.1 5-Oxoprolinuria

Patients with glutathione synthetase deficiency (5-oxoprolinuria) show severe metabolic acidosis secondary to overproduction of 5-oxoproline (pyroglutamic acid) (Njalsson et al., 2005). The reason for such overproduction is that diminished internal glutathione disinhibits the γ-glutamylcysteine synthetase pathway (reaction 1), thereby augmenting the concentration of γ-glutamylcysteine and the subsequent conversion of this dipeptide to cysteine and 5-oxoproline in the cyclotransferase pathway (reaction 4).

Clinical findings include mental retardation, severe metabolic acidosis, hemolysis, and evidence of a spastic quadriparesis and cerebellar disease. In some instances, there has been relatively normal development until late childhood when a progressive loss of intellectual function became appreciated. Pathologic changes include cerebellar atrophy and lesions in the cortex and thalamus. There is no specific therapy.

8.1.2 γ-Glutamylcysteine Synthetase Deficiency

Patients display spinocerebellar degeneration, peripheral neuropathy, psychosis, hemolytic anemia, myopathy, and an aminoaciduria secondary to renal tubular dysfunction.

8.1.3 γ-Glutamyltranspeptidase Deficiency

The absence of γ-glutamyltranspeptidase, which is responsible for glutathione degradation, leads to glutathionuria and varying degrees of mental retardation. The enzyme is present in the brain, primarily in the capillaries. No specific treatment is available.

8.1.4 5-Oxoprolinase Deficiency

Patients with 5-oxoprolinase deficiency excrete increased amounts of oxoproline but they tend not to have neurologic symptoms.

9 Disorders of GABA Metabolism

9.1 GABA Metabolism

Decarboxylation of glutamate via glutamate decarboxylase forms γ-aminobutyric acid (GABA), the most important inhibitory neurotransmitter. GABA is metabolized in neurons and glia, which transaminate it to succinic semialdehyde via γ-aminobutyric acid transaminase (GABA-T). Succinic semialdehyde dehydrogenase oxidizes the semialdehyde to succinate, which then enters the tricarboxylic acid cycle.

9.1.1 Pyridoxine Dependency

Patients with pyridoxine dependency present with severe seizures of early onset, even in utero. They respond dramatically to parenteral pyridoxine (10–100 mg), but the reason for the response is not understood (Baxter, 1999). It may involve faulty binding of pyridoxine, a cofactor in the glutamate decarboxylase reaction, to the enzyme protein, but this has not been proven.

9.1.2 GABA Transaminase Deficiency

Patients with GABA transaminase deficiency have severe psychomotor retardation and hyperreflexia. The concentrations of GABA and β-alanine in the CSF and blood exceed normal, as is the concentration of homocarnosine (γ-aminobutyrylhistidine) in the CSF. GABA-T activity is diminished in blood lymphocytes and liver. There may be increased stature, perhaps reflecting the ability of GABA to evoke the release of growth hormone.

9.1.3 Succinic Semialdehyde Dehydrogenase Deficiency

Patients with succinic semialdehyde dehydrogenase deficiency have mental retardation, cerebellar disease, and hypotonia (Pearl et al., 2003; Gordon, 2004). They excrete large amounts of both succinic semialdehyde and 4-hydroxybutyric acid. There is no curative therapy, although there have been attempts to manage the disease with vigabatrin to inhibit GABA-T, and thereby diminish production of succinic semialdehyde.

10 Disorders of *N*-Acetyl Aspartate Metabolism: Canavan's Disease

Infants are normal at birth, but by 3 months they display developmental delay, increased head circumference (>98th percentile), and rapid deterioration of neurologic function with minimal awareness, spasticity, and seizures. Optic atrophy leads to blindness. Magnetic resonance imaging shows demyelination and brain atrophy with enlargement of the ventricles and widening of the sulci. Pathologic examination shows swelling of the astrocytes with elongation of the mitochondria. Vacuoles appear in the myelin sheets (Madhavarao et al., 2005).

Urinary excretion of *N*-acetylaspartate is elevated. The CSF concentration may be 50 times the control values. The cause is a deficiency of aspartoacylase, which cleaves *N*-acetylaspartate to form aspartate and acetyl-CoA. The enzyme occurs primarily in the white matter, but *N*-acetylaspartate is most abundant in gray matter. The defect is expressed in skin fibroblasts (Surendran et al., 2003).

N-acetylaspartate is among the most abundant amino acids in the brain, although its precise function remains elusive. Putative roles have included osmoregulation and the storage of acetyl groups that subsequently are utilized for myelin synthesis (Baslow, 2003; Madhavarao et al., 2005). The relationship of the enzyme defect to the clinical findings remains problematic.

No specific therapy is yet available.

References

Alemzadeh R, Gammeltoft K, Matteson K. 1996. Efficacy of low-dose dextromethorphan in the treatment of nonketotic hyperglycinemia. Pediatrics 97: 924-926.

Baslow MH. 2003. *N*-acetylaspartate in the vertebrate brain: Metabolism and function. Neurochem Res 2003 Jun; 28(6): 941-953.

Baxter P. 1999. Epidemiology of pyridoxine-dependent and pyridoxine-responsive seizures in the UK. Arch Dis Child 81: 431-433.

Brusilow S, Tinker J, Batshaw ML. 1980. Amino acid acylation: A mechanism of nitrogen excretion in inborn errors of urea synthesis. Science 207: 659.

Caldovic L, Morizono H, Daikhin Y, Nissim I, McCarter RJ, et al. 2004. Restoration of ureagenesis in *N*-acetylglutamate synthase deficiency by *N*-carbamylglutamate. J Pediatr 145: 552-554.

Cedarbaum S. 2002. Phenylketonuria: An update. Curr Opin Pediatr 14: 702-706.

Centerwall SA, Centerwall WR. 2000. The discovery of phenylketonuria: The story of a young couple, two retarded children, and a scientist. Pediatrics 105: 89-103.

Chan H, Butterworth RF. 2003. Cell-selective effects of ammonia on glutamate transporter and receptor function in the mammalian brain. Neurochem Int 43: 525-532.

Chuang DT, Davie JR, Wynn RM, Chuang JL, Koyata H, et al. 1995. Molecular basis of maple syrup urine disease and stable correction by retroviral gene transfer. (Review) J. Nutr 125: 1766S-1772S.

Deutsch SI, Rosse RB, Mastropaolo J. 1998. Current status of NMDA antagonist interventions in the treatment of nonketotic hyperglycinemia. Clin Neuropharmacol 21: 71-79.

Diamond A, Herzberg C. 1996. Impaired sensitivity to visual contrast in children treated early and continuously for phenylketonuria. Brain 119: 523-538.

Dudman NP, Guo XW, Gordon RB, Dawson PA, Wilcken DE. 1996. Human homocysteine catabolism: Three major pathways and their relevance to development of arterial occlusive disease. J Nutr 126(4) (Suppl): 295s-300s.

Eisensmith RC, Woo SL. 1991. Phenylketonuria and the phenylalanine hydroxylase gene. Mol Biol Med 8: 3-10.

Farishian RA, Whittaker JR. 1980. Phenylalanine lowers melanin synthesis in mammalian melanocytes by reducing tyrosine uptake: Implications for pigment reduction in phenylketonuria. J Invest Dermatol 74: 85-89.

Felipo V, Butterworth R. 2002. Neurobiology of ammonia. Prog Neurobiol 67: 259-279.

Flusser H, Korman SH, Sato K, Matsubara Y, Galil A, et al. 2005. Mild glycine encephalopathy (NKH) in a large kindred due to a silent exonic GLDC splice mutation. Neurology 64: 1426-1430.

Folbergrova J, Druga R, Otahal J, Haugvicova R, Mares P, et al. 2005. Seizures induced in immature rats by homocysteic acid and the associated brain damage are prevented by group II metabotropic glutamate receptor agonist (2R,4R)-4-aminopyrrolidine-2,4-dicarboxylate. Exp Neurol 192: 420-436.

Garrod AE. 1923. Inborn Errors of Metabolism. London: Henry Frowde and Hodder & Stoughton.

Gebhardt B, Dittrich S, Parbel S, Vlaho S, Matsika O, et al. 2005. N-carbamylglutamate protects patients with decompensated propionic aciduria from hyperammonemia. J Inherit Metab Dis 28: 241-244.

Gordon N. 2004. Succinic semialdehyde dehydrogenase deficiency (SSADH) (4-hydroxybutyric aciduria, γ-hydroxybutyric aciduria). Eur J Paediatr Neurol 8: 261-265.

Guldberg P, Levy HL, Hanley WB, Koch R, Matalon R, et al. 1996. Phenylalanine hydroxylase gene mutations in the United States: Report from the Maternal PKU Collaborative Study. Am J Hum Genet 59: 84-94.

Huijbregts SC, de Sonneville LM, van Spronsen FJ, Licht R, Sergeant JA. 2002. The neuropsychological profile of early and continuously treated phenylketonuria: Orienting, vigilance, and maintenance versus manipulation-functions of working memory. Neurosci Biobehav Rev 26: 697-712.

Infante JP, Huszagh VA. 2001. Impaired arachidonic (20:4n-6) and docosahexaenoic (22:6n-3) acid synthesis by phenylalanine metabolites as etiological factors in the neuropathology of phenylketonuria. Mol Genet Metab 72: 185-198.

Jan W, Zimmerman RA, Wang ZJ, Berry GT, Kaplan PB, et al. 2003. MR diffusion imaging and MR spectroscopy of maple syrup urine disease during acute metabolic decompensation. Neuroradiology 45: 393-399.

Kahler SG, Fahey MC. 2003. Metabolic disorders and mental retardation. Am J Med Genet 117C: 31-41.

Kaplan P, Mazur A, Field M, Berlin JA, Berry GT, et al. 1991. Intellectual outcome in children with maple syrup urine disease. J Pediatr 119: 46-50.

Kikuchi G. 1973. The glycine cleavage system: Composition, reaction mechanism, and physiological significance. Mol Cell Biochem 1: 169-175.

Kikuchi A, Takeda A, Fujihara K, Kimpara T, Shiga Y, et al. 2004. Arg(184)His mutant GTP cyclohydrolase I, causing recessive hyperphenylalaninemia, is responsible for dopa-responsive dystonia with parkinsonism: A case report. Mov Disord 19: 590-593.

Kluijtmans LA, Boers GH, Stevens EM, Renier WO, Kraus JP, et al. 1996. Defective cystathionine β-synthase regulation by S-adenosylmethionine in a partially pyridoxine responsive homocystinuria patient. J Clin Invest 98: 285-289.

Koch R, Moseley KD, Yano S, Nelson M Jr, Moats RA. 2003. Large neutral amino acid therapy and phenylketonuria: A promising approach to treatment. Mol Genet Metab 79: 110-113.

Kraus JP, Komrower Lecture. 1994. Molecular basis of phenotype expression in homocystinuria. J Inherit Met Dis 17: 383-390.

Levy HL, Ghavami M. 1996. Maternal phenylketonuria: A metabolic teratogen. Teratology 53: 176-184.

Madhavarao CN, Arun P, Moffett JR, Szucs S, Surendran S, et al. 2005. Defective N-acetylaspartate catabolism reduces brain acetate levels and myelin lipid synthesis in Canavan's disease. Proc Natl Acad Sci USA 102: 5221-5226.

Maestri NE, Brusilow SW, Clissold DB, Bassett SS. 1996. Long-term treatment of girls with ornithine transcarbamylase deficiency. N Engl J Med 335: 855-859.

Maestri NE, Hauser ER, Bartholomew D, Brusilow SW. 1991. Prospective treatment of urea cycle disorders. J Pediatr 1119: 923.

Mandel H, Brenner B, Berant M, Rosenberg N, Lanir N, et al. 1996. Coexistence of hereditary homocystinuria and factor V Leiden—effect on thrombosis. N Engl J Med 334: 763-768.

Mueller GM, McKenzie LR, Homanics GE, Watkins SC, Robbins PD, et al. 1995. Complementation of defective leucine decarboxylation in fibroblasts from a maple syrup urine disease patient by retrovirus-mediated gene transfer. Gene Therapy 2: 461-468.

Njalsson R, Ristoff E, Carlsson K, Winkler A, Larsson A, et al. 2005. Genotype, enzyme activity, glutathione level, and clinical phenotype in patients with glutathione synthetase deficiency. Hum Genet 116: 384-389.

Pearl PL, Gibson KM, Acosta MT, Vezina LG, Theodore WH, et al. 2003. Clinical spectrum of succinic semialdehyde dehydrogenase deficiency. Neurology 60: 1413-1417.

Pietz J, Rupp A, Ebinger F, Rating D, Mayatepek E, et al. 2003. Cerebral energy metabolism in phenylketonuria: Findings by quantitative in vivo ^{31}P MR spectroscopy. Pediatr Res 53: 654-662.

Pilla C, Cardozo RFD, Dutra CS, Wyze ATS, Wajner M, et al. 2003. Effect of leucine administration on creatine kinase activity in rat brain. Metab Brain Dis 18: 17-25.

Ristoff E, Larsson A. 1998. Patients with genetic defects in the γ-glutamyl cycle. Chem Biol Interact 111-112: 113-121.

Scriver CR, Beaudet AL, Sly WS, Valle D. 2001. The metabolic and molecular basis of inherited disease. New York: McGraw Hill.

Shefer S, Tint GS, Jean-Guillaume D, Daikhin E, Kendler A, et al. 2000. Is there a relationship between 3-hydroxy-3-methylglutaryl coenzyme a reductase activity and forebrain pathology in the PKU mouse? J Neurosci Res 61: 549-563.

Smith CB, Kang J. 2000. Cerebral protein synthesis in a genetic model of phenylketonuria. Proc Natl Acad Sci USA 97: 11014-11019.

Starkebaum G, Harlan JM. 1986. Endothelial cell injury due to copper-catalyzed hydrogen peroxide generation from homocysteine. J Clin Invest 77: 1370-1376.

Steiner RD, Sweetser DA, Rohrbaugh JR, Dowton SB, Toone JR, et al. 1996. Nonketotic hyperglycinemia: Atypical clinical and biochemical manifestations. J Pediat 128: 243-246.

Surendran S, Matalon KM, Tyring SK, Matalon R. 2003. Molecular basis of Canavan's disease: From human to mouse. J Child Neurol (Sep) 18(9): 604-610.

Toone JR, Applegarth DA, Coulter-Mackie MB, James ER. 2000. Biochemical and molecular investigations of patients with nonketotic hyperglycinemia. Mol Genet Metab 70: 116-121.

Van Hove JL, Kishnani P, Muenzer J, Wenstrup RJ, Summar ML, et al. 1995. Benzoate therapy and carnitine deficiency in nonketotic hyperglycinemia. Am J Med Genet 59: 444-453.

van Spronsen FJ, Smit PG, Koch R. 2001. Phenylketonuria: Tyrosine beyond the phenylalanine-restricted diet. J Inherit Metab Dis 24: 1-4.

Wagner M, Coelho DM, Barschak AG, Araujo PR, Pires RF, et al. 2000. Reduction of large neutral amino acid concentrations in plasma and CSF of patients with maple syrup urine disease. J Inher Met Dis 23: 505-512.

White DA, Nortz MJ, Mandernach T, Huntington K, Steiner RD. 2001. Deficits in memory strategy use related to prefrontal dysfunction during early development: Evidence from children with phenylketonuria. Neuropsychology 15: 221-229.

Ye X, Robinson MB, Batshaw ML, Furth EE, Smith I, et al. 1996. Prolonged metabolic correction in adult ornithine transcarbamylase-deficient mice with adenoviral vectors. J Biol Chem 271: 3639-3646.

Yudkoff M, Daikhin Y, Nissim I, Jawad A, Wilson J, et al. 1996. In vivo nitrogen metabolism in ornithine transcarbamylase deficiency. J Clin Invest 98: 2167-2173.

14 N-Acetylaspartate and N-Acetylaspartylglutamate

M. H. Baslow

1	**Introduction**	307
1.1	N-Acetylaspartate	308
1.2	N-Acetylaspartylglutamate	308
2	**NAA**	309
2.1	Compartmental Distribution	309
2.2	Use of NAA as a Neuron Marker	309
3	**Inborn Errors in NAA Metabolism**	311
3.1	Canavan Disease (Hyperacetylaspartia)	311
3.2	Hypoacetylaspartia	311
4	**NAA Metabolism and Dynamics**	311
4.1	NAA Metabolism Is Coupled to Glc Metabolism	311
4.2	NAA Dynamics and a Metabolic Model	311
4.2.1	NAA Dynamic Model Variable Rate Functions	311
4.2.2	NAA Dynamic Model Functional Constants	312
4.2.3	NAA Dynamic Values in Human Brain	312
5	**NAA Functions**	313
5.1	Proposed Functions	313
5.2	Functions That Fit the Entire Intercompartmental NAA Sequence	313
5.2.1	Myelin Formation Hypothesis	313
5.2.2	NAA Function as an MWP and the Cellular Osmotic Disequilibrium Problem	314
5.3	The NAA Intercellular Cycle as a Water Transporting Mechanism	314
5.3.1	Release of Hydrophilic NAA to the ECF	314
5.3.2	Fate of NAA in the ECF and the Energy Cost of the NAA Cycle	315
5.3.3	The NAA Intercompartmental Cycle and Transport of NAA Obligated Water	315
6	**NAA Function as an MWP and Its Role in the Etiology of Canavan Disease**	315
7	**NAAG**	317
7.1	Similarity to NAA Functions	317
7.2	Relationships Between NAA and NAAG	317
7.2.1	Structural, Metabolic, and Dynamic Relationships Between NAA and NAAG	317
7.3	A Functional Relationship Between NAA and NAAG	318
7.4	A Second Role for NAAG in Maintaining Brain Osmotic Homeostasis	319
8	**Coordinated Functions of NAA and NAAG**	319
8.1	Coupling of Glc, NAA, and NAAG Metabolism	319
8.2	NAAG Is the Major Source of Glu Signal Available to Astrocyte mGluRs	319
8.2.1	The Glu Moiety Is the Neuron Signal to Astrocytes	319

© Springer-Verlag Berlin Heidelberg 2007

8.2.2	Rate of Free Glu Released to the ECF	320
8.2.3	NAAG as a Source of mGluR Signals to Astrocytes	320
8.2.4	Rate and Availability of the Glu Entity to Astrocytes from NAAG	320
8.2.5	NAAG Potential for Neurotoxicity	321
8.2.6	Relationship of NAAG to BOLD Signals	322
8.2.7	Relationship of mGluR Activation and Hyperemia to BOLD Imaging	324
9	***Summary of Brain NAA and NAAG Metabolism and Dynamics***	**325**
9.1	Structural Connections	325
9.2	Target Cell for NAA and Its Obligated Water	325
9.3	Target Cell for NAAG and Its Obligated Water	325
10	***Discussion***	**326**
10.1	The Primary Functions for NAA and NAAG Are Osmoregulatory	326
10.2	NAA and NAAG Functions Are Linked	327
10.3	NAA and Osmoregulation	327
10.4	NAAG and Brain Microcirculation	327
10.5	The Overall Function of the Linked NAA–NAAG System in Brain	327
10.5.1	The NAA–NAAG System Functions as a Unit	327
10.5.2	The NAA–NAAG System and Information Processing	328
10.5.3	Role in Maintaining a Neuron's Minimal Repolarization Period	328
10.6	Temporal Bounds Within Which the NAA–NAAG System Operates	333
10.6.1	Time frame of Information Processing	333
10.6.2	Temporal Interaction of NAAG Dynamics and the fMRI BOLD Response	333
11	***Functional Consequences of Inborn or Engineered Errors in NAA and NAAG Metabolism***	**335**
11.1	Canavan Disease	335
11.2	Hypoacetylaspartia	336
11.3	Genetic Engineering—Absence of NAAG Peptidase Activity in a Mouse	337
12	***Relationship of NAA and NAAG Metabolism to Other Brain Pathologies***	**337**
12.1	Static Measures of Brain NAA as a Neuron Marker	337
12.2	Dynamic Measures of NAA and NAAG	337
12.2.1	Dynamic Measures in Normal Humans	337
12.2.2	Dynamic Measures in a Human Disease	338
13	***Conclusions***	**338**
14	***Epilogue***	**339**

Abstract: *N*-acetyl-L-aspartate (NAA) and *N*-acetylaspartylglutamate (NAAG), a derivative of NAA, are abundant in neurons in the vertebrate brain. However, their physiological roles have remained elusive. In this review, evidence is presented that NAA and NAAG are structurally, metabolically, and dynamically, as well as functionally linked, and their roles are in support of a neuron's continuously changing requirements for energy and removal of products of energy production. Neuronal information-encoded spiking activity requires use of energy in the form of ATP for membrane repolarization, which is replenished primarily by the oxidation of glucose (Glc) that, along with oxygen, is supplied via the vascular system. One product of Glc metabolism is water, which must be removed to extracellular fluid (ECF) against a water gradient. In order to maintain a neuron's ability to transmit meaningful information, water must be removed and Glc supplied in a timely manner. It is proposed that the NAA–NAAG system is a dynamic control mechanism used by neurons to achieve these ends. Both substances have osmoregulatory roles and transport water to ECF for removal via the vascular sink. NAAG also has a second role as a signal molecule for the astrocytic metabotropic glutamate receptor 3, which upon activation initiates an astrocyte–vascular system interaction that results in focal hyperemia providing increased sink capacity and energy supplies. As a result of this homeostatic control mechanism, a neuron's minimum repolarization time and absolute refractory period can be maintained at any level of neurostimulation, thereby retaining its maximal spike frequency-coding capabilities. In this way, the NAA–NAAG system appears to be directly involved in the maintenance of neuronal energy-limited signaling capabilities, and consequently, of global integrated brain functions. As evidenced by inborn errors in metabolism, to the degree that the system fails, there is disorder in these brain functions.

List of Abbreviations: Ac, acetate; AcCoA, acetyl-coenzyme A; ADC, apparent diffusion coefficient; Asp, aspartic acid; ATP, adenosine triphosphate; BOLD, blood-oxygen-level-dependent imaging; CD, Canavan disease; CNS, central nervous system; ECF, extracellular fluid; fMRI, functional magnetic resonance imaging; Glc, glucose; Gln, glutamine; Glu, glutamic acid; HPLC, high-pressure liquid chromatography; mGluR, metabotropic Glu receptor; MRI, magnetic resonance imaging; MRS, magnetic resonance spectroscopy; MWP, molecular water pump; NAA, *N*-acetyl-L-aspartic acid; NAAG, *N*-acetylaspartylglutamic acid; NAH, *N*-acetyl-L-histidine; NMDA, *N*-methyl-D-aspartic acid; PCP, phencyclidine

1 Introduction

The brain is a complex information-processing organ that cannot see, smell, hear, taste, or feel. For these things, it relies on meaningful encoded electrophysiological signals that originate in a variety of environmental neuronal sensors and without which input, normal cognitive abilities cannot be sustained. The brain's metabolic lifeline to the environment is the vascular system, on which it relies for a continuous supply of nutrients, and for removal of waste products. Thus, the well-being of both the brain and the whole organism depends on continuous homeostatic interactions between the brain, its environmental sensors, and the vascular system. Superimposed on basic individual neuronal requirements for energy and waste product removal is a very complex neuronal network that receives, interprets, and responds to brain messages of both external and internal origin. At the most integrated level of neural networking is the realm of higher cognitive functions.

There are two relatively simple substances present in great abundance in the brain of warm-blooded (homeothermic) vertebrates, birds, and mammals, whose possible functions have been the subjects of research efforts over a period of many decades. One is an *N*-acetylated derivative of L-aspartic acid (Asp), *N*-acetyl-L-aspartic acid (NAA), and the other, a dipeptide derivative of NAA, *N*-acetylaspartylglutamic acid (NAAG), in which L-glutamic acid (Glu) is joined to the Asp moiety via a peptide bond.

There is also a strong evolutionary component in the distribution of NAA in vertebrates. This takes the form of a discontinuity between homeotherms and cold-blooded forms (poikilotherms), in that in the brain of poikilothermic vertebrates including bony fishes, amphibians, and reptiles, NAA is present in much less abundance than in homeotherms. However, another *N*-acetylated amino acid, α-*N*-acetyl-L-histidine (α-NAH) is present in these poikilotherms in high abundance, and in one group of amphibians, frogs and toads, NAA is virtually absent and appears to be replaced almost entirely by NAH, which may be a metabolic analog of NAA. This evolutionary component goes even further back in vertebrate evolutionary time where there is a second discontinuity, in that the more ancient elasmobranch fishes, including sharks and rays, have no measurable brain NAH, and NAA is once again present in high abundance and is the major acetylated amino acid in the brain in these groups.

There is yet another phylogenetic discontinuity between vertebrates and invertebrates, in that neither NAA, NAAG, nor NAH is found in high concentration in nerve tissue of invertebrates. However, the reduced levels of these substances do not preclude them from having functional roles in the nervous system of invertebrates, and there is evidence that NAAG may play an important intercellular signaling role in some forms. The phylogenetic relationships between the distribution of NAA and NAH in the vertebrate brain have been described (Baslow, 1997).

In this chapter, evidence is presented based on results of studies in a number of different scientific disciplines, which provides new and unique perspectives from which to evaluate the physiological roles of NAA and NAAG, and allows development of insights into their importance in brain function. This evidence leads to the conclusion that NAA and NAAG operate as a linked metabolic system, which functions as a homeostatic neuronal control mechanism to maintain the ability of neurons to receive and transmit meaningful encoded information. The first international symposium on NAA and NAAG was held in 2004 (Moffett et al., 2006).

1.1 N-Acetylaspartate

NAA is a natural, albeit unusual, amino acid that is present in the vertebrate brain, and in human brain, at about 10 mM; its concentration is among the highest of all free amino acids. Although NAA is synthesized and stored primarily in neurons, it cannot be hydrolyzed in these cells. However, neuronal NAA is dynamic in that it turns over more than once each day by virtue of its continuous efflux down a steep gradient, in a regulated intercompartmental cycling via extracellular fluids (ECF), between neurons and a second compartment, primarily in oligodendrocytes (Madhavarao et al., 2004), where it is rapidly deacetylated. The neuronal membrane transport mechanism for NAA into ECF and its oligodendrocyte docking mechanism are presently unknown. In addition, the specific neuronal sites of NAA efflux are unclear. While the NAA synthetic enzyme has only been partially characterized (Madhavarao et al., 2003), the gene for its hydrolytic enzyme, amidohydrolase II (aspartoacylase), in oligodendrocytes has been cloned. There are no specific inhibitors of either NAA synthase or amidohydrolase II known at present.

A second brain enzyme, amidohydrolase I, whose substrate is NAH, has about 3% of the activity of amidohydrolase II on NAA. In cultured rat brain cells, this enzyme is found in astrocytes, but not in oligodendrocytes (Baslow et al., 2001).

The compartmental metabolism of NAA, between its anabolic compartment in neurons and its catabolic compartment in oligodendrocytes, and its possible physiological role in the brain has been reviewed (Birken and Oldendorf, 1989; Baslow, 1997, 2000, 2003a). Although found primarily in neurons in the brain, NAA is also present in high concentration in several other cells and tissues outside of the brain, and is found in abundance in peritoneal mast cells (Burlina et al., 1997) and in the aneural vertebrate lens (Baslow and Yamada, 1997).

1.2 N-Acetylaspartylglutamate

NAAG, a dipeptide derivative of NAA and Glu, is also present in abundance in neurons as well as in oligodendrocytes and microglia, and at about 1 mM, it is one of the most abundant dipeptides in the vertebrate brain. NAAG is metabolically unusual in that there are three cell types involved in its metabolism. NAAG synthesized in neurons is first exported to astrocytes via an as yet unknown transport mechanism, where it docks with a metabotropic Glu receptor (mGluR), and then an astrocyte-specific enzyme located on the astrocyte surface, NAAG peptidase (GCP II, NAALADase, NAAG peptidase I), hydrolyzes the Glu moiety that is taken up by the astrocytes and converted into glutamine (Gln), which is then transported back to neurons. The residual NAA metabolic product in the ECF diffuses to oligodendrocytes, where the cell-specific enzyme, amidohydrolase II, removes the acetate (Ac) moiety, which is then taken up by the oligodendrocytes. Finally, the ECF-liberated Asp diffuses back to neurons where it is taken up and subsequently used for recycling into NAA and then into NAAG, completing the cycle.

The nature of the NAAG synthase is presently unknown, but NAAG is synthesized in a human neuroblastoma cell line (Arun et al., 2004), and this cell line has been suggested as a model system for the study of NAAG synthesis. The NAAG peptidase (GCP II) has been cloned (Kozikowski et al., 2004), as has been a second membrane-bound brain astrocyte NAAG peptidase, NAAG peptidase II (GCP III), that has about 6% of the activity of NAAG peptidase I on NAAG (Bacich et al., 2002; Bzdega et al., 2004). It is suggested that NAAG peptidase II may be the primary peptidase associated with cerebellar astrocytes rather than cerebral cortical astrocytes, but the physiologically important substrate for NAAG peptidase II remains to be elucidated. No specific inhibitor of NAAG synthase is known, but several specific inhibitors of both NAAG peptidases have been reported. The distribution and unusual metabolism of NAAG in the brain has also been extensively reviewed (Birken and Oldendorf, 1989; Coyle, 1997; Baslow, 2000; Neale et al., 2000; Karelson et al., 2003).

2 NAA

2.1 Compartmental Distribution

Analysis of the content of NAA in various brain compartments indicates that over 98.7% of NAA in the brain is present in a single brain compartment, the neurons. The neuron compartment, which makes up about 50% of vertebrate brain volume, is one of eight selected functional compartments in the brain (Baslow, 2003a). The relative volumes of these compartments and their NAA content are presented in ❷ *Table 14-1*.

2.2 Use of NAA as a Neuron Marker

In addition to the abundant presence of NAA in neurons, its prominent signal in proton nuclear magnetic resonance spectroscopic (MRS) studies, due to its *N*-acetyl methyl group resonance, has led to its use in cognitive research (Jung et al., 1999, 2000; Pfliderer et al., 2004) and in diagnostic human medicine as both an indicator of brain pathology and of disease progression in a variety of central nervous system (CNS) diseases (Tsai and Coyle, 1995; Burlina et al., 2000; Stanley, 2002; Chung et al., 2003).

In cognitive studies of normal humans, positive correlations between a static NAA level and level of performance on some tests of intelligence have been observed, although the precise nature of the relationship between NAA and performance is unknown. Possible explanations offered suggest that NAA may be associated with increased neuron density, increased metabolic efficiency, or increases in synaptic connections to account for the increased performance.

In some pathological conditions, where reduced levels of NAA are observed, the correlations of NAA with performance are still positive. However, interpretation of the MRS data in disease states is also difficult since reduction in NAA levels may indicate either loss of neurons, loss of neuron viability without a loss of neurons, or loss of NAA and several of its functions, without loss of either neuron density or viability (Baslow et al., 2003). In addition, in many MRS studies, NAA levels are often expressed as ratios with respect to other brain metabolites rather than as absolute concentrations. As other metabolites may vary independently of NAA, it is suggested that use of such ratios may confound the quantification of NAA (Li et al., 2003). The association of NAA with normal cognition, and with cognitive losses in many neuropathological conditions has been reviewed (Ross and Sachdev, 2004), and areas of brain research where NAA has been measured and considered as a specific neuronal marker are summarized in ❷ *Table 14-2*.

Static levels of NAA in whole brain of normal humans have been reported to vary by $\pm 13\%$ of the mean, giving a range of 26% in NAA values (Inglese et al., 2004), and reproducible individual differences in such static levels of NAA in whole normal human brain have been observed over a period of several days (Gonen et al., 1998). Stable static levels of NAA in whole brain have also been reported to vary widely within

Table 14-1
Brain compartments and NAA levels[a]

Brain compartment	Volume[b] (%)	NAA[c] (mM)	Total brain NAA[d] (mM)	(%)
C_1 Neurons	50	20.0	10.0	98.7
C_2 ECF	20	0.1	0.02	0.2
C_3 Mast cells	<0.015	75.0	0.01	0.1
C_4 Astrocytes	15	0.0	0.0	0.0
C_5 Microglia	1.5	0.0	0.0	0.0
C_6 Oligodendrocytes	12	0.8	0.1	1.0
C_7 CSF	8	0.0	0.0	0.0
C_8 Vascular	1.5	0.0	0.0	0.0
			10.13	100.0

[a]Provisional values based on a variety of techniques and different organisms
[b]Volume measurements and estimates by various methods are generally based on brain tissues excluding CSF, as is done in this case
C_1, C_8 (Pfeuffer et al., 2000)
C_2 (Nicholson and Sykova, 1998)
C_3 Assumed that mast cells represent no more than 1% of the vascular compartment. It is estimated that fewer than 10,000 mast cells may be present in the brain (Siegel et al., 1981)
C_4 (Williams et al., 1980)
C_5 (Lawson et al., 1990; Long et al., 1998) values derived from numbers and volumes of microglial cells in mouse brain
C_6 Value derived by difference from compartments C_1–C_5 and C_8
C_7 (Guyton, 1966)
[c]NAA content. Reported values of <0.1 mM in C_5, C_7, and C_8 are listed as 0.0
C_1 (Baslow, 2003a)
C_2 (Faull et al., 1999)
C_3 (Burlina et al., 1997)
C_6 (Bhakoo and Pearce, 2000)
[d]Based on weighted averages

Table 14-2
Examples of use of NAA as a neuronal marker, assessed by magnetic resonance techniques, which pervades most areas of brain research

As a marker signal in most NMR metabolic studies
As a metabolite in neuronal metabolic studies
Brain deconstructive diseases such as Alzheimer's disease and multiple sclerosis
Inborn errors in metabolism
Neuropsychiatric disorders such as schizophrenia
Effects of poisons such as exposure to CO
Transient ischemic events
Stroke
Tumors
Age-related dementias
Normal cognition and test performance
Prenatal brain diagnosis

a species as evidenced by observed strain-specific differences in the range of brain NAA found in 6 strains of rats (5.9–9.4 mM) and 14 strains of mice (3.5–9.1 mM) (Marcucci et al., 1969).

3 Inborn Errors in NAA Metabolism

3.1 Canavan Disease (Hyperacetylaspartia)

There are two human inborn errors in metabolism of NAA. One is Canavan disease (CD), where there is a buildup of NAA (hyperacetylaspartia) with associated spongiform leukodystrophy, apparently due to a lack of amidohydrolase II activity in oligodendrocytes (Baslow, 2001; Zeng et al., 2002). A rat with a natural deletion of the amidohydrolase II gene is known that exhibits a CD-like syndrome (Kitada et al., 2000), and a human with a partial deletion of the gene is also known that exhibits the CD phenotype (Tahmaz et al., 2001). In addition, a gene knockout mouse with a CD phenotype has been created (Matalon et al., 2000). The animal CD-like models have been utilized in various biochemical, pharmacological, and genetic engineering studies (Baslow et al., 2002; Surendran et al., 2003).

3.2 Hypoacetylaspartia

The other inborn error in NAA metabolism is a singular human case of lack of brain NAA (hypoacetylaspartia), where the enzyme that synthesizes NAA (NAA synthase) is apparently absent or inactive in neurons and other cells (Martin et al., 2001). A lack of brain NAAG is also observed in this case. While the effects of CD are profound and usually lead to early mortality, the early effects of hypoacetylaspartia in this case were reported to be comparatively mild. However, a follow-up of this patient through 2003 revealed that by age 8, the effects were also profound, with marked abnormalities in motor performance, language ability, and cognition, and with episodic status epilepticus (Boltshauser et al., 2004).

4 NAA Metabolism and Dynamics

4.1 NAA Metabolism Is Coupled to Glc Metabolism

Using MRS involving combined ^{13}C MRS and [1-^{13}C]glucose (Glc) infusion, the rate of NAA synthesis in the human brain has been measured in vivo, and it has been demonstrated that NAA synthesis is both structurally and metabolically coupled to Glc metabolism. The NAA carbon 6 (Ac moiety) is derived from acetyl-coenzyme A (AcCoA), which is derived in turn from [1-^{13}C]Glc metabolism (Moreno et al., 2001). This study connects both the structure and the rate of synthesis of NAA directly with the rate of Glc energy metabolism in the human brain. Similar results have also been reported in rat brain (Choi and Gruetter, 2001; Henry et al., 2003; Karelson et al., 2003). The synthesis of NAA and oxidation of Glc are also physically connected in that their metabolism occurs in the same organelle, the mitochondrion. In rat brain, some NAA synthesis has also been reported to occur in a microsomal fraction (Lu et al., 2004).

4.2 NAA Dynamics and a Metabolic Model

4.2.1 NAA Dynamic Model Variable Rate Functions

On the basis of measurement of the rate of NAA synthesis in humans and other metabolic data, NAA metabolism in the brain has been modeled (Baslow, 2002), and variable rate functions and functional constants determined for each phase in its cellular and intercompartmental metabolism. As Glc is a source of NAA constituent Ac as well as Asp, the model reflects a correlation between the rate of Glc oxidized (G)

and the rate of NAA synthesis (S). The model also indicates a turnover rate (T) for NAA and degree of recycling (R) of NAA and of its hydrolysis products, Ac and Asp.

4.2.2 NAA Dynamic Model Functional Constants

Associated with the dynamics of the model are several functional constants whose level reflects a balance between the inflow and outflow of metabolites, the ratio of Glc oxidized to NAA synthesized $[G/S]=(K_1)$, the ratio of metabolic water produced to Glc oxidized $[G\times 6]=(W_1)$, the ratio of brain metabolic water produced to NAA synthesized $[W_1/S]=(W_2)$, and the mol ratio of metabolic water molecules produced to NAA released to the ECF (K_2). Additional constants include the level of brain NAA (K_3) and the ratio of the rates of oxygen utilized (O) to Glc oxidized $[O/G]=(K_4)$. It also follows that if the level of brain NAA is a functional constant, then the rate of NAA synthesis (S) will also equal the rate of NAA efflux (E) from neurons and the rate of NAA hydrolysis (H). An important element in the model is that it predicts that while the instantaneous level of brain NAA remains constant (K_3), the rate of NAA synthesis and efflux will vary directly with brain metabolism and the rate of Glc utilization. The compartmental locations in which these variables and constants are operational are shown in ● *Figure 14-1*.

◘ **Figure 14-1**
Dynamic model of NAA cycling between neurons and oligodendrocytes. Explanations: G = rate of Glc metabolism, W_1 = rate of metabolic water produced O = rate of oxygen utilitzed, BBB = blood–brain barrier, S = rate of NAA synthesis, K_1 = G/S, R = rates of Ac, Asp recycling, K_2 = molecules of water/NAA, E = rate of NAA efflux, K_3 = NAA level in brain, H = rate of NAA hydrolysis, K_4 = O/G. From Baslow (2003a)

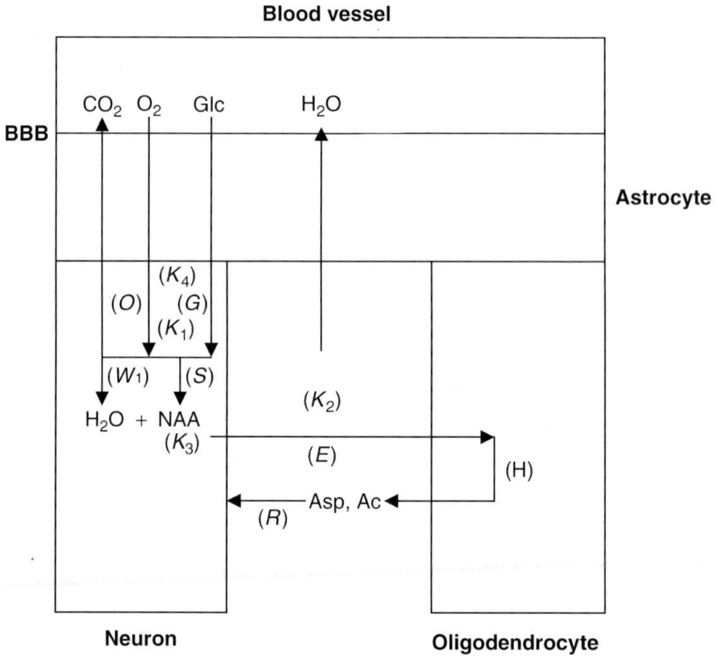

4.2.3 NAA Dynamic Values in Human Brain

On applying this model to humans, it has been calculated that 1 mol of NAA is synthesized for every 40 mol of Glc oxidized in the brain, and each mol of NAA is associated with the transport of 121 or more mol of metabolic water out of neurons. In addition, calculated turnover of NAA is every 16.7 h or 1.4 times per day. Model values for NAA variables and constants are presented in ● *Table 14-3*.

◼ Table 14-3
The NAA dynamic model and measured or derived values for humans[a]

Symbol	Function	Units	Measured/derived
(G)	Rate of Glc metabolism	μmol/g/h	22.20
(O)	Rate of oxygen utilized [G × 6]	μmol/g/h	133
(S)	Rate of NAA synthesis	μmol/g/h	0.55
(T)	NAA turnover [K_3/S]	h	16.7
(R)	Rate of NAA Ac, Asp recycling	%/h	NA[b]
(E)	Rate of NAA efflux [=S]	μmol/g/h	0.55
(H)	Rate of NAA hydrolysis [=S]	μmol/g/h	0.55
(W_1)	Rate of metabolic water produced in brain [=O]	μmol/g/h	133
(W_2)	Metabolic water/NAA [W_1/S]	#	242
(K_1)	Ratio of Glc metabolized NAA synthesized [G/S]	#	40
(K_2)	Mol ratio water/NAA	#	121[c]
(K_3)	NAA in brain	μmol/g	9.20
(K_4)	Ratio of oxygen/Glc used [O/G]	#	4–6

[a]Adapted from Baslow (2003a)
[b]NA, not available for humans. In rats, Kunnecke et al. (1993) report that Asp is highly conserved ($NAA_{Asp} > 95\%/h$), whereas Ac is not ($NAA_{Ac} < 5\%/h$)
[c]Based on neurons producing 100% of NAA while representing only 50% of brain cellular volume and with a metabolic rate equal to surrounding cells

5 NAA Functions

5.1 Proposed Functions

A number of functions for NAA in neurons have been suggested. Initially, proposed functions were focused on its presence in neurons and on intracellular possibilities. As evidence accumulated that NAA metabolism was dynamic and involved two cellular compartments, an anabolic compartment in neurons and a catabolic compartment in oligodendrocytes, additional intercellular functions were also proposed. The proposed functions for NAA in brain are listed in ● *Table 14-4*. From this survey, it is apparent that NAA may participate in several different functions in brain. However, on the basis of the absence of NAA in viable, differentiated cultured neurons and the progressive loss of NAA in cultured organotypic brain slices that still retain viable neurons and glia, it appears that NAA intercompartmental metabolism is not essential for neuron viability or function. Instead, it is associated with a higher order of brain integrated activities that involve neurons, astrocytes, oligodendrocytes, and vascular epithelia—the four cell types that comprise the basic repeating unit of brain structure, as well as intact vascular and ECF systems (Baslow et al., 2003).

5.2 Functions That Fit the Entire Intercompartmental NAA Sequence

5.2.1 Myelin Formation Hypothesis

Of the proposed functions, only the myelin formation hypothesis, based on use of the Ac moiety of NAA by oligodendrocytes (Chakraborty et al., 2001; Madhavarao et al., 2004), and the osmoregulatory hypothesis (Baslow, 2002) incorporate the entire intercompartmental metabolic sequence for NAA. However, a primary myelin-building role does not fit well with the dynamics of NAA metabolism, which at a turnover

◘ Table 14-4
Proposed functions for NAA

Intracellular functions
Cellular osmolyte
Compensates anion deficit
Aspartate storage
Acetate storage
Formation of N-acetylaspartylglutamate
Intercellular functions
Intercellular signaling
Neuroexcitation
Extracellular osmoregulation
Lipid metabolism
Myelin formation
A role in neuroimmune reactions
Molecular water pump

rate of 1.4 times per day is much faster than the 18 day turnover rate for brain myelin (Dunlop, 1983), and no evidence has yet been presented that the release of Ac from NAA and its incorporation into myelin are of a similar order of magnitude. The hypothesis also does not account for the liberation of large amounts of NAA to nonmyelinating oligodendrocytes in gray matter. Nor does it fit with the outcomes of the two rare human genetic inborn errors in NAA metabolism.

In hypoacetylaspartia, NAA is apparently not synthesized in neurons for use by oligodendrocytes, and yet myelin is still formed (Martin et al., 2001; Boltshauser et al., 2004). In hyperacetylaspartia, NAA cannot be catabolized due to a lack of oligodendrocyte amidohydrolase II activity (Baslow, 2000), but the end result is not only hypomyelinization, as observed in other myelin deficiency diseases, but also a profound failure of all neuron–oligodendrocyte connections at their nodal–paranodal junctions and a complete deconstruction of existing neuron–oligodendrocyte associations, the latter being an important clinical characteristic of the CD syndrome that cannot be readily explained by the myelin formation hypothesis. Finally, the myelin hypothesis does not account for the paucity of NAA in normal myelinated amphibian brains, or perhaps most importantly, for the role of the closely linked metabolism of NAAG.

5.2.2 NAA Function as an MWP and the Cellular Osmotic Disequilibrium Problem

On the basis of analyzing these two hypotheses, it has been suggested that although the Ac portion of NAA participates in the formation of lipids and of myelin in myelinating oligodendrocytes, the primary function of the NAA intercompartmental cycle is osmoregulatory. In this hypothesis, it is proposed that NAA functions as a molecular water pump (MWP) to continuously remove metabolic and other water from neurons (Baslow, 2002, 2003a). Previously, NAH, a metabolic analog of NAA in the poikilothermic vertebrate brain (Baslow, 1997), has been demonstrated to be an MWP (Baslow, 1998). Cellular mechanisms to remove water are necessary since all cells have an inherent osmotic disequilibrium problem, and therefore must continuously remove water against a water gradient. The nature of the cellular osmotic disequilibrium problem is postulated in ❷ Table 14-5.

5.3 The NAA Intercellular Cycle as a Water Transporting Mechanism

5.3.1 Release of Hydrophilic NAA to the ECF

Under normal conditions, Asp present in neurons is acetylated by an NAA synthase in mitochondria from AcCoA (Madhavarao et al., 2003). As there is no acylase for hydrolysis of NAA in neurons, its concentration

◘ Table 14-5
Universal nature of the cellular osmotic disequilibrium problem

1. All cells are inherently in osmotic disequilibrium with their extracellular environments by virtue of the presence of nonpermeable osmolytes enclosed within a semipermeable plasma membrane
2. All cells have an innate tendency to swell in an aquatic environment and thus to disrupt the integrity of their plasma membranes
3. All cells have developed homeostatic pump-leak mechanisms to transport water against a water gradient, and thereby establish and remain in dynamic osmotic equilibrium with their external environments
4. All cells require energy to maintain their pump (active transport) and leak (facilitated diffusion) homeostatic systems
5. Interruption of a cell's energy supply for even a short period of time can result in cell swelling, disruption of the plasma membrane, and irreversible cellular pathology

builds, reaching a cytosolic equilibrium level of about 20 mM. As each new NAA molecule is then formed from Glc metabolism at a ratio of approximately 1 NAA for each 40 Glc metabolized in brain (Baslow, 2002, 2003a), an anionic hydrophilic NAA molecule is also released into the ECF, perhaps along with a cationic sodium ion as is the case for the sodium–glucose MWP (Meinild et al., 1998).

5.3.2 Fate of NAA in the ECF and the Energy Cost of the NAA Cycle

In the ECF, NAA diffuses down its gradient to oligodendrocytes where it is rapidly deacetylated to form Asp, most of which is again taken up by neurons from the ECF, via an active transport mechanism, and recycled into NAA (Karelson et al., 2003). The half-life of NAA in the ECF, and that of formed Asp is short due to the presence of high levels of amidohydrolase II activity in oligodendrocytes in both white and gray matter (Klugmann et al., 2003), and a rapid uptake mechanism for Asp in neurons. The enzyme-liberated Ac is primarily taken up by glial cells where it becomes part of the Ac pool, and much of the energy lost by neurons in the synthesis of NAA can be recovered when Ac is eventually oxidized in these cells. Thus, in the overall energy budget of the brain, very little is actually expended to operate the NAA system. At a nominal 1.4 NAA cycles per day, it is estimated that about 0.1% of the brain's daily energy budget is expended on the system (Baslow and Resnik, 1997).

5.3.3 The NAA Intercompartmental Cycle and Transport of NAA Obligated Water

The net result of the NAA cycle is that neuronal water, including metabolic water derived from Glc oxidation at a ratio of about 4–6 molecules of water to 1 molecule of Glc, is transported against a water gradient to the ECF as each molecule of NAA with a minimum of 32 molecules of obligated water (Baslow and Guilfoyle, 2002) is transported down its gradient. From the ECF, after hydrolysis of NAA and recycling of Asp, the released water can then pass into the vascular sink compartment, via astrocyte aquaporin-4 channels and the vascular capillary epithelium, for its eventual excretion (Saadoun et al., 2002). The sink capacity of plasma is about 1.3 mOsmol/l higher than cells in dynamic equilibrium with the ECF, or the ECF itself, so that cellular water pumped into the ECF compartment will therefore diffuse down its gradient into the vascular system (❯ Table 14-6). In addition, from the description of osmolytes present in various compartments as listed in this table, and with neuronal NAA at about 20 mM in humans, it is also clear that neurons are heavily invested with NAA where, of all the thousands of possible organic solutes, this single substance represents approximately 21% of all neuronal organic osmolytes.

6 NAA Function as an MWP and Its Role in the Etiology of Canavan Disease

CD is a usually fatal early-onset autosomal recessive human CNS disease. It is an osmotic disease, and the CD phenotype is characterized very early in its development by a buildup of NAA and a water imbalance in

◘ Table 14-6
Osmolar substances in extracellular and intracellular fluids[a]

Substance	Plasma	Interstitial (mOsmol/l)	Intracellular
Inorganic			
Na^+	144.0	137.0	10.0
K^+	5.0	4.7	141.0
Ca^{2+}	2.5	2.4	
Mg^{2+}	1.5	1.4	31.0
Cl^-	107.0	112.7	4.0
HCO_3^-	27.0	28.3	10.0
$HPO_4^{2-}, H_2PO_4^-$	2.0	2.0	11.0
SO_4^{2-}	0.5	0.5	1.0
Organic			
Phosphocreatine			45.0
Carnosine[b]			14.0
Amino acids	2.0	2.0	8.0
Creatine	0.2	0.2	9.0
Lactate	1.2	1.2	1.5
ATP			5.0
Hexose monophosphate			3.7
Glucose	5.6	5.6	
Protein	1.2	0.2	4.0
Urea	4.0	4.0	4.0
Total mOsmol	303.7	302.2	302.2
Corrected[c]	282.6[d]	281.3	281.3[e]

[a] Adapted from Guyton (1966)
[b] In human brain, this osmotic component would also include homocarnosine, NAA, and NAAG
[c] Based on interactions of molecules, osmotic activity is only 93% of calculated value
[d] The sink capacity of plasma is a positive 1.3 mOsmol/l (osmotic pressure)
[e] Cells are hyperosmotic to interstitial fluid, but osmotic equilibrium is maintained by continuous pumping of water into interstitial fluid, where it is constantly removed via the sink capacity of plasma. ATP is the cells' energy source for pumping water, and when the ATP supply is interrupted, cells immediately begin to swell as a result of the influx of interstitial water (Qiao et al., 2002)

brain, which is evidenced by cellular and ECF edema, and associated megalocephaly. As the disease progresses, additional evidence of water imbalance is observed in continued head enlargement, ventricular enlargement (Breitbach-Faller et al., 2003), and in the formation of large fluid-filled spongiform vacuoles. These vacuoles originate in large part in spaces that result from splitting of the axon-enveloping oligodendrocyte myelin sheaths at their ECF-intraperiod lines, producing a characteristic clinical CD spongiform leukodystrophy.

On the basis of NAA's role as an MWP, an osmotic–hydrostatic mechanism for the etiology of the CD phenotype has been proposed (Baslow, 2003b). In CD, there is a continuous synthesis and efflux of NAA from neurons into the ECF at a rate of 0.22 mOsmol/kg brain/h (Moreno et al., 2001; Baslow, 2003a), coupled with a lack of its hydrolysis and inability of vascular or ventricular epithelial cells to efficiently transport NAA water out of brain. It is the daily addition of 13,375 Pa (equal to 0.132 atm or 1.94 psi) of intractable NAA hydrostatic pressure to the normal equilibrium osmotic–hydrostatic pressure of about 100 psi in brain ECF, on the brain cell side of brain-barrier epithelial membranes, which is considered to be responsible for the brain edema and genesis of the CD syndrome.

With a plasma sink capacity of about 1.3 mOsmol/l (Guyton, 1966), and the daily liberation of 5.28 mOsmol/l of NAA to brain ECF, its rate of removal from the brain and entry into the vascular system is critical since after about 6 h, it would surpass the plasma sink capacity, reversing the flow of water between the brain and the vascular system, further aggravating any edematous brain condition. Use of diuretics to enhance the sink capacity of the vascular system and decrease intracranial pressure is a current symptomatic treatment in CD.

7 NAAG

7.1 Similarity to NAA Functions

Several of the intracellular and extracellular functions of NAA and NAAG parallel one another, and as an MWP it can transport 53 molecules of water per molecule of NAAG liberated to the ECF (Baslow, 1999). The proposed functions of NAAG in the brain are presented in ◉ *Table 14-7*.

◘ Table 14-7
Proposed functions for NAAG

Intracellular functions
Cellular osmolyte
Glutamate storage
Intercellular functions
Neuromodulator
Neurotransmitter
Metabotropic receptor stimulant
Source of glutamate
Source of NAA
Molecular water pump
Cerebral blood flow regulation
Less toxic intercellular delivery system for Glu

7.2 Relationships Between NAA and NAAG

7.2.1 Structural, Metabolic, and Dynamic Relationships Between NAA and NAAG

NAA and NAAG are metabolically and structurally connected (chemometabolic), since Glc metabolism is the source of Ac in both molecules. In rat brains, NAAG is synthesized from NAA and Glu (Tyson and Sutherland, 1998) at a rate of about 0.06 μmol/g/h or about 1 molecule of NAAG for every 10 molecules of NAA synthesized, and under steady-state conditions they are maintained at this 10:1 ratio. Based on its rate of synthesis, and a brain NAAG content of about 1 mM, the turnover time of NAAG at the calculated rate of 6.0%/h is 16.7 h, a value very similar to the turnover rate of NAA (14.2 h) in the rat brain, and identical to the turnover rate in the human brain (◉ *Table 14-8*). Thus, even though brain NAAG content is lower than that of NAA, their turnover rates are similar, suggesting that there is also a dynamic connection. Evidence suggesting an NAA-NAAG precursor-product relationship has also been obtained from chemometabolic studies of cultured human neuroblastoma cells (Arun et al., 2006).

A chemometabolic structure-based relationship between NAA and NAAG is also indicated in the two human inborn errors in NAA metabolism. In hypoacetylaspartia, where it appears that there is little or no synthesis of NAA, its Glu adduct NAAG is also not present in the brain (Martin et al., 2001). In CD, NAA liberated to the ECF cannot be hydrolyzed due to the inactivity of amidohydrolase II in oligodendrocytes, and as a result, there is a buildup of NAA in the ECF. At elevated concentrations, NAA inhibits the action of astrocytic cell surface NAAG peptidase (Berger et al., 1999) by product inhibition (Leshinsky et al., 1997),

◘ Table 14-8
Rat brain NAA and NAAG dynamic values

			Measured/derived	
Symbol	Function	Unit	NAA[a]	NAAG[b]
(S)	Rate of synthesis	μmol/g/h	0.60	0.06
(T)	Turnover [K_3/S]	h	14.2	16.7
		%/h	7.1	6.0
(E)	Rate of efflux [=S]	μmol/g/h	0.60	0.06
(H)	Rate of hydrolysis [=S]	μmol/g/h	0.60	0.06
(K_3)	Brain content	μmol/g	8.50	1.0

[a]Rate of synthesis in whole brain (Choi and Gruetter, 2001). Rate of synthesis of NAA has also been reported in brain homogenates of 30–60-day-old rats, exclusive of cerebella, and in two subcellular components, isolated mitochondrial and microsomal fractions (Lu et al., 2004). In these fractions, the rate of synthesis (S) was 3.5, 6.3, and 29.0 nmol/mg protein/h, respectively. Assuming a nominal value of about 10% protein in whole brain homogenates, S in the portion of brain assayed in this study would be about 0.35 μmol/g/h

[b]Value for NAAG synthesis. Based on NAAG ^{13}C Glu enrichment of 20.0% (6.0%/h) by metabolism of [1-^{13}C]Glc, infused over a period of 200 min (Tyson and Sutherland, 1998), and an average brain NAAG content of about 1 mM for rat and other homeothermic vertebrate species (Robinson et al., 1987)

causing NAAG to build up in the ECF as well. This is reflected in the increased excretion of NAAG as well as NAA in CD (Burlina et al., 1994). In two cases of severe hypomyelination in unrelated children, 20–40-fold elevations in CSF levels of NAAG were observed, and these elevations are considered to be the biochemical hallmark of a new neurometabolic NAAG disorder (Wolf et al., 2004).

In summary, the structures, metabolism, and dynamics of NAA and NAAG are intertwined, as NAAG is synthesized in the normal brain from NAA at a constant rate, maintained at a constant NAA/NAAG ratio, both liberated to ECF at the same rate and also catabolized at the same rate in order to maintain their constant brain metabolite ratio.

7.3 A Functional Relationship Between NAA and NAAG

The evidence from CD and from the dynamics of NAAG synthesis indicates that under normal conditions, NAAG produced in neurons is liberated to the ECF simultaneously with NAA, where they function similarly as MWPs. NAA is specifically targeted to oligodendrocytes, where, upon its hydrolysis it liberates at least 32 molecules of water to ECF. NAAG in the ECF is specifically targeted to astrocytes where Glu is first removed, releasing about 21 molecules of bound water per molecule of NAAG to the ECF. The secondary hydrolysis of NAAG-liberated NAA by oligodendrocytes then releases an additional 32 molecules of bound water, thus completing the NAA–NAAG spatial (geometabolic) and temporal (chronometabolic) MWP cycles. As an MWP, NAAG transports about 14% of metabolic water exported to the ECF. The oligodendrocyte enzyme, amidohydrolase II, is obviously of great importance to this process in the brain in that it is the key component, and an irreplaceable mechanism, for the successful operation of both the NAA and NAAG intercompartmental cycles. A functional relationship between NAA and NAAG, based on their immunocytochemical colocalization in neurons, has also been proposed (Simmons et al., 1991).

Because NAAG is found in oligodendrocytes and microglia as well as in neurons, and NAA is primarily found in most neurons, their distribution in the brain is a function of the relative distribution of these cell types in various areas of the brain. In a study of the distribution of these substances in seven regions of the CNS of the rat, NAA was found to vary from 1.1 to 5.0 mM, and NAAG from 0.1 to 1.6 mM, and the ratios of NAA–NAAG in these regions were found to vary from 0.7 to 48.7 (Battistuta et al., 2001).

7.4 A Second Role for NAAG in Maintaining Brain Osmotic Homeostasis

An additional role in which the astrocyte-targeted NAAG participates is the stimulation of one type of mGluR on the astrocyte surface. NAAG is a selective group II, mGluR3 agonist (Schoepp et al., 1999), one of the several mGluRs that trigger Ca^{2+} oscillations in neuron-associated astrocytic processes which then spread to astrocytic endfeet in contact with arterioles, where release of vasoactive agents induce vascular expansion and increased local blood flow (Zonta et al., 2003). Thus, the rate of release of NAAG can continuously signal astrocytes about the state of focal neurostimulation, the neuronal requirements for vascular Glc or astrocyte lactate (Tsacopoulos, 2002), oxygen, and for increased vascular sink capacity for metabolic water removal.

The distance between any given neuron, or neuronal extensions, and the vascular network is about 1 mm, and via astrocytic endfeet, astrocytes bridge this gap between neuron cell bodies, their axons (or axon nodes where myelinated), dendrites, synapses, and the vascular epithelium. The diffusion path between component neuron regions and the surfaces of astrocyte endfeet used for bidirectional diffusion of substances is also very short, about 100 Å, approximately the width of 25 water molecules. Thus, the combination of minimal diffusion pathways between neuron–astrocyte cellular membranes and the close proximity of all morphological parts of a neuron with the vascular epithelium, bridged by astrocyte cell bodies and extensions in an astrocytic syncytium connected by gap junctions (Bonvento et al., 2002), serve to make the vascular system highly responsive to any neuron initiated astrocytic-vascular signaling required to fulfill local energy and waste removal needs associated with interneuronal communications (Park et al., 2003).

8 Coordinated Functions of NAA and NAAG

8.1 Coupling of Glc, NAA, and NAAG Metabolism

The synthesis and simultaneous release of both NAA and NAAG from neurons appears to be a continuous, compensatory, and coordinated process in brain osmotic and metabolic function. First, metabolic water produced from oxidation of Glc is transported to the ECF against a water gradient by both of these substances, and second, the local blood flow is affected by the NAAG–astrocyte mGluR3 signal, a signal resulting in vasodilation with increased availability of Glc and O_2, and an increased ability of the vascular system to function as a sink for exported water and CO_2. The NAAG signal function appears to be a homeostatic mechanism that can quickly respond to any changes in the rate of neuron stimulation and to the resulting changes in local, regional, or global circulatory requirements. This is because the rate of synthesis and efflux of NAAG is directly and dynamically coupled to changes in the rate of neuron stimulation, Glc utilization, and NAA formation. The dynamics of the coupling of Glc, NAA, and NAAG metabolism is presented in ❷ *Table 14-9*.

8.2 NAAG Is the Major Source of Glu Signal Available to Astrocyte mGluRs

8.2.1 The Glu Moiety Is the Neuron Signal to Astrocytes

Glu is the major signal entity that traffics via the ECF between excited neurons at their synapses and astrocytes (Tsacopoulos, 2002). The excited neurons also liberate ammonia to the ECF, and both of these substances are preferentially taken up by astrocytes where they are transformed into the relatively nontoxic Gln, which is then cycled back to neurons via the ECF where Glu is regenerated. The overall importance of Glu release from neurons in brain is attested to by the fact that more than 80% of neurons are excitatory, and that greater than 90% of synapses release Glu (Raichle and Gusnard, 2002).

◻ Table 14-9
Ratios of molecules of Glc oxidized by neurons to molecules of metabolic products exported to the ECF

Substances	Molecules	Forms	Fate
Imported from the ECF			
Glucose or equivalent	200[a]	Glucose, lactate, amino acids, fatty acids	Oxidized
Oxygen	1000[b]	Dissolved gas	Reduced
Exported to the ECF			
Water	1000[b]	Solute obligated, ionic, hydrates	Excreted
Carbon dioxide	1000[b]	Dissolved gas, carbonates	Excreted
NAA	10[c]	Minimum 32 molecules of water each[d]	Recycled
NAAG	1	Minimum 53 molecules of water each[d]	Recycled

[a]Ratio of Glc to NAA synthesized in whole brain is 40:1. Assumes 50% of Glc or equivalent is oxidized in neuronal compartments
[b]Based on a 5:1 ratio of O_2/Glc
[c]Almost 99% of NAA is synthesized by neurons
[d]Minimums are obligated water in first hydration layer. Water is excreted and carriers metabolized or recycled

8.2.2 Rate of Free Glu Released to the ECF

Chapman et al. (2003) have measured the rate of free Glu release into the ECF using microdialysis in the nucleus accumbens of rat brain. Their baseline value for free Glu delivered to the ECF is calculated to be 0.0026 μmol/ml/h available for the activation of astrocyte mGluRs to trigger Ca^{2+} oscillations leading to vasodilation and functional hyperemia. A similar range of values for the rat striatum ECF Glu has also been reported (Kennedy et al., 2002), and Hillered et al. (1989) reported a 0.0013 μmol/ml/h efflux of free Glu in the rat caudate-putamen. Instantaneous values for Glu in dialysates from the rat frontal cortex and corticostriatal regions were 0.00105 and 0.00126 μmol/ml (Kondrat et al., 2002), respectively. These levels are within an acceptable physiological ECF range for free Glu, and less than toxic levels have been observed to be in the range of 0.01–0.10 μmol/ml when tested against cultured neurons (During and Spencer, 1993).

8.2.3 NAAG as a Source of mGluR Signals to Astrocytes

NAAG is a specific astrocyte mGluR3 agonist of itself, and further, the Glu released by hydrolysis of NAAG by astrocytic NAAG peptidase may also be a significant and important nonspecific signal source for all other astrocytic mGluRs that can be used in their ongoing regulation of vascular flow (Yourick et al., 2003). In cultured rat astrocytes, both mGluR1 and mGluR5 have been shown to be present in addition to mGluR3 (Lea et al., 2003).

In studies of the crayfish giant nerve fiber, the neuron–NAAG–astrocyte connection has been documented. In this preparation, NAAG release is induced by nerve stimulation (Urazaev et al., 2001) that in turn stimulates glial cells. It has also been observed that NAAG by itself mimics the effect of nerve fiber stimulation on glia (Gafurov et al., 2001). From the results of these crayfish studies, it has been suggested that NAAG could be the primary axon-to-glia signaling agent acting on the mGluR3 receptor, and that secondary NAAG-derived Glu could then further contribute to glial hyperpolarization by the activation of many additional mGluRs.

8.2.4 Rate and Availability of the Glu Entity to Astrocytes from NAAG

Based on the brain content of NAAG and its release from cells at a rate of 0.06 μmol/g/h to the ECF in the rat brain, and an ECF compartment size of 20% of brain volume, an estimate of the available Glu entity can be determined. From this, it is calculated that release of the specifically targeted NAAG to astrocytes via the ECF provides a relatively nontoxic bound form of Glu to astrocyte plasma membrane surfaces at a rate of 0.30 µmol/ml/h. An instantaneous basal level of NAAG present in rat brain ECF was 0.0122 mM, a value three times the level of Glu (0.0041 mM) found in this same study (Slusher et al., 1999). As NAAG appears to be produced in all types of neurons, and is synthesized by microglia and oligodendrocytes as well, it is thus available as a specific signal molecule to astrocytes from a wide variety of cell types.

At an efflux rate of 0.30 µmol/ml/h, the availability of NAAG-bound Glu exported to and present in the ECF is also very large when compared to availability of free Glu, and since NAAG-bound Glu is released at a rate of 115–230 times than that of free Glu more Glu is available to astrocytes from NAAG at any given time than from free Glu; it is the major supplier of mGluR3 signal to astrocytes. It has been speculated that the different mGluR receptors on astrocytes may constitute a general mechanism for vasodilation in regions of the brain where Glu is not the main neurotransmitter (Parri and Crunelli, 2003). Thus, as already suggested, astrocyte-targeted NAAG may indeed be the key neuron-to-astrocyte signal central to the dynamic control of brain microcirculation. A comparison of the characteristics of release and availability of free Glu, and of bound Glu from NAAG to astrocytes is presented in ❷ Table 14-10.

◘ Table 14-10
Characteristics associated with the form of Glu exported by brain cells to ECF[a]

Form	Rate of release to ECF (µmol/ml/h)	Release characteristics
Glu	0.0013–0.0026[b]	From glutamatergic neuron synapses
		Not universally distributed in brain
		Not specifically targeted to astrocytes
		Low levels of Glu released[c]
		Targets many mGluR's on many cell types
		Independent of NAA metabolism
		Not coordinated with NAA efflux
		Not specifically tied to Glc metabolism
NAAG	0.30	Widely distributed in all neuron types
		Found in oligodendrocytes and microglia
		Universally distributed in brain
		Specifically targeted to astrocytes
		High levels of bound Glu released[c]
		Targets specifically mGluR3 on astrocytes
		Dependent on NAA metabolism
		Rate of efflux same as that of NAA
		Synthesis tied to rate of Glc metabolism

[a]Rat brain
[b]Neurotoxic level is 0.01–0.10 µmol/ml
[c]Glu is released to the ECF in its bound form (NAAG-Glu) at 115–230 times its rate of release as free Glu

8.2.5 NAAG Potential for Neurotoxicity

NAAG as a Source of Glu At a NAAG efflux rate of 0.30 µmol/ml/h into brain ECF, and toxic levels of free Glu for neurons being 0.01–0.10 µmol/ml, there is also a potential for Glu neurotoxicity from this source. Since Glu is the predominant excitatory neurotransmitter in the vertebrate brain, Glu excitotoxicity is thought to be an important neurotoxic mechanism in many neurological disorders (Aarts and Tymianski, 2003). In a focal or regional ischemic event, the production of ATP is curtailed (Signoretti et al., 2001) and energy-consuming processes such as active transport of Glu are restricted. However, under these same conditions, catabolic processes can continue, and astrocyte surface NAAG peptidase activity could liberate sufficient Glu to the ECF from NAAG to result in both focal and regional neurotoxic effects.

Inhibition of NAAG Peptidase Activity   A number of NAAG peptidase inhibitors have been synthesized, and administration of several of these specific NAAG peptidase inhibitors such as 2-PMPA (2-[phosphonomethyl] pentanedioic acid) and PPDA (4,4'-phosphinicobis[butane-1,3-dicarboxylic acid]) offer a degree of neuroprotection in rat models under ischemic conditions (Thomas et al., 2001; Rong et al., 2002). In addition, administration of inhibitors of NAAG peptidase have also been found to prevent motor neuron cell death, perhaps by reducing Glu excitotoxicity, in an animal model of amyotrophic lateral sclerosis (Ghadge et al., 2003), and showed analgesic properties in a model of rat sensory nerve injury (Kozikowski et al., 2004), where it was demonstrated that there was a significant reduction in pain perception. Another NAAG peptidase inhibitor, an NAAG-analog containing a urea core (ZJ43), has been reported to reverse many of the schizophrenia-associated symptoms in an animal model of this disease induced by phencyclidine (PCP) (Olszewski et al., 2004).

While many of the effects of NAAG peptidase inhibitors have been attributed to a reduction in the rate of Glu formation, it also follows that by reducing the rate of NAAG hydrolysis, the global NAAG–mGluR3 signal effect may be magnified by these inhibitors (Olszewski et al., 2004), and therefore the role of NAAG in regulating focal and regional blood flow may be enhanced. Thus, the neuroprotective effect of NAAG peptidase inhibitors in vivo, under ischemic conditions, may be a function of increased regional collateral blood flow to ischemic affected areas.

On the basis of results using hippocampal slice preparations and the NAAG peptidase inhibitor 2-MPPA (2-[3-mercaptopropyl]pentanedioic acid), it is also suggested that presynaptic NAAG modulation of synaptic efficacies may be essential for preventing pathological processes associated with excessive Glu release (Sanabria et al., 2004). In this regard, the effect of spinal administration of 2-PMPA on electrically evoked responses of an individual spinal cord dorsal horn neuron in a carrageenan-inflamed rat preparation, in vivo, was to significantly reduce the number of action potentials per stimulus, thus modulating synaptic transmission events (Carpenter et al., 2003).

NAAG peptidase inhibition with GPI-5232 (2-hydroxy[2,3,4,5,6-pentafluorophenyl, methyl, phosphinyl, methyl]pentanedioic acid) has been reported to reduce induced aggressiveness in mice, perhaps by the buildup of NAAG in the ECF with its weak agonist action on the NMDA receptor, or by its stronger agonist action on the mGluR3 receptor (Lumley et al., 2004). However, on the basis of a study using whole-cell recordings from cerebellar granule neurons, it is posited that NAAG, even at elevated concentrations, is unlikely to have any physiologically relevant agonist effect on NMDA receptors, and that any interactions are probably mediated by mGluRs, not NMDA receptors (Losi et al., 2004).

The active site of the NAAG peptidase is suggested to have two regions, one of which is sensitive to the structure of the Glu moiety, requiring an intact Glu molecule to maintain the potency of a NAAG peptidase inhibitor (Kozikowski et al., 2004). Finally, a possible relationship between the active sites of the astrocyte mGluR3 receptor and those of the astrocyte surface NAAG peptidase is indicated by the synthesis of an NAAG-based tetraacidic NAAG peptidase inhibitor, that also exhibits the agonist activity of NAAG on the mGluR3 receptor (Nan et al., 2000).

8.2.6 Relationship of NAAG to BOLD Signals

BOLD Measurements   Brain blood oxygen levels can be measured locally using magnetic resonance (MR), based on magnetic properties of blood, which are in turn dependent on the oxygenation state of hemoglobin (Ogawa et al., 1993; Thomas et al., 2000). Blood oxygen exists in two states: a dissolved but freely diffusible gas and a bound form associated with hemoglobin in red blood cells. As oxygen gas diffuses down its gradient out of the vascular system, additional bound oxygen in the vascular compartment is dissociated from hemoglobin.

As deoxygenated hemoglobin is more paramagnetic than oxygenated hemoglobin, it can then act as an intravascular paramagnetic contrast agent. Thus, hemoglobin deoxygenation results in an increased magnetic susceptibility difference resulting in a signal loss in vascular water, caused by changes in proton transverse relaxation time (T_2) and the apparent proton transverse relaxation time (T_2^*). It has also been observed that in addition to a vascular water T_2 signal loss, changes in intravascular magnetic susceptibility can be correlated with changes in the extravascular tissue apparent diffusion coefficient (ADC) of water (Song et al., 2002a; 2003), and can be calibrated to obtain values for focal changes in the cerebral metabolic rate of O_2 consumption (model value O) (Hyder et al., 2002; Smith et al., 2002).

Because the changes in blood oxygenation levels affect the water signal, and are also associated with neurostimulation, this technique is used to image areas of the brain that appear to be stimulated as a function of brain activation tasks. The acronym for these functional MRI (fMRI) measurements is "BOLD" imaging, which stands for blood-oxygen-level-dependent imaging. The basis of the fMRI-induced BOLD signal has been reviewed (Logothetis, 2003). The observed BOLD signals are also reported to consist of both intravascular and extravascular components from both small and large vessels (Duong et al., 2003). In addition, the stimulated areas can also be viewed over time, providing a temporal "fingerprint" of brain areas associated with activation, thus providing a time-based anatomy or "chronoarchitecture" of the brain (Bartels and Zeki, 2004).

Alternate Method for Assessing Neurostimulation An alternate method for observing brain functional activation involves the use of ADC contrast, and it is reported that using ADC contrast, activated areas of the brain appear to be more closely coupled to neuronal activity than those observed using BOLD contrast (Song et al., 2003), and precedes that from BOLD contrast with a timing offset of about 1 s (Gangstead and Song, 2002). While ADC measurements are independent from and different than BOLD measurements, both in tissue distribution and in signal mechanism, it is proposed that by mapping and comparing the two measures in brain that certain inferences regarding various vascular contributions to the BOLD signal can be made from the diffusion values (Song et al., 2002b).

Evaluation of the Extravascular BOLD Effect   Magnetic resonance imaging depends primarily on the proton MR signal elicited by differences in water content, but using MR spectroscopy the spectral resonance peaks of water and individual metabolites can be separated and measured (Stanley, 2002). As the magnetic susceptibility component, which is a function of deoxyhemoglobin in the vascular system, is not only present in the vascular compartment but also to some degree in the extravascular compartment, it has been proposed that the BOLD effect can alter extravascular metabolite signals as well.

On the basis of concurrent MRS and BOLD measurements of signals from water and the extravascular intracellular metabolite NAA during neurostimulation, it was suggested that BOLD effects on individual cerebral metabolites in the extravascular compartment in normal brains could be demonstrated (Zhu and Chen, 2001). In this study, there were significant increases observed in resonance peak height integrals for water and NAA during stimulation, although the exact origins of the increases were unclear. However, the authors proposed that multiple mechanisms might contribute to the signal changes observed, and these mechanisms could include both contributions from T_2 changes as well as changes in proton spin density, a measure of metabolite concentration.

Complex Nature of the BOLD Effect   The BOLD effect can be used as an indirect measure of focal neurostimulation. However, the BOLD effect is complex in nature, and depends on interactions between changes in the metabolic rate of O_2 consumption, blood flow, and blood volume (Song et al., 2003). Since a basic ongoing function of the brain vascular system is to supply oxygen where and as needed, blood oxygen level depletion is also global in nature and exists at a continuous background level against which focal BOLD fMRI measurements are made (Macey et al., 2004), which adds further to the complexity in the interpretation of fMRI studies.

The global nature of the BOLD effect is illustrated in several human experiments. In a study involving repeated challenges of breath holding for 30 s, the result was an initial globally induced increase of blood CO_2 (hypercapnia), decreased blood O_2, and increased deoxyhemoglobin, all leading to vasodilation and a corresponding global rise in BOLD signal intensities of about 3% (Riecker et al., 2003; Macey et al., 2004). In a reverse of this study, involving hyperventilation for 60–180 s, the initial result was an induced decrease in blood CO_2 (hypocapnia), increased blood O_2, decreased deoxyhemoglobin, all leading to vasoconstriction and a corresponding global decrease in BOLD signal intensities (Posse et al., 1997).

The BOLD Effect and NAA

Changes in NAA content under normoxic BOLD effect conditions As the BOLD effect is universal in nature, all brain metabolite measures using MRS are made under BOLD conditions, and therefore it follows that such metabolite measures made under fMRS BOLD conditions in a specific region of interest are equally valid. In this regard, it has been demonstrated in humans, that upon visual stimulation, there is an increase in O_2 consumption (BOLD effect) and a corresponding transient increase in NAA signal in the visual cortex during the stimulation period (Zhu and Chen, 2001).

Changes in NAA content under hypoxic BOLD effect conditions During periods of ischemia caused by either diffuse moderate traumatic brain injury (Signoretti et al., 2001) or focal or global ischemia (Sager et al., 1995) in rats, NAA levels measured by chemical separation using high-pressure liquid chromatography (HPLC), were observed to decline so that within 10 min (brain injury) or 60 min (global ischemia), levels of NAA were reduced by 15–20%. As the catabolic compartment for NAA is in oligodendrocytes, and its anabolic compartment is in neurons, these reductions in NAA indicate the rapid trauma-induced movement of NAA from the intracellular to the extracellular space.

In a similar rat ischemia study, but using a proton MRS methodology instead of a chemical HPLC methodology, it was observed that there was a decrease in the T_2 value for the extravascular metabolite NAA of about 20% after 3 min of forebrain ischemia (Lei et al., 2003). These authors also cautioned that since mechanisms other than BOLD effects may contribute to T_2 decreases, observed changes in T_2 should be taken into account for the quantification of metabolite concentrations during ischemia. However, in this MRS study, the previously reported chemically measured rapid reductions in brain NAA levels associated with ischemia, and their possible effect on the observed T_2 signal of NAA, were not considered. If T_2 measurements may interact with metabolite concentrations as proposed (Lei et al., 2003), when using MRS to determine possible BOLD-related effects involving labile extravascular metabolites, such as NAA under ischemic brain conditions, caution must be exercised in the interpretation of results.

BOLD Signal Phases An initial negative BOLD effect phase occurring in 1–2 s, which is a measure of the decrease in oxygenation level of hemoglobin in the vascular system, especially in capillaries and draining veins, is separate from a second positive BOLD phase within 3–12 s, which is linked to functional hyperemia and increased oxygenation level of hemoglobin due to vascular expansion (Zarahn, 2001; Logothetis, 2003). However, since both are associated with neurostimulation, the interrelationship and balance between these two parameters are incorporated into most BOLD measurements (Zarahn, 2001). On the basis of temporal data obtained using oxygen-dependent phosphorescence quenching of an exogenous indicator, it has been observed that the first event after sensory stimulation was a localized increase in O_2 consumption, and this was followed by hyperemia and a more regional increase in blood flow (Vanzetta and Grinvald, 1999). Therefore, it has been suggested that fMRI focused on the earliest initial phase after stimulation would better colocalize with the actual site of neurostimulation. There is a third BOLD phase, representing oxygenation-blood flow normalization, which is associated with a recovery period after focal neurostimulation is terminated. This phase occurs in about 4–6 s after cessation of stimulation (Zhu and Chen, 2001).

To summarize, the BOLD effect in response to focal neurostimulation, while quite complex in origin, reflects two basic elements: an initial increase in deoxyhemoglobin in response to increased O_2 demand and a subsequent increase in blood flow to meet this demand.

8.2.7 Relationship of mGluR Activation and Hyperemia to BOLD Imaging

It has been proposed, based on the observation of the effect of Glu on initiation of astrocytic Ca^{2+} oscillations and subsequent effects on vascular hyperemia (Zonta et al., 2003), that the activation of astrocyte mGluRs and the ensuing astrocytic control of blood flow is responsible in large part for the measured BOLD response to increased neuronal activity. Although not excluding other factors, it is suggested that mGluR activation of astrocytes is centrally involved in the basic process that underlies most of the fMRI measurements that are made of the BOLD effect.

NAAG mGluR3 Activation and the BOLD Response BOLD imaging signals are associated with changes in the release of Glu from neurons (Hyder et al., 2001). As NAAG release is also a function of the rate of neuronal activity, it is thus a signal molecule specifically targeted to the astrocytic mGluR3 receptor, triggering astrocyte Ca^{2+} oscillations, and delivering 100–200 times more Glu moiety to astrocytes than free Glu in the ECF; it follows that NAAG is also a component in microcirculatory hyperemia, and therefore participates in elicitation of the brain activation-related focal BOLD response. Based on the ratio of NAAG to free Glu availability for mGluR activation, and the suggested importance of hyperemia to the BOLD response, NAAG release may indeed be the primary source of glial hyperpolarization (Gafurov et al., 2001), and therefore of the measurable focal BOLD response.

The putative NAAG-BOLD activation sequence is as follows: neurostimulation; neuron NAAG efflux; astrocyte NAAG-mGluR3 activation; Glu hydrolysis; astrocyte Ca^{2+}-wave initiation; astrocyte-vascular signaling, and resulting vascular hyperemia evident as a focal fMRI BOLD response.

9 Summary of Brain NAA and NAAG Metabolism and Dynamics

9.1 Structural Connections

In the brain, there is a chemometabolic sequence wherein, for every 400 molecules of Glc oxidized in mitochondria, about 10 molecules of NAA are synthesized in neurons from Glc-derived AcCoA, and for every 10 molecules of NAA synthesized, about 1 molecule of NAAG is then synthesized from NAA and Glu. Concurrently, an equivalent number of NAA and NAAG molecules synthesized are also exported to the ECF, carrying a minimum of 32 and 53 molecules of obligated water, respectively, thus maintaining the approximate 10:1 brain NAA/NAAG ratio.

9.2 Target Cell for NAA and Its Obligated Water

The NAA is specifically targeted to oligodendrocytes where the Ac portion of the NAA molecule is removed and incorporated into the general glial Ac pool. The Asp released in the ECF then diffuses down its gradient back to neurons where it is taken up to be recycled into NAA. In the process, at least 32 molecules of obligated water are removed to the ECF for transport to the vascular sink.

9.3 Target Cell for NAAG and Its Obligated Water

NAAG is specifically targeted to astrocytes where it interacts with their mGluR3 surface receptors. To terminate the NAAG-mGluR3 signal, the Glu portion is hydrolyzed by NAAG peptidase, and the Glu then incorporated into the general astrocyte Glu pool. In this mGluR3 stimulation process, a Ca^{2+} oscillatory wave is also initiated in the astrocyte that, on reaching the astrocytic endfeet in contact with the epithelial cell surface of the vascular system, results in release of molecules that signal the vascular system to open, producing local increases in arteriole diameters of 10–20%. In the astrocyte, Glu is processed into Gln, and Gln then returned to neurons via the ECF where it is taken up and formed into Glu once again. In this process, 21 molecules of obligated water are released from NAAG to ECF for transport to the vascular sink.

The NAAG peptidase that terminates the NAAG-mGluR3 signal appears to be an important component in regulation of the intercompartmental NAAG-mGluR3 astrocyte signaling process, and therefore, in the regulation of focal vascular hyperemia. In addition, this enzyme participates in the MWP action of NAAG in that residual NAA produced at the astrocyte surface then diffuses to oligodendrocytes where it joins the NAA stream of the neuron–oligodendrocyte intercompartmental cycle, and liberates an additional 32 molecules of water to the ECF in the process. The metabolic interactions between NAA and NAAG are illustrated in ❯ *Figure 14-2*.

◨ Figure 14-2
Coupling of NAA, NAAG, and Glu metabolism with brain hyperemia, energy availability, and metabolite flux. Enzymes involved in metabolic sequences: (A) NAA synthase, (B) NAAG synthase, (C) NAAG peptidase, (D) Amidohydrolase II, (E) Gln synthase, (F) Glutaminase. Adapted from Baslow (2000)

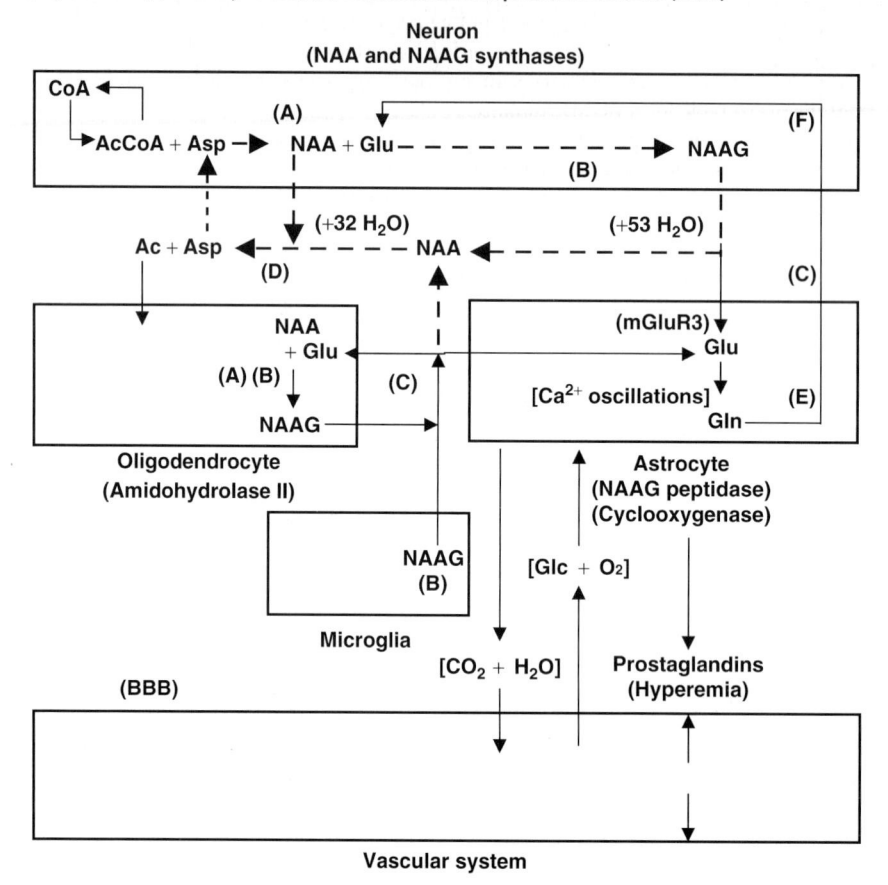

10 Discussion

10.1 The Primary Functions for NAA and NAAG Are Osmoregulatory

One difficulty in understanding the physiological functions of NAA and NAAG in brain has been due to the fact that since the decades these substances were discovered, they have generally been treated as separate entities. Another difficulty is that for most of this period, both the NAA and NAAG metabolic pictures were incompletely known. However, with an increased understanding of their complex intercompartmental relationships, and discovery of the role of astrocyte mGluRs in regulating brain microcirculation, a rationale for primary osmoregulatory functions for both NAA and NAAG is now provided. Nevertheless, it is only when one considers their metabolism and possible physiological functions as linked that a composite picture emerges, and the nature of their complimentary and integrated functions in the brain becomes evident.

10.2 NAA and NAAG Functions Are Linked

In brain, it has been demonstrated that the molecules Glc, NAA, and NAAG are structurally, metabolically, and dynamically linked. In addition, they also appear to be functionally linked in that an important aspect of successful neuronal physiology, under constantly changing conditions of neuronal activity, is a rapid access to energy and a need for dissipation of metabolic products (Park et al., 2003). In this context, an integrated role for the elaborate metabolic and cellular interactions of NAA and NAAG is reasonable.

10.3 NAA and Osmoregulation

Neurons in the brain are continuously firing in the ongoing process of their complex long-term informational storage and retrieval interactions. For this purpose, they require a constant supply of energy-rich molecules, along with sufficient O_2 for their oxidation (Laughlin and Sejnowski, 2003). The intracellular metabolic end products of this energy system are CO_2 and water, the latter of which must be removed to the ECF against a gradient. To serve this specific osmoregulatory requirement, neurons synthesize oligodendrocyte-targeted NAA that functions as a recyclable MWP to transport metabolic water to the ECF.

10.4 NAAG and Brain Microcirculation

To further provide for transient temporal changes in rates of neuron stimulation and continuously changing needs for energy and metabolic product removal, an astrocyte-targeted NAAG signal molecule is also synthesized directly from NAA and Glu. This specific astrocyte mGluR3 signal molecule can then participate in regulating local microcirculation by altering regional blood flow, via astrocyte-vascular interactions, thus increasing or decreasing availability of sources of energy as well as sink capacity for metabolic water and CO_2 as needed. In this role, NAAG serves as a relatively nontoxic intercellular carrier of Glu to the astrocyte mGluR3 receptor, where Glu is regenerated by the action of astrocyte NAAG peptidase.

10.5 The Overall Function of the Linked NAA–NAAG System in Brain

10.5.1 The NAA–NAAG System Functions as a Unit

When viewed from an overall brain perspective, it appears that the NAA–NAAG system functions as a unit, and that these substances play important and complimentary homeostatic and regulatory roles in support of the ever-changing local and global requirements of brain metabolism. Both participate in MWP osmoregulatory activity, and in addition, NAAG plays a role in altering brain microcirculation in order to regulate rates of energy supply and waste removal. Thus, the system appears to play a role in maintaining neurons at a level of their peak efficiency.

While additional important roles for these substances may yet be discovered, their roles as MWPs and in cell signaling are apparent. Special conditions for neurons in the brain, including their metabolic specializations, extensive morphological adaptations, degree of integration, longevity, high rate of metabolic activity and cellular water production, and encapsulation of large portions of their axoplasmic plasma membrane by myelin-producing oligodendrocytes, may have influenced development of these specific osmotic–hydrostatic solutions in the CNS.

10.5.2 The NAA–NAAG System and Information Processing

Attempts to measure whole-brain changes in blood flow during intense mental activity in humans have failed to demonstrate any overall hemodynamic changes. However, measured changes in local blood flow in response to specific cognitive tasks, while relatively small, of the order of 5% or less, are apparently very important to brain function (Raichle and Gusnard, 2002).

Attentiveness and responsive performance are key elements in animal survival, and therefore, specific and rapid changes in energy supply to activated areas of the brain are important to both the processing of input information and performance-related output responses. In this regard, both NAA and NAAG appear to be integral components in the maintenance of normal brain function; they are involved in sustaining the ability of neurons to function at maximum efficiency, and thus to receive, interpret, and transmit meaningful time-coded information. Conversely, failure of one or both components at any level could be expected to result in changes in attentiveness and performance in all animals, with potentially unfavorable biological outcomes. Over the longer term, it is also anticipated that such repeated failures could result in some forms of observable brain pathological syndromes, which would be especially relevant with respect to humans, since we are shielded from many of the unfavorable biological outcomes that impact other animals.

10.5.3 Role in Maintaining a Neuron's Minimal Repolarization Period

The observed focal BOLD response to neurostimulation in specific areas of the brain, and the relationship of hyperemia to the second phase of the BOLD response, clearly indicate that an increase in focal blood flow is an important component in neuron function. As the BOLD response is also correlated with changes in local availability of energy-rich molecules and oxygen, as well as sink capacity for water and carbon dioxide, one question to be asked is how observed temporal changes in the BOLD response reflect neuronal function? Another question to be addressed is how would neuronal function be affected if such focal hyperemic responses were delayed or abolished?

The BOLD Response as an Indicator of Changes in Focal Microcirculation and Neuronal Activity The basic function of neurons is to communicate, and most do this by generating cellular electrophysiological signals in the form of action potentials or "spikes" that are translated into neurochemical signals transmitted to other neurons at synapses, and subsequently interpreted at some level in the CNS neural network (Clifford and Ibbotson, 2000). These energy-driven spike trains are both ephemeral and transient in nature, and a neuron must be able to quickly indicate its needs for increased energy supplies and waste removal in order to sustain this spiking activity. Signal transmission in the mammalian cortex is also considered to be an expensive process that has energy demands tightly coupled to encoding of information by neurons (Smith et al., 2002).

The BOLD response reflects changes in temporal neuronal activity in the form of changes in oxygen utilization, and therefore is an indicator of both changes in focal neuronal energy requirements and the microcirculatory responses to such changes. In addition, based on the association of the BOLD response with changes in energy utilization, along with measured changes in neuronal activity recordings in rat brain at the same time, it has been proposed that the BOLD response is also an indicator of localized changes in neuronal spike frequency (Smith et al., 2002).

Result of Delays in Microcirculatory Responses
Delays Due to Signaling   In the event of delays in astrocytic and/or vascular responses to neuronal chemical signals elicited during neurostimulation, it can be anticipated that the hyperemic response would also be delayed, the indicator BOLD response diminished, and the focal energy supply and vascular sink capacity would not be adequately increased. As a result, the vascular system would become less responsive to a neuron's immediate needs, and the lack of sufficient energy would alter the neuron's ability to send and receive messages. This would be reflected in an increased repolarization time, an increase in the absolute refractive period, and a decrease in the neuron's maximal rate of firing. To the extent that NAAG interacts with astrocyte mGluR receptors and with astrocytic control of focal hyperemia, it would therefore be

directly involved in the maintenance of the maximal rate of neuron firing and as a consequence, with the basic function of neurons, the transmission of interneuronal frequency-encoded communications.

Thus, the NAA–NAAG intercompartmental system in the brain seems to play a specific role as a controlling mechanism in support of neuronal integrative signaling activities. It does this by timely removal of metabolic water, and by increasing the rate of supply of neuronal energy components to be used for resynthesis of ATP, the primary energy source for ongoing metabolic activity and neuron repolarization (Raichle and Gusnard, 2002). For rat brain gray matter, energy consumption depends strongly on the action potential rate, and it has been calculated that an increase in activity of only one action potential per cortical neuron per second will raise oxygen consumption requirements by 21.6% (Attwell and Laughlin, 2001). Focal BOLD responses, which are complex in origin, also reflect any signaling interactions between the NAA–NAAG system and brain hemodynamics.

Delays Due to Vascular Insufficiency   It also follows, that even if the NAA–NAAG system remained intact, and if a state of vascular insufficiency impaired blood–brain barrier function and vasomotor tone, it would affect the ability of focal NAAG-astrocyte-vascular signaling to increase microcirculation when required, and thus could impact neuronal information processing. This appears to be the case in normal aging where it has been reported that the spatial extent of the BOLD response, elicited by a finger-tapping task, is greatly attenuated in old (65 years) versus young (27 years) individuals (Riecker et al., 2003). In addition, the global BOLD signal intensities in response to induced hypercapnia were also much reduced in the old, being only 29.7% of that of the young. On the basis of this study and a review of similar studies, it was suggested that changes in cerebral vasculature could alter the neuronal–vascular coupling so that cerebral vessels might not react as effectively in response to a vasodilating stimulus.

Changes in cerebral vasculature may also be a factor in Alzheimer's disease, where extravascular and intravascular amyloid deposition produces changes in vasomotor control, a specific pathology called cerebral amyloid angiopathy, which has been proposed to explain the chronic state of cerebral ischemia and associated cognitive deficiencies in this disease (Holtzman, 2001; Suhara et al., 2003). Similar outcomes would be expected as a result of any other disease states where a condition of vascular sclerosis occurs that can alter elements of the neuron-vascular signaling process.

Nature of Neuron–Astrocyte Chemical Signals   While it has been reported that the chemical mechanisms responsible for coupling of changes in blood flow in response to focal brain activation are incompletely understood, several possible ECF chemical mediators have been cited (Riecker et al., 2003). These include H^+ ions derived from end products of local energy metabolism, as well as K^+ ions, a product of the neuron depolarization process, and an ATP-related molecule, adenosine. The important role of Glu, and the documented Glu–mGluR neuron–astrocyte interaction with dynamic control of focal hyperemia (Zonta et al., 2003) has already been described, as well as evidence that a specific NAAG-mGluR3 activation may be an important component in this interaction (Gafurov et al., 2001).

Role in Maintaining a Full Range of Neuron Frequency-Coded Information   As synaptic frequency-coded signaling in the brain is a primary mechanism by which information is processed and retrieved (Clifford and Ibbotson, 2000; Chacron et al., 2001; Laughlin and Sejnowski, 2003), any change in the dynamic aspects of the NAA–NAAG intercompartmental metabolic system that significantly alters energy availability and rate of ATP production will be reflected in how the brain processes information and perceives reality. Increases in the repolarization and absolute refractory periods will result in truncation of the maximum rates of neuron firing, with concomitant loss of that portion of coded information that is processed at the highest frequencies. This is a classical phenomenon known as Wedensky inhibition in which high-frequency neural discharges are blocked while low-frequency ones are transmitted, due to an increase in the absolute refractory period of the neuron (Phillips and Mathews, 1993).

Neuron Codes
Nature of Neuron Codes   All neuron codes, whether continuous trains of spikes that are characteristic of information transmitted at low frequencies, or bursts of spikes that are characteristic of information transmitted at high frequencies, are made up of interactions between spikes and relative refractory

periods. At the highest frequencies, the relative refractory periods approach a neuron's absolute action potential refractory period, which may be in the range of 0.5–4 milliseconds (ms), and therefore, at high spike frequencies the effect of lack of adequate energy resources is most critical. While it is difficult to directly investigate how information might be coded within the CNS (Averbeck and Lee, 2004), a study of how some environmental receptors transmit information to the brain is informative. These receptors transmit three kinds of information: the brain-map coordinates of the receptor, the nature of the information transmitted, and the encoded message itself.

Environmental Receptor Coding   As an example, in a study of the isolated salt receptor of the blowfly (Gillary, 1966), it was observed that information about NaCl molarity was transmitted to the CNS using a frequency code in the range of 2–120 Hz, evident as a linear association between the mean frequency response and the logarithm of molarity (❯ Table 14-11). As the relative refractory period is seen to vary inversely with the

◼ Table 14-11
Response of the isolated salt receptor of the blowfly to the molar concentration of NaCl at 25.5°C[a]

NaCl (M)	Signal [Hz] (spikes/s)	Spike interval (ms/spike)	Relative refractiveness[b] (ms between spikes)	Energy cost[c] (relative ATP units)
0.10	2	500	499	1
0.20	22	46	45	11
0.25	36	28	27	18
0.50	52	19	18	26
1.00	80	13	12	40
2.00	104	10	9	52
3.00	120	8	7	60

[a]Adapted from Gillary (1966)
[b]Assumes the total spike period is 1 ms or less (Laughlin and Sejnowski, 2003)
[c]Assumes that the energy cost of repolarization is proportional to spike frequency. At any given frequency, energy can be conserved by shortening the period of information transfer, or by modulating spike periods with relative refractory periods, producing bursts of spikes

mean spike frequency, it is obvious that a small change in the interspike absolute refractory period of only 1 or 2 ms will affect the nature of information transmitted at the highest frequencies. Coding mechanisms of the salt receptor, as well as other taste receptors of the fruit fly are similar to that of the blowfly, and have recently been reviewed (Ishimoto and Tanimura, 2004). Similar responses have been observed in hamster fungiform taste receptor cells exposed to acid stimulation, where the frequency of action currents generated varies directly with increasing acidity between pH 2.5 and 5.5 (Gilbertson et al., 1992). In this case, the frequency was almost linear, with a frequency response of 3.5 Hz at pH 2.5 and 0.25 Hz at pH 5.0.

The effect of such changes in the neuron refractory period on information processing can be tested in a variety of ways, including changing the availability of Glc or oxygen, altering repolarization times with drugs that affect the Na^+/K^+ pump, and especially in poikilotherms, by changing the ambient temperature. In this same study (Gillary, 1966), the effect of alterations in temperature on information processing by the salt receptor was investigated, and it was found that changes in ambient temperature of only ±2–3°C changed the relative refractive periods and strongly influenced the receptor processing and transmission of NaCl molarity information (❯ Table 14-12).

A similar interaction of temperature and action potential duration in humans using an intradermal electrical stimulation technique has also been demonstrated (Mackel and Brink, 2003). In another human study involving changes in ambient temperature and texture perception by a finger placed on a texturized surface, it was observed that afferent coding to the CNS was distorted by cooling the ulnar nerve while maintaining normal hand temperature (Phillips and Matthews, 1993). In this study, there were marked

◘ Table 14-12
Response of the isolated salt receptor of the blowfly to 1 M NaCl at different temperatures[a]

NaCl (M)	Temperature (°C)	Signal (spikes/s)	Spike interval (ms/spike)	Information transmitted[b,c] NaCl (M)
1.00	23	64	16	0.50
1.00	24	68	15	0.70
1.00	25.5	80	13	1.00
1.00	26	82	12	1.10
1.00	27	93	11	1.70
1.00	28	104	10	2.00

[a]Adapted from Gillary (1966)
[b]In this case, relative to 25.5°C, the information transmitted at different temperatures by the isolated salt receptor is incorrect (reality is 1.00 M). However, in a whole organism, concurrent temperature information can be used to integrate the salt receptor data to reconstruct reality
[c]A direct effect of changes in temperature is to change molecular diffusion rates by about 3% per °C, and thus to indirectly alter reaction rates (Guilfoyle et al., 2003). This would include alteration of reactions involving ATP at the plasma membrane, thereby affecting rates of neuron repolarization, and resulting in changes in information processing

changes in both recorded nerve absolute refractory periods and perception of texture by the brain, which were subsequently modeled and then primarily attributed to the phenomenon of Wedensky inhibition.

Coding and Fick's Law   The observed thermal effects in these cases is probably due to the interaction of temperature with the rates of molecular diffusion (Guilfoyle et al., 2003), where according to Fick's law, (Eq. 15-1), the mass movement (J) of any substance is a function of its diffusion coefficient (D), which is in turn dependent on the ambient temperature. In this equation, A is the cross-sectional area, Δx the length of the mixing zone, C the concentrations, and $C_1 - C_2/\Delta x$ the concentration gradient:

$$J = DA\left[\frac{C1 - C2}{\Delta x}\right]. \tag{15-1}$$

Changes in D for water and many other substances with temperature are of the order of 3% per °C, and these changes will alter the mass movement of these molecules, and consequently, the rates of many reactions and of their associated physiological functions. Fick's law has also been proposed to be a component of the physiological basis of the BOLD response in homeotherms, in that the mass movement of oxygen from the vascular system to neuronal mitochondria is determined by the value of the mass transfer coefficient for oxygen and its concentration gradient (Hyder et al., 2001).

Coding Sensitivity to the Absolute Refractory Period   The blowfly studies also show that for an isolated receptor, the response to a continuous and specific stimulus is at the neuron's maximal frequency rate under a given set of conditions, and any changes in those conditions will alter the nature of information transmitted. The importance of maintaining the maximal neuronal response frequency is further demonstrated in a study of a motion-sensitive visual interneuron of the blowfly (Borst, 2003), where it was also observed that the neuron response to any given type of visual stimulus was always with its maximum information rate. At these highest spike frequencies, transmitted information is most sensitive to any changes in the absolute refractory period. This is also illustrated in a study of how visual motion is encoded in the eye of a mammal, the wallaby, where it was found that the absolute refractory period has a greater impact when firing rates are high, since the higher the frequency, the more chance that an individual spike will interact with the absolute refractory period, which in this case was 2 ms (Clifford and Ibbotson, 2000).

Interactions of Fick's Law, Wedensky Inhibition, and Information Processing   Neurons encode and transfer information to other neurons in the form of combinations of spikes and relative refractory periods, the precise nature by which transmissions may be highly modulated in the CNS. However, there is an upper limit to spike frequency that is reflected in an absolute refractory period specific for each neuron, and which is governed in part by limitations imposed by the ability to obtain and use energy. For transmissions at or near this upper limit of spike–refractory period–information codes, a change in availability of energy can truncate high-frequency neuron-coded information transfer by the phenomenon of Wedensky inhibition.

In addition to Wedensky inhibition at high frequencies due to energy constraints, neuron-relative refractory periods at lower frequencies can also interact with any changes in diffusion coefficients (D), substrate concentrations (C), or cell architecture (A, Δx), all of which can affect mass transport (J) of water and other substances (Fick's law, Eq. 15-1) that govern molecular diffusion and reaction rates, and all of which are highly dependent on the changing spatial associations in response to the momentary state of cellular edema. Such changes in mass transport of substances, and consequently in reaction dynamics of these substances, can affect information processing at any spike frequencies, changing the basic nature of information carried in the spike-relative refractory period sequences. This is evident in the response of the salt receptor of the blowfly to a constant salt concentration, with changes in ambient temperature as shown in ❷ *Table 14-12*.

Thus, coded neuron information is subject to constantly changing conditions related to both availability of energy and all other factors affecting molecular diffusion. Clearly, appropriate global control mechanisms must be in place to continuously maintain optimum conditions required for accurate neuron information processing.

Neuron Information Processing and the NAA–NAAG System   As NAAG–astrocyte interactions play a role in the regulation of rates of energy supply to neurons, and consequently in their ability to process information, especially at highest rate-codes, it is of interest that a significant reduction in NAAG peptidase activity, the mechanism by which the NAAG–mGluR3 signal complex is dissociated (Pomper et al., 2002), is reported to modify information processing in mice, as observed in a reduction of induced aggressive behavior (Lumley et al., 2004). This has also been observed in a decreased perception of induced inflammatory pain in a formalin model of rat sensory nerve injury, in response to the inhibition of NAAG peptidase (Kozikowski et al., 2004).

In a biochemical study of schizophrenia, a pathological condition primarily associated with aberrant information processing (Jansma et al., 2004), a reduced NAAG peptidase activity was observed in postmortem samples of the hippocampus and prefrontal cortex (Tsai et al., 1995). In this same study, it was also found that where NAAG peptidase activity levels were reduced, the static levels of NAAG were elevated. These observations suggest that there might be a prolongation of any NAAG–mGluR3 interactions in brains of patients with schizophrenia. This may help to explain in part why, in an fMRI study of monozygotic twins, the identical twin with schizophrenia exhibited a 16-fold increase in the mGluR-related BOLD effect over the healthy twin in response to a similar stimulation of the area of the brain subserving speech (Spaniel et al., 2003). Similar enhanced BOLD effects have been observed in many more, but not in all cases of schizophrenia (Jansma et al., 2004).

Additional support for this explanation is found in a recent study of the levels of NAAG peptidase mRNA in schizophrenia, where it was observed that the mRNA transcripts, located primarily in astrocytes, showed regional abnormalities, and it was suggested that hydrolysis of NAAG is disrupted in patients with this disorder (Ghose et al., 2004). However, in a study of a PCP-induced animal model of schizophrenia, administration of an NAAG peptidase inhibitor was found to reverse many of the PCP-induced symptoms, suggesting that prolongation of NAAG–mGluR3 activation by reducing the rate of NAAG hydrolysis actually decreased the schizophrenic-like symptoms in this case (Olszewski et al., 2004). Alternatively, it may be that the inhibitor itself interacted with the mGluR3–peptidase complex on the astrocyte surface so that NAAG was excluded, and therefore the NAAG–mGluR3 activation inhibited. In any event, there seems to be a general consensus that schizophrenia and aberrant NAAG signal function may be related.

In another biochemical study of postmortem brains of individuals with Huntington's disease or Alzheimer's disease—two well-characterized neurodegenerative disorders associated with cognitive loss—there was evidence of decreased levels of NAAG, NAA, and NAAG peptidase in both diseases. In these cases, the biochemical changes were correlated with neuronal but not glial loss (Passani et al., 1997). In a related fMRI study of patients with mild Alzheimer's disease, using a semantic task it was observed that there was a significant positive correlation between the degree of brain atrophy and of BOLD activation in the left inferior frontal gyrus, a language region (Johnson et al., 2000). Here, it was proposed that in spite of recognized brain atrophy, there might be compensatory mechanisms at work in the hemodynamic response evoked by the stimulus.

As a result of these independent biochemical and fMRI studies, a linkage between NAAG metabolism, the BOLD response, and cognitive loss in humans is observed. In addition, in schizophrenia and mild Alzheimer's disease, two factors are in common: decreased NAAG peptidase activity and increased BOLD responses to stimuli. As NAAG is an agonist of the astrocyte surface mGluR3, associated with hyperemia and BOLD signals, whose action involves the astrocyte surface NAAG peptidase, reduced levels of the enzyme in both diseases could result in the prolongation and regional extension of any NAAG–mGluR3 activation, and lead to the enhanced stimulus BOLD responses observed in these diseases.

A relationship of NAA or NAAG to information processing has also been considered in several recent studies. In a study of attentional performance in relapsing-remitting multiple sclerosis patients, it was concluded that NAA levels provide not only a specific measure of pathology but are also relevant to cognitive functions (Gadea et al., 2003). A role for NAAG in processing information in the spinal cord has also been proposed (Carpenter et al., 2003).

10.6 Temporal Bounds Within Which the NAA–NAAG System Operates

10.6.1 Time frame of Information Processing

The time frame in which normal information processing activities must operate is short and is bounded by the time between neuronal action potentials measured in milliseconds (Clifford and Ibbotson, 2000; Mackel and Brink, 2003) and changes in the rate of microcirculation in seconds (Vanzetta and Grinvald, 1999). Within this time frame, the dynamic metabolic feedback interactions between the NAA–NAAG system and the vascular system, changes in rates of neuronal energy supply and ATP production, and energy-dependent neuronal repolarization all occur. Since the ATP energy required by neurons increases directly with spike frequency (◐ *Table 14-11*), any use of stored ATP to generate an increase in spiking must be replaced by an equivalent increase in Glc oxidation and ATP production. If the number of spikes doubled from 10 to 20 Hz, required energy would also double.

However, since changes in frequency are transient events, the increased energy used only needs to be replaced based on the mean frequency, over a period longer than that of any transient. Thus, a doubling of spike frequency in this example would require an unrealistic reciprocal 100% increase in Glc utilization per second, but only a 10% increase per second over 10 s, the approximate time frame and magnitude of the second phase of the fMRI BOLD response. Thus, the BOLD response seems to represent an afterimage of much shorter neuronal transients in activity and appears to operate in the more realistic time frame of NAA, NAAG, Glc, and oxygen diffusion events.

Under abnormal conditions, such as focal or global anoxia, or hypoglycemia, and at a nominal neuronal information spike frequency of 100 Hz, there could be 30,000 neuronal spikes generated in 5 min, thus depleting existing ATP stores. Without a timely replenishment of ATP supplies, it can be anticipated that neuron repolarization and absolute refractory periods would increase, reducing the maximum spike frequency, and thus altering information processing. On the global level, this could be associated with a loss of cognitive ability and consciousness within minutes.

10.6.2 Temporal Interaction of NAAG Dynamics and the fMRI BOLD Response

In terms of NAA and NAAG dynamic model values (◐ *Tables 14-3* and ◐ *14-8*), the fMRI BOLD response (increased deoxyhemoglobin formation) is a function of changes in the rate of focal oxygen utilization (O), equivalent to approximately 6× the rate of Glc (G) utilization, which is in turn coupled to the rate of NAAG synthesis ($G = S_{NAAG} \cdot K_{1NAAG}$). Thus, the NAAG dynamic model predicts that the BOLD response and NAAG synthesis will vary directly with one another. As S_{NAAG} is equal to the rate of efflux of NAAG (E_{NAAG}) into the ECF, the model also predicts that the BOLD response will vary directly with the release and interaction of NAAG with the mGluR3 receptor on the astrocyte surface, its only known site of agonist action at physiological concentrations, and its only known site of metabolic inactivation (astrocyte surface NAAG peptidase).

The model also indicates that for every 2,400 molecules of oxygen used in the brain to oxidize 400 molecules of Glc, 1 molecule of NAAG is synthesized from Glc and released to the ECF. This drawdown of oxygen from oxyhemoglobin upon neurostimulation would correspond to the initial 1–2 s fMRI BOLD response, and the NAAG-mGluR3 signal-receptor formation would initiate, and correspond to the secondary 3–12 s prolonged astrocyte-mediated hyperemic fMRI BOLD response. The third fMRI BOLD or recovery phase, lasting between 4 and 6 s, would correspond to the hydrolysis of Glu in the astrocyte surface mGluR3–NAAG peptidase complex, with release of NAA to the ECF.

A summary of observed relationships between focal neurostimulation, NAA- and NAAG-linked syntheses, and mGluR signals affecting vascular microcirculation is illustrated in ❯ Figure 14-3, and the likely

Figure 14-3
Interrelationships between focal neurostimulation, NAA and NAAG syntheses, and vascular microcirculation. Stimulation of neurons produces transient increases in water, NAA and BOLD measured by MR (Zhu and Chen, 2001). NAA is synthesized from Glc (Moreno et al., 2001). NAAG is synthesized from NAA (Tyson and Sutherland, 1998). NAAG is one of many mGluR stimulants (Schoepp et al., 1999). NAAG released from stimulated neurons stimulates glial mGluR3 (Urazaev et al., 2001). NAAG alone stimulates glial mGluR3 hyperpolarization (Gafurov et al., 2001). Stimulants of mGluRs produce glial Ca^{2+} oscillatory waves, hyperemia, and participate in BOLD (Zonta et al., 2003)

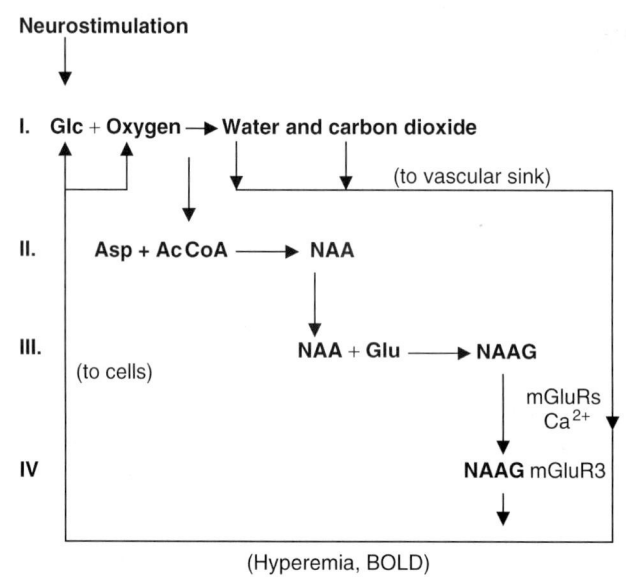

Neurostimulation

I. Glc + Oxygen ⟶ Water and carbon dioxide
(to vascular sink)

II. Asp + AcCoA ⟶ NAA

III. (to cells)   NAA + Glu ⟶ NAAG

mGluRs Ca^{2+}

IV.   NAAG mGluR3

(Hyperemia, BOLD)

interaction of the NAA–NAAG system with neuron repolarization rate and information processing in ❯ Figure 14-4. As illustrated in ❯ Figure 14-4, the rate-limiting step for maintaining a neuron's full capability to communicate is the availability of an adequate ATP supply. To this end, the neuron Glc oxidation-coupled synthesis of NAA and NAAG and their interactions with astrocytes and oligodendrocytes can serve as a feedback mechanism to ensure that energy resources remain adequate at any level of demand. Finally, a proposed temporal sequence and metabolic interactions of NAA and NAAG with a single neuron–neuron signal event is outlined in ❯ Table 14-13.

Although the chain of evidence and weight of the evidence indicate that NAAG plays an important role in regulation of brain microcirculation, only one study linking NAAG and its relative impact on changes in focal blood flow in the brain has been performed. In that study it was observed that the NAAG peptidase

◘ Figure 14-4
Dynamic interactions of the NAA–NAAG system with neuron energy flux, neuron-energy dependent minimal repolarization and refractory periods, and maximal neuron firing rate. Energy flux used in this example is based on complete oxidation of Glc. However, the neuron energy source may be lactate, which has a lower ATP value. Also, Glc is subject to "proton leak in mitochondria," which would reduce the Glc/ATP ratio to 31 (Attwell and Laughlin, 2001). Neurons transmit information in the form of single spikes or bursts of spikes, and in either case, the relative refractoriness is related to interspike intervals, and therefore to information processing (Chacron et al., 2001). Any reduction in the rate of energy supply that increases the minimal refractory period reduces the maximum firing rate, thereby truncating a neuron's signaling repertoire at its highest rates. However, that portion of information transmitted at less than the maximal firing rate would not be affected (Berry and Meister, 1998)

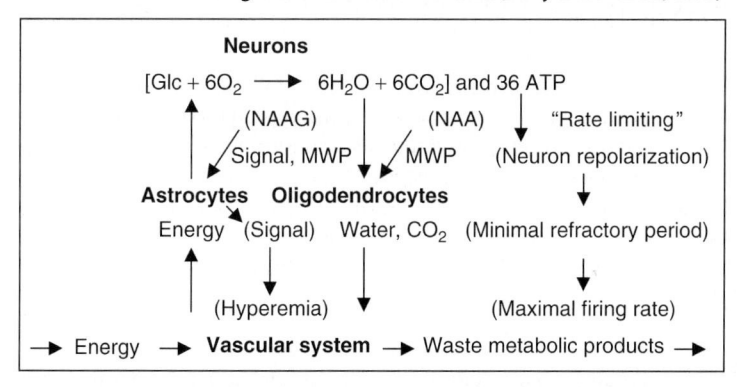

```
                        Neurons
        [Glc + 6O₂  ───▶  6H₂O + 6CO₂] and 36 ATP
              ▲       (NAAG)        (NAA)      "Rate limiting"
              │     Signal, MWP │  / MWP   (Neuron repolarization)
         Astrocytes   Oligodendrocytes          │
           Energy  (Signal)   Water, CO₂   (Minimal refractory period)
              ▲       │             │             │
              │       ▼             ▼             ▼
                  (Hyperemia)            (Maximal firing rate)
   ──▶ Energy  ──▶  Vascular system  ──▶  Waste metabolic products ──▶
```

inhibitor, 2-PMPA, attenuated the BOLD signals in brain of anesthetized mice, suggesting that when NAAG hydrolysis was inhibited, hyperemic responses were also blocked (Baslow et al., 2005).

11 Functional Consequences of Inborn or Engineered Errors in NAA and NAAG Metabolism

11.1 Canavan Disease

In light of the observed osmoregulatory and signal functions of NAA and NAAG, the effects of inborn errors in their metabolism can be further assessed. In CD, in the absence of sufficient amidohydrolase II activity, the MWP activities of both NAA and NAAG, and the microcirculatory activity of NAAG are compromised. With the global buildup of NAAG in ECF due to a product inhibition effect of NAA on NAAG peptidase activity, the astrocyte mGluR3 receptors are continuously exposed to NAAG signal molecules that cannot be readily hydrolyzed and removed. Thus, it is highly likely that any ability of NAAG to control focal or regional hyperemia, via astrocyte-vascular prostaglandin interactions, and resultant feedback regulation of energy availability and metabolite sink capacity, is severely affected. It has also been reported that at elevated NAA concentrations (20 mM), such as found in the ECF in CD, the release of prostaglandins from a human astroglial cell line is completely eliminated (Rael et al., 2004). If this were also the case in CD in vivo, then this would be an additional mechanism that could affect the ability of NAAG to control microcirculation.

As clinically observed, the CD phenotype is that of an osmotic disease characterized by increased brain NAA and NAAG, increased NAA and NAAG in blood and urine, extensive brain water imbalance, a characteristic swelling of astrocytes, and deconstruction of normal neuron–oligodendrocyte and neuron–astrocyte associations. In addition, normal interneuronal communication is affected as evidenced by recurrent seizure and spastic episodes. The known clinical course of this disease is profound, with progressive deterioration in brain function and early mortality.

◘ Table 14-13
Functional and temporal association of NAA and NAAG intercompartmental metabolism with three phases of a neuron–neuron transmitted impulse

Phase I. Steady state (a dynamic homeostatic resting state)
Preformed NAA and NAAG are maintained at high cellular–ECF gradients
Cell membrane is polarized (resting potential)
Plasma membrane surface pump-leak systems are operating
Energy-producing systems are operating
Preformed ATP is available for use
Preformed transmitter is stored at presynaptic efferent membrane
Phase II. Stimulation and signal transmission[a]
Neurotransmitter arrives at postsynaptic afferent receptor
Depolarization is induced and a plasma membrane surface wave (Na^+, K^+) is carried along the axon
Simultaneous efflux and diffusion of preformed NAAG to juxtaposed astrocyte mGluR3
Presynaptic neurotransmitter released to postsynaptic efferent receptor resulting in synaptic signal transmission
Phase III. Recovery (including neuron refractive subperiod)[b]
1. Neurotransmitter removed from postsynaptic afferent membrane receptor
2. ATP depletion is coupled with Na/K membrane repolarization
3. Increased rate of Glc and O_2 use and of supply via NAAG–astrocyte-induced hyperemia
4. Glc oxidation and ATP resynthesis from ADP
5. Glc oxidation products, water and CO_2 formed at increased rate
6. Efflux of preformed NAA with its obligated water to ECF and diffusion to oligodendrocytes
7. Glc coupled NAA resynthesis from oligodendrocyte amidohydrolase II recycled Asp
8. NAA coupled NAAG resynthesis from Glu regenerated from astrocyte surface NAAG-peptidase and astrocyte formed Gln
9. Presynaptic efferent transmitter synthesized and stored
10. Dynamic steady state conditions resumed

[a]In the brain, the volume of signal traffic that can be supported is set by the brain's metabolic rate. In the human cerebral cortex, the permissible traffic is reported to be less than 1 action potential per neuron per second (Laughlin and Sejnowski, 2003)

[b]Availability of energy is the most severe constraint on neural communication (Laughlin and Sejnowski, 2003)

11.2 Hypoacetylaspartia

In hypoacetylaspartia, the singular known case of apparent lack of brain NAA synthase activity, the absence of both NAA and NAAG precludes any of their normal osmoregulatory or microcirculatory functions. While the loss of these functions were correlated with early brain developmental delays associated with this child's inborn error, as well as delays in reaching normal childhood developmental milestones, there was no evidence of any osmotic component. At present, the observed clinical course and outcome of this disease at age 8 is profound. Although there is still no evidence of any osmotic defect, and myelination seems to be normal, this child exhibits motor defects and is reported to be operating at a cognitive level of less than that of a 1-year-old child. (Boltshauser et al., 2004). As there are a variety of cellular osmoregulatory mechanisms available, some may be upregulated to make up for the lack of the NAA intercompartmental system. However, a lack of NAAG and its specific signal effect on the mGluR3 astrocyte receptor suggests that an important neuron microcirculatory control mechanism is inoperative in this case.

As the NAA–NAAG system is not required for individual neuron survival or function but seems only to be important to information processing and higher cognitive function, it follows that any degree of pathology that affects this system will result in a proportionate loss of these functions. In the singular case of hypoacetylaspartia, this could explain why neurons survive (brain structure is normal), are myelinated (unremarkable MRI), and signal one another (normal EEG), but cannot process information well.

To the degree that NAAG–mGluR3 interactions determine local changes in microcirculatory efficiency affecting energy supplies that are key to information processing in response to neuronal stimulation, it could have been anticipated that there would be long-term deficits in brain motor and cognitive functions associated with this disease.

11.3 Genetic Engineering—Absence of NAAG Peptidase Activity in a Mouse

The importance of NAAG and its metabolism in the brain is further supported by the finding that mice made heterozygous for NAAG peptidase by introduction of a null mutation (deletion of exons 9 and 10), appeared normal and expressed normal levels of enzyme activity, but their homozygous null mutant embryos did not survive beyond embryonic day 8 (Tsai et al., 2003). In a similar null mutant experiment, but targeted to exon 1, the homozygous null condition in these embryos was not lethal, although their brain enzyme activity was reduced by about 90% (Bacich et al., 2002). Tsai et al. (2003) suggest that the deletion of exons 9 and 10, but not 1, serves to make it a dominant gene species in the homozygote state, thus preventing expression of a second gene that produces the small residual NAAG peptidase activity.

12 Relationship of NAA and NAAG Metabolism to Other Brain Pathologies

12.1 Static Measures of Brain NAA as a Neuron Marker

NAA has been regarded as a specific neuron marker in studies of almost all aspects of brain function and disease. This has been made possible by its relative ease of measurement and prominent spectroscopic signal using proton MRS measurement techniques. These measurements of brain NAA, and of differences in static levels of NAA (model functional constant K_3) in various human brain conditions, have been universally and routinely used to help in making assessments of the relative degree of impairment, and also in some cases, to make prognoses for a variety of ailments.

The tenfold lower content of brain NAAG, and its very close proximity to the NAA MRS signal peak have precluded separate measures of NAAG in most human studies, but have occasionally been reported (Hajek et al., 2000). As a result, most reported NAA results are considered to be primarily NAA, but with a small NAAG component. However, the inability to measure changes in NAAG levels separately in most cases not only results in less accurate assessments of NAA, but also obscures any possible relationship of changes in static NAAG levels to the pathological condition being investigated. To overcome this problem, it has been proposed that the use of higher magnetic field proton MRS would allow measurement of NAAG in human brains (Di Costanso et al., 2003).

12.2 Dynamic Measures of NAA and NAAG

12.2.1 Dynamic Measures in Normal Humans

In light of the many apparent linkages of NAA and NAAG and their relationship to neurostimulation and Glc utilization (model variable G), a measure of some dynamic aspects of their metabolism may also be informative, and it has been suggested that measurement of such dynamic changes in brain NAA levels may be a useful additional marker of neuronal function (Sigmundsson et al., 2003).

In this regard, it has been demonstrated that upon visual stimulation in normal individuals, there is an induced transient change in NAA resonance peak integrals, perhaps related to NAA increases. The increases in the visual cortex were about 2.5% over a period of 1–2 min, and quickly returned to the prestimulation (K_3) level within 1–2 min after the stimulation is terminated (Zhu and Chen, 2001). If these changes were due in part to changes in NAA content, the increase, and subsequent return to the prestimulus condition could be anticipated based on the NAA model. In this model, any transient increase in rate of NAA synthesis (S) is associated with equivalent increases in its rate of efflux (E) and hydrolysis (H). In this same study it was also shown that there were corresponding and simultaneous 3.0% increases and decreases in the water signal (model variable W_1), and an increased rate of cerebral metabolic rate of O_2 utilization (model variable O) as evidenced by an increase in the BOLD effect. All of these observed transients in response to neurostimulation, both in direction and in comparable magnitude, are predictable by the NAA dynamic model.

Additionally, Moreno et al. (2001) have measured the rate of NAA synthesis (model variable S) in brains of normal adults and children and have provided a composite S value of 9.2 ± 3.9 nmol/min/g (0.55 ± 0.23 μmol/g/h). Because of the much lower brain content of NAAG, and the proximity of its proton MRS signal to that of NAA, dynamic measures of changes in NAAG in humans have yet to be reported.

12.2.2 Dynamic Measures in a Human Disease

A dynamic measure for NAA metabolism has been reported in only one instance of brain pathology thus far, that of CD, where the rate of synthesis (S) of NAA has been measured at 0.22 μmol/g/h, and when compared to its rate of synthesis in normal children's brain (0.58 μmol/g/h), it was found to be significantly reduced in CD (Moreno et al., 2001). In the future it is anticipated that some dynamic measurements of NAA and NAAG metabolism in humans will also become routine, just as static measures of NAA content are at present, and that they will provide new and important insights into normal brain function, as well as into the nature of certain psychopathologies and other brain pathological conditions.

13 Conclusions

As described in this review, there is a complex chain of electrophysiological, metabolic, and diffusion events, related to the degree of neuronal activity in the brain that interact with homeostatic feedback mechanisms to provide neurons with sufficient energy to continuously maintain their full repertoire of signaling capabilities. Maintenance of this signaling ability is the key element in information processing, without which input information cannot be properly interpreted, nor appropriate output responses provided.

Neuronal activity requires use of energy in the form of ATP for repolarization, which is replenished primarily by the oxidation of Glc. It is proposed that the NAA–NAAG system is a neuronal control mechanism that plays an integral role in this process by ensuring both the timely delivery of adequate energy components and the removal of waste products. To reiterate, for every 400 mol of Glc oxidized in the brain (approximately 200 mol by neurons based on 50% neurons and a similar metabolic rate as that of whole brain) to replenish ATP supplies, 10 mol of NAA and 1 mol of NAAG are synthesized by neurons via a chemometabolic process, wherein Glc carbon is used to synthesize the Ac portion of NAA and then NAAG. The NAA is released to the ECF where it diffuses to oligodendrocytes and functions primarily as an MWP. The NAAG is also released to the ECF, where it diffuses to an astrocyte surface mGluR3 receptor and initiates a Ca^{2+} wave that results in a secondary astrocytic messenger sent to the vascular system that results in focal hyperemia, increasing both the availability of energy and the sink capacity for neuron waste metabolites.

It is posited, as a result of the integrated interactions of neurons, the NAA–NAAG system and the vascular system, that ATP supplies are continuously replenished in neurons, thus maintaining the neurons' minimum repolarization times and absolute refractory periods, and thereby retaining their maximal spike frequency-coding capabilities.

Based on the evidence: (1) that NAA and NAAG are not required for individual neuron survival or viability; (2) that NAA and NAAG participate in the intercompartmental transport of water; (3) that NAAG is a specific mGluR3 signal molecule associated with focal hyperemia; (4) of a unique tricellular distribution of NAA and NAAG synthetic and hydrolytic enzymes; and (5) of the phenotypic outcomes of inborn or engineered errors in NAA and NAAG metabolism, the following is a description of how the NAA–NAAG system appears to work, and its apparent role and importance in brain function.

1. NAA and NAAG are structurally and metabolically linked, and their functions involve dynamic cyclic metabolic processes between neurons, astrocytes, and oligodendrocytes. It is a unique metabolic system in which the metabolic components are partitioned between various brain cell types, and therefore is only operational in the intact brain.
2. The primary role of NAA is osmoregulatory in that it functions as an MWP to continuously remove neuronal metabolic water to the ECF against a water gradient. The cells involved in its intercompartmental metabolic cycle are neurons and oligodendrocytes.
3. The primary role of NAAG is communication whereby it continuously signals astrocytes about the state of neurostimulation, and of the neurons' changing requirements for vascular energy supplies and metabolic waste removal. The cells involved in this signaling sequence are neurons, astrocytes, vascular epithelium, and vascular smooth muscle cells, and the product of the stimulatory process is the continuous control of the degree of focal hyperemia.
4. The functions of NAA and NAAG are linked and compensatory in nature. Metabolic water, continuously produced as a result of neuron activity and Glc oxidation to renew ATP energy supplies for repolarization, is rapidly transported to the ECF by both substances for removal via the vascular sink. Concurrently, the state of neuronal activity is also continuously signaled to astrocytes by NAAG, which in turn influences the availability of energy supplies and vascular sink capacity by interacting with cells of the vascular microcirculation. The system is dynamic, homeostatic, and responsive to the focal, regional, or global levels in order to serve the ever-changing and ongoing requirements of the brain.
5. As evidenced by the two human inborn errors in NAA–NAAG metabolism and the results of genetic engineering studies in mice, the system is important to normal brain function, and therefore its failure on various levels may also be involved in the etiology of other known brain cellular pathologies and psychopathological conditions.
6. Cognitive and other information processing aspects of brain function require a high degree of reliability of complex signal pathways and temporal signal sequences at specific synapses. These information-coded ephemeral signal sequences are in turn dependent on a stable synaptic architecture and availability of energy. The NAA–NAAG system provides responsive energy and waste removal mechanisms to maintain these conditions. Thus, the operation of the NAA–NAAG system appears to be essential for normal brain integrative activities and higher brain functions, in that it supports ongoing and meaningful communication between neurons.
7. The function of the NAA–NAAG system is both subtle and powerful. While it is not required for individual neuron survival, it can, by control of access to energy supplies, affect rates and outcomes of neural information processing and retrieval, which are limiting and vital elements in brain function. Therefore, except in the cases of inborn errors in NAA–NAAG metabolism where pathological outcomes may be profound, modest differences in dynamic aspects of the NAA–NAAG system may be an as yet unrecognized, but important factor underlying elements of brain–motor coordination, the ability to learn, and other normal cognitive functions.

14 Epilogue

In writing this review, the author has tried to be as objective as possible. However, a degree of subjectivity arises from the author's interpretation of published research results, and the relative weight given to individual studies. Additional subjectivity derives from the fact that this review is analytical in nature, in that evidence obtained through decades of time, in various research institutions, and in a variety of diverse research disciplines has been selected and pieced together in ways that individual researchers might not have envisioned.

The various research disciplines and subspecialties involved in this evaluation of the role of the NAA–NAAG system in the brain, in alphabetical order, include biochemistry, cell signaling, chemistry, electrophysiology, genetics, information processing, magnetic resonance, metabolism, molecular diffusion, neurophysiology, osmoregulation, pathology, pharmacology, and physics. In addition, it has become clear to the author that any attempt to understand the roles of NAA and NAAG separately, or to try to deduce their linked roles based on analysis of results obtained in only a few disciplines, or to fail to consider their dynamic intercompartmental interactions, would probably not be successful. Several concepts found to be useful in understanding the possible roles of NAA and NAAG that are derived from these disciplines include the following, again in alphabetical order:

Chemometabolism (structure-dependent metabolism): The sequential metabolism of a specific substance used primarily to form another specific substance

Chronoarchitecture (temporal mapping): A time-based mapping of the brain based on areas of neuroactivation in response to a given stimulus

Chronometabolism (temporal metabolism): The sequential metabolism of a substance to form a derivative of that substance measured over time

Dynamic metabolism (rates of metabolism): Metabolic rate functions associated with individual substances

Geometabolism (spatial metabolism): The metabolism of substances in an orderly sequence carried out in different cell types or compartments, based on separation of their metabolic enzymes

Metabolite-linked systems: Physiologic functions requiring the orchestrated actions of chemically related substances

Molecular water pump: Molecular mechanism for removal of intracellular water to extracellular space by release of a hydrated solute down its gradient

Neuron energy-limited information processing: Signal frequency-encoded information restricted by availability of energy

Neuron frequency-encoded information: Information transfer between neurons based on combinations of signaling frequencies and relative refractive periods

Osmotic disequilibrium: Osmotic imbalance, a normal condition requiring continuous removal of water from cells against a water gradient

In this evaluation, attention is focused primarily on two things that are accomplished by the continuous synthesis and release of NAA and NAAG from neurons: (1) transport of large amounts of osmolyte-obligated water to ECF and (2) initiation of mGluR3 signals in astrocytes. Moreover, attention is specifically focused on the logical outcomes that can be associated with these transport and signaling activities.

The overall result of this analysis is the construction of a metabolic puzzle, which when viewed from a technical distance, appears to present a coherent picture of the function of the NAA–NAAG metabolic system in the vertebrate brain. Of course, this picture, while it reflects most of what is currently known about NAA and NAAG, may not as yet be complete, but based on the way the puzzle is constructed the reader should be able to assess both its merits and its faults. If this picture is incorrect, then we have yet to properly organize the existing information, and to obtain additional information, in order to determine the functions of NAA and NAAG and understand how they are related to both normal brain function and various brain pathologies.

On the other hand, if the picture is reasonably accurate, then the NAA–NAAG system is seen to emerge as a keystone metabolic system in support of neuronal information processing, and an important component in cognition and of all other aspects of higher-integrated brain function.

References

Aarts MM, Tymianski M. 2003. Novel treatment of excitotoxicity: Targeted disruption of intracellular signaling from glutamate receptors. Biochem Pharmacol 66: 877-886.

Arun P, Madhavarao CN, Hershfield JR, Moffett JR, Namboodiri MA. 2004. SH-SY5Y neuroblastoma cells: A model system for studying biosynthesis of NAAG. Neuroreport 15: 1167-1170.

Arun P, Madhavarao CN, Moffett JR, Namboodiri AMA. 2006. Regulation of N-acetylaspartate and N-acetylglutamate biosynthesis by protein kinase activators. J Neurochem 98(6): 2034-2042.

Attwell D, Laughlin SB. 2001. An energy budget for signaling in the grey matter of the brain. J Cereb Blood Flow Metab 21: 1133-1145.

Averbeck BB, Lee D. 2004. Coding and transmission of information by neural ensembles. Trends Neurosci 27: 225-230.

Bacich DJ, Ramadan E, O'Keefe DS, Bukhari N, Wegorzewska I, et al. 2002. Deletion of the glutamate carboxypeptidase II gene in mice reveals a second enzyme activity that hydrolyzes N-acetylaspartylglutamate. J Neurochem 83: 20-29.

Bartels A, Zeki S. 2004. The chronoarchitecture of the human brain-natural viewing conditions reveal a time-based anatomy of the brain. Neuroimage 22: 419-433.

Baslow MH. 1997. A review of phylogenetic and metabolic relationships between the acylamino acids, N-acetyl-L-aspartic acid and N-acetyl-L-histidine in the vertebrate nervous system. J Neurochem 68: 1335-1344.

Baslow MH. 1998. Function of the N-acetyl-L-histidine system in the vertebrate eye. Evidence in support of a role as a molecular water pump. J Mol Neurosci 10: 193-208.

Baslow MH. 1999. The existence of molecular water pumps in the nervous system. A review of the evidence. Neurochem Int 34: 77-90.

Baslow MH. 2000. Functions of N-acetyl-L-aspartate and N-acetyl-L-aspartylglutamate in the vertebrate brain. Role in glial cell-specific signaling. J Neurochem 75: 453-459.

Baslow MH. 2001. Canavan's spongiform leukodystrophy: A clinical anatomy of a genetic metabolic CNS disease. An analytical review. J Mol Neurosci 15: 61-69.

Baslow MH. 2002. Evidence supporting a role for N-acetyl-L-aspartate as a molecular water pump in myelinated neurons in the central nervous system. An analytical review. Neurochem Int 40: 295-300.

Baslow MH. 2003a. N-acetylaspartate in the vertebrate brain: Metabolism and function. Neurochem Res 28: 941-953.

Baslow MH. 2003b. Brain N-acetylaspartate as a molecular water pump and its role in the etiology of Canavan disease: A mechanistic explanation. J Mol Neurosci 21: 185-189.

Baslow MH, Guilfoyle DN. 2002. Effect of N-acetylaspartic acid on the diffusion coefficient of water: A proton magnetic resonance phantom method for measurement of osmolyte-obligated water. Anal Biochem 311: 133-138.

Baslow MH, Resnik TR. 1997. Canavan disease: Analysis of the nature of the metabolic lesions responsible for development of the observed clinical symptoms. J Mol Neurosci 9: 109-126.

Baslow MH, Yamada S. 1997. Identification of N-acetylaspartate in the lens of the vertebrate eye: A new model for the investigation of the function of N-acetylated amino acids in vertebrates. Exp Eye Res 64: 283-286.

Baslow MH, Kitada K, Suckow RF, Hungund BL, Serikawa T. 2002. The effects of lithium chloride and other substances on levels of brain N-acetyl-L-aspartic acid in Canavan disease-like rats. Neurochemical Res 27: 403-406.

Baslow MH, Suckow RF, Berg MJ, Marks N, Saito M, et al. 2001. Differential expression of carnosine, homocarnosine and N-acetyl-L-histidine hydrolytic activities in cultured rat macroglial cells. J Mol Neurosci 17: 87-95.

Baslow MH, Suckow RF, Gaynor K, Bhakoo KK, Marks N, et al. 2003. Brain damage results in downregulation of N-acetylaspartate as a neuronal osmolyte. Neuromol Med 3: 95-103.

Baslow MH, Dyakin VV, Nowak K, Hungund BL, Guilfoyle DN. 2005. 2-PMPA, a NAAG peptidase inhibitor, attenuates the BOLD signal in brain of anesthetized mice: Evidence of a link between NAAG release and hyperemia. J Mol Neurosci 26(3): 1-16.

Battistuta J, Bjartmar C, Trapp BD. 2001. Postmortem degradation of N-acetyl aspartate and N-acetyl aspartylglutamate: An HPLC analysis of different rat CNS regions. Neurochem Res 26: 695-702.

Berger UV, Luthi-Carter R, Passani LA, Elkabes S, Black I, et al. 1999. Glutamate carboxypeptidase II is expressed by astrocytes in the adult rat nervous system. J Comp Neurol 415: 52-64.

Berry II, MJ, Meister M. 1998. Refractoriness and neural precision. J Neurosci 18: 2200-2211.

Bhakoo KK, Pearce D. 2000. In vitro expression of N-acetyl aspartate by oligodendrocytes: Implications for proton magnetic resonance spectroscopy signal in vivo. J Neurochem 74: 254-262.

Birken DL, Oldendorf WH. 1989. N-acetyl-L-aspartic acid: A literature review of a compound prominent in H-NMR spectroscopic studies of brain. Neurosci Biobehav Rev 13: 23-31.

Boltshauser E, Schmitt B, Wevers RA, Engelke U, Burlina AB, et al. 2004. Follow-up of a child with hypoacetylaspartia. Neuropediatrics 35: 255-258.

Bonvento G, Giaume C, Lorenceau J. 2002. Neuron-glia interactions: From physiology to behavior. J Physiol (Paris) 96: 167-168.

Borst A. 2003. Noise, not stimulus entropy, determines neuronal information rate. J Comput Neurosci 14: 2331.

Breitbach-Faller N, Schrader K, Rating D, Wunsch R. 2003. Ultrasound findings in follow-up investigations in a case of aspartoacylase deficiency (Canavan Disease). Neuropediatrics 34: 96-99.

Burlina AP, Aureli T, Bracco F, Conti F, Battistin L. 2000. MR spectroscopy: A powerful tool for investigating brain function and neurological diseases. Neurochem Res 25: 1365-1372.

Burlina AP, Corazza A, Ferrari V, Erhard P, Kunnecke B, et al. 1994. Detection of increased urinary N-acetylaspartylglutamate in Canavan disease. Eur J Pediatrics 153: 538-539.

Burlina AP, Ferrari V, Facci L, Skaper SD, Burlina AB. 1997. Mast cells contain large quantities of secretagogue-sensitive N-acetylasparate. J Neurochem 69: 1314-1317.

Bzdega T, Crowe SL, Ramadan ER, Sciarretta KH, Olszewski RT, et al. 2004. The cloning and characterization of a second brain enzyme with NAAG peptidase activity. J Neurochem 89: 627-635.

Carpenter KJ, Sen S, Matthews EA, Flatters SL, Wozniak KM, et al. 2003. Effects of GCP-II inhibition on responses of dorsal horn neurons after inflammation and neuropathy: An electrophysiological study in the rat. Neuropeptides 37: 298-306.

Chacron MJ, Longtin A, Maler L. 2001. Negative interspike interval correlations increase the neuronal capacity for encoding time-dependent stimuli. J Neurosci 21: 5328-5343.

Chakraborty G, Mekala P, Yahya D, Wu G, Ledeen RW. 2001. Intraneural N-acetylaspartate supplies acetyl groups for myelin lipid synthesis: Evidence for myelin-associated aspartoacylase. J Neurochem 78: 736-745.

Chapman MA, Roll JM, Park S, Galloway MP. 2003. Extracellular glutamate decrease in accumbens following cued food delivery. Neuroreport 14: 991-994.

Choi I-Y, Gruetter R. 2001. In vivo ^{13}C NMR measurement of total brain glycogen concentrations in the conscious rat. Proc Intl Soc Magn Reson Med 9: 210.

Chung YL, Barr J, Bhakoo K, Williams SC, Bell JD, et al. 2003. N-acetyl aspartate estimation: A potential method for determining neuronal loss in the transmissible spongiform encephalopathies. Neuropathol Appl Neurobiol 29: 445-450.

Clifford CW, Ibbotson MR. 2000. Response variability and information transfer in directional neurons of the mammalian horizontal optokinetic system. Vis Neurosci 17: 207-215.

Coyle JT. 1997. The nagging question of the function of N-acetylaspartylglutamate. Neurobiol Dis 4: 231-238.

Di Costanzo A, Trojsi F, Tosetti M, Giannatempo GM, Nemore F, et al. 2003. High-field proton MRS of human brain. Eur J Radiol 48: 146-153.

Dunlop DS. 1983. Protein turnover in brain. Synthesis and degradation. Handbook of Neurochemistry, Vol. 5, 2nd edn. Lajtha A, editor. New York: Plenum Press; pp. 25-58.

Duong TO, Yacoub E, Adriany G, Hu X, Ugurbil K, et al. 2003. Microvascular BOLD contribution at 4 and 7 T in the human brain: Gradient-echo and spin-echo fMRI with suppression of blood effects. Magn Reson Med 49: 1019-1027.

During MJ, Spencer DD. 1993. Extracellular hippocampal glutamate and spontaneous seizure in the conscious human brain. Lancet 341: 1607-1610.

Faull KF, Rafie R, Pascoe N, Marsh L, Pfefferbaum A. 1999. N-acetylaspartic acid (NAA) and N-acetyl aspartylglutamic acid (NAAG) in human ventricular, subarachnoid, and lumbar cerebrospinal fluid. Neurochem Res 24: 1249-1261.

Gadea M, Martinez-Bisbal C, Marti-Bonmati L, Espert R, Casanova B, et al. 2003. Spectroscopic axonal damage of the right locus coeruleus relates to selective attention impairment in early stage relapsing-remitting multiple sclerosis. Brain 127: 89-98.

Gafurov B, Urazaev AK, Grossfeld RM, Lieberman EM. 2001. N-acetylaspartylglutamate (NAAG) is the probable mediator of axon-to-glia signaling in the crayfish medial giant nerve fiber. Neuroscience 106: 227-235.

Gangstead SL, Song AW. 2002. On the timing characteristics of the apparent diffusion coefficient contrast in fMRI. Magn Reson Med 48: 385-388.

Ghadge GD, Slusher BS, Bodner A, Canto MD, Wozniak K, et al. 2003. Glutamate carboxypeptidase II inhibition protects motor neurons from death in familial amyotrophic lateral sclerosis models. Proc Natl Acad Sci USA 100: 9554-9559.

Ghose S, Weickert CS, Colvin SM, Coyle JT, Herman MM, et al. 2004. Glutamate carboxypeptidase II gene expression in human frontal and temporal lobe in schizophrenia. Neuropsychopharmacology 29: 117-125.

Gilbertson TA, Avenet P, Kinnamon SC, Roper SD. 1992. Proton currents through amiloride-sensitive Na channels in hamster taste cells. J Gen Physiol 100: 803-824.

Gillary HL. 1966. Stimulation of the salt receptor of the blowfly. II. Temperature. J Gen Physiol 50: 351-357.

Gonen O, Viswanathan AK, Catalaa I, Babb J, Udupa J, et al. 1998. Total brain N-acetylaspartate concentration in normal, age-grouped females: Quantitation with non-echo proton NMR spectroscopy. Magn Reson Med 40: 684-689.

Guilfoyle DN, Suckow RF, Baslow MH. 2003. The apparent dependence of the diffusion coefficient of N-acetylasparate upon magnetic field strength: Evidence of an interaction with NMR methodology. NMR Biomed 16: 468-474.

Guyton AC. 1966. Textbook of Medical Physiology, 3rd edn. Philadelphia: W.B. Saunders; p. 1210.

Hajek M, Burian M, Dezortova M. 2000. Application of LC model for quality control and quantitative in vivo ^{1}H MR spectroscopy by short echo time STEAM sequence. Magn Reson Mater Phys Biol Med 10: 6-17.

Henry P-G, Tkac I, Gruetter R. 2003. ^{1}H-localized broadband ^{13}C NMR spectroscopy of the rat brain in vivo at 9.4 T. Magn Reson Med 50: 684-692.

Hillered L, Hallstrom A, Segersvard S, Persson L, Ungerstedt U. 1989. Dynamics of extracellular metabolites in the striatum after middle cerebral artery occlusion in the rat monitored by intracerebral microdialysis. J Cereb Blood Flow Metab 9: 607-616.

Holtzman DM. 2001. Role of apoE/A-beta interactions in the pathogenesis of Alzheimer's disease and cerebral amyloid angiopathy. J Mol Neurosci 17: 147-155.

Hyder F, Kida I, Behar KL, Kennan RF, Maciejewski PK, et al. 2001. Quantitative functional imaging of the brain: Towards mapping neuronal activity by BOLD fMRI. NMR Biomed 14: 413-431.

Hyder F, Rothman DL, Shulman RG. 2002. Total neuroenergetics support localized brain activity: Implications for the interpretation of fMRI. Proc Natl Acad Sci USA 99: 10771-10776.

Inglese M, Ge Y, Filippi M, Falini A, Grossman RI, et al. 2004. Indirect evidence for early widespread gray matter involvement in relapsing-remitting multiple sclerosis. Neuroimage 21: 1825-1829.

Ishimoto H, Tanimura T. 2004. Molecular neurophysiology of taste in Drosophila. Cell Mol Life Sci 61: 10-18.

Jansma JM, Ramsey NF, ven der Wee NJA, Kahn RS. 2004. Working memory capacity in schizophrenia: A parametric fMRI study. Schizophr Res 68: 159-171.

Johnson SC, Saykin AJ, Baxter LC, Flashman LA, Santulli RB, et al. 2000. The relationship between fMRI activation and cerebral atrophy: Comparison of normal aging and Alzheimer disease. Neuroimage 11: 179-187.

Jung RE, Brooks WM, Yeo RA, Chiulli SJ, Weers DC, et al. 1999. Biochemical markers of intelligence: A proton MR spectroscopy study of normal human brain. Proc R Soc B (Lond) 266: 1375-1379.

Jung RE, Yeo RA, Chiulli SJ, Sibbitt L, Brooks WM. 2000. Myths of neuropsychology: Intelligence, neurometabolism and cognitive ability. Clin Neuropsychol 14: 535-545.

Karelson G, Ziegler A, Kunnecke B, Seelig J. 2003. Feeding versus infusion: A novel approach to study the NAA metabolism in rat brain. NMR Biomed 16: 413-423.

Kennedy RT, Thompson JE, Vickroy TW. 2002. In vivo monitoring of amino acids by direct sampling of brain extracellular fluid at ultralow flow rates and capillary electrophoresis. J Neurosci Methods 114: 39-49.

Kitada K, Akimitsu T, Shigematsu Y, Kondo A, Maihara T, et al. 2000. Accumulation of N-acetyl-L-aspartate in the brain of the tremor rat, a mutant exhibiting absence-like seizure and spongiform degeneration in the central nervous system. J Neurochem 74: 2512-2519.

Klugmann M, Symes CW, Klaussner BK, Leichtlein CB, Serikawa T, et al. 2003. Identification and distribution of aspartoacylase in the postnatal rat brain. Neuroreport 14: 1837-1840.

Kondrat RW, Kananori K, Ross BD. 2002. In vivo microdialysis and gas-chromatography/mass-spectroscopy for ^{13}C-enrichment measurement of extracellular glutamate in rat brain. J Neurosci Methods 120: 179-192.

Kozikowski AP, Zhang K, Nan F, Petukov PA, Grajkowska E, et al. 2004. Synthesis of urea-based inhibitors as active site probes of glutamate carboxypeptidase II: Efficacy as analgesic agents. J Med Chem 47: 1729-1738.

Kunnecke B, Cerdannn S, Seelig J. 1993. Cerebral metabolism of [1,2–^{13}C$_2$]glucose and [U-^{13}C$_4$] 3-hydroxybutyrate in rat brain as detected by ^{13}C NMR spectroscopy. NMR Biomed 6: 264-277.

Laughlin SB, Sejnowski TJ. 2003. Communication in neuronal networks. Science 301: 1870-1874.

Lawson LJ, Perry VH, Dri P, Gordon S. 1990. Heterogeniety in the distribution and morphology of microglia in the normal adult mouse brain. Neuroscience 39: 151-170.

Lea PM, Custer SJ, Stoica BA, Faden AI. 2003. Modulation of strech-induced enhancement of neuronal NMDA receptor current by mGluR1 depends upon presence of glia. J Neurotrauma 20: 1233-1249.

Lei H, Zhang Y, Zhu XH, Chen W. 2003. Changes in proton T_2 relaxation times of cerebral water and metabolites during forebrain ischemia in rat at 9.4 T. Magn Reson Med 49: 979-984.

Leshinsky E, Ribeiro L, Gage DA, Kolodny EH. 1997. N-acetyl aspartoglutamate metabolism in brain: Application to Canavan disease. J Neurochem (Suppl.) 69: 192.

Li BSY, Wang H, Gonen O. 2003. Metabolite ratios to assumed stable creatine level may confound the quantification of proton brain MR spectroscopy. Magn Reson Imaging 21: 923-928.

Logothetis NK. 2003. The underpinnings of the BOLD functional magnetic resonance imaging signal. J Neurosci 23: 3963-3971.

Long JM, Kalehua AN, Muth NJ, Hengemihle JM, Jucker M, et al. 1998. Stereological estimation of total microglia number in mouse hippocampus. J Neurosci Methods 84: 101-108.

Losi G, Vicini S, Neale J. 2004. NAAG fails to antagonize synaptic and extrasynaptic NMDA receptors in cerebellar granule neurons. Neuropharmacology 46: 490-496.

Lu ZH, Chakraborty G, Ledeen RW, Yahya D, Wu G. 2004. N-acetylasparte synthase is bimodally expressed in microsomes and mitochondria in brain. Brain Res Mol Brain Res 122: 71-78.

Lumley LA, Robison CL, Slusher BS, Wozniak K, Dawood M, et al. 2004. Reduced isolation-induced aggressiveness in mice following NAALADase inhibition. Psychopharmacology 171: 375-381.

Macey PM, Macey KE, Kumar R, Harper RM. 2004. A method for removal of global effects from fMRI time series. Neuroimage 22: 360-366.

Mackel R, Brink E. 2003. Conduction of neural impulses in diabetic neuropathy. Clin Neurophysiol 114: 248-255.

Madhavarao CN, Chinopoulos C, Chandrasekaran K, Namboodiri MAA. 2003. Characterization of the N-acetylasparate biosynthetic enzyme from rat brain. J Neurochem 86: 824-835.

Madhavarao CN, Moffett JR, Moore RA, Viola RE, Namboodiri MA, et al. 2004. Immunohistochemical localization of aspartoacylase in the rat central nervous system. J Comp Neurol 472: 318-329.

Marcucci F, Airoldi L, Mussini E. 1969. Brain level of N-acetyl-L-asparate in different strains of mouse and rat. J Neurochem 16: 272-273.

Martin E, Capone A, Schneider J, Hennig J, Thiel T. 2001. Absence of *N*-acetylaspartate in the human brain: Impact on neurospectroscopy? Ann Neurol 49: 518-521.

Matalon R, Rady PL, Platt KA, Skinner HB, Quast MJ, et al. 2000. Knock-out mouse for Canavan disease: A model for gene transfer to the central nervous system. J Gene Med 2: 165-175.

Meinild A-K, Klaerke DA, Loo DDF, Wright EM, Zeuthen T. 1998. The human Na$^+$-glucose cotransporter is a molecular water pump. J Physiol 508: l5-21.

Moffett JR, Tieman SB, Weinberger DR, Coyle JT, Namboodiri AMA, Eds. 2006. *N*-acetylaspartate: A unique neuronal molecule in the central nervous system. Adv in Exp Med and Biol 576, pp 375.

Moreno A, Ross BD, Bluml S. 2001. Direct determination of the *N*-acetyl-L-aspartate synthesis rate in the human brain by ^{13}C MRS and [1-^{13}C]glucose infusion. J Neurochem 77: 347-350.

Nan F, Bzdega T, Pshenichkin S, Wroblewski JT, Wroblewska B, et al. 2000. Dual function glutamate-related ligands: Discovery of a novel, potent inhibitor of glutamate carboxypeptidase II possessing mGluR3 agonist activity. J Med Chem 43: 772-774.

Neale JH, Bzdega T, Wroblewska B. 2000. *N*-acetylaspartylglutamate: The most abundant peptide neurotransmitter in the mammalian central nervous system. J Neurochem 75: 443-452.

Nicholson C, Sykova E. 1998. Extracellular space structure revealed by diffusion analysis. Trends Neurosci 21: 207-215.

Ogawa S, Menon RS, Tank DW, Kim SG, Merkle H, et al. 1993. Functional brain mapping by blood oxygenation level dependent contrast magnetic resonance imaging. Biophys J 64: 803-812.

Olszewski RT, Bukhari N, Zhou J, Kozikowski AP, Wroblewski JT, et al. 2004. NAAG peptidase inhibition reduces locomotor activity and some stereotypes in the PCP model of schizophrenia via group II mGluR. J Neurochem 89: 876-885.

Park JA, Choi KS, Kim SY, Kim KW. 2003. Coordinated interaction of the vascular and nervous systems: From molecule- to cell-based approaches. Biochem Biophys Res Commun 311: 247-253.

Parri R, Crunelli V. 2003. An astrocyte bridge from synapse to blood flow. Nature Neurosci 6: 5-6.

Passani LA, Vonsattel JPG, Carter RE, Coyle JT. 1997. *N*-acetylaspartylglutamate, *N*-acetylaspartate, and *N*-acetylated alpha-linked acidic dipeptidase in human brain and their alterations in Huntington's and Alzheimer's diseases. Mol Chem Neuropathol 31: 97-118.

Pfeuffer J, Tkac I, Gruetter R. 2000. Extracellular-intracellular distribution of glucose and lactate in the rat brain assessed noninvasively by diffusion-weighted ^{1}H nuclear magnetic resonance spectroscopy *in vivo*. J Cereb Blood Flow Metab 20: 736-746.

Pfliderer B, Ohrmann P, Suslow T, Wolgast M, Gerlach AL, et al. 2004. *N*-acetylasparate levels of left frontal cortex are associated with verbal intelligence in women but not in men: A proton magnetic resonance spectroscopy study. Neuroscience 123: 1053-1058.

Phillips JR, Matthews PBC. 1993. Texture perception and afferent coding distorted by cooling the human ulnar nerve. J Neurosci 13: 2332-2341.

Pomper MG, Musachio JL, Zhang J, Scheffel U, Zhou Y, et al. 2002. ^{11}C-MCG: Synthesis, uptake selectivity, and primate PET of a probe for glutamate carboxypeptidase II (NAA-LADase). Mol Imaging 1: 96-101.

Posse S, Olthoff U, Weckesser M, Jancke L, Muller-Gartner H-W, et al. 1997. Regional dynamic signal changes during controlled hyperventilation assessed with blood oxygen level-dependent functional MR imaging. Am J Neuroradiol 18: 1763-1770.

Qiao M, Malisza KL, Del Bigio MR, Tuor UI. 2002. Transient hypoxia-ischemia in rats: Changes in diffusion-sensitive MR imaging findings, extracellular space, and Na$^+$-K$^+$-adenosine triphosphatase and cytochrome oxidase activity. Radiology 223: 65-75.

Rael LT, Thomas GW, Bar-Or R, Craun ML, Bar-Or D. 2004. An anti-inflammatory role for *N*-acetyl aspartate in stimulated human astroglial cells. Biochem Biophys Res Commun 319: 847-853.

Raichle ME, Gusnard DA. 2002. Appraising the brain's energy budget. Proc Natl Acad Sci USA 99: 10237-10239.

Riecker A, Grodd W, Klose U, Schulz JB, Groschel K, et al. 2003. Relation between regional functional MRI activation and vascular reactivity to carbon dioxide during normal aging. J Cereb Blood Flow Metab 23: 565-573.

Robinson MB, Blakely RD, Couto R, Coyle JT. 1987. Hydrolysis of the brain dipeptide *N*-acetyl-L-aspartyl-L-glutamate. Identification and characterization of a novel *N*-acetylated alpha-linked acidic dipeptidase activity from rat brain. J Biol Chem 262: 14498-14506.

Rong SB, Zhang J, Neale JH, Wroblewski JT, Wang S, et al. 2002. Molecular modeling of the interactions of glutamate carboxypeptidase II with its potent NAAG-based inhibitors. J Med Chem 45: 4140-4152.

Ross AJ, Sachdev PS. 2004. Magnetic resonance spectroscopy in cognitive research. Brain Res Rev 44: 83-102.

Saadoun S, Papadopoulos MC, Davies DC, Krishna S, Bell BA. 2002. Aquaporin-4 expression is increased in oedematous human brain tumours. J Neurol Neurosurg Psychiatry 72: 262-265.

Sager TN, Laursen H, Hansen AJ. 1995. Changes in *N*-acetylaspartate content during focal and global brain ischemia of the rat. J Cereb Blood Flow Metab 15: 639-646.

Sanbaria ERG, Wozniak KM, Slusher BS, Keller A. 2004. GCP II (NAALADase) inhibition suppresses mossy fiber-CA3 synaptic transmission by a presynaptic mechanism. J Neurophysiol 19: 182-193.

Schoepp DD, Jane DE, Monn JA. 1999. Pharmacological agents acting at subtypes of metabotropic glutamate receptors. Neuropharmacology 38: 1431-1476.

Siegel GJ, Albers RW, Agranoff BW, Katzman R, editors. 1981. Basic Neurochemistry, 3rd edn. Little, Brown and Co.; p. 857.

Sigmundsson T, Maier M, Toone BK, Williams SCR, Simmons A, et al. 2003. Frontal lobe N-acetylaspartate correlates with psychopathology in schizophrenia: A proton magnetic resonance spectroscopy study. Schizophr Res 64: 63-71.

Signoretti S, Marmarou A, Tavazzi B, Lazzarino G, Beaumont A, et al. 2001. N-acetylaspartate reduction as a measure of injury severity and mitochondrial dysfunction following diffuse traumatic brain injury. J Neurotrauma 18: 977-991.

Simmons ML, Frondoza CG, Coyle JT. 1991. Immunocytochemical localization of N-acetyl-aspartate with monoclonal antibodies. Neuroscience 45: 37-45.

Slusher BS, Vornov JJ, Thomas AG, Hurn PD, Harukuni I, et al. 1999. Selective inhibition of NAALADase, which converts NAAG to glutamate, reduces ischemic brain injury. Nature Med 5: 1396-1402.

Smith AJ, Blumenfeld H, Behar KL, Rothman DL, Shulman RG, et al. 2002. Cerebral energetics and spiking frequency: The neurophysiological basis of fMRI. Proc Natl Acad Sci USA 99: 10765-10770.

Song AW, Woldorff MG, Gangstead S, Mangun GR, McCarthy G. 2002a. Enhanced spatial localization of neuronal activation using simultaneous apparent-diffusion-coefficient and blood-oxygenation functional magnetic resonance imaging. Neuroimage 17: 742-750.

Song AW, Fichtenholtz H, Woldorff M. 2002b. BOLD signal compartmentalization based on apparent diffusion coefficient. Magn Reson Imaging 20: 521-525.

Song AW, Harshbarger T, Li T, Kim KH, Ugurbil K, et al. 2003. Functional activation using apparent diffusion coefficient-dependent contrast allows better spatial localization to the neuronal activity: Evidence using diffusion tensor imaging and fiber tracking. Neuroimage 20: 955-961.

Spaniel F, Hajek T, Tintera J, Harantova P, Dezortova M, et al. 2003. Differences in fMRI and MRS in a monozygotic twin pair discordant for schizophrenia (case report) Acta Psychiatr Scand 107: 155-158.

Stanley JA. 2002. In vivo magnetic resonance spectroscopy and its application to neuropsychiatric disorders. Can J Psychiatry 47: 315-326.

Suhara T, Magrane J, Rosen K, Christensen R, Kim H, et al. 2003. A-beta42 generation is toxic to endothelial cells and inhibits eNOS function through an Akt/GSK-3-beta signaling-dependent mechanism. Neurobiol Aging 24: 437-451.

Surendran S, Rady PL, Michals-Matalon K, Quast MJ, Rassin DK, et al. 2003. Expression of glutamate transporter, GABRA6, serine proteinase inhibitor 2 and low levels of glutamate and GABA in the brain of knock-out mouse for Canavan disease. Brain Res Bull 61: 427-435.

Tahmaz FE, Sam S, Hoganson GE, Quan F. 2001. A partial deletion of the aspartoacylase gene is the cause of Canavan disease in a family from Mexico. J Med Genet (online), 38(3): E9, pp 2.

Thomas AG, Olkowski JL, Slusher BS. 2001. Neuroprotection afforded by NAAG and NAALADase inhibition requires glial cells and metabotropic glutamate receptor activation. Eur J Pharmacol 426: 35-38.

Thomas DL, Lythgoe Pell GS, Calamante F, Ordidge RJ. 2000. The measurement of diffusion and perfusion in biological systems using magnetic resonance imaging. Phys Med Biol 45: R97-R138.

Tsacopoulos M. 2002. Metabolic signaling between neurons and glial cells: A short review. J Physiol 96: 283-288.

Tsai G, Coyle JT. 1995. N-acetylaspartate in neuropsychiatric disorders. Prog Neurobiol 46: 531-540.

Tsai G, Dunham KS, Drager U, Grier A, Anderson C, et al. 2003. Early embryonic death of glutamate carboxypeptidase II (NAALADase) homozygous mutants. Synapse 50: 285-292.

Tsai G, Passani LA, Slusher BS, Carter R, Baer L, et al. 1995. Abnormal excitatory neurotransmitter metabolism in schizophrenic brains. Arch Gen Psychiatry 52: 829-836.

Tyson RL, Sutherland GR. 1998. Labeling of N-acetylaspartate and N-acetylaspartylglutamate in rat neocortex, hippocampus and cerebellum from [1-^{13}C]glucose. Neurosci Lett 251: 181-184.

Urazaev AK, Buttram JG Jr, Deen JP, Gafurov BS, Slusher BS, et al. 2001. Mechanisms for clearance of released N-acetylaspartylglutamate in crayfish nerve fibers: Implications for axon-glia signaling. Neuroscience 107: 697-703.

Vanzetta I, Grinvald A. 1999. Increased cortical oxidative metabolism due to sensory stimulation: Implications for functional brain imaging. Science 286: 1555-1558.

Williams V, Grossman RG, Edmunds SM. 1980. Volume and surface area estimates of astrocytes in the sensorimotor cortex of the cat. Neuroscience 5: 1151-1159.

Wolf NI, Willemsen MAAP, Engelke UF, van der Knapp MS, Pouwels PJW, et al. 2004. Severe hypomyelination associated with increased levels of N-acetylaspartylglutamate in CSF. Neurology 62: 1503-1508.

Yourick DL, Koenig ML, Durden AV, Long JB. 2003. N-acetylaspartylglutamate and beta-NAAG protect against injury induced by NMDA and hypoxia in primary spinal cord cultures. Brain Res 991: 56-64.

Zarahn E. 2001. Spatial localization and resolution of BOLD fMRI. Curr Opin Neurobiol 11: 209-212.

Zeng BJ, Wang ZH, Ribeiro LA, Leone P, De Gasperi R, et al. 2002. Identification and characterization of novel mutations in the aspartoacylase gene in non-Jewish patients with Canavan disease. J Inherit Metab Dis 25: 557-570.

Zhu X-H, Chen W. 2001. Observed BOLD effects on cerebral metabolite resonances in human visual cortex during visual stimulation: A functional ^{1}H MRS study at 4 T. Magn Reson Med 46: 841-847.

Zonta M, Angulo MC, Gobbo S, Rosengarten B, Hossmann K-A, et al. 2003. Neuron-to-astrocyte signaling is central to the dynamic control of brain microcirculation. Nature Neurosci 6: 43-50.

15 Glutathione in the Nervous System: Roles in Neural Function and Health and Implications for Neurological Disease

R. Janáky · R. Cruz-Aguado · S. S. Oja · C. A. Shaw

1	Introduction	349
2	The Biochemistry of Glutathione, Its Transport, and Localization	350
2.1	Synthesis and Degradation	350
2.2	Concentrations In Vivo	352
2.3	Transport in the Brain	352
2.3.1	Transport Across the Blood–Brain Barrier Endothelial Cells	352
2.3.2	Glial Transport	353
2.3.3	Possible Physiological Roles of GSH Transporters	353
2.4	Localization and Distribution	354
2.4.1	Subcellular Localization	354
2.4.2	Cellular Localization	355
2.4.3	Distribution in Different Brain Regions	356
2.5	Intercellular Trafficking of Glutathione, Cystine, and Cysteine	357
3	General Roles of Glutathione in the Central Nervous System	358
3.1	Protection Against Oxidative Stress	358
3.1.1	A Brief Description of Free Radicals and Their Relevance	359
3.1.2	Glutathione as an Antioxidant	360
3.1.3	Glial Versus Neuronal Resistance Against Reactive Oxygen Species	361
3.1.4	Interactions of Glutathione with Other Antioxidants	361
3.2	Protection Against Glutamate Neurotoxicity	362
3.3	Protection Against Cysteine Neurotoxicity	362
3.4	Protection Against Xenobiotics	362
3.5	Transport of Amino Acids	363
3.6	Macromolecule Synthesis, Cell-Cycle Regulation, and Cancer	363
4	Glutathione as a Neuromodulator at Glutamate Receptors	364
4.1	Glutamate Receptors	365
4.2	Glutathione Actions on Ionotropic Glutamate Receptors	365
4.2.1	NMDA Receptors	365
4.2.2	Non-NMDA Receptors	367
4.3	Nitrosoglutathione	367
4.4	Interactions with Glutamate Release	370
4.5	Interactions with Glutamate Uptake	370

Dedication: For István Császár

© Springer-Verlag Berlin Heidelberg 2007

4.6	Interactions with the Neuronal Influx and Intraneuronal Levels of Ca^{2+}	370
4.7	Interactions with the Ca^{2+}-Dependent Evoked Release of Other Neurotransmitters	372
4.7.1	Modulation of the Glutamate-Evoked GABA Release	372
4.7.2	Modulation of the Glutamate-Evoked Dopamine Release	372
5	***Glutathione as a Neurotransmitter***	***373***
5.1	Presence and Characteristics of Glutathione Binding Sites	373
5.2	Pharmacology of Glutathione Binding	374
5.2.1	Effects of GluR Ligands on Binding of Glutathione to Synaptic Membranes	374
5.2.2	Effects of Thiol-Containing Compounds	376
5.3	Effects of Glutathione on Response Properties of Neural Cells	376
5.3.1	Alterations in Evoked Neural Responses	376
5.3.2	Alterations in Neurotransmitter Release and Uptake	377
5.4	Learning and Memory	378
6	***Glutathione in Neuropathology***	***378***
6.1	Brain Aging	379
6.2	Parkinson's Disease	380
6.3	Alzheimer's Disease	380
6.4	Amyotrophic Lateral Sclerosis	381
6.4.1	Etiology and Pathogenesis	381
6.4.2	Therapy	382
6.5	Schizophrenia	383
6.6	Chronic Fatigue Syndrome	385
7	***Conclusions and Perspectives***	***385***

Abstract: Glutathione (γ-L-glutamyl-L-cysteinylglycine, GSH) is the most abundant cellular antioxidant molecule in aerobic cells. In the nervous system, it is found in the millimolar range in both neuronal and nonneuronal cells, arising endogenously by synthesis from its precursor amino acids or taken up from the blood by means of specific transport mechanisms. Glutathione can participate as an antioxidant either as a substrate of glutathione peroxidase (GSHPx) or as an electron donor in spontaneous reactions with reactive oxygen species (ROS). Besides the well-established role of glutathione as an antioxidant, it serves other functions in the nervous system, including neutralization of xenobiotics, participation in amino acid transport, regulation of the activity and conformation of proteins, protection against cysteine and glutamate neurotoxicities, and modulation of glutamatergic transmission. In addition, several lines of evidence suggest that glutathione might act as a neurotransmitter on its own class of receptors. This latter possibility may have significant implications for the possible involvement of GSH in learning and memory processes. Given its ample range of functions in the nervous system, it is not surprising to find significant alterations in glutathione levels and/or glutathione-related enzymes during brain aging and in several neurological conditions including Parkinson's disease (PD), Alzheimer's disease (AD), amyotrophic lateral sclerosis (ALS), schizophrenia, and chronic fatigue syndrome (CFS).

List of Abbreviations: Aβ, amyloid β/A4 protein; AD, Alzheimer's disease; AGE, advanced glycation end product; ALS, amyotrophic lateral sclerosis, Lou Gehrigs's disease; AMPA, (S)-2-amino-3-hydroxy-5-methyl-4-isoxazolepropionate; BBB, blood brain barrier; BSO, buthionine sulfoximine; BUI, brain uptake index; CEE, cysteine monoethyl ester; CFS, chronic fatigue syndrome; CGP 39653, DL-(E)-2-amino-4-propyl-5-phosphono-3-pentanoate; CME, cysteine methyl ester; CNQX, 6-cyano-7-nitroquinoxaline-2,3-dione; CNS, central nervous system; CPP, 3-(2-carboxypiperazin-4-yl)propyl-1-phosphonate; CSF, cerebrospinal fluid; D-AP5, D-(-)-2-amino-7-phosphonopentanoate; DNQX, 6,7-dinitroquinoxaline-2,3-dione; DTNB, 5,5′-dithio-bis(2-nitrobenzoate); DTT, dithiothreitol; eNOS, endothelial nitric oxide synthase; FWD, fluorowillardiine; GABA, γ-aminobutyric acid; GGT, γ-glutamyl transpeptidase; GSA, glutathione sulfonate; GSH, reduced glutathione; GSHMEE, glutathione monoethyl ester; GSHPx, glutathione peroxidase; GSNO, S-nitrosoglutathione; GSSG, oxidized glutathione; GST, glutathione S-transferase; GSTO1, glutathione S-transferase omega-1; HBA, hemoglobin A; ILD, incidental Lewy-body disease; iNOS, inducible nitric oxide synthase; KA, kainate; L-AP3, L-(+)-2-amino-3-phosphonopropionate; L-AP5, L-(+)-2-amino-5-phosphonovalerate; LC, locus ceruleus; LTP, long-term potentiation; MK-801, dizocilpine, 5-methyl-10, 11-dihydro-5H-dibenzo[a,d]cyclohepten-5,10-imine; MPTP, 1-methyl-4-phenyl-1,2,3,6-tetrahydropyridine; NAC, N-acetylcysteine; NBQX, 2,3-dioxo-6-nitro-1,2,3,4-tetrahydrobenzoyl[f]quinoxaline-7-sulfonamide; NFT, neurofibrillary tangles; nNOS, neuronal nitric oxide synthase; NMDA, N-methyl- D-aspartate; NOS, nitric oxide synthase; NO, nitric oxide; PCP, phencyclidine; PD, Parkinson's disease; PLP, pyridoxal 5′-phosphate; QNB, 3-quinuclidinyl benzilate; ROS, reactive oxygen species; SBG, S-butylglutathione; SEG, S-ethylglutathione; SIN-1, 5-amino-3-morpholinyl-1,2,3-oxadiazolium chloride; SMG, S-methylglutathione; SN, substantia nigra; SNAP, S-nitroso-N-acetylpenicillamine; SOD, superoxide dismutase; SPeG, S-pentylglutathione; SPET, single photon emission tomography; SPG, S-propylglutathione; tACPD, *trans*-1-aminocyclopentane-1,3-dicarboxylate

1 Introduction

Glutathione (γ-L-glutamyl-L-cysteinylglycine), a tripeptide of the amino acids glutamate, cysteine, and glycine, has been well known to biochemists for generations. This molecule "celebrated" its one hundred year "anniversary" in 1988, being first described in yeast in 1888 by De Rey-Pailhade (1888a, b). It was then called "philothion," indicating its reactivity with sulfur. Hopkins (1921) further described the reducing properties of the molecule and renamed it glutathione. The molecule was synthesized in 1935. Both the reduced form (GSH) and the oxidized dimer (GSSG) have been implicated in a great variety of molecular reactions (for reviews see Kosower and Kosower, 1978; Deneke and Fanburg, 1989; Meister, 1989, 1991; Lomaestro and Malone, 1995; Anderson, 1997; Dringen and Hirrlinger, 2003).

Many of these reactions to which we return later are crucial to cell survival, so much so that glutathione has been termed "the most important non-protein thiol" (Kosower and Kosower, 1978). Probably, there is no contradiction in further widening its role: glutathione is a key factor for cellular survival. One hypothesis has even suggested that glutathione is responsible for the origin of life (Holt, 1993). While this latter view seems likely to reflect a certain level of scientific hyperbole, it may be difficult to overestimate the central importance of this molecule in the biochemistry of living cells. Some examples of the actions of GSH are given in ❯ *Table 15-1*.

◘ Table 15-1
General multiple functions of GSH

Reducing agent, antioxidant
  Free-radical scavenger (Reed, 1986)
  Protection of cell membranes (Kosower et al., 1982)
  Protection against oxidative stress and lipid peroxidation (elevated level of GSSG) (Boobis et al., 1989; Werns and Lucchesi, 1990)
  Protection against radiation and UV light; DNA repair (Révész and Edgren, 1984; Connor and Wheeler, 1987)
  Maintenance of the SH groups of proteins and other molecules
  Destruction of hydrogen peroxide, other peroxides and free radicals (Kosower and Kosower, 1978; Meister, 1989)

Detoxification of xenobiotics
  Conjugation (Burk et al., 1983)
  Transport of metals between ligands
  Determination of drug resistance (Kramer et al., 1988)
  Reservoir and transfer of cysteine (Meister, 1988; Max, 1989)

Metabolic regulation
  Cofactor and substrate (Kosower and Kosower, 1978)
  Protein and nucleic acid synthesis (Meister, 1988)
  Leukotriene synthesis (Parker et al., 1980)
  Amino acid transport (Meister and Tate, 1976)
  Ca^{2+} homeostasis (protection of –SH in ATPases) (Max, 1989)
  Mitogenesis, cell-cycle regulation (Fischman et al., 1981; Weitberg et al., 1985; Shaw and Chou, 1986; Lu and Ge, 1992)
  Immune functions (Furukawa et al., 1987)
  Thermotolerance (Mitchell et al., 1983)

2 The Biochemistry of Glutathione, Its Transport, and Localization

2.1 Synthesis and Degradation

Glutathione is synthesized from its constituent amino acids and broken down via the γ-glutamyl cycle (❯ *Figure 15-1a* and *b*). The synthesis of GSH is catalyzed by γ-glutamylcysteine synthetase (EC 6.3.2.2.)

◘ Figure 15-1
Metabolism of Glutathione (a) Summary of the γ-glutamyl cycle (modified from Adam et al., 1996). **(b)** Reactions in GSH metabolism (1) γ-glutamyltransferase; (2) γ-glutamylcyclotransferase; (3) 5-oxoprolinase; (4) γ-glutamylcysteine synthetase; (5) glutathione synthetase; (6) dipeptidase; (7) L-cysteine-S-conjugate N-acetyltransferase; (8) glutathione S-transferase; (9) glutathione peroxidase; (10) free-radical quenching (probably nonenzymatic); (11) glutathione transhydrogenases; (12) glutathione disulfide (GSSG) reductase; (13) enzymatic reactions in which GSH is required as a cofactor but not consumed; (14) transport of γ-glutamylcystine and reduction to γ-glutamylcysteine. AA, amino acids; Cys_2, cystine; X, compounds that form conjugates with GSH (From Meister, 1986)

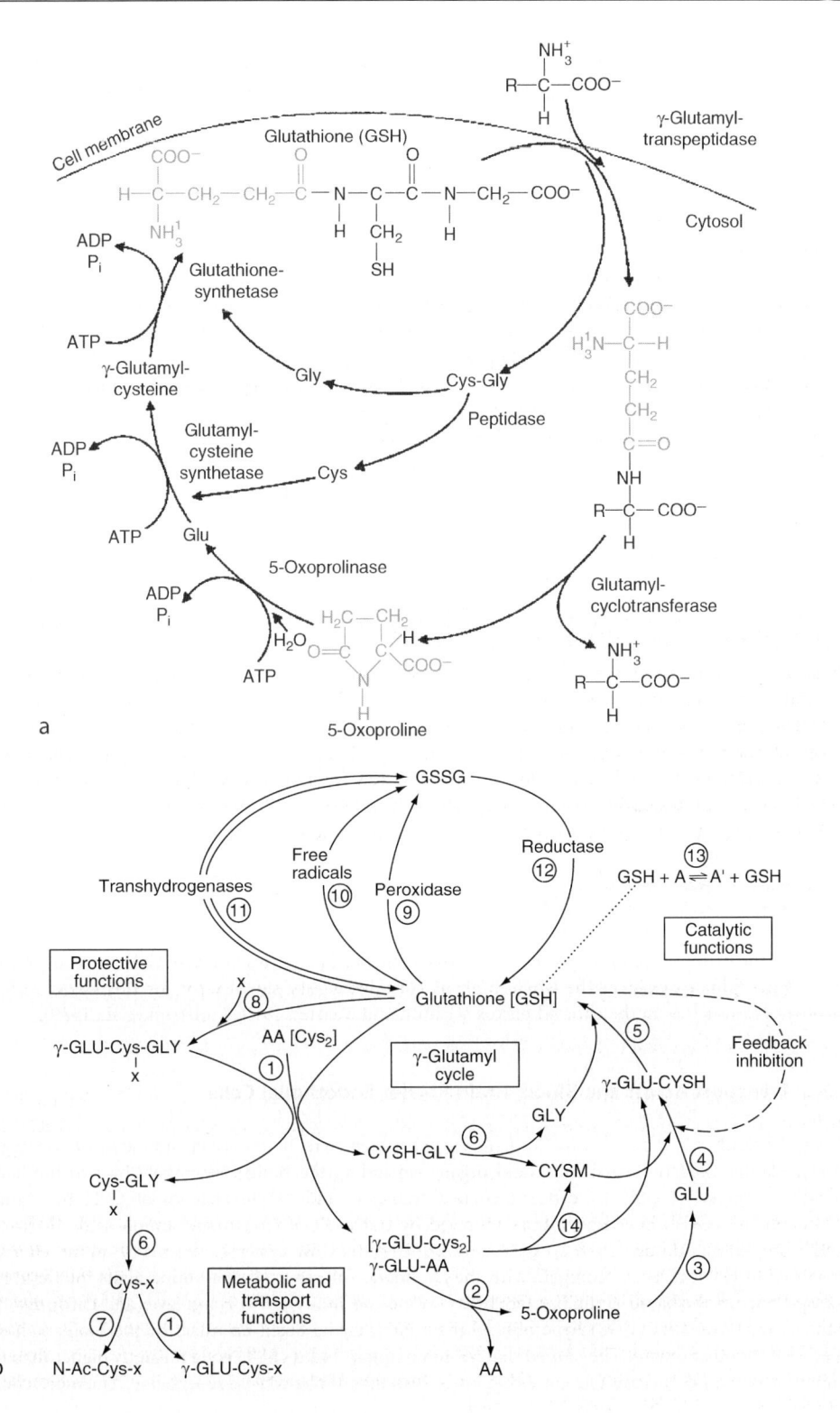

Glutathione in the nervous system

and glutathione synthetase (EC 6.3.2.3.). The activity of both enzymes is under substrate feedback control. The breakdown of glutathione is catalyzed by γ-glutamyl transferase/transpeptidase (EC 2.3.2.2.), a membrane-bound enzyme which catalyzes the transfer of the γ-glutamyl moiety to free amino acids.

2.2 Concentrations In Vivo

The concentration of GSH in the whole brain is 1–3 mM, depending on the species. Within a given mammal species the concentration of GSH in various brain regions tends to be in the following order: forebrain and cortex > cerebellum > brain stem ≥ spinal cord. In the cat, the concentration of GSH in the brain increases from 1.6 to about 2.2 mM during the first month of postnatal development. GSH levels are decreased in the aged rodent brain (Benzi and Moretti, 1995; Cooper, 1998). The concentration of total glutathione (i.e., GSH + 2 × GSSG) in rat plasma is about 15–20 μM, of which at least 80% is accounted for as GSH (Jain et al., 1991). The concentration of total glutathione in rat cerebrospinal fluid (CSF) is 4 μM and the reduced form predominates (Anderson et al., 1989).

Glutathione is easily oxidized to its disulfide form, yet most of the glutathione in the brain is in the reduced form. Maintenance of a high GSH/GSSG ratio requires energy expenditure. Although in the brain only 3%–5% of glucose carbon is converted to CO_2 via the pentose phosphate shunt, the remaining fraction is converted by the tricarboxylate cycle. The pentose phosphate shunt is important to provide NADPH for the reduction of GSSG to GSH (Hotta, 1962; Hotta and Seventko, 1968; Baquer et al., 1988) and is upregulated in cultured astrocytes and neurons exposed to hydrogen peroxide (Ben-Yoseph et al., 1994, 1996a, b).

Although it is technically difficult to measure very low concentrations of GSSG because some nonenzymatic oxidation of GSH to GSSG cannot be avoided during tissue extraction, by taking several precautions, it was possible to show that the concentration of GSSG in rat brain is ≤2 μM, i.e., ≤0.2% of the total glutathione (Cooper et al., 1980). Later, Slivka and coworkers (1987b) showed that in the human and monkey brain, GSSG concentration is ≤1.2% of total glutathione. In oxidative stress, however, the GSSG concentration may dramatically rise (to >1% of the total glutathione) because of the quenching of reactive oxygen species (ROS) and generation of GSSG in the reaction catalyzed by glutathione peroxidase (GSHPx). In these cases, the pentose phosphate pathway–GSSG reductase reaction in brain is unable to help with the oxidation of GSH to GSSG and GSSG accumulates.

2.3 Transport in the Brain

Although the mammalian brain possesses the enzymatic machinery to synthesize GSH from its precursors, the turnover rate of GSH in the whole brain is considered very slow except for tiny pools with rapid turnover rates such as in the choroid plexus (Griffith and Meister, 1979; Anderson et al., 1989).

2.3.1 Transport Across the Blood–Brain Barrier Endothelial Cells

Plasma levels of GSH in rodents and humans range between 10 to 20 μM (Lash and Jones, 1985; Flagg et al., 1993). Plasma GSH is cleared by several organs (including the brain), essentially by two mechanisms: (1) direct uptake of GSH by carrier-mediated transport and (2) breakdown of GSH by γ-glutamyl transpeptidase (GGT) and dipeptidases followed by transport of constituent amino acids (Meister and Anderson, 1983). Although there exists a transport system for cysteine, the major sulfur amino acid precursor of GSH, it has to compete with the abundance of other plasma amino acids for the L-system transport across the blood–brain barrier (BBB) (Wade and Brady, 1981; Hargreaves and Pardridge, 1988). Kannan and coworkers (1998) hypothesized that GSH may be taken up intact by the brain, as has been suggested for other organs. They found that the brain uptake index (BUI) is significantly higher than that of impermeant marker sucrose and somewhat lower than that of phenylalanine–cysteine. The molecular form of uptake was predominantly reduced GSH (>95%), GSSG having BUI similar to that of impermeant sucrose. To exclude the uptake after breakdown, GGT was inhibited either with acivicin or with serine borate. Dual-label experiments further confirmed that GSH molecules were taken up intact. GSH uptake was saturable with an apparent K_m of 5.84 mM. Uptake was by a specific mechanism, since various amino acids, amino acid analogs, GSSG, and γ-glutamyl compounds did not affect GSH uptake. On the other hand, a variety of GSH analogs and organic anions inhibited GSH transport. The inhibition of GSH uptake by GSH monoethyl ester was concentration dependent. In addition to strong inhibition by GSH conjugates and S-alkyl derivatives, GSH uptake by the brain was inhibited by unconjugated bilirubin: unconjugated bilirubin complexed to bovine serum albumin to give a 2:1 ratio caused 60% decrease in GSH uptake. The same concentration of unconjugated bilirubin did not alter the cysteine uptake across the BBB. The uptake of GSH is developmentally regulated: declining from 45% BUI in 2-week-old to 5% in mature adult (6-month-old) rats. Similar results were seen in guinea pigs. The conclusion from these data is that lower GSH uptake may be a reason for decreasing levels of GSH with age (Benzi et al., 1988, 1989a; Ravindranath et al., 1989; Vali Pasha and Vijayan, 1989). The bulk of GSH taken up by the brain capillaries was found in brain parenchyma after 10-min vascular perfusion and appeared in the capillary-depleted supernatant fraction. This finding suggests that GSH is not trapped in the endothelial cells but transcytosis occurs via a luminal transporter for uptake and abluminal transporter for GSH in brain endothelium (Kannan et al., 1998).

Supportive evidence for in vivo findings on GSH transport was obtained from molecular studies in the *Xenopus laevis* oocyte expression system (Fernandez-Checa et al., 1993). *X. laevis* oocytes injected with mRNA from bovine and guinea pig brain capillaries expressed GSH transport even after pretreatment with acivicin to inhibit GGT. This transport was partially (by 49%) dependent on Na^+ removal. When the bovine capillary mRNA was size fractionated and the fractions checked for GSH transport, transport activity was found in three distinct fractions, two of them showing sodium-independent GSH transport, a transcript for GGT being present in one of them, while the third exhibited Na^+-dependent GSH transport and did not contain GGT transcript. This transporter was further characterized by Kannan and coworkers (1998) and showed a striking inhibition by bromosulfophthalein–GSH. The Na^+-dependent GSH transport has two components: a high-affinity component with K_m of 0.40 mM and a sigmoid low-affinity component with a K_m of 10.8 mM (Kannan et al., 1998).

2.3.2 Glial Transport

Glutathione is distributed abundantly in brain endothelial cells and astrocytes, but its level in neurons is very low (Raps et al., 1989; Yudkoff et al., 1990, 1992). The physiological implication of this observation is that neurons have relatively low antioxidant defenses, which may account for their vulnerability to oxidative stress, and that glial cells may protect neurons through release of GSH (Guo et al., 1992; Makar et al., 1994). In 1998, Kannan and coworkers provided evidence that in cultured brain epithelial cells the transport of intact GSH molecules is Na^+ dependent.

In studies of luminal and abluminal membrane vesicles isolated from bovine brain endothelial cells GGT was found on the luminal membrane; the amino acid transport system A was reported to be on the abluminal side (Sanchez del Pino et al., 1995). On the other hand, GSH transport in astrocytes is not Na^+ dependent. Here it is rather a high-K_m bidirectional GSH transporter, present also in other cells (Kannan et al., 1998).

2.3.3 Possible Physiological Roles of GSH Transporters

The exact physiological role of the GSH transporters in the brain is not yet known. According to an hypothetical model, GSH transport trough the BBB occurs by luminal influx that is Na^+-dependent, GGT-independent, and inhibited by bromosulfophthalein–GSH and other organic anions. This system has the possibility of being concentrative, such that low plasma GSH (10–20 μM) can lead to net accumulation in endothelial cells because of coupling to the entry of sodium. There is an efflux via a Na^+-independent GSH transporter from the abluminal surface of the endothelium and from the astrocytes. A second

Na$^+$-independent transporter may also exist both in endothelial and in astroglial cells. These facilitative Na$^+$-independent transporters would have the potential for bidirectional operation as observed in cell culture studies (Lu et al., 1996), but under physiological conditions they would serve as efflux transporters, because net transport would be determined by the concentration gradient (high intra- and low extracellular GSH concentrations). The model also includes the possibility of cysteine transport into the endothelial cells and astrocytes for GSH synthesis followed by GSH efflux into the brain. GGT on the luminal or abluminal pole of endothelial cells or in astrocytes might operate in parallel, hydrolyzing GSH (along with dipeptidase) to cysteine. Cysteine is then taken up and used for intracellular GSH synthesis in neurons. This model is speculative in view of the current state of knowledge, but the evidence for multiple GSH transporters in the brain (capillary endothelial cells and astrocytes) is convincing. It is conceivable that the GSH transporters also transport organic anions, as well as endogenous degradation products (e.g., bilirubin) and drugs (Levine et al., 1982; Ives and Gardiner, 1990). The capacity of newly cloned GSH transporters to transport organic anions will be important to ascertain. Implications of this knowledge would be enormous and, in view of the developmental pattern of expression of transport, might be relevant in the pathophysiology of Kernicterus, the hyperbilirubinemia of the newborn (Odell and Schuetta, 1985).

2.4 Localization and Distribution

Hjelle, Chaudry, and Ottersen characterized the first antibodies against glutathione in 1994 (Hjelle et al., 1994). From that time, GSH has been localized histochemically in the cerebellum, inner ear, retina, spinal cord, and substantia nigra (SN). The small size of the glutathione molecule had to be taken into account when designing the immunization procedure, leading finally to an immunization protocol adapted from one previously used for the generation of antibodies to amino acids. The method is detailed in Hjelle et al. (1998). The antibodies did not show any detectable binding to glutamate, cysteine, or glycine, nor was any labeling observed with γ-glutamylcysteine or cysteinylglycine, or other molecules with structural resemblance to GSH. The antibodies selectively recognized reduced and oxidized glutathione, although reduced glutathione produced almost twice as strong labeling as the oxidized form. The labeling was strongly reduced by pretreating the experimental animals with buthionine sulfoximine (BSO) (Hjelle et al., 1994), which is known to inhibit the synthesis of glutathione. In electron microscopic studies using osmium tetroxide, immunolabeling for glutathione was depressed; later studies used a freeze substitution procedure (Hjelle et al., 1994). With these precautions, cellular and subcellular distribution of glutathione has been determined (Hjelle et al., 1998).

2.4.1 Subcellular Localization

Similarly to most tissues, about 10%–20% of GSH in neural cells can be found in mitochondria, the remaining 80%–90% is cytosolic (Jain et al., 1991; Bonnefoi, 1992; Werner and Cohen, 1993; Huang and Philbert, 1995, 1996). There is GSH also in the nucleus (Hjelle et al., 1994). Mitochondria do not possess enzymes necessary for GSH biosynthesis (Griffith and Meister, 1985) but import GSH and GSSG from the cytosol via at least two transport systems (Kurosawa et al., 1990; Mårtensson et al., 1990). GSSG taken up from the cytosol is reduced to GSH in mitochondria (Kurosawa et al., 1990).

Depletion of the mitochondrial GSH pool by the administration of BSO (an inhibitor of γ-glutamylcysteine synthetase) leads to mitochondrial swelling and cellular degeneration in skeletal muscles, lung capillary endothelium, and gut epithelium, and cataract formation in the eye (Mårtensson and Meister, 1989). Depletion of brain GSH leads to severe mitochondrial damage, reduction of citrate synthase activity in isolated brain mitochondria, disruption of membrane integrity, leakage of enzymes, and death of the cell (Jain et al., 1991; Heales et al., 1995). Treatment of astrocytes in culture with ethacrynic acid (a compound that penetrates mitochondria and depletes GSH by reacting with it to form a glutathione conjugate) leads to a marked reduction in mitochondrial GSSG reductase, and decreases in glyceraldehyde-3-phosphate dehydrogenase activity, ATP levels, and the cell membrane potential. On the basis of these findings,

Huang and Philbert (1996) suggested that loss of mitochondrial GSH renders astrocytes unable to combat oxidative stress (see ❯ *Sect. 3.1*). Similarly, treatment of cerebrocortical cells in culture with methyl iodide lead to the death of these cells because of the depletion of mitochondrial GSH (Bonnefoi, 1992).

2.4.2 Cellular Localization

Immunocytochemical Data   At the outset it needs to be emphasized that our knowledge of the cellular GSH compartmentation in the brain is still far from complete. However, present methods allow the quantitative estimation of GSH concentration by counting immunogold particles even at the subcellular level, and the wider use of glutathione antibodies opens the possibility for a more accurate knowledge (Hjelle et al., 1998). Immunocytochemistry has also the potential of assessing the ratio between reduced and oxidized glutathione. If this could be done by means of the postembedding immunogold procedures, it would provide an indication of the redox status in individual cells and organelles (Viguie et al., 1993; De Mattia et al., 1994; Aukrust et al., 1995). At present, the antibodies raised against reduced glutathione also recognize the oxidized form (Hjelle et al., 1994), and so far there have not been successful attempts to raise antibodies that would react selectively with GSSG. For this reason, the results presented must be taken to represent the total pool of glutathione, both reduced and oxidized. It is likely, however, that the reduced form is responsible for the major part of the immunogold signal as the reduced form is known to predominate over the oxidized form by two orders of magnitude (Meister, 1988).

The structural dissimilarity between oxidized and reduced glutathione should be sufficient to make them immunologically distinguishable, since the present immunization procedure was previously used to raise antibodies that distinguished between two stereoisomers of the same amino acid and between such closely related amino acids as glutamate and glutamine (Ottersen et al., 1992) or homocysteic acid and glutamate (Zhang and Ottersen, 1992). It is likely, therefore, that the failure to produce antibodies specific for the oxidized or reduced form of glutathione reflects the spontaneous oxidization of GSH in the conjugate before or after inoculation, while oxidized glutathione may undergo the converse process (Hjelle et al., 1994). According to this line of reasoning the addition of an antioxidant to the immunogen should favor the formation of antibodies with higher selectivities for the reduced form of glutathione.

Glial Versus Neuronal Glutathione   A central issue is whether glutathione is predominantly neuronal or glial. Numerous claims exist in the literature that glutathione mainly resides in glial cells (Slivka et al., 1987a; Raps et al., 1989; Yudkoff et al., 1990). On the basis of results obtained with a histochemical staining procedure, Slivka and coworkers (1987a) suggested that GSH in the brain is primarily localized in nonneuronal elements, such as glia and axons or nerve terminals. Others have also used histochemical techniques to show that although GSH is well represented throughout the embryonic rat brain, it is largely absent from neurons (except dorsal root ganglia and cerebellar granule cells) in the adult rat brain (Philbert et al., 1991; Lowndes et al., 1994). GSH staining has also been detected in the glomerular layer and granule-cell layer somata of the olfactory bulb (Kirstein et al., 1991). Hjelle and coworkers (1998) discuss in detail the factors, which must be taken into consideration before reaching conclusions about GSH localization. They emphasize the importance of the fact that claims for a dominantly glial GSH loci is drawn mainly from cell culture studies where we should also consider (1) the origin of cells cultured, e.g., cultures from the cerebral cortex show a predominant glial localization, while those from the mesencephalon or striatum show no significant neuronal–glial differences (Langeveld et al., 1996); (2) the sensitivity of the intracellular GSH pool to the extracellular levels of amino acids such as cysteine and glutamate present in the culture media (Kato et al., 1992; Kranich et al., 1996), including the metabolic interaction between neurons and glia that determines the glutathione homeostasis in the brain, which can be perturbed under culture conditions (Sagara et al., 1993); and finally, (3) the presence or absence of growth factors, e.g., nerve growth factor, which stimulates the activity of GSH-synthesizing enzymes in pheochromocytoma PC12 cells, leading to increased levels of GSH (Pan and Perez-Polo, 1993). On the basis of the depolarization-evoked release of glutathione from rat brain slices, it was suggested that glutathione might be present in neuronal elements (Zängerle et al., 1992; Shaw et al., 1996). This suggestion is in accord with the previous finding that GSH is present in isolated nerve terminals (Reichelt and Fonnum, 1969). Moreover, Amara and collaborators (1994) used an immunohistochemical technique to demonstrate the presence of GSH in nerve fiber tracts. After the studies of Hjelle and coworkers (1994) using the above characterized antibodies to GSH and demonstrating the presence of GSH in Golgi epithelial cells, perivascular glia processes, and in Purkinje cells, granule cell bodies and mossy fiber tracts, the present general opinion is that glutathione occurs both in neurons and in glia; moreover, in the cerebellum, substantia nigra (SN), and the spinal cord, the glial labeling does not predominate over that of the neuronal (Hjelle et al., 1998). On the other hand, the labeling of the perivascular glial processes surpasses that of the somata pointing to possible transport functions of glutathione at these sites (see above). Oligodendrocyte viability is dependent on cysteine availability, which suggests an important role for glutathione in this cell type (Yonezawa et al., 1996).

Glutathione in Neurons Considering the fact that a given proportion of brain GSH is present in neurons, and tentatively accepting the idea that GSH may serve as a cascade-neuromodulator and neurotransmitter (detailed later in ❷ *Sect. 5*), it may be of central importance whether or not different neuronal populations show differential glutathione labeling. Hjelle and coworkers in 1998 found significant differences even between two axons in the same cerebellar cortical preparation, one axon showing dense labeling, while the other devoid of immunogold particles (Hjelle et al., 1998). Similar heterogeneity is evident among nerve terminals in the spinal cord (Hjelle et al., 1998; Ramirez-Leon et al., 1999). This sort of GSH heterogeneity may reflect the presence of different neuron populations (i.e., existence of "glutathionergic" neurons) or differences in the actual metabolic and functional status of the respective cells. Correlative studies based on double labeling for glutathione and appropriate transmitters indicate that GSH may be concentrated in terminals with specific transmitters (Ramirez-Leon et al., 1999), a finding which would corroborate the role of GSH in signal transmission (Shaw, 1998) (discussed later in detail in ❷ *Sect. 5*). The dendrosomatic labeling appears to vary among brain regions (e.g., cerebellar compared to spinal) (Hjelle et al., 1998), but there is a debate whether or not the axonal labeling surpasses the dendrosomatic labeling as has been suggested by Amara and coworkers (1994).

Glutathione in Nonneuronal Cells As GSH is thought to play an important role in the detoxification of ototoxic agents (Garetz et al., 1994), it was surprising that hair cells were only weakly labeled (Usami et al., 1996). The strong labeling in the spiral ligament and stria vascularis supports the existence of a perilymphatic GSH transport from the lateral wall tissue to the organ of Corti (Usami et al., 1996). An analogous situation appears to exist in the retina, since the pigment epithelial cells are more densely labeled than the cells of the neuronal part of the retina (Pow and Crook, 1995; Huster et al., 1998). While it is conceivable that the pigment epithelial cells supply retinal neurons and glia with glutathione (Pow and Crook, 1995), it is important to note that a correspondingly high level of glutathione also occurs in the pigment cells of the choroid (Huster et al., 1998). This suggests that the enrichment of GSH is linked to the synthesis of melanin—a function that these two cell types have in common. It is well known that melanin synthesis is associated with the formation of toxic by-products, and GSH might offer protection against such toxins (Hjelle et al., 1998). Endothelial cells were also strongly labeled, although there is a regional heterogeneity (Usami et al., 1996; Hjelle et al., 1998). These differences point to possible regional heterogeneities in transendothelial transport (see above, ❷ *Sect. 2.3*).

2.4.3 Distribution in Different Brain Regions

As detailed so far, the antibodies to GSH have been used to analyze the distribution of glutathione in the cerebellum (Hjelle et al., 1994), inner ear (Usami et al., 1996), retina (Huster et al., 1998), spinal cord, and SN (Hjelle et al., 1998). They have also been applied to cultured neurons and astrocytes (Langeveld et al., 1996).

Cerebellum Gold particles signaling glutathione immunoreactivity occurred both in glial and in neuronal elements. The highest labeling intensities were found in glial processes, particularly in perivascular glia, and in subpopulations of myelinated axons. All neuronal populations exhibited significant immunolabeling.

The labeling intensity of Purkinje cell somata was relatively high and comparable to that of the adjacent glial cell bodies (Golgi epithelial cells). Mossy fiber terminals, which are putatively glutamatergic, also displayed relatively strong glutathione immunoreactivity, whereas basket cell axons (GABAergic) were weakly labeled.

Inner Ear Glutathione is known to protect hair cells of the inner ear against the cytotoxic actions of gentamycin and similar compounds (Garetz et al., 1994), and it was therefore of interest to identify the cells responsible for GSH synthesis in the inner ear. Very low levels of immunoreactivity were found in the hair cells and adjoining supporting cells of the organ of Corti. In contrast, strong immunoreactivity occurred in the stria vascularis and spiral ligament. The immunoreactivity at these sites was concentrated in basal and intermediate cells, and in fibrocytes, respectively. Endothelial cells were also strongly labeled.

Retina The highest concentration of glutathione immunoreactivity occurred in the pigment epithelial cells (Huster et al., 1998); in these cells, mitochondria and cytoplasmic matrix both showed immunolabeling, with mitochondria showing moderate enrichment. In other cell types with significant labeling (Müller cells and receptor inner segments), most of the gold particles were confined to the mitochondria. The outer receptor segments were very weakly labeled.

Spinal Cord Motor neurons were generally weakly labeled for GSH, although some immunoreactivity occurred in the mitochondria. On the other hand, high densities of gold particles were found in a subpopulation of those terminals that were apposed to the dendrites or somata of motoneurons. A large proportion of the glutathione-immunolabeled terminals was enriched in glutamate and is therefore probably glutamatergic. This finding would support the cascade-like neuromodulatory role of GSH (see ❷ Sect. 5).

Substantia Nigra A selective distribution of GSH-like immunoreactivity in the pars reticulata of the SN was found both in rats and in cats. Dendrites with varying caliber showed considerable immunolabeling. There is also considerable immunolabeling of the axoplasm of some myelinated axons, and scattered gold particles were observed over neuronal cell bodies. Interestingly, SN glial elements showed less immunoreactivity than that of myelinated axons. Immunoreactivity was seen preferably over mitochondria. These data may be relevant in view of the profound consequences of GSH loss in the SN pars reticulata, reported to be a probable causative and early pathogenic factor in Parkinson's disease (PD) (see ❷ Sect. 6.2). The present state of the field does not allow us to define whether the immunolabeled dendrites belong to dopaminergic neurons or to other cells. However, it is well known that dopaminergic cells in the pars compacta of the SN send their dendrites into the pars reticulata of the SN, where they are densely covered with GABAergic afferents from the striatum and globus pallidus. In double-labeling experiments, GSH-immunolabeled dendrites were synapsed upon by a large number of GABA-like immunoreactive boutons.

The labeling pattern in the SN is of particular interest because there is a considerable body of evidence suggesting that oxidative stress may play an important role in PD. Reduced glutathione was significantly lower in the SN of patients with PD (Perry et al., 1982). Later, the regional selectivity of this glutathione loss in PD was further noted (Perry and Yong, 1986; Perry et al., 1988). In addition, the severity of the disease correlates with the reduction of total glutathione in the pars compacta of the SN (Riederer et al., 1989). GSH was particularly diminished while GSSG was unaffected or slightly elevated (Sofic et al., 1992; Jenner, 1993a, b; Sian et al., 1994). GSH was diminished already in the pars compacta of the SN in patients with incidental Lewy-body disease (ILD), a condition sometimes considered to be an early form of PD (Jenner et al., 1992a, b; Sofic et al., 1992; Jenner, 1993a, b; Dexter et al., 1994). Supporting the specific pathogenic role of GSH loss was that no similar loss was observed in the pars compacta of the SN in patients with neurodegenerative diseases other than PD.

2.5 Intercellular Trafficking of Glutathione, Cystine, and Cysteine

Yudkoff and coworkers (1990) showed that astrocytes in culture synthesized GSH at a high rate and GSH was released to the medium, an outcome supported by Sagara and coworkers (1996). Microdialysis studies supported GSH release also in vivo (Orwar et al., 1994; Yang et al., 1994), and Li and coworkers (1996) have shown that release of GSH from hippocampal slices was increased in the presence of acivicin (a γ-glutamyltranspeptidase inhibitor). From these results it is tempting to speculate that astrocytes provide nerve endings with GSH under normal physiological conditions and that GSH exchange is yet another example of metabolite trafficking between astrocytes and neurons (Yudkoff et al., 1990, 1992). Such intercellular trafficking of GSH may be physiologically important in the context of (1) GSH protection of neurons against glutamate and cysteine toxicity, (2) GSH modulation of glutamatergic neurotransmission (see ❷ *Sect. 4*), and (3) feedback regulation of glial function as proposed by Guo and Shaw (1992). These authors have presented strong evidence that astrocytes in culture possess high-affinity binding sites for GSH. Therefore, release of GSH from astrocytes or nerves may regulate physiological responses such as G protein coupling, activation of second messengers, regulation of protein kinases, Ca^{2+} release, and gene expression.

Although it is interesting to speculate that neuronal–glial GSH interaction may form a new channel for cellular cross talk, not all studies support this view. For example, Sagara and coworkers (1996) found that neurons in culture do not take up intact GSH. In addition, in the study of Juurlink and collaborators (1996), although astrocytes in culture contained 4.5–5.0 mM GSH, this pool was not released until the cells died following hypoglycemia–hypoxia. Similar extracellular GSH accumulation and increases in glutamate and cysteine concentrations because of the breakdown of released GSH (Orwar et al., 1994; Yang et al., 1994) are considered to be deleterious (Juurlink et al., 1996) (see the roles of GSH against glutamate and cysteine neurotoxicities). The mechanisms of glutamate and cysteine neurotoxicities is detailed below in ❷ *Sects. 3.2* and ❷ *3.3*; here, we only mention that parallel increases in extracellular glutamate and cysteine concentrations during ischemia may be extremely harmful and toxic. Astrocytes possess a Na^+-independent glutamate–cystine antiport system, Xc (Murphy et al., 1989; Cho and Bannai, 1990; O'Connor et al., 1995a). Excess levels of glutamate inhibits cystine uptake, decreasing intracellular GSH synthesis. Such GSH depletion—existing even alone—may prove toxic to cells (Murphy et al., 1990). In addition to this, extracellular accumulation of both glutamate and cysteine may cause overactivation of *N*-methyl-D-aspartate (NMDA) receptors (see ❷ *Sects. 3.2*, ❷ *3.3*, and ❷ *4.2*) with a consequent extreme enhancement in intracellular Ca^{2+} and production of NO•, parallel to a GSH loss and a consequent decrease in the NMDA receptor-mediated protection via relative blockade by nitrosothiols (Hermann et al., 2000, 2002). In this process, the ischemia-induced glial GSH release would represent a "last attempt" to rescue neurons. It is, however, a question of debate whether or not astrocytes (or a subset of astrocytes) may survive such a final loss of their intracellular GSH pool.

In summary, neurons are strongly dependent on exogenous cysteine to maintain cellular GSH levels, and some evidence suggests that this cysteine is derived from astrocytes. In this regard, astrocytes may play an important role in maintaining sulfur homeostasis in the whole brain. Because excess cysteine is neurotoxic (Olney et al., 1990; Janáky et al., 2000b), astrocytes must possess a finely controlled mechanism for releasing the appropriate amount of cysteine necessary for neuronal survival and function. The release of too much cysteine may result in neuronal injury; release of too little cysteine may result in lowered GSH and vulnerability to oxidative damage (see ❷ *Sect. 3.3*).

3 General Roles of Glutathione in the Central Nervous System

The general roles of glutathione are summarized in ❷ *Table 15-1*. Though the general roles of glutathione have remained a focus of research since being first studied, important questions are still open.

3.1 Protection Against Oxidative Stress

Glutathione is synthesized and degraded in many cell types by a series of enzymatic reactions illustrated in ❷ *Figure 15-1* (from Meister, 1986) (see above). As noted above, glutathione exists in both the reduced form (GSH) and the oxidized disulfide form (GSSG), the former by far the larger fraction. GSH and GSSG are interconvertible (❷ *Figure 15-2*).

The ratio of GSH to GSSG is termed the GSH status and is a major determinant of cell health and viability (Kosower and Kosower, 1978). The GSH status is not static, but rather reflects a dynamic system that

Figure 15-2
Interchange between GSH and GSSG. The glutathione peroxidase–glutathione reductase cycle

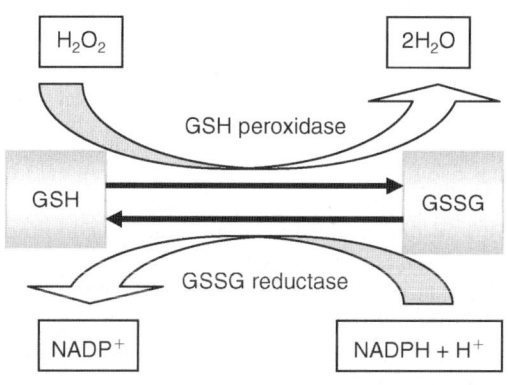

responds to alterations in the cellular environment. As mentioned earlier, in most cells, including those in the nervous system, GSH constitutes approximately 98% of total glutathione (Slivka et al., 1987b); changes in this ratio can affect any or all of the cellular functions of GSH listed in ◉ *Table 15-1*. As we shall discuss in the present section, GSH is crucial for the control of ROS. Generation of ROS and antioxidant defenses are normally in balance. Increases in the former or decreases in the latter change the balance to favor ROS action, and oxidative stress results (Halliwell, 1992a, b, 1996; Cheeseman and Slater, 1993; Evans, 1993).

3.1.1 A Brief Description of Free Radicals and Their Relevance

Reactive oxygen species (ROS) includes nonradicals and radicals centered on oxygen. Nonradicals include oxygen derivatives such as singlet oxygen and hydrogen peroxide. Free radicals are molecules or atoms, capable of independent existence that are often chemically reactive and that possess unpaired electrons in their outer orbitals. Two biologically important free radicals are the superoxide anion radical $O_2^{\bullet-}$ and the hydroxyl radical OH•. $O_2^{\bullet-}$ is not especially reactive, but its conjugate acid, $HO_2\bullet$ is a strong oxidant. OH• is extremely reactive and toxic, producing its effects via a cascade of self-perpetuating free radicals.

The superoxide anion radical and hydrogen peroxide (so called ROS) are generated during normal metabolism, e.g., in the mitochondria via the incomplete reduction of O_2 to H_2O. H_2O_2 is also produced from various oxidase reactions and during the synthesis of prostaglandins. Autooxidation of catecholamines and action of the microsomal cytochrome P-450 system also results in generation of free radicals. In addition to a permanent physiological generation of free radicals, in some pathological situations generation of ROS is perturbed, e.g., during ischemia xanthine dehydrogenase is proteolytically clipped to yield an enzyme with altered catalytic activity. The modified enzyme is xanthine oxidase, which generates H_2O_2 and $O_2^{\bullet-}$ (reviewed in Cooper, 1998). External events, such as radiation, also generate free radicals.

Normally, superoxide dismutase (SOD) is involved in the reduction (by a dismutation reaction) of the superoxide radical, converting the superoxide radical to hydrogen peroxide, which usually breaks down harmlessly to produce water and molecular oxygen

$$O_2^{\bullet-} + H_2 \to H_2O_2$$

$$2H_2O_2 \to 2H_2O + O_2$$

But in the presence of a transition metal (e.g., Fe^{2+}, Cu^{2+}) hydrogen peroxide generates hydroxyl radical in the Fenton reaction

$$H_2O_2 + Fe^{2+} \to OH\bullet + OH^- + Fe^{3+}$$

Fe^{3+} can also react with superoxide in the reaction

$$Fe^{3+} + O_2^{\bullet -} \leftrightarrow (Fe^{3+} - O_2^{\bullet -} \leftrightarrow Fe^{2+} - O_2) \leftrightarrow Fe^{2+} + O_2,$$

leading to the iron-catalyzed Haber–Weiss reaction

$$O_2^{\bullet -} + H_2O_2 \xrightarrow{Fe\ catalyst} OH\bullet + OH^- + O_2$$

These reactions (see Halliwell, 1992a) may be critical components of oxidative stress and neurological damage in cases where iron is released (e.g., stroke, ischemia, head trauma) (Willmore and Triggs, 1991).

In addition to ROS, many cells produce also nitric oxide radical, NO•. Tissues contain three isoforms of nitric oxide synthase (NOS): a calcium-activated inducible form (iNOS) in macrophages and astrocytes, a vascular endothelial form (eNOS), and a calcium-dependent neuronal form (nNOS) (Heales et al., 1996). NO• is a short-lived ($t_{1/2}$ about 1 s) radical, which is now regarded as a neurotransmitter (Lowenstein et al., 1994). NO• diffuses freely out of the place of generation, crosses cell membranes, and its action is terminated by quenching with heme compounds or reaction with cysteine or glutathione, the latter reaction providing longer-lived nitrosothiols (see ❯ Sect. 4.4).

If not quenched, NO• reacts easily with the superoxide anion radical to form peroxynitrite, ONOO⁻, a powerful oxidant, that can damage biological systems (Koppenol et al., 1992)

$$NO\bullet + O_2^{\bullet -} \rightarrow ONOO^-$$

ONOO⁻-mediated conversion of tyrosine to nitrotyrosine may prevent key phosphorylation reactions and disrupt cell-signaling processes. Besides this harmful action, ONOO⁻ is recently considered as the most harmful ROS in the brain also because increased ONOO⁻ production may inhibit the mitochondrial respiratory chain (discussed in Cooper, 1998). In general, free radicals can attack all major groups of biological molecules, but lipids appear to be the most vulnerable to direct and indirect damage as a consequence of lipid peroxidation.

3.1.2 Glutathione as an Antioxidant

Biological systems have evolved a number of antioxidant defenses to protect themselves from the effects of free radicals. These include those defenses designed to prevent the generation of free radicals as well as those whose role is to scavenge free radicals that may be generated. In the first category are various special proteins as well as enzymes such as catalase and GSPHx. Free-radical scavengers include the enzyme superoxide dismutase (see above), reduced glutathione, ascorbic acid (vitamin C), and α tocopherol (a member of the vitamin E family).

GSH can participate as an antioxidant either indirectly or directly. In the first case, glutathione serves as a substrate for GSH peroxidase to reduce hydrogen peroxide and produce GSSG and water

$$2GSH + H_2O_2 \rightarrow 2H_2O + GSSG$$

GSH may also participate directly in three types of oxidation–reduction reactions (for details, see Kosower and Kosower, 1978). In the first, GSH donates a proton to a free radical in a reaction of the form

$$GSH + R\bullet \rightarrow RH + GS\bullet$$

$$2GS\bullet \rightarrow GSSG$$

A typical case of this reaction involves the hydroxyl radical

$$2GSH + 2OH\bullet \rightarrow GSSG + 2H_2O$$

The second type of reaction mediated by GSH is a thiol–disulfide interchange; a third type involves two-electron oxidation. The former may be involved in GSH actions on ionotropic receptor function (see ❯ Sect. 4.2).

3.1.3 Glial Versus Neuronal Resistance Against Reactive Oxygen Species

Neurons are more vulnerable to the toxic action of hydrogen peroxide than are astrocytes (Ben-Yoseph et al., 1994). Astrocytes in culture were completely resistant also to acutely administered ONOO⁻ even at extraordinary high levels (2 mM), whereas neurons in culture were selectively damaged at much lower concentrations (0.1 mM) (Bolaños et al., 1995). It is widely accepted that this difference stems from the different concentrations of GSH in the two cell types and the ability of astrocytes to maintain high GSH levels. Indeed, the concentration of GSH in the neurons was about half of that in the astrocytes, and acute treatment of the neurons with ONOO⁻ resulted in considerable loss of GSH in neurons, but even 2 mM ONOO⁻ had no effect on the pool of GSH in the astrocytes (Bolaños et al., 1995). Although the exact mechanism of maintenance of high GSH levels in astrocytes is not known, it is possible to hypothesize that there is an upregulated synthesis of GSH evoked by administration of ROS. Other studies have shown that there is a strong upregulation of the pentose phosphate shunt in astrocytes exposed to hydrogen peroxide (Ben-Yoseph et al., 1994. 1996a, b) that can match with the strong demand for NADPH to generate GSH through the action of GSSG reductase. The high activity of γ-glutamylcysteine synthase (the first of two steps involved in GSH biosynthesis) in astrocytes compared with neurons (Makar et al., 1994) may also contribute to the resistance of these cells against ROS, and Huang and Philbert (1995) noted higher activities of GSHPx and GSSG reductase in cultured rat cerebellar astrocytes than in cultured cerebellar granule cells. In line with these considerations, Han and coworkers (1996) showed that incubation of glia isolated from newborn rats with L-dopa or with other autooxidizable substances resulted in upregulation of GSH production. Similar results were also observed with fetal rat mesencephalon (which contains glia plus neurons), but not with pure neuronal cultures derived from the mesencephalon. Instead, the pure neuronal cultures were destroyed by application of L-dopa (Han et al., 1996). They also found that upregulation of GSH synthesis provided significant protection against further oxidative challenge. In culture, astrocytes have been shown to protect neurons (Langeveld et al., 1995; Desagher et al., 1996) and oligodendrocytes (Noble et al., 1994) against hydrogen peroxide-induced toxicity.

Some other findings may explain this difference in the resistance of neurons and astrocytes and may underline the protective role of astrocytes in oxidative stress. Desagher and coworkers (1996) pointed out that high levels of catalase contribute to the protective effect of astrocytes. Makar and coworkers (1994) showed that astrocytes in culture have high levels of vitamin E, which also may interact with GSH levels (see later).

Taken together, the available evidence shows that the ability to induce GSH synthesis in astrocytes makes them resistant to oxidative stress; moreover, astrocytes provide cellular defense and protection of neurons against the harmful effects of ROS. We must emphasize, however, that this protection is limited and sensitive for GSH depletion. For example, if cellular GSH is depleted, even astrocytes become susceptible to the action of ONOO⁻, and mitochondrial complexes I and II/III become severely damaged (Barker et al., 1996). Moreover, reduction in GSH induced by buthionine sulfoximine (BSO) (an inhibitor of γ-glutamylcysteine synthetase) (Meister, 1991; Anderson, 1997) is accompanied by increased NOS activity both in whole brain in vivo and in neurons in culture (Heales et al., 1996). The loss of GSH paralleled with increased NOS activity resulted in a rapid neuronal death. Similar results were obtained with cystine depletion (Murphy et al., 1989). In vivo, both the perturbation of hydrogen peroxide generation and release in the brain, e.g., in forebrain ischemia (Hyslop et al., 1995), induced generation of ONOO⁻ in astrocytes which damaged nearby neurons (Barker et al., 1996) and caused depletion of GSH (especially in PD), (Jenner et al., 1992a,b) (see ● *Sect. 6.2*).

3.1.4 Interactions of Glutathione with Other Antioxidants

GSH works in synergy with the antioxidants ascorbic acid and α-tocopherol and plays a crucial role in reactions mediated by SOD. GSH regulates, at least in part, ascorbic acid levels (Meister, 1992)

$$\text{Dehydroascorbate} + 2\text{GSH} \rightarrow \text{ascorbic acid} + \text{GSSG}$$

A decrease in GSH leads to decreased ascorbic acid levels. However, ascorbate can partially compensate for diminution of GSH (Mårtensson and Meister, 1991).

3.2 Protection Against Glutamate Neurotoxicity

Astrocytes contain uptake systems for glutamate and cysteine and a transporter for glycine (Holopainen and Kontro, 1989). Dringen and Hamprecht (1996) proposed that astrocytes protect neurons from the excitotoxic action of glutamate via uptake of extracellular glutamate and GSH synthesis. This assumption is further supported by the fact that the synthesized GSH, which is released from astrocytes and also from neurons (Zängerle et al., 1992), can displace glutamate from its binding sites, thus preventing overactivation and pathological increase of intracellular Ca^{2+} (Janáky et al., 1993a, 1998, 2000a). In addition, a part of glutamate excitotoxic actions stems from the generation of NO• via activation of NMDA receptors. This toxic action of excess of glutamate can also be attenuated via quenching of NO• and forming nitrosoglutathione, the latter being another displacer of glutamate binding (Hermann et al., 2000, 2002).

In summary, there are at least four mechanisms of GSH protection against glutamate neurotoxicity: (1) use of glutamate as a substrate for GSH synthesis, (2) displacing of glutamate from ionotropic receptor binding sites, (3) quenching of excitation-generated NO•, and (4) displacing of glutamate via generation of GSNO.

3.3 Protection Against Cysteine Neurotoxicity

The sulfhydryl of GSH is less reactive than that of free cysteine, and consequently GSH serves as a nontoxic storage form of cysteine. Cysteine can form relatively stable hemithioketals with ketones (e.g., α-keto acids). With aldehydes (e.g., glyoxylate and pyridoxal 5′-phosphate (PLP)), hemithioacetal formation is followed by cyclization to a thiazolidone derivative. Therefore, high concentrations of cysteine inhibit some key PLP-containing enzymes, including glutamate decarboxylase, and this may contribute to the known neurotoxicity of cysteine. This is only one possibility for the mechanism of cysteine neurotoxicity; others have been reviewed (Janáky et al., 2000b). In contrast to cysteine, GSH does not form stable adducts with α-keto acids or with PLP, although it does form a stable adduct with glyoxylate (Cooper, 1998). In addition, GSH has been shown to inhibit the ^{45}Ca influx evoked by overactivation of NMDA receptors with parallel application of excitatory amino acids and reducing agents (e.g., dithiothreitol (DTT)) which acts similarly to cysteine (Janáky et al., 1993a, 2000b). Thus, GSH is a powerful antagonist of cysteine neurotoxicity evoked via reduction of sulfhydryl groups in NMDA receptors and overactivation of NMDA receptor-mediated responses.

3.4 Protection Against Xenobiotics

Many potentially harmful foreign compounds can be detoxified through the mercapturate pathway. The first step of the mercapturate pathway is catalyzed by glutathione S-transferases (GSTs) (❷ Figure 15-1, reaction 8). These enzymes catalyze the conjugation of GSH with a variety of electrophilic substances of endogenous and exogenous origin (Cooper and Tate, 1997). The brain possesses many types of GSTs (reviewed by Johnson et al., 1993; Lowndes et al., 1994; Cooper, 1997). Cytosolic GSTs are present in astrocytes (especially in the end feet), ependymal cells, tanocytes, the subventricular zone, choroid plexus, and oligodendrocytes. Although earlier studies did not show GST activity in neurons, later Makar and coworkers (1994) found GST activity in cultured chick neurons, and Huang and Philbert (1995) found activity in cerebellar granule cells. Microsomal GST is present in Purkinje cells throughout the cerebellar cortex and within neurons of the brain stem and hippocampus of the rat. The brain possesses regional differences in the pattern of α-, μ-, and π-class GST subunit expression (Johnson et al., 1993). These differences may be of clinical significance. For example, in the cerebellar cortex the concentration of the μ-class GST subunit 4 was found to be greatest in the flocculus and lowest in the vermis, a pattern that coincides with the known greater susceptibility of the vermis than of the flocculus to toxic and metabolic insults (Johnson et al., 1993).

Several roles have been suggested for GSTs in the brain. For example, it has been proposed that GSTs protect the myelin sheath against toxic substances (Cammer et al., 1989; Tansey and Cammer, 1991).

A unique µ-GST, found exclusively in the brain and testes, may have a specific role in keeping blood–brain and blood–testis barrier functions (Campbell et al., 1990). GSTs may also remove epoxyeicosatrienoic acid derivatives from cells and participate in the biosynthesis of prostaglandins and leukotrienes in the brain (Cammer et al., 1989). GSTs within ependymal cells and astrocytic end feet may form a first line of defense in the brain against potentially harmful xenobiotics, and participate in the transport of substances into and out of the brain (Cammer et al., 1989). These authors stated that such transport could affect hormonal control over processes such as myelination and neuronal growth and could facilitate removal of endogenous and exogenous toxins from the CNS. Because GSTs are present also in certain types of neurons and oligodendrocytes, they can provide also a second line defense. Because GSTs can act on several eicosanoids in the brain it is also possible that these enzymes participate in local hormonal signaling within the brain. One could speculate that such signaling involves special astrocytic–neuronal interactions.

3.5 Transport of Amino Acids

Orlowski and Meister (1970) pointed out that GSH has a unique metabolic role in the transport of amino acids. They suggested that enzymes involved in GSH metabolism (see ❷ *Figure 15-1*) could be linked to form a process for the recycling of the three amino acid constituents of GSH. The authors termed this process the γ-glutamyl cycle (❷ *Figure 15-1*). It was suggested that the positioning of γ-glutamyltransferase (❷ *Figure 15-1*, reaction 1) on the cell surface and of other enzymes of GSH metabolism within the cell permits the γ-glutamyl cycle to play a role in the translocation of amino acids across cell membranes. While there is strong evidence to suggest that the γ-glutamyl cycle does indeed operate in vivo, it is now recognized that the cycle does not play a key role in amino acid transport (except perhaps in the case of cystine). Rather, the cycle is thought to have a specialized role in metabolism (e.g., in transformations of leukotriene A, estrogens, and prostaglandins and in detoxification of xenobiotics) (see above).

3.6 Macromolecule Synthesis, Cell-Cycle Regulation, and Cancer

Intracellular GSH level rapidly rises in fibroblasts upon mitogenic stimulation of the cells (Shaw and Chou, 1986), and this rapid elevation is followed by a gradual increase as cells enter the S-phase. Prevention of the early increase in intracellular GSH causes a decrease in the percent of cells capable of entering the S-phase. The elevation in GSH content is required for maximum induction of DNA synthesis. The intracellular GSH level decreases as the cells become quiescent by either serum deprivation or high cell density. It is also noteworthy to mention that in plants cellular disulfides decrease during germination.

We can easily explain the above cell-cycle-dependent alterations in the GSH/GSSG status if we consider that GSH is involved in the regulation of the synthesis of macromolecules (Kosower and Kosower, 1978). DNA synthesis is inhibited by the GSH → GSSG conversion. GSH can function as a hydrogen donor for ribonucleotide reduction in *Escherichia coli*. RNA synthesis is similarly inhibited by the GSH→GSSG conversion. Such a conversion inhibits the initiation of protein synthesis and protein elongation and results in cellular lysis in reticulocytes (Kosower and Kosower, 1978). GSSG in these cells may contribute to the production of an interferon similar to a double-stranded RNA. Interferon is an antiviral substance, an altered initiation factor, arising as a by-product of the double-stranded RNA inhibition of protein synthesis. The mechanism of interferon-mediated protein synthesis inhibition may involve the phosphorylation of an initiation factor which prevents its participation in the initiation complex. We note, however, that the normal GSSG level in cells is not sufficient to halt the protein synthesis (Kosower and Kosower, 1978).

In the cell, GSH is present also in mixed disulfides, mostly in a GSS-protein form. The presence or absence of GSH in proteins (the control of cellular protein thiol status) therefore alters the refolding of proteins containing multiple disulfide bridges. In addition, via the above mechanism GSH and GSSG may alter the activity of several enzymes; in general, a relative abundance of intracellular GSSG favors enzyme inhibition, while GSH protects enzymes from such inhibition and reactivates them. Several enzymes are under the control of GSH/GSSG thiol status, e.g., glycogen synthetase D, pyruvic kinase, acid phosphatase,

GSH S-transferases, and glucose-6-phosphate dehydrogenase. This latter enzyme is regulated in the opposite direction: the GSH→GSSG conversion stimulates glycolysis and glycogenolysis. There is also an important regulatory interaction between hemoglobin A (HBA) and GSSG: in the presence of high intracellular GSSG concentrations, HBA-SSG mixed disulfide formation is favored (Kosower and Kosower, 1978).

If we are aware of the above regulatory processes fulfilled by GSH, it is no wonder that significant roles may be attributed to the intracellular GSH/GSSG status in the prevention of carcinogenesis (Flagg et al., 1993). On the other hand, just as the GSH ratio is crucial for the well-being of normal cells, it also protects cancer cells from the effects of oxidative stress. Because chemo- and radiation therapy generates free radicals to kill cancer cells (Meister, 1989, 1991, 1992), normal GSH levels have the potential to compromise anticancer treatments. One strategy to prevent this has been to lower GSH levels by blocking GSH synthesis. BSO, an inhibitor of γ-glutamylcysteine synthetase (Meister, 1991; Anderson, 1997), lowers GSH levels and has been reported to enhance the efficacy of cancer treatments. Such adjuvant GSH depletion has been successfully used in various clinical settings, but must be applied with precautions (reviewed by Estrela et al., 2006). Given the multiple roles of GSH in normal cells (● *Table 15-1*), it is difficult to imagine that BSO treatment, effective as it may be for enhancing the effectiveness of cancer treatments, would not have deleterious effect on healthy cell populations. This is especially the case in the lens, where GSH depletion induces cataract development (Meister, 1991), and in neural cells (Jain et al., 1991; Meister, 1991) that may be vulnerable and sensitive to adjuvant buthionine sulfoximine (BSO) cancer therapy. In addition, GSH depletion is now generally accepted as a pathogenic factor in several neurological and neurodegenerative disorders (discussed in detail in ● *Sect. 6*), and thus prolonged GSH depletion by BSO in cancer therapy, if successful, may come at the expense of later neural function and survival. Indeed, Holt (1993) has reported that an increase in the GSSG/total glutathione ratio following GSSG injections during cancer treatments led to a temporary form of epilepsy. On the other hand, via the GST-catalyzed reactions (see above), GSH may in some case protect normal cells and have neuroprotective effects against cytotoxicity of anticancer drugs such as cisplatin (Cascinu et al., 1995).

4 Glutathione as a Neuromodulator at Glutamate Receptors

Almost 50 years ago, W. F. Loomis, H. L. Lenhoff, and colleagues conducted a classic series of experiments that suggested a role for GSH in signal transduction in the coelenterate *Hydra* (Loomis, 1955; Lenhoff, 1961, 1998; Lenhoff and Bovaird, 1961). The addition of GSH to the media bathing the *Hydra* elicited a feeding response that was highly specific, reliable, and quantifiable. Grosvenor and coworkers (1992) and Bellis and coworkers (1994) later described a class of binding sites for GSH in *Hydra*, perhaps providing the basis for the observed behavioral response. In mammals, GSH was shown to be involved in some aspects of neurotransmitter release (Werman et al., 1971), and binding sites for the tripeptide were found in many areas of the mammalian CNS (Ogita and Yoneda, 1987, 1988).

In spite of such early promising data, the potential for GSH to have a role in neural signal transduction has remained largely unstudied. Recently, several lines of evidence have refocused attention on the possible role of GSH in signal transduction. First, several radioligand binding studies from laboratories of the present authors (Janáky et al., 1993a, 1998, 1999; Ogita et al., 1995, 1998; Pasqualotto et al., 1998; Shaw, 1998) have shown the presence of distinct high-affinity binding sites for GSH and GSH derivatives. The presence of unique binding sites for a putative receptor population might imply that GSH serves a role as neurotransmitter, neuromodulator, or both in various neural circuits (Shaw et al., 1996). Complicating this hypothesis, however, are observations that GSH appears to compete for binding with some glutamate receptor (GluR) populations, a finding that has been taken to suggest that GSH is an endogenous agonist at some GluRs at less than millimolar ligand concentrations (Ogita et al., 1995). We note, however, that antagonist action by GSH may also occur over the same general concentration range (Janáky et al., 1998, 1999), so the matter remains unresolved. We will return to this last point later in this chapter. In addition to such studies, several recent reports have shown that GSH, like other thiol-containing compounds, may mediate reduction–oxidation (redox) reactions to alter the functional characteristics of various ionotropic receptors and ion channels. Notable among the latter are the redox effects on *N*-methyl-D-aspartate (NMDA) receptors (Gozlan and Ben-Ari, 1995). An emerging view is that such reactions are important regulatory mechanisms controlling ionotropic receptor functions (for review, see Janáky et al., 1998).

The studies cited above appear to suggest that GSH may have roles in signal transduction in the nervous system. Whether GSH acts through its own population of receptors, through those for glutamate, or through both remains an open question. In addition, the relationship between GSH as a neurotransmitter versus its role as a neuromodulatory agonist/antagonist and/or redox agent remains unclear.

4.1 Glutamate Receptors

GluRs are thought to play a key role in mechanisms underlying important physiological functions of the brain such as learning and memory (for review, see Cotman et al., 1988; Collingridge and Singer, 1990) and may be involved in acute neurological and chronic neurodegenerative disorders. The former includes hypoxic–ischemic brain injury and trauma; the latter includes epilepsy, Parkinson's disease, Huntington's disease, amyotrophic lateral sclerosis, and AIDS dementia (for reviews, see Choi, 1988; Lipton and Rosenberg, 1994). The most current classification scheme for the GluRs shows them divided into two major categories according to their signal transduction pathway in the mammalian CNS: the metabotropic and ionotropic families (see Seeburg, 1993; Nakanishi and Masu, 1994; Shaw and Ince, 1997). The metabotropic GluR (mGluR) family consists of at least eight independent subunits (mGluR1–8) and is subclassified into three different groups on the basis of differential signal transduction processes: Group I (mGluR1 and 5) stimulates formation of inositol-1,4,5-triphosphate and diacylglycerol; groups II (mGluR2 and 3) and III (mGluR4, 6, 7, and 8) reduce levels of intracellular cyclic AMP or GMP. The ionotropic GluRs are linked to cation channels and are pharmacologically classified on the basis of differential sensitivities to the exogenous agonist NMDA. The NMDA-sensitive receptors are made up of the NMDAR1 subunit together with any one of four different modulatory subunits (NMDAR2A-D). The ionotropic receptors insensitive to NMDA (often referred to as non-NMDA receptors) consist of nine distinct subunits (GluR1–7 and KA1 and 2) in the rat brain and are further divided into two different subtypes, (S)-2-amino-3-hydroxy-5-methyl-4-isoxazolepropionate (AMPA) (GluR1-4) and kainate (KA) (GluR5-7 and KA1 and 2) receptors, according to the differential preference for these agonists.

4.2 Glutathione Actions on Ionotropic Glutamate Receptors

4.2.1 NMDA Receptors

Early pharmacological studies revealed that an ionotropic subclass sensitive to NMDA functions as a receptor ionophore complex consisting of at least four distinct domains: (1) an ion channel domain highly permeable to Ca^{2+}, (2) an NMDA recognition domain with high affinity for the endogenous agonist glutamate, (3) a glycine recognition domain with high affinity for both glycine and D-serine, and (4) a polyamine recognition domain with activating properties (Monaghan et al., 1989; Récasens et al., 1992). To evaluate possible functions of GSH and other thiol-containing compounds in the process involved in opening the channel associated with the NMDA receptor, the effects of these compounds have been investigated on ligand binding to each domain within the NMDA receptor complex in synaptic membranes of the rodent brain (Yoneda et al., 1990; Ogita et al., 1995; Varga et al., 1997; Jenei et al., 1998).

Both GSH and GSSG displace [^{3}H]glutamate binding to rat brain synaptic membranes at temperatures of >30°C in the presence of either Cl^- or Na^+ (Koller and Coyle, 1985; Ogita et al., 1986; Ogita and Yoneda, 1987). However, such temperature-dependent [^{3}H]glutamate binding is now considered to represent Cl^-/Ca^{2+}-dependent transport (Pin et al., 1984) and Na^+-dependent uptake (Vincent and McGeer, 1980; Ogita and Yoneda, 1986) systems for glutamate in the presence of Cl^- and Na^+, respectively. In contrast, GSH is more potent in displacing Na^+-independent binding than Na^+-dependent binding of [^{3}H]glutamate at 2°C (Ogita et al., 1986). Similarly, potent inhibition is seen with GSSG and other peptides containing glutamate as a free amino acid residue (Varga et al., 1989, 1997; Janáky et al., 1998, 1999; Jenei et al., 1998). Moreover, both GSH and GSSG markedly inhibit binding of the ^{3}H-labeled radioligands 3-(2-carboxypiperazin-4-yl)-propyl-1-phosphonate ([^{3}H]CPP) (Yoneda et al., 1990) and DL-(E)-2-amino-4-propyl-5-phosphono-3-pentenoate ([^{3}H]CGP 39653) (Ogita et al., 1995; Pasqualotto et al., 1998; C. A. Shaw et al., unpublished data) to the NMDA recognition domain on the NMDA receptor with potencies greater than those for NMDA. Both [^{3}H]glutamate and [^{3}H]CPP binding have been shown to be inhibited by various S-alkyl derivatives of glutathione, including S-methylglutathione (SMG), S-ethylglutathione (SEG), S-propylglutathione (SPG), S-butylglutathione (SBG), S-pentylglutathione (SPeG), and glutathione sulfonate (GSA) with potencies higher than for NMDA (Jenei et al., 1998). These findings all give rise to a proposal that GSH and its derivatives have some affinity for the NMDA recognition domain in a manner independent of the thiol moiety.

The above data raise the question of whether GSH and its derivatives are agonists or antagonists in their binding at the NMDA receptor. It should be noted that GSH is more potent for inhibiting the binding of [^{3}H]glutamate (agonist) than of [^{3}H]CGP 39653 (antagonist) (Ogita et al., 1995). Under the same experimental conditions, NMDA agonists are more potent in inhibiting [^{3}H]glutamate than [^{3}H]CGP 39653 binding, whereas NMDA antagonists have the opposite actions (Zuo et al., 1993a,b). Because NMDA agonists appear to be more effective for inhibiting binding of the agonist compared with the antagonist, the suggestion has been made that GSH may be an agonist at a recognition domain or a codomain of the NMDA receptor (Ogita et al., 1995). However, this may only be true as a function of ligand concentration because in the micromolar range GSH appears to antagonize NMDA action (Janáky et al., 1993a, 1998; Jenei et al., 1998). Such results may suggest the possibility that GSH functions as an inverse agonist at the NMDA receptor. The effects of GSH derivatives have also been tested on the NMDA channel using ^{3}H-labeled 5-methyl-10, 11-dihydro-5H-dibenzo[a,d]cyclohepten-5,10-imine ([^{3}H]MK-801, dizocilpine), a noncompetitive antagonist that gains access to the open NMDA channel. Both GSH and GSSG at concentrations >1 µM potentiate [^{3}H]MK-801 binding (Ogita et al., 1995), as do S-alkyl derivatives of GSH (Jenei et al., 1998), but again such potentiation is dose-dependent: at high concentrations, GSH decreases [^{3}H]MK-801 binding (Ogita et al., 1995; Varga et al., 1997; Jenei et al., 1998). GSH-mediated potentiation is prevented by NMDA antagonists, e.g., CGP 39653 or L-(+)-2-amino-5-phosphonovalerate (L-AP5) but is not affected by NMDA agonists or glutamate dehydrogenase (Ogita et al., 1995). The latter data suggest that GSH potentiation, when it occurs, is not attributable to free glutamate derived from a potential cleavage of the GSH tripeptide during incubation. Neither GSH nor the above derivatives give potentiation of [^{3}H]MK-801 binding in the presence of glutamate. The implication of these results is that GSH can bind to the NMDA channel as either an agonist or an antagonist in particular circumstances, depending on such action on the γ-glutamyl residue but not on the thiol moiety of GSH (Ogita et al., 1986; Varga et al., 1989; Oja et al., 1995; Janáky et al., 1999). The discrepancy between groups on the agonist versus antagonist actions of GSH may arise because of the presence of distinct subunits consisting of native NMDA receptors on different cell types (see Nakanishi and Masu, 1994).

At the glycine site of the NMDA receptor in the brain, GSH and GSSG at concentrations of >100 µM displace binding of [^{3}H]glycine and 5,7-dichloro[^{3}H]kynurenate (a glycine antagonist) by ~20–30% (Ogita et al., 1995). S-Alkyl derivatives of GSH and GSA (at 1 mM concentrations) also significantly inhibit [^{3}H] glycine binding (Jenei et al., 1998). These data suggest that GSH mediation may be a common, if not equal, regulatory feature at the various ionotropic receptor and receptor subtypes (Jenei et al., 1998) (see also Pan et al., 1995, for a discussion of GSH actions at other ionotropic receptor populations).

The effects of GSH on the response of the NMDA receptor to agonist stimulation have been examined in a series of studies, particularly in relation to redox modulators such as the reducing agent DTT. In cultured rat cortical neurons, NMDA receptor-mediated currents were substantially potentiated after exposure to DTT (Leslie et al., 1992; Lipton, 1993; Lipton and Stamler, 1994). In contrast, the oxidizing agent 5,5′-dithio-bis(2-nitrobenzoate) (DTNB) attenuated NMDA responses and abolished DTT-mediated potentiation (Aizenman et al., 1989). In a related study, DTT also potentiated both NMDA-mediated Ca^{2+} currents and [^{3}H]MK-801 binding; opposite effects were seen for DTNB (Reynolds et al., 1990). The actions of DTT are mimicked by GSH apparently acting on NMDAR1 and 2A subunits in some cell types (Köhr et al., 1994). Both GSH and GSSG have been reported to give rise to a transmembrane Ca^{2+} current, which can be blocked by L-AP5 and MK-801 in dissociated brain cells (Leslie et al., 1992). However, Janáky and coworkers (1993a) and Jenei and coworkers (1998) have reported a clear biphasic effect on NMDA-mediated excitation for GSH that was dependent on concentration in cultured cerebellar granule cells. An antagonist effect of GSSG and S-alkyl GSH derivatives was also noted. These biphasic actions by GSH on NMDA-mediated Ca^{2+} currents may help to resolve the apparently contradictory data of other groups (see, e.g., Levy et al. (1991) versus Leslie et al. (1992)).

Data such as the above lend support to a role for GSH acting at an NMDA modulatory site, perhaps in the following manner: at low concentrations, GSH, GSSG, and S-alkyl derivatives are antagonists or agonists acting as a γ-glutamyl peptide to displace glutamate, with the last being a point of contention among our three laboratories and unresolved as of this writing. At high concentrations, GSH mimics the reducing action of compounds like DTT acting at -SH groups to alter the NMDA receptor as a co-agonist.

4.2.2 Non-NMDA Receptors

Given the data presented above, what is the possibility that GSH and/or GSSG may be ligands at non-NMDA receptors as well? Many γ-glutamyl dipeptides are known to inhibit the Na^+- and temperature-independent binding of [^{3}H]AMPA and [^{3}H]kainate to synaptic membranes isolated from the rodent brain (Varga et al., 1989, 1994a,b). However, because GSH and GSSG are not particularly effective competitors for these radioligands, the cysteinyl and glycyl moieties do not seem to have as crucial a role as the γ-glutamyl moiety. Both GSH and GSSG displace [^{3}H]AMPA and [^{3}H]kainate in a concentration-dependent manner within the concentration range likely to prevail in vivo (Varga et al., 1989; Janáky et al., 1998; Jenei et al., 1998). Both bind preferentially to AMPA receptors; the affinity for the kainate receptor is ~25–30 times lower (Jenei et al., 1998). Similarly, S-alkyl derivatives of GSH (SMG, SEG, SPG, SBG, SPeG, and GSA) bind to the AMPA receptor with relatively high affinity (estimated IC_{50} values are all in the low micromolar range). The similar affinity of GSH and GSSG to displace AMPA binding suggests that redox modulation of sulfhydryl groups in the AMPA receptor complex is not the mechanism of action. This view is in concert with the observed lack of effect of DTT and DTNB for modulating AMPA binding and AMPA-mediated responses (Aizenman et al., 1989). Furthermore, the efficacy of GSH to alter AMPA binding tends to be enhanced when the sulfhydryl group in the cysteine residue is S-alkylated. The alkylation not only enhances affinity but also increases the selectivity of GSH and S-alkyl derivatives for the AMPA versus kainate sites (Jenei et al., 1998).

Related to the issues raised above is the question of whether GSH causes competitive or noncompetitive displacement of ligands from the AMPA sites. GSH and its derivatives significantly reduce B_{max} values and may diminish affinity as well (Jenei et al., 1998). GSSG and GSA also diminish low-affinity binding to AMPA receptors (Oja et al., 1995). The likely mechanism of action is apparently complex and may involve allosteric interactions.

Is the effect of GSH and derivatives on AMPA receptors agonistic or antagonistic? As we discuss below (❯ Sect. 5.3.2), GSH-induced depolarization in cortical slices is not antagonized by 6,7-dinitroquinoxaline-2,3-dione (DNQX) (Shaw et al., 1996; Pasqualotto et al., 1998), suggesting that GSH does not act as an agonist at these sites.

In the light of the above results it appears that physiologically relevant concentrations of GSH displace glutamate mainly from the AMPA subclass of non-NMDA receptors in an allosteric manner owing to the presence of the γ-glutamyl moiety. This effect may play a role in fine tuning glutamatergic neurotransmission by acting to curtail AMPA receptor-mediated current duration. Because AMPA and NMDA receptors are colocalized and cooperate at the postsynaptic membrane (Nicoll et al., 1990; Seeburg, 1993), the co-release of glutamate and GSH from presynaptic terminals (Hjelle et al., 1994, 1998) may have profound consequences for synaptic transmission.

4.3 Nitrosoglutathione

Nitric oxide (NO) is a gaseous odd messenger molecule that is synthesized in the central nervous system by three isoforms (endothelial, glial, and neuronal) of NO synthase (NOS). NMDA receptor agonists activate the calcium/calmodulin-dependent nNOS (Bredt and Snyder, 1994; Knowles and Moncada, 1994). The lifespan of NO generated is only few seconds. It has been suggested that NO requires a carrier system to reach its physiological sites of action, e.g., in blood NO circulates predominantly as an S-nitroso adduct of serum albumin (Stamler et al., 1992). S-NO adducts of cysteine and glutathione are also easily formed in the presence of NO and O_2 (Wink et al., 1994). In the rat cerebellum the concentration of S-nitrosoglutathione (GSNO) is about 15 µmol/kg protein (Kluge et al., 1997). The parent molecule GSH is present intracellularly at millimolar and extracellularly at micromolar concentrations in the rat brain and is released into the extracellular space upon depolarization (Zängerle et al., 1992) (see ❷ Sect. 2.2).

NO itself in excess is neurotoxic (Dawson and Dawson, 1996). This is engendered by the reaction with superoxide anion ($O_2^{•-}$) leading to the formation of peroxynitrite (ONOO⁻), a potent oxidant (see ❷ Sect. 3.1.1) Moreover, NO generates in aerobic aqueous environments several reactive and potentially toxic nitrogen oxide (NOx) metabolites (Wink et al., 1994) (see also ❷ Sect. 3.1.1). Peroxynitrite can also react with thiols, e.g., forming GSNO with GSH. In this manner the cells are protected against oxidative stress (Hogg et al., 1996; Singh et al., 1996). The packaging of NO in the form of nitrosothiols which act as NO donors, may prolong its actions and facilitate delivery to specific target proteins (e.g., enzymes and receptors) (Manzoni et al., 1992, Do et al., 1996). However, liberation of NO may not account for all the biological activities of nitrosothiols. We have for instance shown that GSH (Varga et al., 1997) and S-alkyl derivatives of GSH (Jenei et al., 1998) at micromolar concentrations interact with the binding of ligands to ionotropic GluRs.

GSNO is thus likewise a potential modulator of GluRs. We have studied its effects on the binding of ligands to different ionotropic GluRs in pig cerebral cortical synaptic membranes (Hermann et al., 2000, 2002). GSNO displaced the binding of 20 nM [³H]glutamate, 10 nM [³H]3-(2-carboxypiperazin-4-yl) propyl-1-phosphonate ([³H]CPP), an NMDA antagonist, and 20 nM [³H]kainate to porcine cerebral cortical membranes. GSH and GSSG were also effective but GSNO was always a more potent displacer than GSH or GSSG. In the case of [³H]CPP, GSH and GSSG seem to evoke only partial inhibition of binding, about 70% and 50%, respectively. The estimated IC_{50} values for the displacers were in the low micromolar range. The binding of 20 nM [³H]fluorowillardiine ([³H]FWD), a ligand of AMPA receptors, was not affected by GSNO, GSH, or GSSG. GSNO inhibited [³H]glutamate, [³H]CPP, and [³H]kainate binding more markedly at low than at high ligand concentrations. The estimated kinetic parameters for the binding of [³H]CPP and [³H]kainate demonstrate that GSNO inhibition was indeed competitive in nature. In the case of [³H]glutamate binding, B_{max} was also altered by GSNO. [³H]Dizocilpine (MK-801) binding was enhanced in a concentration-dependent manner by GSNO, GSSG, and GSH. Glycine (10 µM) also activated the binding by 56%. This effect was additive to that of 100 µM GSNO. Similar additive effects were obtained with glycine (10 µM) and GSSG and GSH (both 100 µM). At 50 µM GSH enhanced 1 nM [³H]dizocilpine binding by 72%, GSSG by 88% and GSNO by 123%. Glycine (10 µM) plus GSNO, GSSG, and GSH (all 50 µM) activated the binding by 182%, 160%, and 147%, respectively. The NO donors S-nitroso-N-acetylpenicillamine (SNAP), 5-amino-3-morpholinyl-1,2,3-oxadiazolium chloride (SIN-1), and nitroglycerin failed to affect significantly the binding of [³H]dizocilpine. The present GSNO effects were not significantly affected by the breakdown of GSNO during incubation. In separate experiments in the present incubation media buffered with Tris, 500 µM GSNO yielded 6.7 µM GSH at 25°C in 15 min and 6.4 µM GSH at 0°C in 60 min when incubated with membrane preparations. No measurable production of GSSG, glutamate, or glycine was discernible, nor was any GSH formation detectable when the membranes were omitted. When the incubations were carried out in HEPES–Tris buffer at 23°C with and without membranes, 8.5 µM and 3.7 µM GSH, respectively, were formed in 30 min. No production of GSSG, glutamate, or glycine was detected. The results indicate that synaptic membranes and a high incubation temperature facilitate to some extent NO cleavage from GSNO.

The interaction of GSNO with NMDA and kainate receptors is probably mainly mediated by the γ-glutamyl moiety of the peptide since it was competitive in nature. Such an inference is corroborated by the findings that the NO donors like SNAP, SIN-1, and nitroglycerin with no γ-glutamyl moiety in the molecule failed to interact with binding of specific ligands for GluRs and enhance dizocilpine binding, whereas GSH, GSSG, and GSNO were all effective. The additivity of the enhancing effects of glycine and those of GSH, GSSG, and GSNO in dizocilpine binding provides additional support for this assumption. In view of the fairly similar efficacies of GSH and GSNO on the studied binding processes, the amounts of GSH formed from GSNO breakdown were not large enough to account for all observed effects. The intact molecule of GSNO must thus have retained the properties of GSH (Varga et al., 1997; Jenei et al., 1998), being able to displace glutamate agonists from NMDA and kainate receptors and to enhance dizocilpine binding.

The possible physiological actions of GSNO, however, are apparently more complex. For instance, iontophoretically applied GSNO potentiated the responses of relay neurons to sensory stimuli and excitatory amino acids, while GSH attenuated the responses to NMDA and AMPA in the rat ventrobasal thalamus (Shaw and Salt, 1997). Only glutamatergic transmission is modulated by these peptides since the carbachol effects were not affected by GSH or GSNO. It is difficult, however, to classify the above effects on glutamatergic transmission to be either agonistic or antagonistic in this polysynaptic preparation. The differences in GSH and GSNO actions may be due to liberation of NO from GSNO. Effects of GSNO were partially mimicked by the NO donor SIN-1 and by the precursor of NO synthesis L-arginine (Shaw and Salt, 1997). GSH also possesses high-affinity binding sites, which are not significantly affected by GluR agonists or antagonists (Janáky et al., 2000a). By the agency of these sites, linked to Na^+ ionophores, dose-dependent depolarizing field potentials have been generated by GSH in the rat cerebral cortex (Bains and Shaw, 1998). Moreover, Taguchi and coworkers (1995) report specific binding of [^{3}H]GSNO to synaptic plasma membranes, which is not displaceable by GluR agonists. This finding further supports the hypothesis (see above) that different functional binding sites for GSH and GSNO exist in synaptic plasma membranes (see ❥ Sect. 5).

Nitrosothiols (GSNO, S-nitroso-L-cysteine, S-nitroso-N-acetylcysteine, SNAP) exhibit a wide range of biological activities, including antimicrobial effects (De Groote et al., 1995), vasodilatation (Ignarro et al., 1981), bronchodilatation (Gaston et al., 1994), and inhibition of intestinal motility (Slivka et al., 1994) which are generally thought to be mediated by spontaneous liberation of NO. NO release from S-nitrosothiols is catalyzed by liver cell plasma membranes (Kowaluk and Fung, 1990). Our results show that GSNO is similarly but slowly decomposed in the presence of synaptic membranes. The binding of GSNO to synaptic plasma membranes, e.g., to GluRs, may thus also involve liberation of NO on the site. NO inhibits NMDA receptor-mediated responses (Butler et al., 1995; Hoyt et al., 1992; Manzoni et al., 1992; Fagni et al., 1995). The nitrosylation of cysteine residues (cys-744 and cys-798) in the redox modulatory site in NMDA receptors has been surmised to be responsible for this inhibition (Lei et al., 1992). However, the point mutations of cys-744 and cys-798 for alanine and the exposure of receptors to thiol oxidants (which eliminate free thiols from the nitrosylation targets) do not affect the inhibition produced by NO (Lei et al., 1992; Aizenman and Potthoff, 1999). From these results it seems possible that NO and probably also GSNO affect thiol groups other than those in the redox modulatory sites in NMDA receptors (Tang and Aizenman, 1993a,b; Aizenman and Potthoff, 1999). In vitro, GSNO oxidizes GSH to GSSG under aerobic conditions (Hogg et al., 1996). In vivo, it may also generate disulfide bridges between cysteine side chains in proteins and mixed disulfides with GSH (Lipton and Stamler, 1994; Stamler and Hausladen, 1998). In this manner long-lasting conformational changes in receptor proteins may ensue, which downregulate the receptor activity.

NO and NO donors have been reported to deplete GSH in cultured neurons (Bolaños et al., 1996) and to induce the death of various neural cells (Froissard et al., 1997). Noxious agents enhance cystine uptake and increase GSH synthesis in astrocytes but also subsequently foment the carrier-mediated release of GSH (Sagara et al., 1996). This astrocytic GSH serves as a defense mechanism against the deleterious effects of excess NO and NO• (Hogg et al., 1996). The GSNO formed, together with GSH and GSSG (Varga et al., 1989a; Janáky et al., 1998, 1999, 2000a; Jenei et al., 1998), is an effective displacer of glutamate. They can forestall the activation of NMDA receptors, evoked Ca^{2+} influx via receptor-gated ionophores (Janáky et al., 1993a), and the Ca^{2+}-dependent stimulation of NOS with consequent further generation of NO. Hence, astrocyte-derived GSH may have a double defensive role in nervous tissue, i.e., the capture of excess NO• and prevention of further NO• generation. These actions support the neuroprotective role of GSH and GSNO claimed by Werns and Lucchesi (1990) and by Jain and coworkers (1991).

4.4 Interactions with Glutamate Release

GSH and cysteine (both at 1 mM) evoke release of glutamate from cultured cerebellar granule cells (139% and 316%, respectively). GSH does not interfere with the release evoked by cysteine (R. Janáky et al., unpublished results). This indicates that the mechanism of glutathione action differs from redox modulation, although an inhibition of glutamate uptake may also contribute (Volterra et al., 1994). Given the overall actions of GSH on GluRs, it is plausible that it may also inhibit presynaptic glutamate autoreceptors, although direct evidence is currently lacking.

4.5 Interactions with Glutamate Uptake

In a crude synaptic membrane preparation, GSH and GSSG inhibit the Cl^-/Ca^{2+}-dependent, temperature-sensitive binding of glutamate (Varga et al., 1994a), the event considered to represent glutamate transport (Pin et al., 1984). However, in synaptosomes GSH and GSSG (both at 1 mM) enhance the uptake of L-glutamate (177% and 155%, respectively). One-millimolar GSSG (but not GSH) inhibits by 29% the uptake of D-aspartate. L- and D-aspartate and L-glutamate inhibit both uptake processes, whereas glycine does not have any effect. The enhancement of synaptosomal glutamate uptake is unique for glutathione, since other γ-glutamyl peptides lacking the free thiol group, including γ-L-glutamylglycine, were generally inhibitory (Varga et al., 1994a). D-Aspartate, the unmetabolized analog of L-glutamate, has been thought to mix with the transmitter pool of endogenous glutamate and mimic its behavior (Drejer et al., 1983; Wheeler, 1984), but it is a poor substrate for vesicular transport systems of glutamate in nerve endings (Naito and Ueda, 1985; Maycox et al., 1988). It is likely, therefore, that the vesicular, mitochondrial, and other transport systems for glutamate are differentially modulated by glutathione.

The results on glial uptake of glutamate are likewise contradictory. Volterra and coworkers (1994) report that extracellular GSH prevents inhibition of the glutamate uptake of astrocytes caused by oxygen free radicals. On the other hand, it has been postulated by Albrech et al. (1993) that extracellular GSH does not interfere with glutamate uptake in astrocytes, because the SH groups critical for this process are located within the membrane.

4.6 Interactions with the Neuronal Influx and Intraneuronal Levels of Ca^{2+}

Neuronal excitation elevates cytoplasmic free Ca^{2+}, which triggers a number of physiological responses. A sustained depolarization and an increase in free intracellular Ca^{2+} are also causative factors in neuronal death (Orrenius et al., 1989). Three major mechanisms are involved in the increase in Ca^{2+} (Mayer and Miller, 1990). Ca^{2+} ions are released into the cytosol from intracellular pools emptied by inositol triphosphate (Berridge and Irvine, 1989), extracellular Ca^{2+} may enter into cells through depolarization-activated Ca^{2+} channels located in plasma membranes, and Ca^{2+} entry may also occur via agonist-activated ionotropic GluR-gated channels. Among these, the channels associated with the NMDA receptors exhibit the highest selectivity for Ca^{2+} and a voltage-dependent Mg^{2+} block which is possibly alleviated by other cations or intracellular Ca^{2+} (Récasens et al., 1992).

Since reduction of disulfide bonds activates NMDA receptors (Aizenman et al., 1989, 1990; Lazarewicz et al., 1989; Reynolds et al., 1990; Sucher et al., 1990; Tang and Aizenman, 1993a,b) (see ❷ *Sect. 4.2.1*), it is not surprising that thiol-reducing agents, for example, DTT and L-cysteine, may elicit epileptiform firing in hippocampal neurons (Tolliver and Pellmar, 1987) and long-term potentiation (LTP) (Tauck and Ashbeck, 1990), and aid glutamate in killing cultured retinal ganglion cells (Levy et al., 1990; Olney et al., 1971, 1990). On the other hand, cysteine and DTT have been shown to suppress the cytotoxicity evoked by 10 mM glutamate in a neuronal cell line (Miyamoto et al., 1989; Gilbert et al., 1991). Moreover, if GSH acts as a simple thiol-reducing agent, why does it prevent NMDA receptor-mediated neurotoxicity (Levy et al., 1991)?

We have therefore studied the effects of GSH, GSSG, GSA, and S-alkyl derivatives of glutathione on the influx of Ca^{2+} into cultured cerebellar granule cells induced by glutamate and its agonists (Janáky et al., 1993a,

1998; Jenei et al., 1998). For comparison, the possible interference of DTT and cysteine with the effects of GSH and GSSG was also subjected to study. GSH, GSSG, and the glutathione derivatives tested did not markedly affect the basal influx of $^{45}Ca^{2+}$ into cultured cerebellar granule cells, DTT alone being moderately activating. In agreement with some previous data (Lazarewicz et al., 1989; Levy et al., 1990; Sucher et al., 1990), DTT strongly potentiated the enhancement of Ca^{2+} influx by glutamate and NMDA. The kainate- and quisqualate-evoked influxes were significantly less enhanced. The higher the concentration of glutamate, the greater was the potentiation of the enhanced influx of Ca^{2+} by DTT. This finding indicates that in the activated receptor complexes DTT may preferentially act at accessible disulfide bonds, but probably not located within the binding site for glutamate (Reynolds et al., 1990). Such a synergism between endogenous glutamate agonists and DTT may readily lead to a lethal perturbation of Ca^{2+} influx into neural cells.

The results cited above show that the perturbation of neuronal Ca^{2+} influx can be attenuated by glutathione, although high (millimolar) concentrations of GSH may also slightly increase the influx evoked by millimolar (but not smaller) concentrations of glutamate. On the other hand, micromolar (i.e., the physiological extracellular) concentrations of GSH slightly reduce the influx evoked by 1 mM glutamate and are already sufficient to reduce significantly the potentiation by similar concentrations of DTT. GSSG, SMG, and SpeG inhibit the Ca^{2+} influxes evoked by glutamate. GSSG inhibits the enhancements by glutamate agonists. As predicted by binding studies, S-alkyl derivatives of glutathione have no effect on kainate-evoked Ca^{2+} influx (Jenei et al., 1998). At 0.5 mM concentration, GSH almost completely abolishes the DTT potentiation of the kainate-evoked and considerably attenuates that of the glutamate-, quisqualate-, and NMDA-evoked neuronal Ca^{2+} influxes. The enhancement of NMDA-stimulated Ca^{2+} influx by cysteine is likewise significantly attenuated by GSH: 28% and 23% at 1.0 and 0.1 mM NMDA, respectively (Janáky et al., 1993a). Similar inhibition by GSSG is predictable, but was not tested because of the obvious thiol–disulfide interaction with DTT molecules. Instead, we tested the effect of SMG which significantly attenuates the enhancement of NMDA-evoked responses by DTT and cysteine, 40% and 25%, respectively.

We conclude that GSH can bind to the recognition sites in the NMDA receptors as a γ-glutamyl compound (Varga et al., 1989, 1992; Janáky et al., 1998, 1999), displace glutamate, and hence diminish the number of accessible thiol groups. Such an inference is in keeping with the above results and the assumption that glutathione modulates glutamatergic neurotransmission by displacing glutamate from its binding sites (Oja et al., 1988; Varga et al., 1989; Janáky et al., 1998, 1999). GSH itself also causes a slight activation by reducing disulfide bonds in the vicinity of the γ-glutamyl moiety of the bound GSH molecule where it competes with other more effective disulfide-reducing agents. This is consistent with the finding that GSH mimics the DTT effects in NMDA receptors with NR1-NR2A (but not NR1-NR2B, C, or D) subunit composition, in which DTT rapidly potentiates glutamate-activated whole-cell currents and reduces the time course of desensitization and reactivation (Köhr et al., 1994). GSSG (Oja et al., 1988, 1994; Varga et al., 1989, 1997; Janáky et al., 1998, 1999) and S-alkyl derivatives of glutathione (Oja et al., 1995; Jenei et al., 1998; Janáky et al., 1998, 1999) may also displace glutamate and diminish receptor activation, but have no enhancing effects because of the lacking free SH group. The contribution of such a displacement of agonists to the inhibition of glutamate-evoked influx of Ca^{2+} into neurons is corroborated by the finding that DTNB, a sulfhydryl-oxidizing agent, has only a slight inhibitory effect (Janáky et al., 1993a). Nevertheless, GSSG may also oxidize neighboring thiol groups. This assumption is in agreement with the results of Gilbert and coworkers (1991) obtained on primary cultures of forebrain neurons. In their experiments, GSSG attenuated the NMDA-evoked increase in intracellular free Ca^{2+}. This effect was still apparent after oxidation of the cells by DTNB, but completely reversed only by addition of DTT. Nor can we exclude the possibility of intracellular regulatory actions of GSSG and GSH [e.g., interaction with the intracellular regulatory sites of ion channels (Gilbert et al., 1991; Ruppersberg et al., 1991).

Although the results on the use-dependent binding of dizocilpine may indicate an NMDA agonistic action of GSH, GSSG (Ogita et al., 1995; Varga et al., 1997) (discussed above in ❯ Sect. 4.2), and glutathione derivatives (Jenei et al., 1998), their effects on the neuronal influx and intraneuronal level of Ca^{2+} support rather a biphasic concentration-dependent action of GSH and an inhibitory action of GSSG and S-alkyl derivatives of glutathione. The activation of dizocilpine binding by glutathione derivatives may represent a change in the receptor conformation not necessarily involving any enhanced Ca^{2+} influx and intracellular elevation of free Ca^{2+}. We would be inclined to infer that the tested glutathione analogs are antagonistic in nature.

4.7 Interactions with the Ca^{2+}-Dependent Evoked Release of Other Neurotransmitters

Glutamate affects the release of several other neurotransmitters. This action may also be subject to modulation by glutathione. Although the mechanisms of the redox modulation have not been completely elucidated, the addition of DTT to cortical, hippocampal, and striatal slices results in an increase in the NMDA-evoked release of noradrenaline and dopamine. These effects are associated with an enhancement in the activity of the NMDA receptor through its redox modulatory sites (Woodward and Blair, 1991; Woodward, 1994). In this manner, glutathione may also interact with the effects of DTT by means of competition at the NMDA redox sites.

4.7.1 Modulation of the Glutamate-Evoked GABA Release

Both Ca^{2+}-dependent and seemingly Ca^{2+}-independent releases of GABA can be induced by glutamate and its agonists in the hippocampus (Janáky et al., 1993b) Activation of both ionotropic NMDA and non-NMDA receptors is involved in the transient Ca^{2+}-dependent release of GABA from hippocampal slices, which can contribute to physiological feedback and feedforward inhibition. Quisqualate evokes a sustained release of GABA via activation of metabotropic GluRs (Janáky, 1994; Janáky et al., 1994a), which can contribute to disinhibition and be involved in the generation of LTP.

DTT, GSH, and GSSG have no effect on the basal release of [^{3}H]GABA from hippocampal slices. Coadministration of DTT with glutamate agonists enhances and prolongs the evoked release. The effect of DTT is dose dependent. A 10-min pretreatment with 1 mM DTT followed by washing has no effect on the evoked release. The effect of DTT is also discernible in the absence of external Ca^{2+}. It is insensitive to 2-amino-3-phosphonopropionate, an NMDA antagonist, but blocked by MK-801 and Mg^{2+} (Janáky et al., 1994b). In contrast to DTT, GSH and GSSG do not enhance the release. Moreover, they generally attenuate the enhancement by DTT. GSH attenuates the release evoked by glutamate and glycine, and GSSG the release evoked by quisqualate, glycine, and NMDA together with glycine. The effect of DTT on the kainate-evoked release of GABA is not modified by GSH or GSSG (Janáky et al., 1994b).

4.7.2 Modulation of the Glutamate-Evoked Dopamine Release

The excitatory glutamatergic input from the cerebral cortex to the neostriatum (DiChiara and Morelli, 1993) controls dopamine release (Chéramy et al., 1994; Morari et al., 1996) in a complex manner. A triple regulation has been proposed to exist in vivo (Leviel et al., 1990). The cortical glutamatergic pathway exerts tonic inhibition on GABAergic interneurons mediated by activation of the NMDA receptors. Low concentrations of glutamate temporarily activate dopamine release via AMPA/kainate receptors at dopaminergic nerve endings, and high concentrations of glutamate evoke complex inhibition, possibly via GABAergic interneurons. An indirect facilitation mediated by somatostatinergic and cholinergic interneurons has also been proposed. The glutamatergic regulation of striatal dopamine release thus involves both direct and indirect facilitation and indirect inhibition by GluRs of different types of striatal interneurons and different receptors at nigrostriatal dopaminergic synaptic terminals (Leviel et al., 1990; Krebs et al., 1991; Chéramy et al., 1994; O'Connor et al., 1995b). Such a triple regulation forms the basis for the exact timing of dopamine release from the nigrostriatal nerve endings.

A variety of interactions of glutathione, both direct and indirect, with the striatal dopamine release regulated by glutamate are possible. In our in vitro studies glutamate, kainate, and AMPA evoked [^{3}H] dopamine release in a concentration- and time-dependent manner, kainate and AMPA being significantly more effective than glutamate. NMDA, glycine, and the metabotropic GluR agonist *trans*-1-aminocyclopentane-1,3-dicarboxylate (tACPD) were without effect in standard Krebs–Ringer–HEPES solution. γ-Glutamylcysteine, GSH, GSSG, DTT, DTNB, and L-cysteine were without any marked effect. Likewise, the antagonists of different ionotropic GluRs, for example, 6-cyano-7-nitroquinoxaline-2,3-dione (CNQX),

6,7-dinitroquinoxaline-2,3-dione (DNQX), 2,3-dioxo-6-nitro-1,2,3,4-tetrahydrobenzoyl[f]quinoxaline-7-sulfonamide (NBQX), D-(−)-2-amino-7-phosphonopentanoate (D-AP5), and MK-801 (dizocilpine) failed to affect the basal release (Janáky et al., 1997; Janáky et al., 2007). The release evoked by glutamate was strongly enhanced by GSH and less effectively by L-cysteine. GSSG, glycine, DTT, DTNB, and L-cystine had no effect. γ-Glutamylcysteine was a weak inhibitor (Janáky et al., 1997).

The slight enhancing effect (20%) of NMDA observed in 0.1 mM Mg^{2+} medium in the presence of 50 μM of glycine is not discernible in standard Krebs–Ringer–HEPES solution. This block is relieved in the presence of GSH, but not GSSG (R. Janáky et al., unpublished results). The release evoked by kainate is enhanced by GSSG, but only during the late stimulation phase by GSH. It is inhibited by CNQX and DNQX, NBQX being without effect (Janáky et al., 1997, 2007). The release evoked by AMPA is enhanced by GSSG, GSH being ineffective. tACPD fails to influence the release of dopamine under all conditions (Janáky et al., 1997, 2007).

The studies cited above indicate that striatal dopamine release is regulated by activation of both NMDA and non-NMDA classes of GluRs. This is in concert with both direct and indirect facilitation of dopaminergic nerve terminals or interneurons via GluRs. The results show that striatal dopamine release is enhanced by both reduced and oxidized glutathione, albeit probably by means of different mechanisms and mediated by different GluRs. GSH may act in several ways:

1. Directly via the NMDA receptors on dopamine nerve endings, enhancing the activity solely as a thiol reagent. This is less probable, because DTT fails to facilitate the NMDA effect.
2. Directly as an agonist at the non-NMDA receptors on dopamine nerve endings with a consequent relief of the Mg^{2+} block of NMDA receptors.
3. Indirectly as an agonist at the NMDA receptors on facilitatory somatostatinergic and cholinergic interneurons.
4. Indirectly as a γ-glutamyl competitive antagonist at the NMDA receptors on inhibitory GABAergic interneurons.

Although we cannot at present exclude any of the above mechanisms, the fact that S-ethyl, -propyl, and -butyl derivatives of glutathione are inhibitors of striatal dopamine release indicates the possible importance of both γ-glutamyl and cysteinyl moieties in the enhancement of glutamate-evoked dopamine release. On the other hand, the presence of the γ-glutamyl moiety is necessary for the agonistic action of GSSG exerted at the non-NMDA receptors on the dopaminergic terminals.

5 Glutathione as a Neurotransmitter

5.1 Presence and Characteristics of Glutathione Binding Sites

To evaluate the actions of GSH on its own putative receptors in the nervous system, our groups have used radiolabeled GSH in various preparations (Ogita and Yoneda, 1987, 1988, 1989; Ogita et al., 1988; Guo and Shaw, 1992; Guo et al., 1992; Lanius et al., 1993, 1994; Ogita et al., 1998; Janáky et al., 1999). Binding of [^{3}H]GSH was initially demonstrated in rat brain synaptic membranes by Ogita and Yoneda (1987). The specific binding sites for GSH consisted of at least two components with different affinities: (1) a temperature-independent site of a relatively "high" affinity (K_D = 560 nM) and (b) a temperature-dependent site of "low" affinity (K_D = 13 μM). Most studies since then have confirmed the presence of multiple affinity sites in mammalian CNS preparations. ❶ *Figure 15-3* shows the compounds that gave significant competition for [^{3}H]GSH binding. Note that these are all GSH or cysteine analogues. We discuss the action of these compounds further in ❶ *Sect. 5.2.1*.

In the CNS, the distribution of [^{3}H]GSH binding sites measured in homogenate assays varies for different CNS regions. The highest levels found in decreasing order are as follows: retina, hypothalamus, striatum, spinal cord, midbrain, medulla–pons, hippocampus, cerebellum, and cortex (Ogita and Yoneda, 1987). Regional subvariations revealed by receptor autoradiography show dense binding in retinal photoreceptor and pigment epithelium layers, dentate gyrus of the hippocampus, habenula, hypothalamus, spinal cord, and, depending on the part of the neocortex, either supragranular laminae or lamina 4 (Bains et al., 1997;

Shaw, 1998). At least some of these binding sites appear to be on astrocytes (Guo and Shaw, 1992; Guo et al., 1992). [^{3}H]GSH binding is also detectable in several peripheral tissues, including (in decreasing order) pituitary, adrenal gland, liver, spleen, skeletal muscle, and heart (Ogita et al., 1988).

5.2 Pharmacology of Glutathione Binding

5.2.1 Effects of GluR Ligands on Binding of Glutathione to Synaptic Membranes

As discussed in ● Sect. 4.2, GSH at micromolar concentrations is a ligand at all ionotropic GluRs and can displace glycine from the modulatory site of the NMDA receptor (Janáky et al., 1998). To establish that GSH is not merely binding to glutamatergic receptors, our groups have explored the overlap between GSH and glutamate ligand binding using a great variety of glutamatergic compounds in competition studies of [^{3}H] GSH binding. Of the tested excitatory amino acid analogues and mixed GluR ligands, only L-glutamate and 5-oxoproline demonstrated any competition for [^{3}H]GSH binding and then only in the porcine cortex.

◘ Figure 15-3
Competition curves for various ligands bound to the GSH-binding sites in rat cortical membranes. Membranes were incubated with 2 nM [^{3}H]GSH for 150 min at 37°C. Nonspecific binding was defined in the presence of 1 mM GSH and represented ~5% of total binding. (a) Competition curves are shown for GSH and GSH analogues C-G: cysteinylglycine. Note the similar response profile for all tested compounds and the lower IC$_{50}$ for GSSG (b) Competition with NMDA and various NMDA agonists. Note that with the exception of L-cysteine, most GluR agonists were without significant effect. Data from Pasqualotto et al. (1998) and C. A. Shaw et al. (unpublished data). It is important to note here that the GSH-binding data cited above can be used to distinguish potential binding and nonreceptor sites, e.g., Na$^+$-dependent and -independent transporters for glutathione (or other amino acids or peptides) or various enzymes such as the glutathione S- and γ-glutamyl transferases. Evidence for this is as follows: (a) The binding studies cited above used Tris-acetate buffers, thus excluding ion-dependent transport systems as potential GSH-binding sites; (b) competition studies using inhibitors of GSTs have shown little effect on [^{3}H]GSH binding (Pasqualotto et al., 1998); (c) competition with different substrates of γ-glutamyl transferase has shown to have little effect (Janáky et al., 1999); and (d) γ-glutamyl dipeptides are unable to compete with [^{3}H]GSH binding and have no effect on neural responses (see ● Sect. 5.3) (c) Field potential recordings in the cortical "wedge." Field potentials were recorded as described elsewhere (Harrison and Simmonds, 1985; Shaw et al., 1996) in "wedges" of the adult rat cortex. Here, depolarization is shown as an upward deflection. *Left traces:* Responses to NMDA, GSH, and cysteine. Concentrations (μM) are indicated above each trace. *Center traces*: Effect of coadministration of these compounds (same concentration as in left traces) in the presence of the NMDA antagonist L-AP5 (10 μM). Note that NMDA and cysteine responses are largely suppressed, whereas that of GSH is unaffected. These data support the view that the cysteine response occurs by activation of NMDA receptors, whereas that of GSH does not. *Right traces*: Response to a prolonged "pulse" of GSH. During the GSH-induced depolarization, GABA is applied (50 μM) in concert for 2 min and induces a negative-going potential shift, which is reversible. These data demonstrate that the field potentials to GSH originate in inward-directed currents that can partially be reversed by neural inhibition (Pasqualotto et al., 1998; C. A. Shaw et al., unpublished data) (d) Role of cysteine in the transition from NMDA-like to GSH-like activity. NMDA and various cysteine or GSH analogues were tested in the cortical wedge preparation for their responses alone and in the presence of L-AP5. For each compound, response amplitude in the two conditions was measured and plotted as 1 (peak amplitude of response in the presence of L-AP5/peak amplitude of response alone) × 100 to give the percent L-AP5 inhibition of response. Each compound was tested in several slices ($n = 3$–15). SE values are not shown because the point to be made here is qualitative rather than quantitative. Note that whereas the NMDA response is almost completely inhibited by L-AP5, that of GSH, GSHMEE, and the two cysteine-containing dipeptides (cysteinylglycine (C-G) and glutamycysteine (G-C)) were unaffected. L-Cysteine itself and other cysteine-containing compounds showed partial inhibition by L-AP5 (C. A. Shaw et al., unpublished data)

Glutathione in the nervous system | 15 | 375

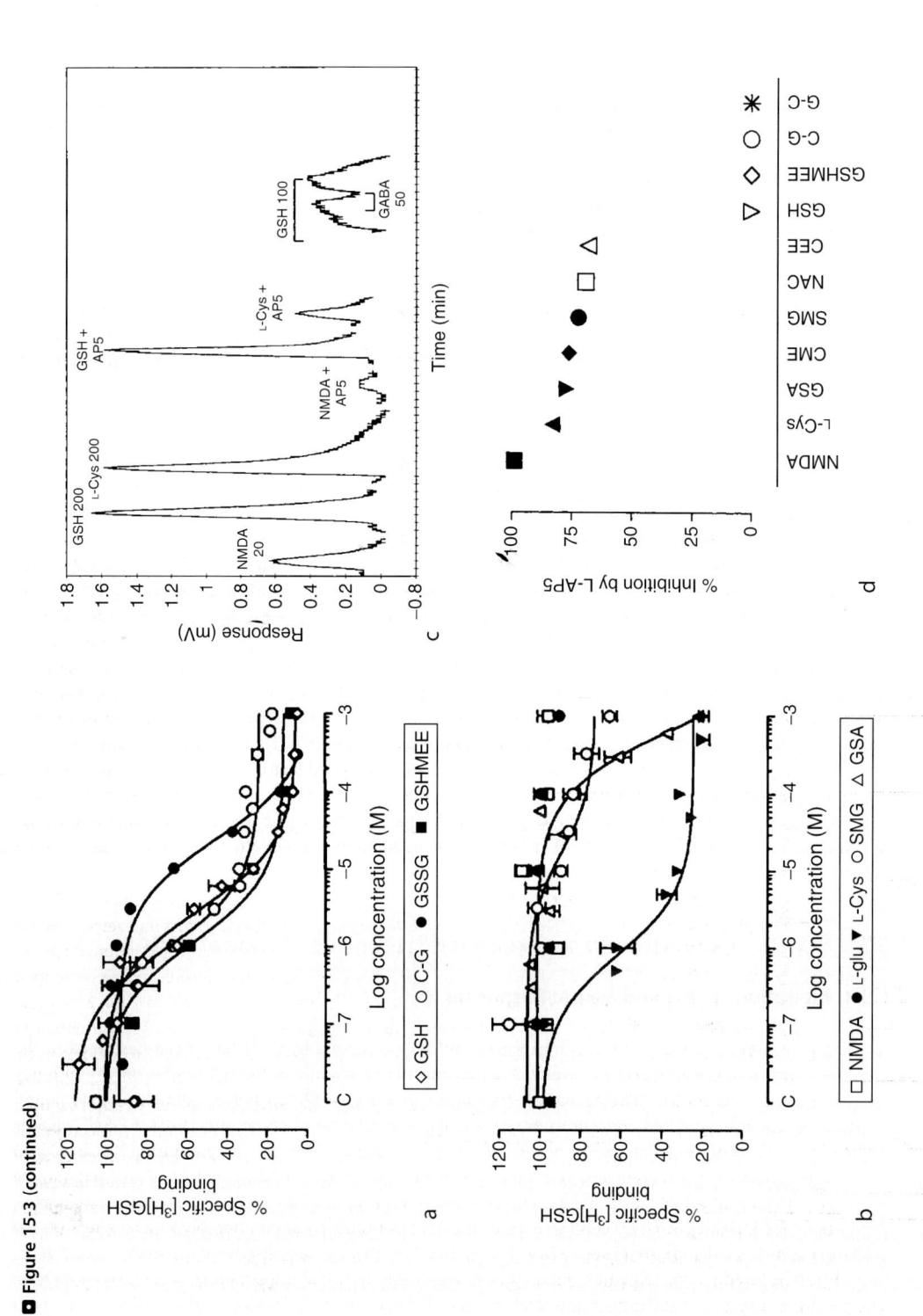

■ Figure 15-3 (continued)

Most NMDA ligands were inactive, with the exception of quinolinate. Glycine and ligands of the NMDA–glycine coactivator site were inactive, except for a slight inhibition by D-serine. Most non-NMDA ligands had no effect, two exceptions being NBQX and fluorowillardiine. Metabotropic glutamate ligands had no effect except L-(+)-2-amino-3-phosphonopropionate (L-AP3). In general, GluR agonists and antagonists had little, if any, effect on [^{3}H]GSH binding. Only thiokynurenate, which can be designated as a cysteine derivative, had a large effect, strongly potentiating binding. (In recording experiments, thiokynurenate greatly potentiated GSH-evoked depolarization.)

These data are striking in that they are quite consistent across our three laboratories, independent of species and, to some extent, binding conditions. Clearly, the characteristics of the binding sites labeled by [^{3}H]GSH do not resemble those of any known ionotropic GluR or mGluR population.

5.2.2 Effects of Thiol-Containing Compounds

We have examined extensively the effects of thiol-containing compounds on [^{3}H]GSH binding and neural responses. The latter is discussed in ◐ *Sect. 5.3.1*. Glutathione, glutathione derivatives, cysteine, cysteamine, and cystamine were the most potent displacers in the pig cortex and in the rat cortex, where dipeptides containing cysteine were also tested. Cysteinylglycine and glutamylcysteine were, along with GSH, the most effective compounds. Replacing either glutamate with aspartate (aspartylcysteine) or glycine with alanine (cysteinylalanine) did not diminish the potency of the dipeptides (Pasqualotto et al., 1998). These results, in addition to the observation that only a dipeptide missing cysteine (glutamylglycine) was unable to displace [^{3}H]GSH binding, lead us to conclude that the cysteine moiety of the GSH molecule is crucial for binding to the putative GSH receptor. We note, however, that the action of cysteine depends on the concentrations of [^{3}H]GSH studied: when the latter are low (<100 nM) cysteine competes for binding (Pasqualotto et al., 1998; Janáky et al., 1999); but above this concentration cysteine potentiates binding (Ogita and Yoneda, 1989).

These data concerning the necessity for the cysteine moiety were further confirmed in electrophysiological studies using tripeptides in which the cysteine was replaced with several neutral amino acids. Such tripeptides showed no response over a wide range of concentrations (C. A. Shaw et al., unpublished data). Because DTT, as a thiol-reducing agent, has only a limited effect on [^{3}H]GSH binding, we believe that another functional group, rather than the free thiol group, is the crucial factor. As oxidation or alkylation of the free thiol group of the GSH molecule diminishes but does not abolish binding, the cysteine moiety itself, and not only the free thiol group, is crucial in GSH binding. The strong activation of [^{3}H]GSH binding by thiokynurenate may indicate that the GSH receptor protein contains a modulatory site to which co-agonists may bind allosterically to activate GSH binding.

5.3 Effects of Glutathione on Response Properties of Neural Cells

5.3.1 Alterations in Evoked Neural Responses

As discussed in ◐ *Sect. 4.2*, redox modifications of NMDA receptors by GSH have been described in the literature. Such redox effects occur at high GSH concentrations (usually in the millimolar range) and may be an important means for altering receptor function in various circumstances. Which, if any, neural response occurs at lower GSH concentrations, particularly in relation to apparently distinct GSH receptor populations? To examine this question we have used a cortical "wedge" preparation (Harrison and Simmonds, 1985) to measure neural response to GSH in the cortex of adult male rats (for details, see Shaw et al., 1996; Pasqualotto et al., 1998). This method is a variation of the grease gap technique which has been widely used in drug discovery and structure–activity relationship studies, notably to screen ligands for excitatory amino acid receptors (see Curry, 1998). It offers the significant advantage that various compounds can be rapidly screened and compared with each other to reveal important structural relationships. The key disadvantage is that the field potentials so recorded can give no information about which cell type is responding and can give only limited information about membrane channels activated by any particular compound We note, however, that the first of these is not necessarily a disadvantage when beginning a survey of previously untested compounds.

In various articles (Shaw et al., 1996; Pasqualotto et al., 1998) we have demonstrated that bath-applied GSH gives a dose-dependent GSH depolarizing potential in the rat cortex (EC_{50} = 4.86 mM). The depolarizing GSH response resembles in time of onset and offset (● *Figure 15-3c*) those evoked by AMPA or NMDA. Antagonists to neither AMPA nor NMDA receptors, DNQX and L-AP5, respectively, were able to diminish the GSH response, nor was it diminished by removal of all external Ca^{2+}. Removal of all Na^+ in the bathing medium reversibly eliminated the GSH-evoked response. Note that in all cases, we have taken care to ensure that the responses observed are not due to electrode redox potentials, e.g., they are not observed in "dead" wedges. In addition, a GSH depolarization during long GSH exposure (~10 min) can be partially reversed by co-application of the inhibitory neurotransmitter GABA. These data suggest that GSH is binding to a site (or sites) that gates membrane depolarization by Na^+.

As stated above, the response to applied GSH does not appear to depend on activation of NMDA channels because L-AP5 and MK-801 were unable to block the GSH response over any part of its response range. These data, taken together with the binding studies cited above in ● *Sect. 5.1*, suggest that GSH binds at two major sites: (1) at its own population of receptor(s) activating an inward Na^+ current and (2) as a redox modulator at the NMDA receptor.

In regard to this last point, it is of interest to compare the responses to GSH, GSH analogues, cysteine, and cysteine analogues. Excitatory field potentials elicited by cysteine, like those after NMDA, were mostly blocked by addition of L-AP5 (● *Figure 15-3d*). However, GSH responses were not affected, nor are responses to GSH monoethyl ester (GSHMEE). In contrast, *N*-acetylcysteine (NAC), cysteine monoethyl ester (CEE), and cysteine methyl ester (CME) responses showed significant inhibition. Dipeptides derived from GSH such as glutamylcysteine and cysteinylglycine as well as other dipeptides, like cysteinylalanine and aspartylcysteine, are more like those of GSH in that they are not inhibited by L-AP5. GSA and SMG responses are like that of cysteine in their sensitivity to inhibition by L-AP5. Non-cysteine-containing dipeptides (glutamylglycine) or tripeptides incorporating neutral amino acids in place of cysteine gave no depolarizing response. These data reveal a dependence on the cysteine moiety for eliciting excitatory activity at the GSH site, i.e., in keeping with the dependence for binding described above in ● *Sect. 5.1*. Nevertheless, the present data reveal clear differences between "GSH-like" versus "cysteine-like" compounds (see also cysteine "transition" and evolution of excitatory receptor populations, below).

To obtain more precise information on GSH action at a neural level, we have also performed several preliminary voltage-clamp experiments on neurons from the rat somatosensory or frontal cortex. As with the data cited above, single-cell responses to GSH show depolarization as well as an increase in spiking. The ionic dependence of these single-cell responses is, at present, not resolved.

5.3.2 Alterations in Neurotransmitter Release and Uptake

Werman and coworkers (1971) described a GSH/GSSG-mediated release of the neurotransmitter acetylcholine at the neuromuscular junction. More recently, various studies have reexamined the possibility that GSH or GSSG can affect neurotransmitter release from neurons. Both GSH and cysteine evoke the release of glutamate from cultured cerebellar granule cells (cited in Janáky et al., 1998). These same investigators have also described an enhancement of both glutamate- and NMDA-induced striatal dopamine release by GSH. In contrast, GSH inhibits potassium-evoked dopamine release in the same preparation.

GSH and GSSG can also inhibit the Cl^-/Ca^{2+}-dependent, temperature-sensitive binding of glutamate in a crude synaptic membrane preparation (Varga et al., 1994a, 1997), with the latter thought to represent glutamate transport (Pin et al., 1984). In synaptosomes, however, GSH and GSSG enhance glutamate uptake (Varga et al., 1994a). Neither GSH nor DTT has an effect on the basal release of [^{3}H]GABA from hippocampal slices. However, coadministration of DTT with glutamate agonists enhances and prolongs the evoked release. The effect of DTT is dose dependent and blocked by MK-801 and Mg^{2+} (Janáky et al., 1999). In contrast to DTT, GSH and GSSG do not enhance the release of GABA and normally act to block

DTT-mediated release. These data suggest that in some cases GSH and GSSG may act to attenuate the potentially pathological release of some neurotransmitters by strong reducing agents. The possible involvement of GSH receptors in the presynaptic regulation of the release of other neurotransmitters, perhaps by the activation of presynaptic GSH autoreceptors, awaits future investigation.

5.4 Learning and Memory

Several proteins with relevant roles in normal brain function are targets for oxidative modifications that modulate their activity, for instance, the glutamatergic NMDA-type receptor (Tang and Aizenman, 1993), protein kinases (Klann and Thiels, 1999), and the transcription factors AP-1 and NF-κB (Tong et al., 1998). In this context, ROS are considered to act not only as deleterious agents but also as cellular messengers. As seen before, aside its roles in keeping the oxidant homeostasis within the cell, GSH may serve functions in intra- and intercellular signaling in the brain. It is noteworthy that genetic deficiencies in GSH-related enzymes lead to marked impairment of nervous system function (Cooper and Meister, 1992). In addition, GSH metabolism shows changes in animal models of neurodegeneration, including decreases in GSH content and alterations in the activity of key GSH-related enzymes, such as GSH peroxidase, GSH reductase, and GSH S-transferase. Some of these changes can be attenuated by neurotrophic treatment (Cruz-Aguado et al., 1998, 1999).

The induction of LTP in the synapse between the perforant pathway and the dentate gyrus granule cells, an electrophysiological model of synaptic plasticity and learning at cellular level, is impaired in GSH-depleted rats (Almaguer-Melian, 2000). Paired-pulse facilitation is also absent in the GSH-depleted animals, suggesting an impairment of short-term synaptic interactions. These findings indicate that a low content of glutathione can impair short-term and long-term mechanisms of synaptic plasticity. In accordance with these evidences, GSH has been found to be crucial for short-term spatial memory, but not inhibitory (avoidance) memory in rats (Cruz-Aguado et al., 2001). The idea of GSH playing a role in learning and memory processes is also supported by the fact that various pharmacological treatments known to enhance cognitive function also induce an increase in GSH levels (Kim et al., 2005; Kiray et al., 2005; Meng et al., 2005; Feng et al., 2006).

6 Glutathione in Neuropathology

As seen before, GSH serves multiple roles in the nervous system. Free radicals can be generated by a variety of normal cellular processes. When their balance is lost, "oxidative stress processes" in neural cells arise. These include oxidative phosphorylation, breakdown products of neurotransmitters such as dopamine and serotonin, overactivation of neurons by calcium or excitatory amino acids, and β-amyloid production. Oxidative stress arising from free-radical formation affects the GSH/GSSG ratio of the cell as GSH is depleted to combat reactive oxygen and nitrogen radicals. The altered GSH status has been linked with normal aging as well as neurodegenerative diseases. On the other hand, GSH and different precursors and GSH derivatives provide neuroprotection against excitatory amino acid toxicity, as in stroke, ischemia, and epilepsy as well as in different neurodegenerative states such as Alzheimer's, Huntington's, and Parkinson's diseases, amyotrophic lateral sclerosis, the Guamian disease ALS–Parkinsonism–dementia complex, prion diseases (transmissible spongiform encephalopathies), AIDS dementia, different stress-related conditions like chronic fatigue syndrome (CFS), mood disorders, and in schizophrenia (for an earlier comprehensive review see Bains and Shaw, 1997) (see ❷ *Sect. 6.6*).

In 1979, Sinet, Lejeune, and Jerome (Sinet et al., 1979) reported a correlation of 0.58 between IQ and erythrocyte GSHPx activity. This original finding generated interest in showing a correlation between the activity and polymorphism of several glutathione-metabolizing enzymes and general IQ as well as in relative incidence of dyslexia and a determination of onset time of different neuropsychiatric diseases with accelerated cognitive decline (e.g., Parkinsonism and Alzheimer's disease). Later all known polymorphisms of GSHPxs and glutathione transferases were tested for any relationship with IQ by five independent research groups and no association was found. (These negative findings were presented in personal communications but were not published.) In 2003, Li and coworkers published a study showing that a single gene (glutathione S-transferase omega-1 or GSTO1) influences the age at which individuals show symptoms of Alzheimer's and Parkinson's diseases. Later, they confirmed their data using gene chips to test probes for 22,000 human gene loci. After filtering and analysis at several levels, a highly significant allelic association was found for the age-at-onset effects in Alzheimer's and Parkinson's diseases for GSTO1 and GSTO2. Such single nucleotide polymorphisms like Ala140Asp, evidenced in GSTO1, may represent therapeutic intervention targets in the above diseases. On the other hand, considering the number of influencing factors (e.g., social status and education) we do not endorse potentially racist hypotheses of some authors (Weiss, 2000).

6.1 Brain Aging

Glutathione maintains the cellular redox potential and has a crucial role in preventing oxidative damage, preserving protein thiol groups, stabilizing cell membranes, and detoxifying xenobiotics (see ❷ *Sect. 3.1*). Because of its function as a cofactor for GSHPx, it is also involved in the detoxification of H_2O_2 and lipid peroxides. Since these functions are implicated in health and longevity, the level of GSH was suggested to affect the health of individuals at molecular, cellular and organ levels. Early studies supported this suggestion: glutathione declined with aging due to decreased synthesis in rodents and mosquitoes (Hazelton and Lang, 1980, 1983); correction of GSH deficiency by administration of its precursors enhanced longevity in mosquitoes (Richie et al., 1987). It was suggested that the blood (e.g., erythrocyte) GSH content could be indicative of the overall glutathione status. Studies following the age-related changes of glutathione in human blood, however, resulted in mixed, inconclusive findings. As a whole, an association of high glutathione levels with good health in aging has been observed (Richie and Lang, 1986; Lang et al., 1992; Julius et al., 1994). It is obvious, however, that the findings obtained in such a complex compartment as serum or blood must be considered with great caution, as well as different species, strains brain regions, cellular and subcellular preparations, methods of assay, etc., must be taken into consideration when examining the total antioxidant capacity of the brain.

Despite conflicting results owing to the above, some conclusions can be drawn. (1) The level of GSH decreases with aging in rodents (Benzi et al., 1988; Chen et al., 1989; Ravindranath et al., 1989; Adams et al., 1993; Desole et al., 1993; Iantomasi et al., 1993). GSSG was generally increased proportionally to GSH decline, and thus the glutathione redox index decreases as well (Benzi et al., 1989a,b; Adams et al., 1993; Iantomasi et al., 1993). (2) Brain mitochondrial total glutathione increased (Martínez et al., 1994) or was not altered in rodents (de la Asuncion et al., 1996) while GSSG was accumulated, resulting in an 110% and 500% increment in mice and rats, respectively, in the GSSG/GSH ratio (de la Asuncion et al., 1996). The rise in the GSSG/GSH ratio correlated with that in the amount of 8-hydroxy-2′-deoxyguanosine (a biomarker of oxidative mitochondrial DNA damage). Moreover, oral antioxidant treatment protected against both glutathione oxidation and mitochondrial DNA damage. We can thus conclude that in spite of significant decreases in cytosolic total glutathione, mitochondria are able to maintain their GSH level, but GSSG can accumulate, which can result in damage of key molecules like DNA. This, in turn, underlines the importance of the GSSG/GSH ratio in determination of cellular health and survival. (3) The brain GSHPx activity does not change with aging in rodents and the GSHPx gene expression is unaffected by aging (Rao et al., 1990; de Haan et al., 1992, 1995). Thus, in the aged brain, enhanced Cu/Zn SOD activity and expression with constant GSHPx activity leads to accumulation of H_2O_2, which may explain the increased lipid peroxidation, a clear index of oxidative stress (Cristiano et al., 1995; de Haan et al., 1995). (4) The activity of glutathione reductase is lower in the brain of old than of adult rodents and the redox index of the glutathione system declines in senescence and the cortex and striatum are affected earlier than other brain structures. The activities of the pentose phosphate pathway, glucose-6-phosphate reductase, and 6-phosphogluconate dehydrogenase are unchanged or slightly increased by aging. Thus, while NADPH is provided to the same extent, reduction of GSSG to GSH decreases in the aging brain (Benzi et al., 1989b). Interestingly, α-tocopherol or propyl gallate feeding increased the GSSG reductase activity, increasing the mean animal life span (Sohal et al., 1984; Harman, 1995) and improving the cognitive performance of aged rats (Socci et al., 1995).

The above findings indicate that with increasing aging the brain becomes less capable of protecting itself against oxidative stress as shown by the increased GSSG/GSH ratio and the decreased glutathione redox index. The oxidation of GSH may result in neuronal damage and death owing to the low GSH levels. The effects of different oxidant stressors are also potentiated in the aged brain. Thus, effects of H_2O_2 and the GSH-depleting agent, cyclohexene-1-one, were more expressed in the aged brain (Benzi et al., 1989, 1991) and the GSH-depleting effect of the dopaminergic neurotoxin, 1-methyl-4-phenyl-1,2,3,6-tetrahydropyridine (MPTP), appeared only in the aged brain (Desole et al., 1993).

6.2 Parkinson's Disease

In 1817, James Parkinson published his *Essay on the Shaking Palsy* in which he described a disorder characterized by three cardinal symptoms: resting tremor, akinesia, and rigidity. The disorder is now known as Parkinson's disease (PD) and is the second most common neurodegenerative disease of the brain after Alzheimer's disease, affecting millions of people worldwide. The prevalence of PD increases dramatically with age and becomes highest in the seventh and eighth decade of life (Schoenberg, 1986).

The diagnostic hallmark of idiopathic PD at autopsy is the frequent presence of Lewy bodies (Gibb and Lees, 1988; Gibb, 1989; Lowe, 1994). Lewy bodies are cytoplasmic inclusions present principally in neurons of the SN and of pigmented noradrenergic cell bodies in the locus ceruleus (LC) of patients with PD (Forno, 1986). These inclusions are composed of a filamentous mesh of neurofilaments that surrounds a dense core. At the cellular level, the major neuropathological feature of PD is severe damage to the neuromelanin-pigmented cell bodies of nigrostriatal dopaminergic neurons in the SN pars compacta. This damage is evidenced by the depigmentation of the parkinsonian SN, loss of neuromelanin-containing tyrosine hydroxylase-immunoreactive neurons, and severe depletion of the catecholaminergic neurotransmitter dopamine in the striatum (Hornykiewicz and Kish, 1986; Hornykiewicz, 1988; Jellinger, 1988, 1990, 1991).

Neuromelanin is produced by the oxidation and polymerization of dopamine, a process in which free-radical species are released. In addition, the catabolism of dopamine by monoamine oxidase also renders ROS. These peculiarities of dopamine metabolism led to the concept of the involvement of oxidative stress in PD which was fully supported by postmortem findings: decreased GSH levels, increased iron concentration, and impaired mitochondrial complex I activity in the SN (Sofic et al., 1992; Schapira et al., 1990; Riederer et al., 1989). Changes in other oxidative stress markers have been also reported (e.g., SOD and GSH peroxidase) (Sian et al., 1994; Yoritaka et al., 1997), but the validity of these findings still remains in controversy. However, the decreased GSH levels in the SN have been consistently demonstrated by both biochemical and immunohistochemical analyses, and it is widely accepted as an evidence of oxidative stress in PD SN. Remarkably, GSH content is also decreased in the SN of patients with incidental Lewy disease (ILD) (Dexter et al., 1994). If the claim of ILD as a form of presymptomatic PD holds true, then the GSH fall might be an early, and perhaps, causal event in the molecular etiopathological cascade of PD. The upstream mechanisms leading to GSH depletion in PD are unknown (reviewed by Jenner, 2003).

6.3 Alzheimer's Disease

Alzheimer's disease (AD) is the most common cause of age-related intellectual impairment. Early onset AD occurs before 65 years, late onset after that age. It is characterized by degeneration of neurons, especially pyramidal neurons in the hippocampus, entorhinal cortex, and neocortical areas with significant synaptic losses. There is also loss of cholinergic neurons in the median forebrain. Two major hallmarks of AD are extracellular deposits of β-amyloid protein in plaques and abnormal intracellular cytoskeletal filaments, the neurofibrillary tangles (NFT).

The amyloid β/A4 protein (Aβ) is a 39–43 amino acid peptide that forms the core of the senile plaques in the AD brain and is thought to play a crucial role in AD pathogenesis (reviewed by Yankner, 1996). Aβ is toxic to neurons in vitro (Yankner et al., 1989) and kills neuronal cultures when aggregated (Pike et al., 1993; Lorenzo and Yankner, 1994). Although the exact mechanisms of Aβ neurotoxicity are not known, it is suggested that Aβ acts both directly via generation of oxidative stress and indirectly through activation of microglia (Yan et al., 1996). These authors identified a receptor for advanced glycation end products (AGEs) that mediates the effects of the β-amyloid peptide on neurons and microglia. The increased expression of these receptors may be relevant to the pathogenesis of neuronal dysfunction and death in AD.

Exposure to Aβ increases neuronal H_2O_2 production (Behl et al., 1992, 1994) and induces peroxidation of membrane lipids. This leads to an influx of Ca^{2+} and activation of Ca^{2+}-dependent degradative enzymes (e.g., phospholipases, proteases and endonucleases) (Okabe et al., 1989; Orrenius et al., 1992). Oxidation of tau, the main component of intracellular NFT, induces dimerization and polymerization of the protein into filaments. The NFT-bearing neurons accumulate iron, a potent catalyst of oxyradical generation, and aluminium (Good et al., 1992). Aluminium salts can induce enhanced release of iron (Gutteridge et al., 1985) which in turn induces lipid peroxidation (Ohyashiki et al., 1993; Oteiza, 1994). Free radicals, released from activated microglia (Colton and Gilbert, 1987) in the presence of iron and aluminium, evoke the aggregation of Aβ and formation of AD plaques (Dyrks et al., 1992). There is a significant decrease in cytochrome oxidase complexes I and III, II, and IV of the electron transport chain in mitochondria of AD platelets and brain (Parker et al., 1990; Kish et al., 1992; Simonian and Hyman, 1993; Chandrasekaran et al., 1994; Mutisya et al., 1994; Parker et al., 1994).

Although Aβ toxicity is directly related to oxidative stress and free-radical damage, little is known about the status of antioxidant defenses in AD. Two findings directed attention toward the possible role of GSH status in the pathogenesis of AD: first, Aβ-caused neuronal damage could be prevented by the free-radical scavenger, vitamin E (Behl et al., 1992); second, both the prevalence of AD and the brain GSH status are strictly age dependent. For this reason, many attempts have been made to link AD prevalence and the cerebral or blood GSH status, but the results are contradictory. The total brain levels of GSH appeared to be unaffected in AD in some studies (Perry et al., 1987; Balazs and Leon, 1994), while in others (Makar et al., 1995) an increase in GSH with aging in the AD brains (but not with normal aging) was found. A compensatory increase of GSH and GSSG was shown in the studies of Adams and Odunze, 1991 and Adams et al. (1993) but the GSH/GSSG ratio was lower in the hippocampus (−39%) and caudate nucleus (−20%) of AD patients than in age-matched healthy subjects. GSH peroxidase and reductase activities were elevated in different brain regions (Lovell et al., 1995). This may also represent a compensatory mechanism. In serum, GSHPx activity was decreased (Jeandel et al., 1989; Perrin et al., 1990) or increased (Ceballos-Picot et al., 1996) in AD.

Interestingly, both GSSG and the thiyl radical GS· inhibit the binding of a specific ligand 3-quinuclidinyl benzilate ([^{3}H]QNB) to muscarinic cholinergic receptors (Frey et al., 1996). Since cholinergic transmission is crucial to memory and learning and its alteration is considered as one of the main causes of cognitive disorders in AD, this finding may have clinical relevance.

6.4 Amyotrophic Lateral Sclerosis

6.4.1 Etiology and Pathogenesis

Amyotrophic lateral sclerosis (Lou Gehrig's disease, ALS) is a degenerative disease of motor neurons. The disease is lethal; the median age of onset is 55, and the median survival is 3–5 years. Only 10%–15% of the cases are inherited as an autosomal dominant trait and in 20% of these cases the primary defects are mutations in the gene encoding the cytosolic copper–zinc SOD1 (Rosen et al., 1993). Thus, motor neuron death is thought to be the consequence of perturbed free-radical generation and oxidative toxicity to motor neurons. Indeed, there are in vitro and in vivo evidences of increased free-radical production in ALS.

SOD1 is a metalloenzyme of about 153 amino acids and its function is to convert the superoxide anion ($O_2^{·-}$) to hydrogen peroxide (H_2O_2) (Halliwell et al., 1995). Hydrogen peroxide is then detoxified by cytosolic GSHPx or peroxisomal catalase to form water. In the case of SOD1 mutation and deficiency accumulated superoxide anions are highly toxic in many ways (see ❷ Sect. 3). Moreover, mutated SOD1 has not only reduced activity but also gains cytotoxic functions, although the mechanisms of neurotoxicity are not known and may include an increased accessibility of the active-site copper to H_2O_2, leading to increased OH· formation (Wiedau-Pazos et al., 1996; Yim et al., 1996) or to peroxynitrite, leading to nitration of critical tyrosine residues in proteins essential to neuronal viability (Beckman et al., 1990, 1993; Ischiropoulos et al., 1992). To examine the gained cytotoxic activity of the mutant SOD1 molecules, Roos and colleagues (1997) expressed mutant SOD1 genes into primary neurons and observed apoptotic cell death which could be prevented by copper chelators, SOD mimics, enhancers of catalase, and agents known to have an antiapoptotic effect. Further evidence for free-radical toxicity in ALS is the toxicity of the CSF of ALS patients in neuronal culture (Couratier et al., 1993). This toxicity is blocked by antioxidant drugs, including vitamin E (Terro et al., 1996).

Besides the presence of mutant SOD1 molecules with deficient or altered activity in familial ALS, oxidative stress can also be explained by significant reduction in GSHPx activity in sporadic ALS. It was suggested that reduction in GSHPx activity may influence even the rate of disease progression (Przedborski et al., 1996), although normal or even elevated GSHPx activities have also been measured in the spinal cord of ALS patients (Ince et al., 1994; Fujita et al., 1996). The CSF GSH level seems not to be altered in ALS (Cudkowicz et al., 1998), but glutathione binding increases significantly (by 16%) in both the dorsal and ventral horns of cervical spinal cords from sporadic ALS patients. Saturation binding studies indicated that the increase in GSH binding is due to an increase in glutathione receptor numbers and not affinity (Lanius et al., 1993).

It seems that by measuring red blood cell SOD1 and GSHPx activities, a given biochemical difference can be made between familial and sporadic ALS patients: generally, red blood cell SOD1 activity is reduced only in familiar ALS with mutated SOD1 while GSHPx activity is reduced only in sporadic ALS patients (Mitchell et al., 1993; Przedborski et al., 1996). A number of indirect biochemical studies in vivo (reviewed in Cudkowicz et al., 1998) exist that underlie the crucial role of neuronal oxidative stress in the development of ALS. Transgenic ALS mice that express a gene-encoding mutant SOD1 protein develop motor neuron degeneration similar to that of the human disease (Gurney et al., 1994). The onset of symptoms is seen at a mean age of 95 days with a mean survival of 134 days. Administration of vitamin E to the mutant SOD mice produced a significant delay (13 days) in time to disease onset without an effect on survival (Gurney et al., 1994). Riluzole and gabapentin, two antiglutamatergic agents, prolonged survival without an effect on the onset of symptoms (Gurney et al., 1996). No similar experiments have been carried out with administration of GSH or GSH derivatives.

6.4.2 Therapy

Antioxidant therapies were among the first to be advocated for the treatment of ALS. The lipid-soluble vitamin E or α-tocopherol was tested first, already in the 1940s (Wechsler, 1940). Because of its lipophilic nature, vitamin E exerts its principal neuroprotective effects at cellular membranes where free radicals generate a tocopheroxyl radical which can be regenerated in the presence of reduced glutathione or possibly vitamin C. Indeed, vitamin E has been reported to arrest the course of ALS (Wechsler, 1940; Quick and Greer, 1967). Along with vitamin C, glutathione contributes to the reducing environment of the cell. GSH exerts its antioxidant activity partly by generating reduced forms of vitamins C and E (see ❯ Sect. 3). For this reason GSH represents a potential treatment target, but owing to the poor intracellular uptake of GSH, it is necessary to utilize other drugs as vehicles to deliver cysteine. NAC, which has been shown to exhibit neuroprotective effects, could be one such possibility. Indeed, orally administered N-acetylcysteine (NAC) to Wobbler mutant mice with muscle degeneration spared motor neurons and diminished changes in the axon diameters of the facial nerve. In addition to this, treatment with NAC enhanced the levels of GSHPx in cervical spinal cord (Henderson et al., 1996). On the other hand, in a placebo-controlled treatment trial of ALS with NAC, where the primary endpoints were the survival and rate of progression (determined by manual muscle testing, forced vital capacity, and the activities of daily living), the observed prolonged survival was not significant and no other significant differences in other outcome measures were observed (Louwerse et al., 1995). However, we should note that different activities of the GSH molecule can be attributed to different constitutive amino acid side chains (Hermann et al., 2000; Janáky et al., 2000), therefore the potential therapeutic activities of different GSH analogues may diverge from that of the GSH precursor NAC. In this respect, we emphasize the potential role of GSH and GSH derivatives in displacing glutamate and aspartate from their binding sites. Although the etiology for sporadic ALS is not known, circumstantial evidence supports the involvement of excitotoxicity. The levels of glutamate and aspartate are elevated in the CSF and glutamate uptake mechanisms are impaired in ALS patients (Rothstein, 1995). Moreover, in the explanted rat spinal cord glutamate selectively destroys motor neurons (Rothstein and Kunel, 1995). This is prevented by the administration of compounds that moderate the release or binding of glutamate or that downregulate the GluR. Antioxidants such as NAC were also protective. Other glutamate-based therapies have also been studied in ALS patients. Gabapentin, an antiepileptic drug that inhibits the release of glutamate, seems to decrease the motor deterioration (Miller et al., 1996). Riluzole, another drug that inhibits glutamate release, slows the progression of weakness in transgenic mice that carry the mutant SOD1 gene (Gurney et al., 1996). Although riluzole did not slow motor deterioration in ALS patients in a multinational controlled trial, the drug increased the probability of survival. The effect was modest, adding about three months to life expectancy (Lacomblez et al., 1996). Similar studies are needed to test the efficacies of GSH derivatives and prodrugs.

6.5 Schizophrenia

Schizophrenia is an endogenous psychosis characterized by positive symptoms, such as delusions, hallucinations, thought disorder, and incoherence, and negative ones such as deficits in cognitive and social abilities and motivation, and poverty of speech accompanied by a loss of emotional content.

Recent neuroimaging studies have shown structural abnormalities in the cerebral cortex (temporal, entorhinal, cingulated, and prefrontal) (Selemon et al., 1995). These cortical areas are richly innervated with dopamine terminals and play an essential role in working memory in primates (Goldman-Rakic et al., 1994). The subcortical regions of the temporal lobe, the basal ganglia, and the thalamus are affected as well. Functional imaging studies also suggest prefrontal cortical hypofunction and a functional impairment of the prefrontal–temporal–limbic cortical connectivity.

At present, two main biological hypotheses have been proposed to explain the pathophysiology of schizophrenia.
1. The dopamine theory proposes that schizophrenia is a manifestation of increased dopaminergic activity in certain brain areas (Matthysse, 1973). This interpretation is supported by the fact that most neuroleptics are antagonists of the dopamine receptors and psychotic symptoms can be induced by drugs that enhance dopaminergic activity such as amphetamine (Seeman et al., 1976). During the past few years, however, evidence has been accumulating that schizophrenic disorders may be characterized by both hypodopaminergia in the mesocortical dopamine neurons and hyperdopaminergia in the mesolimbic dopamine neurons (Davis et al., 1991). D_2-like dopamine receptors were reported to be elevated in postmortem brain tissue of schizophrenic patients (Seeman et al., 1990), supporting the hypothesis that dopaminergic neurotransmission is enhanced in schizophrenia through an overactive response by D_2 receptors. Recently, it was also suggested that the newly characterized D_4 receptors, to which clozapine, a potent neuroleptic, binds with high affinity (van Tol et al., 1992), are involved in the pathogenesis of schizophrenia because expression of these dopamine receptors is elevated in this disease (Seeman et al., 1993).
2. The second major hypothesis, as originally formulated by Kim and coworkers (1980), claimed a role for excitatory amino acids, i.e., glutamate, in the pathophysiology of schizophrenia. On the basis of decreased levels of glutamate in the CSF of schizophrenic patients, these authors postulated that either a hyperactivity of dopaminergic neurons leads to an enhanced inhibition of glutamate release or a hypofunction of GluRs causes the decreased glutamate release. On the basis of the schizophrenia-eliciting activity of the potent noncompetitive NMDA receptor antagonist phencyclidine (PCP), the current hypoglutamatergic theory of schizophrenia proposes diminished glutamatergic neurotransmission at the level of the NMDA receptors (Deutsch et al., 1989; Squires and Saederup, 1991). The glutamate hypothesis also gains support from the increased binding of glutamate to its receptors in postmortem brain tissue which might reflect receptor upregulation owing to decreased glutamatergic activity (Nishikawa et al., 1983; Deakin et al., 1989; Kornhuber et al., 1989; Eastwood et al., 1995).

3. With a very carefully planned, wide and elegant study (Do et al., 1995; Cuénod and Do, 1998), significantly decreased levels were observed for GSH (−42%, $p < 0.05$), its metabolite γ-glutamylglutamine (γ-Glu-Gln) (−16%, $p < 0.01$), and taurine (−15%, $p < 0.001$) in the CSF of drug-naïve schizophrenic patients as compared with age- and sex-matched healthy controls. CSF dopamine and serotonin metabolites were not different from controls and the concentration of glutamate was only slightly lower. The use of GSH, γ-Glu-Gln, glutamate, aspartate, taurine, and isoleucine as discriminating variables allowed Do and colleagues to classify 85% of the subjects correctly, with a specificity of 96%. The results suggest that at least some of these compounds are involved in the pathophysiology of schizophrenia. The above CSF GSH deficit may reflect a GSH deficit in brain areas involved in the pathology of schizophrenia and may originate either from a genetic deficit of enzymes involved in GSH metabolism or from a deficit of their activity (Cuénod and Do, 1998). Indeed, a decrease in the activity of GSHPx in platelets and erythrocytes was found in schizophrenia (Buckman et al., 1987, 1990). The deficit in the level of GSH could have many consequences which alone or in combination could result in functional and morphological changes and in symptoms of schizophrenia.

Since GSH plays a role in detoxification of oxygen radicals resulting from dopamine catabolism (Maker et al., 1981; Spina and Cohen, 1989), Cuénod and Do (1998) suggest that a GSH deficit during development has drastic consequences leading to accumulation of free radicals, subcellular degeneration, and abnormal connectivity of dopaminergic neurons. Moreover, the toxicity of the dopamine metabolites might be restricted to the microenvironment of the terminals of the dopamine fibers innervating the cortex (Goldman-Rakic et al., 1989), leading to the degeneration of spines and dendrites rather than of entire cell bodies. This process would combine the dopaminergic- and GSH-deficiency theories of schizophrenia. This assumption is underlined on one hand by the effectiveness of neuroleptics which not only block D_2 and D_4 receptors but also decrease DA levels with a delay of a few days, paralleling the delay needed for the treatment to be effective, and, on the other hand, by the finding of Shukitt-Hale and coworkers (1997) who were able to produce deficits in psychomotor behavior in rats with BSO-evoked GSH depletion followed by dopamine treatment. The following are speculations concerning the possible mechanisms by which GSH may play a role in schizophrenia:

1. A deficit in GSH would reduce the formation of the nitrosothiol GSNO (see above), thus decreasing the control of the toxicity of NO. Indeed, a reduced density of cells with NOS staining has been reported in the temporal lobe of patients with schizophrenia (Akbarian et al., 1993) which may indicate a permanent loss of such neurons exposed to high NO· levels.
2. GSH displaces excitatory amino acid receptor agonists from their binding sites (Ogita et al., 1995, 1998; Janáky et al., 1998, 2000) (see ● *Sect. 4.2*), and at millimolar concentrations it can also act at the redox modulatory site(s) in the NMDA receptor (Aizenman et al., 1989; for a review see Gozlan and Ben-Ari, 1995; Sucher et al., 1996) and may induce elevated Ca^{2+} influx into neurons (Janáky et al., 1993) and potentiate the L-glutamate-activated currents in cells expressing the NR1-NR2A NMDA receptor channels (Köhr et al., 1994). Because the intracellular concentration of GSH is in the millimolar range, the local concentration available in the synaptic microenvironment following release (Zängerle et al., 1992) can also be in this range under physiological circumstances. On the other hand, at a low level of GSH, the activation of the NMDA receptor might be deficient, while the neurotoxicity of excitatory amino acids may increase because of the loss of GSH-controlled binding of agonists. Such an inadequate activation of the NMDA receptors could be related to some of the symptoms of schizophrenia, as the PCPs induce psychopathological symptoms reminiscent of those observed in schizophrenic patients.
3. Yudkoff and coworkers (1990) reported that a partial depletion of astrocytic GSH evokes a significant reduction of intracellular glutamate. The GTP-mediated transmembrane transport of amino acids, particularly that of glutamine and methionine for which GTP has a particularly high affinity (Tate and Meister, 1974), will be deficient when the GSH availability is below a critical level. This will affect the glutamatergic presynaptic boutons, which rely on a supply of glutamine as precursor of glutamate. Such a decrease in glutamine transport in schizophrenia is supported by the finding of Do and coworkers (1995) who reported that the γ-glutamylglutamine level was decreased in the CSF of patients

with schizophrenia. Additional supporting evidence of this hypothesis is the finding that orally administered methionine which competes glutamine at the active centre of GTP, exacerbates the psychotic symptoms in schizophrenic patients (Park et al., 1965). Alternatively, GSH could be an important reservoir for glutamate (Yudkoff et al., 1990) and GSH depletion would result in glutamate deficit.

The above processes combine the glutamate- and GSH-deficiency hypotheses of schizophrenia because both deficiencies in neurotransmitter pools of the excitatory amino acid glutamate and deficient activation of NMDA receptors by insufficient levels of GSH would lead to a deficit in glutamatergic transmission. A deficit in GSH and GSH-related enzymes might thus play an essential role in the pathophysiology of schizophrenia or at least of some forms of schizophrenia. Such a deficit could form a link between the two main hypotheses created for its pathogenesis and leading to new approaches to its treatment.

6.6 Chronic Fatigue Syndrome

CFS is an illness characterized by the primary complaint of persisting or relapsing fatigue, more prevalent in women and affecting 7–267 cases per 10'000. The disease is more frequent in young adults (Dickinson, 1997) and is often accompanied by numerous neurological symptoms. These include impaired cognition, sleep disturbances, headaches, vestibular and visual disturbances, paresthesias, and gait abnormalities (Komaroff et al., 1996). Moreover, there is a significant overlap between CFS and mood disorders: two-third of CFS patients have been diagnosed with major depression.

There are many morphological abnormalities nuclear magnetic resonance (MRI) in the CNS possibly underlying neurological symptom. Studies have shown that CSF patients have white matter abnormalities more often than healthy volunteers. Subcortical areas of the frontal lobes can be involved and these MRI abnormalities may represent areas of edema or demyelination (Natelson et al., 1993; Lange et al., 1999). It was shown with single-photon emission tomography (SPET) that CFS patients have abnormal perfusion in several brain regions which may result in tissue damage and cellular disfunction (Schwartz et al., 1994). GSH depletion in CSF was first suggested by Droge and Holm (1997). It is now commonly accepted that oxidative stress plays a role in its development but is not determined whether GSH depletion is a cause or a result.

7 Conclusions and Perspectives

Glutathione is an essential molecule in all living cells and nowhere is this more obvious than in the nervous system of animals. In both simple and complex animals, GSH serves a variety of functions ranging from its traditional general roles as antioxidant and xenobiotic to highly specific actions in signal transduction. For the latter, GSH can act both as neuromodulator and as neurotransmitter in specific neuronal circuits. An understanding of the multiple roles of GSH in neural functions has begun to shed light on how any disturbance in GSH synthesis or function can have profound consequences for neural health and survival and it has become apparent that GSH deficits may be critically involved in many aspects of neurological disease.

In view of the colocalization of different GluRs with possible specific sites for glutathione, we are of the opinion that glutathione and its endogenous derivatives may be involved in neuronal information processing and provide a basis for the sharpening of neuronal responses to glutamate. At the same time, these molecules may offer defense against glutamate and cysteine toxicities. Other nontoxic glutathione derivatives, which readily cross the blood-brain barrier (BBB), may also prove to be antiepileptic and neuroprotective in stroke and neurodegenerative diseases involving pathological overexcitation of GluRs. Cloning and localization of glutathione receptors is thus an extremely pressing scientific goal. The possible interaction of glutathione with metabotropic GluRs and the involvement of second messengers other than Ca^{2+} in its action are also provocative questions still awaiting answers. The evaluation of interactions of glutathione as a neuromodulator with the release of other neurotransmitters, ontogenic and phylogenic follow-up studies of glutathione binding, and behavioral studies on glutathione actions should also be carried out.

The above questions are merely some of the myriad ones that must be answered before we have comprehensive understanding of the multiple and overlapping roles of GSH in the nervous system. It would be safe, however, to conclude that future research in this area will have profound implications both for design of new cognitive enhancers which lack neurotoxic properties and for future therapeutics designed to prevent or halt the course of neurological diseases.

Acknowledgments

This work was supported by the competitive funding of Pirkanmaa hospital district (9E112 and 9D099) to R. Janáky and NSERC Canada and the US Army Medical Research and Materiel Command (DAMD17-02-1-0678) to C.A. Shaw.

References

Adam V, Farago A, Machovich R, Mandl J, 1996. *Medical Biochemistry Textbook*. Adam V, editor. Budapest: Semmelweis kiadó; p. 192.

Adams JD Jr., Odunze IN. 1991. Biochemical mechanisms of 1-methyl-4-phenyl-1,2,3,6-tetrahydropyridine toxicity. Could oxidative stress be involved in the brain? Biochem Pharmacol 41: 1099-1105.

Adams JD Jr., Klaidman LK, Odunze IN, Shen HC, Miller CA. 1991. Alzheimer's and Parkinson's disease. Brain levels of glutathione, glutathione disulfide and vitamin E. Mol Chem Neuropathol 14: 213-226.

Adams JD Jr., Wang B, Klaidman LK, LeBel CP, Odunze IN, et al. 1993. New aspects of brain oxidative stress induced by *tert*-butylhydroperoxide. Free Radic Biol Med 15: 195-202.

Aizenman E, Potthoff WK. 1999. Lack of interaction between nitric oxide and the redox modulatory site of the NMDA receptor. Br J Pharmacol 126: 296-300.

Aizenman E, Lipton SA, Loring RH. 1989. Selective modulation of NMDA responses by reduction and oxidation. Neuron 2: 1257-1263.

Aizenman E, Hartnett KA, Reynolds IJ. 1990. Oxygen free radicals regulate NMDA receptor function via a redox modulatory site. Neuron 5: 841-846.

Akbarian S, Vinuela A, Kim JJ, Potkin SG, Bunney WE Jr., et al. 1993. Distorted distribution of nicotinamide-adenine dinucleotide phosphate-diaphorase neurons in temporal lobe of schizophrenics implies anomalous cortical development. Arch Gen Psychiat 50: 178-187.

Albrech J, Talbot M, Kimelberg HK, Ascher M. 1993. The role of sulfhydryl groups and calcium in the mercuric chloride-induced inhibition of glutamate uptake in rat primary astrocyte cultures. Brain Res 607: 249-254.

Almaguer-Melian W, Cruz-Aguado R, Bergado JA. 2000. Synaptic plasticity is impaired in rats with a low glutathione content. Synapse 38: 369-374.

Amara A, Coussemacq M, Geffard M. 1994. Antibodies to reduced glutathione. Brain Res 659: 237-242.

Anderson ME. 1997. Glutathione and glutathione delivery compounds. Adv Pharmacol 38: 65-78.

Anderson ME, Underwood M, Bridges RJ, Meister A. 1989. Glutathione metabolism at the blood–cerebrospinal fluid barrier. FASEB J 3: 2527-2531.

Aukrust P, Svardal AM, Muller F, Lunden B, Berge RK, et al. 1995. Increased levels of oxidized glutathione in CD4+ lymphocytes associated with disturbed intracellular redox balance in human immunodeficiency virus type I infection. Blood 86: 258-267.

Bains JS, Shaw CA. 1997. Neurodegenerative disorders in humans: The role of glutathione in oxidative stress-mediated neuronal death. Brain Res Rev 25: 335-358.

Bains JS, Shaw CA. 1998. Oxidative stress and neurological diseases: Is glutathione depletion a common factor? Glutathione in the Nervous System. Shaw CA, editor. Washington, DC: Taylor & Francis; pp. 355-384.

Bains JS, Pasqualotto BA, Shaw CA, Curry K. 1997. Localization of glutathione binding sites in the central nervous system of rat. Soc Neurosci Abstr 23: 1761.

Balazs L, Leon M. 1994. Evidence of an oxidative challenge in the Alzheimer's brain. Neurochem Res 19: 1131-1137.

Baquer NZ, Hothersall JS, McLean P. 1988. Function and regulation of the pentose phosphate pathway in brain. Curr Top Cell Regul 29: 265-289.

Barker JE, Bolaños JP, Land JM, Clark JB, Heales SJR. 1996. Glutathione protects astrocytes from peroxynitrite-mediated mitochondrial damage: Implications for neuronal/astroglial trafficking and neurodegeneration. Dev Neurosci 18: 391-396.

Beckman JS, Beckman T, Chen J, Marshall PA, Freeman BA. 1990. Apparent hydroxyl radical production by peroxynitrite: Implications for endothelial injury from nitric oxide and superoxide. Proc Natl Acad Sci USA 87: 1620-1624.

Beckman JS, Carson M, Smith CD, Kuppenol WH. 1993. ALS, SOD and peroxynitrite. Nature 364: 584.

Behl C, Davis J, Cole GM, Schubert D. 1992. Vitamin E protects nerve cells from amyloid β protein toxicity. Biochem Biophys Res Commun 186: 944-950.

Behl C, Davis JB, Lesley R, Schubert D. 1994. Hydrogen peroxide mediates amyloid β protein toxicity. Cell 77: 817-827.

Bellis SL, Laux D, Rhoads DE. 1994. Affinity purification of *Hydra* glutathione binding proteins. FEBS Lett 354: 320-324.

Ben-Yoseph O, Boxer PA, Ross BD. 1994. Oxidative stress in the central nervous system: Monitoring the metabolic response using the pentose phosphate pathway. Dev Neurosci 16: 328-336.

Ben-Yoseph O, Boxer PA, Ross BD. 1996a. Assessment of the role of glutathione and pentose phospate pathways in the protection of primary cerebrocortical cultures from oxidative stress. J Neurochem 66: 2329-2337.

Ben-Yoseph O, Boxer PA, Ross BD. 1996b. Noninvasive assessment of the relative roles of cerebral antioxidative enzymes by quantitation of pentose phosphate pathway activity. Neurochem Res 21: 1005-1012.

Benzi G, Moretti A. 1995. Age- and peroxidative stress-related modifications of the cerebral enzymatic activities linked to mitochondria and the glutathione system. Free Radic Biol Med 19: 77-101.

Benzi G, Pastoris O, Marzatico F, Villa RF. 1988. Influence of aging and drug treatment on the cerebral glutathione system. Neurobiol Aging 9: 371-375.

Benzi G, Pastoris O, Marzatico F, Villa RF. 1989a. Cerebral enzyme antioxidant system. Influence of aging and phosphatidylcholine. J Cereb Blood Flow Metab 9: 373-380.

Benzi G, Pastoris O, Marzatico F, Villa RF. 1989b. Age-related effect induced by oxidative stress on the cerebral glutathione system. Neurochem Res 14: 473-481.

Benzi G, Pastoris O, Gorini A, Marzatico F, Villa RF, et al. 1991. Influence of aging on the acute depletion of reduced glutathione induced by electrophilic agents. Neurobiol Aging 12: 227-231.

Berridge MJ, Irvine RF. 1989. Inositol phosphates and cell signalling. Nature 341: 197-205.

Bolaños JP, Heales SJR, Land JM, Clark JB. 1995. Effect of peroxynitrite on the mitochondrial respiratory chain: Differential susceptibility of neurons and astrocytes in primary culture. J Neurochem 64: 1965-1972.

Bolaños JP, Heales SJR, Peuchen S, Barker JE, Land JM, et al. 1996. Nitric oxide mediated mitochondrial damage: A potential neuroprotective role for glutathione. Free Radic Biol Med 21: 995-1001.

Bonnefoi MS. 1992. Mitochondrial glutathione and methyl iodide-induced neurotoxicity in primary neural cell cultures. Neurotoxicology 13: 401-412.

Boobis AR, Fawthrop DJ, Davies DS, 1989. Mechanisms of cell death. Trends Pharmacol Sci 10: 275-280.

Bredt DS, Snyder SH. 1994. Nitric oxide: A physiologic messenger molecule. Annu Rev Biochem 63: 175-195.

Buckman TD, Kling AS, Eiduson S, Sutphin MS, Steinberg A. 1987. Glutathione peroxidase and CT scan abnormalities in schizophrenia. Biol Psychiatry 22: 1349-1356.

Buckman TD, Kling A, Sutphin MS, Steinberg A, Eiduson S. 1990. Platelet glutathione peroxidase and monoamine oxidase activity in schizophrenics with CT scan abnormalities: Relation to psychosocial variables. Psychiat Res 31: 1-14.

Burk RF, Patel K, Lane JM. 1983. Reduced glutathione protection against rat liver microsomal injury by carbon tetrachloride. Dependence on O_2. Biochem J 215: 441-445.

Butler AR, Flitney FW, Williams DL. 1995. NO, nitrosonium ions, nitroxide ions, nitrosothiols and iron-nitrosyls in biology: A chemist's perspective. Trends Pharmacol Sci 16: 18-22.

Cammer W, Tansey F, Abramovitz M, Ishigaki S, Listowsky I. 1989. Differential localization of glutathione-S-transferase Yp and Yb subunits in oligodendrocytes and astrocytes of rat brain. J Neurochem 52: 876-883.

Campbell E, Takahashi Y, Abramovitz M, Peretz M, Listowsky I. 1990. A distinct human testis and brain μ-class glutathione S-transferase. Molecular cloning and characterization of a form present even in individuals lacking hepatic type μ isoenzymes. J Biol Chem 265: 9188-9193.

Cascinu S, Cordella L, Del Ferro E, Fronzoni M, Catalano G. 1995. Neuroprotective effect of reduced glutathione on cisplatin-based chemotherapy in advanced gastric cancer: A randomised double-blind placebo-controlled trial. J Clin Oncol 13: 26-32.

Ceballos-Picot I, Merad-Boudia M, Nicole A, Thevenin M, Hellier G, et al. 1996. Peripheral antioxidant enzyme activities and selenium in elderly subjects and in dementia of Alzheimer's type—place of the extracellular glutathione peroxidase. Free Radic Biol Med 20: 579-587.

Chandrasekaran K, Giordano T, Brady DR, Stoll J, Martin LJ, et al. 1994. Impairment in mitochondrial cytochrome oxidase gene expression in Alzheimer disease. Mol Brain Res 24: 336-340.

Cheeseman KH, Slater TF. 1993. An introduction to free radical biochemistry. Brit Med Bull 49: 481-493.

Chen TS, Richie JP Jr, Lang CA. 1989. The effect of aging on glutathione and cysteine levels in different regions of the mouse brain. Proc Soc Exp Biol Med 190: 399-402.

Chéramy A, Desce JM, Godeheu G, Glowinski J. 1994. Presynaptic control of dopamine synthesis and release by excitatory amino acids in rat striatal synaptosomes. Neurochem Int 25: 145-154.

Cho Y, Bannai S. 1990. Uptake of glutamate and cystine in C-6 glioma cells and in cultured astrocytes. J Neurochem 55: 2091-2097.

Choi DW. 1988. Glutamate neurotoxicity and disease of the nervous system. Neuron 1: 623-634.

Collingridge GL, Singer W. 1990. Excitatory amino acid receptors and synaptic plasticity. Trends Pharmacol Sci 11: 290-296.

Colton CA, Gilbert DL. 1987. Production of superoxide anions by a CNS macrophage, the microglia. FEBS Lett 223: 284-288.

Connor MJ, Wheeler LA. 1987. Depletion of cutaneous glutathione by ultraviolet radiation. Photochem Photobiol 46: 239-245.

Cooper AJL. 1997. Glutathione in the brain: Disorders of glutathione metabolism. The Molecular and Genetic Basis of Neurological Disease. 2nd ed., Rosenberg RN, Prusiner SB, DiMauro S, Barchi RL, editors. Boston: Butterworth-Heinemann; pp. 1195-1230.

Cooper AJL. 1998. Role of astrocytes in maintaining cerebral glutathione homeostasis and in protecting the brain against xenobiotics and oxidative stress. Glutathione in the Nervous System. Shaw CA, editor. Washington DC: Taylor & Francis; pp. 91-115.

Cooper AJL, Meister A. 1992. Glutathione in the brain: Disorders of glutathione metabolism. The Molecular and Genetic Basis of Neurological Disorders. Rosenberg A, editor. New York: Ruttle Graphics; pp. 209-238.

Cooper AJL, Tate SS. 1997. Enzymes involved in processing of glutathione conjugates. Comprehensive Toxicology, Biotransformations, vol. 3. Sipes G, McQueen CA, Gandolfi AJ, editors. Oxford, UK: Elsevier; pp. 329-363.

Cooper AJL, Pulsinelli WA, Duffy TE. 1980. Glutathione and ascorbate during ischemia and postischemic reperfusion in rat brain. J Neurochem 35: 1242-1245.

Cotman CW, Monaghan DT, Ganong AH. 1988. Excitatory amino acid transmission: NMDA receptors and Hebb-type synaptic plasticity. Annu Rev Neurosci 11: 61-80.

Couratier P, Hugon J, Sindou P, Vallat JM, Dumas M. 1993. Cell culture evidence for neuronal degeneration in amyotrophic lateral sclerosis being linked to glutamate AMPA/kainate receptors. Lancet 341: 265-268.

Cristiano F, de Haan JB, Iannello RC, Kola I. 1995. Changes in the levels of enzymes which modulate the antioxidant balance occur during aging and correlate with cellular damage. Mech Ageing Dev 80: 93-105.

Cruz-Aguado R, Fernandez-Verdecia CI, Díaz-Suárez CM, Gonzalez-Monzon O, Antunez-Potashkina I, et al. 1998. Effects of nerve growth factor on brain glutathione-related enzymes from aged rats. Fundam Clin Pharmacol 12: 538-545.

Cruz-Aguado R, Francis-Turner L, Díaz-Suárez CM, Bergado J. 1999. NGF prevents changes in rat brain glutathione-related enzymes following transection of the septohippocampal pathway. Neurochem Int 34: 125-130.

Cruz-Aguado R, Almaguer-Melian W, Díaz CM, Lorigados L, Bergado J. 2001. Behavioral and biochemical effects of glutathione depletion in the rat brain. Brain Res Bull 55: 327-333.

Cudkowicz ME, McKenna-Yasek D, Chen C, Hedley-Whyte ET, Brown RH Jr. 1998. Limited corticospinal tract involvement in amyotrophic lateral sclerosis subjects with the 4V mutation in the copper/zinc superoxide dismutase gene. Ann Neurol 43: 703-710.

Cuénod M, Do KQ. 1998. Glutathione release and nitrosoglutathione presence in the CNS: Implications for schizophrenia. Glutathione in the Nervous System. Shaw CA, editor. Washington, DC: Taylor & Francis; pp. 275-285.

Curry K. 1998. Medicinal chemistry of glutathione and glutathione analogs in the mammalian central nervous system. Glutathione in the Nervous System. Shaw CA, editor. Washington, DC: Taylor & Francis; pp. 217-227.

Davis KL, Kahn RS, Ko G, Davidson M. 1991. Dopamine in schizophrenia: A review and reconceptualization. Am J Psychiat 148: 1474-1486.

Dawson VL, Dawson TM. 1996. Nitric oxide actions in neurochemistry. Neurochem Int 29: 97-110.

Deakin JF, Slater P, Simpson MD, Gilchrist AC, Skan WJ, et al. 1989. Frontal cortical and left temporal glutamatergic dysfunction in schizophrenia. J Neurochem 52: 1781-1786.

De Groote MA, Granger D, Xu Y, Campbell G, Prince R, et al. 1995. Genetic and redox determinants of nitric oxide cytotoxicity in a *Salmonella typhimurium* model. Proc Natl Acad Sci USA 92: 6399-6403.

de Haan JB, Newman JD, Kola I. 1992. Cu/Zn superoxide dismutase mRNA and enzyme activity, and susceptibility to lipid peroxidation, increases with aging in murine brains. Mol Brain Res 13: 179-187.

de Haan JB, Cristiano F, Iannello RC, Kola I. 1995. Cu/Zn-superoxide dismutase and glutathione peroxidase during aging. Biochem Mol Biol Int 35: 1281-1297.

de la Asuncion JG, Millan A, Pla R, Bruseghini L, Esteras A, et al. 1996. Mitochondrial glutathione oxidation correlates with age-associated oxidative damage to mitochondrial DNA. FASEB J 10: 333-338.

De Mattia G, Laurenti O, Bravi C, Ghiselli A, Iuliano L, et al. 1994. Effect of aldose reductase inhibition on glutathione redox status in erythrocytes of diabetic patients. Metab Clin Exp 43: 965-968.

Deneke SM, Fanburg BL, 1989. Regulation of cellular glutathione. Am J Physiol 257: L163-L173.

De Rey-Pailhade J. 1888a. Sur un corps d'origine d'organique hydrogenant de soufre a froid. C. R. Acad Sci 106: 1683-1684.

De Rey-Pailhade J. 1888b. Nouvelle recherche physiologique sur la substance organique hydrogenant le soufre a froid. C. R. Acad Sci 107: 43-44.

Desagher S, Glowinski J, Premont J. 1996. Astrocytes protect neurons from hydrogen peroxide toxicity. J Neurosci 16: 2553-2562.

Desole MS, Esposito G, Enrico P, Miele M, Fresu L, et al. 1993. Effects of ageing on 1-methyl-4-phenyl-1,2,3,6-tetrahydropyridine (MPTP) neurotoxic effects on striatum and brainstem in the rat. Neurosci Lett 159: 143-146.

Deutsch SI, Mastropaolo J, Schwartz BL, Rosse RB, Morihisa JM. 1989. A "glutamatergic hypothesis" of schizophrenia. Rationale for pharmacotherapy with glycine. Clin Neuropharmacol 12: 1-13.

Dexter DT, Sian J, Rose S, Hindmarsh JG, Mann VM, et al. 1994. Indices of oxidative stress and mitochondrial function in individuals with incidental Lewy body disease. Ann Neurol 35: 38-44.

DiChiara G, Morelli M. 1993. Dopamine-acetylcholine-glutamate interactions in the striatum: A working hypothesis. Adv Neurol 6: 102-106.

Dickinson CJ. 1997. Chronic fatigue syndrome—aetiological aspects. Eur J Clin Invest 27: 257-267.

Do KQ, Lauer CJ, Schreiber W, Zollinger M, Gutteck-Amsler U, et al. 1995. γ-Glutamylglutamine and taurine concentrations are decreased in the cerebrospinal fluid of drug-naive patients with schizophrenic disorders. J Neurochem 65: 2652-2662.

Do KQ, Benz B, Grima G, Gutteck-Amsler U, Kluge J, et al. 1996. Nitric oxide precursor arginine and S-nitrosoglutathione in synaptic and glial function. Neurochem Int 29: 213-224.

Drejer J, Larsson OM, Schousboe A. 1983. Characterization of uptake and release processes for D-and L-aspartate in primary cultures of astrocytes and cerebellar granule cells. Neurochem Res 8: 231-243.

Dringen R, Hamprecht B. 1996. Glutathione content as an indicator for the presence of metabolic pathways of amino acids in astroglial cultures. J Neurochem 67: 1375-1382.

Dringen R, Hirrlinger J. 2003. Glutathione pathways in the brain. Biol Chem 384: 505-516.

Droge W, Holm E. 1997. Role of cysteine and glutathione in HIV infection and other diseases associated with muscle wasting and immunological dysfunction. FASEB J 11: 1077-1089.

Dyrks T, Dyrks E, Hartman T, Masters C, Beyreuther K. 1992. Amyloidogenicity of β A4 and β A4-bearing amyloid protein precursor fragments by metal-catalyzed oxidation. J Biol Chem 267: 18210-18217.

Eastwood SL, McDonald B, Burnet PW, Beckwith JP, Kerwin RW, et al. 1995. Decreased expression of mRNAs encoding non-NMDA glutamate receptors GluR1 and GluR2 in medial temporal lobe neurons in schizophrenia. Mol Brain Res 29: 211-223.

Estrela JM, Ortega A, Obrador E. 2006. Glutathione in cancer biology and therapy. Crit Rev Clin Lab Sci 43: 143-181.

Evans PH. 1993. Free radicals in brain metabolism and pathology. Brit Med Bull 49: 577-587.

Fagni L, Olivier M, Lafon-Cazal M, Bockaert J. 1995. Involvement of divalent ions in the nitric oxide-induced blockade of N-methyl-D-aspartate receptors in cerebellar granule cells. Mol Pharmacol 47: 1239-1247.

Feng Z, Qin C, Chang Y, Zhang JT. 2006. Early melatonin supplementation alleviates oxidative stress in a transgenic mouse model of Alzheimer's disease. Free Radic Biol Med 40: 101-109.

Fernandez-Checa JC, Yi J-R, Garcia-Ruiz C, Knezic Z, Tahara SM, et al. 1993. Expression of rat liver glutathione transport in *Xenopus laevis* oocytes. J Biol Chem 268: 2324-2328.

Fischman CM, Udey MC, Kurtz M, Wedner HJ. 1981. Inhibition of lectin-induced lymphocyte activation by 2-cyclohexene-1-one: Decreased intracellular glutathione inhibits an early event in the activation sequence. J Immunol 127: 2257-2262.

Flagg EW, Coates RJ, Jones DP, Eley JW, Gunter EW, et al. 1993. Plasma total glutathione in humans and its association with demographic and health-related factors. Brit J Nutr 70: 797-808.

Forno LS. 1986. The Lewy body in Parkinson's disease. Adv Neurol 45: 35-42.

Frey WH, 2nd, Najarian MM, Kumar KS, Emory CR, Menning PM, Frank JC, Johnson MN, Ala TA. 1996. Endogenous Alzheimer's brain factor and oxidized glutathione inhibit antagonist binding to the muscarinic receptor. Brain Res 714: 87-94.

Froissard P, Monrocq H, Duval D. 1997. Role of glutathione metabolism in the glutamate-induced programmed cell death of neuronal-like PC12 cells. Eur J Pharmacol 326: 93-99.

Fujita K, Yamauchi K, Shibayama K, Ando M, Honda M, et al. 1996. Decreased cytochrome c oxidase activity but unchanged superoxide dismutase and glutathione peroxidase activities in the spinal cords of patients with amyotrophic lateral sclerosis. J Neurosci Res 45: 276-281.

Furukawa T, Meydani SN, Blumberg JB. 1987. Reversal of age-associated decline in immune responsiveness by dietary glutathione supplementation in mice. Mech Aging Dev 38: 107-117.

Garetz SL, Altschuler RA, Schacht J. 1994. Attenuation of gentamicin ototoxicity by glutathione in the guinea pig in vivo. Hearing Res 77: 81-87.

Gaston B, Drazen JM, Jansen A, Sugarbaker DA, Loscalzo J, et al. 1994. Relaxation of human bronchial smooth muscle by S-nitrosothiols in vitro. J Pharmacol Exp Ther 268: 978-984.

Gibb WRG. 1989. The diagnostic relevance of Lewy bodies and other inclusions in Parkinson's disease. Early Diagnosis and Preventive Therapy in Parkinson's Disease. Przuntek H, Riederer P, editors. New York: Springer-Verlag; pp. 171-180.

Gibb WRG, Lees AJ. 1988. The relevance of the Lewy body to the pathogenesis of idiopathic Parkinson's disease. J Neurol Neurosurg Psychiatry 51: 745-752.

Gilbert KR, Aizenman E, Reynolds IJ. 1991. Oxidized glutathione modulates N-methyl-D-aspartate- and depolarization-induced increases in intracellular Ca^{2+} in cultured rat forebrain neurons. Neurosci Lett 133: 11-14.

Goldman-Rakic PS, Leranth C, Williams SM, Mons N, Geffard M. 1989. Dopamine synaptic complex with pyramidal neurons in primate cerebral cortex. Proc Natl Acad Sci USA 86: 9015-9019.

Good PF, Perl DP, Bierer LM, Schmeidler J. 1992. Selective accumulation of aluminium and iron in the neurofibrillar tangles of Alzheimer's disease: A laser microprobe (LAMMA) study. Ann Neurol 31: 286-292.

Gozlan H, Ben-Ari Y. 1995. NMDA receptor redox sites: Are they targets for selective neuronal protection ? Trends Pharmacol Sci 16: 368-374.

Griffith OW, Meister A. 1979. Glutathione: Interorgan translocation, turnover, and metabolism. Proc Natl Acad Sci USA 76: 5606-5610.

Griffith OW, Meister A. 1985. Origin and turnover of mitochondrial glutathione. Proc Natl Acad Sci USA 82: 4668-4672.

Grosvenor W, Bellis SL, Kass-Simon G, Rhoads DE. 1992. Chemoreception in *Hydra*: Specific binding of glutathione to a membrane fraction. Biochim Biophys Acta 1117: 120-125.

Guo N, Shaw C. 1992. Characterization and localization of glutathione binding sites on cultured astrocytes. Mol Brain Res 15: 207-215.

Guo N, McIntosh C, Shaw CA. 1992. Glutathione: New candidate neuropeptide in the central nervous system. Neuroscience 51: 835-842.

Gurney ME, Pu H, Chiu AY, Dal Canto MC, Polchow CY, et al. 1994. Motor neuron degeneration in mice that express a human Cu, Zn superoxide dismutase mutation. Science 264: 1772-1775.

Gurney ME, Cutting FB, Zhai P, Doble A, Taylor CP, Adams PK, Hall ED. 1996. Benefit of vitamin E, riluzole, and gabapentin in a transgenic model of familial amyotrophic lateral sclerosis. Ann Neurol 39: 147-157.

Gutteridge JMC, Quinlan GJ, Clark I, Halliwell B. 1985. Aluminium salts accelerate peroxidation of membrane lipids stimulated by iron salts. Biochim Biophys Acta 835: 441-447.

Halliwell B. 1992a. Reactive oxygen species and the central nervous system. J Neurochem 59: 1609-1623.

Halliwell B. 1992b. Oxygen radicals as key mediators in neurological disease: Fact or fiction? Ann Neurol 32: S10-S15.

Halliwell B. 1996. Cellular stress and protection mechanisms. Biochem Soc T 24: 1023-1027.

Halliwell B, Aeschbach R, Loliger J, Aruoma OI. 1995. The characterization of antioxidants. Food Chem Toxicol 33: 601-617.

Han SK, Mytilineou C, Cohen G. 1996. L-DOPA upregulates glutathione and protects mesencephalic cultures against oxidative stress. J Neurochem 66: 501-510.

Hargreaves KM, Pardridge WM. 1988. Neutral amino acid transport at the blood-brain barrier. J Biol Chem 263: 19392-19397.

Harman D. 1995. Role of antioxidant nutrients in aging: Overview. Age 18: 51-62.

Harrison ML, Simmonds MA. 1985. Quantitative studies on some antagonists of N-methyl-D-asparate in slices of rat cerebral cortex. Brit J Pharmacol 84: 381-391.

Hazelton GA, Lang CA. 1980. Glutathione contents of tissues in the aging mouse. Biochem J 188: 25-30.

Hazelton GA, Lang CA. 1983. Glutathione biosynthesis in the aging adult yellow-fever mosquito [*Aedes aegypti* (Louisville)]. Biochem J 210: 289-295.

Heales SJR, Davies SEC, Bates TE, Clark JB. 1995. Depletion of brain glutathione is accompanied by impaired mitochondrial function and decreased N-acetyl aspartate concentration. Neurochem Res 20: 31-38.

Heales SJR, Bolaños JP, Clark JB. 1996. Glutathione depletion is accompanied by increased neuronal nitric oxide synthase activity. Neurochem Res 21: 35-39.

Henderson JT, Javaheri M, Kopko S, Roder JC. 1996. Reduction of lower motor neuron degeneration in wobbler mice by N-acetyl-L-cysteine. J Neurosci 16: 7574-7582.

Hermann A, Janáky R, Dohovics R, Saransaari P, Oja SS, et al. 2000. Interference of S-nitrosoglutathione with the binding of ligands to ionotropic glutamate receptors in pig cerebral cortical synaptic membranes. Neurochem Res 25: 1119-1124.

Hermann A, Varga V, Oja SS, Saransaari P, Janáky R. 2002. Involvement of the amino-acid side chains in membrane proteins in the binding of glutathione to pig cerebral cortical membranes. Neurochem Res 27: 389-394.

Hjelle OP, Chaudry FA, Ottersen OP. 1994. Antisera to glutathione: Characterization and immunocytochemical application to the rat cerebellum. Eur J Neurosci 6: 793-804.

Hjelle OP, Rinvik E, Huster D, Reichelt W, Ottersen OP. 1998. Antibodies to glutathione: Production, characterization, and immunocytochemical application to the central nervous system. Glutathione in the Nervous System. Shaw CA, editor. Washington, DC: Taylor & Francis; pp. 63-88.

Hogg N, Singh RJ, Kalyanaraman B, 1996. The role of glutathione in the transport and catabolism of nitric oxide. FEBS Lett 382: 223-228.

Holopainen I, Kontro P. 1989. Uptake and release of glycine in cerebellar granule cells and astrocytes in primary culture: Potassium stimulated release from granule cells is calcium dependent. J Neurosci Res 24: 374-383.

Holt JAG. 1993. The glutathione cycle is the creative reaction of life and cancer. Cancer causes oncogenes and not vice versa. Med Hypot 40: 262-266.

Hopkins FG. 1921. On an autooxidisable constituent of the cell. Biochem J 15: 286-305.

Hornykiewicz O. 1988. Neurochemical pathology and etiology of Parkinson's disease: Basic facts and hypothetical possibilities. Mt Sinai J Med 55: 11-20.

Hornykiewicz O, Kish SJ. 1986. Biochemical pathophysiology of Parkinson's disease. Adv Neurol 45: 19-34.

Hotta S. 1962. Glucose metabolism in brain tissue: The hexosemonophosphate shunt and its role in glutathione reduction. J Neurochem 9: 43-51.

Hotta S, Seventko JM Jr. 1968. The hexosemonophosphate shunt and glutathione reduction in guinea pig brain tissue: Changes caused by chlorpromazine, amytal, and malonate. Arch Biochem Biophys123: 104-108.

Hoyt KR, Tang LH, Aizenman E, Reynolds IJ. 1992. Nitric oxide modulates NMDA-induced increases in intracellular Ca^{2+} in cultured rat forebrain neurons. Brain Res 592: 310-316.

Huang J, Philbert MA. 1995. Distribution of glutathione and glutathione-related enzyme systems in mitochondria and cytosol of cultured cerebellar astrocytes and granule cells. Brain Res 680: 16-22.

Huang J, Philbert MA. 1996. Cellular responses of cultured cerebellar astrocytes to ethacrynic acid-induced perturbation of subcellular glutathione homeostasis. Brain Res 711: 184-192.

Huster D, Haug F-M, Nagelhus E, Reichelt W, Ottersen OP. 1998. Subcellular distribution of glutathione in retinal cells of rat and guinea pig: A high resolution, quantitative immunogold analysis. Anat Embryol 198: 277-288.

Hyslop PA, Zhang Z, Pearson DV, Phebus LA. 1995. Measurement of striatal H_2O_2 by microdialysis following global forebrain ischemia and reperfusion in the rat: Correlation with the cytotoxic potential of H_2O_2 in vitro. Brain Res 671: 181-186.

Iantomasi T, Favilli F, Marraccini P, Stio M, Treves C, et al. 1993. Age and GSH metabolism in rat cerebral cortex, as related to oxidative and energy parameters. Mech Ageing Dev 70: 65-82.

Ignarro LJ, Lippton H, Edwards JC, Baricos WH, Hyman AL, et al. 1981. Mechanism of vascular smooth muscle relaxation by organic nitrates, nitrites, nitroprusside and nitric oxide: Evidence for the involvement of S-nitrosothiols as active intermediates. J Pharmacol Exp Ther 218: 739-749.

Ince P, Shaw P, Candy J, Mantle D, Tandon L, et al. 1994. Iron, selenium and glutathione peroxidase activity are elevated in sporadic motor neuron disease. Neurosci Lett 182: 87-90.

Ischiropoulos H, Zhu L, Chen J, Tsai M, Martin JC, et al. 1992. Peroxynitrite-mediated tyrosine nitration catalyzed by superoxide dismutase. Arch Biochem Biophys 16: 149-156.

Ives NK, Gardiner RM. 1990. Blood–brain permeability to bilirubin in the rat studied using intracarotid bolus injection and in situ perfusion techniques. Pediatr Res 27: 436-441.

Jain A, Mårtensson J, Stole E, Auld PAM, Meister A. 1991. Glutathione deficiency leads to mitochondrial damage in brain. Proc Natl Acad Sci USA 88: 1913-1917.

Janáky R. 1994. Glutamate agonist-evoked calcium influx in the cerebellum and release of GABA in the hippocampus. Modulation of the receptor functions by thiol-modulating agents. Acta Univ Tamperensis, Ser. A 403: 1-84.

Janáky R, Ogita K, Pasqualotto BA, Bains JS, Oja SS, et al. 1999. Glutathione and signal transduction in mammalian CNS. J Neurochem 73: 889-902.

Janáky R, Varga V, Saransaari P, Oja SS. 1993a. Glutathione modulates the N-methyl-D-aspartate receptor-activated calcium influx into cultured rat cerebellar granule cells. Neurosci Lett 156: 153-157.

Janáky R, Varga V, Saransaari P, and Oja SS. 1993b. Release of GABA from rat hippocampal slices: Involvement of quisqualate/N-methyl-D-aspartate-gated ionophores and extracellular magnesium. Neuroscience 53: 779-785.

Janáky R, Varga V, Saransaari P, Oja SS. 1994a. Glutamate agonists and [^{3}H]GABA release from rat hippocampal slices: Involvement of metabotropic glutamate receptors in the quisqualate-evoked release. Neurochem Res 19: 729-734.

Janáky R, Varga V, Oja SS, Saransaari P. 1994b. Release of [^{3}H] GABA evoked by glutamate agonists from hippocampal slices: Effects of dithiothreitol and glutathione. Neurochem Int 24: 575-582.

Janáky R, Varga V, Hermann A, Serfözö Z, Dohovics R, et al. 1997. Effect of glutathione on [^{3}H]dopamine release from the mouse striatum evoked by glutamate receptor agonists.

Neurochemistry: Cellular, Molecular and Clinical Aspects. Teelken AW, Korf J, editors. New York: Plenum Press; pp. 733-736.

Janáky R, Varga V, Jenei Zs, Saransaari P, Oja SS. 1998. Glutathione and glutathione derivatives: Possible modulators of ionotropic glutamate receptors. Glutathione in the Nervous System. Shaw CA, editor. Washington, DC: Taylor & Francis; pp. 163-196.

Janáky R, Shaw CA, Varga V, Hermann A, Dohovics R, et al. 2000a. Specific glutathione binding sites in pig cerebral cortical synaptic membranes. Neuroscience 95: 617-624.

Janáky R, Varga V, Hermann A, Saransaari P, Oja SS. 2000b. Mechanisms of L-cysteine neurotoxicity. Neurochem Res 25: 1397-1405.

Janáky R, Dohovics R, Saransaari P, Oja SS. 2007. Modulaton of [^{3}H]dopamine release by glutathione in mouse striatal slices. Neurochem Res, in press.

Jeandel C, Nicolas MB, Dubois F, Nabet-Belleville F, Penin F, et al. 1989. Lipid peroxidation and free radical scavengers in Alzheimer's disease. Gerontology 35: 275-282.

Jellinger K, 1988. Pathology of Parkinson's syndrome. Handbook of Experimental Pharmacology. Calne DB, editor. Berlin: Springer-Verlag; pp. 47-112.

Jellinger K. 1990. New developments in the pathology of Parkinson's disease. Adv Neurol 53: 1-16.

Jellinger KA. 1991. Pathology of Parkinson's disease. Changes other than the nigrostriatal pathway. Mol Chem Neuropathol 14: 153-197.

Jenei Zs, Janáky R, Varga V, Saransaari P, Oja SS. 1998. Interference of S-alkyl derivatives of glutathione with brain ionotropic glutamate receptors. Neurochem Res 23: 1087-1093.

Jenner P. 1993a. Presymptomatic detection of Parkinson's Disease. J Neural Trans (suppl.) 40: 23-36.

Jenner P. 1993b. Altered mitochondrial function, iron metabolism and glutathione levels in Parkinson's disease. Acta Neurol Scand (suppl.) 146: 6-13.

Jenner P. 2003. Oxidative stress in Parkinson's disease. Ann Neurol 53: S26-S38.

Jenner P, Dexter DT, Sian J, Schapira AH, Marsden CD. 1992a. Oxidative stress as a cause of nigral cell death in Parkinson's disease and incidental Lewy body disease. Ann Neurol 32 (suppl.): S82-S87.

Jenner P, Marsden CD, Schapira AH. 1992b. New insights into the cause of Parkinson's disease. Neurology 42: 2241-2250.

Johnson JA, El Barbary A, Kornguth SE, Brugge JF, Siegel FL. 1993. Glutathione S-transferase isoenzymes in rat brain neurons and glia. J Neurosci 13: 2013-2023.

Julius M, Lang CA, Gleiberman L, Harburg E, DiFranceisco W, et al. 1994. Glutathione and morbidity in a community-based sample of elderly. J Clin Epidemiol 47: 1021-1026.

Juurlink BHJ, Schültke E, Hertz L. 1996. Glutathione release and catabolism during energy substrate restriction in astrocytes. Brain Res 710: 229-233.

Kannan R, Yi J-R, Zlokovic BV, Kaplowitz N. 1998. Carrier-mediated GSH transport at the blood–brain barrier and molecular characterization of novel brain GSH transporters. Glutathione in the Nervous System. Shaw CA, editor. Washington, DC: Taylor & Francis; pp. 45-62.

Kato S, Negishi K, Mawatari K, Kuo C-H. 1992. A mechanism for glutamate toxicity in the C6 glioma cells involving inhibition of cysteine uptake leading to glutathione depletion. Neuroscience 48: 903-914.

Kim HC, Shin EJ, Jang CG, Lee MK, Eun JS, et al. 2005. Pharmacological action of Panax ginseng on the behavioral toxicities induced by psychotropic agents. Arch Pharm Res 28: 995-1001.

Kim JS, Kornhuber HH, Schmid-Burgk W, Holzmüller B. 1980. Low cerebrospinal fluid glutamate in schizophrenic patients and a new hypothesis of schizophrenia. Neurosci Lett 20: 379-382.

Kiray M, Bagriyanik HA, Pekcetin C, Ergur BU, Uysal N, et al. 2005. Deprenyl and the relationship between its effects on spatial memory, oxidant stress and hippocampal neurons in aged male rats. Physiol Res 55: 205-212.

Kirstein CL, Coopersmith R, Bridges RJ, Leon M. 1991. Glutathione levels in olfactory and non-olfactory structures of rats. Brain Res 543: 341-346.

Kish SJ, Bergeron C, Rajput A, Dozic S, Mastrogiacomo F, et al. 1992. Brain cytochrome oxidase in Alzheimer's disease. J Neurochem 59: 776-779.

Klann E, Thiels E. 1999. Modulation of protein kinases and protein phosphatases by reactive oxygen species: Implications for hippocampal synaptic plasticity. Prog Neuropsychopharmacol Biol Psychiat 23: 359-376.

Kluge I, Gutteck-Amsler U, Zollinger M, Do K-Q. 1997. S-Nitrosoglutathione in rat cerebellum: Identification and quantification by liquid chromatography–mass spectrometry. J Neurochem 69: 2599-2607.

Knowles RG, Moncada S. 1994. Nitric oxide synthases in mammals. Biochem J 298: 249-258.

Köhr G, Eckardt S, Lüddens M, Monyer H, Seeburg PH. 1994. NMDA receptor channels: Subunit-specific potentiation by reducing agents. Neuron 12: 1031-1040.

Koller KJ, Coyle JT. 1985. The characterization of the specific binding of [^{3}H]-N-acetylaspartylglutamate to rat brain membranes. J Neurosci 5: 2882-2888.

Komaroff AL, Fagioli LR, Geiger AM, Doolittle TH, Lee J, et al. 1996. An examination of the working case definition of chronic fatigue syndrome. Am J Med 100: 56-64.

Koppenol WH, Moreno JJ, Pryor WA, Ischiropoulos H, Beckman JS. 1992. Peroxynitrite, a cloaked oxidant formed by nitric oxide and superoxide. Chem Res Toxicol 5: 834-842.

Kornhuber J, Mack Burkhardt F, Riederer P, Hebenstreit GF, Reynolds GB, et al. 1989. [^{3}H]MK-801 binding sites in postmortem brain regions of schizophrenic patients. J Neural Transm 77: 231-236.

Kosower NS, Kosower EM. 1978. The glutathione status of cells. Int Rev Cytol 54: 109-160.

Kosower NS, Zipser Y, Faltin Z. 1982. Membrane thiol-disulfide status in glucose-6-phosphate dehydrogenase deficient red cells. Relationship to cellular glutathione. Biochim Biophys Acta 691: 345-352.

Kowaluk EA, Fung H-L. 1990. Spontaneous liberation of nitric oxide cannot account for in vitro vascular relaxation by S-nitrosothiols. J Pharmacol Exp Ther 255: 1256-1264.

Kramer RA, Zakher J, Kim G. 1988. Role of the glutathione redox cycle in acquired and de novo multidrug resistance. Science 241: 694-697.

Kranich O, Hamprecht B, Dringen R. 1996. Different preferences in the utilization of amino acids for glutathione synthesis in cultured neurons and astroglial cells derived from rat brain. Neurosci Lett 219: 211-214.

Krebs MO, Desce JM, Kemel ML, Gauchy C, Godeheu G, et al. 1991. Glutamatergic control of dopamine release in the rat striatum: Evidence for presynaptic N-methyl-D-aspartate receptors on dopaminergic nerve terminals. J Neurochem 56: 81-85.

Kurosawa K, Hayashi N, Sato N, Kamada T, Tagawa K. 1990. Transport of glutathione across the mitochondrial membrane. Biochem Biophys Res Commun 167: 367-372.

Lacomblez L, Bensimon G, Leigh PN, Guillet P, Meininger V. 1996. Dose ranging study of riluzole in amyotrophic lateral sclerosis. Lancet 347: 1425-1431.

Lang CA, Naryshkin S, Schneider DL, Mills BJ, Lindeman RD. 1992. Low blood glutathione levels in healthy aging adults. J Lab Clin Med 120: 720-725.

Lange G, DeLuca J, Maldjian JA, Lee H, Tiersky LA, et al. 1999. Brain MRI abnormalities exist in a subset of patients with chronic fatigue syndrome. J Neurol Sci 171: 3-7.

Langeveld CH, Jongenelen CAM, Schepens E, Stoof JC, Bast A, et al. 1995. Cultured rat striatal and cortical astrocytes protect mesencephalic dopaminergic neurons against hydrogen peroxide toxicity independent of their effect on neuronal development. Neurosci Lett 192: 13-16.

Langeveld CH, Schepens E, Jongelen CAM, Stoof JC, Hjelle OP, et al. 1996. Presence of glutathione immunoreactivity in cultured neurones and astrocytes. Neuroreport 7: 1833-1836.

Lanius RA, Krieger C, Wagey R, Shaw CA. 1993. Increased [^{35}S]-glutathione binding sites in spinal cords from patients with sporadic amyotrophic lateral sclerosis. Neurosci Lett 163: 89-92.

Lanius RA, Shaw CA, Wagey R, Kreiger C. 1994. Characterization, distribution, and protein kinase C-mediated regulation of [^{35}S]glutathione binding sites in mouse and human spinal cord. J Neurochem 63: 155-160.

Lash LH, Jones DP. 1985. Distribution of oxidized and reduced forms of glutathione and cysteine in rat plasma. Arch Biochem Biophys 240: 583-592.

Lazarewicz JW, Wroblewski JT, Palmer ME, Costa E. 1989. Reduction of disulfide bonds activates NMDA-sensitive glutamate receptors in primary cultures of cerebellar granule cells. Neurosci Res Commun 4: 91-97.

Lei SZ, Pan ZH, Aggarwal SK, Chen HS, Hartman J, et al. 1992. Effect of nitric oxide production on the redox modulatory site of the NMDA receptor-channel complex. Neuron 8: 1087-1099.

Lenhoff HM. 1961. Activation of the feeding reflex in *Hydra littoralis*. I. Role played by reduced glutathione, and quantitative assay of the feeding reflex. J Gen Physiol 45: 331-344.

Lenhoff HM. 1998. The discovery of the GSH receptor in *Hydra* and its evolutionary significance. Glutathione in the Nervous System. Shaw CA, editor. Washington, DC: Taylor & Francis; pp. 25-43.

Lenhoff HM, Bovaird J. 1961. Action of glutamic acid and glutathione analogs on *Hydra* glutathione receptor. Nature 189: 486-487.

Leslie SW, Brown LM, Trent RD, Lee Y, Morris JL, et al. 1992. Stimulation of N-methyl-D-asparate receptor-mediated calcium entry into dissociated neurons by reduced and oxidized glutathione. Mol Pharmacol 41: 308-314.

Leviel V, Gobert A, Guibert B. 1990. The glutamate-mediated release of dopamine in the rat striatum: Further characterization of the dual excitatory–inhibitory function. Neuroscience 39: 305-312.

Levine RI, Frederick WR, Rapoport SI. 1982. Entry of bilirubin into the brain due to opening of the blood–brain barrier. Pediatrics 69: 255-259.

Levy DI, Sucher NJ, Lipton SA. 1990. Redox modulation of NMDA receptor-mediated toxicity in mammalian central neurons. Neurosci Lett 110: 291-296.

Levy DI, Sucher NJ, Lipton SA. 1991. Glutathione prevents N-methyl-D-asparate receptor-mediated neurotoxicity. Neuroreport 2: 345-347.

Li X, Orwar O, Revesjö C, Sandberg M. 1996. γ-Glutamyl peptides and related amino acids in rat hippocampus in vitro: Effects of depolarisation and γ-glutamyl transpeptidase inhibition. Neurochem Int 29: 121-128.

Lipton SA. 1993. Prospects for clinically tolerated NMDA antagonists: Open-channel blockers and alternative redox states of nitric oxide. Trends Neurosci 16: 527-532.

Lipton SA, Rosenberg PA. 1994. Excitatory amino acids as final common pathway for neurologic disorders. N Engl J Med 330: 613-622.

Lipton SA, Stamler JS. 1994. Actions of redox-related congeners of nitric oxide at the NMDA receptor. Neuropharmacology 33: 1229-1233.

Lomaestro BM, Malone M. 1995. Glutathione in health and disease: Pharmacological issues. Ann Pharmacother 29: 1263-1273.

Loomis WF. 1955. Glutathione control of the specific feeding reactions of *Hydra*. Ann NY Acad Sci 62: 209-228.

Lorenzo A, Yankner BA. 1994. β-Amyloid neurotoxicity requires fibril formation and is inhibited by Congo red. Proc Natl Acad Sci USA 91: 12243-12247.

Louwerse ES, Weverling GJ, Bossuyt PMM, Meyjes FEP, de Jong JMBV. 1995. Randomized, double-blind, controlled trial of acetylcysteine in amyotrophic lateral sclerosis. Arch Neurol 52: 559-564.

Lovell MA, Ehmann WD, Butler SM, Markesbery WR. 1995. Elevated thiobarbituric acid-reactive substances and antioxidant enzyme activity in the brain in Alzheimer's disease. Neurology 45: 1594-1601.

Lowe J. 1994. Lewy bodies. Neurodegenerative Diseases. Calne DB, editor. Philadelphia: Saunders; pp. 51-69.

Lowenstein CJ, Dinerman JL, Snyder SH. 1994. Nitric oxide: A physiological messenger. Ann Intern Med 120: 227-237.

Lowndes HE, Beiswanger CM, Philbert MA, Reuhl KR. 1994. Substrates for neural metabolism of xenobiotics in adult and developing brain. Neurotoxicology 15: 61-74.

Lu SC, Ge JL. 1992. Loss of suppression of GSH synthesis at low cell density in primary cultures of rat hepatocytes. Am J Physiol 263: C1181-C1189.

Lu SC, W-M, Sun Yi J, Ookhtens M, Sze G, et al. 1996. Role of two recently cloned rat liver GSH transporters in the ubiquitous transport of GSH in mammalian cells. J Clin Invest 97: 1488-1496.

Makar TK, Nedergaard M, Preuss A, Gelbard M, Perumal AS, et al. 1994. Vitamin E, ascorbate, glutathione disulfide, and enzymes of glutathione metabolism in cultures of chick astrocytes and neurons: Evidence that astrocytes play an important role in antioxidant processes in the brain. J Neurochem 62: 45-53.

Makar TK, Cooper AJL, Tofel-Grehl B, Thaler HT, Blass JP. 1995. Carnitine, carnitine acetyltransferase, and glutathione in Alzheimer brain. Neurochem Res 20: 705-711.

Maker HS, Weiss C, Silides DJ, Cohen G. 1981. Coupling of dopamine oxidation (monoamine oxidase activity to glutathione oxidation via the generation of hydrogen peroxide in rat brain homogenates. J Neurochem 36: 589-593.

Manzoni O, Prezeau L, Marin P, Deshager S, Bockaert J, et al. 1992. Nitric oxide-induced blockade of NMDA receptors. Neuron 8: 653-662.

Mårtensson J, Meister A. 1989. Mitochondrial damage in muscle occurs after marked depletion of glutathione and is prevented by giving glutathione monoester. Proc Natl Acad Sci USA 86: 471-475.

Mårtensson J, Meister A. 1991. Glutathione deficiency decreases tissue ascorbate levels in newborn rats: Ascorbate spares glutathione and protects. Proc Natl Acad Sci USA 88: 4656-4660.

Mårtensson J, Lai JCK, Meister A. 1990. High-affinity transport of glutathione is part of a multicomponent system essential for mitochondrial function. Proc Natl Acad Sci USA 87: 7185-7189.

Martínez M, Ferrándiz ML, De Juan E, Miquel J. 1994. Age-related changes in glutathione and lipid peroxide content in mouse synaptic mitochondria: Relationship to cytochrome *c* oxidase decline. Neurosci Lett 170: 121-124.

Matthysse S. 1973. Antipsychotic drug actions: A clue to the neuropathology of schizophrenia? Fed Proc 32: 200-205.

Max B. 1989. This and that: The war on drugs and the evolution of sulfur. Trends Pharmacol Sci 10: 483-486.

Maycox PR, Deckwerth T, Hell JW, Jahn R. 1988. Glutamate uptake by brain synaptic vesicles. Energy dependence of transport and functional reconstitution in proteoliposomes. J Biol Chem 263: 15423-15428.

Mayer ML, Miller RJ. 1990. Excitatory amino acid receptors, second messengers and regulation of intracellular Ca^{2+} in mammalian neurons. Trends Pharmacol Sci 11: 254-260.

Meister A. 1986. Glutathione: Metabolism, transport and effects of selective modifications of cellular glutathione levels. Thioredoxin and Glutaredoxin Systems: Structure and Function. Holmgren A, Brändén C-I, Jörnvall H-, Sjöberg B-M, editors. Karolinska Nobel Conference Series. New York: Raven Press; pp. 245-275.

Meister A. 1988. Glutathione metabolism and its selective modification. J Biol Chem 263: 17205-17208.

Meister A. 1989. On the biochemistry of glutathione. Glutathione Centennial: Molecular Perspectives and Clinical Implications. Taneguchi N, Higashi T, Sakamoto Y, Meister A, editors. Academic Press; New York: pp. 3-21.

Meister A. 1991. Glutathione deficiency produced by inhibition of its synthesis, and its reversal: Applications in research and therapy. Pharmacol Ther 51: 155-194.

Meister A. 1992. On the antioxidant effects of ascorbic acid and glutathione. Biochem Pharmacol 44: 1905-1915.

Meister A, Anderson ME. 1983. Glutathione. Annu Rev Biochem 52: 711-760.

Meister A, Tate SS. 1976. Glutathione and related γ-glutamyl compounds: Biosynthesis and utilization. Annu Rev Biochem 45: 559-604.

Meng RS, Li QM, Wei CX, Chen B, Liao HY, et al. 2005. Clinical observation and mechanism study on treatment of senile dementia with Naohuandan. Chin J Integr Med 11: 111-116.

Miller RG, Moore D, Young LA, Armon C, Barohn RJ, et al. 1996. Placebo-controlled trial of gabapentin in patients with amyotrophic lateral sclerosis. Neurology 47: 1383-1388.

Mitchell JD, Gatt JA, Phillips TM, Houghton E, Rostron G, et al. 1993. Cu/Zn superoxide dismutase free radicals, and motoneuron disease. Lancet 342: 1051-1052.

Mitchell JB, Russo A, Biaglow JE, McPherson S. 1983. Cellular glutathione depletion by diethyl maleate or buthionine sulfoximine: No effect of glutathione depletion on the oxygen enhancement ratio. Radiat Res 96: 422-428.

Miyamoto M, Murphy TH, Schnaar RL, Coyle JT. 1989. Antioxidants protect against glutamate-induced cytotoxicity in a neuronal cell line. J Pharmacol Exp Ther 250: 1132-1140.

Monaghan DT, Bridges RJ, Cotman CW. 1989. The excitatory amino acid receptors: Their classes, pharmacology, and distinct properties in the function of the central nervous system. Annu Rev Pharmacol Toxicol 29: 365-402.

Morari M, O'Connor WT, Darvelid M, Ungerstedt U, Bianchi C, et al. 1996. Functional neuroanatomy of the nigrostriatal and striatonigral pathways as studied with dual probe microdialysis in the awake rat. I. Effects of perfusion with tetrodotoxin and low-calcium medium. Neuroscience 72: 79-87.

Murphy TH, Miyamoto M, Satre A, Schnaar AL, Coyle JT. 1989. Glutamate toxicity in a neural cell line involves inhibition of cystine transport leading to oxidative stress. Neuron 2: 1547-1558.

Murphy TH, Schnaar RL, Coyle JT. 1990. Immature cortical neurons are uniquely sensitive to glutamate toxicity by inhibition of cystine uptake. FASEB J 4: 1624-1633.

Mutisya EM, Bowling AC, Beal MF. 1994. Cortical cytochrome oxidase activity is reduced in Alzheimer's disease. J Neurochem 63: 2179-2184.

Naito S, Ueda T. 1985. Characterization of glutamate uptake into synaptic vesicles. J Neurochem 44: 99-109.

Nakanishi S, Masu M. 1994. Molecular diversity and functions of glutamate receptors. Annu Rev Biophys Biomol Struct 23: 319-348.

Natelson BH, Cohen JM, Brasloff I, Lee HJ. 1993. A controlled study of brain magnetic resonance imaging in patients with chronic fatigue syndrome. J Neurol Sci 120: 213-217.

Nicoll RA, Malenka RC, Kauer JA. 1990. Functional comparison of neurotransmitter receptor subtypes in mammalian central nervous system. Physiol Rev 70: 513-565.

Nishikawa T, Takashima M, Toru M. 1983. Increased [^{3}H] kainic acid binding in the prefrontal cortex in schizophrenia. Neurosci Lett 40: 245-250.

Noble PG, Antel JP, Yong VW. 1994. Astrocytes and catalase prevent the toxicity of catecholamines to oligodendrocytes. Brain Res 633: 83-90.

O'Connor E, Devesa A, García C, Puertes IR, Pellín A, et al. 1995a. Biosynthesis and maintenance of GSH in primary astrocyte cultures: Role of cystine and ascorbate. Brain Res 680: 157-163.

O'Connor WT, Drew KL, Ungerstedt U. 1995b. Differential cholinergic regulation of dopamine release in the dorsal and ventral neostriatum of the rat: An in vivo microdialysis study. J Neurosci 15: 8353-8361.

Odell GB, Schuetta HS. 1985. Bilirubin encephalopathy. Neural Energy Metabolism and Metabolic Encephalopathy. McCandless DW, editor. New York: Plenum Press; pp. 229-261.

Ogita K, Yoneda Y. 1986. Characterization of Na$^+$-dependent binding sites of [^{3}H]-glutamate in synaptic membranes from rat brain. Brain Res 397: 137-144.

Ogita K, Yoneda Y. 1987. Possible presence of [^{3}H]-glutathione (GSH) binding sites in synaptic membranes from rat brain. Neurosci Res 4: 486-496.

Ogita K, Yoneda Y. 1988. Temperature-dependent and -independent apparent binding activities of [^{3}H]-glutathione in brain synaptic membranes. Brain Res 463: 37-46.

Ogita K, Yoneda Y. 1989. Selective potentiation by L-cysteine of apparent binding activity of [^{3}H]-glutathione in synaptic membranes of rat brain. Biochem Pharmacol 38: 1499-1505.

Ogita K, Kitago T, Nakamura H, Fukuda Y, Koida M, et al. 1986. Glutathione-induced inhibition of Na$^+$-independent and dependent bindings of L-[^{3}H]-glutamate in rat brain. Life Sci 39: 2411-2418.

Ogita K, Ogawa Y, Yoneda Y. 1988. Apparent binding activity of [^{3}H]glutathione in rat central and peripheral tissues. Neurochem Int 13: 493-497.

Ogita K, Enomoto R, Nakahara F, Ishitsubo N, Yoneda Y. 1995. A possible role of glutathione as an endogenous agonist at the N-methyl-D-aspartate recognition domain in rat brain. J Neurochem 64: 1088-1096.

Ogita K, Shuto M, Maeda H, Minami T, Yoneda Y. 1998. Possible modulation by glutathione of glutamatergic neurotransmission. Glutathione in the Nervous System. Shaw CA, editor. Washington, DC: Taylor & Francis; pp. 137-161.

Ohyashiki T, Karino T, Matsui K. 1993. Stimulation of Fe^{2+}-induced lipid peroxidation in phosphatidylcholine liposomes by aluminium ions at physiological pH. Biochim Biophys Acta 1170: 182-188.

Oja SS, Varga V, Janáky R, Kontro P, Aarnio T, et al. 1988. Glutathione and glutamatergic neurotransmission in the brain. Frontiers in Excitatory Amino Acid Research. Cavalheiro EA, Lehmann J, Turski L, editors. New York: Alan R. Liss; pp. 75-78.

Oja SS, Jenei Zs, Janáky R, Saransaari P, Varga V. 1994. Thiol reagents and brain glutamate receptors. Proc West Pharmacol Soc 37: 59-62.

Oja SS, Janáky R, Saransaari P, Jenei Z, Varga V. 1995. Interaction of glutathione derivatives with brain 2-amino-3-hydroxy-5-methyl-4-isoxazolepropionate (AMPA) receptors. Proc West Pharmacol Soc 38: 9-11.

Okabe E, Sugihara M, Tanaka K, Sasaki H, Ito H. 1989. Calmodulin and free radicals interaction with steady-state calcium accumulation and passive calcium permeability of cardiac sarcoplasmic reticulum. J Pharmacol Exp Ther 250: 286-292.

Olney JW, Ho OL, Rhee V. 1971. Cytotoxic effects of acidic and sulphur containing amino acids on the infant mouse central nervous system. Exp Brain Res 14: 61-70.

Olney JW, Zorumski C, Price MT, Labruyere J. 1990. L-Cysteine, a bicarbonate-sensitive endogenous excitotoxin. Science 248: 596

Orlowski M, Meister A. 1970. The γ-glutamyl cycle: A possible transport system for amino acids. Proc Natl Acad Sci USA 67: 1248-1255.

Orrenius S, McConkey DJ, Bellomo G, Nicotera P. 1989. Role of Ca^{2+} in toxic cell killing. Trends Pharmacol Sci 10: 281-285.

Orrenius S, Burkitt MJ, Kass GEN, Dypbukt JM, Nicotera P. 1992. Calcium ions and oxidative cell injury. Ann Neurol 32 (suppl.): S33-S42.

Orwar O, Li X, Andiné P, Bergström CM, Hagberg H, et al. 1994. Increased intra- and extracellular concentrations of γ-glutamylglutamate and related dipeptides in the ischemic rat striatum: Involvement of γ-glutamyltranspeptidase. J Neurochem 63: 1371-1376.

Ottersen OP, Zhang N, Walberg F. 1992. Metabolic compartmentation of glutamate and glutamine: Morphological evidence obtained by quantitative immunocytochemistry in rat cerebellum. Neuroscience 46: 519-534.

Pan Z, Perez-Polo R. 1993. Role of nerve growth factor in oxidant homeostasis: Glutathione metabolism. J Neurochem 61: 1713-1721.

Pan Z, Bähring R, Grantyn R, Lipton SA. 1995. Differential modulation by sulfhydryl redox agents and glutathione of GABA- and glycine-evoked currents in rat retinal ganglion cells. J Neurosci 15: 1384-1391.

Park LC, Baldessarini RJ, Kety SS. 1965. Methionine effects on chronic schizophrenics. Arch Gen Psychiat 12: 346-351.

Parker CW, Koch D, Huber MM, Falkenhein SF. 1980. Formation of the cysteinyl form of slow reacting substance (leukotriene E4) in human plasma. Biochem Biophys Res Commun 97: 1038-1046.

Parker WD Jr., Filley CM, Parks JK. 1990. Cytochrome oxidase deficiency in Alzheimer's disease. Neurology 40: 1302-1303.

Parker WD Jr., Parks JK, Filley CM, Kleinschmidt-DeMasters BK. 1994. Electron transport chain defects in Alzheimer's disease brain. Neurology 44: 1090-1096.

Pasha KV, Vijayan E. 1989. Glutathione distribution in rat brain at different ages and the effect of intraventricular glutathione on gonadotropin levels in ovariectomized steroid-primed rats. Brain Res Bull 22: 617-619.

Pasqualotto BA, Curry K, Shaw CA. 1998. Excitatory actions of GSH on neocortex. Glutathione in the Nervous System. Sahw CA, editor. Washington, DC: Taylor & Francis; pp. 197-216.

Perrin R, Briancon S, Jeandel C, Artur Y, Minn A, Penin F, Siest G. 1990. Blood activity of Cu/Zn superoxide dismutase, glutathione peroxidase and catalase in Alzheimer's disease: A case-control study. Gerontology 36: 306-313.

Perry TL, Yong VW. 1986. Idiopathic Parkinson's disease, progressive supranuclear palsy and glutathione metabolism in the substantia nigra of patients. Neurosci Lett 67: 269-274.

Perry TL, Godin DV, Hansen S. 1982. Parkinson's disease: A disorder due to nigral glutathione deficiency? Neurosci Lett 33: 305-310.

Perry TL, Yong VW, Bergeron C, Hansen S, Jones K. 1987. Amino acids, glutathione, and glutathione transferase activity in the brains of patients with Alzheimer's disease. Ann Neurol 21: 331-336.

Perry TL, Hansen S, Jones K. 1988. Brain amino acids and glutathione in progressive supranuclear palsy. Neurology 38: 943-946.

Philbert MA, Beiswanger CM, Waters DK, Reuhl KR, Lowndes HE. 1991. Cellular and regional distribution of reduced glutathione in the nervous system of the rat: Histochemical localization by mercury orange and o-phthaldialdehyde-induced histofluorescence. Toxicol Appl Pharmacol 107: 215-227.

Pike CJ, Burdick D, Walencewicz AJ, Glabe CG, Cotman CW. 1993. Neurodegeneration induced by β-amyloid peptides in vitro: The role of peptide assembly state. J Neurosci 13: 1676-1687.

Pin JP, Bockaert J, Récasens M. 1984. The Ca^{2+}/Cl^- dependent L-[^{3}H]glutamate binding: A new receptor or a particular transport process? FEBS Lett 175: 31-36.

Pow DV, Crook DK. 1995. Immunocytochemical evidence for the presence of high levels of reduced glutathione in radial glial cells and horizontal cells in the rabbit retina. Neurosci Lett 193: 25-28.

Przedborski S, Donaldson D, Jakowec M, Kish S, Guttman M, et al. 1996. Brain superoxide dismutase, catalase and glutathione peroxidase activities in amyotrophic lateral sclerosis. Ann Neurol 39: 158-165.

Quick DT, Greer M. 1967. Pancreatic dysfunction in patients with amyotrophic lateral sclerosis. Neurology 17: 112-116.

Ramirez-Leon V, Kullberg S, Hjelle OP, Ottersen OP, Ulfhake B. 1999. Increased glutathione levels in neurochemically identified fibre systems in the aged rat lumbar motor nuclei. Eur J Neurosci 11: 2935-2948.

Rao G, Xia E, Richardson A. 1990. Effect of age on the expression of antioxidant enzymes in male Fischer F344 rats. Mech Ageing Dev 53: 49-60.

Raps SP, Lai JCK, Hertz L, Cooper AJL. 1989. Glutathione is present in high concentrations in cultured astrocytes but not in cultured neurons. Brain Res 493: 398-401.

Ravindranath V, Shivakumar BR, Anandatheerthavarada HK. 1989. Low glutathione levels in brain regions of aged rats. Neurosci Lett 101: 187-190.

Récasens M, Mayat E, Vigres M. 1992. The multiple excitatory amino acid receptors and receptor subtypes and their putative interactions. Mol Neuro Pharmacol 2: 15-31.

Reed DJ. 1986. Regulation of reductive processes by glutathione. Biochem Pharmacol 35: 7-13.

Reichelt KL, Fonnum F. 1969. Subcellular localization of N-acetyl-aspartyl-glutamate, N-acetyl-glutamate and glutathione in brain. J Neurochem 16: 1409-1416.

Révész L, Edgren M. 1984. Glutathione-dependent yield and repair of single-strand DNA breaks in irradiated cells. Br J Cancer (suppl.) 6: 55-60.

Reynolds JJ, Rush EA, Aizenman F. 1990. Reduction of NMDA receptors with dithiothreitol increases [^{3}H]-MK-801 binding and NMDA-induced Ca^{2+} fluxes. Br J Pharmacol 101: 178-182.

Richie JP Jr., Lang CA. 1986. The maintenance of high glutathione levels in healthy, very old women. Gerontologist 26: 80A.

Richie JP Jr., Mills BJ, Lang CA. 1987. Correction of a glutathione deficiency in the aging mosquito increases its longevity. Proc Soc Exp Biol Med 184: 113-117.

Riederer P, Sofic E, Rausch WD, Schmidt B, Reynolds GP, et al. 1989. Transition metals, ferritin, glutathione, and ascorbic acid in parkinsonian brains. J Neurochem 52: 515-520.

Roos R, Lee J, Bindokas V, Jordan J, Miller R, et al. 1997. Gene delivery by replication-deficient recombinant adenoviruses (AdVs) in the study of Cu, Zn superoxide dismutase type 1 (SOD)-linked familial amyotrophic lateral sclerosis (FALS). Neurology 48: A 150.

Rosen DR, Siddique T, Patterson D, Figlewicz DA, Sapp P, et al. 1993. Mutations in Cu/Zn superoxide dismutase gene are associated with familial amyotrophic lateral sclerosis. Nature 362: 59-62.

Rothstein JD. 1995. Excitotoxic mechanisms in the pathogenesis of amyotrophic lateral sclerosis. Adv Neurol 68: 7-20.

Rothstein JD, Kuncl RW. 1995. Neuroprotective strategies in a model of chronic glutamate-mediated motor neuron toxicity. J Neurochem 65: 643-651.

Ruppersberg JP, Stocker M, Pongs O, Heinemann SH, Frank R, et al. 1991. Regulation of fast inactivation of cloned mammalian $I_k(A)$ channels by cysteine oxidation. Nature 352: 711-714.

Sagara JI, Miura K, Bannai S. 1993. Maintenance of neuronal glutathione by glial cells. J Neurochem 61: 1672-1676.

Sagara J, Makino N, Bannai S. 1996. Glutathione efflux from cultured astrocytes. J Neurochem 66: 1876-1881.

Sanchez del Pino MM, Hawkins RA, Peterson DR. 1995. Biochemical discrimination between luminal and abluminal enzyme and transport activities of the blood–brain barrier. J Biol Chem 270: 14907-14912.

Schapira AH, Cooper JM, Dexter D, Clark JB, Jenner P, et al. 1990. Mitochondrial complex I deficiency in Parkinson's disease. J Neurochem 54: 823-827.

Schoenberg BS. 1986. Descriptive epidemiology of Parkinson's disease: Distribution and hypothesis formulation. Adv Neurol 45: 277-283.

Seeburg PH. 1993. The TINS/TIPS Lecture. The molecular biology of mammalian glutamate receptor channels. Trends Neurosci 16: 359-365.

Seeman P, Lee T, Chau-Wong M, Wong K. 1976. Antipsychotic drug doses and neuroleptic/dopamine receptors. Nature 261: 717-719.

Seeman P, Niznik HB, Guan HC. 1990. Elevation of dopamine D2 receptors in schizophrenia is underestimated by radioactive raclopride. Arch Gen Psychiat 47: 1170-1172.

Seeman P, Guan HC, van Tol HHM. 1993. Dopamine D4 receptors elevated in schizophrenia. Nature 365: 441-445.

Selemon LD, Rajkowska G, Goldman-Rakic PS. 1995. Abnormally high neuronal density in the schizophrenic cortex. A morphometric analysis of prefrontal area 9 and occipital area 17. Arch Gen Psychiat 52: 805-818.

Shaw CA. 1998. Multiple roles of glutathione in the nervous system. Glutathione in the Nervous System. Shaw CA, editor. Washington, DC: Taylor & Francis; pp. 3-23.

Shaw CA, Pasqualotto BA, Curry K. 1996. Glutathione-induced sodium currents in neocortex. Neuroreport 7: 1149-1152.

Shaw JP, Chou IN. 1986. Elevation of intracellular glutathione content associated with mitogenic stimulation of quiescent fibroblasts. J Cell Physiol 129: 193-198.

Shaw JP, Ince PG. 1997. Glutamate, excitotoxicity, and amyotrophic lateral sclerosis. J Neurol 244 (Suppl. 2): S3-S14.

Shaw JP, Salt TE. 1997. Modulation of sensory and excitatory amino acid responses by nitric oxide donors and glutathione in the ventrobasal thalamus of the rat. Eur J Neurosci 9: 1507-1513.

Shukitt-Hale B, Denisova NA, Strain JG, Joseph JA. 1997. Psychomotor effects of dopamine infusion under decreased glutathione conditions. Free Radic Biol Med 23: 412-418.

Sian J, Dexter DT, Lees AJ, Daniel S, Jenner P, et al. 1994. Glutathione-related enzymes in brain in Parkinson's disease. Ann Neurol 36: 356-361.

Simonian NA, Hyman BT. 1993. Functional alterations in Alzheimer's disease: Diminution of cytochrome oxidase in hippocampal formation. J Neuropathol Exp Neurol 52: 580-585.

Sinet PM, Lejeune J, Jerome H. 1979. Trisomy 21 (Down's syndrome). Glutathione peroxidase, hexose monophosphate shunt and I.Q. Life Sci 24: 29-33.

Singh SP, Wishnok JS, Keshive M, Deen WM, Tannenbaum SR. 1996. The chemistry of the S-nitrosoglutathione/glutathione system. Proc Natl Acad Sci USA 93: 14428-14433.

Slivka A, Mytilineou C, Cohen G. 1987a. Histochemical evaluation of glutathione in brain. Brain Res 409: 275-284.

Slivka A, Spina MB, Cohen G. 1987b. Reduced and oxidized glutathione in human and monkey brain. Neurosci Lett 74: 112-118.

Slivka A, Chuttani R, Carr-Locke DL, Kobzik L, Bredt DS, et al. 1994. Inhibition of sphincter of Oddi function by the nitric oxide carrier S-nitroso-N-acetylcysteine in rabbits and humans. J Clin Invest 94: 1792-1798.

Socci DJ, Crandall BM, Arendash GW. 1995. Chronic antioxidant treatment improves the cognitive performance of aged rats. Brain Res 693: 88-94.

Sofic E, Lange KW, Jellinger K, Riederer P. 1992. Reduced and oxidized glutathione in the substantia nigra of patients with Parkinson's disease. Neurosci Lett 142: 128-130.

Sohal RS, Farmer KJ, Allen RG, Ragland SS, 1984. Effects of diethyldithiocarbamate on life span, metabolic rate, superoxide dismutase, catalase, inorganic peroxides and glutathione in the adult male housefly *Musca domestica*. Mech Ageing Dev 24: 175-183.

Spina MB, Cohen G. 1989. Dopamine turnover and glutathione oxidation: Implications for Parkinson disease. Proc Natl Acad Sci USA 86: 1398-1400.

Squires RF, Saederup E. 1991. A review of evidence for GABAergic predominance/glutamatergic deficit as a common etiological factor in both schizophrenia and affective psychoses: More support for a continuum hypothesis of "functional" psychosis. Neurochem Res 16: 1099-1111.

Stamler JS, Hausladen A. 1998. Oxidative modifications in nitrosative stress. Nat Struct Biol 5: 247-249.

Stamler SJ, Jaraki O, Osborne J, Simon DI, Keaney J, et al. 1992. Nitric oxide circulates in mammalian plasma primarily as an S-nitroso adduct of serum albumin. Proc Natl Acad Sci USA 89: 7674-7677.

Sucher NJ, Wong LA, Lipton SA. 1990. Redox modulation of NMDA receptor-mediated Ca^{2+} flux in mammalian central nerurons. Neuroreport 1: 29-32.

Sucher NJ, Awobuluyi M, Choi YB, Lipton SA. 1996. NMDA receptors: From genes to channels. Trends Pharmacol Sci 17: 348-355.

Taguchi J, Ohta H, Talman WT. 1995. Identification and pharmacological characterization of an S-nitrosoglutathione binding site in rat brain. Soc Neurosci Abstr 21: 626.

Tang LH, Aizenman E. 1993a. The modulation of *N*-methyl-D-aspartate receptors by redox and alkylating reagents in rat cortical neurones in vitro. J Physiol 465: 303-323.

Tang LH, Aizenman E. 1993b. Long-lasting modification of the *N*-methyl-D-aspartate receptor channel by a voltage-dependent sulfhydryl redox process. Mol Pharmacol 44: 473-478.

Tansey FA, Cammer W. 1991. Depletion of glutathione interferes with induction of glycerolphosphate dehydrogenase in the brains of young rats. Brain Res 564: 31-36.

Tate SS, Meister A. 1974. Interaction of γ-glutamyl transpeptidase with amino acids, dipeptides, and derivatives and analogs of glutathione. J Biol Chem 249: 7593-7602.

Tauck DL, Ashbeck GA. 1990. Glycine synergistically potentiates the enhancement of LTP induced by a sulfhydryl reducing agent. Brain Res 519: 129-132.

Terro F, Lesort M, Viader F, Ludolph A, Hugon J. 1996. Antioxidant drugs block in vitro the neurotoxicity of CSF from patients with amyotrophic lateral sclerosis. Neuroreport 7: 1970-1972.

Tolliver JM, Pellmar TC. 1987. Dithiothreitol elicits epileptiform activity in CA_1 of the guinea pig hippocampal slice. Brain Res 404: 133-141.

Tong L, Toliver-Kinsky T, Taglialatela G, Werrbach-Perez K, Wood T, et al. 1998. Signal transduction in neuronal death. J Neurochem 71: 447-459.

Usami S-I, Hjelle OP, Ottersen OP. 1996. Differential cellular distribution of glutathione—an endogenous antioxidant—in the guinea pig inner ear. Brain Res 743: 337-340.

Van Tol HH, Wu CM, Guan HC, Ohara K, Bunzow JR, et al. 1992. Multiple dopamine D4 receptor variants in the human population. Nature 358: 149-152.

Varga V, Janáky R, Marnela K-M, Gulyás J, Kontro P, et al. 1989. Displacement of excitatory aminoacidergic receptor ligands by acidic oligopeptides. Neurochem Res 14: 1223-1227.

Varga V, Janáky R, Oja SS. 1992. Modulation of glutamate agonist-induced influx of calcium into neurons by γ-L-glutamyl and β-L-aspartyl dipeptides. Neurosci Lett 138: 270-274.

Varga V, Janáky R, Saransaari P, Oja SS. 1994a. Endogenous γ-L-glutamyl and β-L-aspartyl peptides and excitatory aminoacidic neurotransmission in the brain. Neuropeptides 27: 19-26.

Varga V, Janáky R, Saransaari P, Oja SS. 1994b. Interactions of γ-L-glutamyltaurine with excitatory aminoacidergic neurotransmission. Neurochem Res 19: 243-248.

Varga V, Jenei Zs, Janáky R, Saransaari P, Oja SS. 1997. Glutathione is an endogenous ligand of rat brain *N*-methyl-D-aspartate (NMDA) and 2-amino-3-hydroxy-5-methyl-4-isoxazolepropionate (AMPA) receptors. Neurochem Res 22: 1165-1171.

Viguie CA, Frei B, Shigenaga MK, Ames BN, Packer L, et al. 1993. Antioxidant status and indexes of oxidative stress during consecutive days of exercise. J Appl Physiol 75: 566-572.

Vincent SR, McGeer EG. 1980. A comparison of sodium-dependent glutamate binding with high-affinity uptake in rat striatum. Brain Res 184: 99-108.

Volterra A, Trotti D, Tromba C, Floridi S, Racagni G. 1994. Glutamate uptake inhibition by oxygen free radicals in rat cortical astrocytes. J Neurosci 14: 2924-2930.

Wade LA, Brady HM. 1981. Cysteine and cystine transport at the blood–brain barrier. J Neurochem 37: 730-734.

Wechsler IS. 1940. Recovery in amyotrophic lateral sclerosis. Neurology 114: 948-950.

Weitberg AB, Weitzman SA, Clark EP, Stossel TP. 1985. Effects of antioxidants on oxidant-induced sister chromatid exchange formation. J Clin Invest 75: 1835-1841.

Werman R, Carlen PL, Kushnir M, Kosower EM. 1971. Effect of the thiol-oxidizing agent, diamide, on acetylcholine release at the frog endplate. Nat New Biol 233: 120-121.

Werner P, Cohen G. 1993. Glutathione disulfide (GSSG) as a marker of oxidative injury to brain mitochondria. Ann NY Acad Sci 679: 364-369.

Werns SW, Lucchesi BR. 1990. Free radicals and ischemic tissue injury. Trends Pharmacol Sci 11: 161-166.

Wheeler DD. 1984. Kinetics of D-aspartic acid release from rat cortical synaptosomes. Neurochem Res 9: 1599-1614.

Wiedau-Pazos M, Goto J, Rabizadeh S, Gralla E, Roe J, et al. 1996. Altered reactivity of superoxide dismutase in familial amyotrophic lateral sclerosis. Science 271: 515-518.

Willmore LJ, Triggs WJ. 1991. Iron-induced lipid peroxidation and brain injury responses. Int J Dev Neurosci 9: 175-180.

Wink DA, Nims RW, Darbyshire JF, Christodoulou D, Hanbauer I, et al. 1994. Reaction kinetics for nitrosation of cysteine and glutathione in aerobic nitric oxide solutions at neutral pH. Insights into the fate and physiological effects of intermediates generated in the NO/O_2 reaction. Chem Res Toxicol 7: 519-525.

Woodward JJ. 1994. The effects of thiol reduction and oxidation on the inhibition of NMDA-stimulated neurotransmitter release by ethanol. Neuropharmacology 33: 635-640.

Woodward JJ, Blair R. 1991. Redox modulation of N-methyl-D-aspartate-stimulated neurotransmitter release from rat brain slices. J Neurochem 57: 2059-2064.

Yan S-D, Chen X, Fu J, Chen M, Zhu H, et al. 1996. RAGE and amyloid β peptide neurotoxicity in Alzheimer's disease. Nature 382: 685-691.

Yang CS, Chou ST, Lin NN, Liu L, Tsai PJ, et al. 1994. Determination of extracellular glutathione in rat brain by microdialysis and high-performance liquid chromatography with fluorescence detection. J Chromatogr B Biomed Appl 661: 231-235.

Yankner BA. 1996. Mechanisms of neuronal degeneration in Alzheimer's disease. Neuron 16: 921-932.

Yankner BA, Dawes LR, Fisher S, Villa-Komaroff L, Oster-Granite ML, et al. 1989. Neurotoxicity of a fragment of the amyloid precursor associated with Alzheimer's disease. Science 245: 417-420.

Yim MB, Kang JH, Yim HS, Kwak HS, Chock PB, et al. 1996. A gain-of-function mutation of an amyotrophic lateral sclerosis-associated Cu, Zn-superoxide dismutase mutant: An enhancement of free radical formation due to a decrease in K_m for hydrogen peroxide. Proc Natl Acad Sci USA 93: 5709-5714.

Yoneda Y, Ogita K, Kouda T, Ogawa Y. 1990. Radioligand labeling of N-methyl-D-aspartic acid (NMDA) receptors by $[^3H](\pm)$ 3-(2-carboxypiperazin-4-yl)propyl-1-phosphonic acid in brain synaptic membranes treated with Triton X-100. Biochem Pharmacol 39: 225-228.

Yonezawa M, Back SA, Gan Y, Rosenberg PA, Volpe JJ. 1996. Cystine deprivation induces oligodendroglial death: Rescue by free radical scavengers and by a diffusible glial factor. J Neurochem 67: 566-593.

Yoritaka A, Hattori N, Mori H, Kato K, Mizuno Y. 1997. An immunohistochemical study on manganese superoxide dismutase in Parkinson's disease. J Neurol Sci 148: 181-186.

Yudkoff M, Pleasure D, Cregar L, Lin Z, Nissim I, et al. 1990. Glutathione turnover in cultured astrocytes: Studies with ^{15}N-glutamate. J Neurochem 55: 137-145.

Yudkoff M, Nissim I, Hertz L, Pleasure D, Erecińska M. 1992. Nitrogen metabolism: Neuronal-astroglial relationships. Prog Brain Res 94: 213-224.

Zängerle L, Cuénod M, Winterhalter KH, Do KQ. 1992. Screening of thiol compounds: Depolarization-induced release of glutathione and cysteine from rat brain slices. J Neurochem 59: 181-189.

Zhang N, Ottersen OP. 1992. Differential cellular distribution of two sulphur-containing amino acids in rat cerebellum. An immunocytochemical investigation using antisera to taurine and homocystic acid. Exp Brain Res 90: 11-20.

Zuo P, Ogita K, Han D, Yoneda Y. 1993a. Comparative studies on binding of 3 different ligands to the N-methyl-D-aspartate recognition domain in brain synaptic membranes treated with Triton X-100. Brain Res 609: 253-261.

Zuo P, Ogita K, Suzuki T, Han D, Yoneda Y. 1993b. Further evidence for multiple forms of N-methyl-D-aspartate recognition domain in rat brain using membrane binding techniques. J Neurochem 61: 1865-1873.

16 Low Molecular Weight Peptides

K. L. Reichelt

1	Special Aspects of Oligopeptide Biology	402
2	Peptide Formation	403
3	Is There a General Role of Peptides in the Brain?	403
4	Peptide Purification	404
5	Endogenous Peptides	405
5.1	Pyroglutamate Peptides	406
5.2	γ-Glutamyl Peptides	406
5.3	Diketopiperazines	406
5.4	Opioids	407
6	Exogenous Peptides	408
7	Summary	408

© Springer-Verlag Berlin Heidelberg 2007

16 Low molecular weight peptides

Abstract: Peptides active in the central nervous system (CNS) may be endogenous or exogenous. N-Terminally substituted peptides have higher lipid membrane penetration properties, and the main category of such peptides is *N*-acetyl and pyroglu peptides. Though probably also transmitters, most peptides apparently seem to be neuromodulators and are often released differently from traditional transmitters. Peptides are frequently colocated with monoamines, amino acid transmitters, and acetylcholine in the synapse. Peptides show pronounced ability to bind to other molecules, and solely immunoassays (immune-like) do not prove the presence of any given peptide. Mass spectrometry is the method of choice today. Peptides usually show hormetic dose–response curves and are often active in picomolar concentrations.

List of Abbreviations: CCK, Cholecystokinin; DSIP, Delta sleep inducing peptide; GABA, Gamma amino butyric acid; GSH, Glutathione (reduced form); MIF, Melanocyte-stimulating hormone-release inhibiting factor (P-L-GNH$_2$); TFA, Trifluoroacetic acid; TRH, Thyrotropin releasing hormone

1 Special Aspects of Oligopeptide Biology

Peptides are ampholytes and depending on pH their electric charge changes (❯ *Figure 16-1*). Their affinity for lipids increases with the chain length and also with the side groups of aromatic or aliphatic characters. Low-molecular weight peptides are here arbitrarily taken to be peptides up to 12 amino acids in length. The peptides found in brain may be endogenous (formed locally) or exogenous (imported from outside the brain).

◼ Figure 16-1
The different pH-dependent states of ionization of peptides. The R side chains can be aliphatic or aromatic. The latter will increase the HPLC retention time and the affinity for lipid membranes

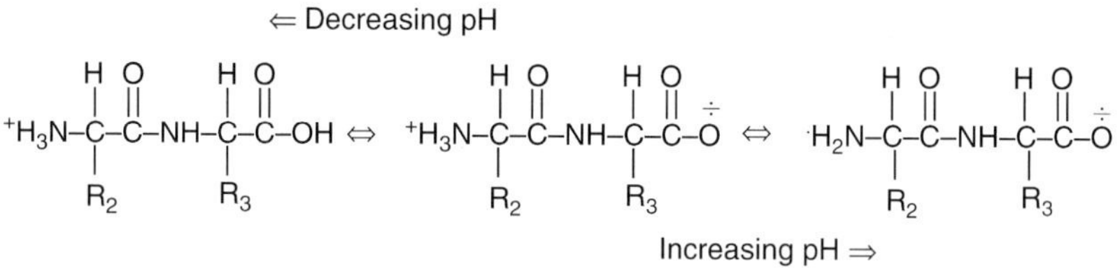

⇐ Decreasing pH

$^+H_3N-\underset{R_2}{\underset{|}{\overset{H}{\overset{|}{C}}}}-\overset{O}{\overset{\|}{C}}-NH-\underset{R_3}{\underset{|}{\overset{H}{\overset{|}{C}}}}-\overset{O}{\overset{\|}{C}}-OH \Leftrightarrow {}^+H_3N-\underset{R_2}{\underset{|}{\overset{H}{\overset{|}{C}}}}-\overset{O}{\overset{\|}{C}}-NH-\underset{R_3}{\underset{|}{\overset{H}{\overset{|}{C}}}}-\overset{O}{\overset{\|}{C}}-O^- \Leftrightarrow H_2N-\underset{R_2}{\underset{|}{\overset{H}{\overset{|}{C}}}}-\overset{O}{\overset{\|}{C}}-NH-\underset{R_3}{\underset{|}{\overset{H}{\overset{|}{C}}}}-\overset{O}{\overset{\|}{C}}-O^-$

Increasing pH ⇒

Unfortunately a lot of work on peptides has been mainly based on immunological techniques and the immune-like prefix is often forgotten. Other techniques to confirm the nature of any given peptide are important: high-pressure liquid chromatography (HPLC) with spiking and mass spectrometry (MS) (the final arbiter with MS/MS techniques causing fragmentation). With MALDI TOF or electron spray MS, the sequences can be obtained directly. The N-terminal amino acid determination with and without sequencing is very useful. Antibodies raised against different parts of the peptide have also been used in an attempt to bypass the limitations of immune techniques, but are not completely safe. "Immuno like" does not guarantee identity. Although peptides often are epitopes (signals binding to Björkman's groove causing antibody formation), the "code" is quite degenerate.

Work with peptides has been difficult for several reasons. First, they show an almost impossible tendency to form aggregates and bind noncovalently to proteins (Walter et al., 1978; Kastin et al., 1984; Burhol et al., 1986; Menezo and Khatchaturian, 1986; Rocetti et al., 1988; Gianfranceschi et al., 1994; Elgjo and Reichelt, 2004) and together with amino acids also to aminoglycans (Stadler and Whittaker, 1978). Interaction with phospholipids and membranes has been described for opioid peptides (Alsina et al., 1991), and demonstration of the presence of a peptide in a membrane could also be an artifact due to this phenomenon. Thus, unexpected high levels of opioid peptides were found in rat membranes (Rothman et al., 1987) and peptide drugs can form high-molecular weight forms in vivo (Hori et al., 1984).

Almost all peptides show bell-shaped dose responses or hormesis (Calabrese, 2001; Calabrese and Baldwin, 2001; 2003). This has proved to be rather exasperating during purification and testing because extensive dilution series must be run from 10^{-6} to 10^{-12} M. In aqueous solutions and at the water–glass interface stability problems remain a problem (Bell, 1997), and polypropanol plastic containers/test tubes and the dried form of the peptides are preferable for storage.

Peptides are also good inhibitors of peptide breakdown and may therefore have indirect effects by potentiating endogenous peptide activity by means of protecting them from breakdown (La Bella et al., 1985). Thus, the effects of adding a peptide may be indirect by prevention of the breakdown of endogenous peptides. Two peptides given simultaneously can result in cooperative effects by complex formation or protection against breakdown or both (Kastin et al., 2002). Peptides that contain SH groups at pH greater than 6 start forming mixed disulfides. At pH values higher than 7 this is quite fast and mixed disulfides may arise with any SH-containing peptides and proteins. At higher pH, sulfonic acids may also be formed and peptide bonds may also move from the α to the β position in aspartic acid. Strongly alkaline conditions may also split some peptide bonds like Asp-Ser.

Finally, chelating different metals may change the activity of peptides as shown for thyroliberin (TRH, thyrotropin-releasing hormone) (Tonoue et al., 1979).

2 Peptide Formation

There are three distinct routes of peptide synthesis in the brain: (1) ribosomal protein synthesis and specific splitting of a peptide, depending on peptidases and glycosylation of the precursor proteins (Loh and Gainer, 1979), (2) direct peptide synthesis by phosphorylation of the involved amino acids, and (3) amino acid or peptide transfer.

(1) An example of the first alternative is TRH and opioids like met-enkephalin and leu-enkephalin as well as cholecystokinin (CCK). Different prepeptides are frequently found to give rise to "families" of peptides with different chain lengths and often with different breakdown patterns in different tissues (Rehfeld et al., 2003). The enzymes involved in cutting out the active peptides from precursor proteins are convertase-1 and -2 (Rehfeld et al., 2003). Peptide fragments also have neuroactive roles (Hallberg and Nyberg, 2003). It is well known that sulfation, phosphorylation, glucosylation, and amidation can give rise to additional derivatives of the main peptide, and intramolecular disulfide bonds can cause considerable conformational specificity. *Trans*-glutaminase can further form pseudopeptide bonds between glutamine and lysine and attachment to different proteins and peptides. Also, esterification of fatty acids to the hydroxyl groups is possible in threonine- and serine-containing peptides and also binding to the N-terminal amine group (Shoji et al., 1986).

(2) Direct synthesis from amino acids: both glutathione (GSH) and peptides such as carnosine (β-alanyl histidine) and acetylaspartylglutamate are apparently synthesized directly by specific enzymes.

(3) Amino acid transfer: GSH can transfer its glutamate to amino acids and peptides and form a series of γ-glutamyl peptides (❷ *Table 16-2*). Because the reduced form GSH is the active peptide involved, this couples the uptake of amino acids and peptides to the redox state in the cells. Peptide leukotrienes are also formed by the group transfer (Shaw and Krell, 1991). Nucleotide peptides have been found in the liver but have not been studied in brain tissues.

3 Is There a General Role of Peptides in the Brain?

It seems that peptides are sometimes transmitters but often also act as modulators of neuronal activity. It is not easy to tell which is which in practice. However, it seems that peptides act more often by modulating transmitter uptake, release, and synthesis and also the firing frequency (Barker, 1976).

Examples are typically represented by the endogenous pentapeptide Q-Y-N-A-D, which blocks Na^+ channels (Meuth et al., 2003), and the tripeptide amide pyroE-W-G-NH_2, which stimulates serotonin uptake into CHO ovarian cells transfected with human gene for the serotonin transporter (Keller, 1997). The peptide also increases serotonin levels in rat brain synapses. This is also reflected in the more global effects of peptides on behavior (Fehm-Wolfdorf and Born, 1991). Thus, CCK 1-4 can induce a panic-like reaction (De Montigny, 1989), facilitate serotonin release in the rat hypothalamus (Voigt et al., 1998), accelerate the habituation rate to a novel environment (Crawley, 1984), affect memory in rats (Harro and Oreland, 1993), and play a role in satiety (Zhang et al., 1986). This seems to confirm the old paradigm that nature is conservative with regard to structure, but radical to function. This is illustrated by vasopressin, which has neuromodulating effects (Brinton et al., 2000), and oxytocin, related to emotional behavior (Pittman and Spencer, 2005; Matsuoka et al., 2005).

Serotonin-moduline is likewise an endogenous peptide involved in control of anxiety (Grimaldi et al., 1999) and serotonin activity (Massot et al., 1996). Tyr W-Mif induces Fos antigen as a potent analogue of Y-P-W-G-NH$_2$ (Liso et al., 1994) in several nuclei of the brain, and complex behavior is induced by the egg-laying hormone in *Aplysia* (Bernheim and Mayeri, 1995), indicating that more modulating roles are probable.

Because peptides are often released by higher-frequency stimulation than most conventional transmitters (Han, 2003; and references therein) and they coexist in the synapse (Lundberg and Hökfelt, 1983; Forloni et al., 1987; Nusbaum et al., 2001), often with presynaptic inhibitory activity (Hökfelt, 1991), all of this strengthens the view that they are more often neuromodulators rather than traditional neurotransmitters. Thus, a series of small dipeptides modulate GABAergic and glutamatergic transmission (Varga et al., 1988, 1994).

4 Peptide Purification

Because immune data need confirmation, a short outline of peptide purification is presented below. Usually the purification is initiated because of biological activity or an antibody-binding compound is found. If one is looking for a known molecular weight (MW) peptide, this may also guide the purification.

Roughly, the steps outlined below can be followed for purification (Pedersen et al., 1999); whenever possible, salt-containing buffers are avoided. They are not easily removed because of the attachment to peptides. It is also of advantage to stick to acidic buffers whenever possible because of alkaline problems, in particular if SH groups are involved, and basic conditions also quickly deamidate peptides. In short, the following steps are common: (1) G-25 gel filtration in 1 M acetic acid to separate low-MW compounds from high-MW compounds (we use 0.9×90 cm columns). (2) Gel filtration in 0.5 M acetic acid on P2 gel columns (1.6×90 cm and flow rate of 4 ml/10 min) to separate small and large peptides. (3) Batch separation on a cation exchanger in the H$^+$ form where nonprotonable compounds (mainly N terminally blocked peptides) pass through and protonable compounds (amino-free peptides) are retained and eluted with 2 M NH$_3$. (4) Anion exchange with a column in the acetate form in batch to separate neutral from acidic peptides, and elution with water followed by 2 M acetic acid. (5) C-18 reverse-phase batch separation of amino acids, ammonia, and salts, etc., from peptides as described in Böhlen et al. 1980. (6) Fractogel MG 2000 in 1 M acetic acid and 40 mM HCl and off-line ninhydrin coloring after hydrolysis for approximate MW determination. (7) C-18 reverse-phase HPLC using gradients of trifluoroacetic acid (TFA) against acetonitrile and the absorption read at 215 nm. When peptides contain aromatic groups, methanol can also be used as an organic phase and monitored at 280 nm. (8) Normal phase with acetonitrile and increasing the water gradient from 5% to 32% over 30 min at a flow rate of 1 ml/min using TSK amide-80 columns (25×0.24 cm) (Tosoh, Japan) is performed. UV monitoring of high-pressure runs is done at 215 and 280 nm.

When we find one N-terminal and/or one C-terminal amino acid, the peptide is considered probably pure. N-Substituted peptides were usually subjected to anion exchange starting with 0.01 M formic acid and increasing the concentration of formic acid to 0.5 M. N-Amino-free peptides separated with reverse phase HPLC as described above.

Off-line peptide monitoring (in particular, after gel filtration) is best done using 2 M alkaline hydrolysis in an aliquot in boiling water for 2 h, neutralizing with HCl, and ninhydrin coloring in an acetate–cyanide buffer where each amino acid has the same molecular absorption coefficient, thereby tryptophan is not destroyed. The reaction is stopped after 15 min boiling with isopropanol/water and read at 570 nm. The method has the advantage of increasing the sensitivity of the assay by splitting many peptide bonds (Reichelt and Kvamme, 1973).

5 Endogenous Peptides

These peptides can roughly be divided into N terminally blocked peptides and amino-free peptides (protonable peptides). Peptides that are N-substituted but with protonable side groups like lysine will behave as ordinary peptides, even if N terminally blocked.

The group of N terminally blocked peptides consists mainly of *N*-acyl, pyroglutamyl and cyclic peptides (diketopiperazines).

N-Acetyl peptides are mostly acetyl-aspartyl derivatives and the most abundant peptide in the brain is acetylaspartylglutamate, which is found up to 0.28 μ moles/g brain tissue. This peptide is probably formed by a nonribosomal mechanism (Reichelt and Kvamme, 1973; Lähdesmäki and Timonen, 1982; Cangro et al., 1987) and has been suspected of being a possible transmitter.

A series of acetyl-aspartyl peptides could be formed from different amino acids (Reichelt and Kvamme, 1973; Lähdesmäki and Timonen, 1982) and the presence of such peptides was confirmed by MS (❯ *Table 16-1*) (Marnela and Lähdesmäki, 1983; Marnela et al., 1984). We proposed that they might be short-term memory

◘ Table 16-1
γ-Glutamyl peptides and peptoids

Monkey brain Reichelt (1970)	Ox brain Sano et al. (1966)	Others
Glu-Glu	Glu-Glu	Glu-Tau[a]
Glu-Gln	Glu-Gln	Glu-Asp[b]
Glu-Ala	Glu-Ala-AIB	Glu-histamine[c]
Glu-Ile	Glu-Ser	Glu-dopamine[d]
Glu-Ser	Glu-Ala	
Glu-Ala-Gly	Glu-Val	
Glu-Cys-Gly	Glu-Met-Cys-Gly	
	Glu-Cys-Gly	

[a]Nakamura et al. (1990)
[b]Cheung and Lim (1979)
[c]Weinreich (1979)
[d]McCaman et al. (1985)
AIB, aminoisobutyric acid

substrates reflecting the sequence of impinging signals on the integrating neuron (Reichelt et al., 1982). Furthermore, the formation of less hydrophilic peptides could ease the reuptake of electrically charged transmitters like amino acids into synaptosomes.

In spite of the slow in vivo metabolism of acetylaspartate (Jacobson, 1959; Reichelt and Kvamme, 1967), the absence of acetylaspartate-splitting acetylase causes a massive loss of *N*-acetylaspartate in the urine (Canavan's syndrome) and also an almost complete absence of synapse formation (Gordon, 2001). This may be indicative of an essential role of *N*-acetylaspartic acid in the formation and/or maintenance of synapses.

5.1 Pyroglutamate Peptides

The most well known of these peptides are TRH and folliliberin (FSH, follicle-stimulating releasing hormone), which has also other functions, affecting the function of different parts of the CNS and being related to depression (Kirkegaard et al., 1979; Banki et al., 1988). Most of the hypothalamic-releasing peptides are N-terminal pyroglutamate compounds, which probably confers some peptidase resistance of these peptides. In ◗ *Table 16-2*, a few pyroglu peptides, recently reviewed by Reichelt (2006), are listed. In many tissues

◘ Table 16-2
N-Substituted peptides

N-Acetyl substituted peptides	
Ac-D-E[a]	Ac-D-E
Ac-D-G-S[a]	Ac-D-tau[b]
Ac-D-E-G[a]	Ac-D-E-tau[b]
Ac-D-E-D[a]	Ac-E-tau[b]
Ac-D-D[a]	Ac-A-A-D-I-S-Q-W-A-G-P-L[c]
Ac-D (tau,tau)[a]	
Pyroglu peptides	Cyclic peptides
PyroE-H-P-NH$_2$ (TRH)	H-P
PyroE-H-W-S-Y-G-L-R-P-NH$_2$ (LHRH)	G-F

[a]Reichelt and Kvamme (1973)
[b]Marnela and Lähdesmäki (1983)
[c]Hippocampal cholinergic neurostimulating peptide
PyroE, pyroglutamyl; TRH, thyroliberin (thyrotropin-releasing hormone); LHRH, luliberin (luteinizing hormone-releasing hormone)

pyroglu peptides are growth-regulating entities acting by means of feedback inhibition (Elgjo and Reichelt, 2004) as demonstrated in hematopoietic (Paukovits and Laerum, 1983) and thymic cells (Gianfranceschi et al., 1994).

5.2 γ-Glutamyl Peptides

These peptides are also relatively abundant and may well reflect amino acid /peptide transport into neurons and/or glia by means of the γ-glutamyl cycle (Griffith et al., 1979). For an overview, see ◗ *Table 16-1* (Sano et al., 1966; Reichelt, 1970).

The γ-glutamyl cycle has the advantage of coupling amino acid uptake to the redox potential of the cell because GSH reflects the level of reductive potential (Sir Hans Krebs, personal communication). The more polar amino acids are also rendered more hydrophobic when their amino groups are blocked, thus facilitating their uptake across membranes, independently of the transport proteins.

5.3 Diketopiperazines

Diketopiperazines or cyclic peptides also show high affinity for other compounds (Gisin et al., 1978). The cyclic dipeptide H-P split from TRH forms such a cyclic peptide, having many different effects in the CNS.

◘ Table 16-3
Some small opioid peptides

Kyotorphin	Y-R
Neo-kyotorphin	T-S-K-Y-R
Met-enkephalin	Y-G-G-F-M
Leu-enkephalin	Y-G-G-F-L
Endorphin 1	Y-P-W-G-NH$_2$
Endorphin 2	Y-P-F-F-NH$_2$

◘ Table 16-4
Some very small peptides with unknown/uncertain function

GABA-H	Pisano et al. (1961)
GABA-K	
GABA-(CH)H	
β-A-K	Kumon et al. (1970)
β-A (CH$_3$)H	
β-A-H (carnosine)	
D-S	
A–F	Versteeg and Witter (1970)
I–L	

◘ Table 16-5
Other small peptides, with presence in the CNS properly demonstrated

Neuropeptide FF	F-L-F-Q-P-Q-R-FNH$_2$
Neuropeptide SF	S-Q-A-F-L-F-Q-P-Q-P-FNH$_2$
Substance P	R-P-K-P-Q-Q-F-F-G-L-MNH$_2$
DSIP	W-A-G-G-D-A-S-G-E
Angiotensin I	D-R-V-Y-I-H-P-F-H-L
Angotensin II	D-R-V-Y-I-H-P-F
Cholecystokinin (CCK) 1–4	D-Y-M-G
Gastrin tetrapeptide	W-M-D-F-NH$_2$
Oxytocin	C-Y-I-Q-N-C-P-L-G-NH$_2$
Vasopressin (arginine)	C-Y-F-Q-N-C-P-R-G-NH$_2$
Vasopressin (lysine)	C-Y-F-Q-N-C-P-K-G-NH$_2$
Melanocyte-stimulating hormone release-inhibiting factor	P-L-G-NH$_2$
5-HT moduline	L-S-A-L
TRH-potentiating peptide	S-F-M-E-S-D-V-T

DSIP, δ-sleep-inducing peptide (Schoenenberger et al., 1978)
In oxytocin and vasopressin, the two cysteines are joined by a disulfide bond
5-HT, serotonin; TRH, thyroliberin

5.4 Opioids

The small opioid peptides like met-enkephalin and leu-enkephalin are both found in the CNS, although MS has been sparingly used to prove their existence. Other low-molecular opioid peptides found are given in

▶ *Table 16-3*. Opioids not only facilitate dopaminergic transmission by inhibiting dopamine uptake (Hole et al., 1979) but also inhibit GABA-mediated transmission (Vaughan et al., 1997). Discussions of β-endorphin-sized peptides are found in other chapters of this handbook.

6 Exogenous Peptides

The uptake of peptides is usually limited, but the N-substitution renders peptides more olefinic, and therefore increases uptake. Furthermore, the formation of glycopeptides, e.g. opioids, increases uptake (Egleton et al., 2005). Low-MW opioids are taken up across the blood–brain barrier (Ermisch et al., 1983, Banks et al., 1996) and interestingly exorphins formed in the gut from casein can penetrate this barrier as well (Nyberg et al., 1989) and induce Fos antigen in several important nuclei of the brain (Sun et al., 1999). Also, behavioral effects (Sun and Cade, 1999) similar to those seen with intracranioventricularly injected opioid from urine of patients with schizophrenia (Hole et al., 1979; Drysdale et al., 1982; Cade et al., 2000) are seen.

The fact that peptides are often active at nanomolar to picomolar concentrations may have vast consequences in neuropsychiatric diseases, where the gut-to-brain axis seems to be important, as described in the autistic syndromes (Reichelt and Knivsberg, 2003). The effect of feeding in infectious bowel disease (Geissler et al., 1995; Hart et al., 1998), causing widespread scattered edema in the white matter seen by means of nuclear magnetic resonance (NMR), points to a fairly intimate relationship of the gut to the brain.

7 Summary

Peptides should be nowadays studied not only using immune techniques but also techniques preferably supplemented with MS or other independent techniques. There is reasonably strong evidence indicating that small peptides are mostly neuromodulators rather than traditional neurotransmitters. The tendency for peptides to form complexes with peptides and proteins as well as membranes may cause problems during identification and subcellular distribution studies.

References

Alsina MA, Sole N, Mestres C, Busquets MA, Haro I, et al. 1991. Physico-chemical interaction of opioid peptides with phospholipids and membranes. Int J Pharm 70: 111-117.

Banki CM, Bisette G, Arato M, Nemeroff CB. 1988. Elevation of immunoreactive CSF TRH in depressed patients. Am J Psychiatry 145: 1526-1531.

Banks WA, Kastin AJ, Harrison LM, Zadina JE. 1996. Perinatal treatment of rats with opiates affects the development of blood–brain barrier transport system PTS-1. Neuroendocrin Teratol 18: 711-715.

Barker JL. 1976. Peptides: Roles in neuronal excitability. Physiol Rev 56: 435-452.

Bell LN. 1997. Peptide stability in solids and solutions. Biotechnol Prog 13: 342-346.

Bernheim SM, Mayeri E. 1995. Complex behavior induced by egg-laying hormone in *Aplysia*. J Comp Physiol A 176: 131-161.

Böhlen P, Castillo F, Ling R, Guillemin R, 1980. Purification of peptides: An efficient procedure for the separation of peptides from amino acids and salts. Int J Pept Prot Res 16: 306-310.

Brinton RD, Thompson RH, Brownson EA. 2000. Spatial, cellular and temporal basis of vasopressin potentiation of norepinephrine-induced cAMP formation. Eur J Pharmacol 405: 73-88.

Burhol K, Jensen TG, Florholmen TG, Jorde H, Vonen B, et al. 1986. Protein-binding and aggregation of somatostatin in human plasma. Ital J Gastroenterol 18: 1-6.

Cade RJ, Privette RM, Fregly M, Rowland N, Sun Z, et al. 2000. Autism and schizophrenia: Intestinal disorders. Nutr Neurosci 2: 57-72.

Calabrese EJ. 2001. Opiates: Biphasic dose responses. Crit Rev Toxicol 31: 585-604.

Calabrese EJ, Baldwin LA. 2001. Hormesis: U-shaped dose responses and their centrality in toxicology. Trends Pharmacol Sci 22: 285-291.

Calabrese EJ, Baldwin LA. 2003. Peptides and hormesis. Crit Rev Toxicol 33: 355-405.

Cangro CB, Namboodiri MA, Sklar LA, Corigliano-Murphy A, Neale JH. 1987. Immunohistochemistry and biosynthesis of N-acetylaspartylglutamate in spinal sensory ganglia. J Neurochem 49: 1579-1588.

Cheung ST, Lim R. 1979. Isolation of γ-glutamylaspartic acid and α-aspartyl-alanine from pig brain. Biochim Biophys Acta 586: 418-424.

Crawley JN. 1984. Cholecystokinin accelerates the rate of habituation to a novel environment. Pharmacol Biochem Behav 20: 23-27.

De Montigny C. 1989. Cholecystokinin tetrapeptide induces panic-like attacks in healthy volunteers. Arch Gen Psychiatry 46: 511-517.

Drysdale A, Deacon R, Lewis P, Olley J, Electricwala A, et al. 1982. A peptide-containing fraction of plasma from schizophrenic patients which binds to opiate receptors and induces hyperreactivity in rats. Neuroscience 7: 1567-1573.

Egleton RD, Bilsky EJ, Tollin G, Bhanasekaranh M, Lowry J, et al. 2005. Biousian glycopeptides penetrate the blood–brain barrier. Tetrahedron-Asymmetr 16: 65-75.

Elgjo K, Reichelt KL. 2004. Chalones: From aqueous extracts to oligopeptides. Cell Cycle 3: 12-15.

Ermisch A, Ruhle HJ, Neubert K, Hartrodt B, Landgraf R. 1983. On the blood–brain barrier to peptides [^{3}H]β-casomorphin-5 uptake by eighteen brain regions in vivo. J Neurochem 41: 1229-1233.

Fehm-Wolfdorf G, Born J. 1991. Behavioral effects of neurohypophyseal peptides in healthy volunteers: 10 years of research. Peptides 12: 1399-1406.

Forloni G, Grzanna R, Blakely RD, Coyle JT. 1987. Colocalization of N-acetyl-aspartyl-glutamate in central cholinergic, noradrenergic, and serotoninergic neurons. Synapse 1: 455-460.

Geissler A, Andus T, Roth M, Kullmann F, Caesar I, et al. 1995. Focal white-matter lesions in brain of patients with inflammatory bowel disease. Lancet 345: 897-898.

Gianfranceschi GL, Czerwinski A, Angiolillo A, Marsili V, Castigli E, et al. 1994. Molecular models of small phosphorylated chromatin peptides. Structure–function relationship and regulatory activity on in vitro transcription and on cell growth and differentiation. Peptides 15: 7-15.

Gisin BF, Ting-Beall HP, Davis DG, Grell E, Tosteson DC. 1978. Selective ion binding and membrane activity of synthetic cyclopeptides. Biochim Biophys Acta 509: 201-217.

Gordon N. 2001. Canavan disease: A review of recent developments. Eur J Pediat Neurol 5: 65-69.

Griffith OW, Bridges RJ, Meister A. 1979. Transport of γ-glutamyl amino acids: Role of glutathione and γ-glutamyl transpeptidase. Proc Natl Acad Sci USA 76: 6319-6322.

Grimaldi B, Bonnin A, Fillion MP, Prudhomme N, Fillion G. 1999. 5-Hydroxytryptamine-moduline: A novel endogenous peptide involved in the control of anxiety. Neuroscience 93: 1223-1225.

Hallberg M, Nyberg F. 2003. Neuropeptide conversion to bioactive fragments—an important pathway in neuromodulation. Curr Protein Pept Sci 4: 31-44.

Han J-S. 2003. Acupuncture: Neuropeptide release produced by electrical stimulation of different frequencies. Trends Neurosci 26: 17-22.

Harro J, Oreland L. 1993. Cholecystokinin receptors and memory: A radial maze study. Pharmacol Biochem Behav 44: 509-517.

Hart PE, Gould SR, MacSweeney JE, Clifton A, Schon F. 1998. Brain white-matter lesions in inflammatory bowel disease. Lancet 351: 1558.

Hökfelt T. 1991. Neuropeptides in perspective: The last ten years. Neuron 7: 867-879.

Hole K, Bergslien H, Jorgensen HA, Berge OG, Reichelt KL, et al. 1979. A peptide-containing fraction in the urine of schizophrenic patients which stimulates opiate receptors and inhibits dopamine uptake. Neuroscience 4: 1883-1893.

Hori R, Suito YS, Yasuhara M, Okumura K. 1984. In vivo conversion of peptide drug into high molecular weight forms. J Pharmacodyn 7: 910-916.

Jacobson KB. 1959. Studies on the role of N-acetylaspartic acid in mammalian brain. J Gen Physiol 43: 323-333.

Kastin AJ, Casillanos PF, Fischman PJ, Profitt JK, Graf MV. 1984. Evidence for peptide aggregates. Pharmacol Biochem Behav 21: 969-974.

Kastin AJ, Pan W, Akerstrom V, Hackler L, Wang C, et al. 2002. Novel peptide–peptide cooperation may transform feeding behavior. Peptides 23: 2189-2196.

Keller J. 1997. Impact of autism-related peptides and 5-HT system manipulations on cortical development and plasticity. The First Annual Report for the EU Project BMH4-CT96-0730, pp. 1-10.

Kirkegaard C, Faber J, Hummer L, Rogowski P. 1979. Increased levels of TRH in cerebrospinal fluid from patients with endogenous depression. Psychoneuroendocrinology 4: 227-235.

Kumon A, Matsuoka Y, Nakajima T, Kakimoto Y, Imaoka N, et al. 1970. Isolation and identification of N-(β-alanyl)lysine and N-(γ-aminobutyryl)lysine from bovine brain. Biochim Biophys Acta 200:170-171.

La Bella FL, Geiger JD, Glavin GB. 1985. Administration of peptides inhibits the degradation of endogenous peptides. The dilemma of distinguishing direct from indirect effects. Peptides 6: 645-660.

Lähdesmäki P, Timonen M. 1982. Non-coded biosynthesis of N-acetyl-aspartyl peptides in mouse brain homogenates. J Liq Chromatogr 5: 989-1002.

Liso JP, Samit JE, Zadina JE, Kenigs V, Kastin AJ, et al. 1994. Induction of Fos immunoreactivity in rat brains by a potent analog of the brain peptide Tyr-W-MIF-1. Regul Peptides 54: 163-164.

Loh P, Gainer H. 1979. The role of the carbohydrate in the stabilization, processing, and packaging of the glycosylated adrenocorticotropin–endorphin common precursor in toad pituitaries. Endocrinology 105: 474-487.

Lundberg JM, Hökfelt T. 1983. Coexistence of peptides and classical neurotransmitters. Trends Neurosci 6: 325-333.

Marnela K-M, Lähdesmäki P. 1983. Mass spectral and hydrolytic determination of amino acid sequences in synaptosomal peptides from calf brain. Neurochem Res 7: 933-941.

Marnela K-M, Timonen M, Lähdesmäki P. 1984. Mass spectrometric analysis of brain synaptic peptides containing taurine. J Neurochem 43: 1650-1653.

Massot O, Rouselle JC, Fillon MP, Grimaldi B, Cloez-Tayarani I, et al. 1996. 5-Hydroxytryptamine-moduline, a new endogenous cerebral peptide, controls the serotoninergic activity via its specific interaction with 5-hydroxytryptamine 1B/1D receptors. Mol Pharmacol 50: 752-762.

Matsuoka T, Sumiyoshi T, Tanaka K, Tsunoda M, Uehara T, et al. 2005. NC-1900, an arginine-vasopressin analogue, ameliorates social behavior deficits and hyperlocomotion in MK-801-treated rats: Therapeutic implications for schizophrenia. Brain Res 1053: 131-161.

McCaman MW, Stetzler J, Clark B. 1985. Synthesis of γ-glutamyldopamine and other peptidoamines in nervous system of *Aplysia californica*. J Neurochem 45: 1828-1835.

Menezo NY, Khatchaturian C. 1986. Peptides bound to albumin. Life Sci 39: 1751-1753.

Meuth SG, Budde T, Duyar H, Landgraf P, Broicher T, et al. 2003. Modulation of neuronal activity by the endogenous pentapeptide QYNAD. Eur J Neurosci 18: 2697-2706.

Nakamura K, Higashiura K, Nishimura N, Yamamoto A, Tooyama I, et al. 1990. Isolation of glutamyltaurine from bovine brains and proof of its γ-linkage by the B/E linked scan SIMS technique. J Neurochem 55: 1064-1066.

Nusbaum MP, Blitz DM, Swensen AM, Wood D, Marder E, 2001. The role of co-transmission in neuronal network modulation. Trends Neurosci 24: 146-154.

Nyberg F, Liberman R, Lindström LH, Lyrenas S, Koch G, et al. 1989. Immunoreactive β-casomorphin-8 in cerebrospinal fluid from pregnant and lactating women: Correlation with plasma level. J Clin Endocrin Metab 68: 283-289.

Paukovits WR, Laerum OD. 1983. Isolation and synthesis of a hemoregulatory peptide. Zschr Zellforsch 37C: 1297-1300.

Pedersen OS, Liu Y, Reichelt KL. 1999. Serotonin uptake stimulating peptide found in plasma of normal individuals and in some autistic urines. J Peptide Res 53: 641-646.

Pisano JJ, Wilson JD, Cohen L, Abraham D, Udenfriend S, 1961. Isolation of γ-aminobutyrylhistidine (homocarnosine) from brain. J Biol Chem 236: 499-502.

Pittman QJ, Spencer SJ. 2005. Neurohypophysial peptides: Gatekeepers in the amygdala. Trends Endocrinol Metab 16: 343-334.

Rehfeld JF, Bundgaard JR, Friis-Hansen L, Goetze JP. 2003. On the tissue-specific processing of procholecystokinin in the brain and gut – a short review. Can J Physiol Pharmacol 54, Suppl 4: 79-79.

Reichelt KL. 1970. The isolation of γ-glutamyl peptides from monkey brain. J Neurochem 17: 19-25.

Reichelt KL, Knivsberg AM. 2003. Can the pathophysiology of autism be explained by the nature of the discovered urine peptides? Nutr Neurosci 6: 19-28.

Reichelt KL, Kvamme E. 1967. Acetylated and peptide bound glutamate and aspartate in brain. J Neurochem 14: 987-996.

Reichelt KL, Kvamme E. 1973. Histamine dependent formation of N-acetyl-aspartyl peptides in mouse brain. J Neurochem 21: 849-859.

Reichelt KL, Edminson PD, Sälid G. 1982. Peptides and memory: A working hypothesis in neuronal plasticity and memory formation. Neuronal Plasticity and Memory Formation. Marsan CA, Matthies H, editors. New York: Raven Press; pp. 63-74.

Reichelt WH. 2006. Naturally occurring small pyroglutamyl peptides acting as signal molecules. Progress in Oncogene Research. Peale LS, editor. New York: Nova Science Publ.; pp. 207-254.

Rocetti G, Venurella F, Roda IG. 1988. Enkephalin binding system in human plasma II. Comparative protection of different peptides. Neurochem Res 13: 221-224.

Rothman RB, Danks JA, Iadrola MJ. 1987. Unexpectedly high level of opioid peptides in rat membranes. Peptides 8: 645-649.

Sano I, Kakimoto Y, Kanazawa A, Nakajima T, Shimizu H. 1966. Identifizierung einiger Glutamylpeptide aus Gehirn. J Neurochem 13: 711-719.

Shaw A, Krell RD. 1991. Peptide leukotrienes: Current status of research. J Med Chem 34: 1235-1242.

Schoenenberger GA, Maier PF, Tobler HJ, Wilson K, Monnier M. 1978. The delta EEG (sleep-inducing peptide, DSIP). Pflügers Arch 376: 119-129.

Shoji S, Hayashi M, Funakoshi T, Kubota Y. 1986. Rapid identification of NH$_2$-terminal myristyl peptides by reversed-phase high-performance liquid chromatography. J Chromatog 356: 179-185.

Stadler H, Whittaker VP. 1978. Identification of vesiculin as a glycosaminoglycan. Brain Res 153: 408-413.

Sun Z, Cade JR. 1999. A peptide found in schizophrenia and autism causes behavioral changes in rats. Autism 3: 85-95.

Sun Z, Cade RJ, Fregly MJ, Privette RM. 1999. β-casomorphin induces Fos-like immunoreactivity in discrete brain regions relevant to schizophrenia and autism. Autism 3: 67-83.

Tonoue T, Minagawa S, Kato N, Kan M, Terao T, et al. 1979. The effect of metal complex of thyrotropin-releasing hormone on locomotor activity of neonatal chicken. Pharmacol Biochem Behav 10: 201-204.

Varga V, Janáky R, Saransaari P, Oja SS. 1994. Endogenous γ-L-glutamyl and β-L-aspartyl peptides and excitatory aminoacidergic neurotransmission in the brain. Neuropeptides 27: 19-26.

Varga V, Kontro P, Oja SS. 1988. Modulation of GABAergic neurotransmission in the brain by dipeptides. Neurochem Res 13: 1027-1034.

Vaughan CW, Ingram SL, Connor MA, Christie MJ. 1997. How opioids inhibit GABA-mediated neurotransmission. Nature 390: 611-614.

Versteeg DH, Witter A. 1970. Isolation and identification of α-aspartyl-serine, alanylphenylalanine, and isoleucylleucine from calf brain stem. J Neurochem 17: 41-52.

Voigt JP, Soht R, Fink H. 1998. CCK-8S facilitates 5-HT release in rat hypothalamus. Pharmacol Biochem Behav 59: 179-182.

Walter R, Deslauriers R, Smith ICP. 1978. Aggregation of Pro-Leu-GlyNH$_2$ in aqueous solution. FEBS Lett 95: 357-360.

Weinreich D. 1979. γ-Glutamylhistamine: A major product of histamine metabolism in ganglia of the marine mollusk, *Aplysia californica*. J Neurochem 32: 363-369.

Zhang D-M, Bula W, Stellar E. 1986. Brain cholecystokinin as a satiety peptide. Physiol Behav 36: 1138-1186.

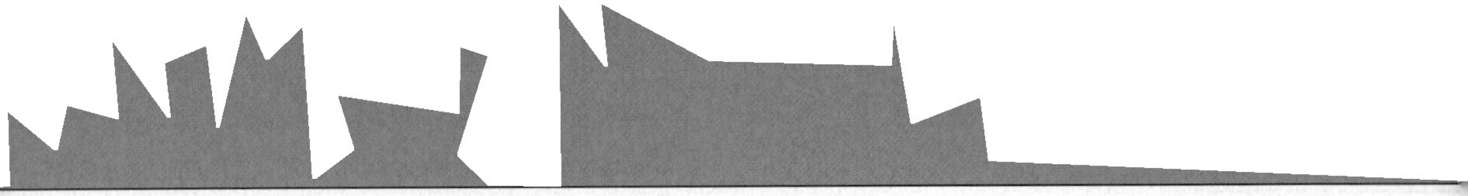

Index

Acetoacetate, 70
Acetylcholine, 75
Acetylcholinesterase, 253
Acetyl-L-carnitine, 256
AD-Amino acids, discovery, 208
Adrenal medulla, 80, 88
Aerobic cells, 349
Aging, 378–381
Agmatine, 100, 106
Agonists glutamate receptors, 179, 180
Alanine aminotransferase, 3, 9–11
Albumin, 64, 66, 69, 75–77, 85
Alcohol, 170, 182
Alzheimer's disease, 310, 329, 333, 378, 380, 381
 – muscarinic sites, 254
 – type II astrocytosis, 251
Amfonelic acid, 87
 – γ-aminobutyrate receptors, 172, 173
 – γ-aminobutyric acid (GABA), 357, 372, 374, 377
Amidohydrolase I, 308
Amidohydrolase II, 309, 311, 314, 315, 317, 318, 326, 335, 336
Amino acid, 2–4, 6–11, 349, 350, 352–355, 362, 363, 369, 374, 376, 378, 381, 383, 384
 – aromatic, 61, 62, 85
 – branched-chain, 62, 64, 73, 76, 85
 – glutamate, 118–127
 – large, neutral, 85
 – isoleucine, 118, 120
 – leucine, 118, 120–122, 124–127
 – neutral, 64, 70, 85
 – valine, 118, 120
 – γ-aminobutyric acid (GABA), 118, 119, 121, 122, 124–127
Aminoacidopathies, 279–289
Aminoacidurias, 281, 182
Aminotransferase, 3, 5, 7, 10, 11, 13, 14
Ammonia effects, 171
Ammonia, 251–253, 255, 256

Amphetamine, 88
Amyotrophic lateral sclerosis, 381–383
Anaplerosis, 2, 8, 10–12
Antiepileptic drugs, 177, 178, 183
Antinociception, 177
Antioxidant, 350, 353, 355, 359–361, 379, 381–383, 385
Apparent diffusion coefficient, 323
Appetite, 82, 86
Arginase
 – immunoreactivity, 105
 – isoforms, 105
Arginine
 – concentration in brain and CSF, 100, 101
 – deimination, 102
 – immunohistochemistry, 101
 – metabolism in astrocytes, 108, 109
 – N-methylated analogues, 100
 – proteinogenic amino acid, 109
 – substrate of NOS, 105
Arginine decarboxylase (ADC), 106, 107
Argininosuccinate lyase
 – localization, 103, 104
Argininosuccinate synthetase
 – immunoreactivity, 103, 104
 – localization cellular/subcellular, 103, 104
Argininosuccinate, 102–104, 109
Aromatic amino acid decarboxylase (AAAD), 74, 77, 78, 82
Aromatic amino acids, 60–62, 85
ASC transport system
 – localisation, 141
 – subtypes, 143
 – transport of SAAs, 141–145
Ascorbic acid, 360, 361
Aspartate aminotransferase, 3, 5, 7
Astrocytes
 – aminotransferase isoform, 126
 – energy metabolism in, 267

 – gliotoxins and, 148
 – glutamate/glutamine cycle, 123, 126
 – glutamate/pyruvate cycle, 123, 124
 – GSH in, 148
 – interaction with neurons, 263, 267, 268
 – lactate, 123, 124
 – localisation of SAAs in, 138
 – morphology of, 265–267, 272
 – oxidative stress in, 262, 265
 – release of SAAs from, 139
 – SAA biosynthesis and metabolism in, 135
 – SAA transport into, 141
 – sodium-calcium exchanger in, 139
Biosynthesis taurine, taurine transporter, 160
Blood-brain barrier (BBB), 48, 66, 352, 353, 385
Bold signals, 322–324, 333, 335
Brain compartments, 309, 310
Brain peptides
 – ampholytes, 402
 – cyclic peptides, 405–407
 – endogenous peptides, 405
 – endorphins, 407
 – exogenous peptides, 408
 – exorphins, 408
 – gamma-glutamyl peptides, 406
 – N-substistuted peptides, 404
 – pepetide "families", 405
Brain, 61 f
Branched-chain amino acids (BCAAs), 62, 66, 76, 85
BCAT nitrogen shuttle in brain, 125
 – catabolic pathway, 118–120, 125, 126
 – glial metabolism, 122
 – nitrogen transfer, 118, 120, 122, 124, 126
 – transport in brain, 121, 125

© Springer-Verlag Berlin Heidelberg 2007

Index

Branched-chain aminotransferase (BCAT), 3, 4
- isoforms, 119
- localization in brain, 119
- nitrogen shuttle in brain, 125

Branched-chain α-keto-acid
- dehydrogenase enzyme complex, 120
- maple syrup urine disease, 120, 127

Breakfast, 86
Calcium, 360, 368, 378
Canavan disease, 301, 311, 315–317, 335, 336
Cancer, 363, 364
Carbamyl phosphate synthetase, 250, 251
"Carbohydrate-craving", 85, 86
Casein, 62
Catecholamine, 60–62, 70, 75, 79–81, 84, 87, 88
Catecholaminergic transmission
- dopamine, 270, 271
- serotonin, 270

Cationic amino acid transporter (CAT), 107, 108
Cell cycle, 350, 363
Cell swelling, 155, 166–170
Cell volume regulation, 160, 165, 167
Cell-damaging-conditions, 163–165, 175, 179
Central nervous system, 25, 29, 31, 34, 35, 39, 40
Cerebellar cell migration, 158
Cerebellum, 352, 354, 356, 368, 373
Cerebral palsy, 250, 251
Chemometabolism, 340
Cholesterol, 61
Choline, 75
Choline acetyltransferase, 253, 254
Choroid plexus, 60, 75
Chronic fatigue syndrome, 378, 385
Chronoarchitecture, 323, 340
Chronometabolism, 340
α-chymotrypsin, 73
Cingulate cortex, 80
Citrulline
- concentration in brain and CSF, 101
- generation within proteins, 105
- immunohistochemistry, 101
- intermediate in citrulline-NO cycle, 105

Citrullinemia, 109, 251, 256, 257
Citrulline-NO cycle, 101, 103–106, 108, 109
Clozapine, 88

Cold stress, 87
Compartmentation, 3, 14
Control, "open loop", 64
Corpus striatum, 64
Creatine, synthesis, 106
Cysteine sulfinate decarboxylase
- biosynthesis of taurine, 135
- metabolism of L-cysteine, 135

Cysteine, 349 f
Cystine-glutamate (x_c^-) exchanger
- glutamate transport and, 144
- GSH synthesis and, 145
- interrelationship with glutamate transporters, 144, 145
- localisation, 144
- neurotoxicity and, 146
- SAA transport and, 141

D-Aspartate
- biosynthesis, 217–219
- localization in the nervous and endocrine system, 217
- metabolism, 219
- neurobiology, 217–220
- precursor for endogenous NMDA, 219
- role of endogenous, 218
- transport and release, 218

DDAH (Dimethylarginine dimethylaminohydrolase), 105
Degeneration retina, 161, 181
Depression, 68, 78, 86, 87
Derivatives taurine, 169, 182–184
Development, 157–159, 161, 167, 174–176, 184
Developmental disability, 283
Dextrin, 85
Dextrose, 85
Diet, high-carbohydrate vs high-protein, 86
Dietary carbohydrates, 64, 78
Dietary proteins, 62, 78, 85
Dietary starches, 62
Diffusion coefficient, 323, 331, 332
Dihydroxyphenylacetic acid (DOPAC), 88
Diseases
- Alzheimer (AD), 101, 104, 107
- multiple sclerosis, 102

Disorders of GABA metabolism
- GABA-transaminase deficiency, 279, 301
- pyridoxine dependency, 279, 300
- succinic semialdehyde dehydrogenase deficiency, 279, 301

Disorders of glutathione metabolism
- γ-glutamyl synthetase deficiency, 279, 300
- γ-glutamyl transpeptidase deficiency, 279, 300
- 5-oxoprolinase deficiency, 279, 300
- 5-oxoprolinuria (glutathione synthetase deficiency), 279, 300

Distribution taurine, 156–159
Disulfide bonds, 370, 371
DL-NAM, 75
DOPA, 61, 72, 77–80, 82, 87
Dopamine, 60 f
D-Serine
- biosynthesis and serine racemase, 211, 212
- metabolism and transport, 212–214
- neuromodulator and, 216, 217
- new roles, 215
- NMDA receptor dysfunction and, 215, 216
- regulation of NMDA receptors, 209–211, 214
- regulation of production, 212
- release, 209, 212, 214–216

Endogenous neuroprotectants
- taurine, 271

End-product inhibition, 79
Energy drinks, 182
Energy metabolism, 123
Enzyme defects, 251, 252
Enzyme phosphorylation, 73
Enzymopathy, 251–257
Eosinophilia-myalgia syndrome (EMS), 69
Ephedrine, 88
Epilepsy, 177, 178, 183, 184
Epinephrine, 60–62, 64, 79
Essential amino acid, 184
Estrogen, 61
Ethanol, 170, 171, 180, 182
Excitotoxicity
- glutamate receptors and, 146
- SAA induced, 146

Facilitated diffusion, 74, 75
"False neurotransmitter", 78
Fatty acids, 64, 75
Fick's law, 331, 332
Free radicals, 359, 360, 364, 370, 378, 381, 382, 384
Fumarate, 70
Functional magnetic resonance imaging, 323–325, 332–334

Index

GABA, 252, 253
– branched chain aminotransferase, 118, 119, 121, 126
– GABA aminotransferase (GABA-T), 126
– GABA shunt, 5, 126
– metabolism, 126, 127
– SAA-induced release of, 141
GABA/glutamate-glutamine cycle, 4–9
GABAergic neurotransmission
– GABA$_A$ receptor complex, 268
– peripheral benzodiazepine receptors, 268
Gabapentin, 124
GABA-T, 5, 10, 13, 14
GAD, 4, 5, 13
Gamma-butyrolactone, 87
Gastric transit, 88
Gene therapy, 250, 257
Geometabolism, 340
Gliotoxicity, 148, 149
Glucocorticoids, 68, 71
Glutamate, 252 f
Glutamate decarboxylase, 4, 5, 13
Glutamate dehydrogenase, 3, 5, 7, 8
Glutamate receptor agonists, 179, 180
Glutamate receptors, 165, 168, 169, 171, 179
– AMPA, 139
– excitotoxicity and, 146
– mGluRs, 140, 141
– NMDA, 139, 140, 146
– SAAs and, 139, 140, 146
Glutamate transporter family
– localisation, 142
– SAAs as substrates, 142, 143
– subtypes of, 142
Glutamate-glutamine cycle, 6, 8, 9
Glutaminase, 5, 6
Glutamine synthetase, 3, 5, 7–10, 12, 118, 119, 123
Glutamine, 2–6, 10, 12–14, 251, 252, 256, 257
Glutaminergic neurotransmission
– nitrosative stress, 268
– NMDA receptor activation, 267
– NMDA receptor/nitric oxide/cGMP pathway, 263
γ-glutamyl cycle
– astrocytes and, 135, 138
– GSH synthesis and 135, 138
– L-cysteine and, 135
γ-glutamylcysteine synthase, 138, 148
Glutamyl cysteine synthetase, 350, 354, 361, 364
Glutamyl transferase, 350, 363, 374

Glutamyl-L-cysteinyl glycine, 349
Glutathione peroxidase, 350, 352, 359, 361, 378, 379
Glutathione reductase, 359, 379
Glutathione S-transferase, 350, 362
Glutathione synthetase, 350, 352
Glutathione, 350–370
Glycemic index, 62
Glycine receptors, 173–175, 177, 179, 184
Glycine, 25 f
GSH
– γ-glutamyl cycle and, 135, 138
– γ-glutamyltranspeptidase and, 138, 147
– L-cysteine/L-cystine as precursors, 134, 135, 141
– non-excitatory neurotoxicity and, 146
Guanidino compounds, 100, 106, 109
Haber-Weiss reaction, 360
Haloperidol, 71, 87, 88
Hemorrhagic shock, 80
High-carbohydrate vs high-protein, 86
Histamine H$_2$ receptor, 53
Histamine, 48, 50–54
Histidase gene, 50–53
Histidine decarboxylase, 50, 53
Histidinemia, 48, 51–54
Homeotherms, 307, 331
Homocystinuria
– cobalamin C disease, 279, 295
– cystathionine synthetase deficiency, 291–294
– hereditary folate malabsorption, 279, 295
– methionine synthase deficiency (Cobalamin E disease), 279, 295
– methylenetetrahydrofolate reductase deficiency, 279, 294
– remethylation deficiency homocystinuria, 279, 294
Homovanillic acid (HVA), 78, 88
Huntington's disease, 332
Hydrogen peroxide, 350, 352, 359–361, 381
5-hydroxyindole acetic acid (5-HIAA), 78, 85, 86
5-hydroxyindoleacetic acid, 254
6-hydroxydopamine, 87
Hydroxyindole-O-methyltransferase (HIOMT), 74
5-hydroxytryptophan (5-HTP), 61, 77, 78, 87
Hyperacetylaspartia, 311, 314
Hyperammonemia, 109, 171

Hyperammonemic brain edema
– cerebral blood flow in, 270
– glutamine in, 269, 270
– organic osmolytes in, 269, 270
Hyperammonemic encephalopathies
– animal models, 264
– clinical characteristics in humans, 263
Hypercapnia, 302, 329
Hyperemia, 320, 324–326, 328, 329, 333–336, 338, 339
Hyperglycemia, 64
Hypernatremia, 239–241
Hypoacetylaspartia, 311, 314, 317, 336, 337
Hypoargininemia, 109
Hyponatremia, 226–228, 232, 236, 239
Hypothalamus, 48, 50–54, 83, 84, 87
Identification of peptides
– aggregation, 402
– Hplc, 402
– immune tests, 404
– mass spectrometry, 402
– mixed disulfides, 403
– non-covalent binding, 402
– sequence, 402
Immobilization stress, 80
In vivo microdialysis, 87
Inborn errors of metabolism, 297
Indoleamine 2,3-dioxygenase (IDO), 67–69, 74
INF-alpha, 69
INF-gamma, 69
Inner ear, 354, 356, 357
Inositol phosphates, enhanced production by SAAs, 140
Insomnia, 69, 86
Insulin, 62–64, 66, 71, 72, 76–78, 85, 86
Ischemia, 163–165, 167, 169, 178–180, 184
Isoleucine, 62, 64, 73, 76
α-ketoglutarate dehydrogenase, effect of ammonia, 255
Kynurenic acid, 68, 69, 74
– regulation of synthesis by L-CSA, 145
α-lactalbumin, 62
Large neutral amino acids (LNAA), 60, 64, 66, 76, 77, 85–88
LAT-1, 75
LAT1–4F2hc heterodimer, 75
Lateral tegmentum, 79
L-carnitine, 256
L-cysteic acid
– biosynthesis and metabolism, 135
– interaction with excitatory receptors, 146

Index

- localisation in the brain, 135, 136
- release, 139
- toxicity, 146
- transport, 143, 144

L-cysteine
- biosynthesis and metabolism, 135
- GSH synthesis and, 135, 138, 141, 149
- interaction with excitatory receptors, 146
- localisation in the brain, 139
- release, 139
- toxicity, 134, 146, 147
- transport, 138, 142, 143, 145

L-cysteine sulphinic acid
- biosynthesis and metabolism, 135
- interaction with excitatory receptors, 146
- localisation in the brain, 135, 136
- release, 139
- toxicity, 146
- transport, 140

L-cystine
- biosynthesis and metabolism, 135
- GSH synthesis and, 135, 138, 141, 145, 146
- localisation in the brain, 135, 136, 141, 142, 145
- release, 135, 138
- toxicity, 146
- transport, 142, 143, 145

Learning and memory, 88, 365, 378, 381
Leucine, 62, 64, 65, 73, 86
L-homocysteic acid
- interaction with excitatory receptors, 146
- localisation in the brain, 136
- release, 139
- toxicity, 146, 148
- transport, 138, 140, 142–144

L-homocysteine
- biosynthesis, 135, 137
- neurotoxicity, 147–148
- potentiation of β-amyloid toxicity, 147
- vitamin B$_{12}$ and, 137, 148

L-homocysteine sulphinic acid
- biosynthesis and metabolism, 135
- interaction with excitatory receptors, 141
- localisation in the brain, 139
- release, 139

L-Homocystine, 135
Lipid peroxidation, 350, 360, 379, 381

Liver transplantation, 256
L-methionine, precursor for L-Hcys, 135
Locus coeruleus, 79
Long-lasting potentiation, 174
Long-term potentiation, 175, 370, 372, 378
L-serine, 25–35
L-serine-O-sulphate
- alanine aminotransferase and, 148
- gliotoxicity, 148
"L-System", 75
L-threonine, 39–41
Lysolecithin, 73
L-α-aminoadipate
- cystine-glutamate exchanger and, 144
- gliotoxicity, 148
Macromolecule synthesis, 363, 364
Magnetic resonance imaging (MRI), 326
Magnetic resonance spectroscopy (MRS), 309, 311, 323, 324, 337, 338
Maple syrup urine disease (MSUD), 120, 127
- classical disease, 279, 289
- intermittent form, 279
- thiamine responsive form, 279
Medial prefrontal cortex, 60, 80, 82–88
Melatonin, 60, 61, 63, 68, 74, 75
Melphalan, 75
Membrane potentials, 172
Memory, 365, 378, 381, 383
Mental retardation, 250, 251, 256, 281, 283, 286, 288, 290, 293–295, 298, 300, 301
Mesocortical tracts, 79, 80, 82, 87, 88
Mesolimbic tracts, 79
Metabolism, 2, 3, 5–7, 9, 11–13
Metabotropic glutamate receptor, 307
Methionine, 65, 73, 76
α-methyldopa, 78
Microdialysis, in vivo, 87
Mitochondria, 3, 5, 14, 354, 357, 359, 379, 381
Mitochondrial oxidative metabolism, 118, 119, 126
Molecular water pump, 314, 317, 340
Monoamine neurotransmitter, 74, 75, 77, 78
Monoamine oxidase (MAO), 74, 254
Multiple sclerosis, 310, 333
Muscarinic cholinergic sites, 254
NAA dynamic model, 311–313, 338
NAA synthase, 308, 311, 314, 326, 336
NAAG peptidase inhibitors, 322

NAAG peptidase, 308, 309, 317, 322, 325–327, 332–337
NAAG synthase, 309, 326
N-acetyl cysteine, 377, 382, 383
N-acetylaspartylglutamate, 308, 309
Neurodegenerative diseases, 120, 357, 378, 385
Neuromodulator, 356, 357, 364, 365, 385
Neuron, branched chain aminotransferase isoform, 126
Neuron codes, 329
Neuron spikes, 328, 329, 331, 333, 335
Neuroprotection, 180
Neurosteroids, 253
Neurotoxicity
- by SAAs, 146
- L-cysteine-induced, 147
- L-homocysteine-induced, 147, 148
- protection against, 149
- S-methylcysteine-induced, 148
Neurotransmission, 358, 367, 371, 383
Neurotransmitter
- acetylcholine, 75
- catecholamine, 60–62, 70, 75, 79, 80, 81, 84, 87, 88
- dopamine, 82, 85, 87, 88
- norepinephrine, 60–62, 70, 73, 79, 80, 83–85, 87, 88
- serotonin, 60–64, 66, 68, 73–78, 80, 82–87
Neurotransmitters
- branched-chain amino acids and, 118
- *de novo* glutamate synthesis, 123–125
- glutamate/glutamine cycle, 123
- glutamate/pyruvate cycle, 125
Neurotransmitters, 158, 161, 162, 164, 169, 176
N-formylkynurenine, 66–68
α–7 nicotinic receptors, 68, 69
Nicotinamide adenine dinucleotide (NAD), 68, 69
Nicotine withdrawal, 85, 86
Nitric oxide synthase(NOS)
- inducible/induction, 101, 104, 105, 109
- neuronal, 101, 105
Nitric oxide(NO), 100, 101, 103–110, 255, 360, 367
Nitrosoglutathione, 362, 367, 370
NMDA receptors, 68, 252, 255
N-methyl-D-aspartate (NMDA) receptor, 358, 362, 365–367, 369–373

Index

Non-esterified fatty acids (NEFA), 64, 66, 75, 76
Non-ketotic hyperglycinemia, 289–291
Non-NMDA receptors, 365, 367, 372, 373
Noradrenaline, stimulation of release by SAAs, 141
Norepinephrine, 60, 62, 70–73, 79, 80, 83–85, 87, 88
Norleucine, 73
Obesity, 66, 86
Olfactory bulb, 79
"Open-loop" control, 64
Ornithine, concentration in brain and CSF, 101
Ornithine decarboxylase, (ODC), 107, 109
Ornithine transcarbamylase (OTC), 251–257
Osmolarity, 161
Osmoregulation, 314, 237, 340
Osmotic disequilibrium, 314, 315, 340
Ovalbumin, 62
Oxidation, 352, 360, 364, 371, 376, 379–381
Oxidative stress, 352, 353, 355, 358–362, 364, 368, 379–382, 385
  – Cystine-glutamate exchanger and, 146
Pain, 69, 86
Parkinson's disease, 78, 84, 87, 88, 357, 365, 380
"Peak E", 69
Peptide formation
  – γ- pseudopeptide bond, 403
  – β-aspartyl transfer, 403
  – γ-glutamyl transfer, 403
  – peptidase effect on proteins, 403
  – phosphate activation and direct synthesis, 403
Peptide function
  – co-transmitter, 402
  – growth control, 403, 406
  – hormesis, 403
  – neuro-modulator, 403, 404
  – transmitter, 403, 405
  – transport-regulator, 406
Peptide purification
  – gel filtration, 404
  – hydrolysis and ninhydrin coloring, 404
  – ion exchange, 404
  – normal phase chromatography, 404

  – reverse phase chromatography, 404
  – UV detection, 404
Peptides and behavior
  – growth control, 403, 406
  – releasing factor, 403, 404
Peptidylarginine deiminase (PAD), 101, 102, 105
Peripheral-type benzodiazepine receptors, 253
Peroxynitrite, 360, 368, 382
Pharmacology, 374–376
Phenylacetate, 255, 256
Phenylalanine hydroxylase (PAH), 61, 62, 72, 73
Phenylalanine, 60–63, 65, 69, 70, 72, 73, 79, 87, 88
Phenylethylamine, 72
Phenylketonuria
  – defects of biopterin metabolism, 279, 280
  – phenylalanine hydroxylase deficiency, 285–287
Phenylketonuria (PKU), 60
Phenylpropanolamine, 88
Phenylpyruvic acid, 72
Phophate activated glutaminase, 6, 12
Phosphorylation, 73, 77, 79, 80
p-hydroxyphenylpyruvate, 70
Pineal gland, 60
Plasma tryptophan ratio, 76, 77, 82, 86
Poikilotherms, 307, 330
Portal vascular system, 72
Potassium stimulation, 167, 168
14–3–3 protein, 77, 78
Premenstrual syndrome, 85, 86
Putrescine, 107, 109
Pyruvate carboxylase
  – anaplerosis, 124
  – glutamate/pyruvate cycle, 123, 124
Pyruvate carboxylase, 7, 8, 11, 12
Pyruvate recycling, 6–8
Quinolinic acid, 68, 255
Raphe nuclei, 82
Receptors γ-aminobutyrate, 172, 173
Receptors glutamate, 165, 168, 169, 171, 179
Receptors glycine, 173–175, 177, 179, 184
Receptors taurine, 172, 175, 176
Receptors, 358 f
Reduction, 350, 352, 354, 357, 359–364, 370, 379, 382, 384
Refractive period, 328, 330, 340
Regulation cell volume, 160, 165, 167

Regulatory volume decrease, 228–231
Release
  – of SAAs, 141
  – SAA-mediated stimulation of, 141
Release taurine, depolarization, 165, 169
Re-polarization, 328–331, 333–336, 338, 339
Reserpine, 87, 88
Retina degeneration, 161, 181
Retina, 79, 80, 88, 180, 181, 183, 184, 354, 356, 357, 373
Schizophrenia, 310, 322, 332, 333, 383–385
Seasonal affective disorder syndrome (SADS), 86
Seizures, 177, 178, 183
Sensors, "variable ratio", 82, 86
Serotonin N-acetyltransferase, 74
Serotonin, 60 f
Signal transduction, 364, 365, 385
Sleep, 64, 86
S-methylcysteine, toxicity of, 148
Sodium benzoate, 255, 256
Sparse-fur mouse (spf mouse), 252, 255, 257
Spinal cord, 352, 354, 356, 357, 373, 382, 383
Splanchnic system, 63
S-sulpho-L-cysteine
  – stimulation of excitatory receptors by, 135, 138
  – stimulation of noradrenaline release by, 141
Substantia nigra, 354, 356, 357
Sucrose, 62, 85
Superoxide dismutase (SOD), 69, 359, 360, 361, 379, 380, 382
Superoxide radical, 359
Swelling, 227–238, 242, 243
Sympathetic nervous system, 79–81, 83, 84
Synapses, 80, 84, 87
Synaptosomes, 72, 75
System L transport, 49
System N transport, 49
System y+
  – induction, 107
Taurine, 156 f
  – biosynthesis, 159, 184
  – derivatives, 169, 182–184
  – distribution, 156–159
  – receptors, 172, 175, 176
TCA cycle, 2–11, 14
Testosterone, 61
Tetrahydrobiopterin (BH$_4$), 72, 77, 79

Thermogenesis, 88
Thermoregulation, 176, 177
Thimerosal, 149
Thiol, 349, 360, 363–366, 368–371, 373, 376, 379
Thyroid gland, 61
TNF-alpha, 69
Tocopherol, 360, 361, 379, 382
Transpeptidase, 352
Transport, 2, 3, 8, 9
  – ASC transporter, 141–143
  – cystine-glutamate exchanger, 141, 142, 144, 145
  – glutamate transporter family, 142, 143
  – into astrocytes, 142, 144, 145
  – of SAAs, 141–145
Transport mechanisms, 349
Transporter taurine, taurine uptake, 160–162, 170, 175, 181, 184
Tricarboxylic acid cycle (TCA cycle)
  – α-ketoglutarate, 120
Tripeptide, 349, 364, 366

Tryptophan dioxygenase (TDO), 62, 66–70
Tryptophan hydroxylase 1 (TPH1), 61, 72, 77, 79
Tryptophan hydroxylase 2 (TPH2), 77, 78, 87
Tryptophan pyrrolase, 68
Tryptophan, 60 f
Tuberohypophyseal neurons, 79
Tuberomamillary neurons, 51
Tyrosine aminotransferase (TAT), 62, 68, 70–72
Tyrosine hydroxylase (TH), 72, 80
Tyrosine, 79–83
Tyrosinemia, 71
Urea cycle, 251–257
  – extramitochondrial enzymes, 103
Urea cycle defects
  – arginase deficiency, 279, 298
  – argininosuccinic aciduria, 279, 298
  – carbamyl phosphate synthetase deficiency, 279, 297

  – citrullinemia, 279, 298
  – hyperammonemia, hyperornithinemia, homocitrullinuria syndrome, 279, 298
  – lysinuric protein intolerance, 279, 298, 299
  – *N*-acetylglutamate synthetase deficiency, 279, 297
  – ornithine transcarbamylase deficiency, 279, 297, 298
Urokanic acid, 50
Valine, 60, 62, 64, 65, 70, 73, 87
"Variable ratio sensors", 82, 86
Vitamin B_{12}, L-Hcys and, 148
Wedensky inhibition, 329, 331, 332
Wedge preparation, 374, 376
X_{AG}^- transporters, *See* glutamate transporter family
Xanthurenic acid, 68
X_c^- exchanger, *See* cystine-glutamate exchanger
Xenobiotics, 350, 362, 363, 379